MODERN OPTICAL METHODS OF ANALYSIS

MODERN OPTICAL METHODS OF ANALYSIS

EUGENE D. OLSEN
Professor of Chemistry
University of South Florida

McGRAW-HILL BOOK COMPANY

New York St. Louis San Francisco Auckland Düsseldorf Johannesburg
Kuala Lumpur London Mexico Montreal New Delhi Panama
Paris São Paulo Singapore Sydney Tokyo Toronto

To W. J. Blaedel, an inspiring teacher and scientist, and to my wonderful family

MODERN OPTICAL METHODS OF ANALYSIS

1 2 3 4 5 6 7 8 9 0 K P K P 7 9 8 7 6 5

This book was set in Modern 8A with 20th Century by Maryland Composition Incorporated. The editors were Robert H. Summersgill and Michael LaBarbera; the designer was Joseph Gillians; the production supervisor was Thomas J. LoPinto. The drawings were done by Reproduction Drawings Ltd.
Kingsport Press, Inc., was printer and binder.

Library of Congress Cataloging in Publication Data

Olsen, Eugene D date
Modern optical methods of analysis.

Includes index.
1. Spectrum analysis. I. Title.
QD95.054 543′.085 74-26885
ISBN 0-07-047697-7

Contents

Preface

The proper use of any instrumental technique for chemical analysis requires a clear understanding of *what* is being measured as well as *how* it is being measured. Only with a balanced knowledge of both the *physicochemical theory* behind the measurement and the *instrumental principles* involved can the scientist extract the maximum information from the instrument. It is the aim of this text to provide this balanced introduction to optical methods of analysis on a level suitable for advanced undergraduates or beginning graduates, as well as for workers in the field.

In broad and modern terms, optical methods of analysis are based on the interaction of matter and radiation, over the entire electromagnetic spectrum, from gamma rays to radio waves. Included are *classical techniques,* such as refractometry, interferometry, polarimetry, turbidimetry, and nephelometry. However, predominant emphasis is given to *spectroscopic methods,* which include the established disciplines of ultraviolet and visible spectrophotometry, atomic absorption spectroscopy, infrared spectrophotometry, and nuclear magnetic resonance spectroscopy. In addition, recent instrumentation improvements are also included, which have increased and extended the relative importance of some of the established methods. In the last few years, for example, commercial instruments have become available for measurements in the far (vacuum) ultraviolet and the far infrared, and the last significant gaps in the electromagnetic spectrum have now been closed. New equipment has also increased the importance of Raman spectroscopy and near-infrared spectrophotometry. Mössbauer spectroscopy has proved to be a very useful adjunct to electron spin resonance and to nuclear spin resonance, and even though it is presently limited in application, its worth will increase with time.

In view of the scope of the spectroscopic techniques that are covered, it is not surprising that the concept of analysis emphasized in this text is also modern (though certainly not new) in that the analytical orientation is balanced between qualitative analysis, quantitative analysis, structural elucidation, and the determination of physical properties of matter.

In presenting background theory, mathematical treatment is not avoided, but derivations and complex operations are included only where they are essential. In some cases, a topic is presented at two levels—in the text proper and in Appendix A. The main body of the text is intended to convey the essential elements necessary for a working knowledge, whereas certain proofs, details, and extensions of some topics are gathered into Appendix A, where they will be available on an optional basis. In studying instrument design, the electronics involved is not avoided but is covered only in sufficient detail to make the operation of an instrument clear. The selected references at the end of each chapter guide the reader to more advanced treatments.

Basic types of instrument design are illustrated with commercially available instruments. The use of a given instrument for illustration should not be taken as an endorsement of the superiority of that instrument. Obviously a complete coverage of available instruments would swell the text to unwieldy proportions. No amount of textbook coverage can obviate the need for studying the instrument manuals accompanying the particular instruments in a given laboratory.

The text is designed to serve either for a separate course in optical methods of analysis (spectrochemical analysis) or to be used in conjunction with a text on electrochemistry for a course in instrumental analysis. Most chapters contain illustrative examples that deal with the principles being covered. At the end of each chapter a collection of exercises is given. Chapter 1 presents general definitions and unifying principles that hold for all optical methods, and thereafter each chapter compartmentalizes a single technique or, in a few cases, two or three closely related techniques. Each chapter is self-contained and does not depend on preceding chapters, except for the first. Thus, the order in which topics are studied is not critical, and topics can be omitted as desired. To facilitate a rapid start on the study of specific instrumental techniques, the first chapter can be scanned rapidly at the outset and then returned to later for a careful reading.

It is the experience of the author that the text contains more material than can be reasonably covered in a one-semester course, but the inclusion of certain subsidiary topics such as turbidimetry, nephelometry, refractometry, etc., should make the text a useful reference book for all practicing chemists and biological scientists.

I would like to acknowledge the helpful comments and suggestions of many students and colleagues, particularly Dr. Walter J. Blaedel, who read Chap. 1; Dr. Marion T. Doig III, who read Chaps. 2 to 5 and 10; Michael G. Heyl, who read Chap. 8; Dr. Jeff C. Davis, Jr., who reviewed parts of Chap. 13; Dr. David S. Wilkinson, who reviewed parts of Chap. 14; and Dr. Galen W. Ewing, who read the entire manuscript. The errors and omissions remain the responsibility of the author.

Eugene D. Olsen

1 Introduction and Unifying Principles

This chapter defines and classifies optical methods of analysis and presents fundamental principles inherent in all optical methods.

1-1 DEFINITION AND CLASSIFICATION OF OPTICAL METHODS OF ANALYSIS

Optical methods of analysis, for the purposes of this text, are defined as those methods which involve the measurement of electromagnetic radiation emanating from matter or interacting with it. All areas of the entire electromagnetic spectrum, from gamma rays to radio waves, are included. Furthermore, all the ways in which electromagnetic radiation can emanate from, or interact with, matter—including emission, absorption, scattering, refraction, reflection, dispersion, interference, diffraction, and polarization—are discussed. Methods based on each of these mechanisms will be described.

It is useful to divide all optical methods of analysis into two fundamental classes, spectroscopic and nonspectroscopic. *Spectroscopic methods* are based on the measurement of the intensity and wavelength of radiative energy. All spectroscopic methods have in common that *spectra* are measured, and furthermore, all spectra are due to transitions between characteristic energy states. *Nonspectroscopic methods*, on the other hand, do not measure spectra, do not involve transitions between characteristic energy states, and instead are based on an interaction between electromagnetic radiation and matter which merely results in a change in *direction* or a change in *physical properties* of the electromagnetic radiation. The specific mechanisms of interaction involved in nonspectroscopic methods are refraction, reflection, dispersion, scattering, interference, diffraction, and polarization. Examples of nonspectroscopic techniques include refractometry, turbidimetry, nephelometry, interferometry, x-ray diffraction, and polarimetry.

The most frequently used optical methods are spectroscopic, and most of them are based on absorption or emission as the mechanism of interaction. Spectroscopic methods are readily identified by the prefix *spectro-* in their name, and the table of contents or the index may be consulted for a list of these methods. Two spectroscopic methods which are exceptional in that neither absorption nor emission is the principal mechanism of interaction are Raman spectroscopy, which is based on a special type of scattering interaction, and spectropolarimetry, which measures polarization as a function of wavelength.

1-2 PROPERTIES OF ELECTROMAGNETIC RADIATION

In simple terms, electromagnetic radiation is energy traveling through space, unaccompanied by any matter. Classically, the behavior of radiation can be attributed to its *wavelike* character, but there are certain phenomena, notably the photoelectric effect, which cannot be explained on the basis of a wave theory and which require that radiation behave like small bundles of energy or particles, called *photons*. Still other phenomena, including the propagation of light in a straight line, reflection, refraction, and Rayleigh scattering, can be explained with *either* waves or photons.

1-2A Wavelike Properties

From classical physics, electromagnetic radiation is basically a force field in space, with characteristic frequency, velocity, and intensity. Radiation is produced whenever a particle possessing an electric and/or magnetic field is accelerated. The field then produces a perturbation which travels in a vacuum with the speed of light and in other media with a somewhat lesser speed. Figure 1-1 pictures the simplest type of electromagnetic wave train, one which is *plane-polarized* and of a single frequency. Being plane-polarized means that the electric field vector **E** vibrates in a single plane and the magnetic field vector **H** vibrates in another plane perpendicular to the electric field. In practice, most electromagnetic radiation is unpolarized, i.e., has electric and magnetic vectors at all orientations perpendicular to the direction of propagation. On passing through matter, the radiation may interact with any particle having an electric charge or magnetic moment, resulting in a transfer

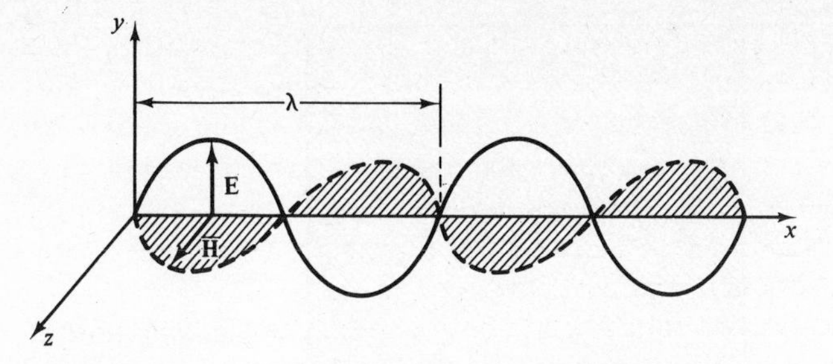

FIGURE 1-1 A plane-polarized electromagnetic wave of a single frequency; **E** = electric vector; **H** = magnetic vector.

of energy between the radiation and matter. Although either the electrical or the magnetic component of radiation can be used to explain the interaction of radiation with matter, the electrical component is perhaps easier to visualize and will be emphasized in this text.

Different kinds of electromagnetic radiation are usually characterized by either the wavelength λ or frequency ν. As shown in Fig. 1-1, the wavelength is defined as the length of a cycle, or the distance between successive maxima or minima. The frequency is the number of cycles occurring per second. The wavelength and frequency are related by

$$\lambda = \frac{v}{\nu} \tag{1-1}$$

where v is the velocity of propagation. All electromagnetic radiation travels through a vacuum with the same velocity c, which has the value of 2.9979×10^{10} cm/s. Thus, for radiation traveling in a vacuum or near vacuum, Eq. (1-1) can be written in the familiar form

$$\lambda = \frac{c}{\nu} \tag{1-2}$$

It should be emphasized that only the frequency is truly characteristic of a particular radiation. Both the velocity v and the wavelength λ depend on the nature of the medium in which the wave travels. Figure 1-2 illustrates this for radiation of a fixed frequency ν passing through the same distance of two different media (medium A and medium B, which, for example, might be air and glass, respectively). The frequency remains the same in the two media, but the wavelengths differ because the velocity of light in the denser medium, medium B, is slower than the velocity of light in medium A. (It is these velocity differences in going from one medium to another that give rise to *refraction*, discussed in Sec. 1-4.) It may be concluded that frequency is a more fundamental characteristic of electromagnetic radiation than wavelength, but either may be

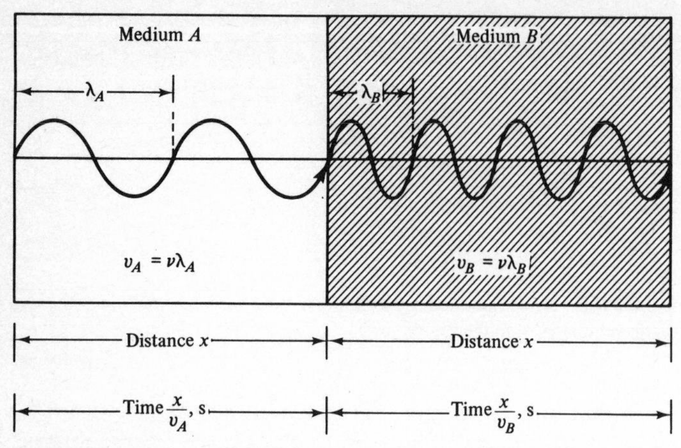

FIGURE 1-2 Effect of density of matter on wavelength and velocity of plane-polarized light of frequency ν. Medium A has low density; medium B has high density.

used in practice to specify radiation, particularly in the ultraviolet, visible, and infrared regions of the electromagnetic spectrum. Fortunately, in these regions of the spectrum, the velocity of radiation in air is within 0.1 percent of the velocity in a vacuum, making it satisfactory to use Eq. (1-2) rather than the more exact Eq. (1-1) to interrelate wavelength and frequency.

Example 1-1 When sodium atoms are excited in a flame, yellow light with a wavelength of 589 nm† is emitted. What is the frequency of this light?

Answer

$$1\text{ nm} = 10^{-9}\text{ m} = 10^{-7}\text{ cm} \qquad 589\text{ nm} = 5.89 \times 10^{-5}\text{ cm}$$

$$\nu = \frac{c}{\lambda} = \frac{3.00 \times 10^{10}\text{ cm/s}}{5.89 \times 10^{-5}\text{ cm}} = 5.09 \times 10^{14}\text{ s}^{-1}$$

Instead of expressing frequency in units of s^{-1} (which is the same as cycles per second, or hertz), *wave numbers* $\bar{\nu}$ are sometimes used, calculated as follows:

$$\bar{\nu} = \frac{1}{\lambda} \tag{1-3}$$

† In the ultraviolet and visible regions of the spectrum the wavelength is most often expressed in units of *nanometers* (nm), equal to 10^{-9} m. The term nanometer was adopted in 1962 by the International Committee on Weights and Measures, replacing the term *millimicrons* (mμ). The reader should know both names, since millimicrons still will be found in the older literature. (See Appendix Table B-1 for metric prefixes.)

where λ is in centimeters and thus $\bar{\nu}$ is in units of cm^{-1}. Wave numbers express the number of waves that occur per centimeter, and this number is directly proportioned to the frequency, as can be seen by combining Eqs. (1-2) and (1-3):

$$\nu = c\bar{\nu} \tag{1-4}$$

Example 1-2 For infrared radiation of 5.00 μm, what is the wave number in cm^{-1}?

Answer

$$1\ \mu m = 10^{-6}\ m = 10^{-4}\ cm \qquad 5.00\ \mu m = 5.00 \times 10^{-4}\ cm$$

$$\bar{\nu} = \frac{1}{\lambda} = \frac{1}{5.00 \times 10^{-4}} = 2000\ cm^{-1}$$

1-2B The Wave-Particle Duality of Radiation

Thus far the wave nature of electromagnetic radiation has been emphasized. Valid though this description is, certain optical phenomena, e.g., the *photoelectric effect,* the *Compton effect,* and the distribution of *spectral intensity* in blackbody radiation, require that radiation be viewed as a flow of particles, or corpuscles. This *corpuscular theory* was originally described by Newton, and the controversy between advocates of the wave theory and the corpuscular theory extended into the beginning of the twentieth century.

Max Planck solved the dilemma by uniting the two separate radiation theories in 1900. Planck took the unprecedented view that the energy of thermally excited, oscillating particles is *quantized,* i.e., that only certain discrete energies are allowed. He further assumed that when the oscillator moves from an allowed *high*-energy level to an allowed *low*-energy level, a *quantum* of energy will be emitted and that this energy is related to the frequency of the emitted radiation by

$$\Delta E = h\nu \tag{1-5}$$

where ΔE = energy of quantum of radiation
ν = frequency of radiation
h = Planck's constant = 6.624×10^{-27} erg-s

Using Eq. (1-5), Planck was able to predict the energy distribution for a hot body correctly.

The unifying feature of Planck's equation is that it relates the *energy* of a quantum of radiation, a corpuscular concept, to the *frequency* of the radiation, a wave-motion concept. Thus, there is no conflict between the two theories. The behavior of radiation can best be described sometimes by one theory and

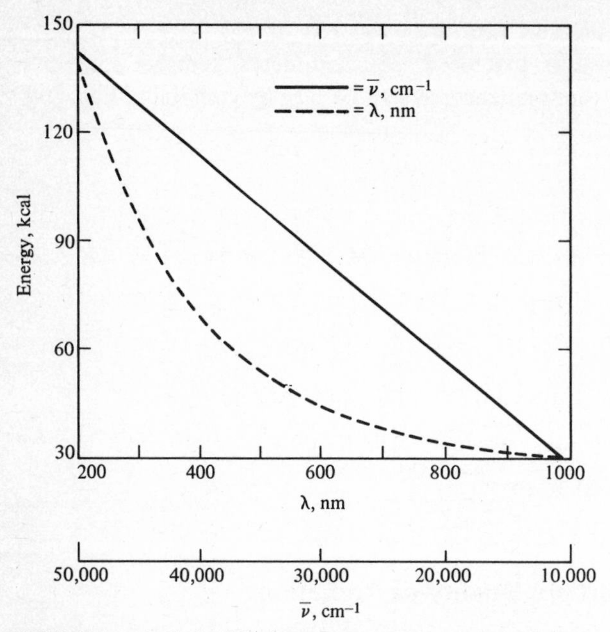

FIGURE 1-3 Relationships between frequency, wavelength, and energy of radiation. (*After H. H. Jaffe, and N. Orchin, "Theory and Applications of Ultraviolet Spectroscopy," p.* 8, *John Wiley & Sons, Inc., New York,* 1962, *by permission.*)

sometimes by the other; only when one dogmatically states that radiation is only a wave motion or is only a flow of particles does a conflict arise.

It should be noted that whereas frequency and wave number are directly proportional to energy, wavelength is inversely proportional. This is illustrated in Fig. 1-3, where both wavelength and wave number are plotted vs. the corresponding energy in kilocalories. The linearity between energy and frequency (or wave number) is a compelling reason for many spectroscopists to use frequency rather than wavelength in characterizing spectra. Nonetheless, the practice of using wavelength is deeply ingrained, and workers in this field should be able to convert readily from one set of units to the other (see Examples 1-1 and 1-2).

1-3 SPECTROSCOPIC METHODS IN EACH REGION OF THE ELECTROMAGNETIC SPECTRUM

Figure 1-4 identifies the various regions of the electromagnetic spectrum and names some spectroscopic methods used in each region. Boundaries between regions are not rigidly defined. The spectrum shown in Fig. 1-4 is complete except for optically unimportant regions at each end of the spectrum, which

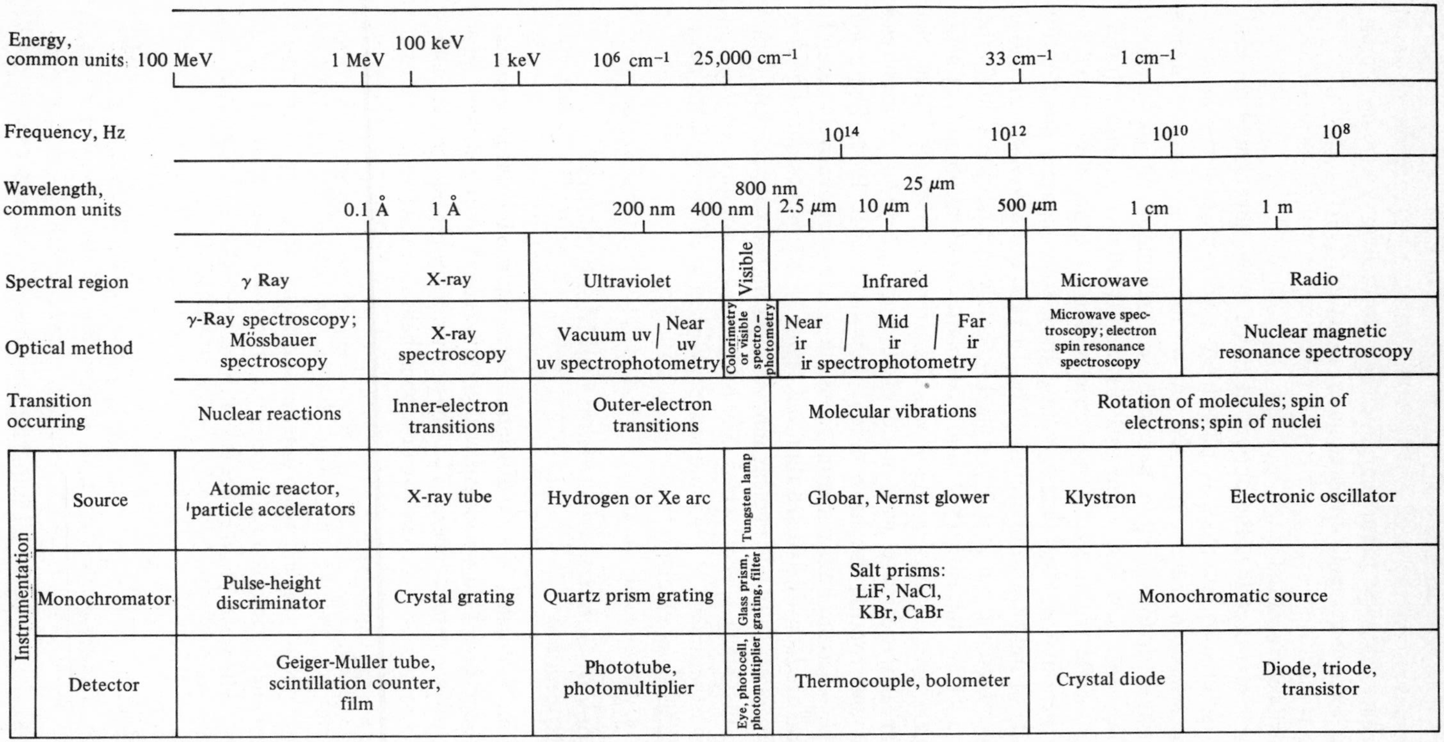

Energy, common units: 100 MeV, 1 MeV, 100 keV, 1 keV, 10^6 cm^{-1}, 25,000 cm^{-1}, 33 cm^{-1}, 1 cm^{-1}

Frequency, Hz: 10^{14}, 10^{12}, 10^{10}, 10^{8}

Wavelength, common units: 0.1 Å, 1 Å, 200 nm, 400 nm, 800 nm, 2.5 μm, 10 μm, 25 μm, 500 μm, 1 cm, 1 m

Spectral region	γ Ray	X-ray	Ultraviolet	Visible	Infrared	Microwave	Radio
Optical method	γ-Ray spectroscopy; Mössbauer spectroscopy	X-ray spectroscopy	Vacuum uv / Near uv uv spectrophotometry	Colorimetry or visible spectro-photometry	Near ir / Mid ir / Far ir ir spectrophotometry	Microwave spectroscopy; electron spin resonance spectroscopy	Nuclear magnetic resonance spectroscopy
Transition occurring	Nuclear reactions	Inner-electron transitions	Outer-electron transitions		Molecular vibrations	Rotation of molecules; spin of electrons; spin of nuclei	
Instrumentation: Source	Atomic reactor, particle accelerators	X-ray tube	Hydrogen or Xe arc	Tungsten lamp	Globar, Nernst glower	Klystron	Electronic oscillator
Instrumentation: Monochromator	Pulse-height discriminator	Crystal grating	Quartz prism grating	Glass prism, grating, filter	Salt prisms: LiF, NaCl, KBr, CaBr	Monochromatic source	
Instrumentation: Detector	Geiger-Muller tube, scintillation counter, film		Phototube, photomultiplier	Eye, photocell, photomultiplier	Thermocouple, bolometer	Crystal diode	Diode, triode, transistor

FIGURE 1-4 The electromagnetic spectrum and some optical methods in each region.

have been omitted. At the *high-energy* end, cosmic rays (originating from interstellar space and even more energetic than gamma rays) have been omitted. At the *low-energy* end, low-frequency radiation (corresponding to ordinary radio transmitters and electric-power generators, which go down to 50 to 60 Hz) have been omitted.

While this text describes the optical methods in use throughout the electromagnetic spectrum, emphasis is placed on a very narrow region in the center of the spectrum from about 200 nm to about 25 μm, which encompasses the near-ultraviolet, visible, near-infrared and middle-infrared regions. (The adjectives *near* and *far*, when used to modify ultraviolet and infrared, relate to the visible region as the reference region.) *Historically*, the *visible region* has been in use the longest, dating back to Bouguer's quantitative measurements of light transmission in 1729. In 1800 the *infrared region* was discovered by Herschel, but it was not used for analytical purposes until after exploratory work by Julius in 1892, and systematic exploitation did not begin until the 1920s. The *ultraviolet region* was discovered in 1801 by Ritter, but no particular analytical use was made of it until the 1920s. Widespread development and routine use have taken place only since 1940. The *other regions* of the spectrum have come into analytical use comparatively recently and in general are less widely used than the region of 200 nm to about 25 μm. The main exception to this is nuclear magnetic resonance spectroscopy (nmr) in the radio-frequency region, which has had an extraordinary development since the first experiments in 1945 and has established itself as one of the major analytical methods for elucidation of organic structure.

1-3A Types of Spectra and Interaction Mechanisms Responsible

All spectra can be divided into three fundamental types, absorption, emission, and Raman spectra. Table 1-1 classifies the various spectroscopic methods according to these three types. The basic principles and distinguishing mechanistic features of each type are described below.

ABSORPTION SPECTRA

As Table 1-1 shows, absorption forms the basis for spectroscopic methods throughout the electromagnetic spectrum, from the gamma-ray region (Mössbauer spectroscopy) through the radio-wave region (nmr and esr spectroscopy). The absorption process can be represented very simply by the reactions

$$X + h\nu \rightarrow X^* \tag{1-6}$$

$$X^* \rightarrow X + \text{heat} \tag{1-7}$$

Equation (1-6) represents the all-important absorption step, whereas Eq. (1-7)

TABLE 1-1 Classification of Spectroscopic Methods According to the Types of Spectra They Exhibit

Absorption spectra	Emission spectra	Raman spectra
Ultraviolet and visible spectrophotometry	Emission spectroscopy	Raman spectroscopy
Infrared spectrophotometry	Flame spectrophotometry	
Atomic absorption	Spectrofluorometry	
Microwave spectroscopy	Spectrophosphorimetry	
Circular dichroism spectrometry	X-ray emission spectroscopy	
X-ray absorption spectroscopy	Gamma-ray spectroscopy	
Nuclear magnetic resonance (nmr) spectroscopy		
Electron spin resonance (esr) spectroscopy		
Mössbauer spectroscopy		

represents the subsequent dissipation of absorbed energy, usually through collisions with other atoms or molecules. The dissipation step [Eq. (1-7)] is generally ignored in considering the absorption process, since the amount of heat liberated is usually negligible, but nonetheless it is important in understanding absorption spectra because it distinguishes absorption spectra from fluorescence and other spectra, as will be seen momentarily.

In order for electromagnetic radiation to be absorbed by matter, two general requirements must be met: (1) It is intuitively obvious that *there must be an interaction between the electric field of the radiation and some electric charge in the material substance.* (Actually, as mentioned earlier, the *magnetic* component of radiation could likewise be invoked in explaining the interaction with matter, and this is particularly advantageous in understanding nmr and esr spectroscopy, as will be seen. However, for all other optical methods it is conceptually preferable to focus on the *electrical* component.) (2) *The energy of the incoming radiation must exactly satisfy the quantized energy requirements of the substance.* Every elementary system, whether nucleus, atom, or molecule, has a number of discrete, quantized energy states, and if the incoming radiation has too little or too much energy to satisfy one of the allowed energy-level transitions, it will be *transmitted* without absorption. The energy or frequency which the incoming photon must have to be absorbed is given by the Bohr equation

$$h\nu = E_f - E_i \qquad (1\text{-}8)$$

where E_f and E_i are the energies of the final and initial states of the substance, respectively. Referring to Eq. (1-6), E_i represents the energy of substance X in the ground state, and E_f represents an allowed higher energy state, denoted by the excited species X*. The average lifetime of an *undisturbed* excited atom or molecule X* is estimated to be of the order of 10^{-8} s, but in ordinary systems (gases at ordinary pressures or condensed phases), the life expectancy is even shorter.

EMISSION SPECTRA

Emission spectra are due to a process which is exactly the reverse of absorption:

$$X^* \rightarrow X + h\nu \qquad (1\text{-}9)$$

Thus, the substance goes from an excited (high-energy) state X* to a lower energy state X by the emission of radiation. In terms of the Bohr equation (1-8), $E_f - E_i$ is negative when emission is occurring.

Three different types of emission processes can be identified, differing in how the substance reaches the excited state before emission.

Emission from Radioactive Nuclei The nuclei of substances which are naturally radioactive or made radioactive (as by neutron bombardment) can spontaneously decay with the emission of energetic gamma rays. The intensity of the radiation as a function of the energy forms a gamma-ray spectrum which is characteristic of the emitting nuclei. Gamma-ray spectroscopy is the only spectroscopic method based on this type of emission.

Emission after the Absorption of Electromagnetic Radiation With absorption spectra, it was pointed out that the substance in the excited state quickly returns to the ground state by giving up its excess energy in the form of heat [Eq. (1-7)]. With some substances, however, one of three other emission mechanisms may deactivate the excited state, namely *resonance* emission, *fluorescence* emission, or *phosphorescence* emission.

Resonance emission is an extremely rare phenomenon, occurring when an atom or molecule which has absorbed incoming radiation returns directly to the ground state with the emission of the same frequency of radiation as that absorbed. This kind of emission is restricted almost exclusively to systems of isolated atoms, where there is little or no chance that the excited species will collide with another substance before emission. Currently there are no spectroscopic methods utilizing this type of emission.

Fluorescence and phosphorescence emissions are delayed reemissions at longer wavelengths than the absorbed radiation. The emissions are at longer wavelengths than the incident radiation because a part of the incident radiation is lost through other modes of deactivation. In general, absorption will result in vibrational as well as electronic excitation, and if the substance has an excited electronic state that is more stable than usual (a necessary requirement of photoluminescent substances), then the excited species has time for vibrational deactivation through collisions with neighboring species; once in the vibrational ground state, it will return to the electronic ground state by the emission of radiation. This process is schematically pictured in Fig. 1-5, which compares the energy-level transitions involved in absorption, emission, and Raman spectra.

Fluoresence and phosphorescence are distinguished from each other by the time between absorption and reemission. In fluorescence the delay between absorption and reemission is only about 10^{-4} to 10^{-8} s, which thus appears to be instantaneous reemission, ceasing as soon as the radiation source is removed. In phosphorescence the delay is much longer, ranging from 10^{-4} to 10 s or more.

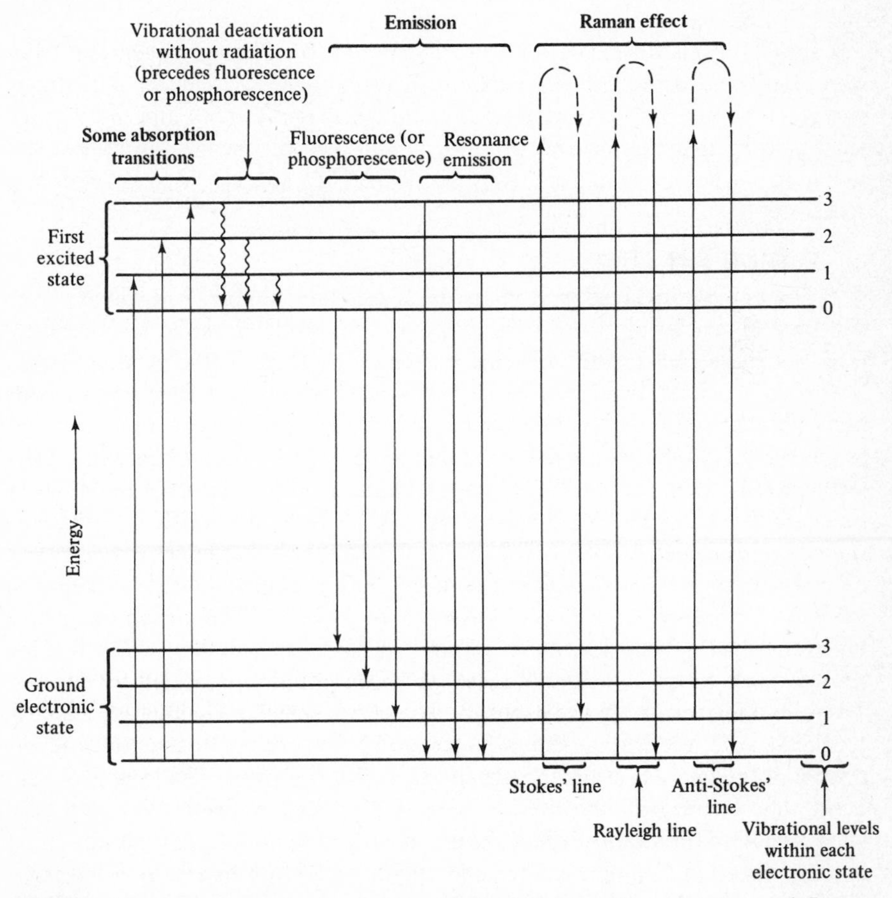

FIGURE 1-5 Some energy-level transitions involved in absorption, emission, and Raman spectra.

As will be shown later, the schematic energy-level diagram in Fig. 1-5 represents electronic energy levels of bonding electrons in a molecule, and the spacing between electronic states corresponds to energies in the ultraviolet or visible regions of the spectrum. The general principles hold throughout most of the electromagnetic spectrum, however. Fluorescent radiation, for example, may lie in the ultraviolet or visible region (spectrofluorometry and spectrophosphorimetry), or it may occur in the x-ray region (x-ray emission spectroscopy).

Emission following Nonelectromagnetic Excitation The most important types of emission spectra are based on using nonelectromagnetic energy to raise an atom or molecule to the excited state, following which the radiation emissions are measured. The process can be portrayed by the two equations

$$X + (\text{electric or thermal energy}) \rightarrow X^* \qquad (1\text{-}10)$$

$$X^* \rightarrow X + h\nu \qquad (1\text{-}9)$$

Equation (1-10) is the excitation step, and Eq. (1-9) is the subsequent emission step. *Electrical* excitation basically involves bombarding matter with high-energy electrons, and this occurs, for example, in x-ray tubes and in ac spark sources used in emission spectroscopy. Examples of thermal excitation are found in flame spectrophotometry and in dc arc sources used in emission spectroscopy.

RAMAN SPECTRA

The process involved in Raman spectroscopy is unique. It involves a special type of scattering process, in which the molecule is raised to an energy level not belonging to the molecule but corresponding to the exciting radiation. This is schematically indicated in Fig. 1-5, where the dotted energy path is not characteristic of the molecule but is of an arbitrary height corresponding to the energy of the exciting radiation that happens to be used. Classically, scattering occurs when the incident radiation induces a dipole moment in the atom or molecule. This dipole oscillates in phase with the incident radiation, thereby producing an emission of the same frequency (*Rayleigh scattering*) or an emission at a *lower* or *higher* frequency than that of the incident radiation (*Stokes' or anti-Stokes' lines*). These three types of Raman emission are illustrated in Fig. 1-5. The most probable event is an emission which returns the substance to the same ground state as it had initially (Rayleigh scattering). The next most probable event occurs when the molecule ends in a higher vibrational state, the Raman emission being of lower energy by the difference in energy of the two vibrational states involved (Stokes' line). The least probable event, but one which nonetheless occurs in enough molecules to give very weak emission lines (anti-Stokes' lines), is where the molecule is initially in an excited vibrational state and after undergoing excitation returns to a lower vibrational state, thereby emitting radiation of higher energy than was used for excitation. Raman spectroscopy is the only spectroscopic method utilizing this effect (see Table 1-1).

1-3B Energy Regions of the Electromagnetic Spectrum

A precise picture of the quantized energy changes occurring in each region of the electromagnetic spectrum will be given in succeeding chapters, but it is worthwhile here to summarize the types of quantized energy changes occurring in each region of the spectrum (Fig. 1-4) and the magnitude of energies involved, to give an overall perspective of the relationships between the various spectroscopic methods.

GAMMA-RAY REGION

Gamma rays are spontaneously emitted from radioactive nuclei with definite, discrete energies, corresponding to quantized energy levels in the nucleus. Because of the tremendous binding forces (millions of electronvolts) required to hold positively charged protons together in the intimate confines of the nucleus, the energy-level differences within nuclei are large, and transitions give rise to radiation of very high energy and frequency. How gamma rays are measured and used will be described in Chap. 14.

X-RAY REGION

X-rays are the shortest-wavelength radiations attributable to reversible changes in the electronic composition of atoms. To generate x-rays efficiently, high-speed electrons are shot at a metal of relatively high atomic number, like tungsten or molybdenum. The high-speed electrons eject inner-shell electrons from the target metal, and electrons from higher-level shells drop into the vacated shell, each transition resulting in an x-ray with an energy corresponding to the difference in energies of the two electron shells involved. As discussed in Chap. 12, these primary x-rays can be used to irradiate and excite other materials, followed by measurement of the amount of *absorption, diffraction,* or *emission.*

Example 1-3 A certain x-ray has a wavelength of 10 Å. Calculate the energy of the x-ray in electronvolts (eV).

Answer (See Appendix Table B-3 for conversions.)

$$1\ \text{Å} = 10^{10}\ \text{m} = 10^{-8}\ \text{cm}$$

$$\nu = \frac{c}{\lambda} = \frac{3.00 \times 10^{10}\ \text{cm/s}}{10 \times 10^{-8}\ \text{cm}} = 3.00 \times 10^{17}\ \text{s}^{-1}$$

$$E = h\nu = (6.62 \times 10^{-27}\ \text{erg-s})\,(3.00 \times 10^{17}\ \text{s}^{-1}) = 19.8 \times 10^{-10}\ \text{erg}$$

$$1\ \text{eV} = 1.60 \times 10^{-19}\ \text{J} \qquad 1\ \text{J} = 10^{7}\ \text{ergs}$$

Therefore,

$$1 \text{ eV} = 1.60 \times 10^{-12} \text{ erg}$$

and

$$E = \frac{19.8 \times 10^{-10} \text{ erg}}{1.60 \times 10^{-12} \text{ erg/eV}} = 1240 \text{ eV}$$

ULTRAVIOLET AND VISIBLE REGION

The energy necessary to cause excitation of outer-valence electrons of atomic and molecular systems is in the range of 8.0 to 1.5 eV. The radiation falls in the ultraviolet to visible region of the electromagnetic spectrum, corresponding to wavelength between 150 and 800 nm.

Example 1-4 Calculate the range of wavelengths, in nanometers, that corresponds to energy transitions of 1.5 to 8.0 eV.

Answer For a 1.5-eV energy transition,

$$E = (1.5 \text{ eV})(1.60 \times 10^{-12} \text{ erg/eV}) = 2.40 \times 10^{-12} \text{ erg}$$

$$E = h\nu = h\frac{c}{\lambda}$$

$$\lambda = \frac{hc}{E} = \frac{(6.62 \times 10^{-27} \text{ erg-s})(3.00 \times 10^{10} \text{ cm/s})}{2.40 \times 10^{-12} \text{ erg}}$$

$$= 8.25 \times 10^{-5} \text{ cm} = 825 \text{ nm}$$

Thus, a 1.5-eV transition corresponds to a wavelength of 825 nm, the nominal edge of the visible region. In similar manner, an 8.0-eV transition can be calculated to correspond to a wavelength of 155 nm.

The excitation of outer electrons is accomplished in a variety of ways, including electron impact, absorption of electromagnetic radiation, or even high temperatures.

Example 1-5 Estimate the energy in electronvolts of a molecule heated to a temperature of 5000 K.

Answer The average kinetic energy of a molecule can be estimated from kT, where k, the Boltzmann constant, is 8.62×10^{-5} eV/K. Thus,

$$kT = (8.62 \times 10^{-5} \text{ eV/K})(5000 \text{ K}) = 0.43 \text{ eV}$$

This is considerably less energy than is required for excitation of an outershell electron, for which about 1.5 eV is needed (see Example 1-4). However, thermally excited molecules have a Maxwell distribution of kinetic energies, and many particles at 5000 K have sufficient energy to cause excitation in the visible and even ultraviolet region of the spectrum.

X-ray emission and absorption spectra are relatively independent of the particular chemical combination of the atoms taking part because the transitions involve inner electrons. However, transitions in the ultraviolet or visible region involve outer electrons, and the spectra are decidedly affected by whether individual *atoms* or molecular *compounds* participate. With individual atoms, both emission and absorption spectra consist of sharp lines, corresponding to transitions between the allowed electronic energy levels of the atom. With molecules, however, both emission and absorption spectra are complicated because electronic transitions are accompanied by simultaneous changes in the vibrational and rotational states of the molecule, and *band* spectra result.

Energy-level diagrams are helpful in visualizing why molecular spectra are band spectra. Figure 1-6 illustrates an energy-level diagram for a diatomic molecule. The major energy levels, of which only two are shown in Fig. 1-6, are the *electronic* energy levels, designated by the *electronic quantum number* n, where $n = 1, 2, 3, \ldots$. Transitions from one to another of these levels involve changes in the energy of one or more electrons in the molecule. A simplified picture of electronic excitation is given in Fig. 1-7*a*. Associated with each electronic energy level are *vibrational* sublevels, designated by a *vibrational quantum number* v, where $v = 0, 1, 2, 3, \ldots$ (see Fig. 1-6). Vibrational transitions involve changes in the energy of the molecule as a whole due to interatomic vibrations of one sort or another, and Fig. 1-7*b* gives a simplified picture of vibrational excitation. Each vibrational level is further subdivided into rotational energy levels, designated by a *rotational quantum number* J, where $J = 0, 1, 2, 3, \ldots$ (see Fig. 1-6). A simplified picture of rotational excitation is given in Fig. 1-7*c*. It should be emphasized again that all these forms of excitation are *quantized,* and thus even with rotation, only certain discrete *speeds* of rotation are allowed.

In the energy-level diagram of Fig. 1-6, only the vertical coordinate has significance, and then only schematically. An accurate energy-level diagram would show three details not included in Fig. 1-6: (1) the energy separation between electronic states is very large; (2) the vibrational levels tend to converge at higher energies; and (3) the rotational levels tend to diverge at higher energies.

Three types of transition are shown in Fig. 1-6: (1) a pure *rotational* transition ($\Delta J = 1$), (2) a *vibrational* transition ($\Delta v = 1$, $\Delta J = 1$), and (3) five different *electronic* transitions, all satisfying the general description $\Delta n = 1$, $\Delta v = 1$, $\Delta J = \pm 1$. Generalizations about which transitions are possible and which are not are called *selection rules.* They are derived in more advanced

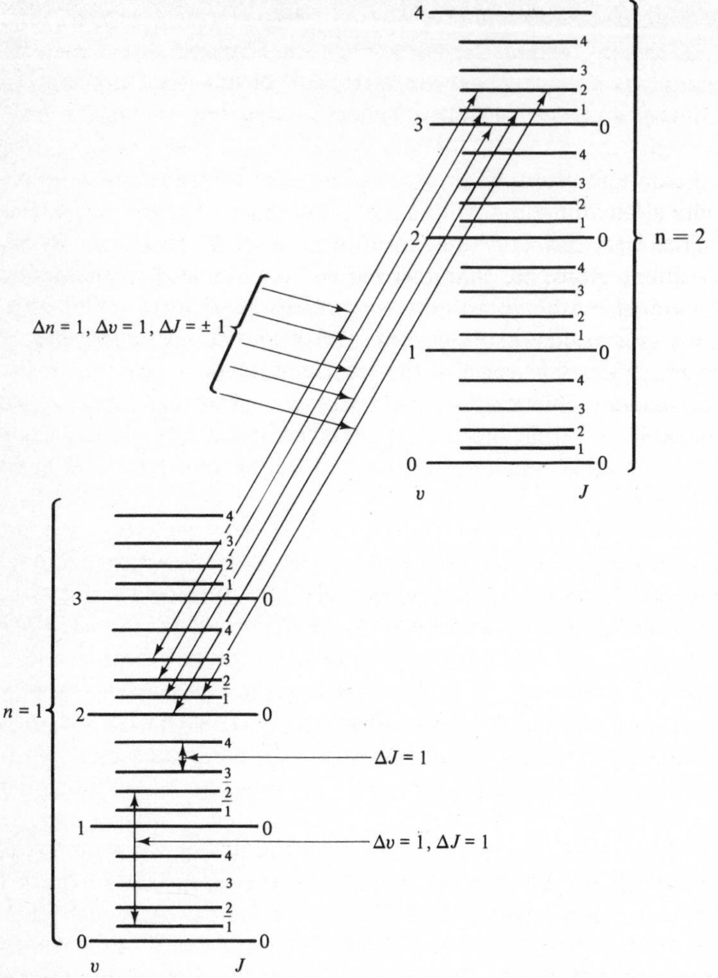

FIGURE 1-6 Energy-level diagram and some allowed transitions for a diatomic molecule.

works on molecular spectroscopy, e.g., Ref. 8, but it will be useful here to list some of the selection rules for simple linear molecules.

1 Pure rotational transitions can take place only between adjacent rotational levels ($\Delta J = \pm 1$).

2 Vibrational transitions *must* be accompanied by simultaneous rotational transitions ($\Delta v = \pm 1$, $\Delta J = \pm 1$). *Note:* Δv can also be ± 2, ± 3, but these occur with much lower probabilities than ± 1.

3 Electronic transitions are generally accompanied by vibrational-rotational transitions but need not be. (Generally, $\Delta n = \pm 1$, $\Delta v = \pm 1$, $\Delta J = \pm 1$.) Further details on electronic transitions will be developed in Chap. 2.

As Fig. 1-6 illustrates, vibrational spectra are legitimately termed vibrational-rotational spectra, and electronic spectra contain vibrational-rotational transitions superimposed on the electronic transitions. Since a large number of molecules in the electronic ground state at room temperature can exist in excited vibrational states, a large number of different transitions of similar energy spacing can take place. Thus, the five electronic transitions shown in Fig. 1-6 are only a few of the many possible transitions of similar energy. For this reason, electronic emission and absorption spectra, found in the ultraviolet and visible regions, are characterized by broad bands for molecules.† Similarly, vibrational-rotational transitions, found in the near- and middle-infrared regions, give rise to band spectra, but the bands are much sharper than for electronic spectra because of the smaller number of allowed transitions of similar energy. Still sharper band spectra can be expected in pure rotational spectra, found in the far-infrared and microwave regions.

INFRARED REGION

The energy necessary to cause *vibrational transitions* (or vibrational-rotational transitions, as described in the previous section) is of the order of 0.05 to 1.2 eV, corresponding to radiation in the wavelength range of 1.0 to 25 μm (where 1 μm = 10^{-6}m). The near-infrared, or *overtone,* region extends from about 0.8 to 2.5 μm, with the middle infrared going from 2.5 to about 25 μm. The far-infrared region extends from approximately 25 to 500 μm, with pure rotational transitions occurring at wavelengths longer than about 100 μm, corresponding to energy spacings of about 0.01 eV or less. Infrared spectrophotometry is discussed in detail in Chap. 3.

MICROWAVE REGION

This region extends from about 500 to 30,000 μm, or 0.05 to 3 cm, with pure rotational transitions having spacings down to about 0.00025 eV (5000 μm). Molecules with a dipole moment can interact with microwave radiation, giving sharply defined absorption spectra.

† Spectra are sharpened at cryogenic temperatures.

(a) O : O

(b) O : O

FIGURE 1-7 Simplified (*a*) electronic, (*b*) vibrational, and (*c*) rotational excitations.

Microwave radiation is in a transition region between the infrared and the radio-wave regions, and as such has some of the properties associated with both. For example, microwaves are generated like radio waves by precise control of electron movements, but unlike radio waves they can be focused by lenses, just as light waves are. As a result of the tremendous resolving power of the electronic systems used in generation and detection, microwave absorption spectra can be more sharply defined than most optical spectra. Microwave spectroscopy will be discussed in detail in Chap. 7.

Microwave radiation of about 1 to 3 cm is also used in *electron spin resonance* (esr) studies, using a technique similar in principle to nmr spectroscopy, discussed in the next section. In brief, esr studies can be used for molecules containing unpaired electrons. First, a homogenous magnetic field is imposed, and then the sample is irradiated with microwave radiation. If the radiation is

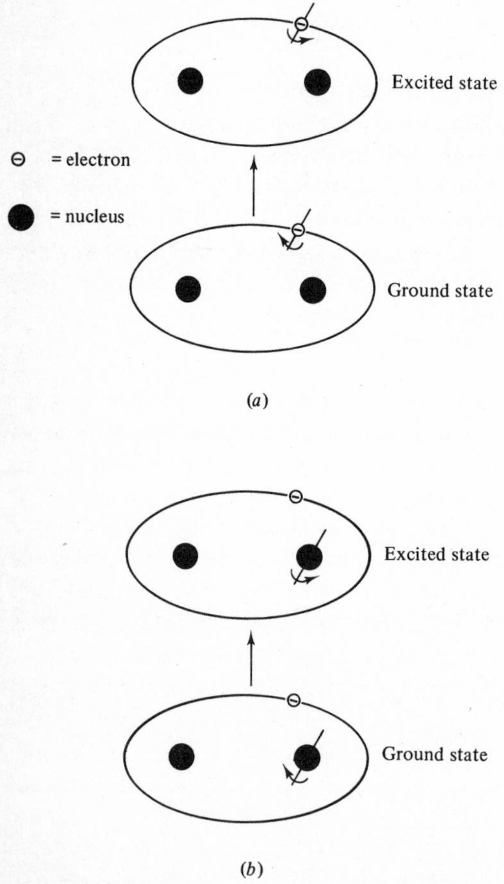

FIGURE 1-8 Spin transitions in a diatomic molecule: (*a*) electron spin excitation; (*b*) nuclear spin excitation.

of precisely the correct frequency, it will be absorbed and cause a transition in electron spin energy, as schematically illustrated in Fig. 1-8*a*. A detailed discussion of esr spectroscopy is given in Chap. 13.

RADIO-FREQUENCY REGION

The portion of the radio-frequency region which has proved most useful for studying interactions with matter extends from about 1 to 300 MHz corresponding to a wavelength of about 1 to 300 m. The analytical technique using this region is known as nuclear magnetic resonance spectroscopy (nmr).

Nuclear magnetic resonance spectroscopy can be used to study nuclei with a spin magnetic moment. Protons are the most widely studied nuclei. First, a homogenous magnetic field is imposed, and then the sample is irradiated with radio-frequency radiation. If the radiation is of precisely the correct frequency, it will be absorbed and cause a transition in nuclear spin energy, as schematically depicted in Fig. 1-8*b*. A detailed discussion of nmr spectroscopy is given in Chap. 13.

1-4 PRINCIPLES OF OPTICS AND NONSPECTROSCOPIC METHODS

In the previous section absorption and emission processes were emphasized, and these concepts are best understood by thinking of radiation as consisting of discrete photons. But other types of interactions that can occur, e.g., *refraction, reflection, scattering, interference, diffraction, polarization,* and *dispersion,* are best explained by considering radiation as waves. Some of these interactions are the basis of a number of analytical methods, and almost all play an essential role in the design of optical instruments.

1-4A Refraction

Refraction is the basis of refractometry, to be discussed in Chap. 9. Refraction might be defined as the bending of radiation as it passes from one material into another, and it can be attributed to a difference in velocity in the two media. Whenever radiation passes from one dielectric material into another, it is partially *reflected* and partially *refracted* (or transmitted), as shown in Fig. 1-9. The reflected energy increases as the change in velocity in the two media increases, and this in turn is the greater the larger the change in density of the two media.

A useful *law of refraction*, called Snell's law, is that the ratio of the sine of the angle of incidence to the sine of the angle of refraction is a constant, called the *index of refraction n*:

$$\frac{\sin\theta}{\sin\theta'} = n = \text{index of refraction} \qquad (1\text{-}11)$$

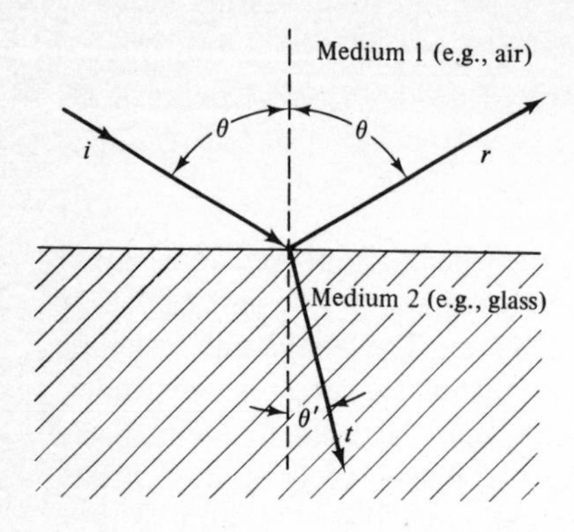

FIGURE 1-9 Refraction of electromagnetic radiation; i = incident beam, r = reflected beam, t = transmitted beam, θ' = angle of refraction.

where θ and θ' are defined in Fig. 1-9. The angular relationships of refraction can also be accounted for on the basis of the relative velocities of the radiation in the two media:

$$\frac{\sin\theta}{\sin\theta'} = \frac{v_1}{v_2} \tag{1-12}$$

where v_1 is the velocity of the radiation in medium 1 and v_2 is the velocity in medium 2. Substituting Eq. (1-12) into Eq. (1-11) gives a useful interpretation of the index of refraction:

$$n = \frac{v_1}{v_2} \tag{1-13}$$

It will be shown later that the velocity of radiation in a given medium depends on its wavelength, and thus in order to put refractive indexes on an absolute basis, both the wavelength of light and the reference medium used must be specified. It is common practice to use the sodium D line (589 nm) as the wavelength, and medium 1 is specified to be a vacuum. Since the velocity of all electromagnetic radiation in a vacuum is a constant, c, the absolute definition of refractive index becomes

$$n_2 = \frac{c}{v_2} \tag{1-14}$$

where the refractive index is given a subscript 2 to emphasize that it is the refractive index of medium 2 which is being defined. (The use of the sodium D line is understood or is specified with a subscript D.) In practice, air is generally used for medium 1, but since the velocity of radiation in air v_{air} is close to that in a vacuum, it is usually sufficiently accurate to approximate the

TABLE 1-2 Index of Refraction of Some Common Substances Using D Line of Sodium

Substance	Index of refraction
Glass†	1.5–1.9
Diamond	2.4173
Fused quartz	1.45843
Quartz crystal	1.544
Glycerin	1.4729
Ethyl alcohol	1.35885 (24°C)
Oleic acid	1.463
Water	1.33262 (24°C)

† The index of glass depends on its composition. Most ordinary glasses have indexes of about 1.5.

absolute refractive index with

$$n_2 = \frac{c}{v_2} \approx \frac{v_{\text{air}}}{v_2} \tag{1-15}$$

The differences between air and a vacuum are usually small, except in high-precision measurements, where the decreased velocity in air must be corrected for. For example, a glass with an index of 1.50000 in air has an index of 1.50044 in a vacuum.

Table 1-2 lists the index of refraction for a few common materials. The index of refraction is often helpful in identifying a substance (Chap. 9) and greatly simplifies our description of refraction. To predict how light will be bent on entering any material from air, only the index of refraction of the material need be known.

Prisms can separate light into its component wavelengths because the velocity of radiation in a given medium depends on its wavelength. From Eq. (1-13), it may be expected that materials with high refractive indexes make the best prisms, which is true as long as the prism material does not significantly *absorb* the wavelengths of interest (see Sec. 1-4*F*).

1-4B Reflection

Whenever radiation is incident upon a boundary between two materials across which there is a change in refractive index, reflection occurs. The quality of the surface has much to do with the nature of the phenomenon. When the surface is smooth, the angle of incidence equals the angle of reflection, as shown in Fig. 1-9. Irregular surfaces give rise to *diffuse reflection*, which is of little interest in most optical work. For most angles of incidence and types of surfaces, the reflected beam differs in total intensity, state of polarization, and phase from the incident beam, and if many frequencies are present, there are usually chromatic differences in the reflected radiation as well.

The amount of reflection, i.e., the ratio of the intensity of the reflected radiation to that which is incident on the interface, is termed the *reflectance* ρ (or sometimes the *reflection power*). For a given angle of incidence θ_i, the reflectance will be greater the larger the refractive index of medium 2. However, the refractive index of medium 1 is also involved. Consider the case where the incident radiation strikes the interface perpendicular to the surface of medium 2, for example, the usual case of radiation passing through a sample. Here the reflectance can be calculated from

$$\rho = \frac{I_R}{I_0} = \frac{(n_2 - n_1)^2}{(n_2 + n_1)^2} \tag{1-16}$$

where I_R = reflected intensity
I_0 = incident intensity
n_1, n_2 = appropriate refractive indexes

Example 1-6 If visible light passes vertically from air into ordinary glass of $n = 1.50$, calculate the percentage of light reflected.

Answer

$$\rho = \frac{(n_2 - n_1)^2}{(n_2 + n_1)^2} = \frac{(1.50 - 1.00)^2}{(1.50 + 1.00)^2} = 0.040, \text{ or } 4\%$$

Example 1-7 Calculate the reflectance at the emerging interface, when visible light passes vertically from glass ($n = 1.50$) to air.

Answer Since the difference between n_2 and n_1 is squared in the numerator of Eq. (1-16), only the absolute difference affects ρ, and thus, from Example 1-6, $\rho = 0.040$, the same value as at the incident air-glass interface.

Example 1-8 Calculate the percentage of visible light transmitted vertically through a block of clear glass with parallel surfaces.

Answer The transmitted light I_t can be calculated from

$$I_t = I_0 - I_R$$

and since $\rho = I_R/I_0$,

$$I_t = I_0(1 - \rho)$$

Therefore, at the first interface,

$$I_t = 1.00(1 - 0.04) = 0.960$$

At the second interface,

$$I_t = 0.96(1 - 0.04) = 0.922$$

Therefore, 92.2 percent of the light is transmitted.

As Eq. (1-16) indicates, the greater the difference in refractive index between two media the larger the reflectance. On the other hand, if $n_2 = n_1$, there can be no reflection. And if no light is reflected from the surface of an object, it will be invisible. Thus, objects are invisible when they are surrounded by a medium of identical refractive index. This principle makes it possible to determine the refractive indexes of glasses and other transparent solids.

1-4C Scattering

Scattering is the basis of turbidimetry and nephelometry, discussed in Chap. 11. In the usual type of scattering, electromagnetic radiation interacts with small particles in its path by inducing oscillations in the electric charges of the matter. The *dipoles* that are induced radiate secondary waves in all directions. In the process, energy is removed from the beam of incident light and emitted without change in wavelength. Although this type of secondary radiation could also be used to explain reflection and refraction, the necessary criteria for the term scattering are that the particles have dimensions of about the same order of magnitude or *smaller* than the incident wavelength and that the particles be distributed in a medium of refractive index *different* from their own. For larger particles (of dimensions greater than approximately 2λ), wave analysis shows that only refraction and reflection occur.

Table 1-3 gives the approximate size of particles that produce scattering in various regions of the optical spectrum. All forms of matter can induce light scattering. For example, the scattering of light by *gases* is observed both in air, where it accounts for the blue color of the sky, and in interstellar dust. Light scattering also occurs in pure *liquids* and in pure *solids*, but its most important analytical applications arise from the scattering in solutions of large molecules or from the scattering of suspended particles. If the radiation the particle scatters is in the visible region, an observer with a microscope or ul-

TABLE 1-3 Particles Producing Scattering in Various Wavelength Regions†

Incident radiation			
Wavelength, μm	Spectral region	Maximum particle size (approximate dimensions), μm	Type of aggregate
10	Infrared	15	Large colloidal particles
0.5	Visible	0.75	Colloidal particles, macromolecules
0.001	X-ray	0.002	Small molecules, atoms

† From H. A. Strobel, "Chemical Instrumentation," 2d ed., p. 255, Addison-Wesley Publishing Company, Inc., Reading, Mass, 1973, by permission.

tramicroscope can "see" the particle, but it appears only as a point; no surface features are resolved.

Within the size range of particles which will cause scattering, two special cases should be discussed. For particles whose longest dimension is no more than 5 to 10 percent of the incident radiation, the scattering is known as *Rayleigh scattering,* whereas particles having dimensions greater that 10 percent of λ are said to give *large-particle scattering.*

RAYLEIGH SCATTERING

This type of scattering produces an intensity pattern of scattered radiation which is relatively symmetrical in all directions around the particles, as illustrated in two dimensions by Fig. 1-10. With the scattering particle visualized as a point at the origin of the axes and the arrow vectors representing the relative intensities of scattering in the various directions, the envelope outline connects the ends of the intensity vectors 360° around the particle. (The three-dimensional picture of Rayleigh scattering may be visualized if the two lobes in Fig. 1-10 are thought of as slightly overlapping spheres.) The intensity of scattering I_s can be calculated from

$$I_s = \frac{8\pi^4\alpha^2}{\lambda^4 r^2}(1 + \cos^2\theta)I_0 \tag{1-17}$$

where α = polarizability of particle
λ = wavelength of incident radiation
I_0 = incident intensity
θ = angle between incident and scattered ray, as shown in Fig. 1-10
r = distance from center of scattering to detector

Polarizability is a measure of the efficiency with which a given frequency of radiation will induce a dipole in the particle. It must be emphasized that the polarizability α varies roughly as the volume of the particle and thus Eq. (1-17) predicts that the scattering will increase strongly as the particles become larger. In any collection of particles of different sizes, the larger ones contribute most heavily to the observed scattering.

The inverse dependency of intensity on the fourth power of the wavelength should also be noted. The intensity increases very strongly as the wavelength becomes shorter, i.e., as it approaches the size of the particle.

Example 1-9 Assuming Rayleigh scattering, estimate the relative efficiency with which violet light of 400 nm is scattered compared with green light of 550 nm.

Answer Assuming the polarizability α of the scattering particles does not vary significantly over this short wavelength interval, all the factors in Eq. (1-17) are constant except I_s and λ. Therefore,

$$\frac{I_{s,1}}{I_{s,2}} = \left(\frac{\lambda_2}{\lambda_1}\right)^4$$

$$\frac{I_{s,400\text{ nm}}}{I_{s,550\text{ nm}}} = \left(\frac{550\text{ nm}}{400\text{ nm}}\right)^4 = 1.375^4 = 3.57$$

In other words, violet light of 400 nm is scattered about 3.57 times more efficiently than green light of 550 nm.

Two everyday examples of this discriminate scattering are the blue color of the sky and the red color of the sun at sunset. The predominant blue color

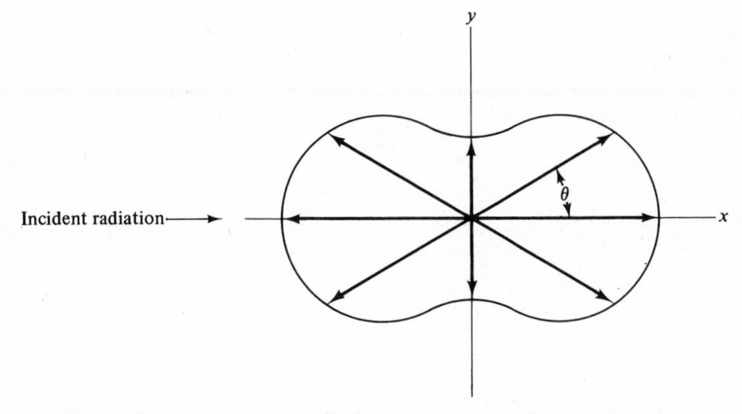

FIGURE 1-10 The intensity pattern for small-particle (Rayleigh) scattering. See text for explanation.

is caused by efficient scattering of short wavelengths of visible light by particles of small dimensions in the atmosphere, such as dust and water vapor. Conversely, in the setting sun, the long wavelengths of visible light (red) are more completely transmitted than the short wavelengths, which are scattered away by smoke, fog, and other small particles.

LARGE-PARTICLE SCATTERING

This type of scattering occurs when the dimensions of the particle are greater than 10 percent of λ, but no larger than about $\frac{3}{2}\lambda$, where the scattering phenomenon terminates. Whereas the characteristic intensity pattern for Rayleigh scattering is fairly uniform in all directions around the scattering particle, as shown by Fig. 1-10, the characteristic intensity pattern for large particles is preponderantly in the forward direction, as illustrated in two dimensions by Fig. 1-11.

The intensity of scattering is inhomogeneously distributed around large particles because of *constructive* and *destructive* interference of waves. In brief, wave analysis shows that two or more waves can interact (or interfere) with each other, and the resulting wave is a *superposition* of the original waves, handled in wave analysis by vectorial addition. The intensity of the resulting radiation strongly depends on the difference in phase of the interaction waves. Figure 1-12 illustrates what happens when two waves of different amplitudes but the same frequency (dotted lines) are superimposed to give the resulting wave (solid line). A maximum in *constructive* interference occurs when the phase difference equals 0, 360, 720°, . . . (Fig. 1-12*a*), whereas the intensity is a minimum due to *destructive* interference when the phase differences have the values of 180, 450°, . . . (Fig. 1-12*b*). In scattering by large particles, in contrast to Rayleigh scattering, the particles can no longer be thought of as point sources, and thus there will be points or centers of scattering distributed over the area of the particle, resulting in some *interference* between secondary rays emitted from separate areas of the particle.

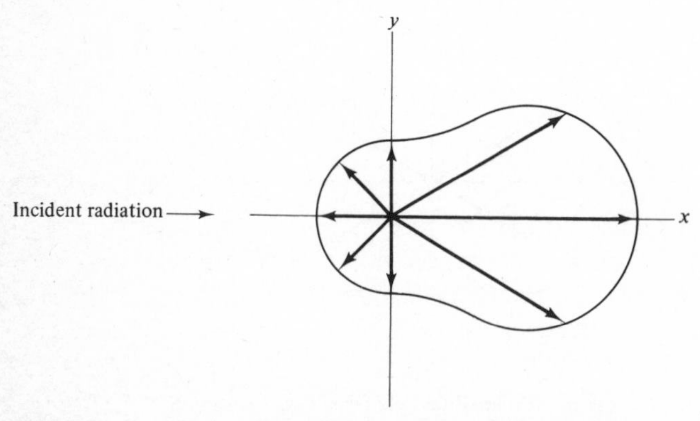

FIGURE 1-11 The intensity pattern for large-particle scattering.

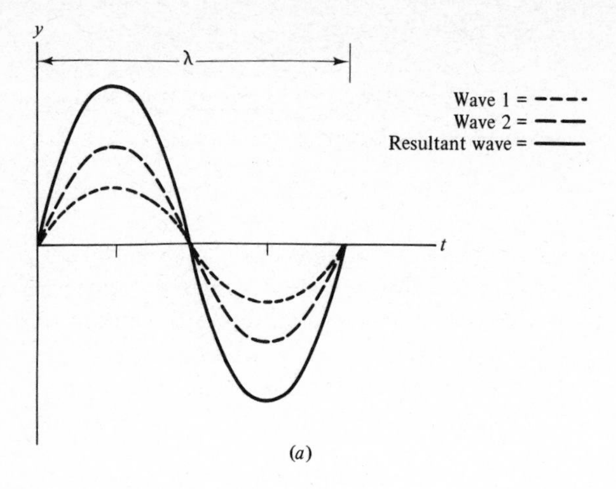

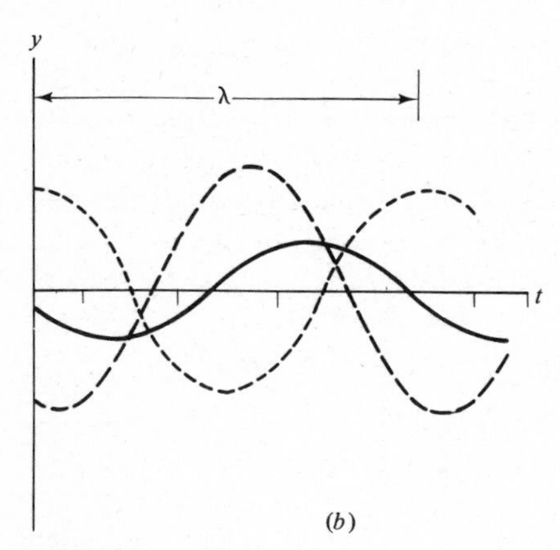

FIGURE 1-12 Superposition of two plane-polarized wave trains of the same frequency: phase difference of (*a*) 0 or 360° and (*b*) about 170°.

This situation is pictured for a large spherical particle in Fig. 1-13. *A* and *B* represent scattering centers on the particle, and points *C* and *D* represent points of interaction between scattered waves. The rays scattered backward are highly susceptible to destructive interference because of the large path *differences* possible, which can be seen, for example, by comparing distances for paths *SAC* and *SBC*. For scattering in the forward direction, on the other hand, the path differences are much less (compare path *SAD* with *SBD*), and thus the interference is usually constructive.

Extension of these arguments to very large particles would show that constructive and destructive interference becomes so predominant that only well-defined refracted and reflected beams remain.

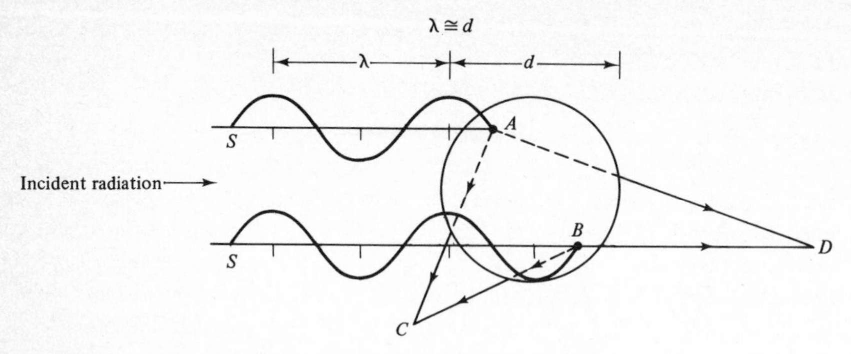

FIGURE 1-13 Interference in scattering from a large molecule.

1-4D Interference

Interference is the basis of interferometry, discussed in Chap. 9. Optical interference can be defined as the modification of the intensity of radiation waves by combining or superimposing two or more waves. The resultant intensity is sometimes decreased, resulting in destructive interference, and sometimes increased, resulting in constructive interference. The phenomenon of interference was described in the previous section. This section will describe optical interference filters which can be used in the near-ultraviolet, visible, and near-infrared regions. Two types have been developed, metal (or *Fabry-Perot*) interference filters and *multilayer dielectric* filters.

THE FABRY-PEROT FILTER

This simple filter is produced by coating both sides of a layer of magnesium fluoride (a transparent dielectric) with semireflecting, thin, parallel films of silver. The thickness of the dielectric is adjusted to one-half the desired wavelength. Radiation entering the dielectric layer is reflected back and forth between the metallic layers, and since the reflecting layers are parallel, a superposition of waves occurs within the dielectric. The ensuing interference causes the obliteration of all but several narrow bands of wavelengths, which are transmitted. The wavelengths of maximum intensity transmitted by the interference filter can be calculated from

$$\frac{\lambda}{n} = \frac{2d}{m} \tag{1-18}$$

where n = refractive index of dielectric material
d = thickness of dielectric layer
m = *order* of interference

Figure 1-14 illustrates the multiple reflections occurring within an interference filter. If the first-order band transmitted by a given filter is centered at λ, the second-order band will appear at $\lambda/2$, the third-order band at $\lambda/3$, etc.

Example 1-10 In a Fabry-Perot interference filter, explain why the thickness of the dielectric should be one-half the desired wavelength for first-order transmission.

Answer For maximum constructive interference the total path of the radiation within the dielectric should be an exact multiple of the wavelength, so that waves interacting at a boundary will have phase differences of 360, 720°,. . . . The first reflected beam leaving the filter will have met *two* reflection surfaces, a thin metallic film on each side of the dielectric. Thus, for maximum constructive interference at the second reflecting surface the wave should have traveled a path of exactly one wavelength (360°), which means that the dielectric thickness should be exactly one-half the wavelength.

Ordinarily, an interference filter gives two transmission bands, e.g., one in the visible (first order) and the other in the ultraviolet (second order). Figure 1-15 gives the transmission curve of a typical Fabry-Perot interference filter. The wavelength lying in the ultraviolet can be eliminated by a glass filter or appropriate coating.

Example 1-11 In a Fabry-Perot interference filter, why are *bands* of radiation transmitted, rather than sharp *lines*?

Answer Although the *maximum* constructive interference occurs at phase differences of 0, 360, 720°, . . . , there is nonetheless *some* constructive interference at phase differences slightly more or less than the optimum values. Thus, wavelengths only slightly out of phase will be transmitted with diminished intensity, resulting in spectral *bands* of radiation.

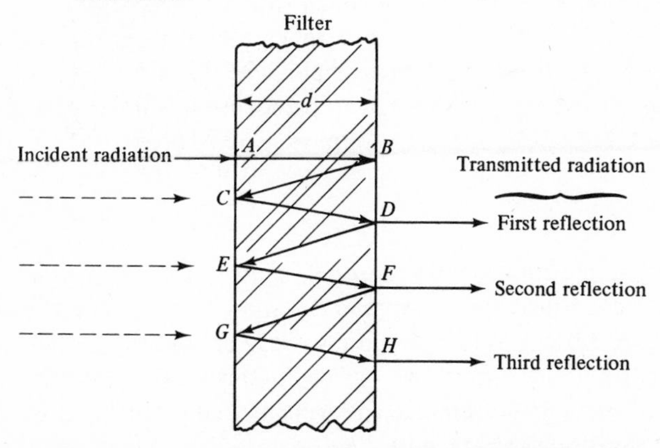

FIGURE 1-14 Multiple reflections within an interference filter. The actual reflections are not separated in space, as shown, but move back and forth upon themselves, in a manner which cannot be drawn clearly.

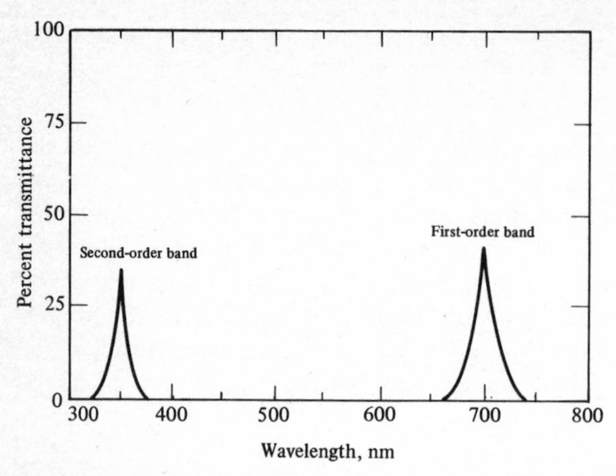

FIGURE 1-15 Transmission curve of a typical Fabry-Perot interference filter.

Example 1-12 If a Fabry-Perot interference filter transmits a first-order band at 600 nm, what other wavelength bands can be expected?

Answer A second-order band at (600 nm)/2 = 300 nm, a third-order band at (600 nm)/3 = 200 nm, etc.

THE MULTILAYER FILTER

This kind of interference filter consists of alternate layers of high- and low-refractive-index materials. Each layer must be of suitable thickness, made possible by a process of vacuum deposition. Some 5 to 25 layers are required for this type of filter, layer thicknesses ranging between one-fourth and one wavelength. Although a detailed discussion is beyond the scope of this book, it should be pointed out that multilayer filters give much sharper peak widths of transmitted light than Fabry-Perot filters, along with greater transmittances (40 to 70 percent). Sharp cutoff filters are also possible. The disadvantages of multilayer filters are greater cost and higher transmittance in unwanted parts of the spectrum. In general, Fabry-Perot filters are more widely used.

1-4E Diffraction

When waves of electromagnetic radiation pass through a slit or travel past the edge of any opaque obstacle, they always spread to some extent into the region which is not directly exposed to the incoming waves. This phenomenon, called *diffraction*, is of paramount importance to the understanding of slits and diffraction gratings in optical instruments, as well as being the basis of the x-ray diffraction patterns discussed in Chap. 12.

Whereas most optical phenomena can be explained by assuming that radiation waves move through space with a linear wavefront, analogous to the crests of water waves coming into shore, diffraction requires a modified view of waves: *each point* on a wavefront can be regarded as a new source of waves.

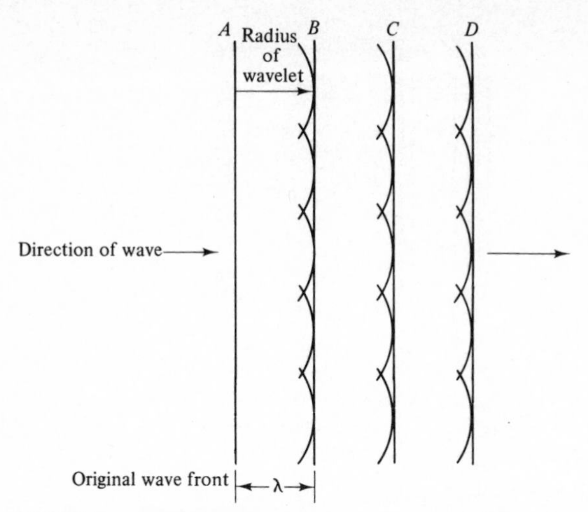

FIGURE 1-16 The construction of wavefronts by Huygens' principle. (*After A. Efron, "Light," p.* 50, *John F. Rider Publisher, Inc., New York,* 1958, *by permission.*)

This is known as *Huygens' principle* and is illustrated in Fig. 1-16. The wavefronts *A, B, C,* and *D* can be visualized as crests of water waves, the distance between each crest being designated as the wavelength, analogous to radiation wavelength λ. For electromagnetic radiation, Huygens' principle requires that we be able to construct secondary spherical wavelets geometrically at every point on a wavefront, and these wavelets, in turn, are propagated in all directions with the velocity of the original wavefront. This principle, in conjunction with the phenomenon of *interference,* helps explain the striking features of diffraction.

DIFFRACTION BY A SLIT

Diffraction of radiation by a slit can be introduced by using the simple and accurate analogy of water waves in a ripple tank. Figure 1-17*a* shows the predominantly linear wavefront generated when water waves pass through a relatively wide opening; the slight curvature at the edges of the wavefront, corresponding to a sidewise spreading around the edge of the barrier should be noted. If the opening in the barrier is large compared with the wavelength, then bending of the waves (or diffraction) becomes insignificant. On the other hand, if the opening is small (of a size comparable to the wavelength), Fig. 1-17*b* shows that completely circular wavelets are formed, analogous to efficient diffraction.

Fraunhofer diffraction by a slit gives rise to very striking effects, as illustrated in Fig. 1-18.† On a screen or photographic plate, the pattern observed would be alternate light and dark bands (called *fringes*), whereas in

† Fraunhofer diffraction implies the use of lenses, which are present in almost all optical work; if lenses are absent, the term *Fresnel diffraction* is used.

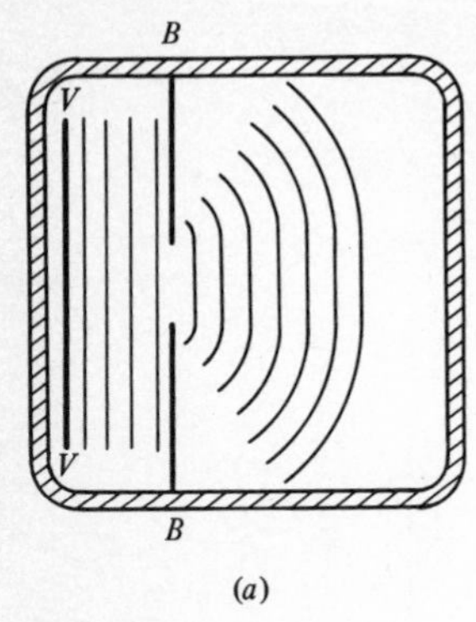

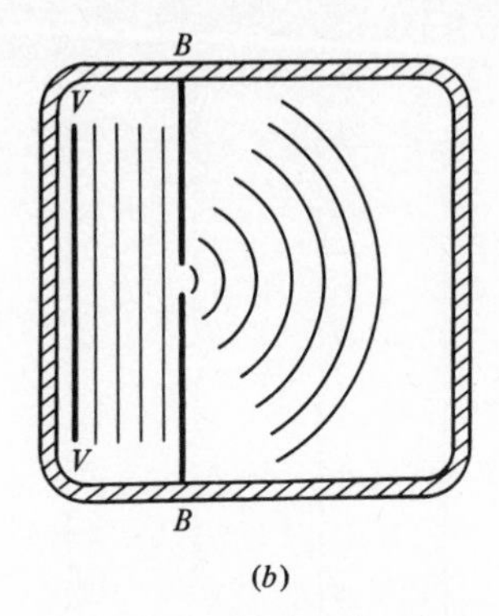

(a) (b)

FIGURE 1-17 Generation of water waves passing through slits of various widths: (a) relatively wide slit; (b) relatively small slit. V is a vibrating bar. (*After W. Bragg, "The Universe of Light," p.* 127, *Dover Publications, Inc., New York,* 1959, *by permission.*)

Fig. 1-18 the relative intensity of the bands of light are shown by the size of the peaks. The central band is of maximum intensity at point P_0 because the interference between wavelets from the slit is *constructive* for the direction directly ahead ($\theta = 0°$), while destructive interference between wavelets causes intensity minima at points P_1, P_2, and P_3.

In Fig. 1-18, the angle θ at which the minimum P_1 occurs can be calculated from the wavelength λ and slit width b as follows:

$$\sin\theta = \frac{\lambda}{b} \tag{1-19}$$

The derivation of Eq. (1-19) is given in Appendix A, Sec. A-1. Equation (1-19) shows that the angle θ becomes smaller when either slit b is widened or the wavelength λ decreases. It follows from Eq. (1-19) that when the slit is large compared with the wavelength, diffraction is no longer important.

The secondary maxima occurring *between* P_1 and P_2 and between P_2 and P_3 are respectively roughly $\frac{1}{22}$ and $\frac{1}{62}$ times as intense as the principal maximum centered at P_0. In optical instruments it is usually advantageous to use only the central (maximum-intensity) band coming through a slit, sacrificing the small amount of energy in the secondary maxima.

OPTIMUM SLIT WIDTH

To take maximum advantage of a prism or diffraction grating for separating a mixture of radiation into its component wavelengths, the entire side of the prism or length of the grating should be illuminated. (This point is further emphasized in Sec. 1-4*F*.) From the previous discussion of diffraction by a slit, it follows that there will be an optimum slit width that will just allow the central (maximum-intensity) band to be focused on the entire length of the prism or grating. Equation (1-19) shows that if the slit width b is too narrow,

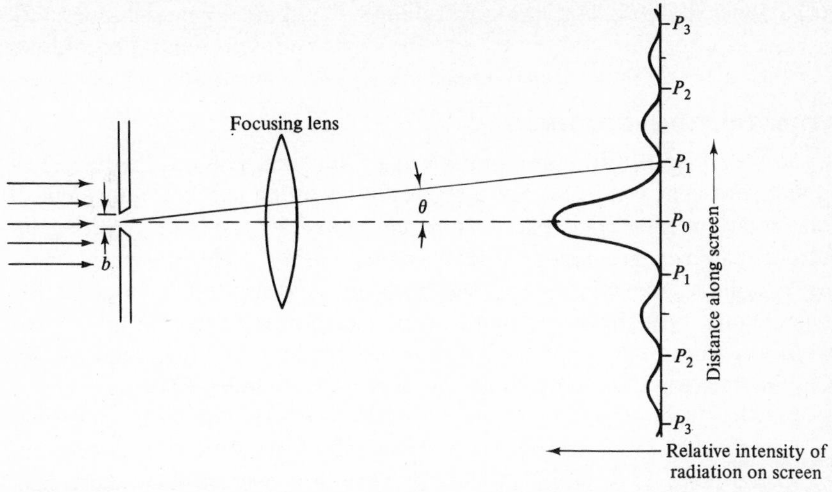

FIGURE 1-18 Fraunhofer diffraction pattern of a narrow slit. The slit width is labeled *b*.

the edges of the central band will miss the prism or grating and be wasted. Similarly, if the slit width *b* is too wide, maximum use of the prism or grating will not be made, since the dark regions adjacent to the central band will also be focused on the dispersion device. The optimum slit width b_{opt} can be calculated from

$$b_{\text{opt}} = \frac{2\lambda d}{w} \tag{1-20}$$

where λ = wavelength to be resolved
d = distance between slit and grating or prism
w = width of grating or prism to be illuminated

Example 1-13 Calculate the optimum slit width for maximum resolution of visible light (400 to 750 nm) using an instrument with a grating 8.0 cm long and a slit-to-grating distance of 0.75 m.

Answer In order to operate at a fixed slit width throughout this wavelength region, the average wavelength should be used in the calculation.

$$\lambda_{\text{av}} = \frac{400 \text{ nm} + 750 \text{ nm}}{2} = 575 \text{ nm}$$

$$b_{\text{opt}} = \frac{2\lambda d}{w}$$

$$= \frac{(2)(575 \text{ nm})(10^{-7} \text{ cm/nm})(0.75 \text{ m})(10^2 \text{ cm/m})}{8.0 \text{ cm}}$$

$$= 1.08 \times 10^{-3} \text{ cm} \approx 0.011 \text{ mm}$$

If a *wider* slit were used, the spectral lines would not be as well separated, whereas a *narrower* slit would give the same optimum separation of lines at the expense of diminished intensity.

DIFFRACTION GRATINGS

The theory of a diffraction grating again involves Huygens' principle of secondary wavelets, but instead of a single slit a grating uses the equivalent of a great many narrow slits spaced at *regular* intervals, and the resulting interference patterns are unique and extremely useful. The arguments that follow assume a *transmission* grating, but they hold as well for a reflection grating. Figure 1-19 shows the maze of intersecting wavelets that are propagated through the various slits of a diffraction grating. Although the diagram looks complicated, it consists simply of sets of semicircles drawn about the centers of the slit openings. For a given slit, the secondary wavelets are propagated directly outward, like the waves through a single slit in Fig. 1-17*b*, but here, where there is a large number of regularly spaced slits, there is a remarkable pattern of constructive interference between the intersecting wavelets. To observe these patterns most easily hold the page of the book nearly on a level with the eye and sight along the following directions:

1 Sight along the imaginary line *EF* and then move the eye slowly toward *AB* to observe the series of wavefronts parallel to *EF* and *AB*. It can be seen that the wave systems emerging through each slit recombine in a short time to re-form the *original linear wavefront*. This parallel wavefront is termed the *zero-order wavefront*.

2 Sight along the imaginary line *BC*, and notice the uniformly spaced wavefronts as the eye moves away from the grating but parallel to *BC*. This parallel wavefront makes up the *first-order wavefront*. (An additional first-order wavefront, having the same angle with the grating as *BC* but moving to the opposite side, can be seen by sighting along the imaginary line *AD* and moving the eye away from the grating but parallel to *AD*.)

3 Sight along the imaginary line *BE* and parallel wavefronts. These make up the *second-order wavefront* (with a symmetrical second-order wavefront moving to the opposite side along line *AF* and parallel fronts).

4 Sight along *BG* (and the symmetrical front *AG*) to see the *third-order wavefronts*. There are no higher-order wavefronts in the figure.

Figure 1-20 identifies the diffraction angles for the first-, second-, and third-order wavefronts in two different ways. The upper quadrant of Fig. 1-20 was extracted directly from Fig. 1-19. The angles which these wavefronts make with the diffraction grating are the diffraction angles θ_1, θ_2, and θ_3, respectively. In the lower quadrant of Fig. 1-20 these same diffraction angles are represented in the more conventional way, i.e., as angles made by the di-

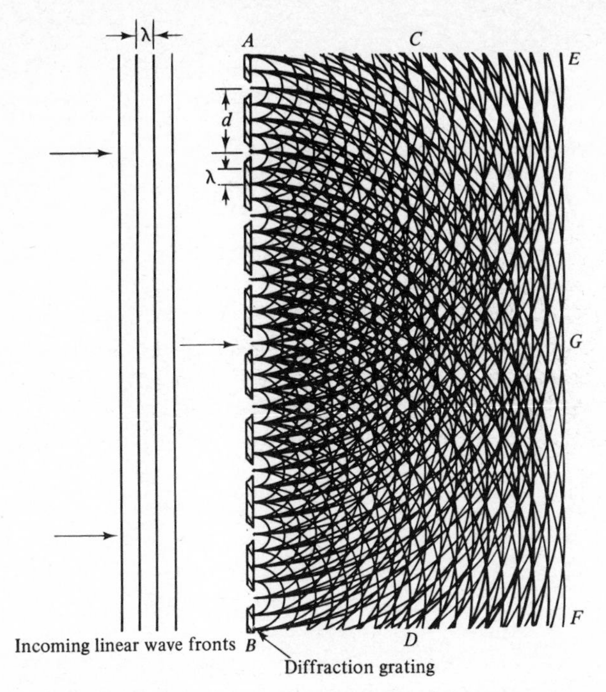

FIGURE 1-19 Diffraction at a grating. (*After W. Bragg, "The Universe of Light," p.* 131, *Dover Publication, Inc., New York,* 1959, *by permission.*)

rection of the wavefronts from the direction of the zero-order wavefront. Since lines *BI, BJ,* and *BK* are perpendicular to lines *BC, BE,* and *BG*, respectively, it follows that the two methods of representing the diffraction angles are equivalent.

The semicircle wavelet representation can be used to illustrate two other important grating principles. Figure 1-21 illustrates the destructive interference that results when the spacing between slits is irregular. Though the zero-order wavefront is unaffected by the irregular spacing, mutual destruction prevents higher-order wavefronts from forming, which can be seen when straight lines are drawn tangent to some of the wavelets. Although a given straight line may be tangent to a *few* wavelets, it goes between most, and in the main there will be mutual destruction of all higher-order wavefronts. Thus, it is a requirement of a good diffraction grating that the spacing between slits (or rulings) be equal along the barrier. This regular spacing, identified in Fig. 1-19 as the distance *d*, is called the *grating constant*.

In the discussions thus far it has been implicitly assumed that the radiation incident on the gratings is *monochromatic*. The more important case of polychromatic radiation is illustrated in Fig. 1-22, where two different wavelengths, λ_1 and λ_2, are shown impinging upon a diffraction grating. As can be

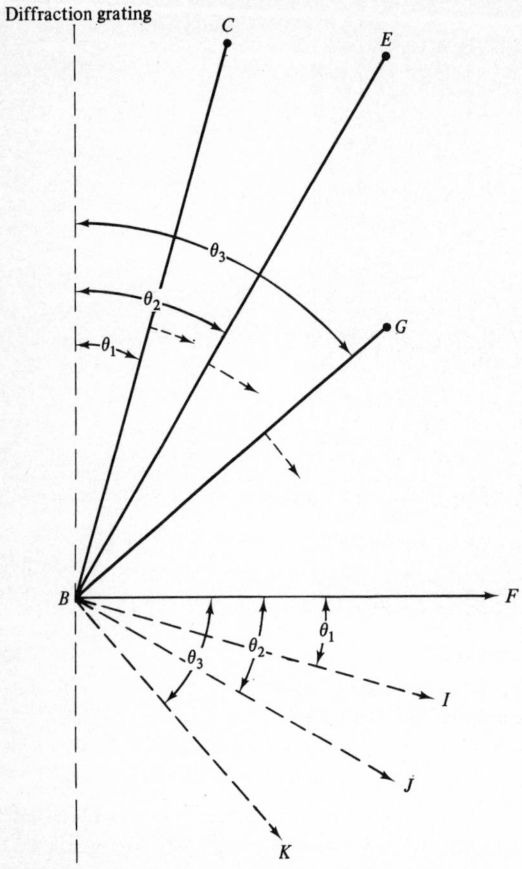

FIGURE 1-20 Diffraction angles for first-, second-, and third-order diffraction waves.

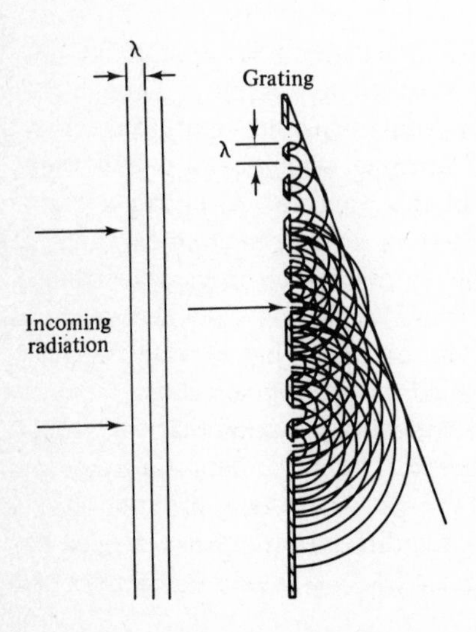

FIGURE 1-21 Diffraction grating with irregularly spaced slits. (*After W. Bragg "The Universe of Light," p.* 133, *Dover Publications, Inc., New York,* 1959, *by permission.*)

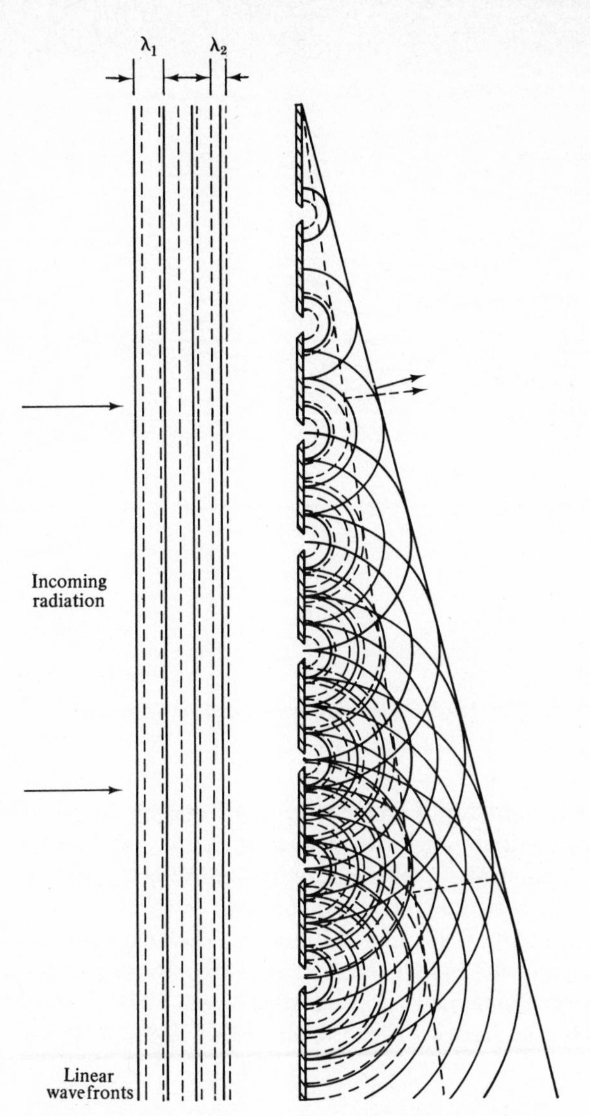

FIGURE 1-22 Radiation of two different wavelengths passing through a diffraction grating. (*After W. Bragg, "The Universe of Light," p.* 133, *Dover Publications, Inc., New York,* 1959, *by permission.*)

seen, longer wavelengths, as exemplified by λ_1, are diffracted through wider angles than shorter wavelengths, exemplified by λ_2. This separation of mixed wavelengths is the basis of the most important applications of diffraction gratings.

Grating Equation Figure 1-23 illustrates the usual optical arrangement of a grating, giving Fraunhofer diffraction. The diagram is simplified by

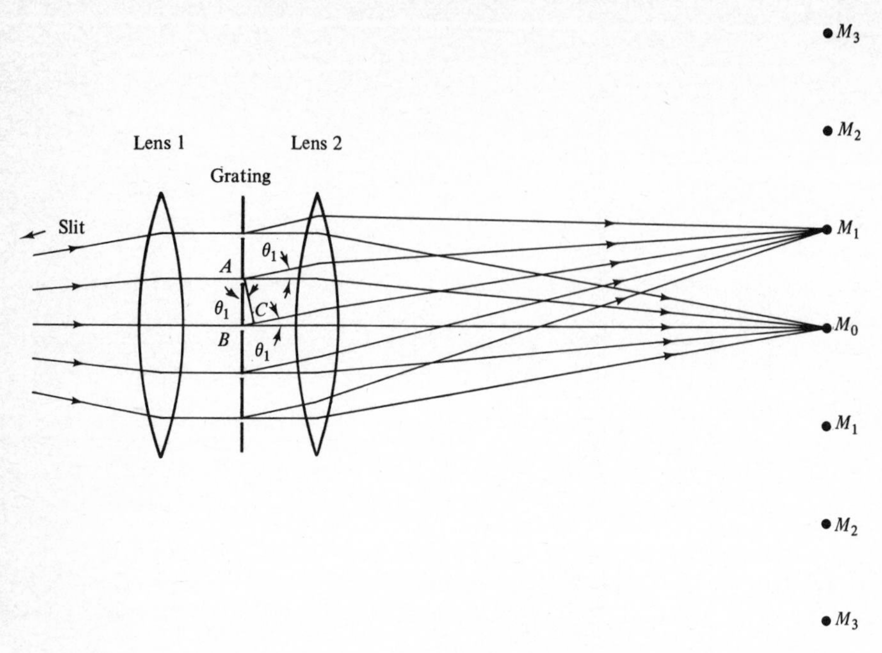

FIGURE 1-23 Arrangement for Fraunhofer diffraction at a grating. (*After A. C. Hardy and F. H. Perrin, "The Principles of Optics," p. 559, McGraw-Hill Book Company, New York, 1932, by permission.*)

assuming that monochromatic radiation is incident upon the grating. Some of this radiation will of course continue on in its original direction and will be brought to focus by lens 2 at point M_0. Additional radiation will be diffracted at an angle θ_1 with the original direction and will be brought to focus at point M_1. These latter rays represent the first-order diffraction. The condition for *constructive interference* at point M_1 is that the difference in path lengths for two diffracted rays be an *integral* number of wavelengths.

For paths AM_1 and BM_1, let length BC represent *one* wavelength. From the geometry of Fig. 1-23,

$$BC = AB \sin \theta_1 \tag{1-21}$$

and since AB is the grating constant spacing d, it follows that

$$\lambda = d \sin \theta_1 \tag{1-22}$$

Constructive interference would also occur if the distance BC were equal to 2λ, 3λ, or any other integral number of wavelengths, leading to second-, third-, or higher-order reinforcement at points M_2, M_3, and other points not shown.

Thus, the general condition for reinforcement is

$$m\lambda = d \sin \theta \tag{1-23}$$

where m is any integer and is the *order* of the diffraction.

If polychromatic radiation is incident upon the grating, it is apparent that radiation of all wavelengths will be combined at M_0 to form an image identical with the source. For first-order diffraction ($m = 1$) it is clear that since θ varies with λ, two spectra will be formed on opposite sides of M_0, with short wavelengths focused closer to M_0 than longer wavelengths. For $m = 2$, a second pair of spectra will be formed, the corresponding portions of which lie at greater distances from M_0 than those of first order. Figure 1-24 illustrates this for a visible (white) light source. Four representative wavelengths from the visible region are shown, focused on the screen, whereas the actual spectrum of white light would contain all the wavelengths in between those shown.

Example 1-4 A grating 2.0 cm long contains 1000 uniformly spaced slits. Calculate the arc, in degrees, subtended by the first-order spectrum of visible light from 400 to 750 nm.

Answer

$$d = \frac{2.0 \text{ cm}}{1000} = 0.0020 \text{ cm}$$

For 400-nm light, Eq. (1-23) gives

$$\sin \theta = \frac{400 \times 10^{-7} \text{ cm}}{0.0020 \text{ cm}} = 0.020$$

$$\theta = 1°9'$$

For 750 nm,

$$\sin \theta = \frac{750 \times 10^{-7} \text{ cm}}{0.0020 \text{ cm}} = 0.0375$$

$$\theta = 2°9'$$

Therefore, the entire first-order spectrum covers approximately 1° of arc. Through a telescope with a magnifying power of 10, it subtends an angle of 10°, which would amount to a length of about 3 in as viewed from a distance of 20 in.

If the grating contains 10,000 slits instead of the specified 1000, it can be shown with the same sort of calculation that the first-order visible

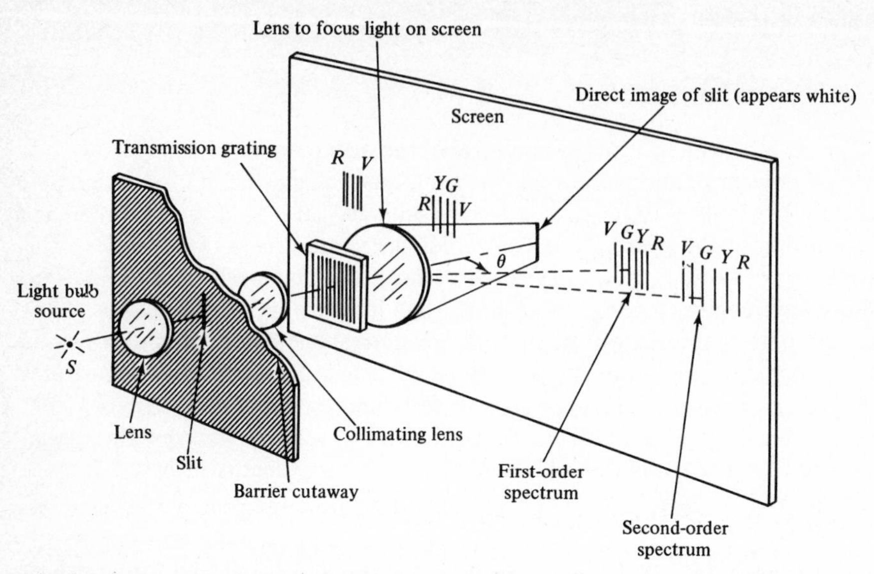

FIGURE 1-24 Experimental setup for producing diffraction spectra of white light; V = violet, G = green, Y = yellow, R = red. (*After K. N. Ogle, "Optics,"* p. 15, 1961. *Courtesy of Charles C Thomas, Publisher, Springfield, Illinois.*)

spectrum extends from 11°32′ to 22°0′, an arc of 10°28′. Thus, a tenfold decrease in the grating constant spreads the spectrum out more than tenfold.

The *overlapping of orders* can be a problem with diffraction gratings if the wavelength range to be separated is large. For example, if our source in Fig. 1-24 gives off ultraviolet as well as visible light, the grating equation (1-23) predicts that the yellow line at 600 nm in the first order coincides exactly with the line at 300 nm in the second order, 200 nm in the third order, etc., i.e., has the same diffraction angle θ.

Example 1-15 Suppose a diffraction grating is being used with a visible light source and a certain red line of 700 nm is being observed in the third-order spectrum. Is there any danger that other orders of visible light will overlap?

Answer Equation (1-23) predicts that for a given grating the angle of diffraction will be the same for wavelengths satisfying the general equation

$$m_1\lambda_1 = m_2\lambda_2$$

Therefore,

$$(3)\,(700\text{ nm}) = m_2\lambda_2$$

$$\lambda_2 = \begin{cases} \dfrac{2100\text{ nm}}{4} = 525\text{ nm} & m_2 = 4 \\[2ex] \dfrac{2100\text{ nm}}{5} = 420\text{ nm} & m_2 = 5 \\[2ex] \dfrac{2100\text{ nm}}{6} = 350\text{ nm} & m_2 = 6 \end{cases}$$

(350 nm is not in the visible region.) Therefore, there would be some overlap of the 700-nm line by light of 525- and 420-nm wavelength.

Although the overlap of spectra can be troublesome, it is easily avoided in practice by using suitable filters to absorb from the incident light those wavelengths which would overlap the region under study. For example, ordinary glass can be used to filter out ultraviolet radiation if only visible light is of interest. Also reducing the seriousness of overlap is the fact that the amount of energy distributed into higher orders is smaller than that going into the first. Furthermore, it is possible to rule gratings to greatly enhance the distribution of energy in the low orders. For example, the *echelette ruling* concentrates 80 percent of the diffracted light into the first two orders.

General Grating Equation For completeness, it should be pointed out that it is possible to make allowance for radiation striking a grating at an angle other than perpendicular. If radiation strikes the grating at an angle i from the perpendicular, then Eq. (1-23) is converted to the more general grating equation

$$m\lambda = d(\sin\theta + \sin i) \tag{1-24}$$

In the more usual case, $i = 0°$, and Eq. (1-24) reduces to Eq. (1-23).

Types of Gratings Gratings for use in (or near) the visible region are ruled with from 10,000 to 30,000 lines per inch, and their construction is no simple matter. The two basic types are *transmission* gratings, of the type we have been discussing, and *reflection* gratings, which are more widely used. Transmission gratings are made by ruling fine lines on a transparent surface (such as glass, in the visible region) with a diamond point. The roughened spaces produced by the ruling form the opaque regions of the grating.

Reflection gratings are made in the same way by ruling on a polished, opaque surface with a diamond point. The light reflected from the narrow rulings (or the spaces *between* them) interferes to produce maxima and minima in exactly the same way as the light transmitted by a transmission grating. (Since Huygens' principle states that secondary wavelets, forming at every point on a wavefront, are propagated in *all directions*, the previous arguments for transmission gratings hold as well for reflection gratings.)

If a grating is ruled on a concave mirror, the grating can serve as its own collimating mirror, producing sharp spectral lines without the use of lenses. This can be an important advantage, e.g., in the ultraviolet region, where glass and some other common optical materials absorb strongly.

Because of the many difficulties attendant upon the production of a perfect, ruled grating (slight defects in rulings cause "ghost" spectra, which are faint, false lines on each side of an intense one), there are only a relatively small number of *original* ruled gratings in existence. For many purposes, *replica* gratings, the manufacture of which has been brought to a high degree of perfection, are perfectly satisfactory. Both transmission and reflection replica gratings are available. For example, a collodion impression of the rulings of an original grating can be made by pouring a thin layer of collodion solution over the surface of the ruled grating and allowing it to harden. The hardened collodion film is then stripped from the grating and placed between glass plates (to form a *transmission* grating) or mounted against a silvered surface (to form a *reflection* grating). Any number of replicas can be made from a single original.

X-RAY DIFFRACTION

The theory of diffraction can easily be extended to x-ray diffraction studies of crystals by viewing atoms in a crystal as apertures or scattering centers in a two- or three-dimensional grating array. This will be done in Chap. 12. The important point to realize here is that with *any* grating, if the spacing or grating constant d becomes equal to or less than the wavelength of light incident upon it, the grating will act like a semitransparent plate without any structure. Under these conditions the grating cannot be used as a device for separating wavelengths, and in fact we have reached the limit of our ability to view the structure of matter (the so-called *microscopic limit*).

Example 1-16 Show that for radiation normal to a diffraction grating the smallest grating spacing that can be used is a spacing equal to the wavelength.

Answer For incident radiation that is normal to a diffraction grating, the grating equation is

$$m\lambda = d \sin \theta \tag{1-23}$$

If $d = \lambda$, $\sin\theta = 1$ for first-order diffraction and is greater than 1 (making θ indeterminate) for higher orders. When $\sin\theta = 1$, $\theta = 90°$, which means that the diffracted radiation is totally deflected, or glances along the surface of the grating. Only the zero-order diffraction (which according to the grating equation is $\sin\theta = 0$, or $\theta = 0°$) is transmitted, and without deviation. Obviously, for d less than λ, θ is indeterminate even for first order, and again only zero order is transmitted. Thus, an object ceases to act like a grating and becomes transparent when wavelengths equal to or larger than the spacing are used.

Example 1-17 A block of rock salt has a lattice spacing (distance between sodium ions or chloride ions) of about 5×10^{-8} cm. (*a*) Will the rock salt act as a diffraction grating for ultraviolet light of 200 nm? (*b*) What wavelength of light must be used to study the lattice spacings in rock salt?

Answer (*a*)

$$\lambda = 200 \text{ nm} = 200 \times 10^{-7} \text{ cm}$$

$$\text{Atomic spacing } d = 5 \times 10^{-8} \text{ cm} = 0.5 \times 10^{-7} \text{ cm}$$

$$m\lambda = d \sin\theta$$

$$\sin\theta = \frac{m\lambda}{d} = \frac{(1)(200 \times 10^{-7} \text{ cm})}{0.5 \times 10^{-7} \text{ cm}} = 400$$

However, a sine value greater than 1 is meaningless. Since λ is larger than d, $\sin\theta$ will be indeterminate for all but zero order. Therefore, the rock salt will be transparent to the radiation and will not act as a diffraction grating.

(*b*) To study the lattice spacings, the wavelength used must be smaller than the spacings. Radiation smaller than 5×10^{-8} cm (5 Å) is in the x-ray region (see Fig. 1-4).

1-4F Dispersion

Dispersion can be defined as breaking apart or separating a mixture of wavelengths into its component wavelengths. This can be accomplished with a *prism* using the phenomenon of refraction or with a *grating* using the phenomenon of diffraction. Each of these two methods will be described. An optical method based on dispersion, called *optical rotatory dispersion*, will be discussed in Chap. 10.

DISPERSION BY A PRISM

From the principles of refraction discussed in Sec. 1-4*A*, it follows that if we send radiation through a block of dielectric material having parallel faces, the rays emerging will be parallel to the incident rays. On the other hand, if the block of material has nonparallel faces, the direction of the refracted beam will be changed. Furthermore, since the velocity of light in a medium other than a vacuum is different for different wavelengths, i.e., the refractive index of a substance varies with wavelength, each component in a mixture of wavelengths will follow its own path and thus the degree of bending will vary with wavelength, allowing *dispersion* of a mixture of wavelengths. This is illustrated in Fig. 1-25 with a mixture of red and blue light incident upon a glass prism, and Table 1-4 illustrates how the index of refraction varies with the wavelength of visible light. Notice in Fig. 1-25 that shorter wavelengths (like blue) are refracted more than longer wavelengths (like red).

The ability of a prism to disperse radiation is often characterized by the derivative $d\theta/d\lambda$, called the *angular dispersion*, which is a measure of the angular separation of two light rays which differ in wavelength by an amount $d\lambda$. This term can be visualized from Fig. 1-25 as

$$\frac{d\theta}{d\lambda} \approx \frac{\Delta\theta}{\Delta\lambda} = \frac{\theta_2 - \theta_1}{\lambda_2 - \lambda_1} \tag{1-25}$$

which of course is valid only if the increments $\Delta\theta$ and $\Delta\lambda$ are small. It is informative to resolve the angular dispersion into the product of two factors

$$\frac{d\theta}{d\lambda} = \frac{d\theta}{dn}\frac{dn}{d\lambda} \tag{1-26}$$

where n is the refractive index of the prism material. The first factor, $d\theta/dn$, can be evaluated from geometrical considerations alone (either in terms of the angles of the prism faces or their lengths), and no more need be said of it. Of much greater interest is the factor $dn/d\lambda$, which is a characteristic property of the prism material and is usually referred to simply as the *dispersion* of the prism. The variation of n with λ is shown in Fig. 1-26 for several different materials commonly used for lenses and prisms. The refractive index increases as

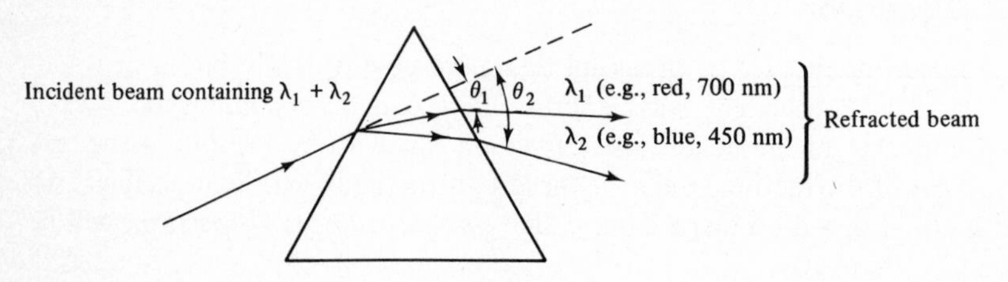

FIGURE 1-25 Dispersion of light by a glass prism.

TABLE 1-4 Variation of Index of Refraction of Crown Glass with Wavelength of Visible Light

Wavelength, nm	Color	Index of refraction
400	Violet	1.532
450	Blue	1.528
550	Green	1.519
590	Yellow	1.517
620	Orange	1.514
750	Red	1.513

the wavelength decreases for all these materials, which explains why short wavelengths are refracted to a greater degree than longer wavelengths. The *rate* of increase becomes greater at shorter wavelengths, which is another way of saying that the dispersion $dn/d\lambda$, which is the slope of the curve, neglecting the negative sign, increases with decreasing wavelength. Thus, the spectrum produced by a prism will spread out on a much larger scale at the violet end of the spectrum than at the red end. A prism, then, has the disadvantage of giving *nonlinear* dispersion; i.e., the dispersion varies nonlinearly with wavelength.

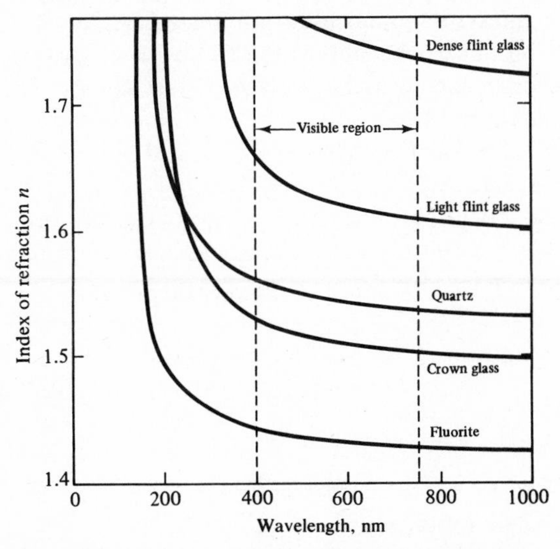

FIGURE 1-26 Variation of refractive index with wavelength for some common optical materials. (*After F. A. Jenkins and H. E. White, "Fundamentals of Optics,"* 3d *ed., p.* 466, *McGraw-Hill Book Company, New York,* 1957, *by permission.*)

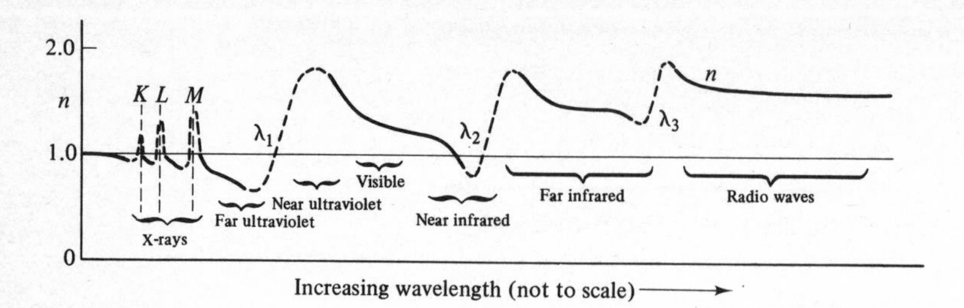

FIGURE 1-27 Variation of refractive index with wavelength over the entire electromagnetic spectrum. The material is glass. (*After F. A. Jenkins and H. E. White, "Fundamentals of Optics," 3d ed. p.* 478, *McGraw-Hill Book Company, New York,* 1957, *by permission.*)

The variation of refractive index with wavelength over the entire electromagnetic spectrum is shown in Fig. 1-27 for a substance (such as glass) which is transparent in the visible region. The dotted portions of Fig. 1-27 show several discontinuities in the dispersion curve, where the dispersion is *anomalous*, in contrast to the *normal dispersion* elsewhere. Anomalous dispersion occurs in regions where the substance shows *absorption* bands. For example, the K, L, and M regions of Fig. 1-27 correspond to absorption of x-radiation of sufficient energy to cause ejection of K-, L-, and M-shell electrons in the optical material. Similarly, the λ_1 region of Fig. 1-27 corresponds to a broad region of strong absorption and anomalous dispersion in the ultraviolet region. Somewhere in the near infrared another absorption band will be encountered at λ_2; for example, with quartz this absorption begins to be strong at 4 or 5 μm. At longer wavelengths, narrow and weak absorption regions are found in the radio frequencies, as exemplified by λ_3.

A quantitative relationship between anomalous dispersion and absorption bands is discussed in Appendix A, Sec. A-2.

Resolving Power of a Prism A dispersion device is said to resolve two wavelengths when they appear as clearly separated lines. A useful working definition of resolving power is given by

$$R = \frac{\lambda}{\Delta\lambda} \tag{1-27}$$

where λ is the average wavelength of the closest pair of lines that are resolvable and $\Delta\lambda$ is their wavelength difference.

Example 1-18 What is the resolving power of a prism that will just resolve the sodium D lines at 589.0 and 589.6 nm?

Answer

$$R = \frac{\lambda}{\Delta\lambda} = \frac{589.3 \text{ nm}}{0.6 \text{ nm}} = 980$$

Example 1-19 A certain prism has a resolving power of 6000 at 3500 Å. How close together can a pair of spectral lines be at this wavelength and still be resolved?

Answer

$$R = \frac{\lambda}{\Delta\lambda}$$

$$\Delta\lambda = \frac{\lambda}{R} = \frac{3500 \text{ Å}}{6000} \approx 0.6 \text{ Å}$$

Thus, a pair of lines at 3499.7 and 3500.3 Å can be resolved.

The ultimate limit of resolution is governed by the dispersing element itself. For a prism, the resolving power can be predicted by the equation

$$R_{\text{prism}} = b\frac{dn}{d\lambda} \tag{1-28}$$

where b is the thickness of the base of the prism and $dn/d\lambda$ is the dispersion of the prism material. It should be noted that the above equation assumes that the entire side face of the prism is illuminated by radiation from the slit. If this is not the case, the thickness b of the prism base must be replaced by $b - t$, the *effective thickness*, where b and t are the thicknesses of the prism material through which the two extreme rays of the beam pass. This modification is consistent with Eq. (1-28), since when the entire side of the prism face is illuminated, b is then the actual base thickness and t goes to zero since it is at the apex of the prism.

Equation (1-28) serves as a guide for selecting a prism which will give optimum resolution. From Eq. (1-28) it is clear that larger prisms (larger b) will provide better resolution, but it turns out that the choice of prism material (because of the factor $dn/d\lambda$) is even more important. The material selected should have a large dispersion in the wavelength region of interest, but the *paradox* of prism materials is that the dispersion is greatest as absorption by the material increases, and thus a compromise must often be made between the amount of dispersion and the amount of absorption, which results in loss of energy.

Example 1-20 A certain glass prism with a 3.0-cm base has a dispersion $dn/d\lambda$ of 2.70×10^{-4} per nanometer in the blue region (450 nm) of the visible spectrum. Calculate the resolving power of the prism at 450 nm.

Answer

$$R = b\frac{dn}{d\lambda}$$

$$= (3.0\ \text{cm})\,(2.70 \times 10^{-4}\ \text{nm}^{-1})\,(10^{7}\ \text{nm/cm}) = 8100$$

Example 1-21 The dispersion $dn/d\lambda$ of a certain light flint glass is 5000 cm^{-1} at 400 nm (violet light) and 1000 cm^{-1} at 589 nm (yellow light), whereas the dispersion of a certain quartz at those same wavelengths is 1200 and 400 cm^{-1}, respectively. How much better is the resolving power of the flint glass than the quartz when used as prism materials at each of the two wavelengths?

Answer For prisms the same size

$$\frac{R_1}{R_2} = \frac{(dn/d\lambda)_1}{(dn/d\lambda)_2}$$

$$\text{At 400 nm:}\quad \frac{R_{\text{glass}}}{R_{\text{quartz}}} = \frac{5000\ \text{cm}^{-1}}{1200\ \text{cm}^{-1}} = 4.2$$

$$\text{At 589 m}\mu\text{:}\quad \frac{R_{\text{glass}}}{R_{\text{quartz}}} = \frac{1000\ \text{cm}^{-1}}{400\ \text{cm}^{-1}} = 2.5$$

Example 1-22 As Example 1-21 indicates, light flint glass would appear superior to quartz as a prism dispersion material, and the advantage of flint glass seems to increase toward shorter wavelengths. However, quartz is always chosen for prisms to be used in the ultraviolet region down to 200 nm. Explain.

Answer As Fig. 1-26 indicates, the dispersion (and more importantly, the *absorption*) by glass increases greatly in the ultraviolet region, whereas the absorption by quartz is not too severe as low as 200 nm. Therefore, glass is not used in the ultraviolet region because it would block too much of the radiant energy by absorption. As Fig. 1-26 shows, quartz is a very good prism material for the ultraviolet region down to 200 nm because its dispersion and resolving power are high in this region.

DISPERSION BY A GRATING

As for a prism, the ability of a diffraction grating to disperse radiation is often characterized by the *angular dispersion* $d\theta/d\lambda$. Although the angular dispersion for a prism is difficult to calculate from purely theoretical considerations, the angular dispersion for a grating is readily evaluated using the general grating equation

$$m\lambda = d(\sin\theta + \sin i) \tag{1-24}$$

If Eq. (1-24) is differentiated with respect to λ (realizing that i is independent of wavelength), an expression for the angular dispersion results:

$$\frac{d\theta}{d\lambda} = \frac{m}{d\cos\theta} \tag{1-29}$$

Equation (1-29) reveals several important features of a diffraction grating. (1) For a given small wavelength difference $d\lambda$ the angular separation $d\theta$ is directly proportional to the order m, and hence the second-order spectrum will be twice as wide as the first order, the third 3 times as wide as the first, etc. (2) The angular separation $d\theta$ is inversely proportional to the grating constant d, and thus the more rulings (or the smaller the spacing) the more widely spread the spectra will be. (3) Most important, the angular dispersion will be constant for a given order, assuming only that $\cos\theta$ is reasonably constant. Since $\cos\theta$ is 1 for $\theta = 0°$ and decreases only slowly as θ increases, $\cos\theta$ will be substantially constant and close to unity for the small range of diffraction angles used in practice. Thus, neglecting this very small change in $\cos\theta$, the spectral dispersion for a grating will be *linear*; i.e., the spacing between spectral lines in a given order will be directly proportional to the difference in wavelength. This feature constitutes one of the most important advantages of gratings over prisms. The comparative spectral-line spacings for a prism and a grating are shown in Fig. 1-28. Whereas the spectral lines from a grating are uniformly spaced with wavelength, the lines from a prism become more crowded together toward longer wavelengths.

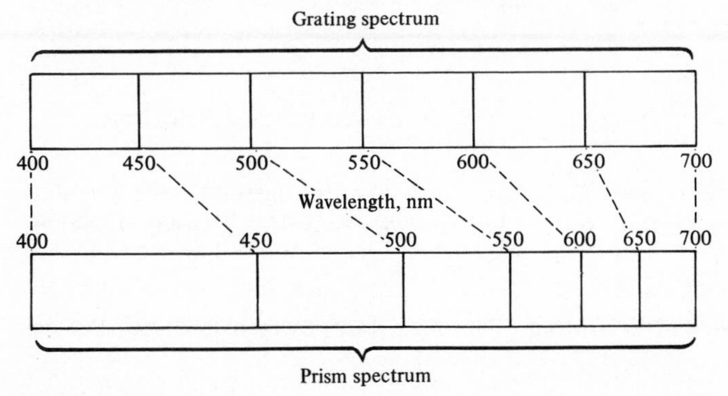

FIGURE 1-28 Comparison of the spacing of spectral lines for a prism and a grating.

Example 1-23 The first-order spectrum produced by a grating having 5000 lines distributed uniformly over its 1-in length has a certain spectral line falling at an angle 4° from the zero-order image of the slit. Identify the wavelength of that line. Assume that the radiation is incident normally on the grating.

Answer

$$m\lambda = d(\sin\theta + \sin i)$$

but since $i = 90°$,

$$m\lambda = d\sin\theta$$

$$d = \frac{2.54\ \text{cm}}{5000} = 5.08 \times 10^{-4}\ \text{cm}$$

$$\sin 4° = 0.0698$$

$$\lambda = d\sin\theta = (5.08 \times 10^{-4}\ \text{cm})(\sin 4°)$$

$$= 0.356 \times 10^{-4}\ \text{cm} = 356\ \text{nm}$$

With a grating, all that is needed to identify a given wavelength is the grating constant d and the angle of diffraction θ. With a prism, no such simple relationship exists, since the angle of deviation is not related in any simple way to the wavelength but depends on the varying dispersion characteristics of the material from which the prism is constructed and its geometrical characteristics.

Example 1-24 Using the data in Example 1-23, calculate the angular dispersion for the first-order spectrum from this grating.

Answer

$$\text{Angular dispersion} = \frac{d\theta}{d\lambda} = \frac{m}{d\cos\theta}$$

$$\frac{d\theta}{d\lambda} = \frac{1}{5.08 \times 10^{-4}(\cos 4°)} = 0.197 \times 10^{4}\ \text{deg/cm}$$

Again, note how few data are necessary to calculate the angular dispersion of a grating. A comparable calculation for a prism would be very involved.

Resolving Power of a Grating The resolving power of a dispersion device was defined as

$$\text{Resolving power} = R = \frac{\lambda}{\Delta\lambda} \tag{1-27}$$

where λ is the average wavelength of the closest pair of lines that are resolvable and $\Delta\lambda$ is their wavelength difference. For a grating, the resolving power can be predicted by the simple equation

$$R_{\text{grating}} = mr \tag{1-30}$$

where m is the spectral order and r is the number of rulings on the grating. As with prisms, it should be noted that Eq. (1-30) assumes that the entire length of the grating is illuminated by radiation from the slit. If this is not the case, the *effective* number of rulings r' should be substituted for r, where r' is the number of rulings actually under illumination from the slit. Thus, the resolving power of a grating depends only on the product of the order and the number of rulings, and not on the wavelength or grating spacing.

Example 1-25 How many rulings must a grating have to resolve two lines at 742.2 and 742.4 nm in the first-order spectrum?

Answer

$$R = \frac{\lambda}{\Delta\lambda} = mr$$

$$r = \frac{742.3 \text{ nm}}{0.2 \text{ nm}} = 3710$$

Thus, a grating with only 3710 rulings will theoretically have sufficient resolving power.

COMPARISON OF RESOLVING POWER OF PRISMS AND GRATINGS

Although gratings have been ruled with more than 80,000 lines per inch, the upper limit of commercially available gratings is about 30,000 lines per inch and a commonly used ruling is about 15,000 lines per inch. Thus, a typical grating might be 4 in long and [from Eq. (1-30)] have a resolving power R of (4 in)(15,000 in^{-1})m, or $6.0 \times 10^4 m$ at any wavelength that can be diffracted. (Higher-order spectra can be used to improve resolution.) For comparison, a large 60° crystalline-quartz prism 4 in on a side would have a resolving power which varies with wavelength, being about 1.5×10^5 at 200 nm, 1.3×10^4 at 400 nm, and 2.0×10^3 at 800 nm. Thus, in general, gratings tend to give superior resolution, except perhaps in regions where prisms have their highest dispersion. Finally, it should be noted that the resolving powers for prisms and gratings, as calculated from Eqs. (1-28) and (1-30), are theoretical *maximum* resolving powers, corresponding to perfect alignment and (for gratings) perfectly regular spacing of rulings. The resolving power attained in practice may be significantly less.

1-4G Polarization

Most radiation sources produce electromagnetic disturbances in which the vibrations of electric and magnetic vectors occur at *all* orientations perpendicular to the direction of travel, as represented by the end-on view in Fig. 1-29*a*. Such *nonpolarized* radiation can be represented more simply by hypothetically resolving the infinite number of vibrational planes into two major planes by vector addition, with the result illustrated in Fig. 1-29*b*. If the amplitude of the vectors becomes unsymmetrical about the direction of travel, the radiation is polarized and the elimination of all but one plane results in *plane polarization*.

It is important to understand the concept of polarization because it results from numerous types of interactions between matter and ordinary unpolarized radiation. Furthermore, these interactions are involved in polarimetry and optical rotatory dispersion, discussed in Chap. 10. The principal ways of producing plane-polarized radiation are reflection, dichroism, double refraction, and scattering.

POLARIZATION BY REFLECTION

Both reflected and refracted rays are partially polarized, except when radiation is perpendicular to the material medium. Figure 1-30*a* schematically illustrates this for ordinary visible light incident at an angle θ. In the simplified side view of radiation shown in Fig. 1-30*a* and *b*, dots present vibrations in and out of the plane of the paper, and line segments represent vibrations up and down in the plane of the paper. The size of the dots and the length of the lines represent the relative amplitude of the respective vibrations;

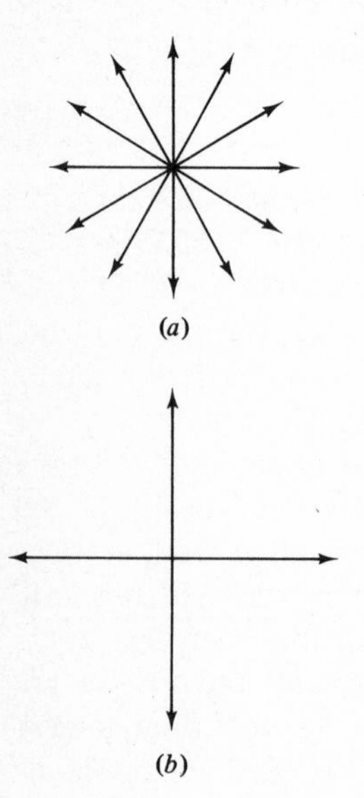

FIGURE 1-29 Vector representation of unpolarized radiation: (*a*) end view illustrating vectors in all directions; (*b*) simplified end view after resolution of vectors into the two major planes.

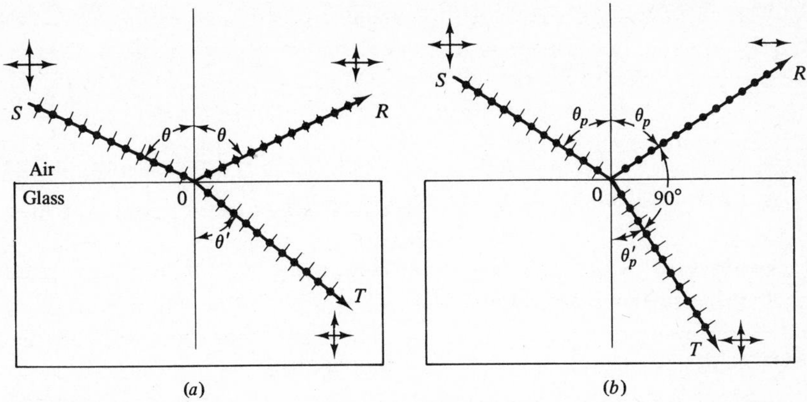

FIGURE 1-30 Polarization by reflection and refraction. (*a*) Light incident at any angle θ (except 0, 90°, and θ_p). (*b*) Light incident at angle θ_p (about 56°). (*After F. A. Jenkins and H. E. White, "Fundaments of Optics," 3d ed., p.* 491, *McGraw-Hill Book Company, New York,* 1957, *by permission.*)

these amplitudes are further clarified by the end-on views inserted near each of the three rays. As Fig. 1-30*a* indicates, when ordinary unpolarized light is incident at an angle θ to glass, there will always be a reflected ray *OR* and a refracted ray *OT*, both partially polarized. (The component of light which is vibrating in and out of the plane of the paper is the one that is more completely *reflected*, whereas the component vibrating up and down on the plane of the paper is selectively refracted.) Polarization by reflection is always more efficient than polarization by refraction, and there is an angle of incidence θ_p, called the *polarizing angle*, at which the reflected ray is completely plane-polarized, as shown in Fig. 1-30*b*. The polarizing angle for a given material can be calculated from its refractive index, which can be shown by a simple derivation based on Fig. 1-30*b*. First, from the definition of refractive index given in Sec. 1-4*A*,

$$n = \frac{\sin \theta}{\sin \theta'} \qquad (1\text{-}11)$$

where n is the refractive index and θ is any angle of incidence (Fig. 1-30*a*). The condition necessary to make $\theta = \theta_p$ is that the reflected and refracted rays be 90° apart or, as shown by Fig. 1-30*b*, that the angle *ROT* be 90°. Under these conditions θ_p and θ_p' are complementary angles; that is, $\theta_p + \theta_p' = 90°$, which means that $\sin \theta_p' = \cos \theta_p$, giving

$$\frac{\sin \theta_p}{\sin \theta_p'} = \frac{\sin \theta_p}{\cos \theta_p} = \tan \theta_p$$

or

$$n = \tan \theta_p \tag{1-31}$$

Thus, the angle of incidence for maximum polarization depends only on the refractive index. Angle θ_p is called the *Brewster angle.*

Example 1-26 Calculate the polarizing angle θ_p for ordinary glass having a refractive index n_D of 1.50.

Answer

$$n = \tan \theta_p$$

$$1.50 = \tan \theta_p$$

$$\theta_p = 56.3°$$

Although this polarizing angle holds exactly only with yellow light (the sodium D line having been used to measure the refractive index), for ordinary glass the refractive index does not change much over the visible region (Table 1-4), and thus the polarizing angle for glass is relatively constant at about 56° throughout the visible region. (See Exercise 1-8 at the end of the chapter.)

In reflection off smooth metals there is also partial polarization, but the polarizing effect is not as great as with dielectrics, and there is no angle at which complete plane polarization occurs. Thus, there is no polarizing angle for metals.

POLARIZATION BY DICHROISM

Dichroism, a property exhibited by a number of crystalline materials, consists of a selective *absorption* of one vibrational plane of radiation. This phenomenon is illustrated in Fig. 1-31 for a crystal which absorbs horizontal vibrations strongly and vertical vibrations only slightly. Thus, if the crystal is cut to the proper thickness, complete plane polarization is possible.

The selective absorption is due to parallel alignment of absorbing groups. One of the best known examples of a manufactured dichroic polarizer is Polaroid, developed by Land in 1934. Although the early types contained embedded crystals, the later types are made from polyvinyl alcohol film, which has exceedingly long molecules. When a sheet of this film is stretched

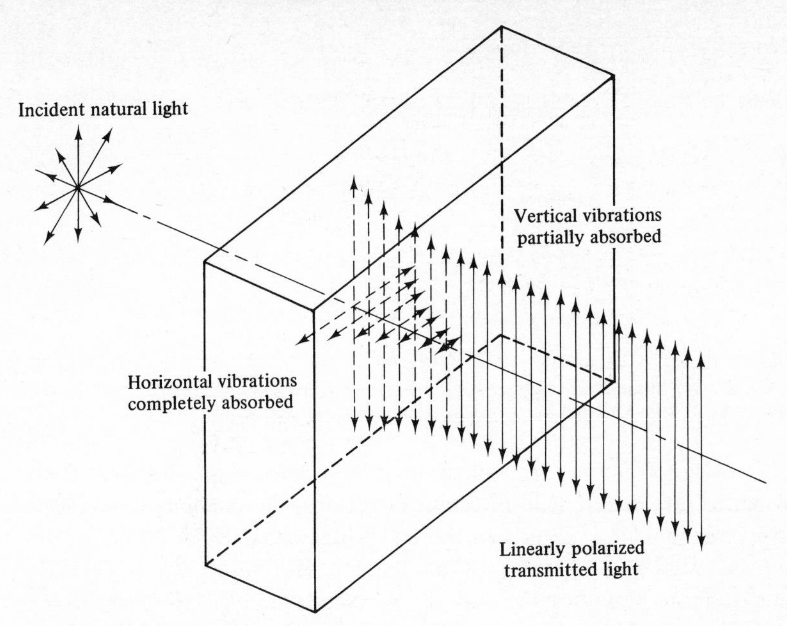

FIGURE 1-31 Polarization of light by a dichroic crystal. (*After F. W. Sears, "Optics," 3d ed., p.* 182, *Addison-Wesley Publishing Company, Reading, Mass.*, 1949, *by permission.*)

mechanically in one direction, the long, tangled molecules straighten out and align themselves in the direction of the stretch. After impregnation with iodine the iodine molecules are aligned with the polymer molecules, where they serve as absorption sites for components of light vibrating parallel to the bond direction.

POLARIZATION BY DOUBLE REFRACTION

From the standpoint of optical instruments, this way of producing polarization is one of the most important. It affects the design of prisms and lenses and permits the design of precision instruments for measuring polarized light, e.g., using double-refracting nicol prisms.

Double refraction is a phenomenon shown by a number of crystalline materials, two of the most important being calcite, $CaCO_3$, and quartz crystals, SiO_2. When ordinary unpolarized radiation passes through a crystal with this property, the beam of radiation is *split* into two separate diverging beams of equal power and the two beams are plane-polarized at right angles to each other. This effect is illustrated in Fig. 1-32 for ordinary unpolarized light perpendicularly incident upon a block of calcite. One beam obeys the ordinary laws of refraction and is called the *ordinary ray,* whereas the second beam has an index of refraction that varies with the direction it travels, violating the ordinary laws of refraction, and is called the *extraordinary ray.* Substances

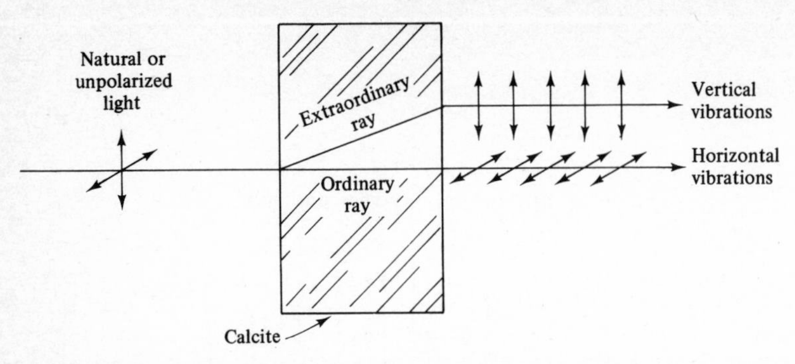

FIGURE 1-32 Double refraction of light passing through calcite. (*After A. Efron, "Light," p.* 60, *John F. Rider Publisher, Inc., New York,* 1958, *by permission.*)

exhibiting double refraction are called *optically anisotropic* because their optical properties are different in different directions; the phenomenon results because the arrangement of atoms in the crystalline lattice is different in one direction from another. On the other hand, amorphous solids, certain crystalline materials, most liquids, and all gases are *optically isotropic,* which means that their optical properties are the same in all directions and they obey the ordinary laws of refraction.

Another term for double refraction is *birefringence,* which literally means "breaking off twice." Thus, if a block of transparent calcite (sometimes called Iceland spar) were laid over the page of this book, the reader would see a double image of the printed material. Obviously, this phenomenon must be recognized in designing focusing lenses in optical instruments.

The usefulness of birefringent materials for optical instruments is illustrated by quartz for prisms. Quartz is a widely used prism material because it transmits throughout the visible and near ultraviolet (down to about 200 nm) and gives good dispersion, particularly in the ultraviolet. However, the birefringence of quartz is a disadvantage. This troublesome feature can be compensated for, however, by substituting two prisms with 30° angles for the usual 60° prism, one cut from right-handed quartz, the other left-handed (where the terms indicate that the crystal structures are mirror images of each other), so that the birefringence of one is exactly counteracted by the other. This type of prism was devised by Cornu, and its principle is illustrated in Fig. 1-33 for monochromatic radiation. In Fig. 1-33*a*, a right-handed crystal of quartz (*R*) is used, and the double refraction (somewhat exaggerated in Fig. 1-33) results in divergent beams even though their wavelength is the same as the incident beam. In Fig. 1-33*b*, this double refraction is corrected for by joining right- and left-handed 30° quartz prisms. Another method of eliminating the effect of bifringence is to employ a single 30° prism with a reflecting coat of silver on its back surface so that the rays traverse the prism twice, in opposite directions. Both arrangements are widely used, particularly the latter, and examples will be seen in Chap. 2.

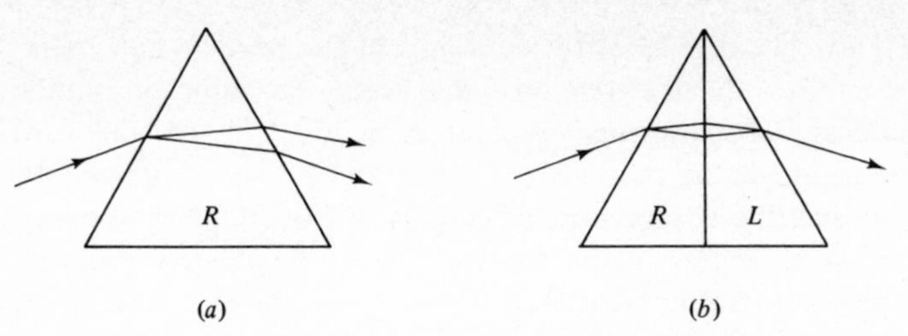

FIGURE 1-33 Principle of a Cornu prism to eliminate double refraction: (*a*) Simple quartz crystal prism; (*b*) Cornu prism. (*After F. A. Jenkins and H. E. White, "Principles of Optics,"* 3*d ed., p.* 581, *McGraw-Hill Book Company, New York,* 1957, *by permission.*)

POLARIZATION BY SCATTERING

Scattering causes polarization, but it usually is not of great importance because only when the scattering particles are much smaller than a wavelength, giving Rayleigh scattering (see Sec. 1-4*C*), does complete plane polarization result, and then with relatively low intensity. Figure 1-34 shows the two-dimensional vector outline of Rayleigh scattering given in Fig. 1-10, modified only by the addition of a z axis to show the directions of plane polarization caused by scattering. In Fig. 1-34 ordinary unpolarized radiation is incident (along the x axis) upon a small particle located at the origin of the axes. The conclusions of Sec. 1-4*C* were that interaction of the radiation with

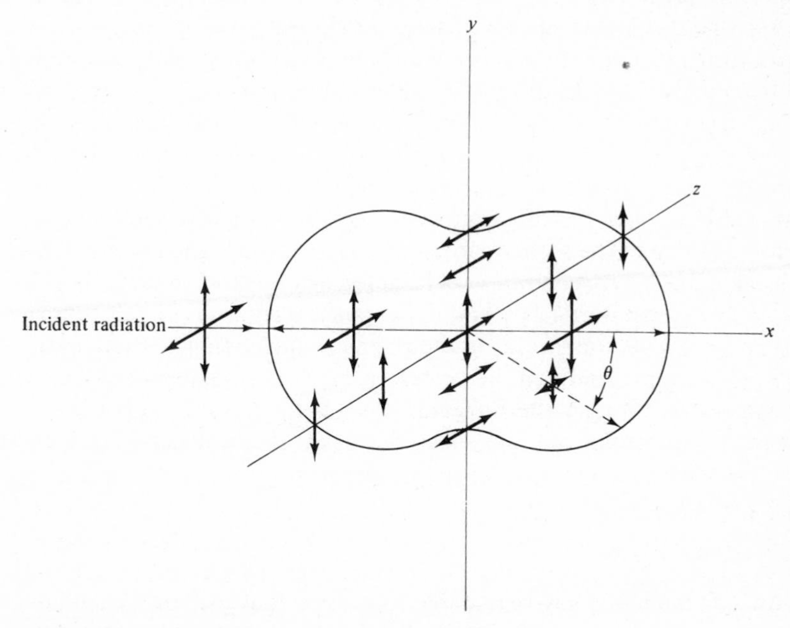

FIGURE 1-34 Polarization caused by Rayleigh scattering.

the charges in the particle cause forced oscillations of the particle and radiation is emitted (scattered) in all directions, the vector envelope or outline showing that the intensity of scattering is greater in the forward and backward directions than at right angles to the incident radiation (instead of the two-dimensional vector outline, the reader should imagine two overlapping spheres, the outer surface of which represent the relative lengths of intensity vectors). As Fig. 1-34 illustrates, radiation scattered 90° to the incident radiation is plane-polarized, whereas radiation scattered forward and backward is completely unpolarized; at all other angles in between (as exemplified by scattering angle θ) the radiation will be partially polarized. The relative sizes of the small arrow vectors give an approximate indication of the relative intensity of various vibrations. The single plane of polarization is the yz plane of Fig. 1-34, and thus, looking down the z axis, e.g., through a sheet of Polaroid, one would see only up and down, or vertical, vibrations, and looking down the y axis one would see only vibrations along the z axis (or in and out of the plane of the paper).

When particles are larger than about 10 percent of the wavelength of the incident radiation, only partial polarization can occur. The degree of polarization steadily decreases as the particles increase in size.

1-5 LAWS OF RADIATION ABSORPTION

When electromagnetic radiation interacts with matter, absorption occurs if the frequency of the radiation corresponds precisely to the energy required to raise the system to a higher allowed energy level. The types of energy-level transitions which can occur in various regions of the electromagnetic spectrum were described in Sec. 1-3. In this section, we shall be concerned with the two fundamental laws governing the *fraction* of incident radiation absorbed on passing through a given sample medium. The first law, formulated by Bouguer in 1729 and restated by Lambert in 1768, is called *Lambert's law* and predicts the effect of *thickness* of a sample medium upon the fraction of radiation which is absorbed. The second law, stated in 1852 and called *Beer's law,* deals with the effect of *concentration.* The combined form of these two laws has come to be called Beer's law since the concentration of the absorbing species is the value most important in the chemical application of these laws. While Beer's law is used most in the center of the electromagnetic spectrum, from the ultraviolet through the infrared, it holds for absorption processes anywhere in the spectrum and is in fact the basis of quantitative analysis throughout the spectrum.

1-5A Lambert's Law

Lambert reached the intuitively reasonable conclusion that each unit length of material through which radiation passes absorbs the same fraction of radiation. Given a parallel monochromatic beam of intensity I passing through a thickness db of absorbing material, the reduction in intensity can be stated mathematically as

$$dI = -kI\,db \tag{1-32}$$

where dI is the change in intensity, k is a proportionality constant, and the negative sign arises from the fact that I becomes smaller when b becomes larger. Rearranging Eq. (1-32), we obtain

$$\frac{dI}{I} = -k\,db \tag{1-33}$$

which is a mathematical statement of the fact that the fraction of radiation absorbed is proportional to the thickness traversed. If I_0 is the incident intensity when $b = 0$, we integrate

$$\int_{I_0}^{I} \frac{dI}{I} = -k \int_0^b db \tag{1-34}$$

obtaining

$$\ln \frac{I}{I_0} = -kb \tag{1-35}$$

Removing the negative sign and converting to logarithms to the base 10, we obtain

$$\log \frac{I_0}{I} = \frac{k}{2.303}\,b \tag{1-36}$$

Equation (1-36), the final form of Lambert's law, is an exact law and applies to any homogeneous, nonscattering medium, regardless of whether it is a gas, liquid, solid, or solution. In the majority of cases of analytical interest one deals with solutions. The proportionality constant k depends on the wavelength and temperature for a given absorbing substance, and for a solution the concentration must remain constant. (I_0 and I are sometimes symbolized by P_0 and P, since the intensity to which we refer has units of energy per unit time, or power.)

1-5B Beer's Law

Lambert's law, though exact, is not of very great application in chemistry. Of much greater interest is the dependence of intensity on the concentration of absorbing solutes in solution. Beer found that increasing the concentration of radiation-absorbing solute in solution had the same effect as a proportional increase in the radiation-absorbing path. This should be obvious, since an increase of solute in the same volume of solution will increase the effective thickness of the light-absorbing layer by the same factor. Thus, the proportionality constant k in Eq. (1-36) is in turn proportional to the concentration C of absorbing solute, which can be expressed as

$$\frac{k}{2.303} = aC \tag{1-37}$$

where a is the new proportionality constant. Substituting Eq. (1-37) into Eq. (1-36) gives the logarithmic form of Beer's law:

$$\log \frac{I_0}{I} = abC \tag{1-38}$$

Beer's law is fundamental to optical methods of analysis, forming the basis for estimating the concentration of a substance by measuring the radiation absorbed by a solution of that substance.

Equation (1-38) is so widely applied that the terms must be discussed from the standpoint of practical use. The ratio I/I_0, called the *transmittance*, is dimensionless. The combination abC, being a logarithm, must also be a pure number, and it is essential that the units of each quantity be carefully specified. The cell width b is usually expressed in centimeters and is a constant in most work. When C is expressed in grams per liter, a is called the *absorptivity* (or, less desirably, the extinction coefficient) and has units of liters per gram-centimeter. The absorptivity is a constant, characteristic of the absorbing substance and of the particular wavelength of radiation used. When the concentration C is expressed in moles per liter, the proportionality constant a is changed to ϵ and called the molar *absorptivity* (or molar extinction coefficient).

Most often, in experimental work, the terms *percent transmittance* ($\%T = I/I_0 \times 100$) and *absorbance* [$A = \log(I_0/I) = 2 - \log \%T$] are used. Absorbance is also called (less desirably) *optical density* (OD). Substitution of A into Eq. (1-38) gives the shortest statement of Beer's law

$$A = abC \tag{1-39}$$

or

$$A = \epsilon bC \tag{1-40}$$

according as C is in grams per liter or moles per liter. A summary of the notation and synonyms used in Beer's law is given in Table 1-5. The Beer's law symbols and definitions used in this text are those endorsed by the advi-

TABLE 1-5 Beer's Law Nomenclature†

Accepted symbol	Meaning	Accepted name	Synonym‡: Abbreviation or symbol	Synonym‡: Name
T	$\frac{I}{I_0}$	Transmittance		Transmission
A	$\log \frac{I_0}{I}$	Absorbance	OD, D, E	Optical density, extinction
a	$\frac{A}{bc}$	Absorptivity	k	Extinction coefficient, absorbancy index
ϵ	$\frac{A}{bM}$ $M = \text{mol/l}$	Molar absorptivity	a_M	Molar extinction coefficient, molar absorbancy index
b	Path length of radiation through sample, cm	Cell path	l or d	

† Based on recommendations in *Anal. Chem.*, **45**: 2449 (1973).
‡ Not recommended but occasionally seen in the literature.

sory board of *Analytical Chemistry,* which has attempted to establish some consistency in terminology.

Example 1-27 The intensity of a light beam is reduced 20 percent in passing through 1.00 cm of an absorbing medium. What will be the reduction after going through 5.00 cm of the same medium?

Answer Since the percentage reduction is 20 percent, the percent transmittance is

$$100\% - 20\% = 80\% = \%T$$

From Eq. (1-38)

$$\log \frac{I_0}{I} = -\log T = abC$$

If $b = 1.00$ cm,

$$aC = -\log 0.80 = 0.096$$

When $b = 5.00$ cm,

$$-\log T = aCb = (0.096)(5.00) = 0.48$$

$$T = 0.331 \qquad \%T = 33.1\%$$

$$\text{Percent reduction} = 100\% - 33\% = 67\%$$

Thus, the intensity is only 33 percent of what it was originally. The same answer could have been obtained using the Lambert law principle that each unit length of material through which radiation passes absorbs the same fraction of radiation. Thus, if 80 percent of the radiation is transmitted on passing 1 cm, 0.8×0.8 would be transmitted after 2 cm, $0.8 \times 0.8 \times 0.8$ after 3 cm, etc., the fraction transmitted after 5 cm being $0.80^5 = 0.33$, the same answer as above.

Example 1-28 The molar absorptivity of a solute is 1.10×10^4. Find the absorbance and percent transmittance through a 0.50-cm thickness of a 3.00×10^{-5} *M* solution.

Answer

$$A = \epsilon bC = (1.10 \times 10^4)(0.50)(3.00 \times 10^{-5})$$

$$= 0.165 = -\log T$$

$$T = 0.685, \text{ or } 68.5\%$$

SPECTRAL CURVES

Beer's law is of great importance in quantitative analyses because it shows that absorbance A is directly proportional to the concentration C of a solute. To apply Beer's law, the optimum wavelength must be selected, and for this purpose a spectral curve is determined. In practice, either percent transmittance (percent T) or absorbance A is plotted as a function of wavelength, holding the concentration C and path length b constant. The resulting curves, illustrated in Fig. 1-35 for potassium permanganate and potassium dichromate, are called the *transmittance* spectra (Fig. 1-35*a*) and *absorption* spectra (Fig. 1-35*b*), respectively. As we shall see in Chap. 2, most instruments give linear readout in percent transmittance rather than absorbance, because detectors respond to the amount of light transmitted rather than to the light absorbed, or missing. Spectral curves are characteristic of the absorbing substance, as can be seen from the distinctive spectra of potassium permanganate and potassium dichromate in Fig. 1-35, and thus these curves can be used for *qualitative analysis* of solute species. Some substances have

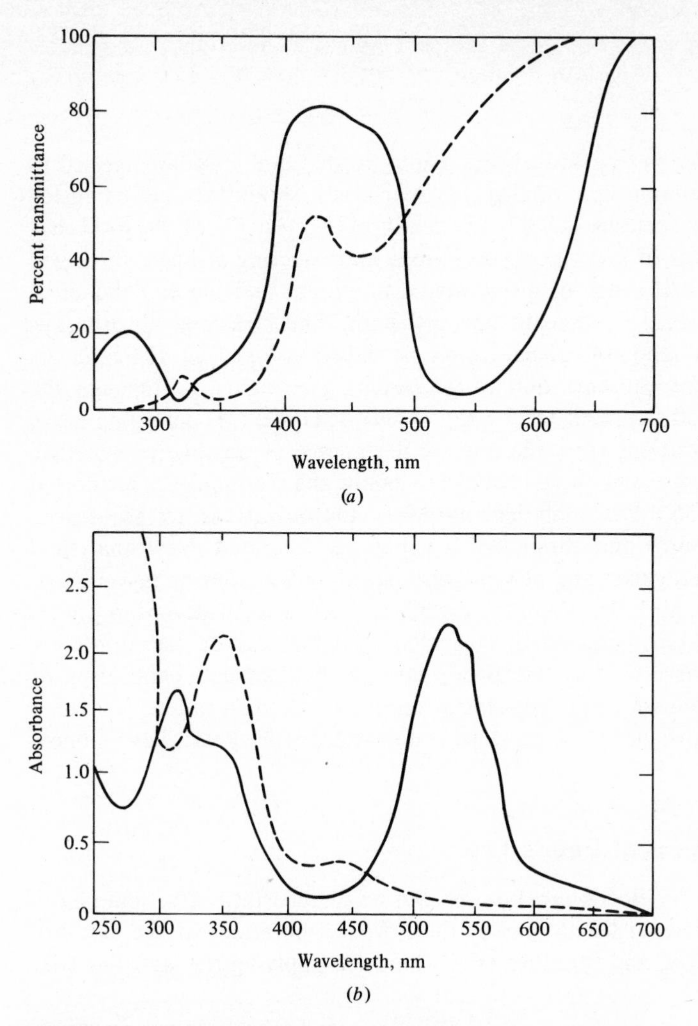

FIGURE 1-35 Spectral curves for $KMnO_4$ (*solid curve*) and $K_2Cr_2O_7$ (*dotted curve*) each 0.001 *M*, 1 *M* in H_2SO_4: (*a*) transmittance spectra; (*b*) absorption spectra.

unusual spectra with many peaks. For example, some types of didymium glass have 10 distinguishable peaks between 400 and 700 nm and are used to calibrate wavelength scales on spectrophotometers. Organic substances usually have more complicated spectral curves than inorganic ones. Spectra in the infrared region are highly discrete and are therefore especially useful for qualitative analysis.

For *quantitative* analysis one selects a wavelength where the desired solute absorbs strongly and interfering substances absorb minimally. For example, to analyze for $KMnO_4$ in the presence of $K_2Cr_2O_7$, Fig. 1-35 indicates that 550 nm would be a good choice. Other considerations in the choice of a wavelength will be described below.

TESTING BEER'S LAW

Whereas there are no known deviations from Lambert's law applied to homogeneous material, the validity of Beer's law should always be tested before using it for accurate quantitative analyses. To test Eq. (1-39) or (1-40), a series of solutions of known concentrations and covering a range which will bracket the expected range of unknowns is prepared, and the absorbance A determined at a fixed wavelength and cell path. The measured absorbances are plotted as a function of concentration, as shown in Fig. 1-36. If Beer's law is obeyed over the concentration range tested, a straight line through the origin should result, as predicted by Eq. (1-39) or (1-40) (see solid-line curve in Fig. 1-36). Deviations from the law are designated as positive or negative, according as the observed curve is above or below the straight line (see dotted curves in Fig. 1-36). For analytical purposes, the deviations are sometimes traced to their source and eliminated, or they are tolerated. New analytical procedures, new reagents, and new instruments should always be investigated as sources of deviation. If, after investigation, the source of deviation proves to be something which is not easily corrected but which can be maintained reproducibly, the curve used to test Beer's law can be used as a calibration or working curve from which the concentration of an unknown can be read after measuring its absorbance. The reasons for deviations from Beer's law will now be examined.

1-5C Limitations of Beer's Law

The many causes for Beer's law failures can be divided into three main categories: (1) fundamental limitations of the law, (2) violations of the assumptions behind the law, and (3) other errors made in applying the law. The first

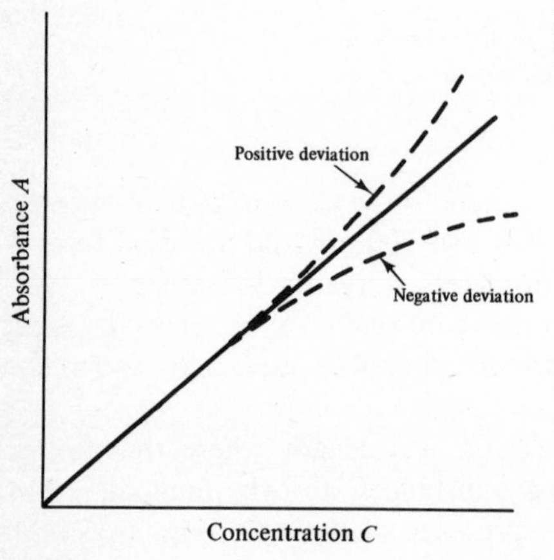

FIGURE 1-36 Calibration curves for quantitative analysis.

category is due to an inherent weakness in Beer's law which causes it to be inexact at relatively high concentrations. All other causes are more *apparent* than *real,* since if it were feasible to satisfy the assumptions behind the law or to avoid other errors, no deviations would occur.

A FUNDAMENTAL LIMITATION OF BEER'S LAW

Although the derivation of Beer's law does not take it into account, the absorptivity a is more correctly a function of the true absorptivity a_{true} and the refractive index†

$$a = a_{true} \frac{n}{(n + 2)^2}$$

Since the refractive index varies with concentration, so does a. In practice, however, n is essentially constant at concentrations of about 0.01 M or lower, and thus Beer's law may be regarded as exact at these low concentrations. It does suffer from this fundamental limitation at higher concentrations, however, and will be shown in the next section that high concentrations may cause apparent deviations for still another reason. Thus Beer's law should be thought of as a *limiting law,* strictly valid only at low concentrations. This does not rule out quantitative analyses at high concentrations, since bracketing standard solutions and a calibration curve can provide sufficient accuracy.

VIOLATIONS OF BEER'S LAW ASSUMPTIONS

The four assumptions implicit in the derivation of Beer's law are (1) that the only mechanism of interaction between the electromagnetic radiation and the solute species in question is *absorption*; (2) that monochromatic radiation is used; (3) that the absorbing solute species (molecules and/or ions) act independently of one another, regardless of number and kind; and (4) that the absorption is limited to a volume of uniform cross section. Some of these assumptions will now be discussed.

Interactions between Solute and Radiation by Mechanisms Other Than Absorption There are three types of interactions which cause attenuation of the light intensity and which therefore appear similar to absorption, *resonance emission, fluorescence* (or *phosphorescence*), and *scattering.*

"True" absorption was defined in Sec. 1-3*A* by

$$X + h\nu \rightarrow X^* \tag{1-6}$$

$$X^* \rightarrow X + \text{heat} \tag{1-7}$$

where $h\nu$ represents the energy of the incident radiation. *Resonance emission*

† G. Kortum and M. Seiler, *Angew. Chem.*, **52**:687 (1939).

involves Eq. (1-6) as a first step, but the important difference is in the decay of the excited state

$$X^* \rightarrow X + h\nu \tag{1-41}$$

Equation (1-41) may appear to be the exact reverse of Eq. (1-6), but the two steps are not reversible, since the *emitted* radiation $h\nu$ in Eq. (1-41) is emitted in *all directions,* whereas the incident radiation $h\nu$ in Eq. (1-6) is a collimated parallel beam of radiation. The apparent absorbance is less than the true absorbance by the amount of resonance emission emitted in the forward direction. Fortunately, resonance emission is restricted almost exclusively to systems of isolated atoms (Sec. 1-3*A*).

Fluorescence (or phosphorescence) again involves a first step identical to true absorption [see Eq. (1-6)], but the subsequent deactivation of the excited state may be written

$$X^* \rightarrow X + \text{heat} + h\nu' \tag{1-42}$$

where ν' represents the reduced frequency of the fluorescent (or phosphorescent) emission. Radiation $h\nu'$ is emitted in all directions and again causes the apparent absorbance to be less than the true absorbance.

Scattering occurs for incident frequencies not corresponding to absorption frequencies. The scattering process can be visualized by equations analogous to the preceding:

$$X + h\nu \rightarrow X^* \tag{1-6}$$

and

$$X^* \rightarrow X + h\nu_{scat} \tag{1-43}$$

Here ν is *any* incident frequency (though it will be recalled from Sec. 1-4*C* that the probability of scattering increases as the incident frequency increases), and ν_{scat} is the same frequency as the incident radiation but emitted in all directions from the scattering particle, thus causing deviations from Beer's law. Scattering is generally insignificant for clear solutions devoid of dust or other suspended particulate matter. Raman scattering, a special type of scattering (see Sec. 1-3*A* and Chap. 6), is always insignificant in making absorption measurements and need not be considered here.

The Use of Nonmonochromatic Radiation In the derivation of the Lambert and Beer laws monochromatic radiation was assumed. In practice, spectrophotometers pass a *band* or range of wavelengths through the sample cell, and the narrowness of the band depends on the particular instrument and the circumstances of the analysis. For a constant cell path *b*, Beer's law will fail if *a* is not constant over the concentration range of interest. More specifically, *Beer's law will fail if the absorptivity a varies over the band of wavelengths used in the analysis.* Proof of this statement is given in Sec. A-3.

Example 1-29 The absorption spectrum for a certain solute is determined and shown in Fig. 1-37, where $\Delta\lambda$ represents the *bandpass* for the instrument used (in other words, the range of wavelengths simultaneously passed while the instrument is set at the nominal wavelengths, λ_1 or λ_2). Which wavelength should be selected for the quantitative analysis of this substance, λ_1 or λ_2? Assume no interfering substances at either wavelength.

Answer The wavelength λ_2 is the best one to use since the absorptivity a is nearly constant over the bandpass at λ_2. Ordinarily, the strongest absorption peak (in this case λ_1) would be chosen, since this would allow the lowest concentration of the analyzed substance to be determined. However, Beer's law would be expected to fail at λ_1.

Example 1-30 Figure 1-38 represents the working curves (slightly exaggerated) that might result if Beer's law were tested at both λ_1 and λ_2 of Fig. 1-37. Suppose that a fixed error ΔA is made in measuring the absorbance. Which working curve will give the least error in concentration ΔC?

Answer As can be seen from Fig. 1-38, the concentration error is least when the calibration curve for λ_2 is used, or in general when Beer's law is most closely obeyed.

Interactions between Absorbing Solute Species For Beer's law to be followed, all absorption centers (molecules and ions) must act independently of one another, regardless of number and kind, and this in turn causes Beer's

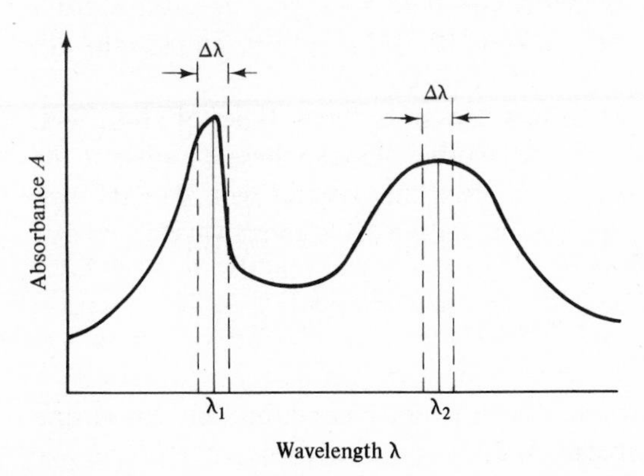

FIGURE 1-37 Absorption spectrum of a single solute.

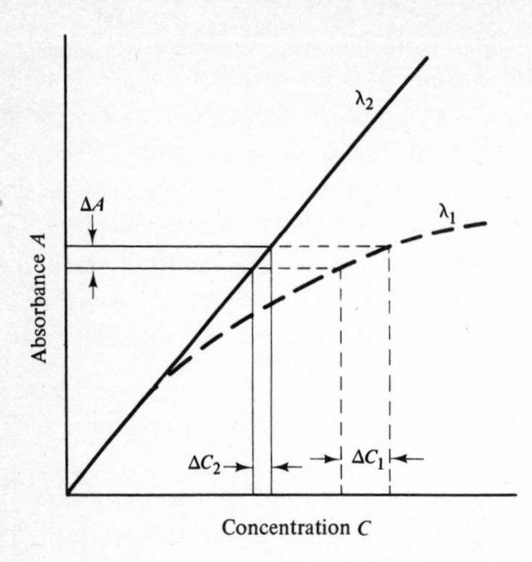

FIGURE 1-38 Test of Beer's law at wavelengths 1 and 2 of Fig. 1-37.

law to be a limiting law, applicable mainly in dilute solutions (concentrations $< 10^{-2}$ *M*). (In this sense, Beer's law is analogous to the Debye-Hückel limiting law for calculating activity coefficients of electrolytes, and for much the same reasons.) Interactions between absorption centers will have the effect of altering the charge distribution either in the absorbing species or the excited species, or both, and thus will have the effect of changing the energy requirements for absorption of the incident radiation. As a result, the absorption peaks may be altered in position, shape, and height as the concentration increases.

It should also be pointed out that interfering interactions are not just limited to those between the absorption centers themselves; some absorbing species are influenced by the electrostatic forces of so-called inert salts that are present, and these interactions may cause Beer's law deviations even at very low concentrations of the absorbing species. For example, some complex organic molecules like eosin and methylene blue may give deviations from Beer's law at concentrations as low as about 10^{-5} *M* if certain simple salts are present.

Beer's law can also be applied to mixtures of different types of absorbers, as long as they act independently of one another. Under these conditions the absorbances are additive, and since each absorbing species will have its own characteristic absorptivity *a*, the law can be written for a given wavelength as

$$A_{\text{tot}} = a_1bc_1 + a_2bc_2 + \cdots = b \sum_i a_iC_i \tag{1-44}$$

where A_{tot} is the total absorbance at the given wavelength and the subscripts $1, 2, \ldots, i$ designate absorbing species $1, 2, \ldots$.

Nonuniform Absorption Cross Section Beer's law assumes that all rays pass through the same number of absorbing centers, i.e., that square or rectangular sample cells are used. However, cylindrical cells, e.g., test tubes, are often used for expedience or economy. It has been shown [5, pp. 2773–2775] that approximate agreement with Beer's law is obtained when cylindrical cells are used, but obviously the absorbance A is less than with a uniform path equal to the diameter of the cell. The deviation is minimized by restricting the aperture. Even when half the diameter of the curved cell is illuminated, the deviation amounts to only about 2.5 percent [5], and usually narrower beams of radiation are used.

1-5D Errors Made in Applying Beer's Law

A number of common errors can occur in making quantitative absorption measurements, apart from the Beer's law errors just described. They may be classified as chemical, instrumental, and personal.

CHEMICAL ERRORS

Equilibria Effects A common but often overlooked source of error in quantitative absorption measurements occurs when the substance being sought is in equilibrium with other chemical species. It is important to realize that the value of C specified in Beer's law is the concentration of the *absorbing species*. It is not necessarily true that C is equal to or even directly proportional to the total or analytical concentration. Several examples will make this clear.

Dimerization equilibria Potassium dichromate absorbs in the visible region at about 450 nm. Suppose that a 0.1000 M solution of potassium dichromate is diluted two-, three-, and fourfold to give apparently 0.0500, 0.0333, and 0.0250 M solutions respectively. If the absorbance of these solutions is measured at 450 nm and a calibration curve prepared, it will deviate strongly from Beer's law. The deviation can be attributed to the equilibrium

$$\underset{\text{(Orange)}}{Cr_2O_7^{2-}} + H_2O \rightleftharpoons 2HCrO_4^- \rightleftharpoons 2H^+ + \underset{\text{(Yellow)}}{2CrO_4^{2-}} \qquad (1\text{-}45)$$

The $Cr_2O_7^{2-}$ species absorbs differently from $HCrO_4^-$ and CrO_4^{2-}, and thus any shift in equilibrium (1-45) will affect the measured absorbance. When 0.1000 M $Cr_2O_7^{2-}$ is diluted exactly twofold, the concentration of $Cr_2O_7^{2-}$ species is not 0.0500 M but something appreciably less because equilibrium (1-45) shifts to the right upon dilution. Thus, equilibria involving the absorbing species must be controlled. The best way of controlling equilibrium (1-45) is to convert practically all the Cr(VI) species to CrO_4^{2-} by making the solution 0.05 M in potassium hydroxide. Beer's law will then be followed, and the solution is sufficiently reliable to be used for checking other possible Beer's law deviations [7].

Acid-base equilibria If an absorbing species is involved in an acid-base equilibrium, Beer's law will fail unless the pH and ionic strength are kept constant or a wavelength corresponding to an *isobestic* point is used (an *isobestic point* is a wavelength where two species in equilibrium with each other show the same absorptivities).

Complexation equilibria When the absorbing species is a complex ion, e.g., complex of Cu(II) and ammonia or Fe(III) plus thiocyanate, the concentration of free ligand, ammonia or thiocyanate, in these examples, must be kept constant in order for Beer's law to hold. This is usually accomplished by adding an amount of ligand which is large in comparison to the amount of the metal being sought.

Solvent Effects The effect of changing the solvent on the absorption of a given solute cannot be predicted in a general way, but solvent-solute interactions often give rise to spectral shifts, band broadening, and/or deviations in Beer's law. For example, acetic acid in hexane exists almost completely as molecular HA, with its characteristic absorbance, but acetic acid in water exists appreciably in the ionic state (A^-), giving a different absorption characteristic. In any case, it is desirable to choose a solvent that does not absorb appreciably in the wavelength region to be examined.

Absorbing Impurities in Reagents Errors due to absorbing impurities in distilled water and other reagents may be serious, since many spectrophotometric methods are sensitive enough to measure trace amounts of absorbing substances. Even reagent-grade chemicals (with impurity levels that are insignificant in most methods of analysis) may prove unsuitable. A *blank determination*, which involves measuring the absorption due to cell, solvent, and reagents, should be determined and subtracted from the sample absorbance to obtain the net absorbance due to the substance being sought. However, if the absorbance of the blank is large, a small relative error in its measurement can introduce a large relative error in the final result.

Absorbing Interferences in the Sample If the sample contains a substance which absorbs appreciably over the wavelength interval chosen for measurement of the material being analyzed for, special precautions must be taken to avoid error. The principal techniques for eliminating or minimizing errors of this type can be summarized as follows:

1 Remove the interference. In practice, this alternative is often avoided because of the time, effort, and increased chances of error involved. However, removal of the interferences may become feasible if a rapid or simple separation such as liquid-liquid extraction or columnar adsorption is applicable.

2 Convert the interfering substance into a noninterfering form. This approach is often used, since it can often be accomplished in situ with rapidity and simplicity. For example, in determining manganese as

permanganate, Fe(III) interferes by absorbing at the same wavelength as permanganate. The interference can be simply eliminated by the addition of H_3PO_4, which forms nonabsorbing iron(III)–phosphate complexes.

3 Convert the substance sought into a nonabsorbing form. This approach is a modification of the preceding one and can be used when the substance sought can be selectively converted into a nonabsorbing form without altering the absorption of the interfering substance. For example, if manganese in steel is to be spectrophotometrically determined as permanganate and chromium is present as dichromate (which would interfere), the sample solution can be divided in two and the permangate in one portion selectively reduced with potassium nitrite to nonabsorbing Mn(II), without altering the dichromate. The reduced portion can then be used as the reference blank, against which the absorbance of the untreated portion is measured.

4 Analyze the sample as a mixture. If it is not possible or convenient to eliminate the interference by a separation or other chemical treatment, it may be possible to measure both the interfering and sought substances in the presence of each other. For example, suppose the two absorbing interfering species are substances X and Y. By measuring the total absorbances at each of two wavelengths, λ_1 and λ_2, two simultaneous Beer's law equations can be set up similar to general equation (1-44):

$$A_1 = a_{X,1}bC_X + a_{Y,1}bC_Y \tag{1-46}$$

$$A_2 = a_{X,2}bC_X + a_{Y,2}bC_Y \tag{1-47}$$

where subscripts 1 and 2 refer to wavelengths 1 and 2 and subscripts X and Y refer to substances X and Y. The absorptivities $a_{X,1}$ and $a_{X,2}$ are determined in a separate experiment by measuring the absorbance of a known concentration of X at wavelengths 1 and 2, and similarly $a_{Y,1}$ and $a_{Y,2}$ are measured with a known concentration of Y. The absorbances A_1 and A_2 of the mixture at wavelengths 1 and 2 are measured, and coupled with the known cell path b, Eqs. (1-46) and (1-47) can be solved simultaneously for C_X and C_Y.

This approach assumes that Beer's law holds for the system. Where Beer's law fails, multicomponent analysis becomes very complex, although approaches have been developed [5, 6].

INSTRUMENTAL ERRORS

The pertinent instrumental errors are reading errors and stray radiation. Other instrumental errors, including cell reflections and inaccurate wavelength calibration, are simple to avoid with proper technique and need not be discussed. The problem of nonmonochromatic radiation, which might be considered an instrumental error, was discussed in Sec. 1-5*C*.

Reading Errors The random error involved in reading the transmittance or absorbance scales is an instrumental error that is always present, and every user of spectrophotometers for quantitative measurements should be aware of it. It will be shown in this section and Sec. A-4 that a small, random error in reading a percent transmittance scale causes a large relative error in the concentration C when percent T is either small or large. There is an optimum intermediate transmittance range of about 20 to 60 percent, and wherever possible the sample concentration or cell thickness should be chosen to fall within this range.

The errors at the extremes of the transmittance scale stem from the exponential nature of Beer's law, expressed in the form

$$T = \text{transmittance} = \frac{I}{I_0} = 10^{-abC} \tag{1-48}$$

where all terms were defined in Sec. 1-5*B*. To demonstrate the reading error problem, a plot of percent transmittance vs. concentration is given in Fig. 1-39 showing graphically the uncertainty in concentration caused by a given uncertainty in the transmittance reading. In Fig. 1-39, an arbitrary 1 percent error in the transmittance reading is plotted at three positions along the ordinate, at 10, 37, and 90 percent transmittance. The corresponding uncertainty in concentration is largest at the 10 percent transmittance point, resulting in poor precision in measuring concentration. At the other end of the transmittance scale, as illustrated by 90 percent T, the absolute uncertainty in concentration is much smaller, but the *relative* error (absolute error divided by the concentration being determined) will be large, again giving poor precision. It is intuitively reasonable that intermediate values of transmittance will give optimum precision, and it is proved mathematically in Sec. A-4 that the relative error reaches a minimum at about 37 percent T for most instruments.

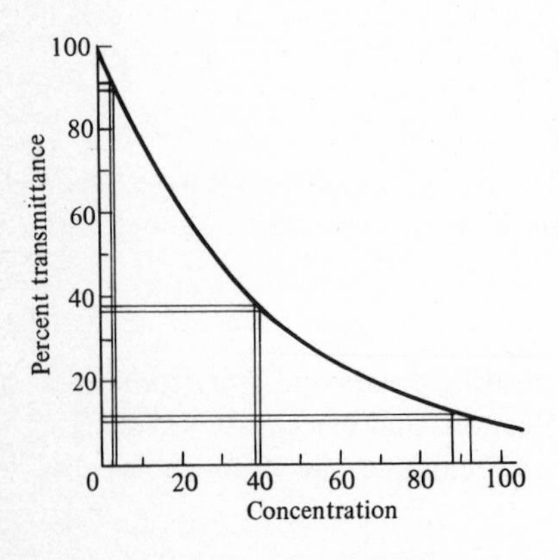

FIGURE 1-39 A plot of exponential form of Beer's law. (*From G. W. Ewing, "Instrumental Methods of Analysis,"* 3d *ed., p.* 70, *McGraw-Hill Book Company, New York,* 1969, *by permission.*)

Stray Radiation Stray radiation is extraneous radiation which strikes the detector but does not go through the sample. It can be caused by dust, reflections, or defects in the optical system such as scratches or leaks. Although some stray radiation is nearly always present, the effects are most serious at high absorbances (low transmittances), as can be seen from the equation

$$T_{obs} = \frac{T_{true} + \rho}{1 + \rho} \tag{1-49}$$

where T_{obs} = observed transmittance
T_{true} = true transmittance, observed in the absence of stray light
ρ = fraction of stray light present compared with incident intensity transmitted by reference cell

The derivation of Eq. (1-49) is given in Sec. A-5.

Example 1-31 Under a fixed set of conditions a certain sample transmits 10.0 percent of the incident radiation with no stray light present. What error in concentration would result if 1.0 percent stray radiation were present?

Answer

$$T_{obs} = \frac{T_{true} + \rho}{1 + \rho}$$

For $T_{true} = 0.100$ and $\rho = 0.010$,

$$T_{obs} = \frac{0.100 + 0.010}{1 + 0.010} = 0.109$$

Since $A = -\log T = abC$, the relative error in the concentration can be evaluated from the absorbances involved. Since $T_{true} = 0.100$, $A_{true} = 1.000$; and since $T_{obs} = 0.109$, $A_{obs} = 0.962$. Therefore,

$$\% \text{ error in } C = \frac{A_{true} - A_{obs}}{A_{true}} \times 100$$

$$= \frac{1.000 - 0.962}{1.000} \times 100 = 3.8\%$$

Example 1-32 What error in concentration would 1.0 percent stray radiation cause if the sample had transmitted 40.0 percent of the incident radiation with no stray light present? Compare this error with that obtained in Example 1-31.

Answer

$$T_{obs} = \frac{0.400 + 0.010}{1 + 0.010} = 0.406$$

For $T_{true} = 0.400$, $A_{true} = 0.398$; for $T_{obs} = 0.406$, $A_{obs} = 0.392$. Therefore,

$$\% \text{ error in } C = \frac{0.398 - 0.392}{0.398} \times 100 = 1.5\%$$

A comparison of Examples 1-31 and 1-32 shows that the effect of stray light becomes less serious for samples which transmit more light. This is another reason, in addition to reading error problems, for avoiding measuring samples which transmit less than about 20 percent of the radiation. A simple dilution will often help.

PERSONAL ERRORS

No attempt will be made to list and classify the variety of handling and technique errors an operator can make. Discussion will be limited to two areas of great importance to absorption measurements: (1) the care and use of absorption cells and (2) temperature control.

Care and Use of Absorption Cells All absorption cells should be kept scrupulously clean and free of scratches, fingerprints, and lint. Glass and quartz cells used in the visible and ultraviolet regions should be cleaned with cold concentrated nitric acid or aqua regia rather than with dichromate solutions, which have a tendency to be absorbed by these cells. Hot concentrated acids should be avoided, since they may attack some cells. The cells should be rinsed with distilled water and then with several small portions of the solution whose absorption is to be measured, rather than attempting to dry the inside. The outsides of cells should be carefully dried with paper tissues and the solutions inspected for air bells both before and after measurement. Solutions should also be examined for light-scattering inhomogeneities such as dust or undissolved particles of the sample. One cell is usually reserved exclusively for the solvent, but the equivalence of the cells used should be checked by placing the same solution in each cell and measuring the transmittance.

Cells for use in the infrared region present a particular problem because the path length must be very short (of the order of 0.05 mm or less) and the cell windows are made of salt. Such windows may be ruined by moisture or other polar solvents. The care and use of infrared cells is discussed in detail in Sec. 3-3.

Temperature Control Temperature usually has only secondary effects on absorption measurements, and for most analytical work precise temperature control is unnecessary. In most quantitative absorption measurements it is sufficient simply to equilibrate sample and standards at room temperature. However, if the absorbing solute is involved in an equilibrium, temperature control may be critical and must be checked for specific cases.

If absorption measurements are made over an appreciable range of temperatures, as in determining equilibrium constants or in making kinetic measurements, the temperature change may result in a shift in the position of the absorption band. In general, absorption bands in the ultraviolet and visible regions are shifted to longer wavelengths as the temperature increases, since the vibrational energy level of the absorbing species will be raised in accordance with the Boltzmann distribution law and thus less energy will be required of the incoming radiation to raise the absorbing species to a higher electronic state. In the infrared region, absorption bands are generally shifted slightly to shorter wavelengths as the temperature increases, since the energy separation between vibrational states increases slightly at higher vibrational levels. Consequently, if absorption measurements are made over a range of temperatures and at a fixed wavelength, the absorbance will decrease whether the temperature increases or decreases since all measurements after the first will be made on one side of the band or the other, rather than on the absorption maximum.

1-5E Differential Techniques to Increase Precision

The precision of quantitative absorption measurements is ordinarily limited to about 1 percent, and then only if the sample transmits between the limits of about 20 to 60 percent. However, three differential or scale-expansion techniques can be used to increase precision, the high-absorbance method, the trace-analysis method, and the maximum-precision method.

HIGH-ABSORBANCE METHOD

This method is useful for sample solutions of high concentration (more properly, high absorbance), particularly when their transmittances lie in the 0 to 10 percent range, where high reading errors occur (Sec. 1-5*D*). Although a dilution will often suffice to improve the precision, the high-absorbance method may be more convenient since it involves only a simple scale expansion. To explain this scale-expansion technique, the ordinary method of making quantitative absorption measurements must be outlined first.

In *ordinary* absorption measurements, the transmission scale of the instrument is calibrated by adjusting the scale (1) to read zero with no light reaching the detector (shutter closed) and (2) to read 100 (full scale) with pure solvent in the beam of radiation. To complete the analysis the transmittance of at least one solution of known concentration is measured to establish the proportionality between absorbance A and concentration C; then the transmittance of the unknown solution is measured.

In the *high-absorbance method*, pure solvent to adjust the scale to read

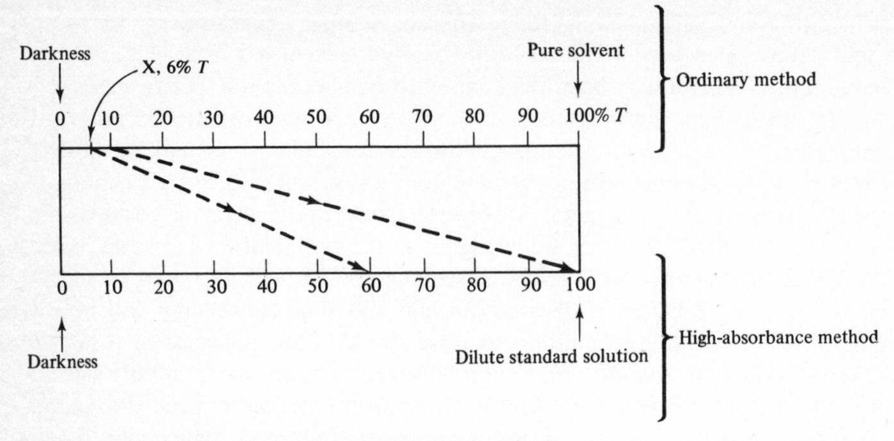

FIGURE 1-40 Scale expansion in high-absorbance method.

100 is not used; it is replaced by a standard solution that is only slightly more dilute than the unknown. For example, suppose in the ordinary method of analysis a standard solution shows 10 percent transmittance and the unknown solution reads 6 percent. The two readings are being compared within only 10 percent of the total transmission scale, and the unknown lies in the region of very high reading error. By setting the instrument to read 100 (full scale) with a standard solution that shows 10 percent T, a tenfold scale expansion is accomplished, as schematically illustrated in Fig. 1-40, and the unknown solution of 6 percent T now reads 60 percent of full scale.

To accomplish this scale expansion, the instrument used must have sufficient reserve sensitivity. There are two practical ways of increasing the sensitivity. The first is by increasing the amplification of the detector output, and the second is by increasing the slit width (a third, increasing the illumination by the source, is impractical with most instruments). There is a practical limit to the first method because the stability of all measuring devices decreases with increasing sensitivity and beyond a certain sensitivity *noise* becomes the limiting factor. Attempts to increase the sensitivity beyond this limit also increase the noise, with no net improvement in the signal-to-noise ratio. Increasing the slit width will cause decreased spectral purity (a wider band of wavelengths will be incident upon the sample), and the magnitude of the resulting error will depend on the sharpness of the absorption band being measured and on whether any interferences are present which absorb at a nearby wavelength. Assuming no interferences, no adverse effects will result if the absorption band is sufficiently broad, but sharp absorption bands will lead to serious deviations from Beer's law (see Sec. 1-5C) and decreasing adsorptivities (a's). The presence of stray radiation in the instrument can also limit the usefulness of the high-absorbance method. In summary, a significant increase in precision can be expected using the high-absorbance method, but to achieve the greatest gain the chemical system should exhibit a broad absorption band and the instrument should be stable and have low amounts of stray radiation. It is recommended that a calibration curve be used for the high-absorbance method, even if it is known that Beer's law will hold for the system under normal operating conditions, since increasing the sensitivity may cause some Beer's law deviations.

TRACE-ANALYSIS METHOD

This method is analogous to the high-absorbance method but is used for very dilute sample solutions, particularly those with transmittances in the 90 to 100 percent range. In this method, instead of adjusting the instrument scale to read zero with no light reaching the detector, a standard solution having a concentration somewhat higher than the unknown is used to set the instrument to read zero. For example, suppose in the ordinary method of analysis a standard solution shows 90 percent transmittance and the unknown solution reads 94 percent. Figure 1-41 illustrates such a case and shows how the scale expansion allows the unknown transmittance to be read with better precision. Whereas only 10 percent of the scale is utilized in the ordinary method, setting the instrument to read zero with a standard solution of 90 percent T results in a tenfold scale expansion, and the unknown solution of 94 percent T now reads 40 percent of full scale.

In order to use this method, the instrument must have a zero or dark-current control or a *bucking circuit*, with which the response obtained through a fairly dilute solution can be made to read zero. Most recorders have a zero adjustment sufficiently variable for this purpose. Again, there will be some sacrifice in stability due to a higher background noise level, but a significant increase in sensitivity and precision can be obtained. A calibration curve is a necessity with this method, since positive deviations from Beer's law linearity

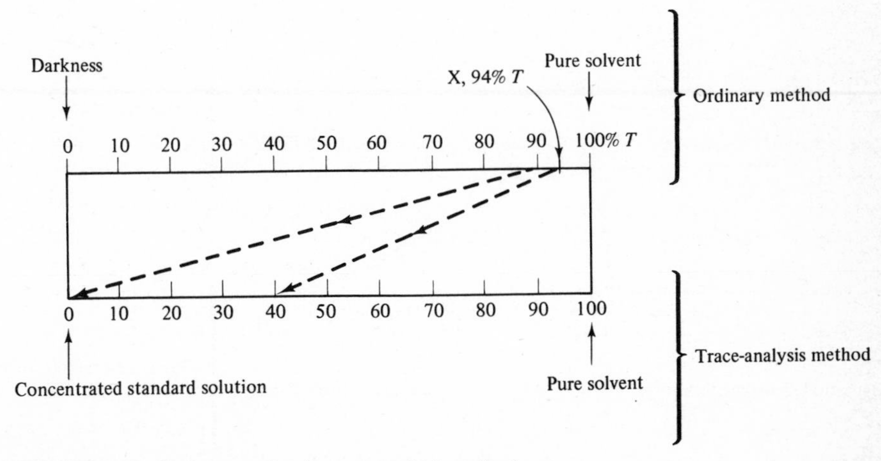

FIGURE 1-41 Scale expansion in trace-analysis method.

are the basis of the increased sensitivity. The path length of the cells used must be known with at least the precision expected in the analysis, or this will limit the precision obtainable. One answer is to use the same cell for all absorbance measurements.

MAXIMUM-PRECISION METHOD

This method, also known as the *ultimate-precision* procedure, is simply a combination of the preceding two methods. Both ends of the scale are calibrated with standard reference solutions, with one solution slightly more dilute and one slightly more concentrated than the unknown solution. The closer together the three concentrations are, the greater the scale expansion and the reading precision. Figure 1-42 illustrates this case for an unknown that reads 44 percent T by the ordinary method and 40 percent of full scale after a scale expansion of tenfold. In this example, although the scale expansion is no greater than the expansion illustrated for the two previous methods, the maximum-precision method improves an already favorable transmission reading (one in the 20 to 60 percent transmission region by the ordinary method) by making the reading 10 times more precise.

The instrument requirements for this method are the combined requirements of the previous two methods, and the limitations likewise tend to be a combination, though stray light is usually not a problem unless the method is used on a highly absorbing sample. Again, a calibration curve will be necessary, as positive deviations from Beer's law linearity occur. Assuming the availability of an instrument with very high sensitivity and stability (characteristics which tend to be incompatible), the precision of this method can be increased at will by making the difference between the concentrations of the two references sufficiently small, and the limiting source of error would be the accuracy of preparation of the standard solutions. In practice, this

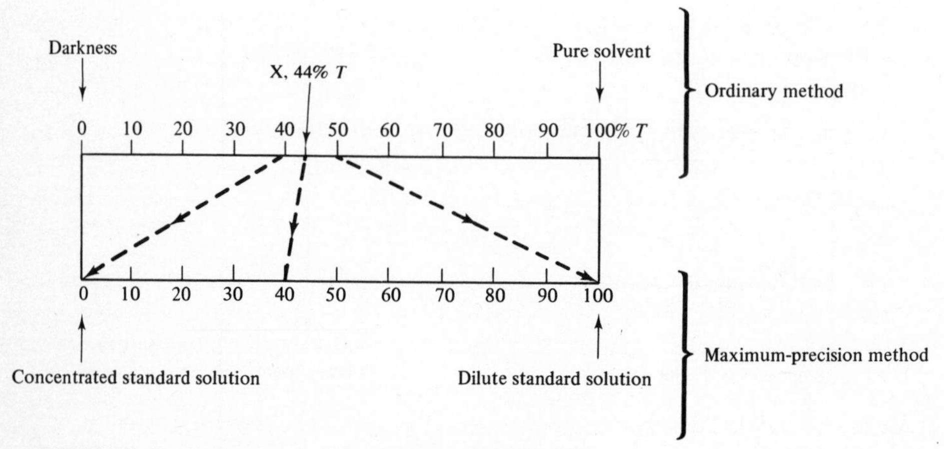

FIGURE 1-42 Scale expansion in maximum-precision method.

method has not proved as useful as the other two, though a tenfold gain in precision, to the 0.1 percent error level in favorable cases, has sometimes been achieved.

EXERCISES

1-1 To acquire facility with the various units used to characterize electromagnetic reduction, mentally convert the following wavelengths to frequency and wave numbers: 200 nm, 250 nm, 500 nm, 1.0 μm, 2.5 μm, 3.0 μm, 10 μm, 25 μm.

1-2 Convert the following wave numbers to wavelengths: 15,000 cm^{-1}, 8750 cm^{-1}, 6667 cm^{-1}, 5000 cm^{-1}, 3000 cm^{-1}, 2500 cm^{-1}, 2200 cm^{-1}, 2000 cm^{-1}, 1000 cm^{-1}, 200 cm^{-1}.

1-3 Assume that forced vibrations or oscillations which occur in particles during the process of scattering exist for the period of the incident radiation. Calculate the period (or time interval of one cycle) for visible light with a wavelength of 600 nm. *Ans:* 2×10^{-15} s

1-4 A spectrophotometric quantitative analysis at 400 nm is to be performed on a certain solute species known to absorb at 400 nm and to fluoresce at 390 nm. Will the deviations from Beer's law caused by the fluorescence be positive or negative?

1-5 A glass prism with a 2.5-cm base has a dispersion $dn/d\lambda$ of 2.50×10^{-4} per nanometer at 500 nm. Calculate the minimum separation between spectral lines that can be resolved in the region of 500 nm with this prism.

Ans: 0.1 nm

1-6 The grating constant for a certain grating is 2×10^{-3} cm. Calculate the arc in degrees subtended for the second-order visible spectrum from 400 to 750 nm.

1-7 What slit width, in millimeters, should be used to focus radiation of 15 μm on a screen in a Fraunhofer diffraction pattern so that the edge of the central (maximum-intensity) band is at an angle of 5° from the center of the slit? *Ans:* 0.17 nm

1-8 Calculate the variation of the polarizing angle through the visible spectrum (400 to 750 nm) for the borosilicate crown glass listed in Table 1-4. Give only the values at the extreme limits and their difference.

Ans: 56.75°; 56.43°; 19′15′′

1-9 Calculate the critical angle for radiation passing from water to air.

Ans: 49°

1-10 A certain sample transmits 10 percent of the incident radiation with no stray light present. The measurement is repeated with 1.5 percent stray light present. (*a*) What is the new percent transmittance? (*b*) What relative error in concentration will the stray light cause? *Ans:* (*a*) 11.3 percent; (*b*) 5 percent

1-11 A 0.1000 *M* aqueous solution reagent-grade potassium dichromate is prepared, and then a series of accurate dilutions are made. Predict whether a test of Beer's law, made at a wavelength where $Cr_2O_7^{2-}$ absorbs, will result in positive or negative deviations from Beer's law.

1-12 The absorptivity a for $K_2Cr_2O_7$ at 450 nm is 12.0 l/(g)(cm). What is the molar absorptivity ϵ?

1-13 With an ordinary method of spectrophotometric analysis, an unknown solution gives a transmittance reading of 38 percent. Then with bracketing standards that read 37 and 40 percent T by the ordinary method, the method of maximum precision is used. (*a*) What percent T reading can be expected for the unknown on the expanded scale? (*b*) Assuming that the instrument is as stable under the conditions used for the maximum-precision method as for the ordinary method and that the preparation of the standard solutions does not limit the accuracy, estimate the maximum precision obtainable with the maximum-precision method.

Ans: (*a*) 33.3 percent T; (*b*) 0.033 percent relative deviation

1-14 (*a*) In the differential high-absorbance method, by what factor is the scale expanded if a solution that reads 6 percent T by the old method is made to read 100 percent T? (*b*) Estimate the increase in precision accomplished by the scale-expansion method in part (*a*). Assume a broad absorption peak, a stable instrument with no stray radiation, and a nominal reading error in reading the transmission scale of 1 percent.

1-15 Derive Eq. (1-24) from principles like those used to derive Eq. (1-23).

1-16 What length of grating having 10,000 lines per inch would be required to accomplish the resolution of two lines at 742.2 and 742.4 nm?

REFERENCES

INTRODUCTORY

1 EWING, G. W.: "Instrumental Methods of Analysis," 3d ed., McGraw-Hill, New York, 1969.

2 WILLARD, H. H., L. L. MERRITT, JR., and J. A. DEAN: "Instrumental Methods of Analysis," 4th ed., Van Nostrand, New York, 1965.

3 BLAEDEL, W. J., and V. W. MELOCHE: "Elementary Quantitative Analysis," 2d ed., Harper & Row, New York, 1963.

INTERMEDIATE AND ADVANCED

4 STROBEL, H. A.: "Chemical Instrumentation," Addison-Wesley, Reading, Mass., 1960.
5 KOLTHOFF, I. M., and P. J. ELVING (eds.): "Treatise on Analytical Chemistry," pt. I, vol. 5, Wiley, New York, 1964.
6 FRED, M., and F. W. PORSCHE: *Ind. Eng. Chem., Anal. Ed.,* **18:** 603 (1946).
7 HAUPT, G. W.: *J. Res. Natl. Bur. Stand.,* **48:** 414 (1952).
8 BANWELL, C. N.: "Fundamentals of Molecular Spectroscopy," McGraw-Hill, New York, 1966.

OPTICS

9 JENKINS, F. A., and H. E. WHITE: "Fundamentals of Optics," 3d ed., McGraw-Hill, New York, 1957.
10 HARDY, A. C., and F. H. PERRIN: "The Principles of Optics," McGraw-Hill, New York, 1932.
11 PHYSICAL SCIENCE STUDY COMMITTEE: "Physics," Heath, Lexington, Mass., 1971.
12 BRAGG, W.: "The Universe of Light," Dover, New York, 1959.
13 ROSSI, B.: "Optics," Addison-Wesley, Reading, Mass., 1957.
14 REUCHARDT, E.: "Light, Visible and Invisible," University of Michigan Press, Ann Arbor, 1960.
15 SEARS, F. W.: "Optics," 3d ed., Addison-Wesley, Reading, Mass., 1949.

2
Ultraviolet and Visible Spectrophotometry

Absorption spectrophotometry in the ultraviolet and visible region constitutes one of the oldest of the physical methods for quantitative analysis and structural elucidation. While infrared and nmr techniques are more powerful tools for structural elucidation and qualitative identifications, ultraviolet and visible spectrophotometry still surpasses all other optical methods of analysis as a means for quantitative analysis and serves as a useful auxiliary tool for structural elucidation.

The wavelengths corresponding to the ultraviolet and visible regions of the electromagnetic spectrum are identified in Fig. 2-1. For convenience, the ultraviolet region is divided into the *far ultraviolet* (10 to 200 nm) and the *near ultraviolet* (200 to 400 nm). The far ultraviolet is also known as the *vacuum ultraviolet,* since oxygen absorbs strongly just below 200 nm and evacuation of the spectrophotometer is the best remedy (flushing with nitrogen suffices down to about 150 nm, where it begins to absorb strongly). The division between the near ultraviolet and the *visible* region is mainly a physiological one, radiation below about 400 nm being invisible to the human eye.

The unifying feature of the entire ultraviolet and visible region is that absorption in this region causes *excitation of electrons* to higher energy levels. In general, tightly held electrons will require energetic photons (of short wave-

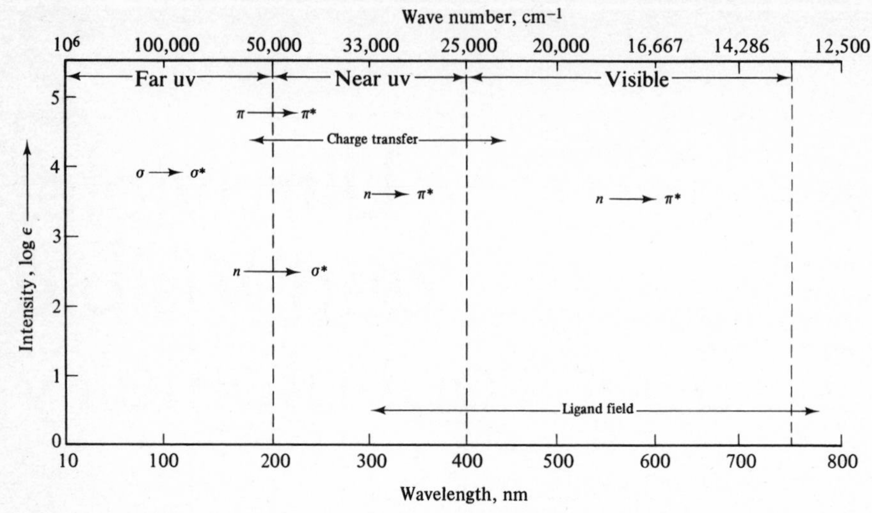

FIGURE 2-1 Ultraviolet and visible regions of the spectrum and the types of absorption bands that most often occur. (*After C. N. Banwell, "Fundamentals of Molecular Spectroscopy," McGraw-Hill Book Company, New York,* 1966, *by permission.*)

length) to accomplish absorption, whereas more loosely held (*delocalized*) electrons can be excited with longer-wavelength radiation. The theory of the absorption process will be discussed in the next section, and subsequent sections will cover instrumentation, sample handling, and applications. A final section deals specifically with the vacuum-ultraviolet region.

2-1 THEORY OF ABSORPTION

The total energy of a molecule can be broken down into electronic, vibrational, rotational, and translational contributions. The *change* ΔE in the energy of a molecule upon absorption of electromagnetic radiation is given by

$$\Delta E = \Delta E_{\text{elec}} + \Delta E_{\text{vib}} + \Delta E_{\text{rot}} + \Delta E_{\text{trans}} \tag{2-1}$$

where ΔE_{elec} is the spacing between allowed electronic energy levels, ΔE_{vib} is the spacing between allowed vibrational energy levels, etc. As the energy-level diagram of Fig. 1-6 indicated, the excitation of electrons to higher energy levels requires the greatest amount of energy ΔE_{elec}, whereas the vibrational energy-level spacings ΔE_{vib} are generally about 10 times less, and rotational energy spacings ΔE_{rot} are about 10 or 100 less than vibrational energy spacings. The translational energy of a molecule, which is its energy due to linear motion through space, is also quantized, but the translational energy spacings ΔE_{trans} are extremely small and are unimportant in all absorption regions of interest.†

2-1A Vibrational-Rotational Fine Structure of Electronic Spectra

As pointed out in Sec. 1-3, in connection with the energy-level diagram of Fig. 1-6, the excitation of electrons is accompanied by changes in vibrational and rotational energy; thus what would otherwise be an absorption *line* in the ultraviolet and visible regions becomes a broad *band* containing vibrational and rotational fine structure. In the vapor phase, the spectrum observed is due to isolated molecules, and often the vibrational and rotational structure can be resolved. In solution, molecules are not isolated but are solvated by solvent molecules, which restrict free rotation and obliterate rotational structure. Rotational restriction also occurs at fairly high pressure in the spectra of vapors and is called *pressure broadening*. Furthermore, the force fields of solvents modify the vibrational energy levels in an irregular manner, and when absorption is averaged over a large number of molecules with indefinite energy levels, broad bands result. In general, the more polar the solvent the greater the loss of vibrational detail. These effects are illustrated in Fig. 2-2 with the visible absorption spectra of *sym*-tetrazine in the vapor state and two different solvents.

† For neutral molecules, changes in translational energy ΔE_{trans} brought about by electromagnetic radiation are *disallowed*, since there can be no electric or magnetic interaction with a neutral molecule. No such restriction exists for changes in electronic energy ΔE_{elec}, since the electromagnetic fields of incoming radiation will readily interact with the electric fields of electrons.

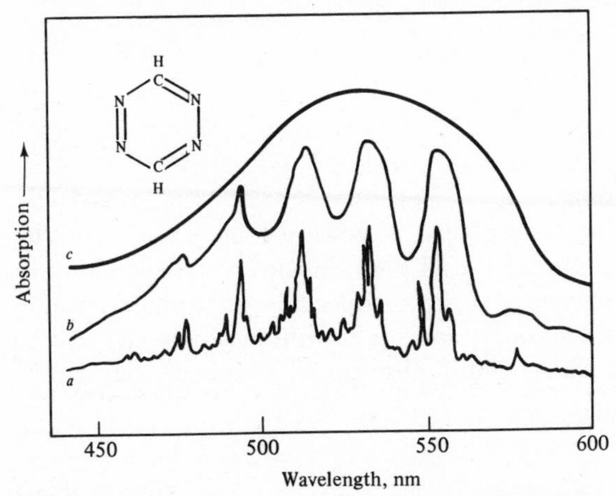

FIGURE 2-2 Absorption spectra of *sym*-tetrazine in two solvents and in the vapor state: (*a*) vapor state; (*b*) cyclohexane solvent; (*c*) water solvent. (*After S. F. Mason, J. Chem. Soc.*, **1959:** 1265.)

The vibrational resolution of solution spectra is usually enhanced by working at *lower temperature,* since fewer vibrational levels are occupied and solute-solvent interaction is decreased. If an absorption band lies on the edge of the visible spectrum, temperature-dependent changes in the shape of the band can also cause *thermochromism,* i.e., the reversible appearance of visible color on heating or cooling.

Though *rotational* fine structure is not ordinarily observable in the ultraviolet and visible regions, the rotational structure in the vapor state can sometimes be observed with specially designed spectrographs of very high resolution.†

2-1B Types of Absorption Bands

In order to understand and be able to interpret electronic absorption spectra, one must understand the specific types of electronic transitions that can take place when ultraviolet or visible radiation is absorbed. The six types of absorption bands that can be recognized in the ultraviolet and visible region (not counting the *charge-resonance* and *electron-resonance* spectra of unstable organic ions and radicals) are due to the following electron transitions: $\pi \rightarrow \pi^*$, $n \rightarrow \pi^*$, $\sigma \rightarrow \sigma^*$, $n \rightarrow \sigma^*$, charge-transfer, and ligand-field transitions. The types of transition can be identified fairly definitely by the spectral regions in which they are generally found, their relative absorption intensities, and in some cases by spectral shifts when the solvent is varied.

Absorption intensities are generally specified in terms of molar absorptivities, defined in Chap. 1 as

$$\epsilon = \frac{A}{bC} \tag{1-40}$$

where A = absorbance
b = path length, cm
C = molar concentration

The magnitude of the molar absorptivity is governed by the *size* of the absorbing species and by the *probability* of the transition. For interaction to take place, a photon must obviously strike a molecule approximately within the space of the molecular dimensions, and the transition probability will be the proportion of target hits which lead to absorption. These concepts can be expressed mathematically‡ as

$$\epsilon = 0.87 \times 10^{20} PA \tag{2-2}$$

† See S. F. Mason, *J. Chem. Soc.*, **1959:** 1269, who obtained the rotational structure of *sym*-tetrazine in the vapor state at about 551 nm.

‡ E. A. Braude, *Nature,* **155:** 753 (1945); *J. Chem. Soc.*, **1950:** 379.

where P is the transition probability and A is the cross-sectional target area in square centimeters. The area A can be estimated from x-ray data (Chap. 12) or electron-diffraction data and is the order of 10^{-15} cm^2 for ordinary organic molecules. Thus, for a transition of unit probability, $\epsilon \approx 10^5$. The highest molar absorptivities are of this order, and hence absorption with ϵ values of about 10^4 or greater is considered high-intensity, or very strong absorption, and is due to transitions of high probability (quantum-mechanically allowed transitions, with $P = 0.1$ to 1). Absorption with ϵ values from about 10^3 to 10^4 ($P = 0.01$ to 0.1) is considered strong absorption, and absorption with ϵ values less than about 10^3 is considered low-intensity, or weak, absorption. The last is usually due to transitions of low probability (forbidden transition, with $P = 0.01$ or less).

Absorption energies (or wavelength regions) in which the various types of absorption bands usually occur are indicated schematically in Fig. 2-1. The approximate intensities of the various types of absorption bands are indicated along the ordinate. It should be emphasized that both the wavelength regions and the intensities in Fig. 2-1 generalize a mass of data in which there are many exceptions. Each type of transition will now be discussed, and empirical rules will be given for predicting where in the absorption spectrum a given chemical structure will have an absorption maximum.

$\pi \rightarrow \pi^*$ AND $n \rightarrow \pi^*$ ABSORPTION BANDS

It is convenient to consider $\pi \rightarrow \pi^*$ and $n \rightarrow \pi^*$ transitions together, since many chemical groups contain both π electrons and n electrons and both may contribute to the ultraviolet and visible spectrum.† Any multiple bond, such as C═C, C≡C, C═N, contains π electrons, and atoms to the right of carbon in the periodic table, notably nitrogen, oxygen, sulfur, and the halogens, possess n electrons. The carbonyl group is a good example of a group showing both $\pi \rightarrow \pi^*$ and $n \rightarrow \pi^*$ transitions, as schematically illustrated with electron-dot formulas in Fig. 2-3.

† Outershell electrons not directly involved in bonding are called *nonbonding* or n electrons.

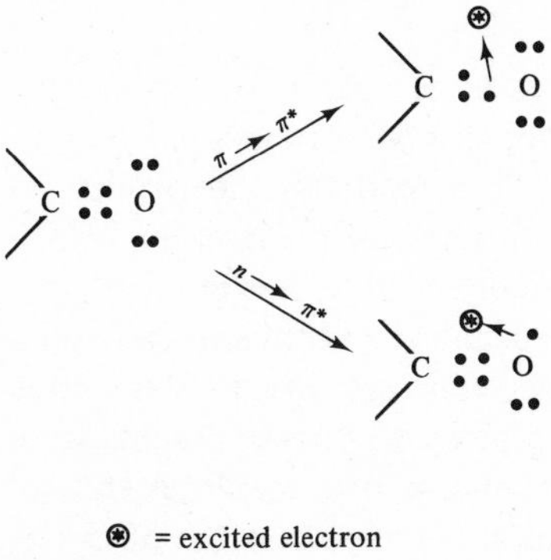

FIGURE 2-3 Schematic representation of $\pi \rightarrow \pi^*$ and $n \rightarrow \pi^*$ transitions in a carbonyl group.

⊛ = excited electron

TABLE 2-1 Absorption Maxima for $\pi \rightarrow \pi^*$ and $n \rightarrow \pi^*$ Transitions of Some Isolated Functional Groups†

	Transition, nm	
Functional group	$\pi \rightarrow \pi^*$ (intense absorption)	$n \rightarrow \pi^*$ (weak absorption)
>C=C<	170	None
—C≡C—	170	None
>C=O	166	280
>C=N—	190	300
—N=N—		340
>C=S		500
—N=O		665

† Data from Ref. 5, p. 229, and Ref. 4, p. 9.

Table 2-1 lists the approximate wavelengths of maximum absorption λ_{max} for $\pi \rightarrow \pi^*$ and $n \rightarrow \pi^*$ transitions of various isolated functional groups. It should be noted that for $\pi \rightarrow \pi^*$ transitions λ_{max} is relatively independent of the atoms making up the double or triple bond, whereas for $n \rightarrow \pi^*$ transitions the location of the λ_{max} is strongly dependent on the atoms bonded. This behavior is reasonable, since π electrons are delocalized and thus relatively independent of the atoms involved, whereas n electrons are directly associated with the parent atoms and increasing electronegativity will cause the unshared electrons (or n electrons) to be more tightly bound, with a resulting shift of λ_{max} to shorter wavelengths; compare, for example, the wavelengths of C=S,

C=N, and C=O. These data are approximate since different substituents on the functional groups will produce some variation in the wavelength of the $\pi \rightarrow \pi^*$ and especially the $n \rightarrow \pi^*$ transitions. For example, the λ_{max} for the $n \rightarrow \pi^*$ transition of CH_3COCH_3 is 272 nm, while for CH_3CHO it is 294 nm. These wavelength shifts can be useful for obtaining information about the substituents on a particular group, though the data are less definitive, in general, than comparable infrared or nmr data. Finally, it should be noted that though the $\pi \rightarrow \pi^*$ transitions of various isolated functional groups give intense absorption bands, the absorption maxima are usually in the far ultraviolet, which is at present accessible only with difficulty. On the other hand, *conjugated* π-electron systems are much more easily excited, giving $\pi \rightarrow \pi^*$ transitions in the easily accessible near ultraviolet, where their absorption forms the basis of some of the most important applications of ultraviolet spectrophotometry.

Conjugated π-Electron Systems Conjugated π-electron systems are those which have alternate single and multiple bonds, for example, —C=C—C=C or —C=C—C=O. Because of conjugation both the $\pi \rightarrow \pi^*$ and $n \rightarrow \pi^*$ transitions shift to longer wavelengths and increase in intensity as the extent of conjugation increases. For example, Table 2-2 gives the approximate λ_{max} and ϵ of the $\pi \rightarrow \pi^*$ transitions in various carbon-carbon bonds, and Table 2-3 gives the approximate λ_{max} for both the $\pi \rightarrow \pi^*$ and $n \rightarrow \pi^*$ transitions of some oxygen-containing molecules. Tables 2-2 and 2-3 both show that conjugation immediately shifts the very intense $\pi \rightarrow \pi^*$ transition into the near ultraviolet, making this region very useful for studying conjugated and aromatic systems. It should also be noted from Table 2-3 that the $n \rightarrow \pi^*$ absorption of *p*-benzoquinone has entered the blue region of the *visible* spectrum at 435 nm. Organic compounds that are colored invariably contain considerable conjugation, a fact that will be reemphasized in Sec. 2-1*C*.

Some Spectral Terms Defined Before continuing, it is necessary to define several terms which are frequently used in discussing electronic spectra.

TABLE 2-2 Approximate Absorption Maxima and Molar Absorptivity of $\pi \rightarrow \pi^*$ Transitions in Various Carbon–Carbon Bonds†

Structure	λ_{max}, nm	ϵ
—C=C—	170	16,000
—C=C—C=C—	220	21,000
—C=C—C=C—C=C—	260	35,000

† Data from Ref. 5, p. 230.

Chromophores Groups of atoms responsible for absorption in the ultraviolet or visible region. Thus, to be very general, π-electron systems are chromophores for visible and ultraviolet spectra, and σ electrons are chromophores for the far (vacuum) ultraviolet.

Auxochromes Saturated groups which when added to chromophores cause a shift in the absorption to a longer wavelength and an increased intensity in the absorption peak. Some common auxochromic groups are hydroxyl, amino, sulfhydryl (and their derivatives), and some of the halogens. These groups all contain nonbonding electrons, and $n \rightarrow \pi^*$ transitions are usually responsible for these effects. To illustrate the auxochromic effect, the benzene ring can be considered a chromophore, with a λ_{max} at 255 nm and ϵ = 230 (a forbidden $\pi \rightarrow \pi^*$ transition), whereas an OH group substituted for one of its hydrogens causes λ_{max} to shift to 270 nm, and ϵ becomes 1450.

Shifts in absorption maxima and changes in absorption intensities are very important in studying the effect of solvents and temperature on spectra, as well as the effects of substituent groups, and thus the following commonly used terms should also be defined:

Bathochromic Effect, or *red shift* A shift of an absorption maximum toward longer wavelength. Incidentally, ions and free radicals usually have spectra that are shifted toward longer wavelengths (bathochromic shift) with respect to the related neutral and undissociated molecules. Often, therefore, ions and radicals are colored while the corresponding neutral (undissociated) molecules are not.

TABLE 2-3 Approximate Absorption Maxima for $\pi \rightarrow \pi^*$ and $n \rightarrow \pi^*$ Transitions of Some Oxygen-containing Molecules†

Structure	Transition, nm	
	$\pi \rightarrow \pi^*$ (strong absorption)	$n \rightarrow \pi^*$ (weak absorption)
—C=O	166	280
—C=C—C=O	240	320
—C=C—C=C—C=O	270	350
O=⟨cyclohexadiene ring⟩=O	245	435

† Data from Ref. 5, p. 230

Hypsochromic Effect, or *blue shift* A shift toward shorter wavelength.

Hyperchromic Effect An effect leading to increased absorption intensity.

Hypochromic Effect An effect leading to decreased absorption intensity.

Effects of Solvents on $\pi \rightarrow \pi^*$ and $n \rightarrow \pi^*$ Transitions The $\pi \rightarrow \pi^*$ and $n \rightarrow \pi^*$ transitions can be differentiated by varying the polarity of the solvent. In general, as the polarity of the solvent increases, $n \rightarrow \pi^*$ (and also $n \rightarrow \sigma^*$) bands undergo a hypsochromic (blue) shift, whereas $\pi \rightarrow \pi^*$ bands undergo a gradual bathochromic (red) shift. These effects can be understood by realizing that in $\pi \rightarrow \pi^*$ transitions the excited state is more polar than the ground state, and therefore dipole-dipole interactions with polar solvents will lower the energy of the excited state more than that of the ground state. On the other hand, nonbonding electrons in the ground state are stabilized, relative to the excited state, by hydrogen bonding or electrostatic interaction with polar solvents. In the extreme of acid solutions, for example, $n \rightarrow \pi^*$ transitions "disappear" completely (shift to very high energies), because the lone-pair electrons become involved in bonding with a proton. In general, $\pi \rightarrow \pi^*$ bands shift about 10 to 20 nm bathochromic in going from hexane to ethanol, whereas in contrast the $n \rightarrow \pi^*$ band of acetone shifts about 7 nm hypsochromic in going from hexane to ethanol and another 8 nm in going to water.

Isolation of Chromophores From the standpoint of their effect on absorption spectra, there are three ways in which chromophores containing π electrons can be linked together in a molecule:

1 If two chromophores, A and B, are *directly* linked together, the spectrum of molecule A—B will usually be very different from that of either A or B, and in fact A—B must be considered a new chromophore. For example, the ethylene group (C=C) has a λ_{max} at about 170 nm and an ϵ of about 10,000, but the allene group (C=C=C) has a λ_{max} at about 225 nm and an ϵ of about 500.

2 If two chromophores are *conjugated,* i.e., separated by one single bond, each chromphore tends to retain its individuality, except that both $\pi \rightarrow \pi^*$ and $n \rightarrow \pi^*$ transitions undergo a bathochromic (red) shift, accompanied by increased (hyperchromic) intensity, as was illustrated in Tables 2-2 and 2-3.

3 If two chromophores, A and B, are separated by more than one single bond, the π electrons are said to be *isolated* and the spectrum is practically the sum of those of A and B. Hence, a chromophore which is isolated can be expected to absorb at a characteristic wavelength, as illustrated in Table 2-1 and the accompanying discussion.

Empirical Rules for Predicting the Wavelength of Absorption Peaks At present, the physical nature of the electronic excited state is understood only semiquantitatively, at best. Furthermore, molecules in excited states interact in various ways with solvent molecules. Hence, it is normally impossible to calculate a complete ultraviolet spectrum from a theoretical model. Nonetheless, from the mass of empirical data available, certain trends and rules can be formulated for predicting the approximate wavelength at which a given structure will absorb. Like most trends, they are not without exceptions, and they should be viewed accordingly.

Four types of compounds particulary lend themselves to this type of numerical correlation: (1) conjugated dienes and cyclic trienes, (2) unsaturated ketones and aldehydes, (3) substituted benzene rings, and (4) unsaturated acids, esters, nitriles, and amides.

Conjugated dienes and cyclic trienes The observed absorption properties of conjugated dienes can be qualitatively explained with energy-level diagrams of the molecular orbitals. One of the best-known systems is that of butadiene, and Fig. 2-4 illustrates how the π electrons of two ethylenic double bonds interact to generate four new π orbitals in 1,3-butadiene. Since each double bond contains a pair of electrons, four $\pi \rightarrow \pi^*$ transitions are theoretically possible in butadiene, but by far the most probable transition is the one between

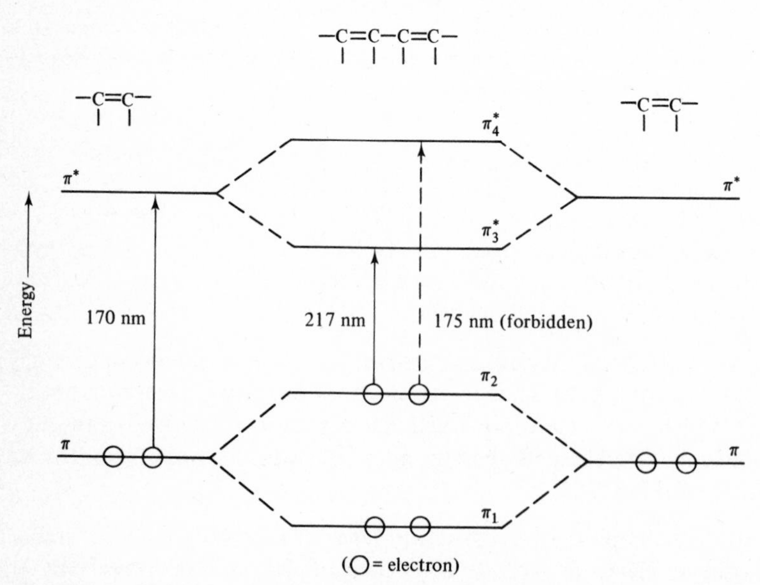

FIGURE 2-4 Energy-level diagram and electronic transitions for ethylene and butadiene.

the $\pi_2 \rightarrow \pi_3{}^*$ molecular orbitals, which occurs at about 217 nm (see Fig. 2-4). Hence, though the simple ethylene chromophore does not absorb in the near ultraviolet, the conjugation of two such chromophores give rise to strong absorption at 217 nm ($\epsilon \approx 21{,}000$). It should be pointed out here that alkyl substitution on a simple chromophore like ethylene will give a small bathochromic shift because of a small interaction between the σ-bonded electrons of the alkyl group and the π-bond system, but the shift is small compared with an increase in conjugation.

Fortunately, the effect of alkyl substitution in dienes is additive, and a study of a large number of open-chain dienes and dienes in six-membered rings has led to the following rules (sometimes known as *Woodward's rules*) for predicting the absorption wavelength:†

1 Assign a base value of 214 nm to an open-chain diene or to conjugated double bonds in separate but fused six-membered rings (a heteroannular diene); if the conjugated double bonds are contained in the same ring (a homoannular diene), assign a base value of 253 nm.

2 Add 30 for each additional double bond in conjugation.

3 Add 5 for each alkyl substituent or residue of ring attached to the double-bond system.

4 Add 5 for each double bond which is *exocyclic*, i.e., any double bond which goes off a ring (symbolized >C=, where the single bonds are part of a ring).

Example 2-1 Calculate the absorption wavelength for cholesta-2,4,6-triene:

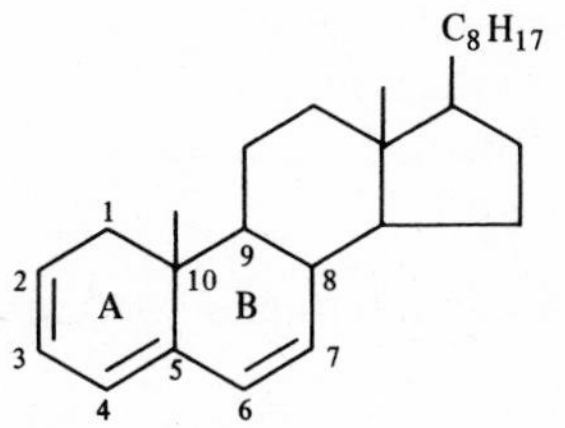

† R. B. Woodward, *J. Am. Chem. Soc.*, **64:** 72, 76 (1942); L. Fieser, *J. Org. Chem.*, **15:** 930 (1950); and A. I. Scott, "Interpretation of the Ultraviolet Spectra of Natural Products," Pergamon, Oxford, 1964.

Answer The compound could be considered either a homo- or a heteroannular diene with an additional double bond; the presence of a homoannular diene (ring A in this case) takes precedence in such cases.

Parent homoannular diene		253 nm
Extra double bond (6,7) in conjugation		30
Three substituents, attached at positions 2, 5, and 7	$3 \times 5 =$	15
One exocyclic double bond (the 4,5 bond is exocyclic to ring B)		5
Total		303 nm

The experimentally observed absorption is 306 nm.

These rules hold well only for conjugated dienes and trienes in cyclic systems. Fair agreement can be obtained in cyclic systems with four conjugated double bonds, but the rules are unsatisfactory for systems of more extended conjugation, for which other empirical rules have been formulated.†

Unsaturated ketones and aldehydes The $\pi \rightarrow \pi^*$ absorption of an isolated carbonyl group lies in the far ultraviolet, but conjugation with a double bond moves the absorption into the region of about 215 to 280 nm (the weak $n \rightarrow \pi^*$ bond is shifted above 310 nm in most cases, but this particular transition is not amenable to numerical correlation).

Woodward's rules‡ for α,β-unsaturated carbonyl compounds of the general formula

$$\begin{matrix} \delta & & \gamma & & \beta & & \alpha & & 1 & & \\ C & = & C & - & C & = & C & - & C & = & O \end{matrix}$$

are as follows:

1 Assign a base value of 215 nm to a parent α,β-unsaturated ketone or 210 nm to a parent α, β-unsaturated aldehyde. (One exception: if the α,β-unsaturated ketone is contained in a *five*-membered ring system, assign a base value of only 200 nm.)

2 Add 30 for each additional double bond in conjugation.

3 Add 10 for an α-alkyl substituent.

4 Add 12 for a β-alkyl substituent.

5 Add 18 for an α- or higher-alkyl substituent.

† Fieser, *loc. cit.*

‡ R. B. Woodward, *J. Am. Chem. Soc.*, **63:** 1123 (1941).

6 Add 5 for each exocyclic double bond.

7 Add 39 if a homoannular diene is present.

In addition to these principal rules, there are a large number of rules for auxochromic substituents in the α, β, and γ positions. In general, these substituents cause appreciable bathochromic shifts, but the interested reader is referred elsewhere.†

In addition to substituent rules, solvent corrections are necessary since the spectra of these compounds are affected significantly by the polarity of the solvent. For the above rules, ethanol is assumed to be the standard solvent, and Table 2-4 gives the wavelength correction for other solvents. Note that the solvents are listed in the order of increasing polarity and that increasing polarity causes a bathochromic shift.

Example 2-2 Calculate the absorption wavelength for ergosta-4,6,8,(9), 22-tetraene-3-one (8):

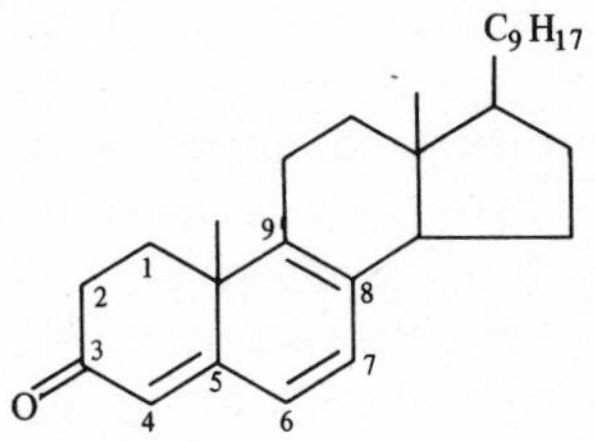

Answer

Parent ketone	215 nm
Two extra double bonds (6,7 and 8,9) in conjugation	60
One β-alkyl substituent (on C_5)	12
One substituent on C_8	18
Two substituents on C_9	36
One exocyclic double bond (4,5)	5
Homoannular diene component	39
Total	385 nm

The observed absorption is at 388 nm, with ϵ = 12,300.

Substituted benzene rings Benzene absorbs strongly in the far ultraviolet at 184 nm (ϵ = 50,000) and in the near ultraviolet at 204 nm (ϵ = 7000) and 254 nm (ϵ = 200). All these bands are ascribed to $\pi \rightarrow \pi^*$ transitions, the third band at 254 nm showing extensive vibrational fine structure in the vapor state and lesser structure in solution.

† Scott, *op. cit.*, or Ref. 8, p. 23.

Monosubstituted Benzene When benzene is substituted with a single functional group, the fine-structure bands generally tend to be diminished in complexity, the intensity is increased, and there is a shift to longer wavelength. Though empirical rules are not available, some of the trends can be seen from Table 2-5. Substitution by halogen or alkyl causes only a slight shift, with a small increase in molar absorptivity, but substitution by groups carrying nonbonding or π electrons such as —OH, —NH_2, or —CHO causes a very pronounced shift and a greatly intensified absorption. Note that the aniline cation, —NH_3^+, has almost the same spectrum as benzene, whereas the spectrum of aniline, —NH_2, is much different. This is because in aniline there is conjugation between the lone pair of electrons on the nitrogen and the π electrons of the benzene ring (n-π conjugation), whereas in the aniline cation there is no longer a free pair of electrons and thus no n-π conjugation.

Disubstituted Benzenes In disubstituted benzenes, the trends can be understood by dividing the compounds into two classes. In class I, the substituents are *electronically complementary,* i.e., one group is *electron-donating* and the other is *electron-withdrawing*, and are situated para to each other. An example is *p*-nitroaniline, which absorbs at 375 nm with ϵ = 16,000. Compounds of this class absorb at appreciably longer wavelength than would be predicted from either constituent considered separately because of the extension of the chromophore from the electron-donating group through the benzene ring to the electron-withdrawing group.

In class II are all other disubstituted benzenes, including groups situated ortho or meta to each other, and para groups of similar electron-donating properties. An example is *p*-dinitrobenzene which absorbs at 260 nm with ϵ = 13,000. In compounds of this class, the observed spectrum is usually closer tc that of the separate noninteracting chromophores.

A special case of class I compounds that yield to empirical rules for predicting the absorption maximum are compounds of the type R—C_6H_4—COX,

TABLE 2-4 Solvent Corrections for α,β-Unsaturated Ketones

Solvent	Correction, nm
Hexane	−11
Ether	−7
Dioxane	−5
Chloroform	−1
Ethanol, methanol	0
Water	+8

TABLE 2-5 Ultraviolet Absorption Maxima of Some Monosubstituted Benzenes in Water†

C_6H_5X, where —X is	Primary band		Secondary band	
	λ_{max}, nm	ϵ	λ_{max}, nm	ϵ
—H	203.5	7,400	254	204
$—NH_3^+$	203	7,500	254	169
$—CH_3$	206.5	7,000	261	225
—I	207	7,000	257	700
—Cl	209.5	7,400	263.5	190
—Br	210	7,900	261	192
—OH	210.5	6,200	270	1,450
$—OCH_3$	217	6,400	269	1,480
$—CO_2H$	230	11,600	273	970
$—NH_2$	230	8,600	280	1,430
$—COCH_3$	245.5	9,800		
—CHO	249.5	11,400		
$—NO_2$	268.5	7,800		

† From J. R. Dyer, "Applications of Absorption Spectroscopy of Organic Compounds," p. 18, Prentice-Hall, Inc., Englewood Cliffs, N.J., 1965, by permission.

where R is an electron-donating (ortho-para-directing) group, —COX is an electron-withdrawing carbonyl group, and X is alkyl, —H,—OH, or —OAlk. The following rules allow the calculation of the strongest absorption band in the near ultraviolet. For compounds of the formula $R—C_6H_4—COX$:

1 Assign a base value of 250 nm when X = H; when X = alkyl or a ring residue, assign a base value of 246 nm; when X = —OH or —OAlk, assign a base value of 230 nm.

2 Add 3 when R is an ortho- or meta-substituted alkyl or ring residue.

3 Add 10 when R is a para-substituted alkyl or ring residue.

4 Add 7 when R is an ortho- or meta-substituted —OH, —OMe, or —OAlk.

5 Add 25 when R is a para-substituted —OH, —OMe, or —OAlk.

Rules for other substituents are available.†

† Scott, *op. cit.*, or Ref. 8, p. 31.

Example 2-3 Calculate the absorption wavelength for 6-methoxytetralone:

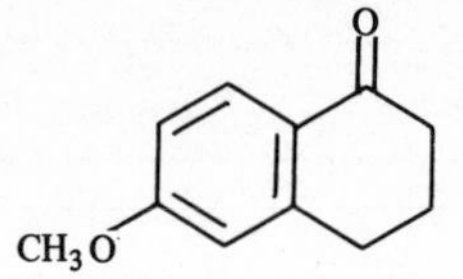

Answer

Parent chromophore with X = ring residue	246 nm
Ortho ring residue	3
Para methoxyl	25
Total	274 nm

The observed absorbance is at 276 nm, with ϵ = 16,500.

Trends with polysubstituted benzene rings are complicated, particularly when steric hindrance is involved.

Unsaturated acids, esters, nitriles, and amides α,β-Unsaturated acids, esters, nitriles, and amides generally show a high-intensity (ϵ = 10,000 to 20,000) absorption peak in the region 205 to 225 nm. The empirical rules for α,β-unsaturated acids and esters are as follows:

1 Assign a base value of 197 nm to the acid or ester.

2 Add 30 for a double bond extending the conjugation.

3 Add 10 for each alkyl substituent present.

4 Add 5 for an exocyclic double bond.

5 Add 5 for an endocyclic double bond in a five- or seven-membered ring.

The shift in going from an acid to an ester is usually less than 2 nm.

α,β-Unsaturated nitriles have been little studied, but usually they absorb at slightly shorter wavelengths than the corresponding acids.

α,β-Unsaturated amines also absorb at shorter wavelengths than the corresponding acids, and usually near 200 nm ($\epsilon \approx$ 8000).

Example 2-4 Calculate the absorption wavelength for $(CH_3)_2C{=}CH{-}COOH$.

Answer

Parent acid		197 nm
2 alkyl substituents	2×10	20
Total		217 nm

The observed absorbance is 216 nm, with ϵ = 12,000.

$\sigma \rightarrow \sigma^*$ ABSORPTION BANDS

Sigma electrons are the bonding electrons in single bonds, and $\sigma \rightarrow \sigma^*$ transitions occur as a result of absorption in the far ultraviolet. Relatively little work has been done in this region of the spectrum, because of instrumental difficulties.

$n \rightarrow \sigma^*$ ABSORPTION BANDS

A substituent with unshared electron pairs introduces the possibility of $n \rightarrow \sigma^*$ transitions. The most common examples are saturated compounds containing heteroatoms such as sulfur, nitrogen, bromine, or iodine (which absorb in the wavelength region around 200 nm) and compounds containing oxygen and chlorine (which absorb at shorter wavelengths). Table 2-6 lists the absorption wavelength and intensity of some simple compounds containing the common heteroatoms. Heteroatoms of increasing electronegativity cause the absorption maximum to be shifted to shorter wavelengths, consistent with the higher binding energy of the ground-state electrons. That most saturated oxygen-containing compounds absorb well below 200 nm makes it possible for compounds like water, alcohol, and ethers to be used as solvents in the near ultraviolet.

If two heteroatoms have overlapping lone-pair orbitals, the absorption maxima are shifted to longer wavelengths, as illustrated in Table 2-6 with some polyhalides and disulfides.

CHARGE-TRANSFER ABSORPTION BANDS

Charge-transfer absorption is a very important type of absorption process, being responsible for the intense ultraviolet (and sometimes visible) absorption of a large number of inorganic and organic species. In this process, an electron is transferred from one part of the system to another. This can be exemplified for a complex ion as

$$M—L + h\nu \rightarrow M^+—L^- \tag{2-3}$$

where M is a metal ion and L is a ligand. The opposite case, of an electron being transferred from the ligand to the metal, also occurs. In all such spectra, two components can be identified, an electron donor (D) and an electron acceptor (A). (Alternatively, these components can be respectively identified as a Lewis base and a Lewis acid or as a reductant and an oxidant.

TABLE 2-6 Absorption Wavelength and Intensity Due to $n \rightarrow \sigma^*$ Transitions of Some Saturated Molecules Containing Heteroatoms†

Heteroatom	Compound	λ_{max}, nm	ϵ_{max}
—Cl	CH_3Cl	173	200
—O/	CH_3OH	184	150
—Br	CH_3Br	204	200
	n-PrBr‡	208	320
—I	CH_3I	259	365
	CH_2I_2	292	1320
	CHI_3	349	230
—N<	CH_3NH_2	215	600
	$(CH_3)_3N$	227	820
—S/	$(CH_3)_2S$	210	1020
	Et_2S‡	215	1600
—S—S—			
Acyclic	$C_2H_5SSC_2H_5$	202	2100
In six-membered ring	Tetramethyl disulfide	295	300
In five-membered ring	Trimethyl disulfide	334	160

† Data from J. C. D. Brand and G. Eglinton, "Applications of Spectroscopy to Organic Chemistry," Oldbourne, London, 1965, and Ref. 6.
‡ In n-hexane solution; all other compounds were in the vapor state.

Some common examples of charge-transfer absorption are the intense red color of the complex formed between chloranil (yellow) and hexamethylbenzene (colorless), the deep brown color due to iodine and aromatic hydrocarbons, the strong visible absorption of quinhydrone, and the strong visible absorption of polynitro- and picric acid components with aromatic hydrocarbons, amines, and phenols. Charge transfer also explains the ultraviolet absorption of many hydrated inorganic ions, such as

$$Cl^-(H_2O)_n \rightarrow Cl(H_2O)_n^- \quad (2\text{-}4)$$

$$Fe^{2+}(H_2O)_n \rightarrow Fe^{3+}(H_2O)_n^- \quad (2\text{-}5)$$

$$Fe^{3+}OH^- \rightarrow Fe^{2+}OH \quad (2\text{-}6)$$

The color produced when transition-metal ions react with chromogenic analytical reagents is likewise usually due to charge-transfer bands. Examples are the thiocyanate test for Fe^{3+} (forming Fe^{3+} CNS^-), the use of peroxide in estimating Ti^{4+} (forming $Ti^{4+}—O_2H^-$), and the use of phenols in detecting Fe^{3+}, Cu^{2+}, and Ti^{4+} (forming complexes of the type $M^{n+}—O—R^-$).

There is also intramolecular charge-transfer absorption in substituted organic molecules, as exemplified by

$$C_6H_5—NR_2 \xrightarrow{h\nu} {}^{-}C_6H_5{=}\overset{+}{N}R_2 \tag{2-7}$$

$$C_6H_5—C(R){=}O \xrightarrow{h\nu} {}^{+}C_6H_5{=}C(R)—O^- \tag{2-8}$$

Calculation of Absorption Wavelength The energy (and thus the wavelength) of the electronic transition, generalized as

$$D—A \xrightarrow{h\nu} D^+—A^- \tag{2-9}$$

can be estimated by the equation

$$h\nu = I_D - E_A - C \tag{2-10}$$

where I_D = ionization potential of donor
E_A = electron affinity of acceptor
C = mutual electrostatic energy of D^+ and A^- relative to that of D and A

If only one component of the complex carries a charge, as in $Cl^-(H_2O)_n$ and other hydrated inorganic ions, and if the charge is simply redistributed in going from the ground to the excited state, the coulombic term C will be zero. Actually, as in $Cl^-(H_2O)_n$, there will be a small coulombic contribution arising from the different polarization energy of the H_2O in the field of the Cl^-, compared with that of Cl in the field of H_2O^-, but this can usually be neglected.

In the charge-transfer bands of bivalent transition-metal ions, the energy of the absorption transition is directly proportional to the redox potential of the system

$$M^{2+} \rightleftarrows M^{3+} + e$$

This suggests that the charge-transfer transition is

$$M^{2+}(H_2O)_n \xrightarrow{h\nu} M^{3+}(H_2O)_n^-$$

rather than

$$M^{2+}(H_2O)_n \xrightarrow{h\nu} M^+(H_2O)_n{}^+$$

***Example** 2-5* Gaseous sodium chloride will absorb in the ultraviolet due to a charge-transfer transition. Calculate the wavelength at which absorption may be expected. In the gas phase, $I_{Na} = 5.14$ eV, $E_{Cl} = 3.82$ eV, and $C \approx 6.2$ eV.

Answer

$$h\nu = I_D - E_A - C$$

$$= 5.14 - 3.82 - 6.2 = -4.9 \text{ eV}\dagger$$

To convert the energy (in electronvolts) to wavelength (in nanometers), the following conversion factor may be used:

$$\Delta E \text{ (eV)} \lambda \text{ (nm)} = 1240$$

$$\lambda = \frac{1240}{4.9} = 253 \text{ nm}$$

(The observed absorption of gaseous NaCl is about 248 nm.)

***Example** 2-6* The benzene–Br_2 complex absorbs strongly at about 292 nm (ϵ = 6880). The ionization potential of bromine has been established at 9.24 eV, and the coulombic energy of the complex is about 3.22 eV. Calculate the electron affinity for Br_2.

Answer

$$h\nu = I_D - E_A - C$$

$$h\nu = \Delta E = \frac{1240}{\lambda \text{ (nm)}} = \frac{1240}{292 \text{ nm}} = 4.24 \text{ eV}$$

$$4.24 \text{ eV} = 9.24 \text{ eV} - E_A - 3.22 \text{ eV}$$

$$E_A = 1.78 \text{ eV}$$

† The minus sign implies that the ionic state, Na^+Cl^-, is more stable (has a lower energy) than the covalent state, NaCl, and therefore the charge-transfer reaction is

$$Na^+Cl^- \xrightarrow{h\nu} NaCl$$

Hence, only the absolute value of $h\nu$ has quantitative significance, the sign merely giving the direction of the reaction.

Since electron affinities are often difficult to measure by other means, charge-transfer spectra are useful for this purpose.

LIGAND-FIELD ABSORPTION BANDS

Transition-metal ions and compounds exhibit two kinds of electronic spectra. Besides the intense ($\epsilon \approx 10^3$ to 10^4) charge-transfer absorption bands described above, they exhibit weak ($\epsilon \approx 10^{-1}$ to 10^2) ligand-field absorption bands. Whereas charge-transfer bands occur chiefly in the ultraviolet (occasionally extending into the visible region), ligand-field bands usually are found in the visible region, though they may occasionally extend into the near infrared or near ultraviolet regions (see Fig. 2-1). The characteristic color of transition-metal ions and complexes can almost always be attributed to ligand-field absorption.

The name ligand field arises because a transition-metal atom or ion in a vacuum has *d* orbitals which are degenerate, or equal in energy, whereas the presence of ligands produces an electrostatic field which splits the various orbitals into levels of different energy. If the *d* or *f* subshell is unfilled, radiation of the proper wavelength can cause electronic transitions between these levels of different energy. An introductory picture of ligand-field splitting can be obtained by considering the important case of an *octahedral* complex, like that of $Ti(H_2O)_6{}^{3+}$. Ti^{3+} contains one electron in a *d* orbital ($3d^1$), and the octahedral shape of the hydrated ion can be visualized by bringing six H_2O molecules in along the coordinate axes toward the metal ion at the origin. The five *d* atomic orbitals are pictured on the *x, y, z* axes in Fig. 2-5, and in the free state all have equivalent energy (in other words, the $3d^1$ electron in Ti^{3+} would have an equal probability of being in any one of the five *d* orbitals). However, once the six H_2O ligands have been brought close to the Ti^{3+} ion, electrical repulsion between the negative end of H_2O (at the oxygen atom) and

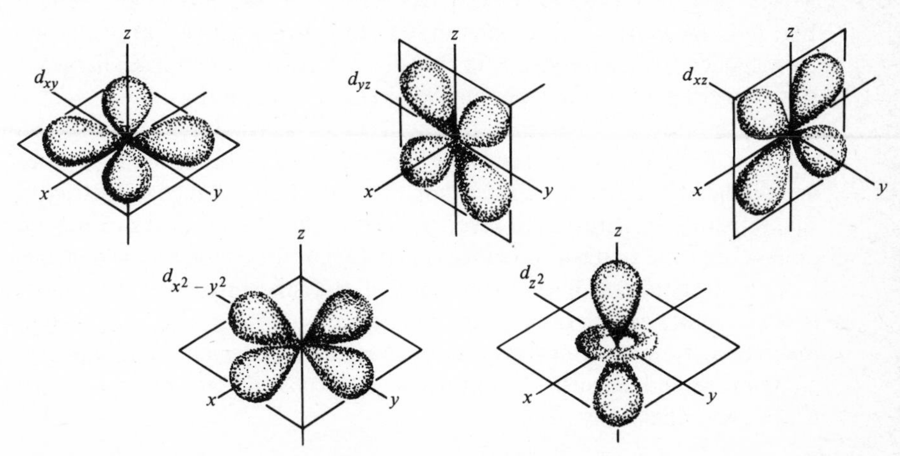

FIGURE 2-5 Schematic diagram of the five *d* atomic orbitals. (*From G. M. Barrow, "Physical Chemistry," 2d ed., p.* 326, *McGraw-Hill Book Company, New York,* 1966, *by permission.*)

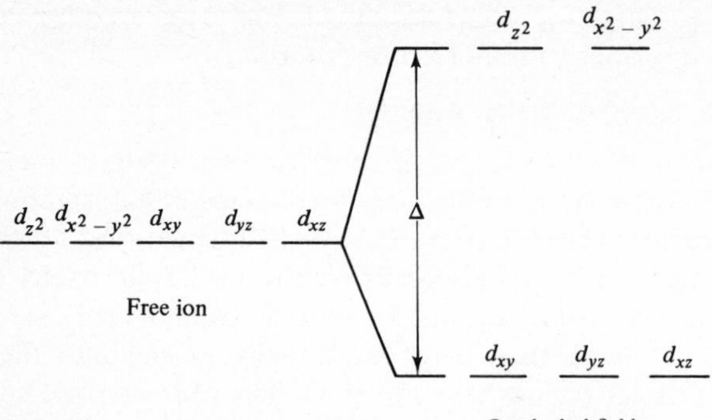

FIGURE 2-6 Splitting of *d* orbitals in an octahedral field.

the *d* electron in Ti^{3+} will raise the energy of a *d* orbital directed along the axes and lower the energy of a *d* orbital concentrated between the axes. Thus, in an octahedral complex, two *d* orbitals, the d_{z^2} and the $d_{x^2-y^2}$, will be raised in energy, and the remaining 3*d* orbitals, d_{xy}, d_{yz}, and d_{xz}, will be lowered. This splitting process is diagrammed in Fig. 2-6, where the extent of the splitting is designated by Δ (sometimes the arbitrary term $10Dq$ is used), where Δ is a measure of the strength of the ligand field. For $Ti(H_2O)_6^{3+}$, $\Delta = 20{,}400\ \text{cm}^{-1}$, corresponding to a wavelength of 490 nm, and its absorption spectrum is shown in Fig. 2-7. The weakness of the absorption band ($\epsilon \approx 5$) results from the fact that transitions of this type are forbidden by the rules of quantum mechanics, and the breadth of the band is caused by the fact that the electronic transition is accompanied by a host of vibrational excitations, spread over a range of almost 300 nm. These features of intensity and breadth are fairly general for *d-d* transitions (or ligand-field spectra), although in certain cases, for example, Mn(II) and Cr(III) spectra, relatively narrow absorption bands are found. In Fig. 2-7, the sharply rising band in the near-ultraviolet region (below 400 nm) corresponds to a charge-transfer transition. Sometimes intense charge-transfer bands overlap and largely obscure weaker *d-d* transitions.

It should be noted that the *pattern* of energy-level splittings brought about by a ligand field depends on the symmetry of the complex formed, i.e., on whether the complex formed has a tetrahedral, octahedral, tetragonal, or square-planar structure. However, it suffices for the purposes of this book to discuss only the pattern of octahedral field splitting, which is one of the commonest. Regardless of the symmetry of the structure, it is not possible to make absolute calculations of the ligand-field splitting, and thus from purely theoretical considerations we cannot predict the wavelength at which a given complex should absorb. Nevertheless, it is possible to make the following useful generalizations:

1 For a given ligand, the ligand-field splitting Δ is greater the higher the ionic charge on the central metal ion because a higher charge increases the electrostatic attraction between metal ion and ligand. In general, for

divalent ions, Δ is in the order of 10,000 cm^{-1} (1000 nm), and for trivalents this splitting is increased to about 20,000 cm^{-1} (500 nm).

2 For a given ligand, the ligand-field splitting Δ is greater the larger the *d* orbitals of the metal ion. Larger *d* orbitals extend farther into space and therefore interact more strongly with the ligands. For example, the splitting is about 40 to 80 percent larger for the 4*d* and 5*d* orbitals than for the 3*d* orbitals.

3 For a given metal ion, the ligand-field splitting Δ depends on the nature of the ligand, being dependent on its charge density distribution and polarizability. With few exceptions, the order of increasing Δ (sometimes called the *spectrochemical series*) is I^-, Br^-, Cl^-, F^-, OH^-, $C_2O_4^{2-}$, C_2H_5OH, H_2O, NCS^-, NH_3, ethylenediamine, NO_2^-, 1,10-phenanthroline, CN^-, CO.

A good example of the visible effects of the spectrochemical series on ligand-field splitting is the effect of various ligands on the color of copper salts and solutions. For example, $CuSO_4 \cdot 5H_2O$ and aqueous solutions of the Cu^{2+} ion, $Cu(H_2O)_4^{2+}$, appear pale blue, but cuprammine salts and ammoniacal cupric solutions, $Cu(NH_3)_4^{2+}$, show a much more intense blue-violet color due to the increase in ligand field which occurs when H_2O is replaced by NH_3. On the other hand, anhydrous $CuSO_4$ is colorless because the SO_4^{2-} group provides so small a ligand field that the *d-d* absorption band moves into the infrared region.

Although the fields of most ligands can be explained largely with electrostatic forces (crystal-field theory), some form of covalent bonding between ligands and the central metal atom is usually present also. With most ligands there is a σ-bond contribution to the bonding, and in the particularly strong fields of ligands like CO, CN^-, 1,10-phenanthroline, and NO_2^-, the extra large splittings are caused by π-bond formation with the central metal atom, which markedly increases the magnitude of the ligand-field splitting.

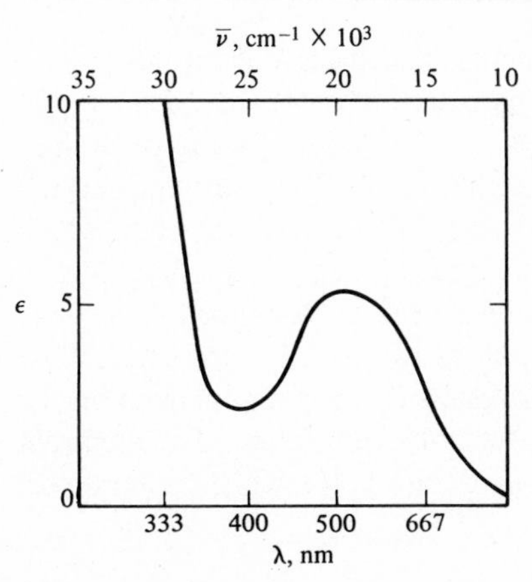

FIGURE 2-7 Absorption spectrum of $Ti(H_2O)_6^{3+}$. (*From F. A. Cotton and G. Wilkinson, "Advanced Inorganic Chemistry," 2d ed., p. 573, John Wiley & Sons, Inc., New York, 1966, by permission.*)

TABLE 2-7 Some Representative Ligand-Field Splittings Δ in Wave Numbers (cm^{-1})†,‡

Electron configuration	Metal ion	$6Br^-$	$6Cl^-$	$6F^-$	$6H_2O$	$6NH_3$	$6CN^-$
$3d^1$	Ti^{3+}				20,300		
$3d^3$	V^{2+}				12,600		
	Cr^{3+}		13,600		17,400	21,600	26,300
$4d^3$	Mo^{3+}		19,200				
$3d^5$	Mn^{2+}	7,000	7,500	7,800	7,800		
	Fe^{3+}			10,000	13,700		30,000
$3d^6$	Fe^{2+}				10,400		33,000
	Co^{3+}				18,600	23,000	34,000
$4d^6$	Rh^{3+}	18,900	20,300		27,000	33,900	
$5d^6$	Ir^{3+}	23,100	24,900				
$3d^8$	Ni^{2+}	7,000	7,300		8,500	10,800	
$3d^9$	Cu^{2+}		6,500		12,600	15,100	

† To convert wave numbers in cm^{-1} to nanometers use 1/wave numbers (cm^{-1}) $\times 10^7$ = nm.

‡ Data from R. A. Plane and M. J. Sienko, "Physical Inorganic Chemistry," p. 56, Benjamin, New York, 1963, and Ref. 11, p. 523.

Table 2-7 gives some representative values of ligand-field splittings for various transition-metal ions and ligands. The three generalizations given above fit these spectral data fairly well.

In addition to the ligand-field (*d-d*) spectra of transition metals, the lanthanides and actinides (with unfilled 4*f* and 5*f* subshells) likewise have ligand-field (*f-f*) bands. These bands tend to be rather narrow, and good progress has been made in interpreting them by ligand-field theory.

2-1C The Basis of Color and Colorimetry

The phenomenon of color deserves to be discussed separately from the general theory of visible radiation because of the importance of color judgments to the scientific observer. Color has both a physical and a physiological basis in that, on one hand, it can be correlated mathematically and absolutely and, on the other, the human response to color varies from person to person. Light which is visible to the average eye is generally considered to range from about 400 to 700 nm, though the extremes may be as wide as 380 to 780 nm. If the eye is stimulated by light containing all wavelengths of the visible region, the effect is that of *white* light. The sensation of color results if wavelengths from one or more portions of the visible region are appreciably diminished. The apparent color that results is always the complement of the color which is removed.

Table 2-8 gives the colors of radiation of successive wavelength regions, together with their complementary colors. The wavelength ranges are only approximate, as they differ slightly from observer to observer.

The complementary relationship given in Table 2-8 is useful for predicting absorption wavelengths for compounds of known color. For example, a yellow solution can be predicted to absorb blue light of about 450 to 480 nm. Similarly, to analyze for a MnO_4^- (purple) solution, green light of about 500 to 560 nm should be used, or if a filter is to be used, it should be green, since it will then transmit green radiation.

Colorimetry is an optical method of analysis based on *color comparison*, using the eye as the detector. An accuracy of ± 5 percent is about the best that can be expected. In general, whatever may be done with a *colorimeter* or *color comparator* (such as the Duboscq colorimeter) can be done more accurately with a spectrophotometer. Therefore, although colorimetric procedures are still occasionally used, the reader is referred elsewhere for specific details.† A spectrophotometer has the following advantages over a colorimeter:

1 Since a spectrophotometer can restrict the light to a narrow wavelength region of maximum absorption, maximum *sensitivity* is obtained. (Light that is used at wavelengths where absorption does not occur is wasted, and in fact lowers the sensitivity of the detector for the wavelengths of interest.) In addition, Beer's law has a better chance of holding, thus improving accuracy and convenience (see Sec. 1-5).

† For example, E. B. Sandell, "Colorimetric Determination of Traces of Metals," 3d ed., Wiley, New York, 1959.

TABLE 2-8 Relation between Absorption and Observed Color†

Wavelength region removed by absorption, nm	Color absorbed	Complementary color of the residual light, as seen by the eye
400–450	Violet	Yellow-green
450–480	Blue	Yellow
480–490	Green-blue	Orange
490–500	Blue-green	Red
500–560	Green	Purple
560–580	Yellow-green	Violet
580–600	Yellow	Blue
600–650	Orange	Green-blue
650–750	Red	Blue-green

† Data from W. J. Blaedel and V. W. Meloche, "Elementary Quantitative Analysis," 2d ed., p. 508, Harper & Row, New York, 1963, and Ref. 1, p. 49.

2 Since a spectrophotometer can avoid using wavelengths where other substances interfere, *higher selectivity* can be obtained. A colorimeter is generally useful for only one colored component, whereas a spectrophotometer allows analysis of multicomponent colored systems.

3 *Eye fatigue* and partial *color blindness* severely limit routine colorimetric matching procedures but are of no consequence when a spectrophotometer with electronic detection is used.

4 With color-matching procedures it is necessary to make continuous use of standard solutions or artificial standards, whereas in spectrophotometric analysis, once a working curve is established, it is only necessary to recheck points on the working curve occasionally.

2-2 INSTRUMENTATION

2-2A Nomenclature of Instruments

The two main types of instruments in use for measuring the absorbance of ultraviolet or visible radiation by a solution are *photometers* and *spectrophotometers*.

Photometers Technically, the name photometer is applied to any device for measuring the intensity of radiation, but by common usage the term is restricted to a relatively inexpensive instrument using a *filter* to isolate a narrow-wavelength region and a *photocell* or *phototube* to measure the intensity of radiation.

Spectrophotometers A spectrophotometer, on the other hand, is a more elaborate instrument containing a *monochromator* instead of a filter. The monochromator allows a continuous variation in the selection of wavelength, and thus allows a large wavelength region to be scanned. In addition, a spectrophotometer uses the most sensitive detectors available, usually *phototubes* or *photomultipliers*.

Since some confusion exists in instrument terminology, for the purposes of this book it will be useful to define some other instrument names and types. These definitions are not universally accepted, but they are in reasonably general agreement with popular usage.

Colorimeters A colorimeter is a simple instrument, *using the human eye as a detector*, for comparing the *color* of a substance being sought with that of a standard solution. This type of instrument was discussed briefly in Sec. 2-1*D* and will not be discussed further. (It should be noted that *any* instrument used for making absorption measurements in the visible region could be called a colorimeter, and, in fact, some commercial filter photometers are called colorimeters.)

Spectroscopes A spectroscope is an instrument that permits *spectral lines* to be viewed with the *eye*. (The suffix *-scope* comes from the Greek verb meaning to view or examine.) A spectroscope is sometimes useful for qualitative and semiquantitative analyses of elements having emission lines in the visible region of the spectrum.

Spectrographs A spectrograph is an instrument that records spectral lines on a *photographic plate*. (The suffix *-graph* comes from the Greek verb meaning to write.) A spectrograph is generally used only to record *emission* or *Raman* spectra. For examples, see Chaps. 4 and 6.

Spectrometers The name spectrometer is a *general* name that can be applied to any spectral instrument using *electrical* methods of detection. Spectrometer results are usually presented on a graphical recorder, though the readout may range all the way from a simple meter to a digital display. A spectro*photo*meter is thus a spectrometer that measures *photons*, in contrast to a spectrometer, which measures particles of finite mass, e.g., a *mass* spectrometer or a *photoelectron* spectrometer, both of which fall outside the category of optical methods described in this book. Whereas spectrometer is used throughout the electromagnetic spectrum, the use of spectrophotometer is generally restricted to the region from the ultraviolet through the infrared.

2-2B Basic Components of Photometers and Spectrophotometers

All photometers and spectrophotometers, no matter how simple or how sophisticated, contain three basic components: (1) a *source* of radiant energy, (2) a *filter* or *monochromator* for isolating a narrow band of radiant energy, and (3) a *detector* for measuring the radiant energy transmitted through the sample. The sample solution is usually placed after the monochromator and just before the detector. While these three components are the fundamental units, the filter or monochromator has associated with it an optical system for producing a parallel beam of radiation and guiding it to the detector. Associated with the detector will be a readout system for presentation of the detector response.

RADIATION SOURCES

Ideally, a radiation source should be continuous over a wide range of spectrum and have high intensity, and the intensity should not vary appreciably with the wavelength. In practice, most sources have an output intensity that varies with wavelength.

Sources of radiation in the ultraviolet and visible regions of the spectrum may be divided into *thermal* sources, in which the radiation is the result of high temperature, and sources depending on *electric discharge* through gases.

Thermal Sources In the visible region, a *tungsten-filament* lamp (most widely used) is a thermal source. Unfortunately, the tungsten lamp emits the major portion of its energy in the near-infrared region of the spectrum, as shown in Fig. 2-8. At the usual operating temperature of about 3000 K, only

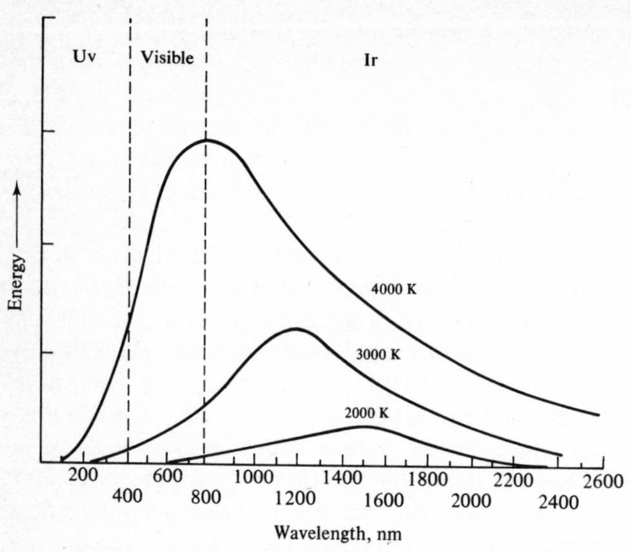

FIGURE 2-8 Spectral distribution curves of a tungsten (blackbody) lamp. (*After H. A. Strobel, "Chemical Instrumentation," 2d ed., p.* 47, *Addison-Wesley Publishing Company, Reading, Mass.*, 1973, *by permission.*)

about 15 percent of the total radiant energy falls in the visible region, and at 2000 K it is only 1 percent. Increasing the operating temperature above 3000 K greatly increases the total energy output and shifts the wavelength of maximum intensity to shorter wavelengths, but the life expectancy of the lamp is drastically shortened (the melting point of tungsten is about 3600 K). To remove the unwanted infrared radiation, a heat-absorbing filter is often inserted between the lamp and sample holder.

The spectral distribution shown in Fig. 2-8 is typical of blackbody or incandescent sources (both names implying that their radiation is due to the *temperature* of the solid material, with only small dependence on the chemical nature of the solid). Tungsten has proved to be the most satisfactory material for lamp filaments, though the *carbon arc* (a thermal source operating near 4000 K) is occasionally used when a more intense source of visible light is needed. The continuous-emission spectrum from heated solids results from atoms and molecules being thermally excited to virtually a continuum of closely spaced electronic and vibrational energy levels. The energy levels are smeared together by the mutual interaction of atoms that are closely packed together. Qualitatively, an increase in temperature should cause a greater output intensity as well as a shift to shorter wavelengths, since a greater number of atoms are excited, and to higher energy levels. The wavelength of maximum emission (λ_{max}, in nanometers) can be estimated from the Wien displacement law

$$\lambda_{max} T \approx 3.0 \times 10^6 \qquad (2\text{-}11)$$

where T is the absolute temperature in kelvins. Thus, the wavelength of maximum emission varies inversely with the absolute temperature, which agrees with the data in Fig. 2-8.

Electric-Discharge Sources In the ultraviolet region, various kinds of electric-discharge sources are used, including the hydrogen or deuterium lamp, the xenon-discharge lamp, and the mercury arc. In all cases the excitation is caused by the passage of electrons through a gas, and the collisions between electrons and gas molecules cause electronic, vibrational, and rotational excitations in the gas molecules. At very low pressures of gases, only line spectra are emitted, but at higher pressures band spectra and continuous spectra result, since again the mutual interaction between atoms leads to a spread in energy levels.

Hydrogen-discharge lamps are commonly used for a continuous radiation source in the ultraviolet region. Although high-voltage lamps (about 2500 volts ac) have been developed, low-voltage types (about 40 V dc) are most often used. The pressure of hydrogen gas is usually about 0.2 to 5 mm. Figure 2-9 gives a typical near-ultraviolet spectral distribution curve for a hydrogen-discharge lamp. The hydrogen continuum extends down to about 165 nm, but the envelope or window material generally limits the transmission. Quartz windows absorb below 200 nm, and fused silica absorb below 185 nm. The lamp's upper limit of usefulness is about 375 nm, above which the output energy of the continuum is too small (see Fig. 2-9), but the energy from a tungsten lamp can be used above 375 nm.

Deuterium lamps produce a wavelength range similar to that of hydrogen, but the intensity is 3 to 5 times the intensity of a hydrogen lamp of comparable design and wattage.

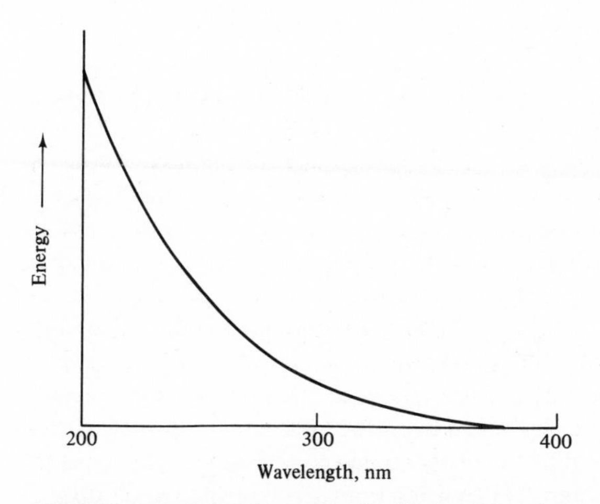

FIGURE 2-9 Spectral energy distribution of a low-voltage hydrogen-arc lamp. (*After L. R. Koller, "Ultraviolet Radiation,"* 2d *ed., p.* 67, *John Wiley & Sons, Inc., New York,* 1965, *by permission.*)

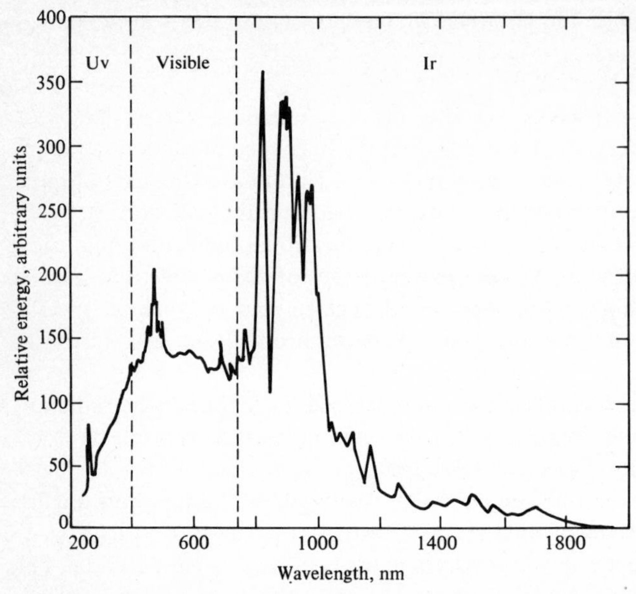

FIGURE 2-10 Spectral energy distribution of a typical high-pressure xenon lamp.

The *xenon-discharge lamp* has shown considerable promise as a source for the ultraviolet region, operating from a low-voltage dc source similar to that of the hydrogen lamp but at xenon pressures in the range of 10 to 30 atm. An intense arc is formed between two tungsten electrodes separated by about 8 mm. A typical spectral-energy-distribution curve is shown in Fig. 2-10. The intensity in the near ultraviolet is actually much greater than that of the hydrogen lamp, but the even greater intensity in the visible region (coupled with some very intense lines in the near infrared) poses potential stray-radiation problems. Another difficulty is the spatial stability of the extremely small, intense spots of light produced.

The *mercury arc*, a standard source for much ultraviolet work, is generally not suitable for continuous spectral studies because of the presence of sharp lines or bands superimposed on a continuous background, even at very high pressures, for example, 110 atm. A typical spectral energy distribution is shown in Fig. 2-11. The low-pressure mercury arc, however, is very valuable for calibration. Mercury emits 35 separate lines from 127 to 405 nm and 24 lines in the region of 200 to 400 nm.

Source *stability* is important, particularly with single-beam instruments. The energy output of these sources depends on the voltage applied across the lamp, and the detector photocurrent is proportional to the lamp voltage raised to some power that is larger than 1 (3 to 4 for tungsten lamps). Therefore the source *voltage* must be very stable; stability is achieved by using storage batteries or constant-voltage transformers and electronic voltage regulators.

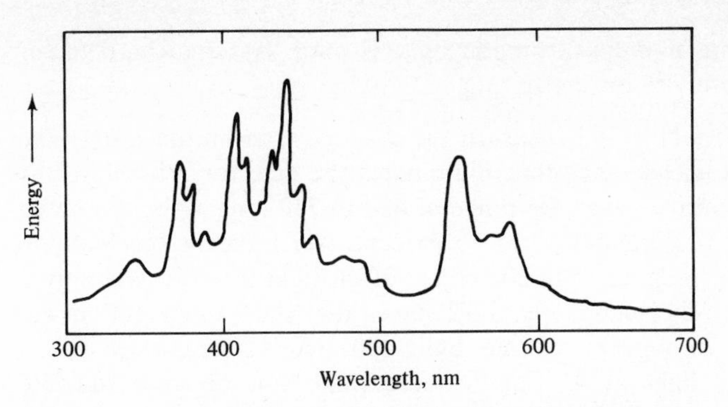

FIGURE 2-11 Spectral energy distribution of high-pressure mercury-arc lamp. (*After L. R. Koller, "Ultraviolet Radiation," 2d ed., p. 43, John Wiley & Sons, Inc., New York,* 1965, *by permission.*)

FILTERS AND MONOCHROMATORS

It is vitally important to the selectivity, accuracy, and sensitivity of all absorbance measurements to isolate a narrow band of wavelengths emanating from the broad-spectrum source. The importance of spectral isolation on *selectivity* and *accuracy* was discussed in Sec. 1-5. The importance of spectral isolation to the *sensitivity* of absorption measurements can be illustrated with a simple example.

Example 2-7 Figure 2-12 gives the transmission spectrum of a 5×10^{-4} *M* solution of *m*-nitroaniline-*N,N*-diacetic acid as measured between 340 and 750 nm with a 1-cm cell. Estimate the transmittance of this solution if a white-light source with a uniform output of radiation from 400 to 750 nm is used without a monochromator, and compare this to the transmit-

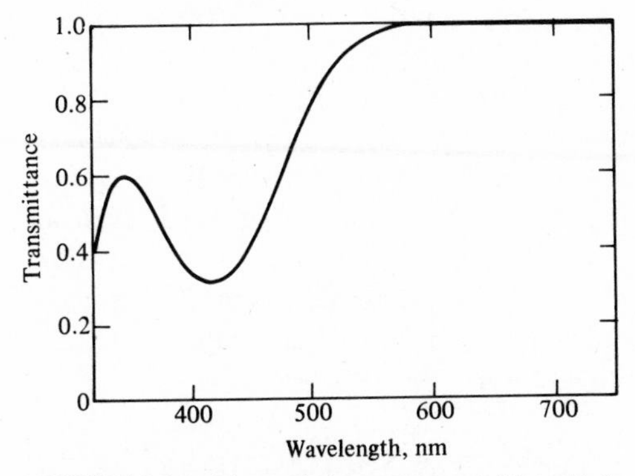

FIGURE 2-12 Transmission spectrum of 5×10^{-4} *M* *m*-nitroaniline-*N,N*-diacetic acid (1-cm cell). (*After L. Meites and H. C. Thomas, "Advanced Analytical Chemistry," p.* 252, *McGraw-Hill Book Company, New York,* 1958, *by permission.*)

tance at 415 nm if monochromatic light is used. Assume the detector responds uniformly to the entire range of visible light.

Answer The detector will measure the *average* transmittance over the entire range of wavelengths that illuminates the detector. Therefore the average transmittance over the range of 400 to 750 nm can be estimated from Fig. 2-12 to be about 80 or 85 percent, which means this solution would *absorb* only about 15 to 20 percent of the light from the white-light source. On the other hand, Fig. 2-12 shows that this same solution will absorb about 70 percent of the light (30 percent *transmittance*) if *monochromatic* light at 415 nm is used, and thus about a fourfold increase in sensitivity is achieved through spectral isolation.

To isolate a narrow band of wavelengths *filters* or *monochromators* or both are used.

Filters The two basic types of filters in use are absorption filters and interference filters. *Absorption filters* function by selective absorption of unwanted wavelengths and are generally made from some form of glass tinted with a pigment which is dissolved or dispersed in the glass. Filters are also made from liquids or gelatin.

Absorption filters are classed as either *cutoff* or *bandpass* filters. Figure 2-13 shows the transmission characteristics of three typical cutoff filters, and Fig. 2-14 shows the transmission of a whole series of bandpass filters. Figure 2-13 also illustrates the fact that two sharp-cutoff filters (C-9780 and C-3482 in Fig. 2-13) can be combined to give the effect of a bandpass filter. (A pair of

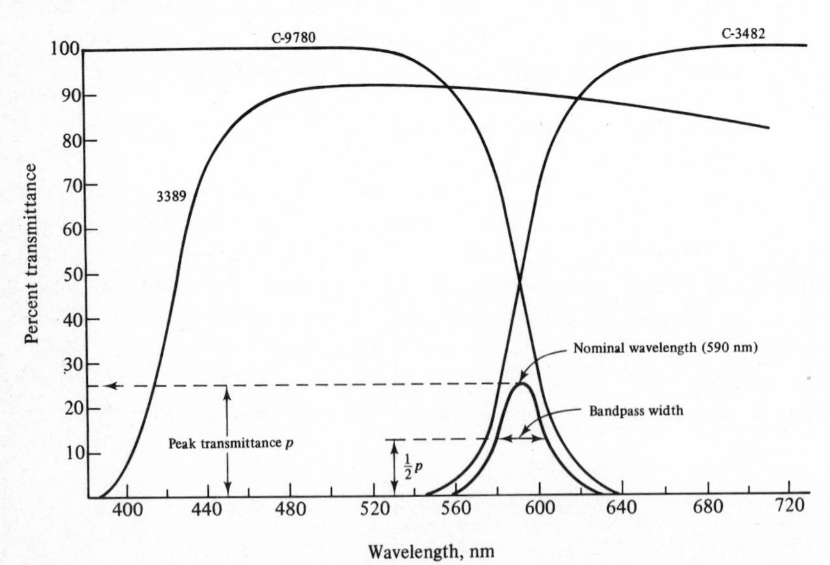

FIGURE 2-13 Transmission characteristics of some sharp-cutoff filters. (*Corning Glass Works.*)

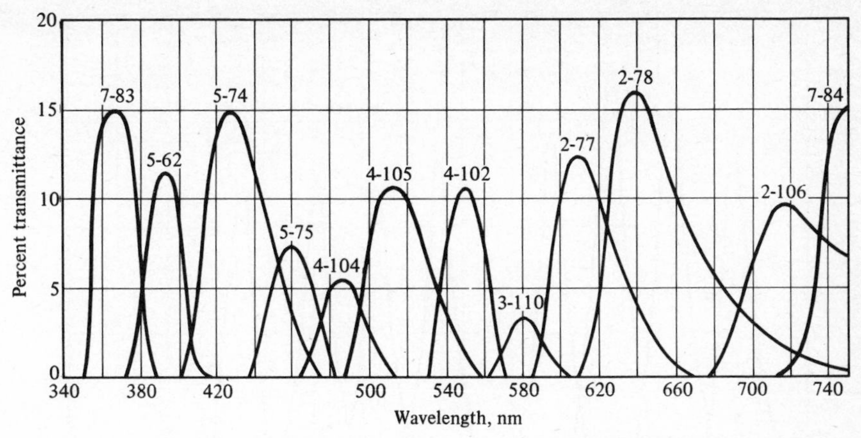

FIGURE 2-14 Transmission characteristics of some glass bandpass filters. (*Corning Glass Works.*)

bandpass filters can likewise be combined to provide a further reduction in the width of the band of wavelengths transmitted.) Bandpass filters are characterized by their *bandpass width* (also called *spectral bandwidth* and *effective bandwidth*), the *nominal wavelength* at the center of the band, and *peak transmittance*. Figure 2-13 illustrates these terms. Most glass filters have relatively wide bandpass widths of about 35 to 50 nm, and their peak transmittance is only about 5 to 20 percent, decreasing with improved spectral isolation.

Interference filters provide narrower bandwidths and greater peak transmittances, as illustrated in Fig. 2-15. The composition and general theory of interference filters was discussed in Sec. 1-4*D*. A wide selection of interference filters is now commercially available.

Monochromators A monochromator is a device for producing a beam of radiation of high spectral purity (narrow bandwidth) while allowing the wavelength to be varied at will. The essential elements of a monochromator are an *entrance slit*, which sharply defines the incoming beam of heterochromic radiation, a *dispersing element*, which may be a *prism* or *grating*, and an *exit slit*. The prism (or grating) disperses the heterochromic radiation into its component wavelengths, and the exit slit transmits the nominal wavelength together with a band of wavelengths on either side of it. The prism (or grating) is rotated to vary the nominal wavelength passing through the exit slit.

The theory of dispersion by prisms and gratings was discussed in Sec. 1-4*F*. The essential features reiterated here are that prisms bend shorter wavelengths more than longer wavelengths and thus have the disadvantage of giving *nonlinear dispersion*, as illustrated in Table 2-9. On the other hand, prisms have the advantage of giving spectral purity, which means that wave-

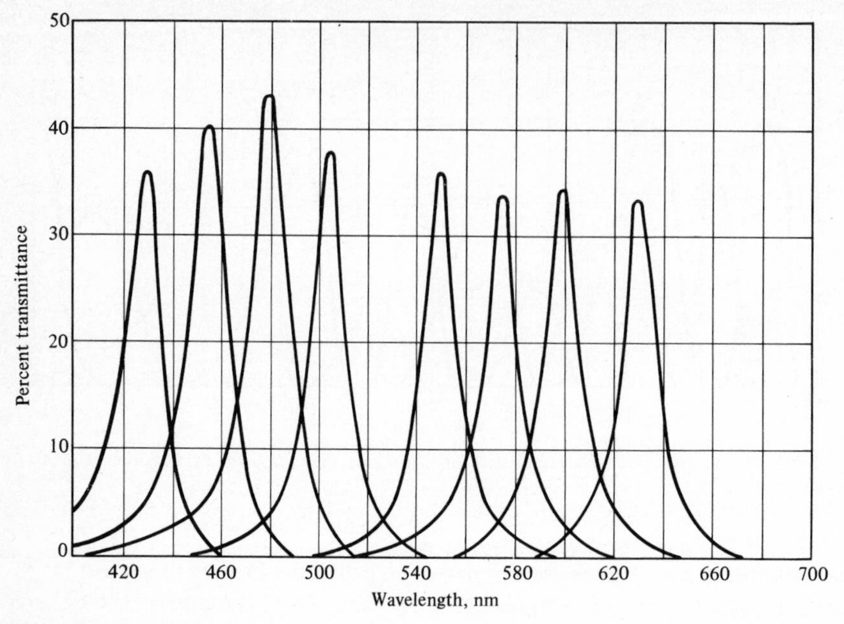

FIGURE 2-15 Transmission characteristics of some interference filters. (*Vacuum Coating Division, Bausch & Lomb.*)

lengths do not overlap. Gratings offer higher resolution and give linear dispersion throughout the spectrum. Also, since the light is usually reflected (some gratings transmit), there are no absorption losses within a reflection grating. On the other hand, gratings suffer from an overlap of spectral orders, although an echelette grating concentrates most of the dispersed radiation into the first two orders. Filters can be used to reduce the radiation of different orders and stray radiation. A prism is sometimes used in conjunction with a grating to sort out the first-order spectrum.

The role of the exit slit in a monochromator deserves discussion. Narrowing the slit reduces the intensity of the beam of radiation and at the same time narrows its bandwidth. Hence, it is possible to increase resolution by decreasing slit width, but a limit is reached where the diffraction pattern from the slit becomes appreciable (see Sec. 1-4*E*). With a good monochromator this slit width may be 0.01 mm or less. The sensitivity of the detector may also place a limit on narrowness of the slit width. Most instruments provide variable slit widths, to give the instrument flexibility, and the entrance and exit slits are adjusted simultaneously to the same width. Since the dispersion of an instrument with a prism monochromator changes with wavelength (see Table 2-9), such instruments *require* a slit aperture that is continuously variable if it is to provide radiation of approximately equal bandwidth over its wavelength range.

For a monochromator the spectral bandwidth is defined in terms of the relative distribution of energy passing through the exit slit. This distribution is assumed to have a triangular distribution of intensity with wavelength. The middle wavelength (peak transmittance), called the *nominal wavelength*, is the value read on the wavelength scale of the instrument. The effective bandwidth is taken as the bandwidth at one-half the peak transmittance and contains approximately 75 percent of the transmitted radiant energy going through the slit. Since the energy distribution is assumed to be an isosceles triangle, the spectral slit width will be exactly twice the effective bandwidth. The relationship between the *mechanical* slit width (usually given in millimeters) and *spectral* slit width (usually given in nanometers in the ultraviolet and visible regions) depends on the design and focal length of the monochromator and the nature of the dispersing unit. In a grating monochromator effective bandwidth for a given slit is constant throughout the spectrum, and thus the slit control can be calibrated in effective-bandwidth units. In a prism monochromator, however, the nonlinear dispersion of the prism makes it necessary for the manufacturer to supply a graph showing how the effective bandwidth varies with wavelength and mechanical slit width. An example of such a graph is shown in Fig. 2-16. As a useful rule of thumb, two sharp absorption bands or two emission lines can be resolved with a spectrophotometer if their wavelengths of maximum intensities (nominal wavelengths) differ by an amount equal to the spectral slit width (or twice the effective bandwidth) or more.

Material used for prisms and transmitting optical components like windows, lenses, and sample containers (cuvettes) should not be used at wavelengths where its absorbance is greater than about 0.2 (transmittance less than 60 percent). Ordinary silicate glasses are transparent from about 350 nm to

TABLE 2-9 Effective Bandwidths at Various Nominal Wavelengths with a Beckman DU (Prism) Monochromator (Exit Slit = 1 mm)†

Nominal wavelength, nm	Bandwidth, nm
200	1
400	10
600	35
800	75

† Data from Beckman Instruments, Inc.

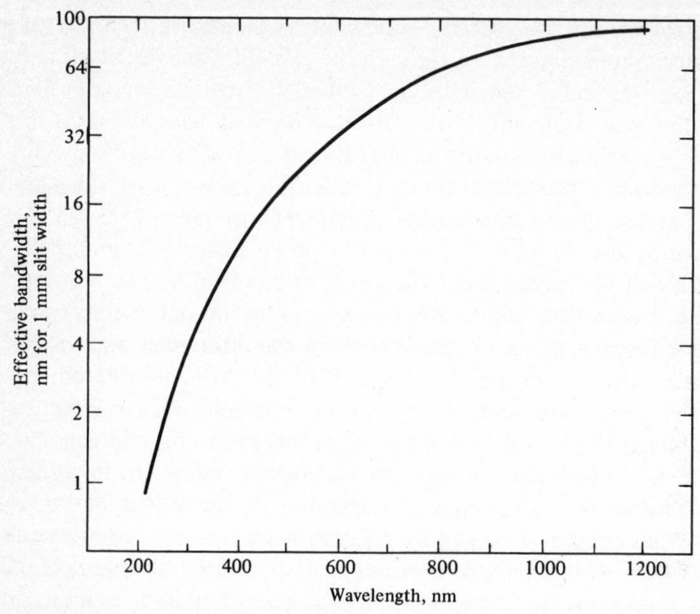

FIGURE 2-16 Dispersion data for Beckman Model DU (quartz-prism) spectrophotometer. (*Beckman Instruments, Inc.*)

2.2 μm, and special Corex (Pyrex) glass can extend the ultraviolet range to about 300 nm. For all work below 300 nm quartz or fused silica must be used. The limit for quartz (optical silica) is about 190 nm, and some samples of fused silica can be used down to 185 nm.

DETECTORS

The three types of photosensitive devices used in the ultraviolet and visible regions, photovoltaic cells (also called barrier-layer cells), phototubes (also called photoemissive tubes), and photomultiplier tubes, will be described and evaluated in terms of the following important criteria: sensitivity (minimum intensity level detectable), linearity of response with radiation intensity, response time, dependence of response on wavelength, amenability of output to amplification, and stability.

Photovoltaic Cells Photovoltaic cells are simple, require no external power supply, and can be hooked directly to a microammeter or galvanometer to read their output. However, their sensitivity is only moderate, and their use is generally restricted to instruments like filter photometers that permit a wide band of radiation to strike the detector. Figure 2-17 diagrams the construction of a photovoltaic cell. A metal like iron (or aluminum) is used for one electrode, and a thin layer of a semiconductor material, like selenium, is deposited on this base electrode. Then a very thin layer of silver (or gold) is sputtered over the surface of the semiconductor to act as a second collector electrode.

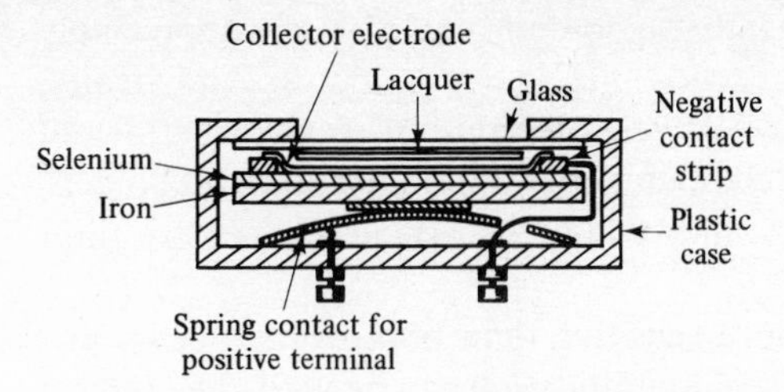

FIGURE 2-17 Construction of a photovoltaic cell.

Radiation striking the surface of the selenium semiconductor excites electrons at the silver-selenium interface and causes electrons from the selenium surface to pass to the silver collector electrode. Although electrons readily pass from selenium to the silver collector electrode, there is a resistance to electron flow in the reverse direction. Consequently, the cell generates an electromotive force between the base electrode and the collector electrode, and if the external circuit has a low resistance (about 400 Ω or less), a photocurrent will flow that is very nearly directly proportional to the intensity of the incident radiation beam.

Figure 2-18 shows the spectral-response curve of a typical selenium photovoltaic cell with a glass protective cover and compares it with the sensitivity characteristics of the human eye. Photovoltaic cells are sensitive over the whole visible range, and the response is somewhat more uniform with wavelength than the human eye. Nevertheless a photovoltaic cell is less sensitive in the blue region than in the green and yellow.

The response time of photovoltaic cells is only fair, and thus the light beam cannot be chopped or modulated, as is sometimes done with other detectors, for amplification or reduction of noise. In addition, photovoltaic cells

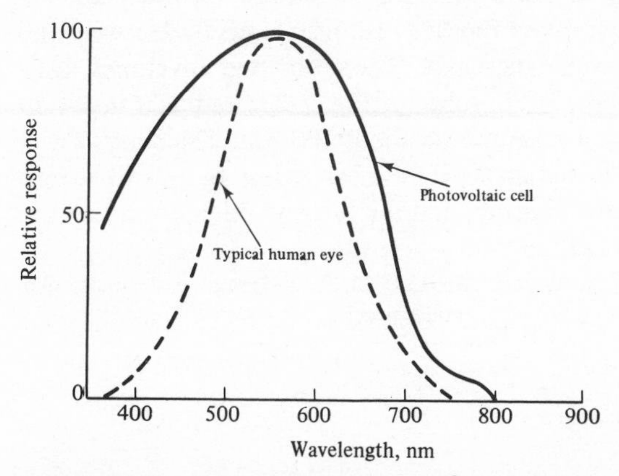

FIGURE 2-18 Spectral-response curve of a typical selenium photovoltaic cell with protective glass cover (response of the human eye is shown for comparison).

show fatigue effects, whereby the initial photocurrent may be appreciably higher than the steady-state value reached after a few minutes. And finally, the cells age very slowly over a period of years because of transformations of the selenium layer. Although these characteristics rule out photovoltaic cells for sophisticated spectrophotometers, they still serve well in inexpensive filter photometers.

Phototubes Phototubes are more sensitive than photovoltaic cells, primarily because a high degree of external amplification can be used. The typical vacuum phototube contains a light-sensitive cathode in the form of a half cylinder of metal, coated on its inside surface with a light-sensitive layer, and an anode wire located more or less along the axis of the cylinder. Figure 2-19 diagrams a simple phototube circuit. When radiation strikes the cathode K, photoelectrons are emitted and are drawn to the positive anode A, where they are collected and returned via the external circuit. The photoelectric current results in an iR drop across the resistor R_L which is proportional to the current. Resistor R_L will normally be a grid load resistor in an amplifier input circuit, and thus the iR drop will be amplified proportionately. Phototube signals as small as 10 pA (produced at low levels of illumination) can easily be amplified. For an accuracy of 1 percent or better, a calibration is required to overcome the slight nonlinearity in the photocurrent-vs.-illumination power curve unless a suitable null method can be employed.

The spectral sensitivity of the phototube will depend on the nature of the substance coating the cathode and can be varied by using different alkali metals or by changing the method of preparing the cathode surface. Most commercial phototubes use a composite coating on the cathode consisting of a mixture of an alkali metal, alkali metal oxide, silver metal, and silver oxide. Figure 2-20 shows spectral-response curves for photocathodes made from different alkali metals. The heavier, more easily ionized alkali metals emit electrons at longer wavelengths than the lighter alkali metals, and in progressing from sodium to cesium the relative response tends to decrease and the breadth of the spectral response increases. The evacuated envelopes, used to seal in the electrodes, are glass in the tubes of Fig. 2-20, and as a result all the phototubes show a maximum response at about 400 nm. The short-wavelength maximum has been attributed to light absorption by alkali atoms located beneath the surface of the coating, and the long-wavelength maximum

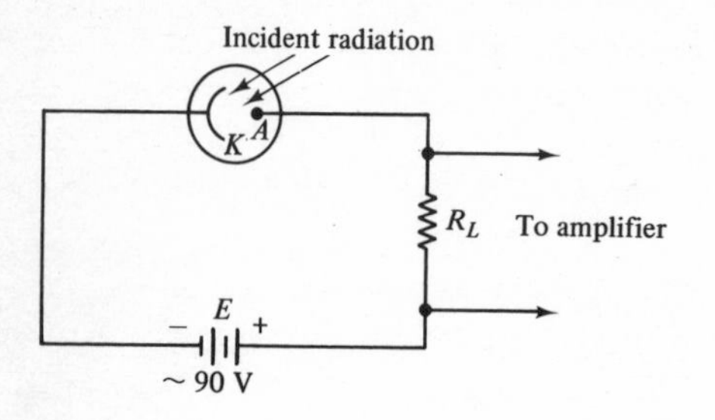

FIGURE 2-19 Diagram of a simple phototube circuit.

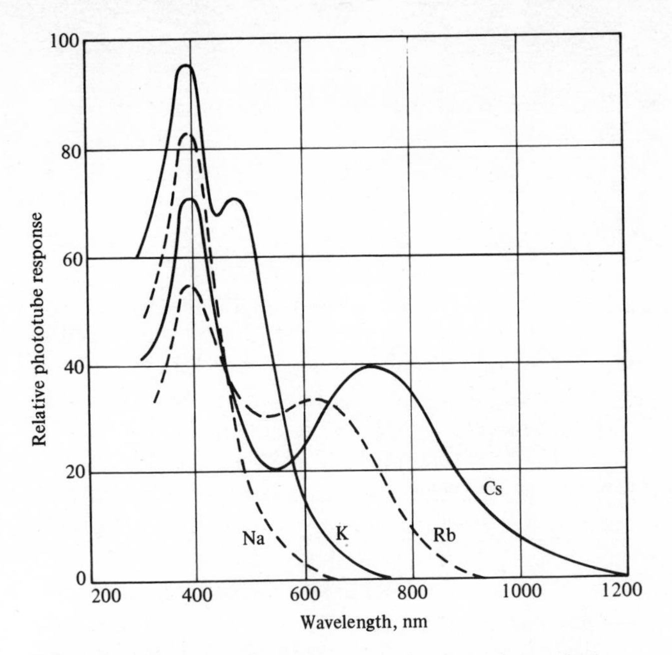

FIGURE 2-20 Spectral-response curves for some photocathodes consisting of silver, oxygen, and various alkali metals. (*After V. K. Zworykin and E. G. Ramberg, "Photoelectricity," p.* 48, *John Wiley & Sons, Inc., New York,* 1949, *by permission.*)

is ascribed to light absorption by alkali atoms located on the surface of the oxide coating. If quartz or fused-silica windows are used, the range of these phototubes can be extended through the near-ultraviolet and into the far-ultraviolet region.

Photomultiplier Tubes Photomultiplier tubes are extremely sensitive, fast-responding types of vacuum phototube, designed so that an amplification of several millionfold is achieved within one tube by the emission of *secondary electrons*. Radiation striking a photocathode causes the ejection of primary electrons, as in an ordinary vacuum phototube, but in the photomultiplier tube these electrons are accelerated by a positive potential to a second sensitive surface, where each electron striking it causes the release of four to five secondary electrons. These electrons in turn are accelerated to another sensitive surface, where the number of electrons is again increased by a factor of 4 or 5. This practice can be repeated as many times as desired, though most commercial photomultipliers have about 10 target electrodes, or *dynodes*. Figure 2-21 schematically illustrates the cross section of the widely used *circular-cage* photomultiplier design.

The amplification factors achieved depend critically on the voltage applied to each dynode, and a very stable high-voltage power supply is required. Typically, each dynode is made 75 to 100 V more positive than the preceding

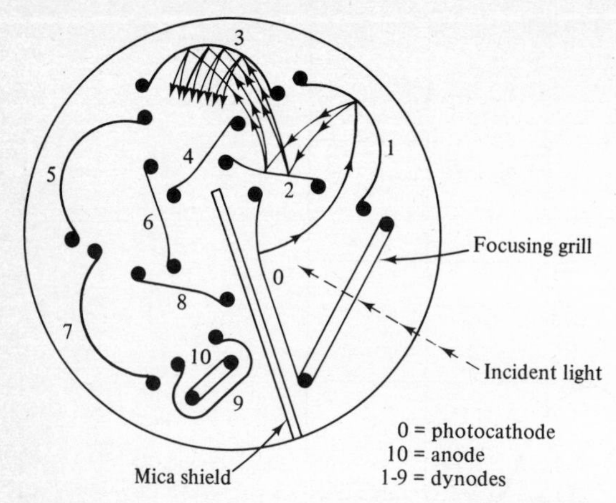

FIGURE 2-21 Schematic diagram of a photomultiplier tube.

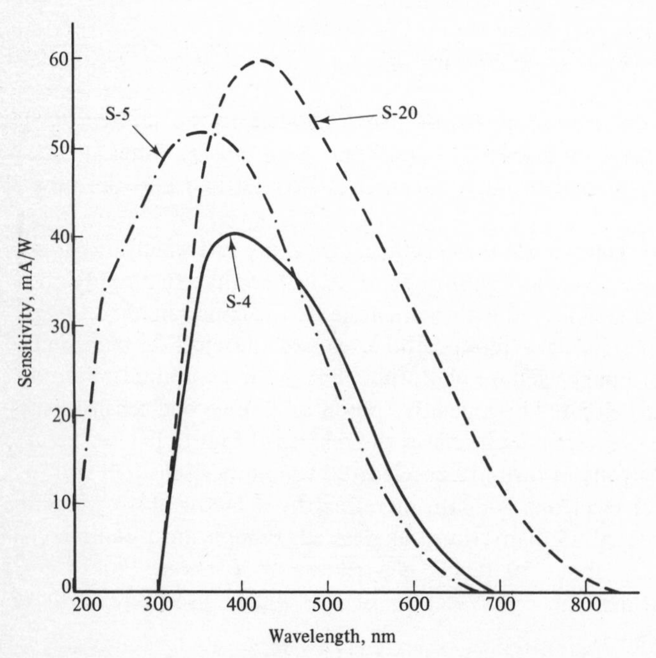

FIGURE 2-22 Relative spectral response of some photomultiplier tubes.

dynode, and overall amplification factors of about 10^6 are common. In addition, the output of the photomultiplier can be further amplified. The photomultiplier tube can measure intensities about 200 times weaker than those measurable with an ordinary phototube and amplifier. The limit of detection is set by the inherent dark-current noise, which is due to thermionic emission and other random noise. The dark current can be reduced by lowering the temperature of the tube, but this is usually not feasible in spectrophotometry. The response time is extremely short, responding to light pulses as brief as 10^{-9} s. The output current is linearly dependent on illumination over a fairly wide range but becomes nonlinear at high levels of illumination.

To prevent serious *deterioration* of electrode surfaces due to local heating effects and to prevent tube *fatigue*, which shows up as drifting signals, the anode current must be kept below 1 mA. This in turn requires that voltage between the last dynode and the anode be restricted to 50 V or less and high-intensity radiant energy should be avoided. Photomultiplier tubes must be carefully shielded from stray light.

Figure 2-22 shows the relative spectral response of three types of photomultiplier tubes having particularly good sensitivity in the ultraviolet and visible regions. The S numbers correspond to a standard classification code adopted by tube manufacturers to identify the spectral-response characteristics of all types of phototubes.

Reflectance-Detector Attachments Special reflectance attachments, which are available as accessories for many spectrophotometers and replace standard detectors, are useful for quantitatively measuring the color of opaque materials like paints, textiles, plastics, inks, dyes, glass, tile, and a wide variety of other products. Figure 2-23 gives a schematic optical diagram of a typical reflectance attachment. Light from a monochromator enters the reflectance attachment and is directed onto the opaque colored sample, which absorbs some

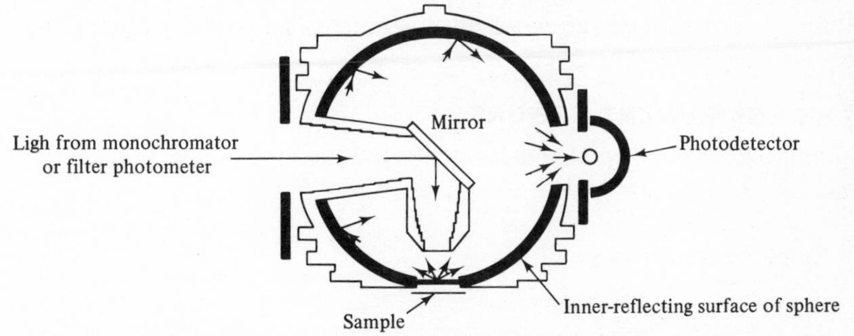

FIGURE 2-23 Schematic optical diagram of a reflectance attachment. (*From H. H. Willard, L. L. Merritt, Jr., and J. A. Dean, "Instrumental Methods of Analysis,"* 4*th ed., p.* 103, *Van Nostrand Reinhold Company, New York,* 1965, *by permission.*)

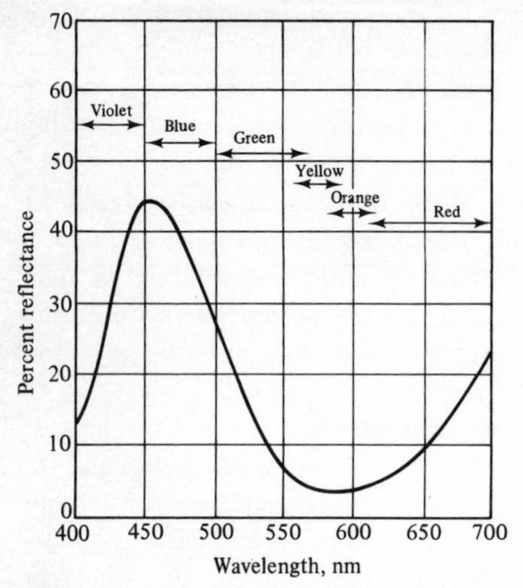

FIGURE 2-24 Spectral reflectance curve of an opaque blue material. (*After H. H. Willard, L. L. Merritt, Jr., and J. A. Dean, "Instrumental Methods of Analysis,"* 4*th ed., p.* 103, *Van Nostrand Reinhold Company, New York,* 1965, *by permission.*)

light and reflects the rest. An integrating sphere directs the reflected light through the photodetector aperture, where it is measured by a photomultiplier tube. Figure 2-24 shows a typical reflectance curve for an opaque blue material.

2-2C Principles of Instrument Design

Now that the basic components of photometers and spectrophotometers have been described, the fundamental principles of modern instrument design can be outlined and described, after which a summary of some commercially available instruments will be given.

BASIC INSTRUMENT DESIGNS

All photometers and spectrophotometers fall into one of four basic design patterns:

1 Single beam, direct reading

2 Single beam, null balance

3 Double beam, direct reading

4 Double beam, null balance

Each of these designs has advantages and disadvantages, and the relative merits of the various options will be described briefly.

Single Beam vs. Double Beam In single-beam instruments, radiation is passed through only one solution at a time, whereas in double-beam instruments, one beam of radiation passes through a sample solution and a second beam passes through a blank solution to provide a reference signal. With the *single-beam* instrument, the following minimum number of steps are necessary to make an absorption (or transmittance) measurement at a given wavelength:

1 With no radiation reaching the detector (shutter closed), the meter is zeroed (at zero transmittance or infinite absorbance).

2 With the solvent blank in the measuring position, the shutter is opened, and the amplifier gain or source intensity is adjusted to make the instrument read 100 percent T (zero absorbance).

3 The sample solution is moved into the measuring position, and its transmittance (or absorbance) is read from the meter.

These three steps must be repeated for every wavelength of interest, since both the source output and the detector response change with wavelength (see Sec. 2-2*B*). Furthermore, the absorbance of the solvent blank may change with wavelength. The sequential nature of these three steps also means that both the source output and the detector sensitivity must remain constant during the course of the three steps, which in turn means that the voltage supply for the source and detector must be very stable.

In a *double-beam* instrument, the operation is simplified, and most of these problems are overcome. To calibrate the instrument before use it is only necessary to carry out the following steps:

1 With solvent blank in *both* beams, the recorder pen is adjusted to zero (zero absorbance or 100 percent transmittance).

2 With solvent blank in the reference beam and no radiation reaching the sample detector (a shutter or opaque object is used to block the sample beam), the recorder is adjusted to full-scale deflection (infinite absorbance or zero transmittance).

Once the instrument is calibrated by these steps, the transmittance of the sample can be measured at any wavelength within the range of the instrument without repeating the calibration, since the signal from the sample beam is continually being referred to the signal from the solvent-blank reference beam and only the *difference* in the signals from the two beams is amplified and recorded. This amounts to measuring the ratio of the intensity of the beams from the two cells, and this ratio will be virtually independent of the fact that the source output and detector response change with wavelength. Furthermore, any absorbance by the solvent blank is automatically subtracted out as long as the absorption cells are properly matched. It follows that fluctuations in source output and detector sensitivity due to small fluctuations in the voltage supply are likewise automatically compensated for, assuming that the detectors are properly matched. (In another, more widely used type of double-beam instrument a *single* detector is used to measure both beams, thereby eliminating the problem of trying to match detectors.)

In summary, double-beam operation offers great advantages for automatic recording and tends to minimize errors that arise from fluctuations and drift in the applied voltage, source intensity, detector response, amplifier gain, and other irregularities. Of course, rapid fluctuations, e.g., those arising from a faulty component, cannot be smoothed or easily allowed for. Inherently, double-beam spectrophotometers are more complex and thus more expensive to construct and maintain. On the other hand, a quality single-beam instrument may surpass the accuracy and reliability of a given double-beam instrument if higher-quality components are used and operated well under their rated capacity, or if a superior measuring system, e.g., null balance rather than direct reading (see next section), is used. In general, single-beam spectrophotometers are well suited for quantitative analysis of single substances and of simple mixtures but are not as well suited for qualitative analysis where complex absorption curves must be obtained over a wide wavelength range.

Direct Reading vs. Null Balance A direct-reading instrument is one that uses an output meter to read out the transmittance or absorbance. A null-balance instrument, on the other hand, uses a balancing device in the computation stage, which allows a *comparison* of the measurement signal with a standard signal. The potentiometric servo recorder is a good example of a null-balance device. The null-balance method of measurement is inherently more accurate than the direct-reading method.†

The essence of the two methods and the relative advantages of the null-balance approach can be seen by comparing the direct and null-balance methods of measuring the voltage of an unknown battery, as shown in Fig. 2-25. The voltmeter M used in the direct method may be thought of as a simple laboratory (moving-coil) meter, whereas the circuit used in the null-balance method is a simple slide-wire potentiometer. Although the unknown voltage E_X is pictured in Fig. 2-25 as a dry cell, it is completely analogous to an

† Direct-reading digital voltmeters, now coming into widespread use, are much more precise than direct-reading meters and in certain instances may be more accurate than null-balance instruments. Digital voltmeters should be put in a category separate from meters and recorders, since, for example, digital voltmeters avoid the mechanical lag that may characterize both galvanometers and potentiometric recorders.

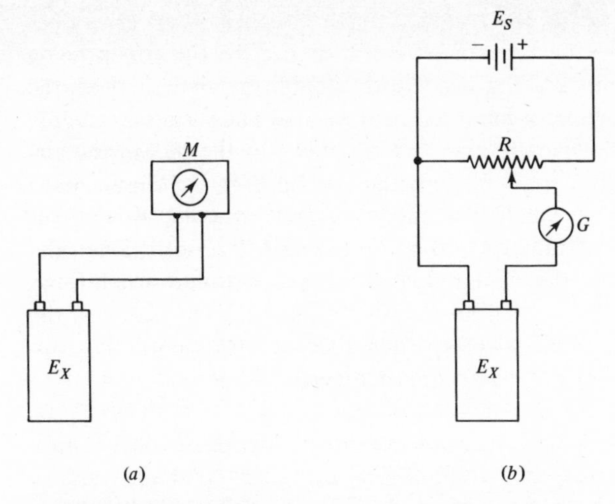

FIGURE 2-25 Comparison of direct and null-balance methods of measuring the voltage of an unknown battery: (*a*) direct method; (*b*) null-balance method. E_X = unknown voltage to be measured, E_S = known reference voltage of magnitude greater than E_X, M = voltmeter; R = calibrated slide-wire of uniform resistance, and G = galvanometer (null-detector).

output signal E_X from a detector in a spectrophotometer. There are two fundamental reasons why the null-balance method will give a more accurate reading than the direct method:

1 The accuracy of a direct-reading meter is limited by the shortness of the scale (readability) and the inherent lack of operational accuracy. Most good meters are accurate only to ± 1 percent of full-scale deflection, and the very best are seldom reliable to more than ± 0.5 percent of full-scale deflection. On the other hand, in the null-balance method, the accuracy is limited only by the linearity of a slide-wire and the accuracy with which the sliding tap can be positioned. With a good slide-wire it is possible to achieve an overall accuracy of ± 0.01 percent or better. In practice, an accuracy of ± 0.1 percent is aimed for in the slide-wire potentiometers of most null-balance spectrophotometers. It should be mentioned that the galvanometer shown in Fig. 2-25 is used simply as an indication of the point of balance and is chosen sensitive enough not to limit the accuracy of the measurement. In automatic-recording instruments, a servomotor takes over the function of finding the point of balance.

2 Direct-reading meters draw appreciable current and by so doing disturb the circuit being measured and cause changes in the voltage being

measured. This change in the voltage (which constitutes an error) becomes more serious as the resistance (or impedance) of the circuit being measured goes up. On the other hand, with the null-balance method, the unknown voltage is precisely balanced against a standard-voltage supply until, at null balance, virtually *zero* current flows in the measuring circuit. Thus, the measurement is made without disturbing the circuit being measured. (Actually, a small but finite current may be flowing at balance, the exact magnitude depending on the sensitivity of the null-detector used; in any case, the current will be far less than that which flows in the direct-reading meter.) In an automatic-recording instrument, the servomotor moves the slide-wire tap (and recorder pen) until the condition of zero current (or null balance) is achieved.

In summary, direct-reading instruments have advantages over null-balance instruments with respect to simplicity, cost, and perhaps ease of maintenance. Furthermore, in comparison with a *manual* null-balance instrument, the direct-reading meter is faster. However, null-balance instruments are capable of higher photometric accuracy, or accuracy in measuring the transmittance scale. In general, direct-reading photometers and spectrophotometers give accuracy in the range of ±1 to 3 percent in transmittance, though exceptionally good direct-reading instruments may approach ±0.5 percent in transmittance readings. Null-balance spectrophotometers can be obtained with accuracies as good as ±0.2 percent in transmittance and may range to about ±1 percent in the more inexpensive instruments.

In direct-reading instruments that provide output terminals for hooking up a servo recorder across the meter, the accuracy may still be limited by the accuracy of the meter, no matter how good the recorder.

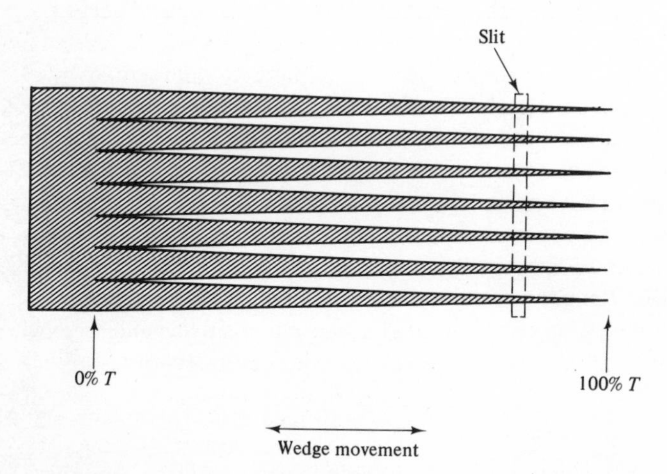

FIGURE 2-26 Schematic drawing of an optical wedge giving linear attenuation of light beam. Note the relationship of the wedge to the slit image.

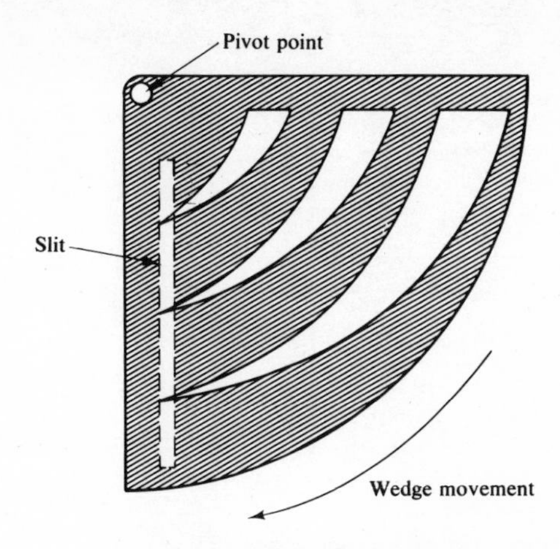

FIGURE 2-27 Schematic drawing of an optical wedge giving logarithmic attenuation of light beam. (*Perkin-Elmer Corp.*)

Optical-Null vs. Potentiometric-Null Balance In the preceding discussion of null-balance instruments, only potentiometric-null balance was considered. However, it is possible to use an optical-null-balance system in double-beam spectrophotometers. In the optical-null method, a wedge is pushed into the reference beam until the sample and reference beams give signals of equal strength. The wedge may be of the type shown in Fig. 2-26, in which case the movement of the wedge will be linearly proportional to the percent *transmission*, or the wedge may be shaped like that of Fig. 2-27, in which case the light will be attenuated logarithmically and the movement will be directly proportional to the *absorbance*. The optical-null system loses accuracy as solutions become highly absorbing,† although it is possible to compensate for this partially by electronics. More ultraviolet and visible spectrophotometers use the potentiometric-null than the optical-null principle, but both types will be illustrated in the next section.

2-2D Some Commercially Available Instruments

The commercially available instruments for absorption measurements in the ultraviolet and visible regions will be divided into filter photometer, manual (nonrecording) spectrophotometers, and automatic-recording spectrophotometers. In each category representative instruments, chosen because they clearly illustrate certain design features, will be described. No attempt has been made to give a comprehensive coverage of the multitude of instruments available, nor is there any intention of appearing to endorse one instrument over another.

FILTER PHOTOMETERS

Filter photometers have a place in the array of modern instruments because of their inherent simplicity and low cost. It is generally true that

† This assumes that a wedge is inserted in the reference beam; in the Hilger-Spekker absorptiometer, the wedge is *removed* from the sample beam, thereby maintaining a constant radiation intensity and overcoming this objection.

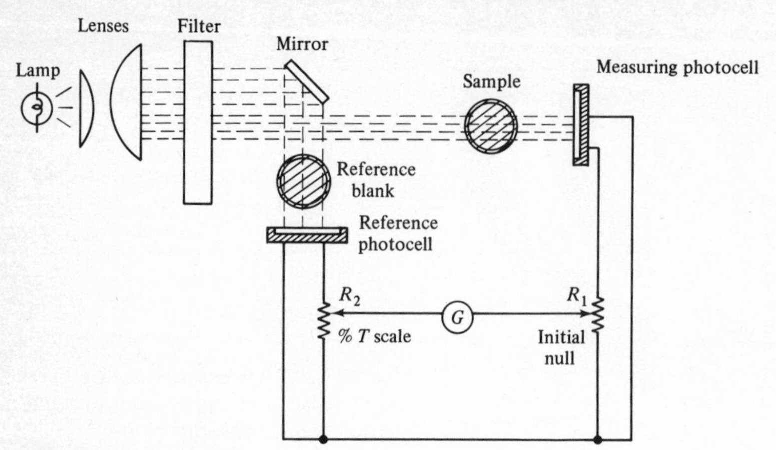

FIGURE 2-28 Schematic optical and electrical diagram of the Fisher Electrophotometer II. (*Fisher Scientific Co.*)

anything a filter photometer can do can be done more accurately and with greater sensitivity with a spectrophotometer, but, on the other hand, where a filter photometer can achieve the necessary sensitivity and accuracy, as for certain routine analyses, it can be the instrument of choice.

Filter photometers are often classified as being of single- or double-beam design. Single-beam instruments are usually direct-reading and suffer the disadvantage of giving readings which fluctuate with variations in the source intensity. This effect is sometimes minimized by using a storage battery as a current supply or by using a more expensive constant-voltage transformer.

A better means of minimizing the effect of fluctuations in the intensity of the light source is to use two photocells in a double-beam circuit, so that the fluctuations are observed equally by both cells and are canceled out. Almost all double-beam filter photometers use the null-balance method of measurement. Figure 2-28 gives a schematic (combined optical and electrical) diagram of the Fisher Electrophotometer II, a double-beam instrument employing a potentiometric null-balance measuring system. To operate this type of instrument, the null-balance galvanometer is adjusted mechanically to position the needle at midscale with the lamp off (which in effect calibrates the instrument to read 0 percent transmittance with no light striking the detectors). Then, with the lamp on and blank solution in both light beams, the potentiometer dial R_2 is set to read 100 percent T and the slide-wire contact R_1 is adjusted to null the galvanometer. Standards and unknowns can now be introduced into the measurement beam and the slide-wire contact R_2 adjusted to renull the meter. The transmittance (linear) or absorbance (nonlinear) can then be read off the potentiometer dial for each sample.

Most other double-beam filter photometers, i.e., the Klett-Summerson photoelectric colorimeter, Photovolt's Lumetron model 402, and the Hilger-Spekker absorptiometer, do not provide for the insertion of a solvent blank in the reference beam and instead simply use the reference beam to compensate for normal variations in the lamp supply voltage. Thus, these instruments are not "true" double-beam instruments by our definition since their only purpose in splitting the beam into two segments is to add stability and allow ordinary line voltage to be used for the lamp supply voltage instead of automatically compensating for absorption by the solvent blank.

MANUAL (NONRECORDING) SPECTROPHOTOMETERS

Commercially available manual *spectro*photometers may be classified according to the four basic design patterns described earlier.

Single-Beam Direct-reading Spectrophotometers A good example of a relatively inexpensive direct-reading grating spectrophotometer is shown in Fig. 2-29, and the optical system is schematically illustrated in Fig. 2-30. The range of this instrument is 340 and 650 nm with a blue-sensitive phototube, and the range can be extended to 950 nm by the use of a red-sensitive phototube and a red-transmitting filter. A small replica diffraction grating and fixed slits provide a constant bandpass of 20 nm. The phototube output is amplified by a stable dc difference amplifier arranged in a Wheatstone bridge circuit, and the meter in the bridge circuit is graduated in linear scale divisions from 0 to 100 percent transmittance, as well as in nonlinear absorbance units.

Single-Beam Null-Balance Spectrophotometers A sensitive, reliable, and

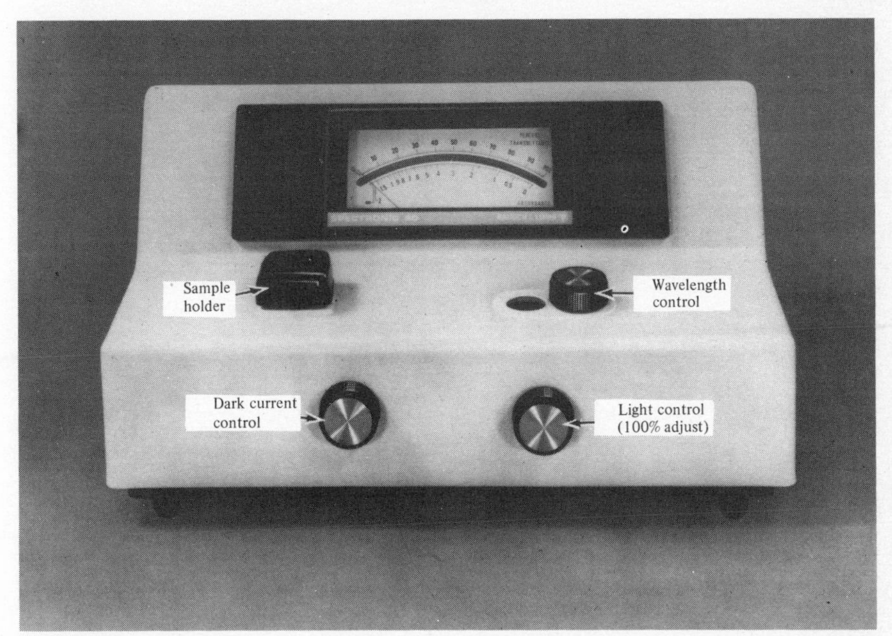

FIGURE 2-29 Bausch & Lomb Spectronic† 20, with controls labeled. (*Analytical Systems Division, Bausch & Lomb.*)

† Registered trademark of Bausch & Lomb, Inc.

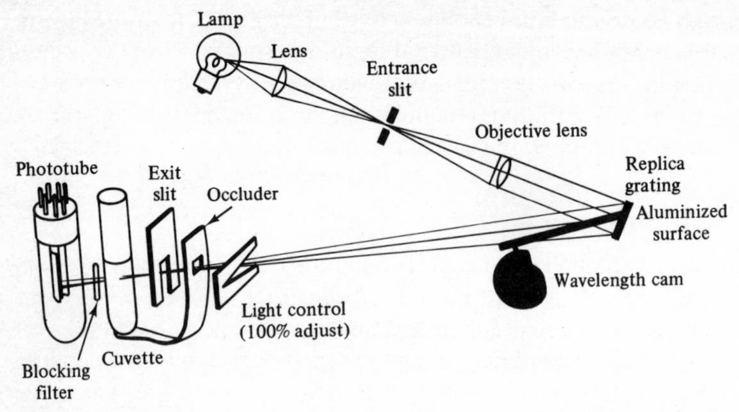

FIGURE 2-30 Schematic diagram of optical system of Bausch & Lomb Spectronic 20. (*Analytical Systems Division, Bausch & Lomb.*)

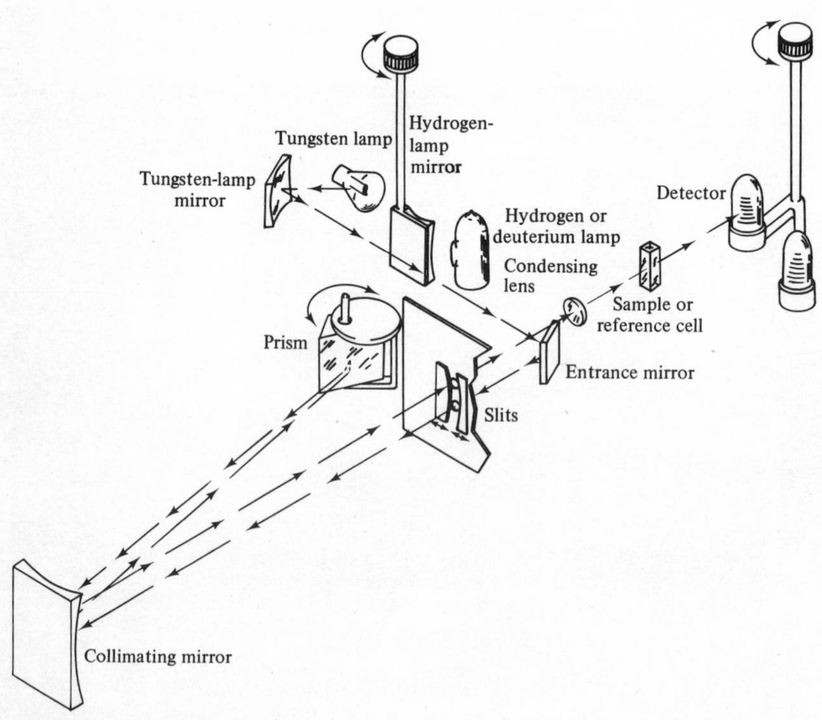

FIGURE 2-31 Schematic diagram of optical system in Beckman model DU-2 spectrophotometer. (*Beckman Instruments, Inc.*)

accurate instrument suitable for both research and routine quantitative determinations is the Beckman model DU-2 spectrophotometer, schematically diagramed in Fig. 2-31. The model DU-2 is a modernized version of the popular model DU. The range of the Beckman model DU is 210 to 1000 nm, while the DU-2 goes down to 190 nm. Both models use a 30° quartz prism with a mirror backing (called a *Littrow prism*) and have slits continually adjustable from 0.01 to 2.0 mm. The resolution (or bandpass) of these instruments is approximately 0.01 nm in the ultraviolet region and 0.5 nm in the visible region (see Fig. 2-16, for example). A blue-sensitive photomultiplier (optional in the model DU) is used below 625 nm, and a red-sensitive phototube is used above 600 nm. A hydrogen or deuterium source provides the ultraviolet radiation, and a tungsten lamp provides the visible light. The detector signal is amplified by a direct-coupled dc amplifier incorporated in a null-balance circuit. The photometric accuracy is better than 0.3 percent in transmittance.

Double-Beam Null-Balance Spectrophotometers In 1966, the first manual double-beam null-balance spectrophotometer was put on the market by Applied Physics Corporation, the Cary model 16, shown in Fig. 2-32. Figure 2-33 gives its schematic optical diagram. The wavelength range is 186 to 800 nm, with an optional phototube to extend the range down to 170 nm. The photometric accuracy is within ± 0.06 percent transmittance unit over the entire range (equivalent to 0.00024 absorbance unit near 0 absorbance and 0.001

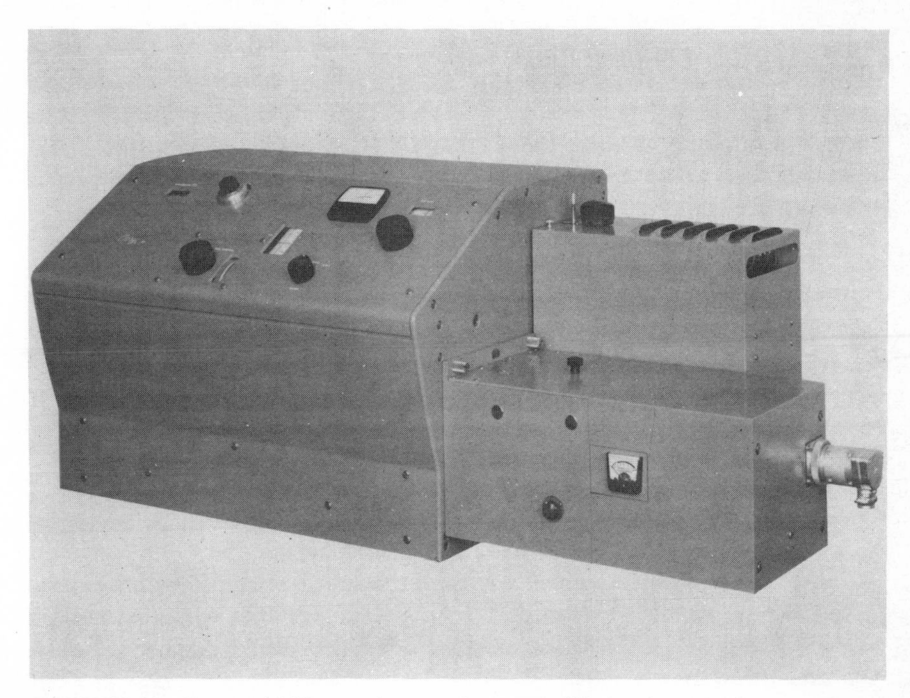

FIGURE 2-32 Cary model 16 manual spectrophotometer. (*Varian Associates.*)

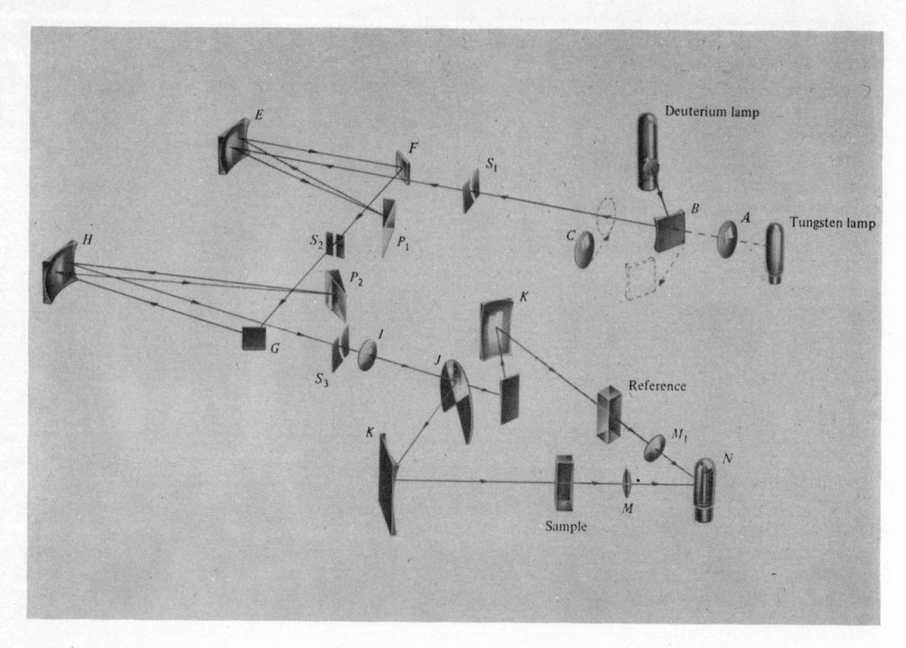

FIGURE 2-33 Schematic optical diagram of Cary model 16 manual spectrophotometer. B = mirror, which is withdrawn when the tungsten lamp is used; P_1, P_2 = fused-silica Littrow prisms; E, F, G, H, K, and K_1 = mirrors; J = rotating mirror-chopper. (*Varian Associates.*)

absorbance unit near absorbance of 1). This high accuracy is achieved to a large extent through the use of a 100-in null-balance slide-wire for the percent transmittance scale.

RECORDING SPECTROPHOTOMETERS

Almost all recording spectrophotometers use the double-beam principle. As explained in Sec. 2-2*C*, a wavelength scan with a single-beam instrument is subject to variations in source output, detector response, and solvent absorption, all of which are automatically corrected for with the double-beam design. Commercially available recording double-beam spectrophotometers will be divided into direct-reading, optical-null, and potentiometric-null types.

Direct-reading Double-Beam Recording Spectrophotometers There are a number of this type available. Many of these instruments can be purchased without a recorder and used as manual instruments, but they are best classified as recording instruments because they come equipped with a built-in motor for automatic wavelength scanning and terminals for connecting a recorder, and the simple addition of a recorder makes them a much more useful *recording* spectrophotometer.

Figure 2-34 gives a schematic optical and electrical diagram of the

Beckman model DB-G. The double beam is achieved through the use of a vibrating mirror assembly in the sample compartment, which alternately directs the monochromatic beam first through the sample solution and then the reference solution, at the rate of about 35 Hz. The beam striking the photomultiplier tube thus consists of alternating pulses of sample and reference energy, and a high-speed magnetic switch, operating synchronously with the vibrating-mirror assembly, diverts the reference signal to a regulating circuit, where it is compared with a constant voltage from the power supply. Any difference in voltage causes the voltage applied to the dynodes of the photomultiplier to be changed until the reference signal matches the constant voltage from the power supply. In this way, the reference signal will be stabi-

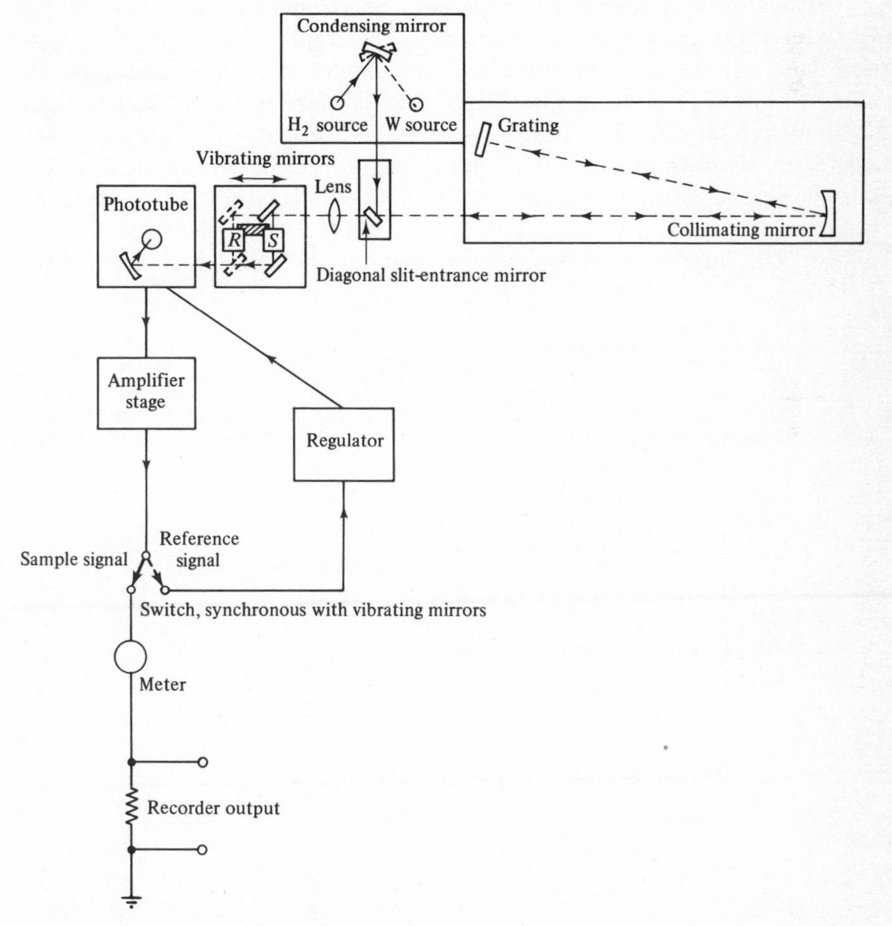

FIGURE 2-34 Schematic optical and electrical diagram of Beckman model DB-G spectrophotometer. (*Beckman Instruments, Inc.*)

lized at a constant value. Meanwhile, every half-cycle, the signal from the sample solution is sent to the output meter where its amplitude is indicated directly on the meter as percent *T*. This system gives true double-beam operation, since as the wavelength changes, the reference-regulator circuit compensates for changes in solvent-blank absorption as well as for source output and detector sensitivity.

The model DB-G covers the wavelength range from 190 to 700 nm, with optional photomultipliers that extend the range somwhat. For manual operation the slits are adjustable from 0.01 to 10 mm, and for automatic scanning they are programmed to compensate for variations in energy from the source output and detector response, thereby keeping the energy level approximately constant. Resolution is about 0.2 nm throughout the wavelength range. The sample compartment will hold cells up to 40 mm long.

Optical-Null Double-Beam Recording Spectrophotometers One of the few ultraviolet- and visible-region recording spectrophotometers to use the optical-null principle is the Perkin-Elmer model 202 spectrophotometer, schematically diagramed in Fig. 2-35. Radiation leaving the monochromator is alternately split into two beams by means of a rotating mirror or chopper, rotating at the rate of 26 Hz. One beam passes through the sample solution, and the second beam passes through the reference solution. The alternating beams are then recombined on the detector, and the resulting ac signal is amplified. The larger the imbalance between the reference and the sample

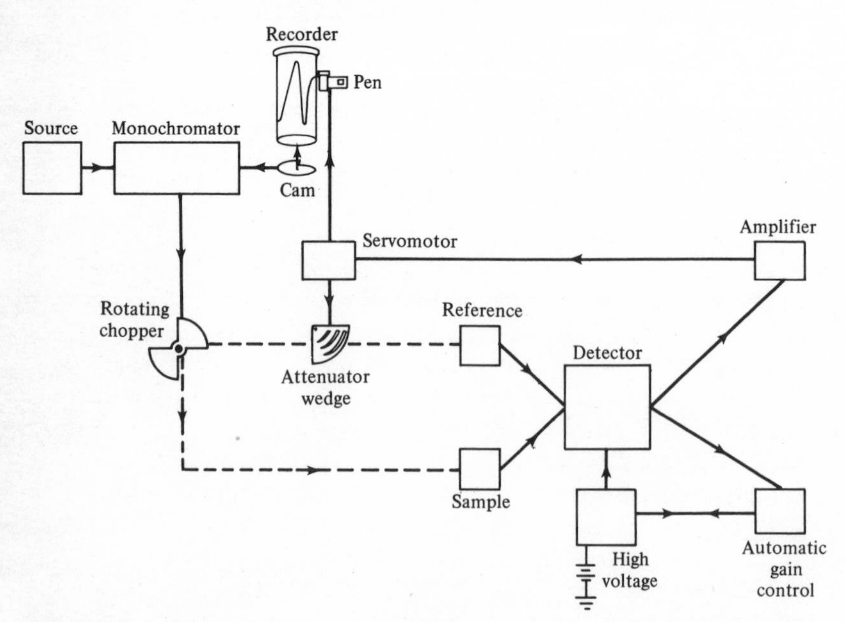

FIGURE 2-35 Schematic block diagram of the optical-null system of the Perkin-Elmer model 202 spectrophotometer. (*Perkin-Elmer Corp.*)

signals, the larger the amplitude of the amplified ac signal. This ac signal is used to drive a servomotor, which moves the attenuator wedge into the reference beam until the intensities of the two beams are equalized (at which time the ac signal from the amplifier will be zero). The movement of the recorder pen is linked to the movement of the attenuator wedge, and since the slits in the attenuator wedge open logarithmically (see Fig. 2-27), the recorder readings are linear in absorbance units.

With this optical-null system, the intensity of radiation striking the photomultiplier decreases rapidly as the absorbance of the sample solution increases, and if nothing were done about this, the measurement signal would get weak at high absorbances, with concomitant large errors. To rectify this problem, an automatic-gain-control system monitors the detector signal and automatically increases the high-voltage supply to the photomultiplier as the detector signal decreases, thereby increasing the gain of the photomultiplier and keeping the output of the multiplier anode constant.

The Perkin-Elmer model 202 covers the wavelength range from 190 to 390 nm on one chart and from 350 to 750 nm on a second chart. A deuterium-arc source is used for the ultraviolet range and a tungsten lamp for the visible region. Photometric accuracy is ± 0.01 absorbance unit between an absorbance of 0 to 1.0 (which on a percent transmittance scale would amount to ± 1 percent T) and ± 1 percent of the reading between an absorbance of 1.0 and 1.5. The monochromator uses a 60° fused-silica prism and automatically programmed slits, which give a resolution of 0.2 nm at 210 nm, 0.4 nm at 390 nm, and 1.5 nm at 700 nm. The slit-override control provides a means of widening the slit program to accommodate absorbing solvent blanks, at the expense of resolution.

Potentiometric-Null Double-Beam Recording Spectrophotometers Most ultraviolet- and visible-region recording spectrophotometers are of this type. A few of these instruments, e.g., the Cary models 10, 11, and 15 and the Perkin-Elmer model 450, use a double beam which is separated *in space*. In other words, these instruments use a beam splitter and mirrors to create two separate paths, and then *separate detectors* measure the radiant power of each beam. This type of design requires that the two detectors be carefully matched. The problem of matching detectors can be avoided, however, by using a double beam which is separated *in time*, whereby a single beam leaving the monochromator is alternately switched between reference and sample paths. After leaving the sample compartment, the two beams, separated in time, are diverted back to a *single* detector. This double-beam-in-time system was used in the previously described recording spectrophotometers and will be further illustrated in this section with the Cary model 14 spectrophotometer.

The Cary model 14 spectrophotometer is an example of a high-resolution high-accuracy spectrophotometer with a wavelength range that extends through the near infrared. Figure 2-36 shows the Cary model 14, and Fig.

2-37 gives a schematic optical diagram. The model 14 covers the wavelength range of 185 nm to 265 μm. Fittings are provided for flushing with inert gas in the far ultraviolet. A deuterium lamp (*A* in Fig. 2-37) is used for the ultraviolet, a tungsten lamp (*C*) is used for the visible region, and another tungsten lamp (*Y*) is used in the near infrared, where the direction of the radiation is reversed from that used in the ultraviolet and visible regions. Radiation is dispersed by the prism (*F*) and grating (*J*). Three slits (*D, H*, and *L*) are simultaneously adjustable in width over a range of 0 to 3.0 mm. A semicircular mirror (*O*) is rotated at the rate of 30 Hz, sending radiation alternately through the reference cell (T') and the sample cell (*T*). In this way, reference and sample energy alternately strike the photomultiplier detector (*X*), separated in time by a short interval of no light caused by a chopper disk (*N*) rotating on the same shaft as the semicircular mirror. If there is a difference in energy between the sample and reference beams, the resulting pulsating ac signal from the photomultiplier is amplified and fed to a servomotor, which drives a slide contact on the measuring slide-wire (servo recorder) until null balance is achieved with an accurate comparison circuit.

When the instrument is used in the near-infrared region, the tungsten lamp (*Y*) is slid into the place of the photomultiplier (*X*) and radiation traces a reverse path back to a lead sulfide detector, *f* (see Sec. 3-2 for a discussion of infrared detectors). The reverse direction is used in the near-infrared region to

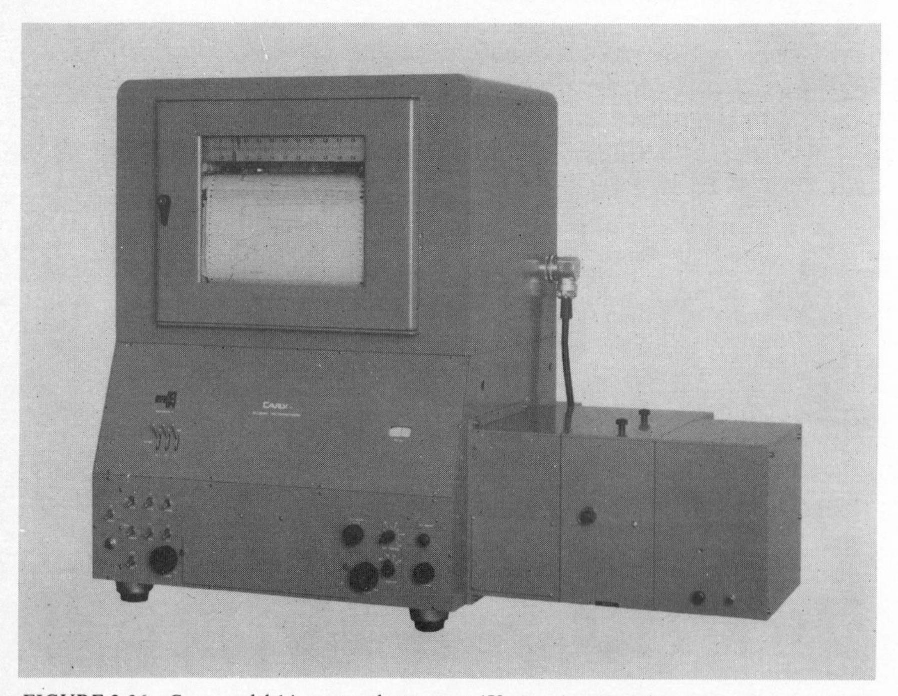

FIGURE 2-36 Cary model 14 spectrophotometer. (*Varian Associates.*)

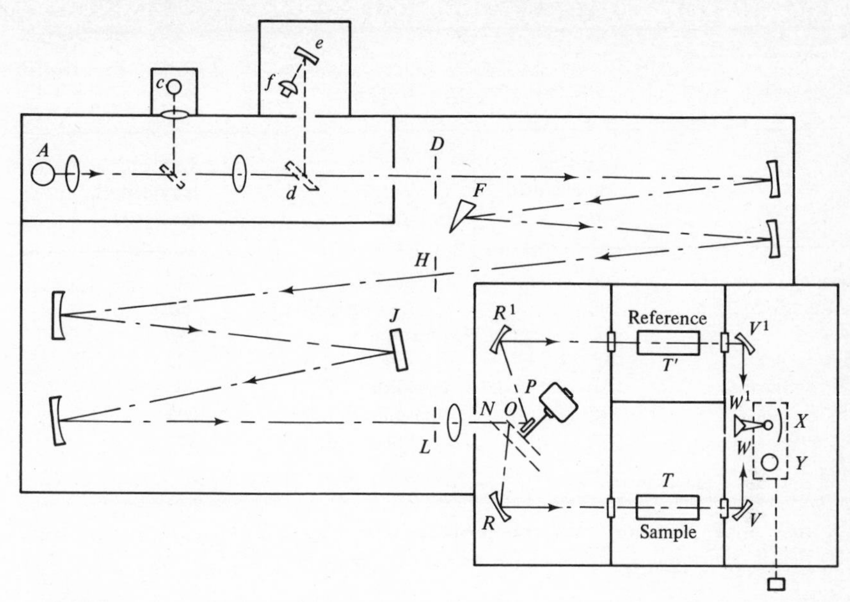

FIGURE 2-37 Schematic optical diagram of Cary model 14 spectrophotometer. (*Varian Associates.*)

eliminate stray thermal energy emitted from the chopper unit. If the direction were not reversed, polychromatic infrared radiation thermally emitted from the chopper system would be measured, whereas by reversing the direction of the radiation, this stray thermal radiation is eliminated by the monochromator before detection.

The resolution of the model 14 is 0.1 nm in the ultraviolet and visible regions and 0.3 nm in the near infrared. The photometric accuracy is 0.002 absorbance unit near 1.0 absorbance and 0.005 absorbance unit near 2.0 absorbance. The standard slide-wire gives linear readout on 0 to 1 and 1 to 2 absorbance scales, with automatic range change between the two scales, which effectively gives a 20-in-wide chart readout for the 0 to 2 to absorbance range. Wavelength accuracy is ± 0.4 nm over most of the range. Stray light is less than 0.0001 percent over much of the range and less than 0.1 percent at the ends of the range.

2-3 SAMPLE HANDLING

Cleanliness of the *cells*, inside and out, is of utmost importance. Between successive determinations cells should be rinsed with solvent, then filled with solvent, dried on the outside, and checked for absorption. Detergent or hot nitric acid can be used for cleaning.

Many *solvents* are available for use in the ultraviolet and visible regions. Table 2-10 lists the approximate transmission limits of some common

TABLE 2-10 Approximate Ultraviolet Transmission Limits of Some Common Solvents Based on 1-cm Path Length†

Solvent	Limiting wavelength, nm	Solvent	Limiting wavelength, nm
Water	200	Chloroform	245
Ethanol, 95 or 100%	195	Carbon tetrachloride	262
Methanol	195	Benzene	280
Diethyl ether	205	Xylene	290
Isopropyl alcohol	210	Pyridine	305
Cyclohexane	212	Acetone	328
Isooctane	215	Carbon disulfide	375
1,4-Dioxane	220		

† Data from R. P. Bauman, "Absorption Spectroscopy," p. 173, Wiley, New York, 1962, and H. B. Klevens and J. R. Plah, *J. Am. Chem. Soc.*, **69:** 3055 (1947).

solvents, based on a 1-cm path length. Probably the most widely used solvents are water, 95% ethanol, cyclohexane, and 1,4-dioxane.

Water needs to be distilled or deionized, but the small amount of residual impurities do not normally cause significant absorption in the accessible region of the ultraviolet. A 95% solution of ethanol is suitable without further purification and is generally a good choice when a polar organic solvent is required. Absolute ethanol is often found to contain benzene, which can be checked by measuring its transmittance in the 250-nm region against water as a blank.

A spectrographic grade of cyclohexane is available. Cyclohexane can be freed of aromatic impurities by passage through an activated silica-gel column. It is a good solvent for aromatic compounds, particularly the polynuclear aromatics. The spectra of aromatic compounds tend to retain their fine-line structure when determined in cyclohexane, whereas it is often lost in more polar solvents.

Dioxane is a good low-dielectric-constant solvent for certain classes of compounds, but it may contain benzene or develop peroxides. Dioxane can be purified by distillation from sodium. Benzene contamination can be removed by the addition of methanol followed by distillation to remove the benzene-methanol azeotrope.†

Care should be exercised to choose a solvent which will not interact with the solute. For example, the spectra of aldehydes should not be determined in alcohols.

† R. A. Friedel and M. Orchin, "Ultraviolet Spectra of Aromatic Compounds," 2d ed., Wiley, New York, 1958.

Finally, one should be alert to the possibility of *photochemical reactions* in the ultraviolet beam of the instrument. This can be checked by measuring the change in absorbance with time while the sample is exposed to the ultraviolet beam, compared with any change with time that may occur in an unexposed sample.

2-4 APPLICATIONS

In this brief list of important applications of ultraviolet and visible spectrophotometry, along with short descriptions of each, no attempt is made at an exhaustive survey or detailed descriptions. Further details will be found in the references.

2-4A Quantitative Analysis

Undoubtedly the largest single application of ultraviolet and visible spectrophotometry is for quantitative analysis. Substances that would not ordinarily absorb in this region can often be determined by forming suitable derivatives, e.g., by combining an organic compound with a metallic ion in solution. In general, the broad absorption peaks in this region contribute to *accurate* analyses, and the intensity of the absorption bands contributes to *sensitive* methods of analysis (the molar absorptivity ϵ often has values of the order of 10^4 or 10^5, allowing concentrations in the 10^{-4} to 10^{-6} M range to be determined). Compared with classical gravimetric and volumetric analysis, the accuracy of spectrophotometric methods is usually inferior, but the sensitivity and speed often make spectrophotometry the method of choice. The accuracy of a spectrophotometric analysis varies with the instrument being used (as well as with the procedures); while the instrumental error is usually about ± 1 percent, under favorable conditions the error may be as low as ± 0.2 percent.

Quantitative analysis by spectrophotometric methods is based on Beer's law, details of which were given in Sec. 1-5. A prerequisite for quantitative analysis of a mixture is a knowledge of the spectra of all of the pure components. The observed spectrum of the mixture must be accounted for completely in terms of bands of the pure components. If a band is present which cannot be identified, a component is probably present which can contribute in an unknown way to any of the bands selected for measurement.

Occasional checks on the accuracy of a spectrophotometer should be carried out, particularly before publication of a spectrum or spectrophotometric method. Both wavelength calibration and transmittance (or absorbance) scales should be checked, the accuracy of the latter being subject to aging of the photodetectors. The best way of performing such a check is to determine the spectrum of a standard solution, preferably at more than one concentration. A number of standards have been studied. Probably the best is potassium chromate; 4 mg potassium chromate in 100 ml of 0.05 M potassium hydroxide has maxima at 275 and 375 nm, giving ϵ of 3675 and 4815, respectively.

2-4B Qualitative Analysis

While absorption spectra in the ultraviolet and visible region are sometimes useful for qualitative identification, they are generally much less useful for qualitative purposes than infrared spectra because the ultraviolet and visible spectra of most substances, both organic and inorganic, are relatively broad and featureless, whereas infrared spectra are much more detailed. The certainty of identification is directly related to the total number of spectral identities, e.g., inflection points, maxima, and minima, which can be made between the unknown and a comparison substance of known composition. Additional spectral detail can be obtained in the vapor state (from vibrational and rotational features in the spectra), but this is usually much more inconvenient than obtaining solution spectra.

Benzene is one of the relatively few organic substances with a highly characteristic ultraviolet spectrum. Around 250 nm a dilute solution of benzene has four strong, sharp absorption maxima, plus several weaker maxima (see Fig. 2-38). Thus the ultraviolet spectrum is of considerable use in testing purity specifications where the benzene content must be carefully controlled, e.g., in absolute alcohol or in cyclohexane. Benzene is also easily identified in mixtures with substituted benzenes, e.g., toluene and ethylbenzene.

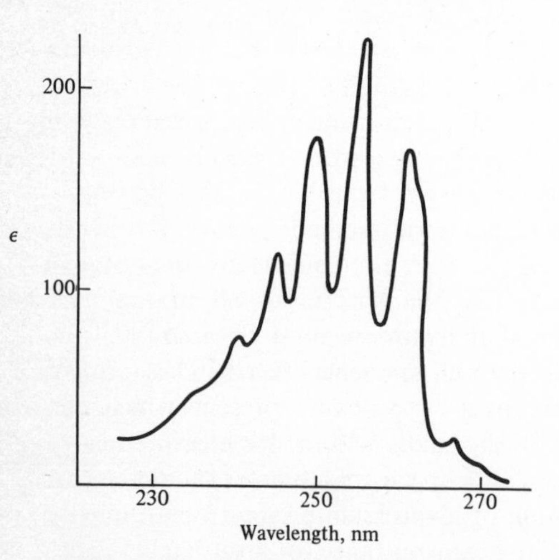

FIGURE 2-38 Ultraviolet absorption spectrum of benzene in 95% ethanol. (*From D. H. Williams and I. Fleming, "Spectroscopic Methods in Organic Chemistry," p. 28, McGraw-Hill Book Company, New York, 1966, by permission.*)

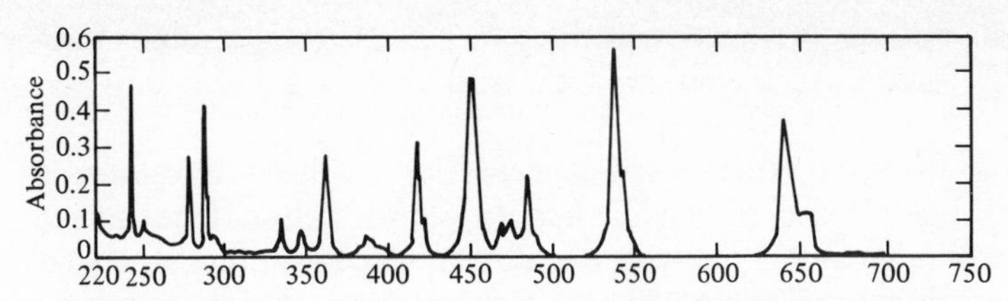

FIGURE 2-39 Absorption spectrum of 0.2 g Ho_2O_3 dissolved in 10 ml of 1 *M* $HClO_4$, measured in a 1-cm cell with a Cary model 14 spectrophotometer. [*By permission from C. V. Banks and D. W. Klingman, Anal. Chim. Acta,* **15:** 356 (1956).]

The intensity of absorption of a compound in solution, expressed in terms of its molar absorptivity ϵ, is sometimes a useful additional criterion for identification, particularly if the intensity is exceptionally high. However, the compound must be available in high purity if misleading results are to be avoided. The range of intensities of absorption of organic compounds is very wide, as may be seen from Tables 2-2, 2-5, and 2-6. Thus, acetone absorbs at 272 nm and phenol at 274 nm, but the absorption bands cannot be confused since acetone has an ϵ_{max} of about 17 whereas phenol has an ϵ_{max} of about 1450.

The trivalent rare-earth ions in solution are among the relatively few inorganic substances having highly characteristic ultraviolet and visible absorption spectra. Figure 2-39 gives the absorption spectrum of Ho_2O_3 dissolved in 1 *M* $HClO_4$. Other rare earths have spectra sufficiently different often to permit qualitative and quantitative analysis in mixtures containing several rare earths.

2-4C Structure Determination

Ultraviolet and visible absorption spectra provide a valuable tool in the identification of chromophoric groups and the elucidation of the structure of organic compounds; however, these spectra alone cannot provide conclusive evidence of a particular compound or chromophore but should be used *in conjunction* with infrared and nmr spectra and any other physical, chemical, and analytical information available. For example, in confirming the presence of a carbonyl group, a band in the 270 to 290 nm region (with ϵ equal to about 50) might indeed be due to the carbonyl group, but, on the other hand, it might also be due to an impurity having a short conjugated system. Thus, the presence of only 0.25 percent of an impurity with high-intensity absorption ($\epsilon \approx 20{,}000$) will simulate the carbonyl absorption band and may lead to erroneous deductions.

One widespread use of ultraviolet and visible absorption spectra for structural elucidation is in deciding between possible structures after the main functional groups and skeleton have been found by other means or in deciding between relative positions of functional groups in the molecule, rather than their presence or absence. An example will illustrate this.

Example 2-8 Mesityl oxide, a self-condensation product of acetone, is known to consist of two isomers, I and II, shown below. One isomer exhibits a maximum at 235 nm with $\epsilon = 12{,}000$, while the other shows no high-intensity absorption above 220 nm. Which of the two isomers absorbs at 235 nm?

$$\underset{\text{I}}{CH_2{=}\underset{\substack{|\\ CH_3}}{C}{-}CH_2{-}COCH_3} \qquad \underset{\text{II}}{CH_3{-}CH{=}\underset{\substack{|\\ CH_3}}{C}{-}COCH_3}$$

Answer Isomer II absorbs at 235 nm, since its double-bond system is conjugated, whereas isomer I has unconjugated double bonds and would be expected to absorb only at shorter wavelengths.

2-4D Kinetic Studies

Ultraviolet and visible spectrophotometry has found wide application as an analytical tool for studying reaction rates and kinetics. Spectrophotometry is particularly well suited to the basic problem of kinetics—determining concentration as a function of time. Reactants and products must, of course, have sufficiently different spectra for the change to be followed readily, and the absorbing substance should absorb strongly.

Most recording spectrophotometers can be used to produce a continuous recording as a function of time by simply disconnecting or deactivating the wavelength drive and allowing the chart to move while the wavelength is set at some predetermined value. This is particularly valuable for studying moderately fast reactions.

Kinetic studies make it possible to analyze certain substances when no other approach is feasible; e.g., substances that exert a measurable catalytic influence on a chemical reaction can be determinable at extremely low concentrations because of this influence, and two or more substances with similar physical and chemical properties may display sufficient difference in their reaction rates to make rate measurements a feasible method of distinguishing the substances and performing a quantitative analysis.

2-4E Measurements of Dissociation Constants of Acids and Bases

If the acidic or basic function of a compound is part of a chromophore, the spectrum of the substance will vary with the pH of the solution and the dissociation constant can be measured from spectra obtained at different pHs. For example, the dissociation of an acid, HA, in a solvent, SH, can be represented by the equilibrium

$$HA + SH \leftrightarrows A^- + SH_2^+ \tag{2-12}$$

where SH_2^+ is the solvated proton. The acid dissociation constant K_a is thus given by

$$K_a = \frac{[A^-][SH_2^+]}{[HA]} \tag{2-13}$$

To determine K_a it is only necessary to measure the concentration of the acid, HA, and its conjugate base, A^-, by spectrophotometric measurement while determining the concentration (or activity) of the species SH_2^+ with a pH meter. At least three spectra must be run to determine K_a: one at a pH near to the pK_a of the acid, a second on the acid side at least 2 pH units away from pK_a, and the third on the alkaline side at least 2 pH units away from pK_a. Separating the spectra by at least 2 pH units ensures that the acid species (HA) can be measured with less than 1 percent interference by the salt species (A^-), and vice versa. The presence of an *isobestic* point (see Sec. 1-5*D*) is evidence that the equilibrium in question involves only two absorbing species. Further details can be found in the references.

2-4F Determination of Molecular Weight

The spectrophotometric determination of molecular weight is another useful application. If a compound forms a derivative with a reagent which has a characteristic absorption in another wavelength region, the molar absorptivity ϵ of the derivative is often nearly the same as that of the reagent. For example, many amine picrates exhibit maxima at 380 nm with $\epsilon = 13{,}400$, which is constant (within 1 percent) and identical with the value of picric acid itself because the absorption originates in the picryl group. Thus, the molecular weight of a given amine picrate can be determined by dissolving a weighed amount of the solid derivative in a known volume of solvent (usually ethanol) and measuring the absorbance at 380 nm. From Beer's law,

$$A = abC \tag{1-39}$$

where C is the concentration in grams per liter, the absorptivity a is calculated in liters per gram-centimeter. The molecular weight can be calculated from

$$\epsilon = a \times \text{molecular weight} \tag{2-14}$$

since ϵ is known to have a value of 13,400. This method assumes that only one amino group is present and that the amine itself has no appreciable absorption at 380 nm. If this is not the case, appropriate corrections must be applied.

2-4G Determination of Empirical Formulas

Several spectrophotometric methods have been used for the determination of the mole ratio of the components in a complex compound. One of the best known methods is Job's *method of continuous variations*, whereby the values of the coefficients m and n in the equilibrium

$$mA + nB \rightleftarrows A_mB_n \tag{2-15}$$

can be determined. In brief, the total number of moles of A and B is kept constant, and absorbance measurements are made at an appropriate wavelength while varying the *ratio* of the moles of A to moles of B. A plot of absorbance vs. mole fraction of A or B will pass through a maximum, making it possible to establish the stoichiometry.

2-4H Photometric Titrations

Photometric or spectrophotometric methods can be employed to locate the end point in a titration where a color change occurs and may even be used in the ultraviolet region with appropriate equipment. In this technique, the titration vessel is placed directly in the light path of the instrument (this often requires modification of the cell compartment). The absorbance of the solution is determined after increment additions of titrant, and a plot of absorbance as a function of volume of titrant is prepared. Typical titration curves are shown in Fig. 2-40. If the reaction is complete, the titration curve will consist of two straight lines intersecting at the end point, similar to amperometric and conductometric titrations. If the titration reaction is appreciably incomplete, there will be appreciable curvature in the end-point region, but extrapolation of the two linear segments of the titration curve to their intersection establishes the end-point volume. This method has the decided advantage of making small additions in the vicinity of the end point unnecessary. Figure 2-40*a* is characteristic of a case where only the titrant absorbs, as in the titration of arsenic(III) with bromate-bromide, measured at a wavelength where bromine absorbs. Figure 2-40*b* is typical of systems where only the product of the reaction absorbs, as in the titration of copper (II) with ethylenediaminetetraacetic acid (EDTA). Figure 2-40*c* is illustrative of a case where only the substance being titrated absorbs, the titrant and product being nonabsorbing. An example is the titration of *p*-toluidine in butanol with perchloric acid at 290 nm.

Photometric titrations have several advantages over direct spectrophotometric analysis. (1) Nonabsorbing substances can often be determined, since only one absorber need be present among the reactant, the titrant, or the reaction products. (2) The presence of other substances that absorb at the analytical wavelength does not necessarily cause interference, since only the *change* in absorbance is significant. However, the interference must not be too intense, or absorbance measurements will be crowded into a narrow region on one end of the scale. (3) An accuracy of about 0.5 percent or better can be obtained, because the absorbance readings of a number of points are averaged in constructing the straight-line segments of the titration curve.

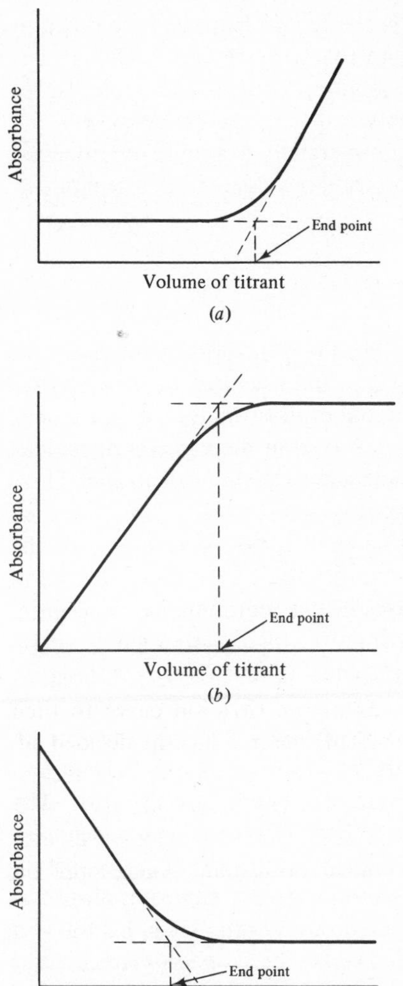

FIGURE 2-40 Typical photometric titration curves.

2-4I Other Applications

Other useful applications of ultraviolet and visible spectroscopy include studies of hydrogen bonding and other associations, the determination of steric configuration, distinguishing between cis and trans isomers, the detection of tautomerism, and the study of ions and free radicals. Further details can be found in the references.

2-5 VACUUM-ULTRAVIOLET SPECTROPHOTOMETRY

The vacuum ultraviolet, or far ultraviolet, is the region from 200 nm down to 0.1 nm (1 Å), or from the near ultraviolet to the x-ray region. Although numerous studies have been made in this region, vacuum-ultraviolet spectrophotometry has not yet arrived as an analytical tool, largely because of the instrumentation problems and the lack of commercially available instruments. Since all molecules absorb in this region, molecular studies in this region can be expected to increase in importance.

2-5A Theory

Whereas the electronic transitions responsible for absorption in the near ultraviolet are of the order of 4 eV, transitions in the vacuum ultraviolet are of the order of 6 eV. Since the ionization potential of most molecules lies around 6 eV, the electronic transitions most often observed in the vacuum ultraviolet are from the ground state of the neutral molecule to the molecular *ion*. These transitions are known as *Rydberg transitions*.

The two main instrumental difficulties in the far ultraviolet are the absorbance of air and the unavailability of transparent substances for use in optical systems. The absorption of most gases in this region makes it essential to work with equipment that has been evacuated or filled with a gas having a relatively low absorbance. Of the atmospheric constituents, oxygen is the most troublesome. Oxygen begins to absorb near 195 nm, goes through an absorption maximum at about 145 nm, and exhibits varying amounts of absorption down to 0.2 nm. Nitrogen begins to absorb at about 145 nm and has been used to flush spectrophotometers at longer wavelengths. The nitrogen must be free of oxygen and water vapor, however. Water begins to absorb at about 178 nm. The rare gases have comparatively little absorption except at relatively short wavelengths. For example, argon begins to absorb at about 107 nm, neon at about 74 nm, and helium at 60 nm. Thus, helium is a particularly good gas for flushing out far-ultraviolet spectrophotometers.

The other major instrumentation problem in this region is the scarcity of transparent substances for use in optical systems. Quartz, fluorite, and lithium fluoride have been used for optical accessories in the far ultraviolet.

Crystalline quartz absorbs strongly below about 185 nm and cannot be used as a prism or lens material below this wavelength. Fused quartz, which is amorphous, can be used down to about 170 nm. Calcium fluoride (fluorite) and lithium fluoride have been almost the only materials suitable for lenses and prisms at shorter wavelengths, though their transmittance limit has been about 100 nm. Below 100 nm some work has been done with thin films of silicon monoxide, aluminum, and beryllium, but the techniques necessary with such thin films have not yet been fully developed.

2-5B Instrumentation

VACUUM-ULTRAVIOLET SOURCES

The availability of good light sources has long been a problem in the vacuum ultraviolet. There is no single source of intense, continuous radiation extending over the entire vacuum-ultraviolet region, but there are a variety of sources useful in different parts of the spectrum. The most successful is the Lyman type, which uses a high-voltage discharge through a capillary of about 1-mm bore filled with hydrogen or a rare gas. The hydrogen-discharge lamp gives a usable continuum down to about 165 nm, below which it gives a multilined molecular spectrum of little use. Figure 2-41 shows emission spectra of each of the rare gases and hydrogen. The helium continuum from 60 to 100 nm is too unstable to be used for spectrophotometric studies, but the other rare gases have been successfully used. The argon capillary-discharge tube covers the fairly broad range of 107 and 165 nm, meeting the lower limit of the hydrogen-discharge lamp, but krypton and xenon give more intense outputs in the intermediate region.

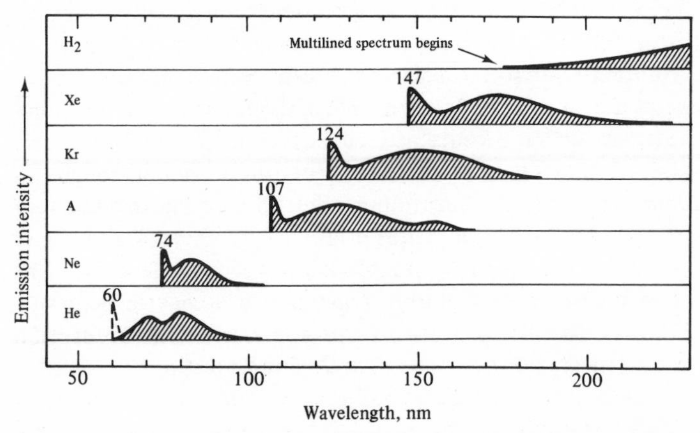

FIGURE 2-41 Continuous-emission spectra of the rare gases and hydrogen. (*From W. C. Price, p.* 59 *in H. W. Thompson (ed.), "Advances in Spectroscopy," Vol. I, John Wiley & Sons, Inc., New York,* 1959, *by permission.*)

VACUUM-ULTRAVIOLET MONOCHROMATORS

In the longer wavelength region of the vacuum ultraviolet, down to perhaps 125 nm, prism instruments have been widely used because of their low cost and ease of construction. However, scattered light is a serious problem in the vacuum ultraviolet, and radiation scattered from the face of a prism may be of greater intensity than the dispersed radiation. One successful method of eliminating this scattered radiation consists of using a vibrating mirror and tuning the detector to respond to the vibration frequency of the mirror, thereby discriminating against the scattered radiation.

Concave gratings are being increasingly used. One problem with the efficiency of gratings is caused by a decrease in reflecting power as the wavelength decreases. Platinum and aluminum seem to be the best reflectors in this region, and successful gratings have been prepared by platinizing aluminum surfaces. A quarter-wave thickness of MgF_2 greatly improves the reflectivity of mirrors.

The Beckman model DK-A far-ultraviolet spectrophotometer is an example of a commercially available prism instrument, and the Baird vacuum-ultraviolet spectrophotometer is a commercially available grating instrument. Both are useful to about 160 nm. The McPherson Instrument Corporation markets several vacuum-ultraviolet monochromators.

VACUUM-ULTRAVIOLET DETECTORS

The most convenient detector in the vacuum-ultraviolet region is an end-on photomultiplier which can be used down to 160 nm with a fused-quartz window. For shorter wavelengths it is customary to coat the window with a fluorescent material, like sodium salicylate. The sodium salicylate phosphor absorbs the short-wavelength radiation and reemits radiation of a longer wavelength, which can be detected by the photomultiplier.

A photosensitive Geiger counter is now available which can be made sensitive in the region between 105 to 250 nm. A central wire anode is made about 1 kV positive of a cylindrical photosensitive cathode, and a lithium fluoride window is used to seal the cylinder after it has been filled with ionizing gases, e.g., an ethyl formate–helium mixture. Photons entering the active volume of the Geiger tube cause ionization of the filling gas, and the ions produced are accelerated toward the appropriate electrode, colliding with other gas molecules and producing more ionization along the way. This internal amplification means that the output pulse is a relatively large, easily measured signal. Such tubes have been used to study the solar spectrum from rockets and should find increasing application in the laboratory.

2-5C Applications

Very little has yet been done in the vacuum ultraviolet in the way of useful analytical applications, but once instruments and convenient techniques become available, the applications will grow. A brief list of promising applications may be cited.

1 The analysis of substances that cannot otherwise be determined spectrophotometrically. For example, methane absorbs only at about 135 nm, octane at about 170 nm, and cyclopropane at about 190 nm. There are even some unsaturated compounds that give electronic transitions only in the far ultraviolet; some outstanding examples are the aliphatic cyanides (nitriles) and sulfones, in addition to ethylene and unconjugated polyenes.

2 The determination of the total olefin content of a sample. Jones and Taylor† have published one of the few analytical studies in the far ultraviolet, and they found, among other things, that the spectra of *unsaturated* hydrocarbons give intense, broad, almost structureless absorption bands, ideally suited for determining the total olefin content. Other substances give characteristic spectra suitable for specific identification.

3 Electronic structure studies. In addition to the general problem of correlating spectra with structure, specific problems, e.g., photochemical cis-trans isomerizations, can be studied. Also, the Rydberg transitions of nonbonding electrons and the effects of various substituents should prove interesting.

4 Determination of ionization and excitation potentials and dissociation energies.

5 Photoelectric studies of metals in the vapor state.

6 Determination of photoionization cross sections (absorptivities) for correlation with theoretical quantum-mechanical predictions.

The future of the vacuum-ultraviolet region is promising, and the research connected with space exploration should speed the technology necessary for laboratory applications.

EXERCISES

2-1 (*a*) Explain why spectra of solutions in the ultraviolet and visible regions tend to consist of a few broad peaks, whereas the same solutions measured in the infrared tend to show many sharp peaks. (*b*) Explain why the ultraviolet spectrum of benzene vapor contains many sharp bands whereas the same spectrum of benzene solution in alcohol lacks fine detail and consists only of broad peaks.

† L. C. Jones, Jr., and J. W. Taylor, *Anal. Chem.*, **27**:228 (1955).

2-2 Calculate the lowest molar concentration of solute that can be determined spectrophotometrically. Assume that the minimum absorbance accurately measurable is 0.01 and that a cell with a 10-cm light path is available.

Ans: 10^{-8} *M*

2-3 Using Woodward's rules, predict the wavelength of maximum absorption for each of the following compounds:

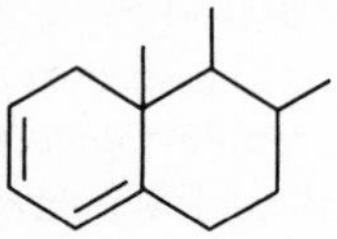

C_8H_{17}

AcO

H

(*a*) 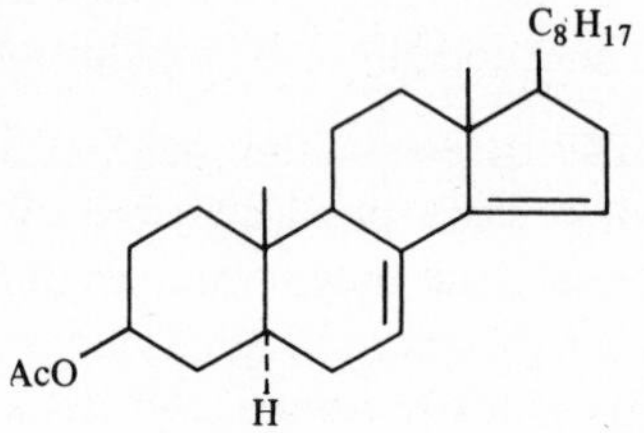 (*b*)

C_8H_{17}

AcO

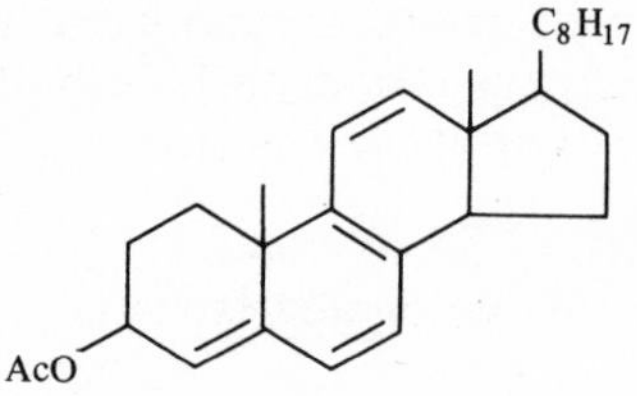

(*c*)

Ans: (*a*) 273 nm (observed, 275 nm); (*b*) 244 nm (observed, 242, nm); (*c*) 343 nm (observed, 355 nm)

2-4 α-Cyperone is known to have one of the following two structures:

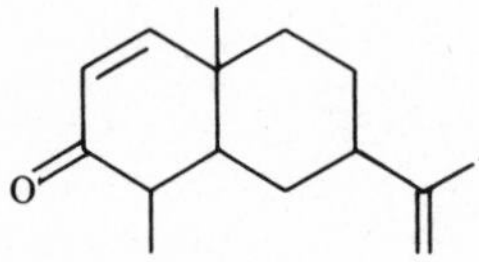

O or O

(*a*) (*b*)

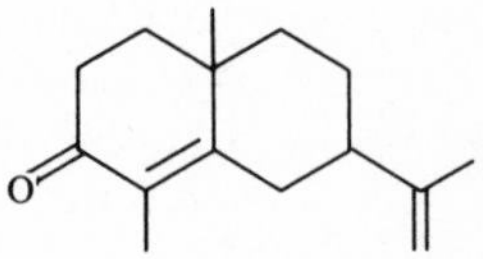

α-Cyperone is found to have an absorption maximum in ethanol at 252 nm. Using Woodward's rules, decide whether structure (*a*) or (*b*) is consistent with the observed spectrum.

Ans: (*b*) is the correct structure, with a calculated absorption at 254 nm. Structure (*a*) has a calculated absorption at 227 nm.

2-5 Using Woodward's rules, predict the absorption maximum for the following compound:

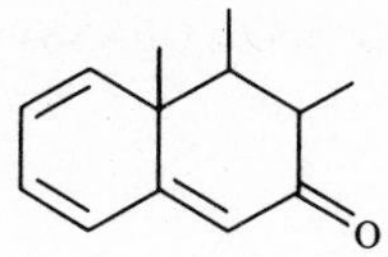

Ans: 349 nm

2-6 A 10^{-3} *M* solution of Cu(II) in concentrated HCl is a pale green-blue, whereas Cu(II) in concentrated ammonia is an intense blue-violet color. (*a*) What is the mechanism of absorption which is responsible for cupric-ion solutions being colored? (*b*) Explain why ammonia gives Cu(II) solutions an intense blue-violet color whereas HCl imparts a pale green-blue color. (*c*) For a quantitative analysis of cupric ion in ammoniacal solution with a spectrophotometer, approximately what wavelength of visible length would you use? *Ans:* (*c*) About 580 nm

2-7 Hemoglobin, an iron-containing porphyrin in blood, is reddish-purple in color. Predict the wavelength region in the visible spectrum where hemoglobin should absorb.

Ans: About 490 to 560 nm. In practice, 555 nm is often used for analysis.

2-8 When analyzed spectrophotometrically by the Lieberman-Burchard reaction, cholesterol gives a green solution. Predict the approximate wavelength for spectrophotometric analysis.

Ans: About 650 nm. In practice, 625 nm is often used.

2-9 Why do most photometers and spectrophotometers give linear readout in transmittance rather than a linear readout in absorbance?

2-10 Discuss the advantages and disadvantages of spectrophotometry in the vacuum-ultraviolet region of the spectrum.

2-11 When 5 mg of a certain amine picrate is dissolved in 100 ml of ethanol, its absorbance is measured in a 2.00-cm cell at 380 nm as 0.415. What is the molecular weight of the amine picrate?

2-12 Discuss the advantages of double-beam vs. single-beam spectrophotometers and illustrate the two main ways of achieving double-beam operation.

REFERENCES

INTRODUCTORY

1 EWING, G. W.: "Instrumental Methods of Chemical Analysis," 3d ed., McGraw-Hill, New York, 1969.

2 WILLARD, H. H., L. L. MERRITT, JR., and J. A. DEAN: "Instrumental Methods of Analysis," 4th ed., Van Nostrand, New York, 1965.
3 SILVERSTEIN, R. M., and G. C. BASSLER: "Spectrometric Identification of Organic Compounds," 2d ed., Wiley, New York, 1967.

INTERMEDIATE

4 DYER, J. R.: "Applications of Absorption Spectroscopy of Organic Compounds," Prentice-Hall, Englewood Cliffs, N.J., 1965.
5 BANWELL, C. N.: "Fundamentals of Molecular Spectroscopy," McGraw-Hill, New York, 1966.
6 SCHWARZ, J. C. P. (ed.): "Physical Methods in Organic Chemistry," Oliver & Boyd, Edinburgh, 1964.
7 STROBEL, H. A.: "Chemical Instrumentation," Addison-Wesley, Reading, Mass., 1960.
8 WILLIAMS, D. H., and I. FLEMING, "Spectroscopic Methods in Organic Chemistry," McGraw-Hill, New York, 1966.
9 KOLTHOFF, I. M., and P. J. ELVING (eds.): "Treatise on Analytical Chem.," pt. I, vol. 5, Wiley, New York, 1964. A chapter by A. A. Schilt and B. Jaselskis is particularly good on instrumentation.
10 FREEMAN, S. K.: "Interpretative Spectroscopy," Reinhold, New York, 1965.

ADVANCED

11 JAFFÉ, H. H., and M. ORCHIN: "Theory and Applications of Ultraviolet Spectroscopy," Wiley, New York, 1962.
12 SANDORFY, C.: "Electronic Spectra and Quantum Chemistry," Prentice-Hall, Englewood Cliffs, N.J., 1964.

3 Infrared Spectrophotometry

Infrared spectrophotometry (also called infrared spectroscopy) is one of the most powerful tools available for solving problems of molecular structure and chemical identification. Although organic compounds are usually studied, inorganic compounds containing polyatomic cations or anions also give useful infrared spectra.

The infrared region lies between the visible and the microwave regions, extending from about 750 nm (0.75 μm) to about 1000 μm. Both wavelength (μm) and wave-number (cm^{-1}) units are used to characterize infrared radiation, and the student must become adept at using both. The relationship between the two is

$$\bar{\nu}\ (\text{cm}^{-1}) = \frac{10^4}{\lambda\ (\mu\text{m})} \tag{3-1}$$

Example 3-1 What is the frequency in wave numbers corresponding to 2.75 μm?

Answer

$$\bar{\nu} = \frac{10^4}{2.75} = 3636\ \text{cm}^{-1}$$

Since $\bar{\nu} = \nu/c$, where ν is the frequency and c is the speed of light, both $\bar{\nu}$

TABLE 3-1 Common Subdivisions of the Infrared Spectrum

Region	Type of energy-level transition	Wavelength range, μm	Wave-number range, cm^{-1}
Near infrared	Overtones	0.75–2.5	13,300–4000
Fundamental	Vibrations, rotations	2.5–25	4000–400
Far infrared	Skeletal vibrations, rotations	25–1000	400–10

and ν are directly proportional to the energy of the radiation E, as given by Planck's equation

$$E = h\nu = h\bar{\nu}c \tag{3-2}$$

where h is Planck's constant $= 6.62 \times 10^{-27}$ erg-s.

Table 3-1 gives the common subdivisions of the infrared spectrum. The spectrum is divided into three regions to show the different kinds of molecular information obtainable and to emphasize basic differences in the instruments of each region. In the *near infrared* the majority of the absorption bands are due to overtones of hydrogen-stretching vibrations, which are particularly useful for the quantitative analysis of various functional groups. Instruments used in the near infrared are similar to those used in the visible region.

In the *fundamental*-infrared region, a vast amount of qualitative and quantitative information about functional groups and molecular structure is obtainable. The fundamental region is generally meant when the term infrared is unqualified. The optical materials and other instrumental components used in this region differ from those used in the visible and near-infrared regions. The majority of infrared instruments cover the limited range from 2.5 to 15 μm (or 4000 to 667 cm^{-1}).

The *far-infrared* region gives information mainly about rotational transitions, vibrational modes of crystal lattices, and skeletal vibrations of large molecules. The instrumentation requirements are considerably different in the far-infrared region, and suitable instruments have only recently become commercially available.

In marked contrast to the visible and near-ultraviolet regions, almost all substances show absorption in the infrared region, the only exceptions being monatomic and homopolar molecules such as Ne, He, O_2, N_2, and H_2. Furthermore, no two compounds with different structures have the same infrared spectrum, except for optical isomers and certain high-molecular-weight polymers differing only slightly in molecular weight, which may not be differentiable with the instruments available. Hence infrared spectrophotometry is unexcelled as a general tool for molecular identification.

On the other hand, the detection limits and sensitivity of infrared spectrophotometry are appreciably less than those of ultraviolet and visible spectrophotometry. In general, components in a sample that are present below about 1 percent are not detectable. Furthermore, the low energy of infrared radiation creates a number of other instrumentation problems. For example, the signals measured are often of the same order of magnitude as electronic noise in the detecting circuit. In addition, warm components of the spectrophotometer radiate infrared energy, and this false radiation must be distinguished from the true signal. These and other problems can be minimized to a large degree, as we shall see, but nonetheless they tend to limit the usefulness of infrared spectrophotometry for quantitative analysis.

In the sections that follow, a brief discussion of the theory of infrared absorption will be followed by sections on instrumentation and applications. Most of this discussion will deal with the fundamental-infrared region, from 2.5 to 25 μm, but at the end of the chapter will be found brief discussions of the near- and the far-infrared regions.

3-1 THEORY OF INFRARED ABSORPTION

3-1A Requirements for Absorption

There are two requirements for the absorption of electromagnetic radiation by matter: (1) the radiation must have precisely the correct energy to satisfy the energy requirements of the material, and (2) there must be a coupling (or interaction) between the radiation and matter. Radiation in the infrared region has the proper magnitude of energy to cause *vibrational* transitions in molecules, and the first requirement for absorption is satisfied if a given frequency of infrared radiation corresponds exactly to a fundamental vibrational frequency of a given molecule. To satisfy the second requirement for absorption, the molecule must undergo a *change in dipole moment* when the fundamental vibration occurs. If *no* change in dipole moment occurs when the molecule vibrates, there will be no interaction between the electromagnetic radiation and the nonelectromagnetic molecule and no absorption will take place regardless of the energy compatibility. Such a vibration is said to be *infrared-inactive*.

The *dipole moment* μ of two equal and opposite charges is defined as the product of the charge q and the distance r separating them

$$\mu = qr \tag{3-3}$$

For a molecule, it is the *effective center* of the positive and negative charges which is important, r being the distance between those centers. Figure 3-1 illustrates how the separation of charge in a polar molecule like water may be visualized. As will be shown momentarily, all the possible modes of vibration of a water molecule involve a change in dipole moment, and thus all its vibrational modes are *infrared-active*. With a linear molecule like CO_2, however,

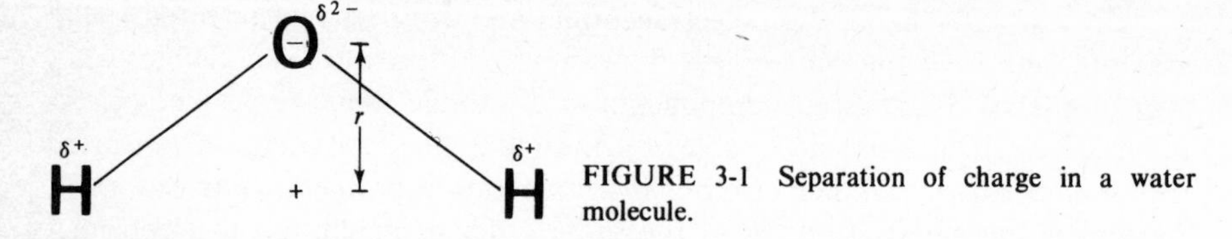

FIGURE 3-1 Separation of charge in a water molecule.

Fig. 3-2*a* shows that its dipole moment is zero. More important is that the *change* in dipole moment is zero when the molecule is symmetrically stretched (Fig. 3-2*b*). Thus, the symmetrical stretch of CO_2 is infrared-inactive. On the other hand, as shown in Fig. 3-2*c*, an antisymmetrical (or asymmetrical) stretch involves a change in dipole moment, and thus this mode of vibration is infrared-active.

3-1B Types of Vibrations

NAMES OF VIBRATIONAL MODES

It is necessary to know the descriptive names given to the various modes of vibration, since these names are used to identify absorption peaks in infrared spectra. Vibrations of the methylene group are given in Fig. 3-3. In Fig. 3-3, the carbon atom may be imagined to be rigidly fixed to the remainder of the molecule (not shown), and the motions indicated reverse their direction every half-cycle. Two basic types of vibration, stretching and bending, can be distinguished in Fig. 3-3. Modes *a* and *b* in Fig. 3-3 are *stretching* vibrations, whereas the remaining four modes are *bending* (sometimes called deformation) vibrations. Ring compounds can undergo a stretching mode involving the entire ring, called *breathing*, illustrated in Fig. 3-4.

THEORETICAL NUMBER OF FUNDAMENTAL MODES

It is useful and fairly accurate to visualize a molecule as an assembly of balls and springs in constant motion, the balls representing nuclei and the springs representing chemical bonds. If this model is suspended in space and struck a blow with a solid object, the balls will appear to undergo random

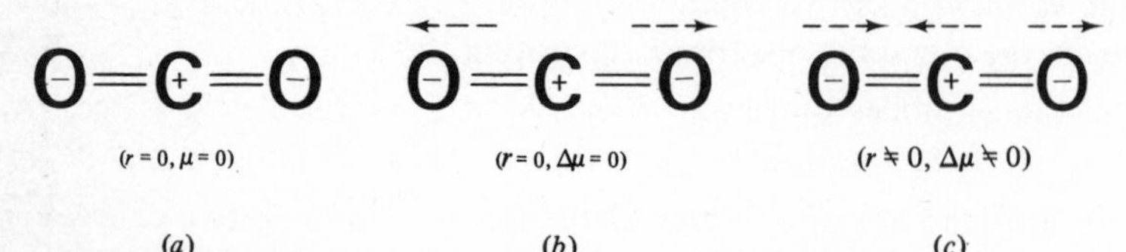

FIGURE 3-2 Carbon dioxide and its stretching vibrations. (*a*) Normal shape (the center of the carbon atom is the center of gravity for both positive and negative charges). (*b*) Symmetrical stretch (infrared-*inactive*); *r* remains zero during the stretch. (*c*) Antisymmetric stretch (infrared-*active*).

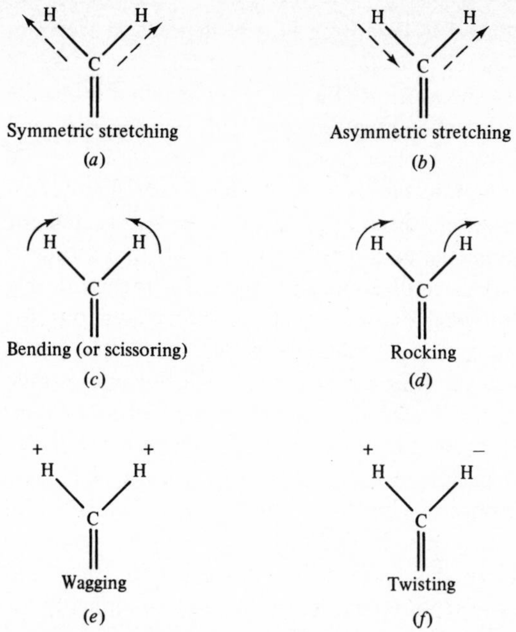

Symmetric stretching
(*a*)

Asymmetric stretching
(*b*)

Bending (or scissoring)
(*c*)

Rocking
(*d*)

Wagging
(*e*)

Twisting
(*f*)

FIGURE 3-3 Names given to modes of vibration; + and − represent movements at right angles to the surface of the page.

chaotic motions, but if the vibrating model is observed with a variable-frequency stroboscopic light, at certain light frequencies some of the balls appear to remain stationary. These represent the *fundamental modes of vibration and their corresponding frequencies*.

In dealing with actual molecules instead of large mechanical models, one additional refinement must be made. Whereas with a mechanical model the vibration can be initiated with *any* amount of impact energy, with a molecule only certain, fixed amounts of radiation energy are effective. This is because the vibrational energy levels of a molecule are quantized; i.e., only radiation with an energy corresponding precisely to an allowed vibrational frequency of the molecule will interact with the molecule. Radiation of any other energy will not interact, i.e., will not be absorbed.

The theoretical number of fundamental modes of vibration of any molecule are easily predicted. A molecule containing N atoms can have only $3N - 6$ fundamental modes of vibration or $3N - 5$ modes if it is a linear

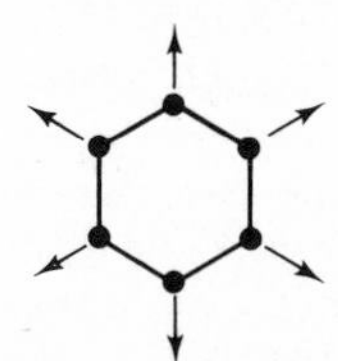

FIGURE 3-4 Skeletal breathing vibrations.

molecule. These rules can be deduced by counting all the degrees of freedom of a molecule and subtracting out all motions which are not vibrational in character. Specifically, to describe the motion of a given nucleus in a molecule completely, the three coordinates in space must be specified, i.e., the x, y, and z cartesian coordinates. Hence, for a molecule with N atoms, a total of $3N$ coordinates is required, and the molecule is said to have $3N$ degrees of freedom of motion. Of these, however, three degrees must be subtracted for *translational* motion, since the molecule as a whole can drift off in a straight-line (transitional) motion in any of three directions in space. Furthermore, for a nonlinear molecule, three additional degrees must be subtracted out for *rotational* motion about the three cartesian coordinates, leaving $3N - 6$ fundamental modes of vibration for a nonlinear molecule. For a linear molecule only two degrees of rotational motion need be subtracted out, since a linear molecule can be placed along one cartesian coordinate and rotation about this axis cannot be recognized as a spatial movement. Hence, a linear molecule will have $3N - 5$ fundamental modes of vibration. Thus, a bent molecule like water has three fundamental modes of vibration, whereas a linear molecule like HCN has four fundamental modes of vibration. In practice, three and four fundamental frequencies of absorption can be found spectroscopically for H_2O and HCN, respectively.

OBSERVED NUMBER OF MODES OF VIBRATION

The correlation between the expected and observed number of absorption bands does not hold in every case. Occasionally *more* absorption bands than would be predicted are found, and more often *fewer* absorption bands are identified. First we consider the reasons why more absorption bands may be found.

The number of observed absorption bands may be increased by the presence of bands which are not fundamentals but *combination tones, overtones,* or *difference tones*. A *combination tone* is the sum of two or more different frequencies, such as ν_1 and ν_2; that is, vibrations 1 and 2 in a molecule are simultaneously excited by the absorbed photon. An *overtone* is a multiple of a given frequency, for example, 2ν (first overtone) and 3ν (second overtone). [The first overtone may occur, for example, because of a transition from the $v = 0$ state to the $v = 2$ excited state (where v is the vibrational quantum number, as explained in Sec. 1-3) instead of the more usual $v = 0$ to $v = 1$ fundamental transition.]† Finally, a *difference tone* is the difference between two frequencies, say $\nu_1 - \nu_2$. A difference tone is relatively rare and occurs when a molecule already in one excited state ν_2 absorbs enough addi-

† It should be noted that observed overtones often do not fall at precisely two, three, etc., times the fundamental frequency because vibrational energy levels are often not equally spaced but tend to become closer together at higher and higher vibrational levels; as a result, overtones usually appear at a frequency slightly less than the predicted multiple of the fundamental frequency.

tional radiant energy to raise it to another excited vibrational state ν_1. All three of these additional (nonfundamental) transitions have much less probability of occurring than the fundamental transitions, and thus nonfundamental absorption bands usually are much less intense than fundamental bands, often by a factor of 10 to 100. Most absorption bands in the near infrared (0.75 to 2.5 μm) are *overtones* or *combinations* of hydrogen-stretching vibrations. The 5- to 6-μm region often has absorption bands due to *overtones* or *combination tones* of bending vibrations.

Now consider the more common case where *fewer* absorption bands can be found than are predicted by the $3N - 6$ or $3N - 5$ rule. The number of observed bands may be less than the predicted number for the following reasons:

1 A given vibration may be *infrared-inactive* if a change in dipole moment does not accompany the vibration (see Sec. 3-1*A*; the examples to be considered shortly will also illustrate this).

2 Because of symmetry, some vibrations may occur at precisely the same fundamental frequency as other vibrations. Such vibrations are said to be *degenerate* (have the same energy) and will be observed as only one band. (This phenomenon will be illustrated in the examples that follow.)

3 Some vibrations may have frequencies so nearly alike that they are not resolved by the spectrophotometer.

4 Some fundamental vibrations may absorb so weakly that the bands are not observed or are overlooked.

5 Some of the fundamental vibrations may occur at frequencies outside the range of the instrument being used. Most instruments go down to 2.5 μm (4000 cm^{-1}), which essentially covers all of the most *energetic* fundamental vibrations; on the other hand, most instruments go out to only about 15 to 25 μm (667 to 400 cm^{-1}), and a fair number of bending vibrations and especially skeletal vibrations occur at longer wavelengths.

It should be clear that it is rarely possible to locate all the fundamental absorption frequencies predicted from the geometry of a molecule; nor is it necessary to attempt to do so. Nonetheless, an understanding of the fundamental modes of vibration is helpful in interpreting infrared spectra, especially for relatively simple molecules.

Example 3-2 Predict the number and give the names of the fundamental modes of vibration of HCl.

Answer With only two atoms, HCl is obviously linear and has $3N - 5 =$ 1 fundamental mode of vibration, which is the *symmetrical stretch*. This vibration is observed to occur at 2886 cm^{-1}, or 3.46 μm.

Example 3-3 How many fundamental vibrational frequencies would you expect to observe in the infrared absorption spectrum of H_2O?

Answer The bent H_2O molecule will have $3N - 6 = 3$ fundamental modes of vibration. To decide whether all three are infrared-active and whether any degeneracy exists, it is helpful to diagram the vibrational modes, as follows:

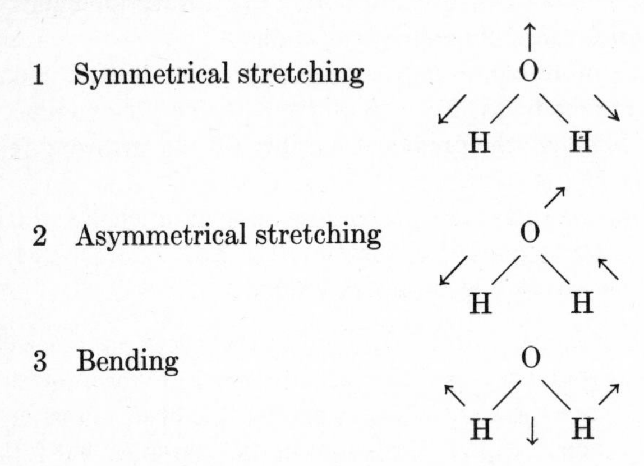

No other fundamental modes of vibration can be diagramed. If, for example, an attempt were made to portray rocking, wagging, or twisting modes, these would in reality be *rotations*, and not vibrations, since the oxygen atom is not anchored but free to rotate.

All three modes of vibration are infrared-active (undergo a change in dipole moment) and are nondegenerate (vibrate with different energies). The observed absorptions occur at 3652 cm^{-1} (2.74 μm), 3756 cm^{-1} (2.66 μm), and 1595 cm^{-1} (6.27 μm), respectively. Note that an asymmetrical stretch takes slightly more energy than a symmetrical stretch, whereas a bending vibration requires appreciably less energy.

Example 3-4 How many fundamental vibrational frequencies would you expect to observe in the infrared absorption spectrum of CO_2?

Answer Since CO_2 is linear, it has $3N - 5 = 4$ fundamental vibrational modes, which may be diagramed as follows:

1 Symmetrical stretching $\overleftarrow{O}{=}C{=}\overrightarrow{O}$

2 Asymmetrical stretching $\overrightarrow{O}{=}\overleftarrow{C}{=}\overrightarrow{O}$

3 Bending (on the plane of the paper) $\uparrow O{=}\underset{\downarrow}{C}{=}O \uparrow$

4 Bending (perpendicular to the plane of the paper) $\underset{+}{O}{=}\underset{-}{C}{=}\underset{+}{O}$

Of these four vibrational modes, the symmetrical stretch (1) will be infrared-inactive, since no change in dipole moment occurs during the vibration.

Furthermore, it is easy to see that the two bending modes (3 and 4) are degenerate and will absorb at the same frequency. Thus, the actual vibration will be a combination of modes 3 and 4, amounting to a wobble, or precession of each oxygen atom around the normal bond axis.

We conclude, then, that CO_2 should exhibit *two* fundamental absorption bands, one due to the asymmetrical stretch (2) and another due to the doubly degenerate bending modes (3 and 4). In practice, these two absorption bands are observed at 2349 cm^{-1} (4.26 μm) and 667 cm^{-1} (14.99 μm), respectively.

Example 3-5 Nitrogen dioxide, NO_2, could exist in a linear configuration, O—N—O, or a bent configuration. The infrared spectrum of gaseous NO_2 reveals three strong bands, at 1616 cm^{-1} (6.19 μm), 1323 cm^{-1} (7.56 μm), and 750 cm^{-1} (13.33 μm). On the basis of this spectrum, what is the probable configuration of NO_2? What vibrations can be assigned to these bands?

Answer If NO_2 were linear, it would have $3N - 5 = 4$ fundamental modes of vibration, which we can diagram as follows:

1 Symmetrical stretching $\overleftarrow{O}$—N—$\overrightarrow{O}$

2 Asymmetrical stretching $\overleftarrow{O}$—$\overrightarrow{N}$—$\overleftarrow{O}$

3 Bending (in plane) ↑O—N—O↑ (N ↓)

4 Bending (out of plane) O—N—O (+ − +)

By the same reasoning used for CO_2 (Example 3-4) a linear NO_2 should exhibit only two fundamental absorption bands, and therefore NO_2 is probably not linear.

If NO_2 were *bent*, it would have $3N - 6 = 3$ fundamental modes of vibration, which we can diagram as follows:

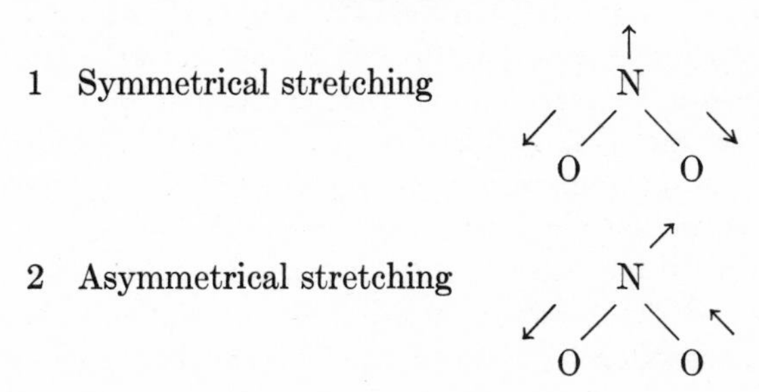

3 Bending

```
       N
  ↖  / \  ↗
   O  ↓  O
```

All three of these modes of vibration are infrared-active, and there is no degeneracy among them; therefore these three vibrations must correspond to the three absorption bands. By analogy to the assignments made for the H_2O molecule (Example 3-3) we would predict that the asymmetrical stretch is responsible for the 6.19-μm band, the symmetrical stretch for the 7.56-μm band, and the bending mode for the 13.33-μm band.

3-1C Intensity of Absorption Bands

The intensity of an infrared absorption band is a measure of the probability of the vibrational transition, and this in turn depends on the magnitude of the *change of dipole moment*. Bands with only small dipole-moment changes absorb weakly; e.g., the C=C stretching vibration in 2-methyl-but-2-ene has a molar absorptivity ϵ of only about 5. Similarly, C=N, C—C, and C—H stretching vibrations tend to give weak absorption bands. On the other hand, carbonyl and similar highly polar groupings like Si—O, C—Cl, or C—F tend to give very intense infrared bands, with molar absorptivities in the range 100 to 1000. The precise theory governing the intensity of absorption bands is not well established, however, and qualitative generalizations frequently have exceptions. For example, whereas C—H stretches tend to give weak bands, many C—H bending vibrations such as out-of-plane bending in aromatic or olefinic compounds may give rise to strong absorption bands. Similarly, whereas combination bands are usually rather weak, it is not too uncommon to find some combination bands which are more intense than some of the fundamentals.

It should be emphasized that the intensity of infrared vibrational absorption bands of even the most polar groups is two or three orders of magnitude smaller than that of the most intense electronic transitions found in the ultraviolet and visible regions. Furthermore, the rather low energies and relatively wide slit widths that must be tolerated in infrared spectrophotometry make molar absorptivities less useful in qualitative identifications. In the ultraviolet and visible regions the slit widths can be made very much smaller than the true bandwidth, whereas in infrared spectrophotometry, the slit widths must often be of the same width as the true absorption bandwidth. As a result the measured absorption maximum and bandwidth of infrared peaks are strongly influenced by the slit width used. Thus, the molar absorptivity ϵ of a substance will vary from instrument to instrument. There is some trend toward using *integrated absorption* intensities, or absorptivities integrated over the area of the absorbtion band, since these have fundamental value, but widespread use of this approach must await instruments that give integrated absorptions directly. In the meantime, the usual practice is to describe the appearance of absorption bands qualitatively, using notations like *vs* (very strong), *s* (strong), *m* (medium), *w* (weak), *vw* (very weak), *wm* weak to medium), etc.

3-1D Theoretical Group Frequencies

With many fundamental modes of vibration, the main participants in the vibration are just two atoms and the chemical bond in between. Such modes have frequencies which depend primarily on the weight of the two vibrating atoms and the strength of the bond joining them. The frequencies are only slightly affected by the other atoms attached to the two atoms concerned. Thus, these vibrational modes are characteristic of the groups in the molecule and are extremely useful in identifying a compound. These characteristic frequencies are known as *group frequencies*.

As a first approximation, the frequency of vibration can be calculated from the equation for a harmonic oscillator

$$\bar{\nu} = \frac{1}{2\pi c}\sqrt{\frac{k}{m_r}} \tag{3-4}$$

where k is the force constant in units of dynes per centimeter and m_r is the reduced mass, which for two atoms is defined as

$$m_r = \frac{m_1 m_2}{m_1 + m_2} \tag{3-5}$$

where m_1 and m_2 are the masses of atoms 1 and 2.

The force constant k is a measure of the strength or stiffness of a chemical bond when it is in its normal equilibrium position. A number of theoretical and empirical relationships have been proposed for calculating force constants, taking into account the bond length and valence forces or electronegativities of the two atoms involved. The accuracy of these estimates is, at best, about ± 5 percent. Table 3-2 shows the estimated stretching-force constants of some chemical bonds. Notice that for carbon-carbon single, double, and triple bonds, the force constants are approximately 5, 10, and 15 $\times\ 10^5$ dyn/cm, respectively. For bonds of the type H—X, the force constant increases with the electronegativity of atom X. Bending-force constants can be estimated with less accuracy than stretching-force constants, but in general they have values of the order of one-tenth those of the stretching constants.

The reduced mass m_r may be thought of as a weighted average of the two atoms. Notice particularly that if $m_1 \gg m_2$, then $m_r \approx m_2$ [see Eq. (3-5)]. In other words, if there is a great disparity in the weights of atoms 1 and 2, it is the *lighter* atom that determines the reduced mass, and the vibrational frequency will be independent of the mass of the heavier atom.

TABLE 3-2 Approximate Stretching-Force Constants of Some Chemical Bonds†

Bond	Molecule	k, dyn/cm
H—F	HF	9.7×10^5
H—Cl	HCl	4.8
H—Br	HBr	4.1
H—I	HI	3.2
H—O	H_2O	7.8
H—S	H_2S	4.3
H—N	NH_3	6.5
H—C	CH_3X	4.7–5.0
H—C	C_2H_4	5.1
H—C	C_2H_2	5.9
Cl—C	CH_3Cl	3.4
C—C		4.5–5.6
C=C		9.5–9.9
C≡C		15–17
N—N		3.5–5.5
N=N		13
N≡N		23
C≡N		16–18
C—O		5.0–5.8
C=O		12–13

† From E. Bright Wilson, Jr., J. C. Decius, and Paul C. Cross, "Molecular Vibrations," p. 175, McGraw-Hill Book Company, New York, 1955, by permission.

Equation (3-4) assumes that the vibrating atoms execute simple harmonic motion about an equilibrium point, and linkages with the remaining atoms are ignored. In practice, molecular vibrations tend to be *anharmonic*, i.e., the vibrations are not symmetrical about an equilibrium point, and quantum theory predicts that additional higher-order terms must be added to Eq. (3-4) to obtain good agreement between theory and experiment. However, reasonable agreement can be obtained without the higher-order terms, and they are usually neglected. To use Eq. (3-4), k must be in dynes per centimeter and m_1 and m_2 must be the weight (in grams) of one atom of substance 1 and 2, respectively. Calculations can be simplified, however, if we use k in units of 10^5 dyn/cm and m_1 and m_2 in gram atomic weight units. With these new units for m_1 and m_2, we obtain

$$\bar{\nu} = 1307 \sqrt{\frac{k}{m_r}} \tag{3-6}$$

which is much more convenient than Eq. (3-4).

Example 3-6 Estimate the frequency of absorption of the C—H stretching vibration.

Answer

$$m_r = \frac{m_1 m_2}{m_1 + m_2} = \frac{(1)(12)}{1 + 12} = 0.92$$

$$k_{C-H} \approx 5.0 \times 10^5 \text{ dyn/cm} \qquad \text{(see Table 3-2)}$$

$$\bar{\nu} = 1307 \sqrt{\frac{5.0}{0.92}} = 3000 \text{ cm}^{-1}$$

In agreement with this estimate, experimentally observed C—H stretching vibrations are found in this frequency region. For example, the C—H stretch in chloroform, $CHCl_3$, is 2915 cm^{-1} (3.43 μm).

Example 3-7 Estimate the frequency of absorption of the C—D stretching vibration where D is deuterium.

Answer

$$m_r = \frac{(2)(12)}{2 + 12} \doteq 1.71$$

Since the only difference between hydrogen and deuterium is the extra neutron in the nucleus of deuterium, it may be assumed that the C—D force constant is approximately the same as the C—H force constant, or

$$k_{C-D} \approx 5.0 \times 10^5 \text{ dyn/cm}$$

$$\bar{\nu} = 1307 \sqrt{\frac{5.0}{1.71}} \approx 2200 \text{ cm}^{-1}$$

(Note that substituting deuterium for hydrogen lowers the absorption frequency by about $\sqrt{2}$, which is a useful rule of thumb; isotopic substitution is often a useful means of proving the assignment of spectral absorption bands.)

In agreement with this estimate, the observed C—D stretching frequency in deuterochloroform is 2256 cm^{-1} (4.43 μm).

Evaluation of Eq. (3-6) for various possible combinations of atoms would show that fundamental stretching vibrations fall in a frequency range of about 4000 to 800 cm^{-1} (2.5 to 12.5 μm) and fundamental bending modes extend from about 1700 to 400 cm^{-1} (6.0 to 25 μm). The region beyond 400 cm^{-1} (25 μm) is considered to be the far infrared, and certain skeletal vibrations and lattice vibrations of crystals can occur in this region, in addition to pure rotational transitions of small molecules.

Some simple correlations of functional-group vibrations in various regions of the fundamental infrared region are shown in Fig. 3-5. The funda-

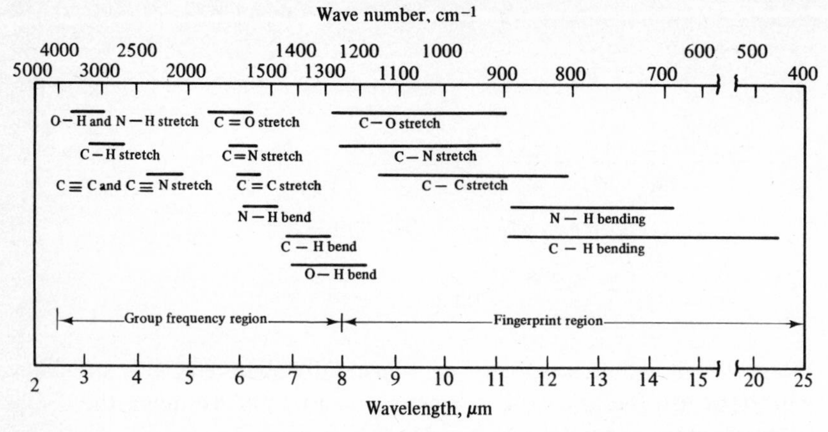

FIGURE 3-5 Simple correlations of group vibrations in the fundamental-infrared region. (*After R. T. Conley, "Infrared Spectroscopy," 2d ed., p.* 88, *Allyn and Bacon, Inc., Boston, 1972, by permission.*)

mental-infrared region is conveniently divided into a *group-frequency* region from 2.5 to 8 μm and a *fingerprint* region from 8 to 25 μm. In the group-frequency region, absorption bands tend to be characteristic of specific groups of atoms and relatively independent of the composition of the rest of the molecule. In the fingerprint region, however, vibrational frequencies are profoundly affected by the molecular structure as a whole, and therefore bands in this region are considered specific for a particular molecule rather than for a particular functional group. The uniqueness of a given molecule's spectrum in this region justifies calling this the fingerprint region. It should be acknowledged, however, that there is some limitation on this specificity. It is doubtful, for example, whether one could detect any difference in the spectra of the molecules $CH_3(CH_2)_xCH_3$ if x were varied from, say 16 to 18. The added methylene groups would not add any new absorption bands or significantly shift the frequency of existing bands.

In the fingerprint region, as Fig. 3-5 shows, a given vibrational mode may be found anywhere over a large range of frequencies, depending on the structure of the molecule as a whole. In the group-frequency region, however, the vibrational frequency of a given functional group can be pinpointed to a much narrower range of frequencies, and thus the group-frequency region is extremely useful for diagnosing the presence or absence of certain functional groups. The region from 2.5 to 4 μm (4000 to 2500 cm^{-1}) is restricted to hydrogen-stretching vibrations, the region from about 4 to 6.3 μm (2500 to 1600 cm^{-1}) emphasizes double- and triple-bond stretching, and the region from about 6.3 to 8 μm contains mainly hydrogen-bending vibrations.

The hydrogen-stretching region is fairly independent of the molecular configuration as a whole, although the presence of hydrogen bonding, for example, tends to cause a broadening of absorption bands coupled with a shift toward lower frequencies. At longer wavelengths (lower frequencies), vibrations become more sensitive to overall molecular structure. Hence, the region from 4 to 6.3 μm may be thought of as a transitional region, since the hydrogen-bending modes beyond 6.3 μm are fairly dependent on the structure as a whole. Within the transition region from 4 to 6.3 μm, it is usually still possible to assign an absorption at a given wavelength to a particular atomic configuration. Notice the regularity in shifts to longer wavelength in going from triple to double to single bonds. In the region from 6.3 to 8 μm assignments can sometimes be made to methyl and methylene groups, as well as to the OH bending vibrations of alcohols and acids, but it will rarely be possible to assign all the observed bands in a complex spectrum. Beyond 8 μm, in the fingerprint region, there is little point in trying to make specific band assignments, and the absorption bands are mainly useful for comparison with known spectra.

3-1E Factors Affecting Group Frequencies and Band Shapes

PHYSICAL STATE OF THE SAMPLE

Infrared spectra depend on the physical state of the sample. When spectra of gas, liquid, and solid states are compared, shifts in group absorption frequencies are generally small (1 to 2 percent) unless specific interactions like hydrogen bonding or solvent effects occur. However, the general appearance and complexity of the spectra may be markedly affected. Figure 3-6 shows the spectra of *p*-dichlorobenzene, run as a vapor, solution, and solid. In the *gaseous* state, individual molecules are free to vibrate and rotate with little interference from other molecules. As a result, absorption bands tend to be a little broader and stubbier than in the liquid state, because the rotational transitions are superimposed on the vibrational transitions (compare Fig. 3-6*a* and *b*). In general, the absorption spectra of smaller molecules show more rotational fine detail than large molecules (see Sec. 3-1*F*).

In the *liquid* state, individual molecules are confined to a "molecular cage," where they are continually buffeted by other particles, and as a result they can no longer execute quantized rotational motion. Consequently, any rotational fine structure disappears, and absorption bands appear narrower and more symmetrical, somewhat similar to a normal error curve (see Fig. 3-6*b*).

Absorption spectra in the *solid* state tend to be more complex than those for the liquid state (compare Fig. 3-6*b* and *c*). Vibrational absorption bands in solid-state spectra are often split, new bands may appear, and some bands may become sharper. There are a variety of reasons for these modifications. For example, in the unit cell of a crystal, the vibrations of some molecules

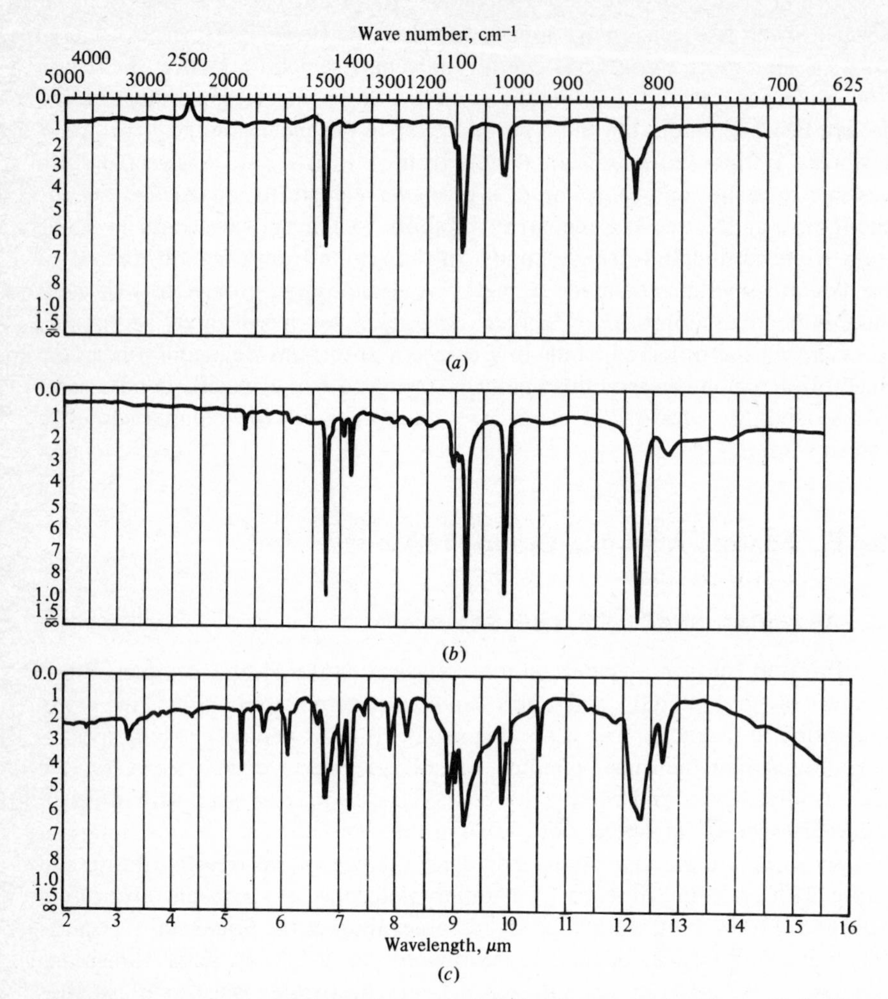

FIGURE 3-6 Comparison of infrared spectra of dichlorobenzene in the (*a*) gas, (*b*) liquid, and (*c*) solid states. (*From L. A. Smith, in I. M. Kolthoff and P. J. Elving (eds.), "Treatise on Analytical Chemistry," pt. I, vol.* 6, *p.* 3647, *John Wiley & Sons, Inc., New York,* 1965, *by permission.*)

may be out of phase with other molecules, and the interactions of these motions may lead to different frequencies. In addition, *lattice vibrations* of low frequency may be superimposed on fundamental absorption frequencies. Furthermore, strong local fields may change the frequency of a vibrational band or may even make a normally inactive mode infrared-active. When dealing with *flexible* molecules, e.g., long-chain acids or esters, the solid-state spectra tend to be sharper and more distinctive than the liquid-state spectra, particularly in the fingerprint region, because the liquid state allows more numerous conformations, which in turn result in broader bands. When dealing with *rigid* systems, e.g., the steroids, on the other hand, spectra are almost equally sharp in the liquid and solid states.

If one is given a choice of phases, the liquid state is probably to be preferred for most applications, since the spectra tend to be easier to interpret. However, the increased complexity of solid-state spectra may sometimes be an advantage for qualitative analyses. With the liquid state, dilute solutions of the solute in relatively inert, nonpolar solvents tend to give the most invariant and reproducible group frequencies, intensities, and band shapes. Polar solvents tend to shift absorption bands to lower frequencies (longer wavelengths), but the extent of the shift is difficult to generalize. Pure liquids are often analyzed without a solvent, but it should not be overlooked that strong *intermolecular* attractions are likely to occur. As a result the absorption frequency may be significantly different from those found in dilute solution, where molecules are well separated by relatively inert solvent molecules.

The effect of *temperature* changes on spectra depends on whether the solid, liquid, or gaseous state is being studied. However, the effect of temperature on absorption *frequencies* is generally small for all phases. With the *gaseous* state, increasing temperature causes a pronounced broadening of spectral bands due to increased collisions and also causes other changes in band shapes by changing the number of molecules in various rotational (and vibrational) energy levels. With molecules in the *liquid* state, bands often sharpen at higher temperatures, due to decreased intermolecular interactions. With *solids*, bands often sharpen with decreasing temperature, probably because of decreased lattice vibrations.

VIBRATIONAL COUPLING

Vibrating groups that have approximately the same frequency and are located in adjoining portions of the molecule sometimes interact with each other to give a mixed vibration in which both groups take part. Such vibrational coupling is fairly common and does not prevent the resulting absorption frequency from being useful for qualitative identification of the participating groups. For example, in substituted ethylenes the well-known C=C frequency at about 1620 cm^{-1} (6.2 μm) is in reality about 60 percent due to C=C stretching and 35 percent due to C—H bending. When the interaction is removed by the substitution of deuterium for hydrogen, the C=C frequency drops to its "normal" position at about 1520 cm^{-1} (6.6 μm). The coupling is mechanical and can be observed in ball-and-spring models of the molecule. Similar mechanical interaction is known to occur between the C—O—H bending motion and the C—O stretching motion in alcohols, between the C—N—H bend and C—N stretch in amides, and between the out-of-plane bending motions of adjacent hydrogens on aromatic rings.

A special case of vibrational coupling occurs when an overtone or combination frequency interacts with a fundamental vibration. This type of interaction, called *Fermi resonance*, results in *two* bands in the region where *one* fundamental band was expected. For example, the CH stretch in aldehydes might be expected to occur at around 2800 cm^{-1} (3.57 μm), but instead a characteristic doublet in the region of 2700 to 2900 cm^{-1} (3.70 to 3.45 μm) is found because the CH stretch interacts with the first *overtone* of the 1400-cm^{-1} in-plane CH bending mode. Another example is the doublet observed in the benzene spectrum at 3099 and 3045 cm^{-1} (3.23 and 3.28 μm). A single CH stretching vibration would be expected to absorb at about 3070 cm^{-1} (3.26 μm), but this stretch coincides with a *combination* of two fundamental CH bending vibrations at 1485 and 1585 cm^{-1}, and thus Fermi resonance occurs to give the observed doublet. Fermi resonance is not always easy to diagnose, but it should be suspected whenever a band that is normally a singlet is split into a doublet.

Finally, a third type of vibrational coupling, involving two identical bands located in close proximity to each other, also gives rise to closely spaced doublets. For example, equivalent carbonyl groups in simple anhydrides or imides produce the doublets listed in Table 3-3, instead of exhibiting a single carbonyl absorption band in the region of 1700 cm^{-1}. The carbonyl stretching frequency is split into two modes (an asymmetrical stretching mode at a slightly higher frequency than normal and a symmetrical stretch at a slightly lower frequency) because carbonyl groups in close proximity are subject to mechanical interaction.

ELECTRICAL EFFECTS

The force constants between atoms depend on the distribution of electrons in the molecule, and this in turn is altered by conjugation, resonance, and inductive effects.

Conjugation Conjugation refers to alternate multiple and single bonds, compared with isolated multiple bonds. Hybridization occurs with the result that the multiple bonds transfer some of their π-electron character to the intervening single bond. As a result of conjugation, C═C stretching frequencies are shifted to lower frequencies by about 20 to 40 cm^{-1}, compared with isolated double bonds, and similarily the C—C stretching frequencies are shifted to higher frequencies. A *splitting* of absorption bands also results, corresponding to in-phase and out-of-phase stretching of alternate C═C bonds. Conjugation can also occur between such groups as C═C and C═O, with the result that both bonds are shifted to lower frequencies. An increase in *intensity* also occurs, which may be visualized as being due to a resonance contribution of the type

$$\mathrm{C{=}C{-}C{=}O} \leftrightarrow \overset{+}{\mathrm{C}}\mathrm{{-}C{=}C{-}}\overset{-}{\mathrm{O}}$$

with more efficient absorption caused by the increased polarity of the vibrating atoms.

Resonance Resonance refers to the quantum-mechanical phenomenon whereby a molecular structure is stabilized by contributions from several hypothetical stationary states in such a way that the total energy of the system is minimized. The concept of resonance can be used, for example, to explain the effect of adjacent groups on the carbonyl stretching frequency. Table 3-4 gives the frequency of absorption of the carbonyl stretch for five different

TABLE 3-3 Vibrational Coupling of Carbonyl Groups in Anhydrides and Imides to Produce Doublets

Compound	C═O stretching wave number, cm^{-1}	
	Asymmetrical	Symmetrical
Succinic anhydride (CH_2—C(═O)—O—C(═O)—CH_2, five-membered ring)	1872	1796
Succinimide (CH_2—C(═O)—NH—C(═O)—CH_2, five-membered ring)	1770	1689
N-methylglutarimide (CH_2—CH_2—C(═O)—N—CH_3—C(═O)—CH_2, six-membered ring)	1729	1686

TABLE 3-4 Effect of Adjacent Groups on the C=O Stretching Frequency

		C=O absorption	
Group	Formula	Wave number, cm^{-1}	Wavelength, μm
Acids	RCOOH	1650	6.06
Amides	$RCONH_2$	1670	5.99
Ketones	RCOR′	1700	5.88
Esters	RCOOR″	1735	5.76
Acid chlorides	RCOCl	1800	5.56

compounds of the type

```
R'
  \
   C=O
  /
R
```

where R′ is being varied. The effect of R′ can be understood if we consider the contribution of three simple resonance structures:

```
R'              R'               R'
  \               \                \
   C=O   ↔         C—O⁻   ↔         C⁺—O⁻
  /               //               /
R               R                R
                +

   I               II               III
```

The actual contribution of any particular structure (I, II, or III) will depend on the ability of groups R and R′ to attract or repel electrons, which in turn depends on the relative electronegativities of R and R′. If the contribution of structure I is significantly greater than that of structures II and III, a "normal" carbonyl stretching frequency of around 1700 cm^{-1} is to be expected, like that observed with most dialkyl ketones; e.g., the C=O band is at 1705 cm^{-1} in acetone and 1715 cm^{-1} in diethyl ketone. If, on the other hand, structure II is a significant factor in the electron distribution within the carbonyl group, we can expect a shift of the observed band position to lower frequencies as a result of the increased single-bond character of the C=O bond system. Structure II becomes important when R is a basic (electron-donating) group, and this is the case in acids and amides where R is hydroxyl or amido (see Table 3-4). Conversely, when R is an electron-withdrawing group like OR or Cl, structure III becomes significant, and the observed band position shifts to higher frequencies since the highly electrostatic nature of the $C^+—O^-$ bond increases the strength (raises the force constant) of the bond. Thus, the concept of resonance qualitatively explains the effects of shifts in the carbonyl stretching frequency, but it is difficult to be more quantitative about the predictions because of the difficulty in quantitating resonance-structure contributions.

Inductive Effects The ability of an atom or group of atoms to attract or repel electrons is responsible for inductive effects, just as it is for resonance effects. However, unlike resonance effects, inductive effects can be made more quantitative with the use of physical data like electronegativities, Hammett- or Taft-substituent constants (sigma values), or acid-dissociation constants. As a single example of the use of such data to correlate absorption frequencies, Fig. 3-7 illustrates how the OH stretching frequency of various substituted carboxylic acids can be correlated with the pK_a of the acid. Points on straight line *A* are for substituted acetic acids, whereas line *B* is for substituted unsaturated acids of the type R═CHCOOH. For a given type of acid, it is reasonable that there should be a direct correlation between the OH stretching frequency and the pK_a, since the pK_a is a measure of the ease of ionization of the acid hydrogen. The slope is different for an unsaturated acid because of the contribution of resonance to the stretching frequency.

Taft-substituent constants give a direct correlation with stretching frequencies for a large variety of substituent groups, as long as the substituent is reasonably well removed from the vibrating group.

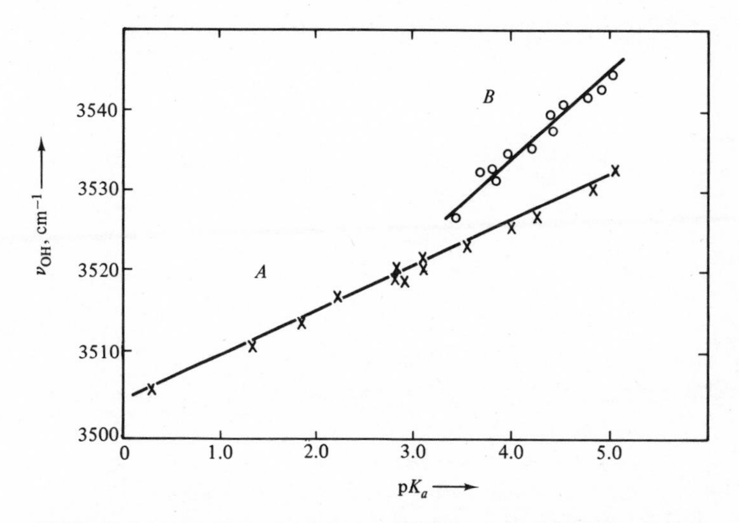

FIGURE 3-7 Correlation of OH stretching frequencies with pK_a of substituted carboxylic acids (see text for explanation). (*From J. D. S. Goulden, Spectrochim. Acta,* **6**:129 (1954), *by permission from Pergamon Press.*)

HYDROGEN BONDING

Hydrogen bonding is a special type of interaction between a proton-donor group XH and a proton-acceptor atom, whereby the hydrogen atom serves as a bridge between the two electronegative atoms. In infrared studies X is usually oxygen or nitrogen, although it may be fluorine.

The effect of hydrogen bonding on infrared spectra is pronounced, since it both shifts and broadens the XH stretching band. The shift in frequency is to *lower* energy (longer wavelength), since the hydrogen-bonding association will tend to *weaken* the X—H bond. The broadening is generally due to the random degree of association between various molecules, though there may be additional factors such as the initiation of a resonance condition.

Of the two types of hydrogen bonding, *intermolecular* (between different molecules) and *intramolecular* (within one molecule), intermolecular hydrogen bonding is by far the more widespread. The essential difference between intermolecular and intramolecular hydrogen bonding is that the spectral shifts for intermolecular hydrogen bonding show a strong concentration dependence, whereas spectral shifts for intramolecular hydrogen are independent of concentration. Figure 3-8 shows the absorbance spectra in the region of the C—H stretch for three different concentrations of *n*-butanol in carbon tetrachloride. To explain the changing spectra, we may assume that equilibria are set up between monomers, dimers, trimers, and other polymeric species. Thus, at very low concentrations of alcohols, we may assume that mainly monomer is present, giving a sharp absorption band at about 2.76 μm (the so-called free-OH stretching frequency). At higher concentrations, higher pro-

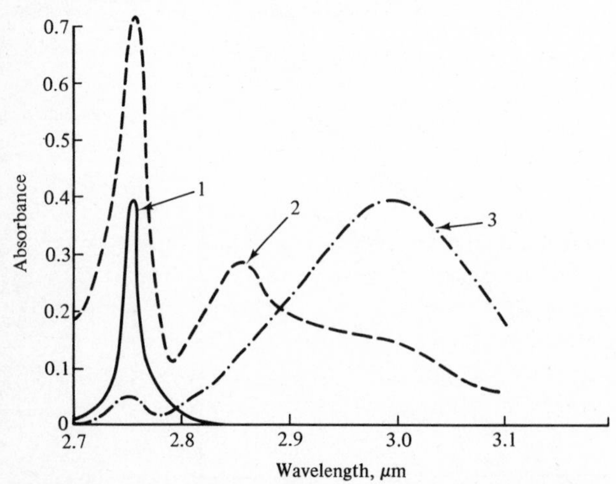

FIGURE 3-8 Infrared absorbance spectra for three different concentrations of *n*-butanol in CCl_4. Curve 1, 0.005 *M* in 20-mm cell; curve 2, 0.0075 *M* in 3-mm cell; and curve 3, 1.7 *M* in 0.025-mm cell. (*After L. A. Smith in I. M. Kolthoff and P. J. Elving (eds.), "Treatise on Analytical Chemistry," pt. I. vol. 6, p.* 3640. *John Wiley & Sons, Inc., New York,* 1965 *by permission.*)

TABLE 3-5 Effect of Ring Strain on C=C and C—H Stretching Frequencies in Some Cycloalkenes†

Hydrocarbon	Bond angle, deg	Absorption frequency, cm^{-1} C=C stretch	C—H stretch
Cyclohexene	120	1646	3017
Cyclopentene	110	1611	3045
Cyclobutene	94	1571	3070

† Data from Ref. 8, p. 3644.

portions of the associated species form, with the result that their absorption maxima shift to longer wavelengths (lower frequencies) and broaden considerably.

Similar but less pronounced behavior occurs for amines and amides. Because of the strong dependence of the intensity at a given wavelength on hydrogen bonding, OH and NH absorptions are not suitable for quantitative analyses, except in special instances.

RING STRAIN

The effect of bond-angle strain on absorption frequencies has been studied in detail. In general, for ring systems, stretching frequencies between constrained atoms *decrease* with increasing strain, whereas frequencies of attached atoms *increase*. This is illustrated in Table 3-5 for six-, five-, and four-membered cycloalkenes. Evidently the C =C bonds in the ring are weakened as the ring is strained, but it is not understood whether the C—H bonds involving the attached hydrogens are concomitantly *strengthened* or the frequencies are raised due to some type of mechanical effect, like vibrational coupling. At any rate, it is a well-known fact that vibrations involving attached atoms are shifted to higher frequencies with increasing ring strain, and this correlation can be very useful to the chemist trying to establish the ring size of complex molecules. To illustrate this point, Table 3-6 shows the effect of ring size on the position of the carbonyl stretching frequency in ketones, lactones, and lactams.

3-1F Vibrational-Rotational Fine Structure

The importance of rotational transitions depends on the physical state of the sample. With gases at low pressure, each molecule exists free and uninfluenced by neighboring molecules. As a result, rotational transitions are an important part of the observed infrared spectra, often resulting in clearly de-

TABLE 3-6 Effect of Ring Size on the Position of the Carbonyl Stretching Frequency in Ketones, Lactones, and Lactams†

	Carbonyl stretching frequency, cm^{-1}		
Ring size	Ketones (>C=O)	Lactones (–C(=O)–O–)	Lactams (–C(=O)–NH–)
4	1780	1818	1745
5	1745	1770	1700
6	1715	1735	1677
7	1705	1727	1675

† Data from Ref. 6, p. 152.

fined rotational fine structure. This is exemplified by the near-infrared (first-overtone) spectrum of gaseous HCl, shown in Fig. 3-9. When the gas is at high pressure (it being immaterial whether the high pressure is achieved by using the absorbing gas itself or by adding an inert, nonaborbing gas), the rotational structure of the vibrational bands becomes more and more smeared out, as shown in Fig. 3-9*b* to *d*. The loss of rotational fine structure and the accompanying band broadening come from an increasing number of collisions, which restrict or dampen the amount of rotation and also cause a more

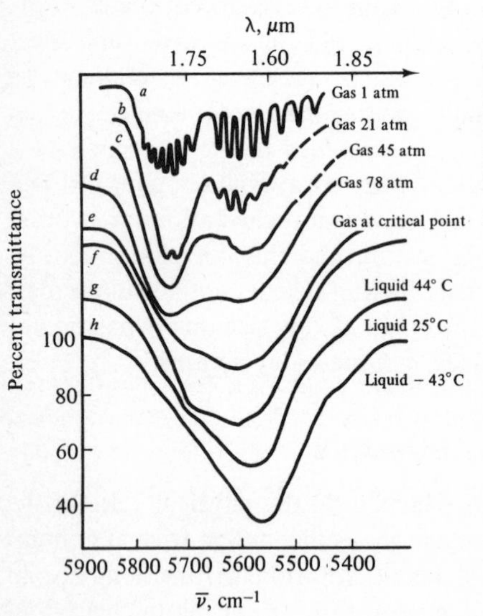

FIGURE 3-9 Importance of rotational transitions on the near-infrared spectra of HCl in the gaseous state at different pressures and in the liquid state at different temperatures. The individual curves are displaced on the ordinate. (*From W. West, J. Chem. Plys.*, 7:795 (1939), *by permission.*)

random distribution of rotational energy levels among the molecules. Even before the liquid state is reached, the rotational detail may have disappeared almost completely, as shown by Fig. 3-9*d*. When the condensed, or liquid, state has been reached, intermolecular interactions so disturb and restrict the rotational energy levels that they become merged into a continuum, as shown in Fig. 3-9*f* and *g*. Cooling the liquid serves to restrict the very limited amount of rotation further, and the bands tend to become sharper and more prominent.

For molecules larger than HCl, the rotational fine structure of the gaseous state is more difficult to resolve (the spacing between peaks becomes smaller as the molecular weight increases). In the case of liquids, only a few molecules heavier than HCl (including H_2O, NH_3, and CH_4, all in inert solvents), show any significant amount of rotational detail. The spectra of pure liquids, in general, are essentially free from signs of quantized free rotation. This also applies to solids, for which the fine structure sometimes found can be explained by vibrational interactions and force fields, as discussed in Sec. 3-1*E*.

The frequency of bands making up a vibrational-rotational spectrum can be estimated by assuming that vibrational and rotational contributions can be added directly without correcting for interaction effects, leading to

$$\bar{\nu} = \bar{\nu}_0 + \frac{hm}{4\pi^2 I} \tag{3-7}$$

where $\bar{\nu}_0$ = center of vibrational-rotational band and corresponds to *vibrational* frequency which can be estimated from Eq. (3-4)

h = Planck's constant

m = molecular rotational quantum number = $J + 1$ (J being the original rotational quantum number, defined in Chap. 1)

I = moment of inertia of molecule

I may be defined for a molecule containing i number of atoms by

$$I = \sum_i m_i r_i^2 \tag{3-8}$$

where r_i is the distance of the particle i of mass m_i from the center of gravity of the system.

The second term in Eq. (3-7) determines the rotational fine-structure separation in the band, and since the moment of inertia I is in the denominator, the rotational band spacing becomes smaller as the molecular weight of the molecule increases.

The molecular rotational quantum number m can undergo *changes* of ± 1, and thus m in Eq. (3-7) can take on values such as 1, 2, 3, . . . , giving a series of lines, called the *R branch*, on the high-frequency side of ν_0. Likewise, m can take on negative values such as -1, -2, -3, . . . ; the resulting series of

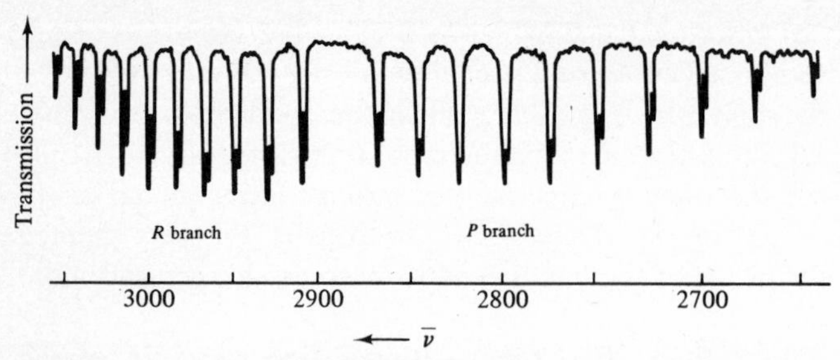

FIGURE 3-10 Vibrational-rotational band structure of HCl vapor at high resolution. [*After Pure and Applied Chemistry,* **1**:572(1961), *by permission.*]

lines on the low-frequency side of ν_0 are called the *P branch*. Figure 3-10 shows the vibrational-rotational fine structure of HCl vapor at high resolution in the region of its fundamental vibrational transition. The separation in the center of the spectrum, around 2890 cm^{-1}, is caused by the absence of a pure vibrational transition, since with linear molecules like HCl a rotational transition *must* accompany a vibrational transition. The spacing between lines is, according to Eq. (3-7), $h/4\pi^2 I$, and thus by measuring the frequency spacing from such spectra the moment of inertia I can be calculated. The moment of inertia, in turn, allows bond distances r to be calculated from equations like (3-8). (The very small splittings that show up are caused by the presence of both ^{35}Cl and ^{37}Cl in HCl.) In more complicated, polyatomic molecules, it is possible to have vibrational transitions *without* rotational changes [m in Eq. (3-7) is 0], and then a sharp vibrational band called the *Q branch* appears between the *P* and *R* branches.

Equation (3-7) holds fairly well for linear molecules and simple molecules like methane, but for the majority of polyatomic molecules more complex equations are necessary to describe their complicated spectra, and such equations are beyond the scope of this book.

3-2 INSTRUMENTATION

3-2A Basic Components of Infrared Spectrophotometers

As mentioned in Sec. 2-2*B*, all spectrophotometers contain three basic components: (1) a *source* of radiant energy, (2) a *monochromator* for isolating a narrow band of radiant energy, and (3) a *detector* for measuring the radiant energy transmitted through the sample. One fundamental difference between infrared spectrophotometers and ultraviolet and visible spectrophotometers is in the location of the sample with reference to the monochromator. In ultraviolet and visible spectrophotometers the sample is placed *after* the monochromator, in order to minimize the amount of exposure of the sample to energetic ultraviolet and visible radiation. Infrared spectrophotometers have the sample placed *before* the monochromator, so that the monochromator can dispose of stray radiation (emanating from the sample and cell) before it reaches the detector.

RADIATION SOURCES

Ideally, radiation sources should be continuous over a wide range and have high intensity and the intensity should not vary appreciably with wavelength. In practice this is not attainable, and in fact the best sources of infrared radiation are blackbody radiators, which have emission spectra in the near and fundamental infrared, as shown in Fig. 3-11. As discussed in Sec. 2-2*B*, blackbody radiators have a spectral distribution which depends only on the temperature of the solid material, and the Wien displacement law can be used to estimate the wavelength of maximum emission λ_{max}, as follows:

$$\lambda_{max} T \approx 3000 \qquad (3\text{-}9)$$

where T is the absolute temperature in kelvins and λ_{max} is in micrometers. [Note that, in Eq. (2-11), λ_{max} was in nanometers.] Thus, the wavelength of maximum emission varies inversely with the absolute temperature, in accordance with the data in Fig. 3-11. Figure 3-11 also implies that higher temperatures will increase the source-output intensity, but unfortunately temperatures in excess of about 2000 K result in too much radiation in the near-infrared and visible regions. This short-wavelength radiation is particularly undesirable in infrared instruments, since short-wavelength radiation is readily scattered [see Sec. 1-4*C* and Eq. (1-17)]. This scattered radiation, being more energetic than infrared radiation, may overwhelm the low-energy signal being measured. As a result, it is necessary to limit the source temperature to below about 2000 K and use filters and baffles to minimize stray radiation.

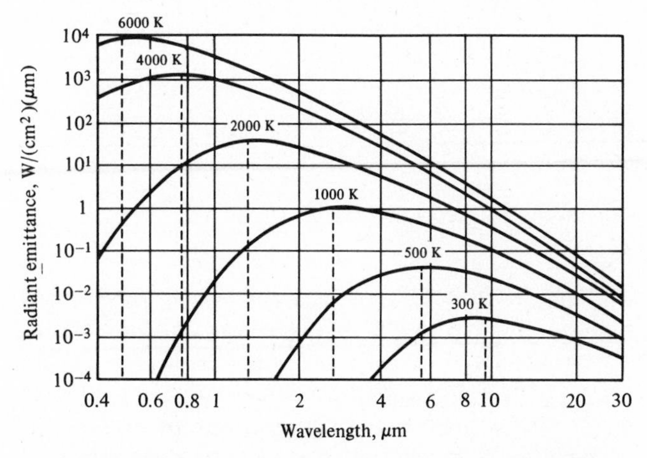

FIGURE 3-11 Blackbody (thermal radiation curves in the near-infrared and fundamental-infrared regions. (*Adopted from H. L. Hackforth, "Infrared Radiation," p.* 16, *McGraw-Hill Book Company, New York,* 1960, *by permission.*)

Ideal blackbody radiators are not available, but the sources in use approach the ideal, and might be termed "graybody" radiators. The three most common sources in use are the *Nernst glower,* the *Globar,* and *incandescent wires.*

Nernst Glower The Nernst glower is probably the most widely used source of radiation for the fundamental-infrared region. It is composed of a fused mixture of the oxides of zirconium, yttrium, and thorium, which is molded in the shape of a hollow rod 1 to 3 mm in diameter and 2 to 5 cm in length. The ends of the rod are attached to ceramic tubes for mounting and have platinum leads for the electrical connections. The source is nonconducting at room temperature and must be preheated to bring it to a conducting state. The preheating is usually accomplished with a separate heater and reflector. After preheating, a final heating to about 1900°C is accomplished using about 75 V ac at about 1.2 A (giving a moderate power consumption of about 90 W). The spectral energy of the emission, shown in Fig. 3-12, is about 50 to 60 percent that of a perfect blackbody radiator of the same temperature. The energy output is predominantly concentrated between 1 and 10 μm, with relatively low energy beyond 10 μm; nonetheless the Nernst glower is sometimes used out to about 40 μm. The deviations from a smooth curve noticeable in Fig. 3-12 near 2.7, 4.3, and 6.7 μm result not from any peculiarity of the glower itself but from the absorption of water vapor and carbon dioxide in the atmosphere (see Exercise 3-9).

The *advantages* of the Nernst glower are its intense radiation up to about 10 μm, its low power consumption (less than 100 W), its ability to operate in air (though its stability is improved if it is enclosed or shielded from drafts), and its relatively long life (several hundred hours if it is not abused by turning it on and off frequently). Its main *disadvantages* are that it is very fragile, its

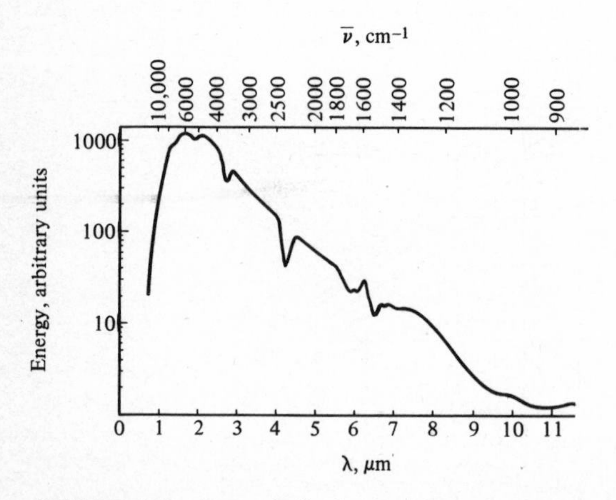

FIGURE 3-12 Spectral distribution of energy output of Nernst glower at 1900°C.

lead connections often fail through breakage or melting, and it requires a current stabilizer because of its large negative temperature coefficient. Current stabilization is easily achieved, however, by using a ballast resistor with positive temperature coefficient.

Globar The Globar is perhaps the second most widely used infrared source. It consists of a cylindrical rod of silicon carbide, SiC, usually about 6 to 8 mm in diameter by about 5 cm long. It is often encased in a water-cooled brass tube, with a slot provided for the emission of radiation. Aluminum electrodes are firmly mounted to the ends to provide electrical connections. An advantage over the Nernst glower is that SiC conducts in the cold, so that the Globar can be heated up simply by turning on the current. The power consumption varies with the size of the Globar but ranges from about 200 to 400 W, using voltages from 50 to 115 V. This relatively high power consumption necessitates efficient water (or sometimes air) cooling to avoid damage to the surroundings. The usual operating temperature is 1200 to 1400°C, with oxidation and loss of binding material becoming significant above 1400°C. The absolute upper temperature limit is 2000°C, since this is the point where silicon carbide vaporizes. The spectral output of the Globar is about 80 percent that of a blackbody radiator, and thus its spectral distribution can be estimated from Fig. 3-12. In comparison with a Nernst glower, the Globar is a less intense source below 10 μm, the two sources are comparable out to about 15 μm, and the Globar is the better of the two beyond 15 μm. The Globar is sometimes used in commercial instruments out to about 50 μm.

In summary, the Globar has the advantage of being a more stable source than a Nernst glower and more rugged. In addition, it is a slightly more intense source beyond 15 μm. The main disadvantages are that it requires more power and must be cooled, and a variable transformer must be used as a voltage source since after long use its resistance increases and the applied voltage must be increased. The life of the Globar is probably comparable to that of the Nernst glower, but the Globar requires maintenance at the electrode contacts because of oxidation.

Incandescent-Wire Sources Many of the newer, lower-cost spectrophotometers employ one of two types of incandescent-wire sources. One type is a closely wound coil of nichrome (a nickel-chromium alloy), electrically heated to incandescence at about 800 or 900°C. The spectral output is roughly comparable to that of a Globar. The nichrome coil requires no water cooling and little or no maintenance and gives long service.

Another type of incandescent-wire source uses a coil of rhodium wire packed in a sealed aluminum oxide tube. The rhodium wire serves as a high-temperature heating coil, and the aluminum oxide becomes the infrared radiator. Electric contacts are soldered onto the rhodium wire, and solder also caps the aluminum oxide tube at both ends. The rhodium coil would deteriorate in the atmosphere. In use the source is largely enclosed in a ceramic shield to minimize power requirements and prevent air currents. The source consumes only 30 W at about 2.8 V and develops an external (ceramic) temperature of about 1200°C. It has a useful output in the region of about 2.5 to 25 μm and requires no ballasting or water cooling. The useful life is approximately 1000 h.

MONOCHROMATORS

The function and design of monochromators was discussed in Sec. 2-2*B*; the only features of infrared monochromators that need special discussion here are the optical materials in use and the relative merits of prisms and gratings.

Optical Materials In contrast to the convenient glasslike materials used in the ultraviolet and visible regions, ionic crystals must generally be used for optical components in the infrared region. Table 3-7 gives a list of some optical materials commonly used in the infrared, along with the longest wavelength (lowest frequency) at which each is usable. Glass and quartz absorb very strongly beyond 4 μm and thus cannot be used for the fundamental-infrared region. The alkali and alkaline-earth halides absorb at frequencies corresponding to the frequency of their lattice vibrations, and heavier atoms absorb at lower frequencies, consistent with a heavier reduced mass in Eq. (3-4). Unfortunately, a number of the best optical materials are hygroscopic (see Table 3-7), which means that great care must be exerted to protect them from moisture.

Whereas high transmittance is one of the primary requirements for optical materials used in windows and cells, with prisms one must compromise between high transparency and high dispersion, since dispersion of a prism increases with increasing absorption (see Sec. 1-4*F*). Therefore, to

TABLE 3-7 Some Common Infrared Optical Materials†

Material	Limiting transmission wavelength, μm	Limiting transmission frequency, cm^{-1}
Glass	2	5000
Quartz	4	2500
LiF	6	1667
CaF_2	9	1111
NaCl‡	15.5	645
KBr‡	25	400
CsBr‡	40	250
CsI‡	50	200

† Data from P. Delahay, "Instrumental Analysis," Macmillan, New York, 1957, p. 221, and Ref. 5, p. 113.
‡ Hygroscopic.

achieve high resolution over a wide wavelength range with prisms, it is necessary to change prisms. For example, NaCl is a widely used prism material for the wavelength range of 2.5 to 15 μm, but if higher resolution is needed in the short-wavelength (hydrogen-stretching) region, LiF or CaF_2 prisms must be used.

In general, lenses are avoided in infrared instruments, and instead focusing is accomplished with variously shaped mirrors. This is dictated by the availability of very efficient, rugged mirrors and the corresponding inefficiency and fragility of lens materials. Infrared mirrors are usually made from glass or fused quartz, with gold or aluminum sublimed onto the front face. An aluminum-faced mirror will reflect up to about 96 percent of infrared radiation and gold up to about 99 percent. On the other hand, lenses fabricated from ionic crystals would absorb about 10 percent of the already weak-intensity infrared radiation.

Grating materials are similar to those used in the ultraviolet and visible regions. Original reflection gratings are made by ruling some soft metal of high infrared reflectivity (such as aluminum), and replica gratings are now being produced from molded plastic.

Prisms vs. Gratings Gratings suitable for the infrared region have not long been available in commercial quantities, and thus the majority of instruments in established laboratories are prism instruments. Grating instruments are inherently superior because they give higher resolution, higher transmission (less energy loss, especially through absorption), and resolution which is independent of wavelength.

In comparison with prisms, gratings also have a few disadvantages. With gratings it is necessary to eliminate other spectral orders (see Sec. 1-4*E*), gratings give more scattered radiation, and each grating has only a limited wavelength range. In general, gratings of the echelette type are preferred because they concentrate radiation into the first two orders, but nonetheless it is always necessary to eliminate at least the predominant overlapping wavelengths. To eliminate overlapping orders (as well as scattered radiation), either a double monochromator with a foreprism (prism preceding the grating) or bandpass filters are used. Bandpass filters are selected which have a very low transmission at unwanted frequencies and very high transmission in the region of interest. To cover an appreciable wavelength region, two or more gratings are generally used, in conjunction with several filters.

Prism monochromators have the advantage of being simpler than gratings, and therefore cheaper. The entire infrared region over which the prism is transparent can be scanned in one continuous sweep of the scanning arm, without the need for any auxiliary spectral separating devices. Disadvantages of prisms are that they are easily damaged by moisture, abrasion, or mishandling; resolution is limited and varies with wavelength; and dispersion is temperature-sensitive.

Both grating and prism infrared monochromators are usually provided with an automatic slit-width program, so that the slit width is increased as the source energy decreases. This is desirable because of the large decrease in energy of infrared sources as longer wavelengths are approached. The slit program is usually controlled by a cam which operates synchronously with the turning of the prism or grating.

DETECTORS

For detecting radiation in the fundamental-infrared region, nearly all commercially available infrared spectrophotometers currently utilize *thermal* detectors of which three types are in widespread use: (1) *thermocouples*, which use a heat-sensitive junction of dissimilar metals to generate a small voltage when the temperature changes; (2) *bolometers*, which are resistors that change in resistance as the temperature changes; and (3) *Golay pneumatic detectors*, which use a gas chamber that expands in volume as the temperature increases. These detectors will be briefly described and compared, particularly in terms of their response time and wavelength dependency.

Response time is important with infrared detectors since with all modern infrared spectrophotometers some type of flickered or interrupted radiation (usually in the 6- to 30-Hz range) must be used to minimize the effect of stray radiation signals and to look alternately at the sample and reference in double-beam arrangements. The response time of all three types of infrared detectors is largely limited by the rate of heat transfer by the absorber and is markedly inferior to the response time of ultraviolet and visible detectors.

It is of only moderate interest to compare the sensitivity of these three detectors, since all use a small absorbing receiver which converts radiant energy to thermal energy, and though the resulting small temperature change is measured in a different way with each detector, the sensitivities turn out to be roughly equivalent. However, it should be emphasized that the sensitivity of these infrared detectors is also far inferior to the sensitivity of the detectors used in the ultraviolet and visible regions.

Thermocouples At present thermocouples are the most widely used detectors in commercially available infrared spectrophotometers. Two dissimilar metals are placed together, and the infrared radiation to be measured is focused on one of the junctions, to which a small receiver plate is attached. The dissimilar metals are often antimony and bismuth or silver and platinum, and the receiver plate is commonly gold or platinum which has been blackened with, for example, gold soot deposited in a vacuum. Black soot is quite effective in absorbing the incident radiation and converting it to heat, and the temperature difference $T_2 - T_1$ between the hot junction and the isolated cold junction sets up a thermoelectric voltage between the two junctions. The resulting small current can be measured with an impedance-matching transformer, which couples the thermocouple circuit to an electronic amplifier circuit, whereby the small current is greatly amplified. Transformer coupling is possible because the radiation signal reaching the receiver plate is a pulsating signal coming from the sample during one half-cycle and from the reference during the other half-cycle. Therefore, a pulsating, or alternating, current is generated by the thermocouple, suitable for transformer coupling.

The optimum design of a thermocouple detector involves a compromise between sensitivity and speed of response. For high sensitivity, the rate of heat transfer from the receiver to its surroundings must be low, so that the incident radiation can cause as large a temperature differential as possible. Unfortunately, however, a low rate of heat transfer results in a very slow response time. Three mechanisms of heat transfer are involved: conduction, convection, and reradiation. Reradiation is not a design variable, since it depends on the blackness of the receiver plate and it is necessary to have as black a target receiver as possible for efficient absorption of the incident radiation. In the interests of acceptable sensitivity, convection must be minimized by isolating the thermocouple in an evacuated case (usually steel) with a KBr (or CsI, etc.) window. Optimum conductivity is arrived at by proper choice of wire material and the size, for both the thermocouple and connecting wires. The receiver plate is made as thin and small as possible, consistent with mechanical strength and the area of the exit slit. Too large a receiver plate would minimize the temperature rise of the receiver.

Most thermocouple detectors used in infrared spectrophotometers have a response time of about 60 ms, the longest of the infrared detectors. This response time dictates the chopping frequency that can be employed in the instrument, which is generally 10 to 13 Hz. If a faster chopping rate were used, the detector would not be able to keep up. On the other hand, if a slower chopping speed were used, amplifier noise would increase, along with a necessity for slower spectral scans.

Thermocouples have a response which is almost constant and independent of wavelength out to about 30 μm, where the efficiency begins to fall off. In order to give a smooth, nonspecific response as a function of wavelength, the detector must, in effect, appear black at all wavelengths. For wavelengths beyond 30 μm, the dimensions of radiation waves are large compared with the dimensions of the rough ridges in the black surface, and the surface begins to look smooth and shiny to the incident radiation. This results in a higher and higher percentage of radiation which is reflected rather than absorbed.

Recently a more sensitive version of a thermocouple, called a *thermopile*, has been incorporated in some infrared instruments, e.g., the Perkin-Elmer models 180, 325, and 727. A thermopile consists of a number of thermocouple junctions joined to a common receiver plate, accentuating the sensitivity of the thermocouple.

Bolometers A bolometer is a device which changes resistance as its temperature changes. Figure 3-13 schematically illustrates the Wheatstone bridge circuit commonly used to measure the bolometer resistance. R_d represents the resistance of the detector bolometer, and R_r represents the

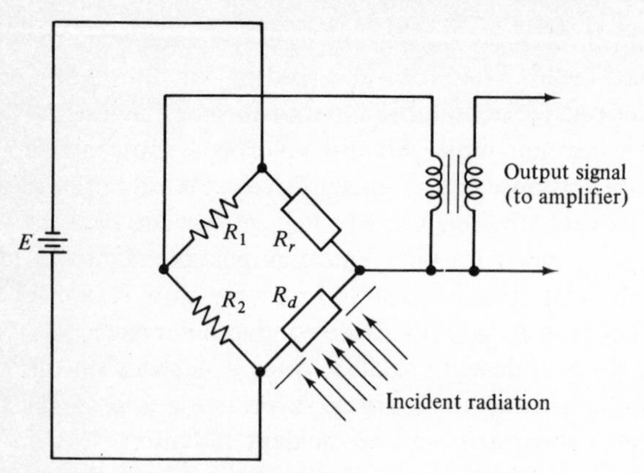

FIGURE 3-13 Schematic diagram of a typical bolometer circuit. (See text for explanation.)

resistance of a reference bolometer. The reference bolometer has thermal properties similar to the detector bolometer, but it is isolated from the incident radiation to compensate for changes in background temperature. This circuit can be analyzed in a conventional manner with Kirchhoff's laws, but a simplification may be invoked in order to visualize how the output signal is proportional to the radiation intensity. Assume that the reference bolometer is chosen to have a resistance R_r which is at least 10 times the resistance of the detector bolometer R_d. Since the bridge is supplied with a constant voltage E, which may be either ac or dc, the detector bolometer will be supplied with virtually constant current. Therefore, any change in resistance R_d will manifest itself in a voltage change which is directly proportional to the resistance change. Transformer coupling to an amplifier can be used, since the radiation signal is chopped and pulsating.

Metals like nickel and platinum have been used for the sensing element in bolometers, but in recent years *thermistor* bolometers have become prominent. Thermistors have a negative thermal coefficient of resistivity of about 4 percent per Celsius degree, which is about 10 times that of metal bolometers. Thermistors are composed of sintered oxides of nickel, cobalt, and manganese and are used in the form of thin flakes about 1 to 5 mm long by 0.1 to 1 mm wide. Most of the design compromises discussed in connection with thermocouples apply as well to bolometers. The fragile thermistor flakes are mounted on a glass or quartz backing plate, metal contacts are fastened to each end, and the whole assembly is placed in an evacuated metal case with an infrared-transparent window.

Response times for metal bolometers and thermistor bolometers are about 4 ms. Thermistor bolometers respond well over the 1- to 15-μm range, whereas metal bolometers with a blackened surface respond out to about 30 μm, very similar to the spectral response of thermocouples. The sensitivity of bolometers is somewhat better than that of thermocouples, but the need for a very stable power supply with bolometers probably explains why they are less frequently used in commercially available spectrophotometers.

The detection limit of bolometers and thermocouples is set by *Johnson noise*, which is thermal motion of electrons in metals. Johnson noise can be reduced by cooling the detector, and *superconducting bolometers* can be made by cooling to cryogenic temperatures. Superconducting bolometers have extremely high sensitivity and very fast response. For example, tantalum nitride at liquid-helium temperature (4 K) and niobium nitride at 15 K have sensitivities about 20 to 30 times better than the best room-temperature bolometers. Although these extreme temperatures discourage widespread use, recent work with nickel-ribbon bolometers at liquid-nitrogen temperature (77 K) indicates that the sensitivity may be four- or fivefold that of room-temperature bolometers. At the relatively low cost of liquid nitrogen, this may become a feasible approach to increased sensitivity.

Golay Pneumatic Detectors The Golay detector, shown in Fig. 3-14, utilizes the expansion of a gas as the measuring device. The small pneumatic chamber is filled with xenon gas and sealed at one end with an infrared-transparent window and at the other end with a metallized plastic film. The film is coated with evaporated metal, which causes the film to offer an electric impedance to the electromagnetic wave and the absorbance to be independent of wavelength as long as the wavelength is small with respect to the area of the film. This detector is reported to have uniform absorptivity for radiation from the ultraviolet to microwaves. As the gas in the pneumatic chamber expands or contracts, it moves through a short duct and in turn moves a mirrored membrane (labeled "flexible mirror" in Fig. 3-14). A beam of light directed at the mirrored membrane is reflected and focused on a phototube detector. A slight movement of the mirrored membrane causes a large change in the intensity of light reaching the phototube detector, and this is the basis of the detection signal. By amplifying the signal, extremely small displace-

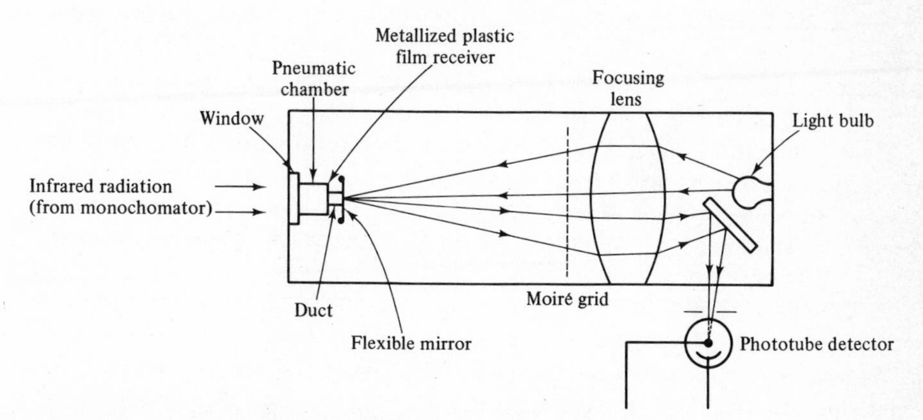

FIGURE 3-14 Simplified schematic diagram of a Golay pneumatic detector.

ments (10^{-9} cm) of the mirror can be measured, giving this type of detector a relatively high sensitivity, slightly superior to that of thermocouples and bolometers. The response time is about 4 ms. The large size of the detector unit is a disadvantage in many spectrophotometer designs, and relatively few commercial instruments employ the Golay detector at present.

3-2B Principles of Infrared Instrument Design

There are three major differences in the type of instruments available in the infrared region compared with the ultraviolet and visible regions. (1) Whereas ultraviolet and visible spectrophotometers are readily available in all four categories of instruments outlined in Sec. 2-2*C*, almost all infrared instruments are of the double-beam null-balance type. (2) Whereas almost all ultraviolet and visible instruments of the null-balance type use a *potentiometric-null* system, almost all infrared instruments use an *optical-null* system. (3) Instruments in the infrared region are almost exclusively of the automatic recording type.

There are several reasons for these differences. First, single-beam instruments tend to be impractical in the infrared, except for specific purposes, because of the interference of atmospheric water vapor and carbon dioxide. A number of commercially available infrared spectrophotometers are made to be operated in either single- or double-beam modes, and Fig. 3-15 illustrates the dramatic differences in the spectra obtained in the two modes of operation when scans are made with no sample present. The lower curve is for single-beam operation and clearly shows the strongly interfering bands due to water and carbon dioxide absorption (the 2.67- and 5.5–7.5-μm bands are due to H_2O, and the 4.25- and 15-μm bands are due to CO_2). In double-beam

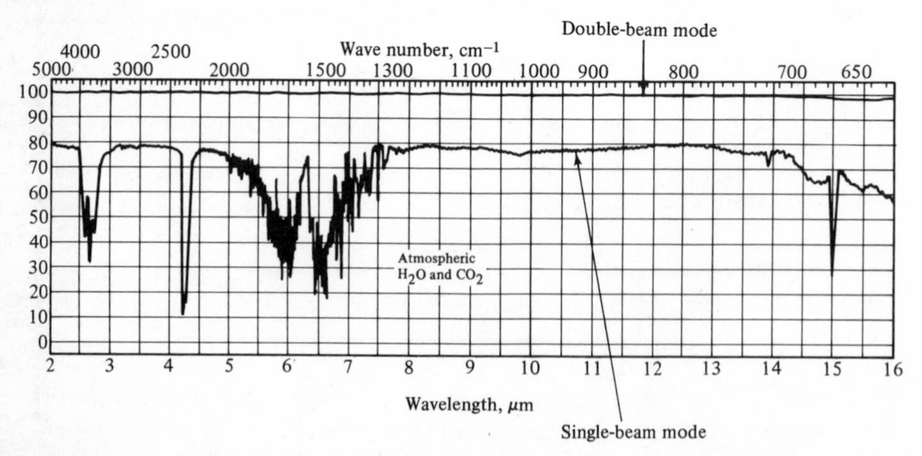

FIGURE 3-15 Effect of atmospheric water vapor and carbon dioxide on infrared spectra obtained in single- and double-beam modes. (*Beckman Instruments, Inc.*)

operation, as shown by the upper curve, the reference beam exactly compensates for these interferents, and a stable base line is achieved.

Second, the relatively low amount of energy available in the infrared, plus the relatively low sensitivity and slow response of infrared detectors, make it difficult to design a potentiometric-null infrared spectrophotometer with a low signal-to-noise ratio. The cost of successful potentiometric-null instruments tends to be commensurate with the complexity involved.

Finally, nonrecording instruments are almost nonexistent in the infrared region because of the sharpness of most infrared bands and the complexity of most infrared spectra. Obviously, such spectra would be extremely tedious to determine point by point manually. Furthermore, in contrast to the great value of manual ultraviolet and visible spectrophotometers for carrying out standard quantitative-analysis procedures, in which the wavelength for analysis is specified and fixed, it is seldom possible to have such rigid standard procedures for quantitative analysis in the infrared. Not only do infrared absorption bands tend to be sharp, but their sharpness depends on the resolution of the particular instrument being used, and further, small changes in experimental conditions may cause a small but significant shift in the wavelength of maximum absorption. Finally, even small inaccuracies in the wavelength calibration of the instrument being used would make a fixed-wavelength quantitative-analysis procedure impractical.

3-2C Some Commercially Available Instruments

The commercially available infrared spectrophotometers can be classified according to the type of monochromator system used, including variable interference filters, prisms, gratings, and double monochromators. With some minor exceptions, this classification system presents the instruments in order of increasing complexity, precision, and cost. In each category representative instruments illustrating significant design features will be described. No attempt has been made to give a comprehensive coverage of the many instruments available, nor is there any intention of endorsing one instrument over another.

INSTRUMENTS USING VARIABLE INTERFERENCE FILTERS

After wedgeshaped interference filters were developed, Beckman Instruments, Inc., marketed low-cost instruments using these filters in a new type of monochromator system. Figure 3-16 illustrates the optical diagram of the Microspec infrared spectrophotometer. Light from the nichrome wire source is divided into two beams, which enter the monochromator through variable-opening entrance slits. One beam passes through the sample, and the second beam passes through the reference, after which the beams are alternately chopped. First, the sample beam and then the reference beam passes through

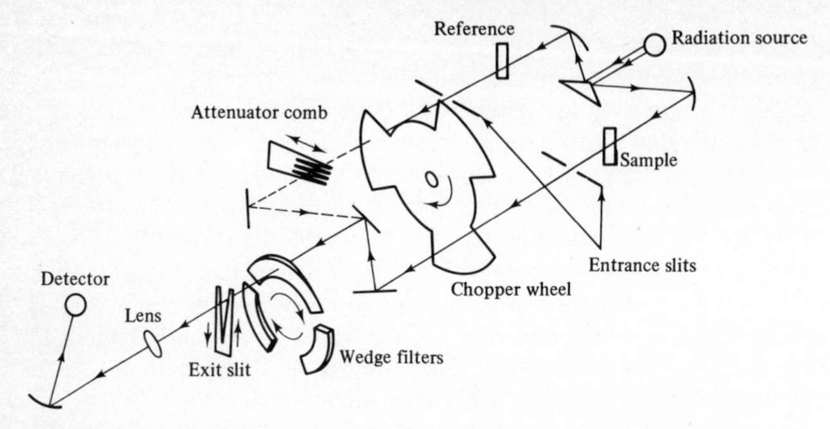

FIGURE 3-16 Schematic optical diagram of Beckman Microspec infrared spectrophotometer, using a variable-interference-filter monochromator. (*Beckman Instruments, Inc.*)

the interference filter then through the V-shaped exit slit, finally reaching the thermocouple detector. Absorbance of the sample is recorded by the optical-null principle. If the sample-beam signal is the same as the reference-beam signal, there will be no unbalance signal and the recorder will read 100 percent *T*. If, on the other hand, the sample absorbs more strongly than the reference material, the unbalance signal which arises at the detector is amplified and fed to a servomotor, which drives the attenuator comb into the reference beam to decrease the transmission of that beam until the two signals are equal (nulled). The recorder pen is mechanically linked to the attenuator comb to record the absorption. (See Sec. 2-2*C* for a discussion of attenuator combs.)

The monochromator consists of a wheel on which are mounted three segments of a circular variable filter. The wavelength of radiation which is passed varies as a function of the angle of rotation. Longer wavelengths are passed as the wedge becomes thicker because the spacing between layers in the filter becomes greater (see the discussion of interference filters in Sec. 1-4*D*). The first filter segment passes radiation in the range of 2.5 to 4.5 μm, the second from 4.4 to 8.0 μm, and the third from 7.9 to 14.5 μm. A cam-operated slit program operates synchronously with the filter wheel; the V-shaped slit becomes wider as a filter segment becomes thicker, closes between filter segments, and then reopens with the start of the next filter segment. The spectrum is presented in three segments, a space between segments representing the time necessary to go from one filter segment to another.

INSTRUMENTS USING PRISMS

A good example of a relatively low-cost double-beam instrument using an NaCl prism is the Perkin-Elmer model 137 spectrophotometer pictured in Fig. 3-17. The schematic optical diagram of the instrument is shown in Fig. 3-18. Radiation from the source (ceramic tubing heated from within by a coil of rhodium wire) is split into sample and reference beams, and after passing through the sampling area the two beams are recombined into a common path by the semicircular-sector mirror rotating at 13 Hz. The rotating sector alternately *passes* the sample beam and *reflects* the reference beam, so that the two beams are separated in time. After passing through the prism monochromator the pulsing beams are detected with an evacuated thermocouple.

If the energy in both sample and reference beams is equal, a dc voltage is produced by the thermocouple which is not amplified by the ac amplifier of the instrument. When the sample absorbs radiation at its characteristic wavelengths, the intensity of the sample beam is reduced and a pulsating voltage is produced by the thermocouple, which is amplified and used to activate a servomotor. The servo moves the optical wedge into the reference beam until the intensities of the sample and reference beams are equalized, or nulled. Since the recorder pen is mechanically linked to the optical wedge, the absorbance is simultaneously recorded.

The Perkin-Elmer 137-B has scanning speeds of 3 and 12 min for the complete spectrum of 2.5 to 15 μm. The wavelength accuracy is ± 0.03 μm, and photometric accuracy is ± 1 percent T. Stray light is less than 1 percent from 2.5 to 14 μm, and less than 3 percent at 14.7 μm. The maximum resolution is 0.04 μm at 10 μm. The slit is automatically programmed to compensate for variations in source intensity and prism dispersion throughout the

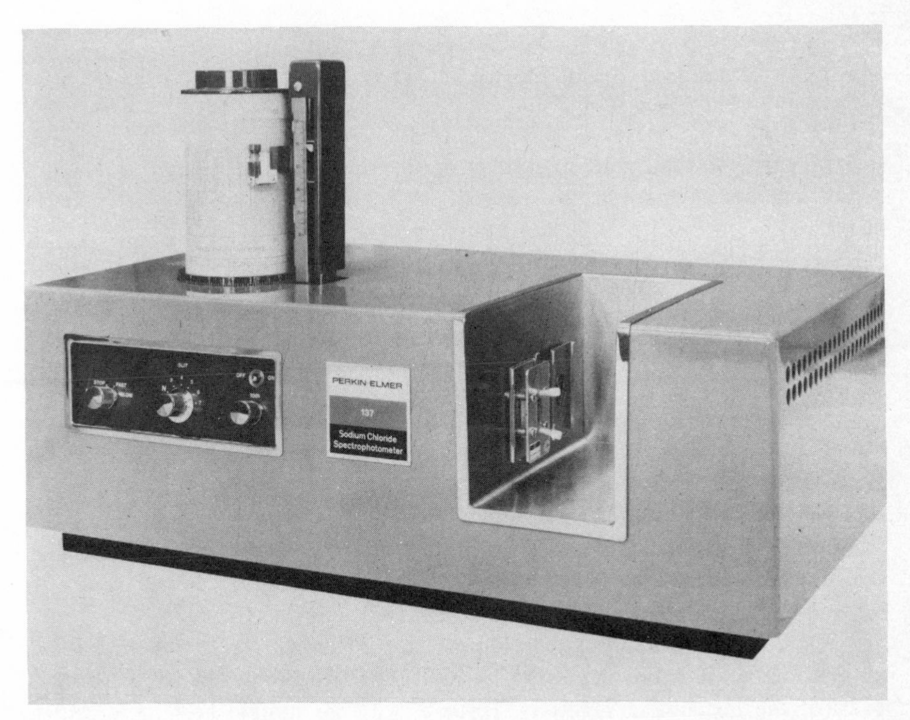

FIGURE 3-17 Perkin-Elmer model 137 infrared spectrophotometer. (*Perkin-Elmer Corp.*)

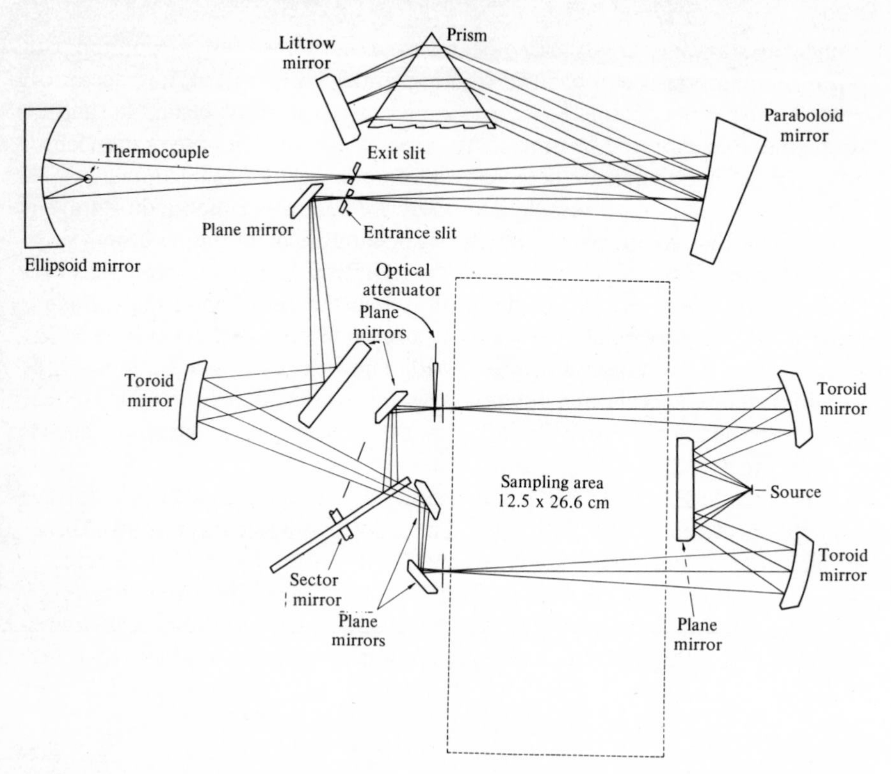

FIGURE 3-18 Schematic optical diagram of Perkin-Elmer model 137 infrared spectrophotometer. (*Perkin-Elmer Corp.*)

spectral range so that, with no absorption in either beam, the energy at the detector remains constant while scanning. A manual adjustment of the slit is possible.

To operate optical-null instruments of this type, the *gain* of the amplifier is first checked to see that the pen neither overshoots an equilibrium value (gain too high) nor responds too slowly (gain too low). In essence the response can be quickly checked by moving a small opaque object in and out of the sample beam.

Next the *balance* or stability of the amplifier is checked by blocking both the sample and reference beams with an opaque material like thick cardboard to see if the pen remains in a stable position. A slow drift toward 100 percent may be tolerated, but any drift toward 0 percent transmittance is unacceptable because the instrument may go dead during a subsequent zero measurement. Drift is easily eliminated with a balance control provided, which electronically balances out any background signals that may be causing the drift.

For qualitative analysis, the largest single application of infrared spectrophotometry, it is not necessary to adjust the 0 percent and 100 percent T levels of the instrument precisely. Instead, with no sample or sample cell in either beam, the recorder pen is arbitrarily adjusted to 100 percent T (it may be adjusted to some value like 95 percent T) with the 100 percent T control, which moves a small optical wedge in or out of the sample beam. Then the spectrum of a standard sample, like a thin sheet of polystyrene or liquid indene, is scanned, checking the spectral bands against the accepted values. Figure 3-19 gives the spectrum of polystyrene film and shows various reference peaks marked with their accepted wavelength values.

For quantitative analysis it would be desirable to adjust both the 100 and 0 percent T settings precisely, but unfortunately the inherent limitations of infrared sampling methods make a meaningful 100 percent T setting difficult to achieve, and limitations of the optical-null method preclude a precise 0 percent T setting. In brief, the 100 percent T level *should* correspond to zero absorbance of the sample constituent at the wavelength of analysis. To calibrate the 100 percent T level precisely would require exact compensation for all other sources of absorption, including sample cell, solvent, or any other material in the beam, as well as for reflection losses in the cell window. In practice, such compensation is extremely difficult to achieve, since it is virtually impossible to produce and maintain accurately matched infrared cells. Furthermore, at high concentrations of solute even matched cells do not provide equal thickness of *solvent*. (For that matter, even a single-beam instrument using the same cell for both the sample and solvent would not provide equal thickness of solvent at high solute concentrations; furthermore, with a single-beam instrument there is always a danger of changing the cell characteristics between scans by virtue of fogging from water vapor or other cell deterioration.)

The optical-null method precludes a precise 0 percent T setting since the more absorbing the substance in the sample beam the more the reference beam is blocked by the optical wedge. Thus, the closer the pen gets to 0 percent T the smaller the signal upon which the instrument operates, and

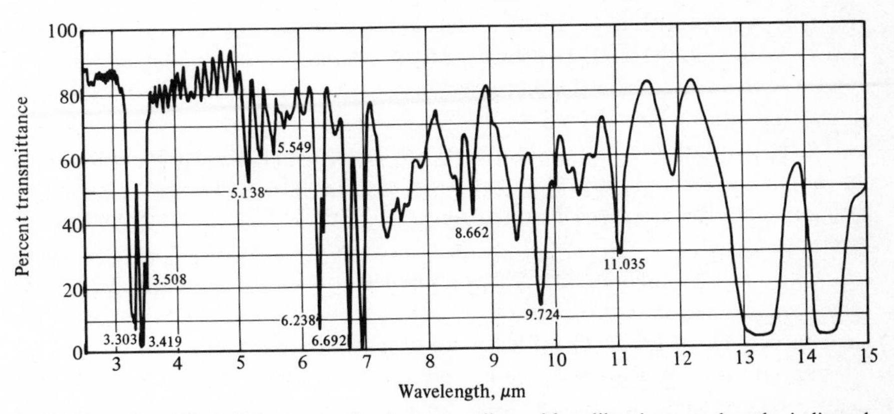

FIGURE 3-19 Infrared spectrum of polystyrene film, with calibration wavelengths indicated. (*Perkin-Elmer Corp.*)

eventually, somewhere in the vicinity of 1 percent transmittance, the servomotor will no longer have sufficient power to drive the pen. There are various electronic devices for preventing the servo from completely going dead, but nonetheless they limit the accuracy of the instrument in the region of 0 percent *T*. This source of error can be minimized by limiting analyses to bands which do not absorb too strongly.

In order to check the 0 percent *T* reading of the instrument and correct it if necessary, an opaque object is slowly moved into the sample beam, in increments, until the sample beam is completely blocked (the reference beam should remain unobstructed), allowing the pen to come to rest at or near 0 percent *T*. If the pen zero does not correspond with true zero, a screwdriver readjustment of the pen location is usually necessary.

INSTRUMENTS USING GRATINGS

Grating instruments require overlapping wavelength orders to be removed, and the most common method is a series of interference filters which operate on a synchronous program with the turning of the grating. In general, filters are inserted near a slit or slit image, where the required size of the filter is not excessive.

A good example of a low-cost grating instrument is the Perkin-Elmer model 337. The outward appearance of model 337 is very much like that of the model 137-B, shown in Fig. 3-17, and the optical diagram is likewise like Fig. 3-18, except that back-to-back gratings replace the prism and a filter wheel is added just beyond the exit slit.

The Perkin-Elmer 337 covers the wavelength range of 2.5 to 25 μm in two separate ranges, the first range going from 2.5 to 8.3 μm and the second from 7.5 to 25 μm. The instrument has a resolution of about 2 cm^{-1} (0.02 μm) at 1000 cm^{-1} (10 μm), and the photometric accuracy is about ± 1 percent *T*. The wavelength accuracy is within ± 0.012 μm on range 1, and within ± 0.036 μm on range 2.

An example of a sophisticated, higher-cost grating instrument is the Perkin-Elmer model 180. It uses a *potentiometric* instead of an optical-null system, and furthermore (as discussed in Sec. 3-6) it can readily be equipped for analysis in the far infrared.

A potentiometric-null system does not have the fundamental problem of optical-null systems of losing all response as a sample approaches zero transmission since the reference beam is never blocked and serves as a continuous reference signal.

The Perkin-Elmer model 180 is shown in Fig. 3-20, and Fig. 3-21 is its schematic optical diagram. The optical path may be followed by starting with the source (*A* in Fig. 3-21), which is an exceptionally large, forced-air-cooled Globar. (*B* and *I* in Fig. 3-21 are a mercury-arc source and a pyroelectric detector, respectively, which are used only in the far-infrared version of the instrument.) Radiation from the Globar is directed by a series of flat and

FIGURE 3-20 Perkin-Elmer model 180 infrared spectrophotometer. (*Perkin-Elmer Corp.*)

toroidal mirrors to a 15-Hz rotating chopper (*C* in Fig. 3-21). Chopper *C* alternately directs the beam through the sample and reference areas. After passing through the sample and reference, the two beams are recombined by a second chopper (*D*).

The combined time-shared beam is transferred by a series of flat and toroidal mirrors onto the entrance slit (*E*) of a single-pass Ebert monochromator. The heart of the monochromator is a series of diffraction gratings located on a carousel mount (*F*). In the normal operating range of the instrument (4000 to 180 cm^{-1}, or 2.5 to 55.6 μm), five different gratings are sequentially and automatically moved into the beam, each grating covering a selected portion of the overall range. When the instrument is equipped to cover the far-infrared region, two more gratings are added (extending the range to 31.25 cm^{-1}, or 317 μm), making a total of seven in the carousel. (The far-infrared version of the instrument will be discussed in Sec. 3-6.)

Energy from the monochromator exit slit then passes through a 14-position filter wheel (*G*), which contains a series of long-pass filters to remove unwanted orders. Thereafter, the beam is directed to the detector (*H*), which is a permanently evacuated thermopile fitted with a flat CsI window. The ac signal (first from the sample and then from the reference) is amplified and demodulated, and the ratio of the sample signal to the reference signal is recorded.

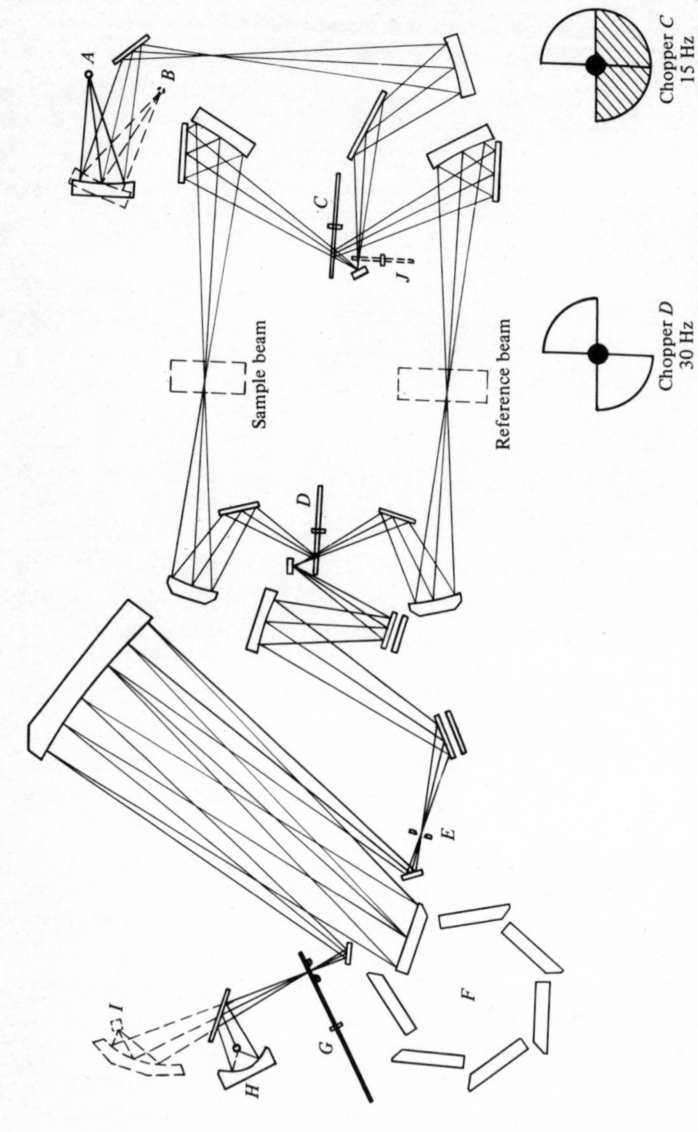

FIGURE 3-21 Schematic optical diagram of Perkin-Elmer model 180 infrared spectrophotometer. (*Perkin-Elmer Corp.*)

The resolution of the Perkin-Elmer model 180 is better than 0.2 cm^{-1} at 1030 cm^{-1} (9.7 μm), and better than 0.4 cm^{-1} at 3000 cm^{-1} (3.3 μm). Stray light is less than 2 percent over the entire instrument range and less than 0.4 percent from 2.5 to 25 μm. The abscissa is linear in wave numbers, and the ordinate can be selected to be linear in absorbance or transmittance with a switch. Photometric accuracy is better than ± 0.4 percent T. The abscissa can be expanded or compressed in steps up to a hundredfold, and the ordinate can be expanded up to twentyfold or compressed fourfold.

The double-chopper system used in the model 180 deserves discussion. The infrared spectrophotometers discussed thus far have only one chopper, which must necessarily be located between the sample and monochromator (see, for example, Fig. 3-18), since this chopper is the means of recombining the sample and reference beams into one beam. The disadvantage of chopping *after* the sample, rather than before, is that any radiation coming from the sample (as from warm samples) will be chopped and counted with the transmitted signal, causing errors. The two-chopper system shown in Fig. 3-21 is able to discriminate against any radiation originating from the sample or reference because the detector system is tuned to the 15-Hz chopper (C), while the 30-Hz chopper (D) serves only as a convenient means of recombining the sample and reference beams into one beam. In the detailed views of choppers C and D shown near the bottom of Fig. 3-21, the shaded half of chopper C represents a masked or blocking portion, whereas the outlined quadrants represent mirrors. (The other quadrants are open, allowing radiation to pass; note that chopper C allows the sample to receive energy from the source during only one-quarter of the cycle, thus significantly reducing the degree of sample heating due to the source.) Actually, both sectors C and D rotate at 15 rps, but because of the mirror design, chopper C chops the source energy at a frequency of 15 Hz while chopper D chops any radiation from the sample or reference at a frequency of 30 Hz. The 30-Hz signal is electronically rejected, whereas the 15-Hz signal is accepted and used. (See Exercise 3-11.)

Model 180 can also be used as a single-beam instrument. For this purpose choppers C and D in Fig. 3-21 are stopped so that they are out of the optical beam, and an auxiliary built-in 180° 15-Hz chopper (J in Fig. 3-21) is turned on.

DOUBLE-MONOCHROMATOR INSTRUMENTS

All the instruments discussed previously were single-monochromator instruments, inasmuch as only one prism or grating (or variable interference filter) was in use at a time. A number of precision double-monochromator instruments have been developed, to give higher resolution and lower stray light than single-monochromator instruments. Most of these instruments use a prism coupled with a grating, although two prisms are sometimes used. The prism-grating combination takes advantage of the high resolution obtainable with a grating and the spectral purity obtainable with a prism. Thus, by placing the prism ahead of the grating (in which case it is called a *foreprism*) the prism can be used to isolate a narrow band of wavelengths, and only a single order remains to be diffracted by the follow-up grating monochromator.

The Perkin-Elmer model 325 infrared spectrophotometer is shown in Fig. 3-22, and its schematic optical diagram is given in Fig. 3-23. The model 325 covers the range of 5000 to 200 cm^{-1} (2 to 50 μm), using a thermostated KBr prism (*P* in Fig. 3-23) as a foremonochromator in the spectral region 5000 to 450 cm^{-1} (2 to 22 μm) and interference filters (*F* in Fig. 3-23) for the region 450 to 200 cm^{-1} (22 to 50 μm). Potassium bromide begins to absorb strongly beyond 22 μm (the limiting transmission wavelength is 25 μm; see Table 3-7). Therefore, at 22 μm a scatter plate (*SP*) drops into the optical beam, blocking the path to the prism, and simultaneously interference filters (*F*) are inserted to isolate a narrow bandpass before entrance to the grating monochromator (compartment 6). Two echelette back-to-back gratings are used in their second and first order, with an automatic program positioning the proper grating for a given wavelength region.

Compartments 1 to 7 in Fig. 3-23 correspond closely to the physical arrangement of the instrument as pictured in Fig. 3-22. The source (*G* in compartment 1) is an air-cooled Globar. Compartment 2 contains chopper 1, which rotates at 13 Hz and splits the source energy into sample and reference beams. In the sample compartment (3), these beams pass alternately through the sample (*S*) and reference (*R*). In compartment 4 the two beams are optically balanced by means of optical wedge (*A*) and recombined into a single beam by chopper 2. Movement of the optical wedge is recorded in this optical-null system. The radiation then passes into the foremonochromator compartment (5), thence to the main monochromator (6), and finally to a thermopile detector with a CsI window (*D*, in compartment 7). In principle the double-chopper system used here is similar to the chopper system already described for the Perkin-Elmer model 180, and will not be elaborated upon.

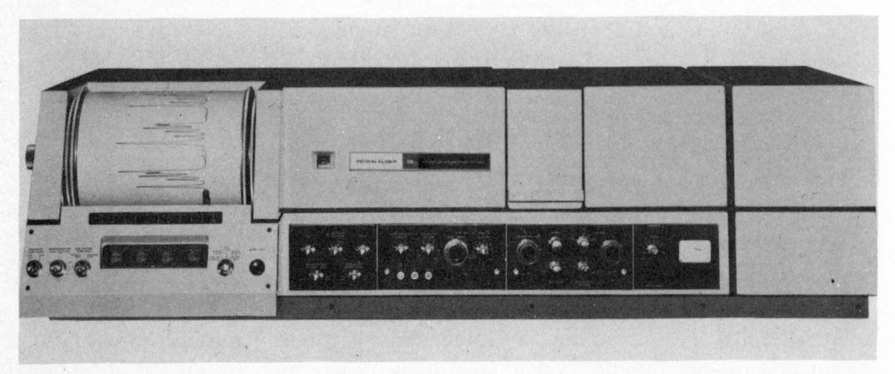

FIGURE 3-22 Perkin-Elmer model 325 infrared spectrophotometer. (*Bodenseewerk Perkin-Elmer & Co. GmbH.*)

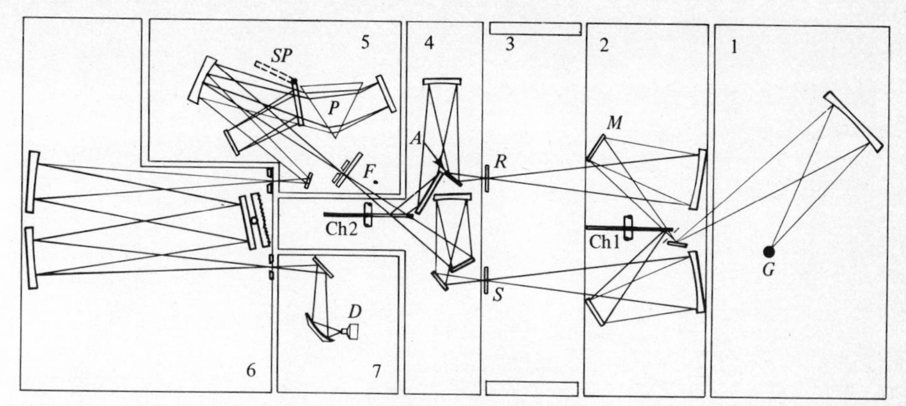

FIGURE 3-23 Schematic optical diagram of Perkin-Elmer model 325 infrared spectrophotometer. (*Bodenseewerk Perkin-Elmer & Co. GmbH.*)

The net effect of the double-chopper system is exactly as described before, namely, the elimination of any radiation originating from the sample or reference.

The resolution of the model 325 is better than 0.5 cm^{-1} over the entire range (being better than 0.33 cm^{-1} at 2400 cm^{-1}, 0.16 cm^{-1} at 1034 cm^{-1}, and 0.41 cm^{-1} at 286 cm^{-1}). Stray light is less than 0.2 percent from 2 to 37 μm and less than 2.5 percent from 37 to 50 μm. Six wavelength ranges are furnished, with two survey ranges (from 2.5 to 50 μm or 5 to 50 μm) and four expanded ranges (from 2 to 5, 4 to 10, 10 to 25, or 24 to 50 μm). The abscissa is linear in wave number and accurate to $\pm$0.02 percent. The ordinate is linear in transmittance or absorbance, with a photometric accuracy of $\pm$0.5 percent T.

3-3 SAMPLE HANDLING

Sample handling in the infrared region presents a number of problems. The majority arise from the fact that almost all substances absorb in the infrared, which in turn poses restrictions on the choice of cell materials, thickness of the sample path, and the solvents that can be used. On the other hand, an almost infinite variety of infrared techniques have been devised to examine almost all types of materials. Instead of having a rigid set of sampling rules, the analyst is often left to his own ingenuity to solve his particular sampling problems. In this section, some of the most useful sampling techniques for gases, liquids, and solids will be described briefly. Further details and variations can be found in the references.

3-3A Gases

Gaseous samples are often measured in a cell about 10 cm in length, an example of which is shown in Fig. 3-24. The cell is evacuated, and then the

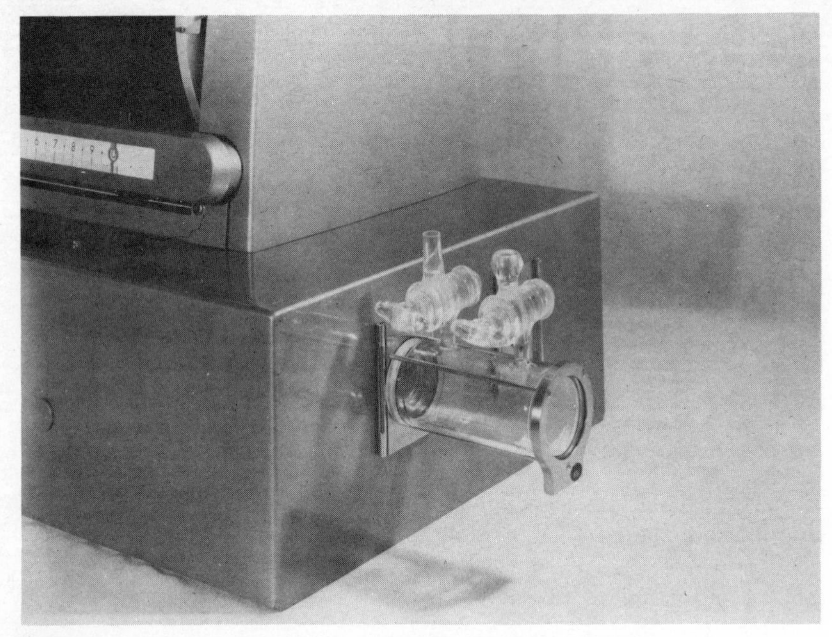

FIGURE 3-24 Typical gas absorption cell (10-cm path). (*Perkin-Elmer Corp.*)

sample is drawn in to the desired pressure. Gas pressures may range from 5 mm or even less for strongly absorbing substances like fluorocarbons, to ½ atm for weak absorbers such as hydrogen chloride or water vapor. Where very long path lengths are required, as in air-pollution work or breath analysis, paths up to 40 m can be obtained using multiple-reflection cells. Multiple-reflection cells are usually placed 90° from the normal sample-beam direction and have a mirror at each end of the cell that repeatedly reflects the beam between the ends of the long cell before the beam resumes its normal direction.

3-3B Liquids

Samples that are liquid at room temperature are usually scanned in their pure (neat) form or as a solution in conjunction with a solvent. Liquids of *low volatility* can most readily be scanned for qualitative analysis by placing a drop between rock-salt plates which are squeezed together and clamped between two plates of a demountable cell. Figure 3-25 shows an example of a demountable cell. When no spacer is placed between the salt plates, the thickness is generally designated as capillary thickness. To increase the sample path, spacers (0.005 to 0.5 mm thick) may be used.

For *volatile liquids* (liquids boiling below about 70°C), a sealed cell is required. (Demountable cells are not tight enough to hold volatile samples, and

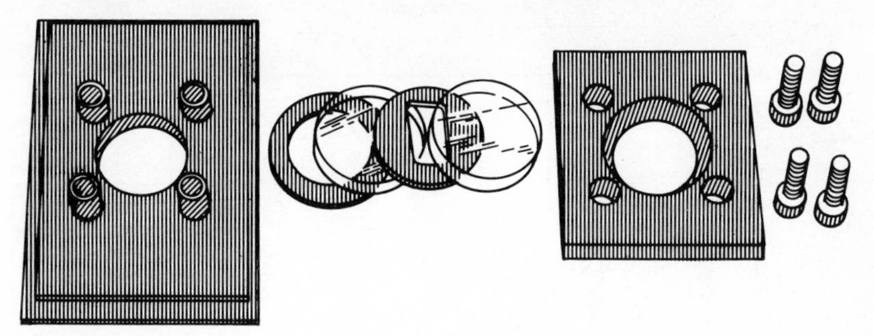

FIGURE 3-25 Demountable infrared cell. (*Perkin-Elmer Corp.*)

some evaporation may occur before completion of the scan.) Figure 3-26 shows sealed cells of both the fixed-thickness and variable-space type.

Solutions are generally handled in cells with paths from 0.1 to 1 mm. Concentrations are commonly between 0.05 and 10 percent, which usually requires about 1 to 15 mg of solute. A compensating cell of fixed or variable thickness is usually placed in the reference beam.

FIGURE 3-26 Infrared sample cells of (*a*) fixed-thickness and (*b*) variable-space type. (*Perkin-Elmer Corp.*)

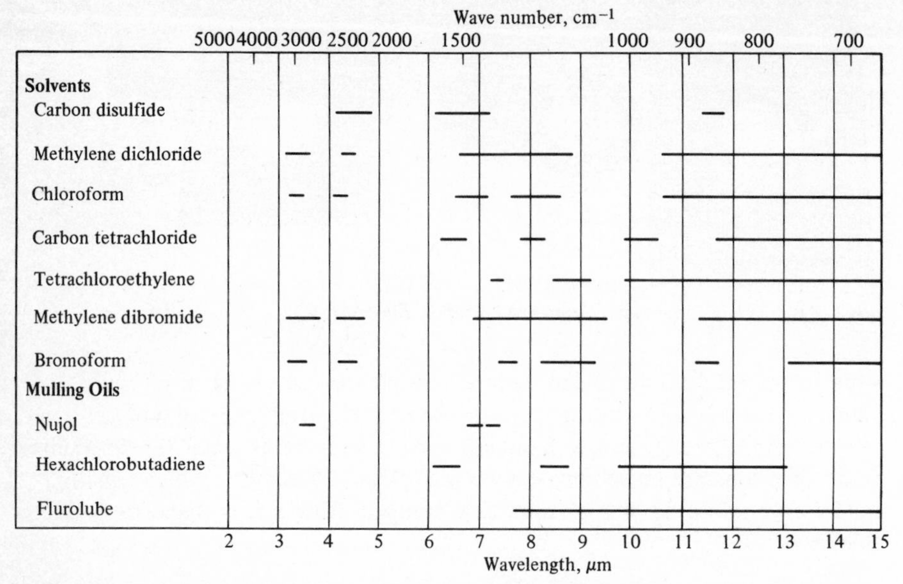

FIGURE 3-27 Transparent regions of some infrared solvents and mulling oils (open regions are those in which a 1-mm thickness of solvent or thin film of oil transmits more than 25 percent). (*From R. M. Silverstein and G. C. Bassler, "Spectrometric Identification of Organic Compounds,"* 2d *ed., p.* 71, *John Wiley & Sons, Inc., New York,* 1967 *by permission.*)

Solvents used must be dry and transparent in the wavelength region of interest. Figure 3-27 gives the transparent regions of a few selected solvents and mulling oils (the latter are used for solid samples, as will be discussed). All common solvents absorb in the infrared, but carbon tetrachloride, carbon disulfide, and chloroform are the three solvents most commonly used. In addition to transparency and solubility considerations, inertness toward the sample may be important. For example, carbon disulfide should not be used with primary and secondary amines, since they will react to give thiocarbamates. Similarly, aliphatic amines undergo a slow photochemical reaction with carbon tetrachloride to form the amine hydrochloride.

Although it is often more convenient to scan a liquid in its pure form than to dilute it with a solvent, there are certain disadvantages to the latter procedure for quantitative analysis. (1) It is easier to reproduce the longer path lengths used with dilute solutions. (2) The concentration can be conveniently adjusted for optimum viewing of key absorption bands. (3) Although the solvent may absorb somewhat, some compensation for the solvent absorption can be achieved by using a matched cell in the reference beam. With optical-null instruments, however, compensation can never be highly precise in regions of strong solvent absorption, since in these regions little or no energy is reaching the detector.

DETERMINATION OF CELL THICKNESS

The small dimension of infrared cells makes it difficult to measure and reproduce cell thicknesses accurately. Thicknesses greater than about 0.6 mm can be determined by measuring the gasket thickness with a micrometer, but thicknesses smaller than this are determined most accurately by the method of interference fringes. This method takes advantage of the maxima and minima produced in the spectrum of an empty cell by the reinforcement and destructive interference of radiation reflected from the internal surfaces of cell windows. With spectrophotometers that are linear in wave numbers, the thickness b (in centimeters) can be calculated from

$$b = \frac{n}{2} \frac{1}{\bar{\nu}_1 - \bar{\nu}_2} \tag{3-10}$$

where n is the number of maxima between frequencies $\bar{\nu}_1$ and $\bar{\nu}_2$. For spectrophotometers that are linear in wavelength, the corresponding equation is

$$b = \frac{n}{2} \frac{\lambda_1 \lambda_2}{\lambda_2 - \lambda_1} \tag{3-11}$$

where b, λ_1, and λ_2 are all in micrometers.

Example 3-8 Figure 3-28 shows the interference pattern for an empty cell. Use this pattern to calculate the path length in centimeters for the cell.

Answer Either Eq. (3-10) or (3-11) can be used. With Eq. (3-11),

$$b = \frac{10}{2} \frac{(7.0)(11.0)}{11.0 - 7.0} = 96\ \mu\text{m}$$

Since 1 μm = 10^{-4} cm,

$$b = 0.0096 \text{ cm}$$

To obtain distinct interference fringes, the cell windows must be flat, and measurements are best made immediately following repolishing of the win-

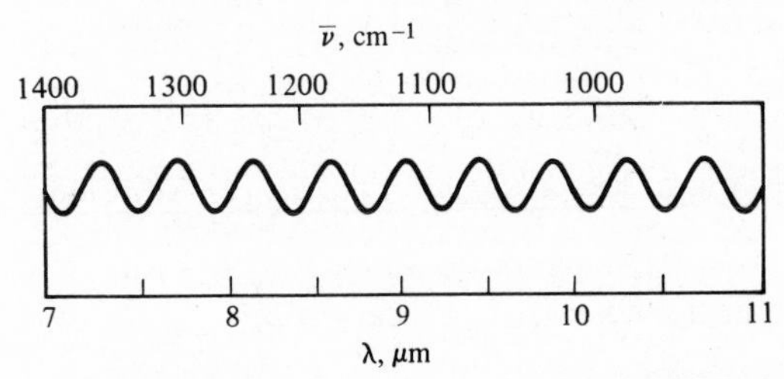

FIGURE 3-28 Infrared spectrum of an empty cell for Example 3-8.

dows. Since cell thicknesses change with time because of gradual compression of the spacer or erosion of the windows, they must be remeasured periodically.

3-3C Solids

There are more infrared techniques for examining solids than for gases and liquids. The two most important techniques for powdered samples involve mineral-oil mulls and KBr pellets. Thin films or other smooth surfaces can be examined by attenuated total reflectance (ATR).

MINERAL-OIL MULLS

Mineral-oil mulls, in which a finely ground sample is suspended in mineral oil, are extensively used for solid samples. Three kinds of mulling oils are commonly used (see Fig. 3-27). Nujol, a highly refined mineral oil consisting of a mixture of various straight-chain saturated hydrocarbons (averaging about C_{25}), has the disadvantage of strong absorption due to CH stretching vibrations at 3.3 to 3.5 μm and CH bending modes at 6.81 and 7.25 μm. Thus, Nujol cannot be used to examine aliphatic CH vibrations in a sample. On the other hand, since Nujol is transparent in most other regions, for aromatic compounds and samples in which only functional-group analysis is desired, this medium is very useful. For samples requiring aliphatic C—H analysis, chlorinated or fluorinated oils, e.g., hexachlorobutadiene or Fluorolube, are used.

To prepare a mineral-oil mull, the general procedure is first to grind the dry sample to a fine powder using an agate mortar, glass plates, or automatic pulverizing equipment. A few drops of the mulling oil are added and the grinding continued until a smooth paste is obtained. A few drops of this paste are sandwiched between polished sodium chloride disks, clamped with a cell holder, and scanned. Amorphous materials can be ground with dry ice to make them brittle.

It is important that the sample be finely ground. Specifically, the particle size should be less than the wavelengths of the incident radiation in order to minimize scattering. Refraction and reflection do not occur with smaller particles because a wavefront of coherent radiation cannot be formed (see Sec. 1-4*C*). Scattering at phase boundaries, i.e., Rayleigh scattering (see Sec. 1-4*C*), is not a serious problem since the refractive index of mineral oil is compatible with most solid particles. The majority of infrared spectral studies begin at about 2 μm; it is desirable but generally impossible to reduce all particles in a sample to below the 2-μm size. Therefore, the transmission spectrum of solid samples often shows a sharp decrease from 2 to 4 μm. The steepness of this slope can be used as a measure of the scattering.

POTASSIUM BROMIDE PELLETS

The KBr-pellet technique involves mixing a finely divided sample with KBr powder and then pressing the powder in a die to form a transparent pellet. Specifically, about 1 mg of sample is usually ground and mixed with, say, 300 mg of high-purity KBr. After transferring to a die, the die assembly should be evacuated to remove entrapped and occluded air in the mixture. The mixture is then pressed for several minutes in a laboratory press, applying pressures between 8 and 20 tons/in^2. Properly formed pellets will be completely clear, the pressed KBr being transparent out to about 25 μm (and in fact transparent throughout the visible and ultraviolet regions).

The main disadvantages of the potassium bromide technique are the hygroscopic nature of potassium bromide and the difficulty of obtaining reproducible spectra. A properly pressed pellet will remain clear for a long time when stored in a dry atmosphere, but the presence of moisture or the lack of evacuation to remove entrapped air before pressing will show up as visibly opaque pellets, rendering analysis in the —OH and —NH region difficult. Lack of reproducibility results largely from the problem of obtaining uniform particle size in grinding, although variations in pressure and occasional reactions between the sample and KBr can contribute. If a pellet contains too much sample, it may be possible to reduce its thickness by rubbing it on fine sandpaper, for example, 320-A, and very fine emery paper, for example 4/0.

ATTENUATED TOTAL REFLECTANCE

The technique of attenuated total reflectance (ATR) has become available fairly recently for obtaining spectra of solid plastics, elastomers, fabrics, adhesives, powders, foams, and inorganics. The method is based on the fact that when a beam of radiation is passed into a prism with a high index of refraction, total reflection will take place at the back face of the prism. If a sample is placed on the back face of the prism, the radiation will penetrate a few micrometers beyond the interface into the low-index sample material before being totally reflected. While it is in the sample material, the radiation can be selectively absorbed, and the resulting spectrum is very similar to an absorption spectrum.

Accessories for ATR are commercially available, and Fig. 3-29 gives a schematic optical diagram of a typical ATR unit. The prism is a high-

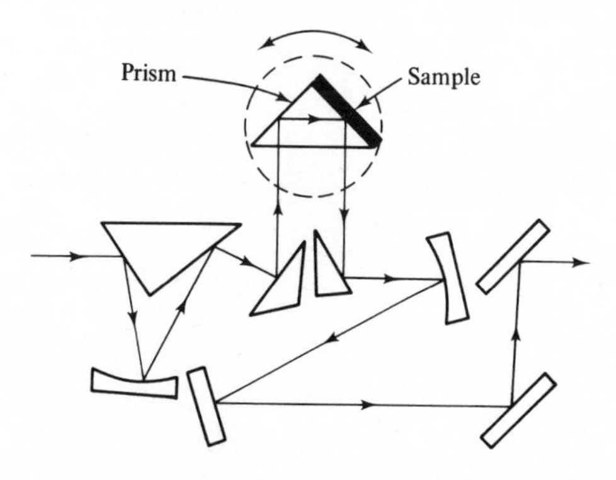

FIGURE 3-29 Schematic optical diagram of a typical ATR unit. (*Beckman Instruments, Inc.*)

refractive-index medium such as AgCl, Ge, or KRS-5 (thallium bromide–thallium iodide). The sample must make a firm, intimate contact with the prism face, and thus the best samples have very smooth surfaces, such as films. The most important operating variables are the prism material (in order to get a proper ratio between the refractive index of the prism and that of the sample) and the angle of incidence which the radiation makes with prism-sample interface. The angle of incidence is generally easily adjustable, and it is advantageous to have available an assortment of prisms made from various materials.

3-4 APPLICATIONS

3-4A Structure Determination

Since the principles of structure determinations form the background for organic qualitative analysis, it will be useful to discuss structure determination first, after which the approach to qualitative analysis will follow readily. The inherent difference between these two applications is that whereas samples for qualitative analysis are usually mixtures or, at best, impure, the samples for structure determinations *must* be pure. Wasted effort and/or invalid conclusions may result from attempting a structure determination on a sample that has not been carefully purified.

No fixed procedure can be given for determining the structure of a compound, nor is it possible for infrared information alone to reveal the exact structure of a complex molecule unequivocally. On the other hand, in conjunction with chemical and other physical data, a routine infrared analysis provides much useful information, and it is worthwhile to run infrared spectra early in a structure study.

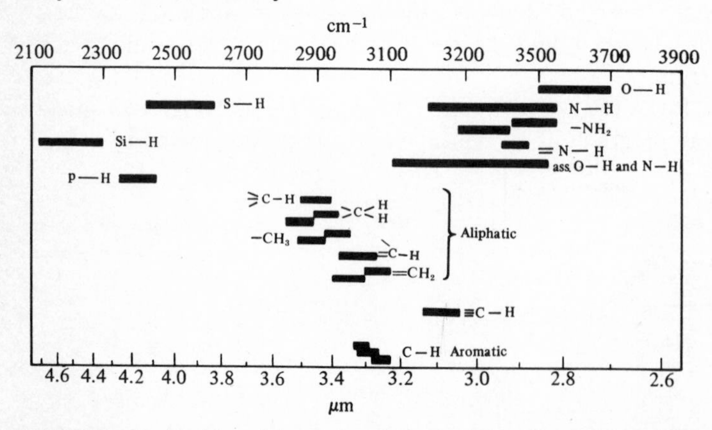

FIGURE 3-30 Position of the characteristic stretching vibrational bands of XH groups.

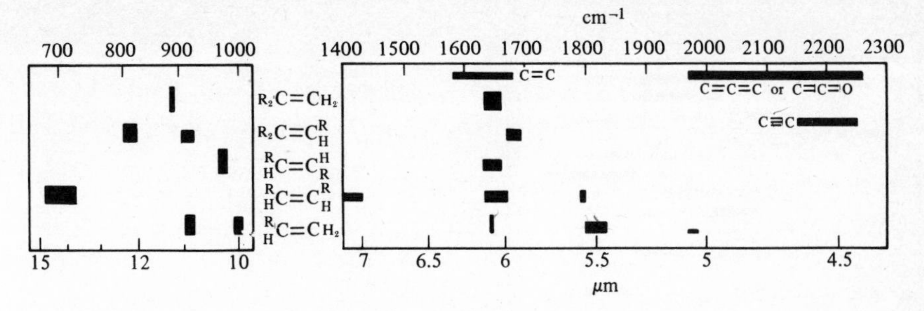

FIGURE 3-31 Position of the characteristic bands of various carbon-carbon bonds.

A knowledge of the previous reaction conditions and treatment prior to spectral analysis may be very helpful in identifying spurious peaks; e.g., moisture may show up in the 2.7- to 2.9-μm and 6.14-μm region; bottle-cap liners may have been leached to contribute carbonyl absorption at 5.7 to 5.9 μm; silicone oil or grease may contribute at 7.9 μm ($SiCH_3$), etc. Infrared examination of the starting materials is well worth the effort.

The most useful region of the infrared for structure determinations is the characteristic *group-frequency* region, from about 2 to 7 μm (see Fig. 3-5). There are several ways in which group-frequency data can be represented and tabulated. Two approaches will be given here. Figures 3-30 to 3-35 present graphs of wavelength (or frequency) of absorption for various functional groups, whereas Fig. 3-36 is a comprehensive chart of wavelength of absorption organized according to the class of compound. Each approach has merit, the first simplifying the initial spectral evaluation and the second (Fig. 3-36) giving more comprehensive information. The monographs listed in the references should be consulted for a wealth of additional spectral data.

After the sample has been run, using a suitable sampling technique

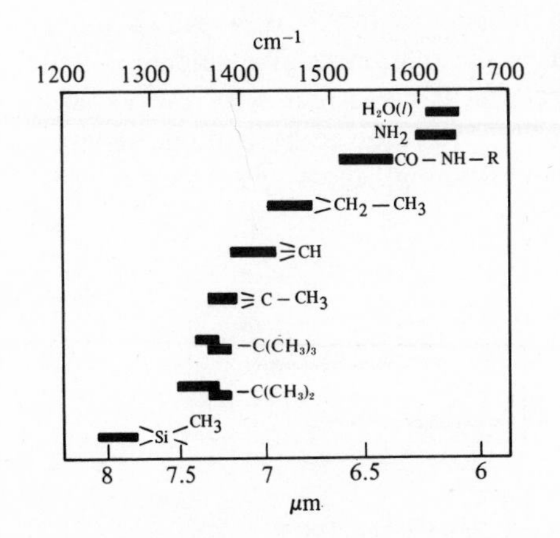

FIGURE 3-32 Position of the characteristic hydrogen-bending vibration bands.

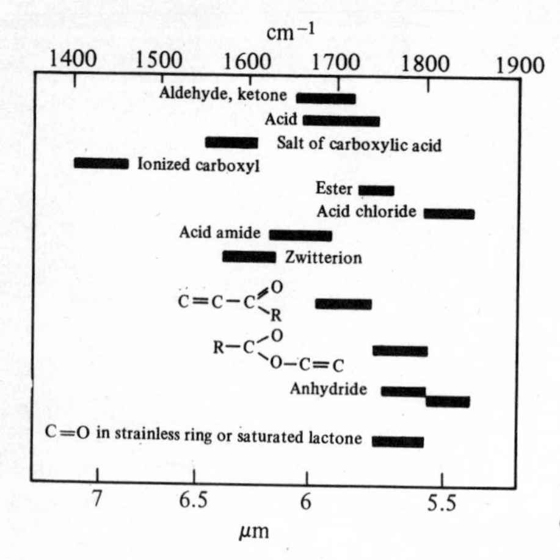

FIGURE 3-33 Position of the characteristic C=O stretching vibration bands.

described earlier (Sec. 3-3), the spectrum is critically examined. The strongest bands are selected, and an attempt is made to identify the functional groups corresponding to those bands. This can seldom be carried out unambiguously, since characteristic frequencies often overlap. Therefore other information must be used. It may include the relative intensities of various bands (see Fig. 3-36 and Refs. 5, 7, and 16), other spectroscopic data, or any other chemical or physical data that are available. After the strongest bands have been assigned, an attempt should be made to assign the bands of moderate or even weak intensity. This is much more difficult, and it is usually impossible to make a complete assignment, even if the compound is pure, because of such complications as overtones and combination vibrations. It is useful to examine spectra of related compounds. Finally, a series of possible models should be assumed, and, when possible, the number and type of frequencies should be predicted by selection rules and compared with those observed. A complete vibrational analysis is often difficult and laborious, but it is worthwhile for reasonably simple molecules, and several of the references give guidelines and examples. The model (or models) most nearly consistent with the experimental data is selected, and any ambiguities are resolved by isotopic separations, polarization studies, or other techniques, e.g., mass spectrometry, nuclear magnetic resonance, and x-ray methods.

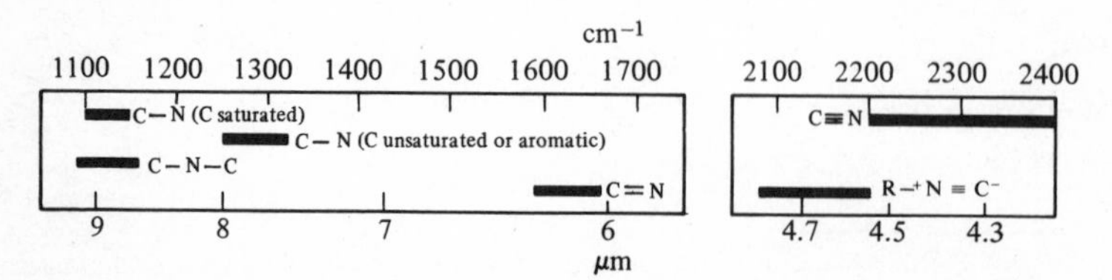

FIGURE 3-34 Position of the characteristic bands of carbon-nitrogen bonds.

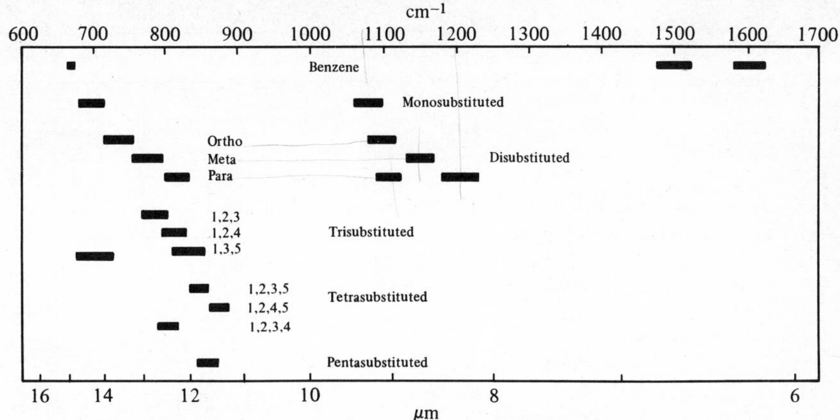

FIGURE 3-35 Position of the characteristic bands of benzene and benzene derivatives.

Example 3-9 Thioacetic acid is acetic acid in which one oxygen atom has been replaced by sulfur. From the infrared spectrum of thioacetic acid given in Fig. 3-37, determine whether the structure of thioacetic acid is

$$\underset{\text{I}}{CH_3C(=O)SH} \qquad \text{or} \qquad \underset{\text{II}}{CH_3C(=S)OH}$$

Answer The very strong absorption at 5.78 μm (1730 cm^{-1}) is consistent with that of an acid C=O stretch (see Fig. 3-33), and the band at about 4.0 μm (2500 cm^{-1}) corresponds to an S—H stretch (see Fig. 3-30). The peaks in the 3.3-μm (3000-cm^{-1}) region and 7- to 7.2-μm (1430- to 1380-cm^{-1}) regions can be attributed to a CH_3 group attached to a C=O group, according to Fig. 3-36. All these data point toward I as the correct structure. Furthermore, C=S can be ruled out because it should give a very strong band in the vicinity of 7.5 μm (1340 cm^{-1}) and an acid OH group should absorb roughly in the region of 3.0 to 3.3 μm (3333 to 3030 cm^{-1}) (see Fig. 3-36). The spectrum of Fig. 3-37 allows an unequivocal distinction of the two structures, since if structure II is assumed, there is no way of explaining the most intense peak of all, the 5.78-μm peak, which is very characteristic of C=O. Thus, without question, the structure is I.

Example 3-10 The product of a certain reaction gave white crystals in the form of sheets. The product melted at 100°C and was shown to have

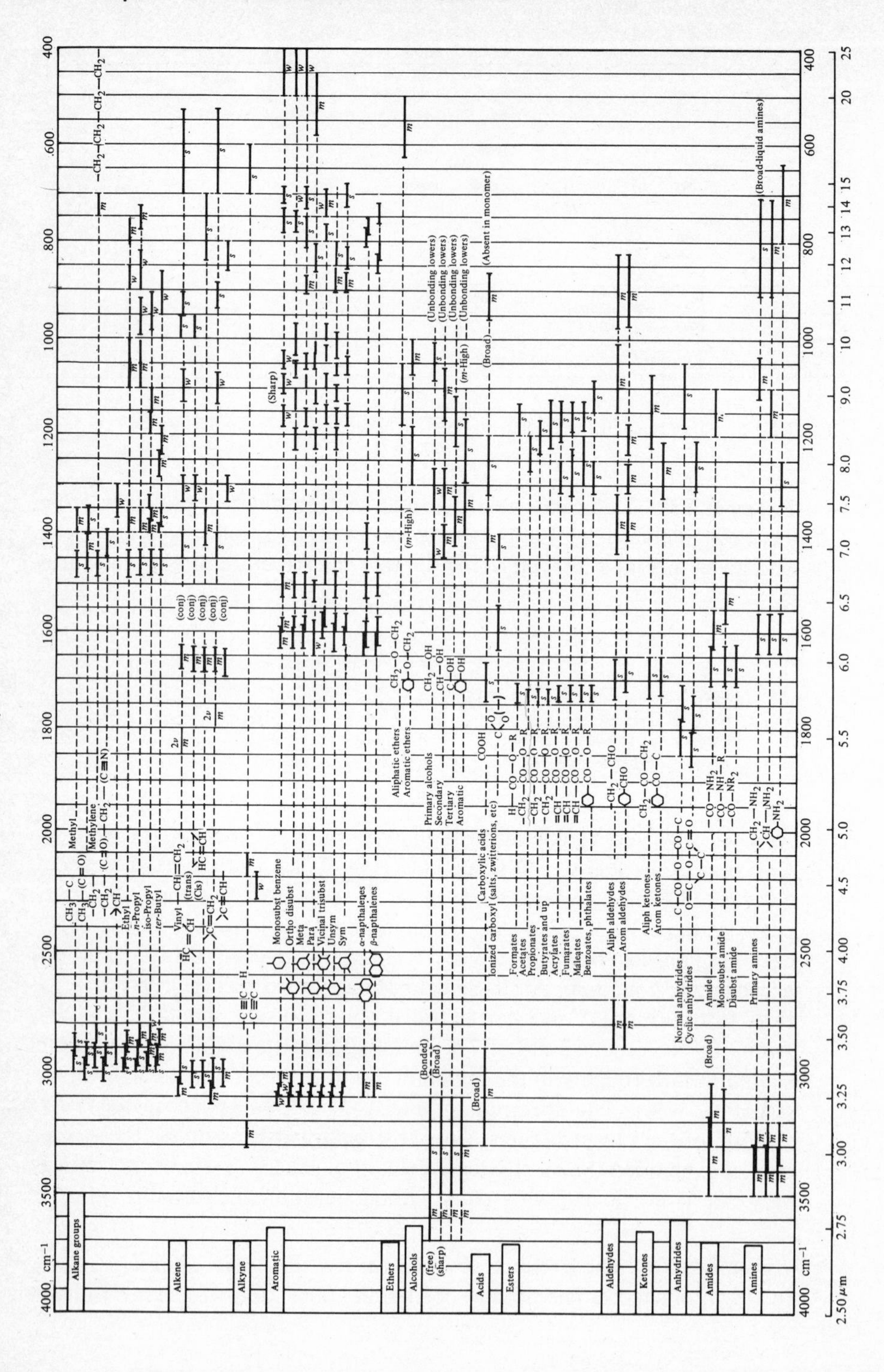
Alkane groups
Alkene
Alkyne
Aromatic
Ethers
Alcohols
Acids
Esters
Aldehydes
Ketones
Anhydrides
Amides
Amines

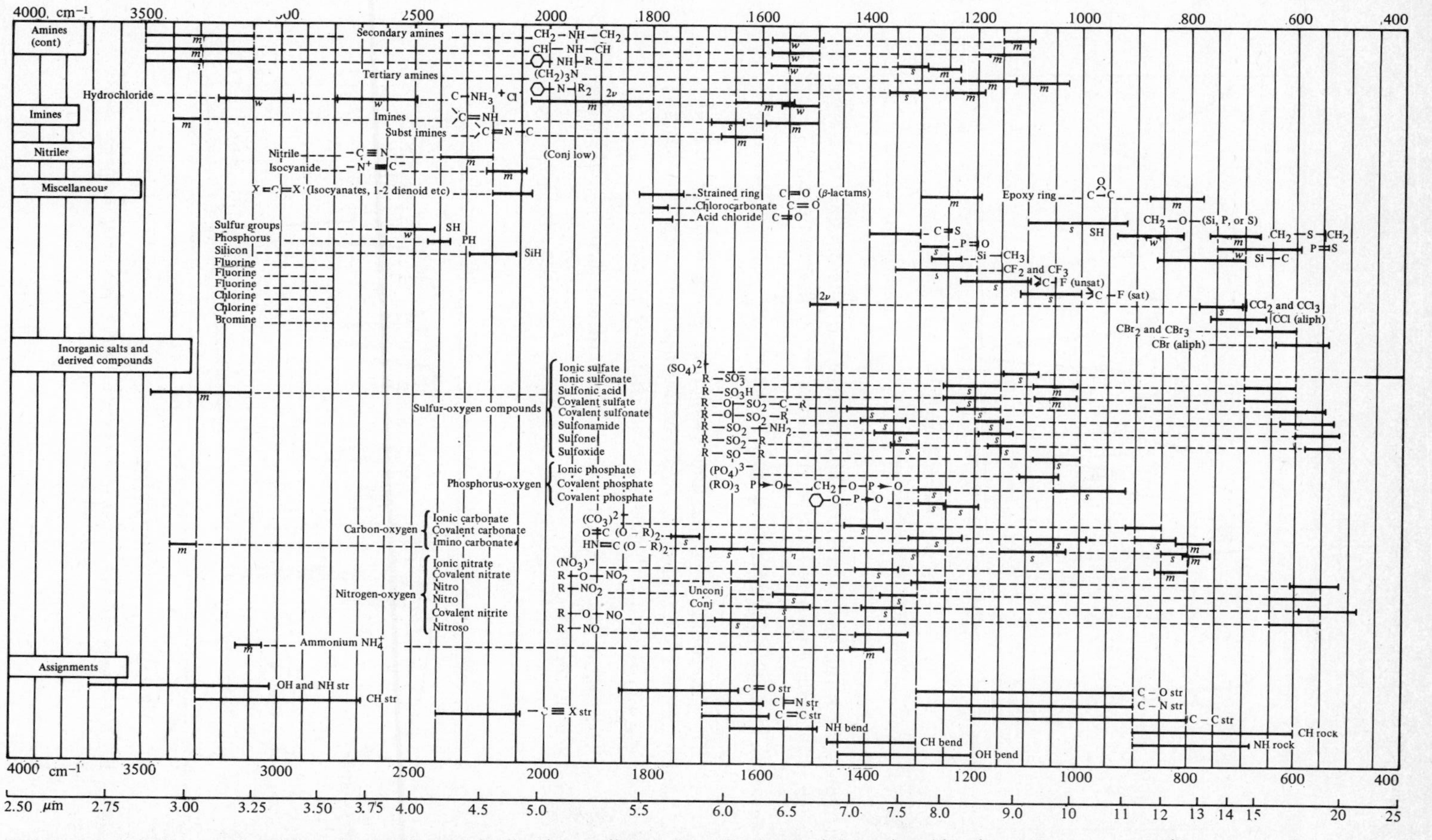

FIGURE 3-36 Characteristic infrared group frequencies listed according to class of compound. Intensity of band: s = strong; m = medium; w = weak. Overtone bands are marked 2ν. (*From N. B. Colthup, J. Opt. Soc. Am.*, **40**:397–400 (1950), *by permission.*)

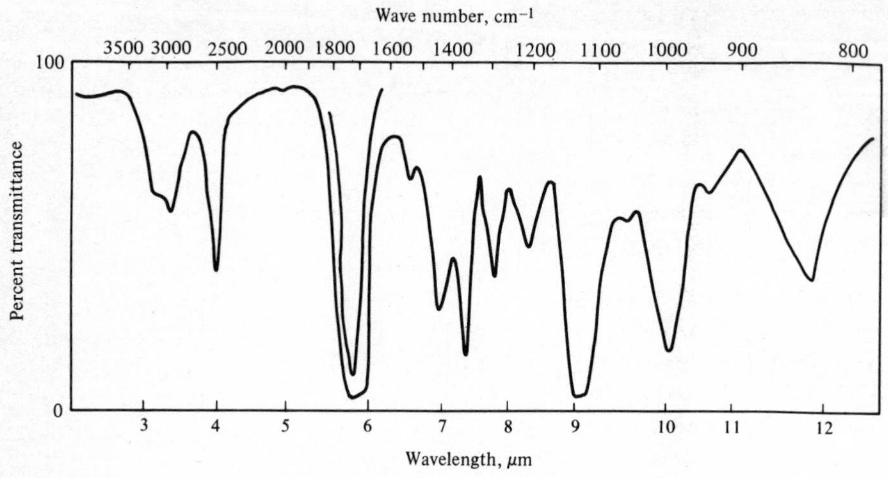

FIGURE 3-37 Infrared spectrum of thioacetic acid.

a molecular formula of $C_{13}H_{10}O$. The spectrum of the crystals in a KBr pellet is shown in Fig. 3-38. Deduce the structure of this compound.

Answer The following reasoning may be used:

1 The molecular formula and properties given indicate that the substance may be aromatic. This is confirmed by the ═CH— band at 3.3 μm and aromatic-ring bands at 6.2, 6.3, and 6.7 μm.

2 The ring is disubstituted in the ortho position, as indicated by the band at 13.3 μm.

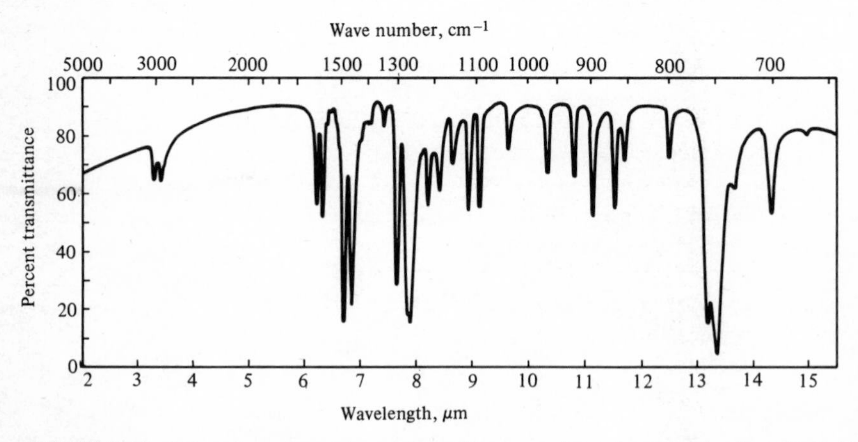

FIGURE 3-38 Infrared spectrum of a compound known to have the molecular formula $C_{13}H_{10}O$ (KBr pellet).

3 At least one carbon atom is in the form of a saturated methylene group, as indicated by the bands at 3.42, 3.5, and 6.85 μm. However, the methylene bands are so weak compared to the aromatic bands that there is probably no more than one nonaromatic methylene group. Two benzene rings connected by a methylene bridge might be suspected.

4 Though the compound contains oxygen, the spectrum gives no indication of an O—H or C=O group. However, the band at 7.9 μm is probably indicative of an unsaturated ether linkage.

5 Though it is suspected that two benzene rings are connected by a methylene bridge (conclusion 3), there is no evidence of a monosubstituted benzene ring, as indicated by the absence of strong bands in the region of 9 to 9.5 μm, and 14 to 14.7 μm. Therefore, *two* ortho-substituted benzene rings are probably present (from conclusion 2). This indicates a second bridge between the two benezene rings, the second bridge being an ether linkage (from conclusion 4). Accordingly, the structure is probably

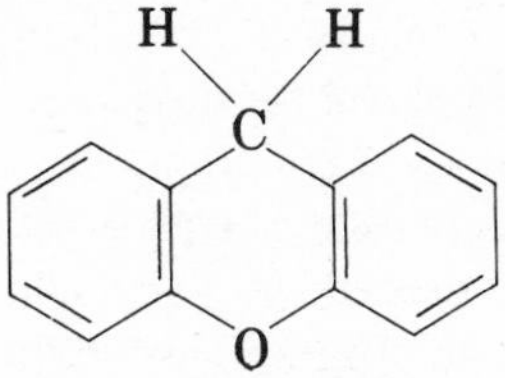

H H
C
O

Xanthene

This structure was confirmed by running a KBr pellet spectrum of known xanthene and comparing the two spectra.

It should be reemphasized that only rarely will it be possible to assign all observed bands in a spectrum to particular chemical groups or modes of vibration. Correlation charts have ambiguities due to overlapping absorption ranges, and conclusions are complicated by overtones, combination modes, and sample conditions. For example, carbonyl groups show a weak overtone that is easily mistaken for a trace of OH. Sometimes the experienced spectroscopist can use the characteristic shape and intensity of absorption bands to clarify ambiguities. For example, the C=C stretching vibration and the C=O stretching vibration of amides both occur at 1650 cm^{-1} (6.06 μm), but the two vibrations can be distinguished because the C=C band is sharp whereas the C=O band is broader and blunter. Sample conditions are also very important, because changes in state can cause significant changes in band positions and intensities. Spectral comparisons should be made only when the known and unknown are run using the same method of sample preparation.

3-4B Qualitative Analysis

Infrared spectrophotometry is probably the most widely used technique available for the identification of unknown substances and mixtures. The basis of qualitative analysis by infrared is that no two substances which absorb infrared radiation absorb it at the same frequencies to the same extent. Thus, an infrared absorption spectrum is a fingerprint of a substance as unique as the fingerprint of a person. Furthermore, the spectrum of a mixture of components is generally a simple addition or overlap of individual spectra, except in the relatively few cases where intermolecular reactions take place.

Both organic and inorganic materials can be analyzed by infrared spectrophotometry. All organic materials absorb infrared, and the multiplicity of bands, differing in position, intensity, and shape, allows great specificity of characterization. Whereas most physical data like melting points, boiling points, etc., are single-valued, an infrared spectrum presents some 30 to 50 physical values which characterize the substance.

All inorganic substances containing polyatomic cations and/or anions absorb infrared radiation because of the presence of covalent bonds. Examples are ammonium, sulfate, carbonate, and borate ions. In general, however, inorganic spectra are characterized by fewer and broader absorption bands than organic spectra and thus are somewhat less useful for qualitative analysis.

Infrared analyses are rapid, require only a small amount of sample, and are not destructive of the sample. These attributes are very fortunate, since many isolation procedures and synthetic procedures in chemical, biological, and pharmaceutical fields result in only micrograms or milligrams of final product. It is often possible to obtain spectrum on as little as 10 μg of material.

The main criterion for qualitative identification of an unknown single substance or mixture of substances is that the spectrum of the unknown match precisely the spectrum of the known (or knowns) in terms of both band positions and intensities. As with structural determinations, the correspondence of the strongest bands is first checked, then the moderately strong ones, and finally the weakest bands. It is advantageous if the different spectra are obtained with the same apparatus or at least with instruments having similar resolution. Punched-card methods of comparison and indexing are very useful for systematic analyses.

3-4C Quantitative Analysis

Several difficulties commonly occur in quantitative infrared analysis. First, deviations from Beer's law are common, since absorption peaks are usually sharp and slit widths must generally be relatively wide because of the small energy available in the infrared. Further, stray radiation or scattered radiation is often present. Mull methods of sampling are usually unsatisfactory for quantitative analyses because of both scattering and sample-preparation problems. However, KBr-pellet techniques are sometimes sufficiently accurate to be of value. The problem of Beer's law deviations can often be partially overcome by using calibration curves (see Sec. 1-5*B*).

A second difficulty, inherent in infrared instruments using the optical-null principle, is the uncertainty in setting the 0 percent transmittance reading (see Sec. 3-2*C*). Fortunately, modern instruments are sufficiently sensitive to keep the error in this setting small (usually about 1 percent or less). Furthermore, the effect of this uncertainty can be minimized by limiting analyses to bands which do not absorb too strongly. The error in absorbance ΔA due to an error δ in measuring the position of the 0 percent line is given by

$$\Delta A = \log \frac{I_0 + \delta}{I + \delta} - \log \frac{I_0}{I}$$

$$= \log \frac{1 + \delta/I_0}{1 + \delta/I} \tag{3-12}$$

The error in position δ will be small compared with I_0, and if I is not too small, δ/I will also be small compared with unity.

Example 3-11 The uncertainty in setting the 0 percent T line for a certain instrument is 0.5 percent. If a sample has a transmittance of 37 percent at the analytical wavelength, estimate the percentage error in the concentration of the sample due to the zero-setting error.

Answer

$$\Delta A = \log \frac{1 - 0.005/1}{1 - 0.005/0.37} = \log \frac{0.995}{0.9865}$$

$$= \log 1.01 = 0.004$$

Since $A = -\log T = abC$,

$$A = -\log 0.37 = 0.432$$

$$\frac{\Delta A}{A} \times 100 = \frac{0.004}{0.432} \times 100 \approx 1\%$$

A third difficulty in infrared-quantitative analysis is uncertainty in determining the position of the 100 percent transmittance level. The 100 percent transmittance setting is made uncertain by the problem of very thin cells used in the infrared and the variability of their path length with use. Whether the same cell is used for both the sample and reference or matched cells are used, there is always some uncertainty in setting the 100 percent transmittance line. Because of this uncertainty, the empirical *base-line method* is preferred for quantitative analysis in the infrared. In the base-line method, I_0 is approximated (and called I_0') by drawing a straight base line between points of minimum absorption (called *end points*) on either side of the absorption maximum. Figure 3-39 illustrates how the base-line method can be used to calculate the empirical transmittance ratio for chart paper calibrated in linear transmittance units and in nonlinear absorbance units. In Fig. 3-39*a* it is convenient to use a millimeter ruler to measure the distances I and I_0' followed by a calculation of log (I_0'/I); in Fig. 3-39*b* the absorbance can be read directly from the chart, as shown. Usually, the most difficult part of the base-line method is trying to determine the end points. The only guideline that can be given is to *try to draw the base line as nearly as possible where the pen tracing would go if the band were not present*. This may be difficult to decide if interfering bands are nearby, but as long as the same end points are used consistently, any systematic error will tend to be compensated for by the calibration curve. The calibration curve consists of a plot of log (I_0'/I) vs. concentration.

The base-line method assumes that the background is linear over the width of the band, and to the extent that this assumption is unjustified the technique will be in error. Analysis bands lying on the side of a band due to another substance are particularly likely to give nonlinear background, and thus, where possible, it is preferable to choose an analysis band having a

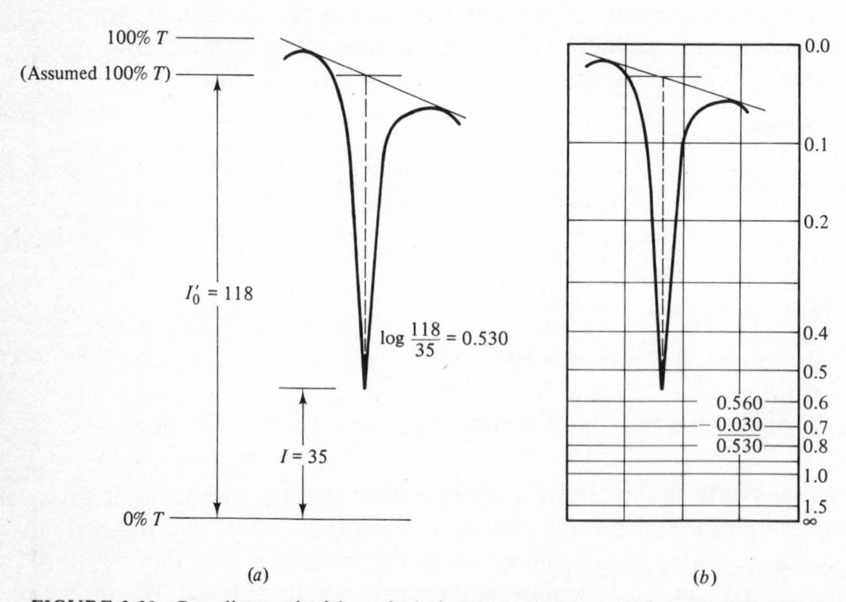

FIGURE 3-39 Base-line method for calculating transmittance ratio in quantitative analysis: (*a*) chart paper calibrated in linear transmittance units, measurement made using a millimeter scale; (*b*) chart paper calibrated in nonlinear absorbance units, absorbance read directly from chart. [*After L. A. Smith in I. M. Kolthoff and P. J. Elving (eds.), "Treatise on Analytical Chemistry," pt. I, vol.* 6, *p.* 3689, *John Wiley & Sons, Inc.*, 1965, *by permission.*]

nearly horizontal base line. Similarly, base lines falling at low absorbance values are to be preferred over those at higher absorbance values.

The base-line method is extensively used in quantitative infrared analyses and has several advantages over some of the conventional approaches used in the ultraviolet and visible regions.

1 The same cell is used for all determinations, and thus absorption due to the cell and other constituents tend to be compensated for.

2 All measurements are made at points on the spectrum which are sharply defined by the spectrum itself, and thus there is no critical dependence on the wavelength settings of the instrument.

3 Use of this type of transmission ratio eliminates changes in instrument sensitivity, source intensity, or changes in the optical system, and thus calibration curves remain valid for long periods of time, subject to routine verification of one or two points. With proper care, the base-line method can yield accuracies of ± 1 percent.

3-4D Detection of Impurities

In general, the presence of impurities tends to cause reduced sharpness of individual bands, a loss of resolution in the spectrum, and the appearance of extra bands. High concentrations may be necessary to see the extra impurity bands clearly. If the pure compound is available, the *difference method* can be very useful in revealing the actual spectrum of the impurity. In the difference method, the pure compound is placed in the reference beam, and the same thickness of impure sample is placed in the sample beam.

3-4E Control of Reactions and Separations

Infrared spectrophotometry can be used to follow the progress of a reaction. Often examination of a small portion of the spectrum is sufficient to indicate whether the desired product is formed and when the reaction has gone to completion. As an example, the Clemmensen reduction of a ketone ($C{=}O \rightarrow CH_2$) is accompanied by the disappearance of the carbonyl peak. The separation of a mixture can be followed similarly.

3-4F Association Studies

Infrared spectrophotometry is a powerful and widely used tool for studying hydrogen bonding. Hydrogen bonding through an O—H or N—H group always alters the characteristic vibrational frequency of that group. This occurs whether the bonding is intramolecular (as in salicylaldehyde) or between molecules (in a crystal or in concentrated solutions). In general, the stronger the hydrogen bonding, the greater the lowering of the fundamental N—H or O—H vibration frequency (bathochromic shift). An infrared spectrum is a useful adjunct to the study of crystalline compounds by x-ray diffraction, since x-ray does not easily yield the positions of the hydrogen atoms (see Chap. 12). The infrared data on crystals will often make it possible to decide between what groups hydrogen bonding occurs. Similarly, with solutions, it is often possible to tell to what groups the solvent is hydrogen-bonded. Intermolecular and intramolecular hydrogen bonding can be differentiated by obtaining a series of spectra at different dilutions. Intermolecular hydrogen bonding decreases as molecules are separated by dilution, whereas intramolecular hydrogen bonding shows no comparable dilution effect. An example of the effect of dilution on intermolecular hydrogen bonding was given in Sec. 3-1*E*.

3-4G Other Applications

Other useful applications of infrared spectrophotometry include kinetic studies, determining the structure of crystals using polarized radiation, the study of biological activity, and fundamental studies of molecules, e.g., determining molecular geometry and bonding, especially for small molecules in the gaseous state. Further details may be found in Refs. 1 and 9.

3-5 NEAR-INFRARED SPECTROPHOTOMETRY

The near infrared is the region between the visible and the fundamental infrared, comprising a wavelength range of about 0.75 to 3.0 μm. Below 1.0 μm a few electronic transitions can occur, and above 2.7 μm a few fundamental stretching vibrations appear, but between 1.0 and 2.7 μm the absorption bands are, almost without exception, due to overtones or combinations of hydrogen-stretching vibrations. Since the most important and most intense near-infrared bands are found in the region of 1.0 to 3.0 μm, this region will be emphasized here.

3-5A Spectral Correlations in the Near Infrared

Near-infrared spectra are not so uniquely characteristic of a molecule as the spectra obtained in the fundamental fingerprint region of the infrared. Consequently, the near-infrared region lacks the general usefulness for qualitative analysis of the fundamental-infrared region. Nevertheless, for functional groups containing unique hydrogen atoms, the near-infrared region is extremely valuable for identifying and then quantitatively determining those unique hydrogen atoms.

Figure 3-40 summarizes the functional-group correlations available in the near-infrared region. Since intensity data are relatively reliable and easy to obtain in the near-infrared region (compared with the fundamental-infrared region), molar-absorptivity data are a useful secondary source of information, for both qualitative and quantitative analysis. In Fig. 3-40, the molar absorptivities that are known are listed beside the absorption-band correlations. From the spectra-structure correlations given in Fig. 3-40 it is usually possible to predict reliably whether the near-infrared region is appropriate for the particular problem.

The near-infrared spectra of primary, secondary, and tertiary amines are quite different and constitute one of the clearest ways of distinguishing between these three types in dilute solution. As can be seen from Fig. 3-40, primary amines have two bands between 2.85 and 3.05 μm, due to fundamental N—H stretching vibrations, and a corresponding doublet in the first overtone region between 1.45 and 1.55 μm. In addition, combination modes give one or two peaks in the region around 2.0 μm and a single peak near 1.0 μm. In contrast, secondary amines have no peaks in the 2.0-μm region and only singlet bands in the other regions. Tertiary amines, since they have no N—H constituents, have none of the N—H absorption bands just mentioned. Of course, the near-infrared spectrum of any amine will also contain bands due to C—H groups or other functional groups in the molecule.

Example 3-12 A certain compound known to be an amine gives the near-infrared spectum shown in Fig. 3-41. Is the compound a primary, secondary, or tertiary amine?

Answer The relatively strong peak in the 2.0-μm region indicates that the compound is a primary amine. Further, the very strong absorption in the 2.8-μm region suggests an aromatic amine. The singlet at 1.0 μm and the doublet (barely resolved) at 1.45 and 1.5 μm likewise substantiate a primary amine (note that the 1.45-μm peak is much less intense than the 1.5-μm peak, consistent with an aromatic primary amine; see Fig. 3-40). The unknown compound is aniline.

3-5B Near-Infrared Instrumentation

SOURCES

The source is no problem in the near infrared, since an ordinary tungsten lamp (blackbody radiator) radiates a maximum amount of energy in this region (see Fig. 3-11). With an abundance of energy at the proper wavelength, stray radiation is a relatively minor problem in the near infrared.

MONOCHROMATORS

Near-infrared monochromators are unique only in the optical materials used. It is possible to modify many conventional infrared spectrophotometers for work in the near-infrared region simply by replacing the conventional

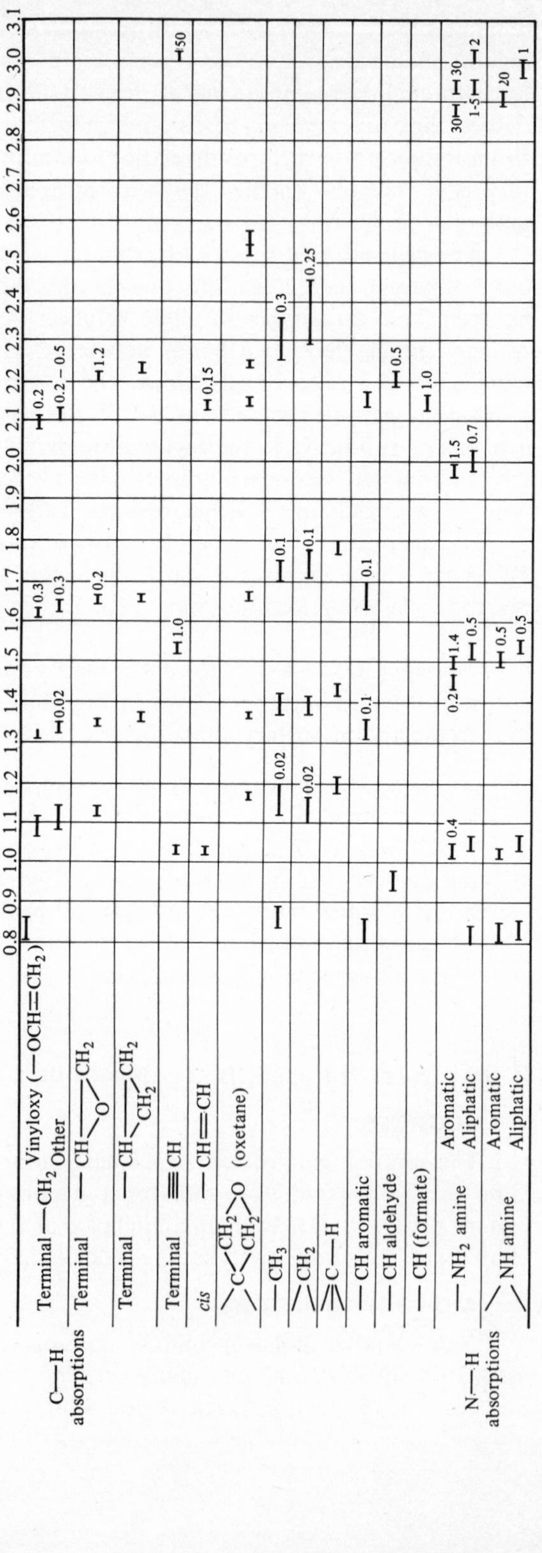

μm

0.8 0.9 1.0 1.1 1.2 1.3 1.4 1.5 1.6 1.7 1.8 1.9 2.0 2.1 2.2 2.3 2.4 2.5 2.6 2.7 2.8 2.9 3.0 3.1

C—H absorptions

Terminal —CH_2 Vinyloxy (—$OCH{=}CH_2$) 0.3 0.2
Other 0.02 0.3 0.2 – 0.5
Terminal —CH$\langle$O$\rangle$$CH_2$ 0.2 1.2
Terminal —CH$\langle CH_2 \rangle$$CH_2$
Terminal ≡CH 1.0 50
cis —CH=CH 0.15
$>C<(CH_2)_2>O$ (oxetane)
—CH_3 0.02 0.1 0.3
$>CH_2$ 0.02 0.1 0.25
$\geq$C—H
—CH aromatic 0.1 0.1
—CH aldehyde 0.5
—CH (formate) 1.0

N—H absorptions

—NH_2 amine Aromatic 0.4 0.2 1.4 1.5 30 30
Aliphatic 0.5 0.7 1-5 2
$>$NH amine Aromatic 0.5 20
Aliphatic 0.5 1

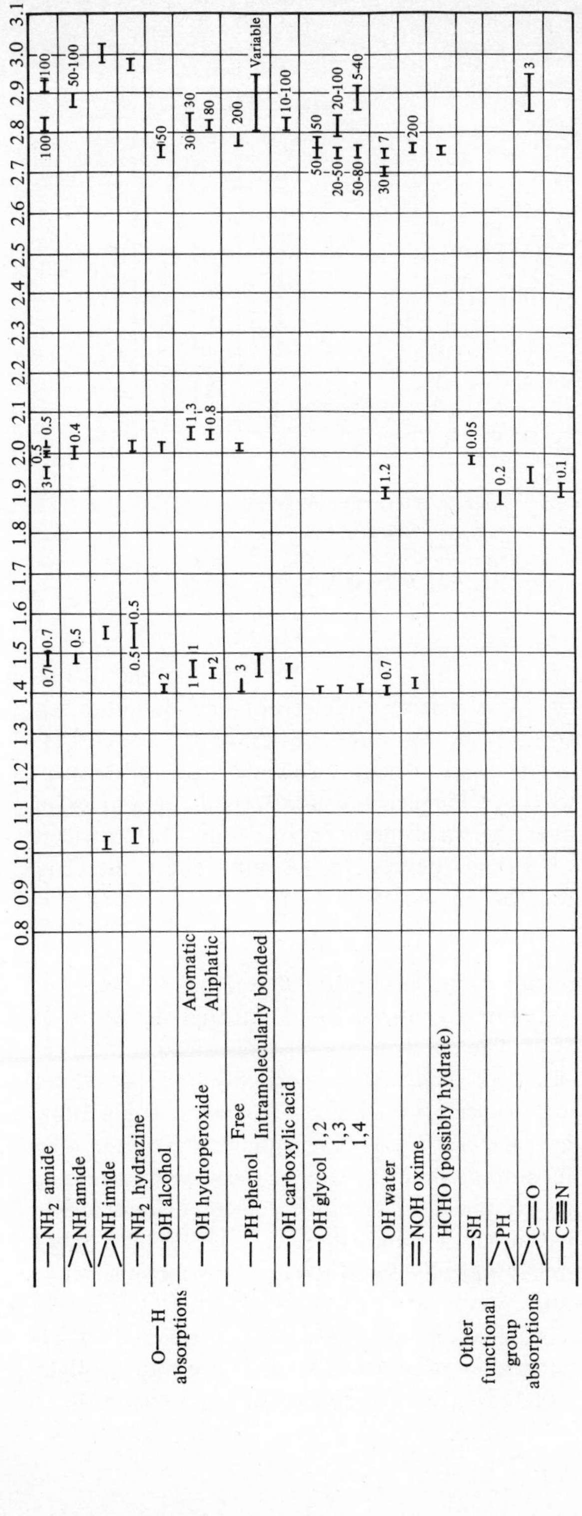

μm

0.8 0.9 1.0 1.1 1.2 1.3 1.4 1.5 1.6 1.7 1.8 1.9 2.0 2.1 2.2 2.3 2.4 2.5 2.6 2.7 2.8 2.9 3.0 3.1

—NH_2 amide — 0.7, 0.7; 3, 0.5, 0.5; 100, 100

>NH amide — 0.5; 0.4; 50-100

>NH imide

—NH_2 hydrazine — 0.5, 0.5

O—H absorptions

—OH alcohol — 2; 50

—OH hydroperoxide, Aromatic — 1, 1; 1,3; 30, 30

—OH hydroperoxide, Aliphatic — 2; 0.8; 80

—PH phenol, Free — 3; 200

—PH phenol, Intramolecularly bonded — Variable

—OH carboxylic acid — 10-100

—OH glycol 1,2 — 50, 50

—OH glycol 1,3 — 20-50, 20-100

—OH glycol 1,4 — 50-80, 5-40

OH water — 0.7; 1.2; 30, 7

=NOH oxime — 200

HCHO (possibly hydrate)

Other functional group absorptions

—SH — 0.05

>PH — 0.2

>C=O — 3

—C≡N — 0.1

FIGURE 3-40 Summary of functional-group correlations in the near-infrared region (numbers alongside band locations are molar absorptivities). (*From R. F. Goddu, "Advances in Analytical Chemistry and Instrumentation,"* vol. I, pp. 360–361, *John Wiley & Sons, Inc., New York,* 1960, *by permission.*)

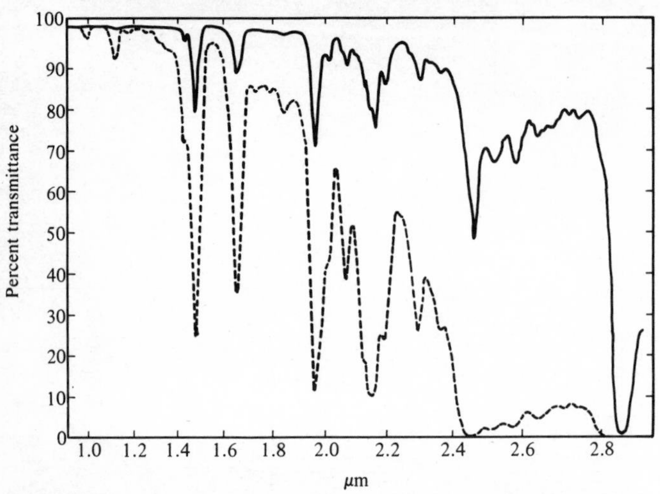

FIGURE 3-41 Near-infrared absorption spectrum of an amine in carbon tetrachloride. Solid line, 1% solution in a 1-cm cell; dotted line, 10% solution in a 1-cm cell. (*From R. F. Goddu, "Advances in Analytical Chemistry and Instrumentation," vol. I, p.* 358, *John Wiley & Sons, Inc., New York,* 1960, *by permission.*)

prism with a prism made from fused silica, quartz, lithium fluoride, or calcium fluoride and adding a more sensitive detector. Few laboratories, however, can spare the instrument or the time necessary to make these conversions, and the common trend is to design ultraviolet and visible spectrophotometers specifically to cover the near-infrared region. Examples of precision instruments that cover the wavelength region from the ultraviolet through the near infrared are Beckman models DK-1A and DK-2A, the Cary model 14, and Perkin-Elmer models 323 and 450.

DETECTORS

The most widely used detector in the near-infrared region is the lead sulfide detector, which is far superior to conventional infrared detectors in sensitivity and speed of response. The spectral response of a lead sulfide detector is shown in Fig. 3-42. This detector is normally used at room temperature (293 K), but, as can be seen from Fig. 3-42, cooling would allow it to be used at slightly longer wavelengths. At room temperature, it can be used out to about 3.2 μm, with cutoff at about 3.6 μm. In the wavelength range of about 1 to 2.5 μm, it is by far the most sensitive detector commercially available, having a sensitivity about a hundredfold greater than a good thermocouple. The response time is typically about 100 μs, again far superior to all conventional infrared detectors.

The lead sulfide detector is an example of a *photoconductive cell,* which means that its resistance changes in proportion to the intensity of light striking it. Specifically, lead sulfide is a semiconductor, having a resistance of

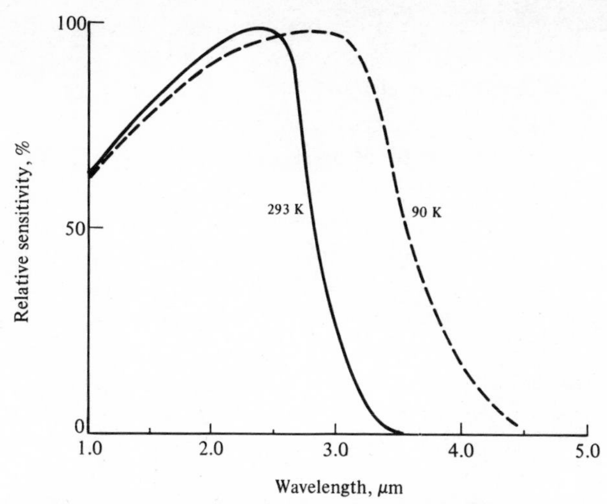

FIGURE 3-42 Spectral response of PbS photoconductive cell. [*After Wilbur Kaye, Spectrochim. Acta,* 7:181 (1955), *by permission from Pergamon Press.*]

about 0.1 to 10 MΩ in the dark and a linearly decreasing resistance with an increase in luminous flux. Like other semiconductors, it has electronic energy levels, or *bands*, within the solid crystal, and the electrons in the lower level are bound to individual atoms, giving the material a normally high resistance. If an energy sufficient to elevate electrons from the ground state to a higher-energy conduction band is supplied, the conductivity of the crystal will increase, and in proportion to the number of electrons elevated. The threshold energy of incoming photons corresponds to the energy spacing between bands. With lead sulfide at room temperature, it is about 3.6 μm.

To use the lead sulfide detector, a small voltage is imposed across the cell, allowing only an extremely small current to flow when the cell is not under illumination. Radiant energy striking the cell causes a very large drop in resistance, with a correspondingly large rise in current through the cell. The current is amplified and the resultant signal fed to a meter or recorder.

3-5C Near-Infrared Techniques

The techniques used in the near-infrared region are much less troublesome than those required in the normal-infrared region and are quite similar to those used in the ultraviolet and visible regions. *Sample cells* are generally a special grade of silica or quartz which transmit almost 90 percent of the incident light throughout the near infrared (and in fact from about 300 nm through 3.5 μm). Ordinary quartz or silica cells are transparent except for moderate-intensity absorption bands in the 2.8- to 3.1-μm region, and glass or Corex may be used out to about 2.4 μm. Cell paths are generally between 0.1

and 10 cm in length, and there is no problem in obtaining matched cells with such dimensions.

In choosing a *solvent*, it is generally best to avoid those containing O—H or N—H and even C—H groups if possible. Table 3-8 summarizes the transmittance characteristics of some near-infrared solvents. Carbon tetrachloride is the ideal near-infrared solvent, since it is transparent throughout this region. Carbon disulfide is also a good solvent, but it has a weak absorption band at 2.22 μm which may interfere in a 10-cm cell although it is completely transparent in a 1-cm cell. Methylene chloride is a good solvent for the 2.7- to 2.9-μm region when samples do not dissolve in carbon tetrachloride or carbon disulfide.

The *slit-width* adjustment is particularly important in near-infrared spectrophotometry, since absorption bands tend to be very sharp, usually having half-height bandwidths of 0.01 μm or less. Thus, for proper resolution, it is

TABLE 3-8 Transmittance Characteristics of Some Near-Infrared Solvents†

Solid Lines Indicate Usable Regions; Numbers Indicate Maximum Desirable Lengths in Path Centimeters

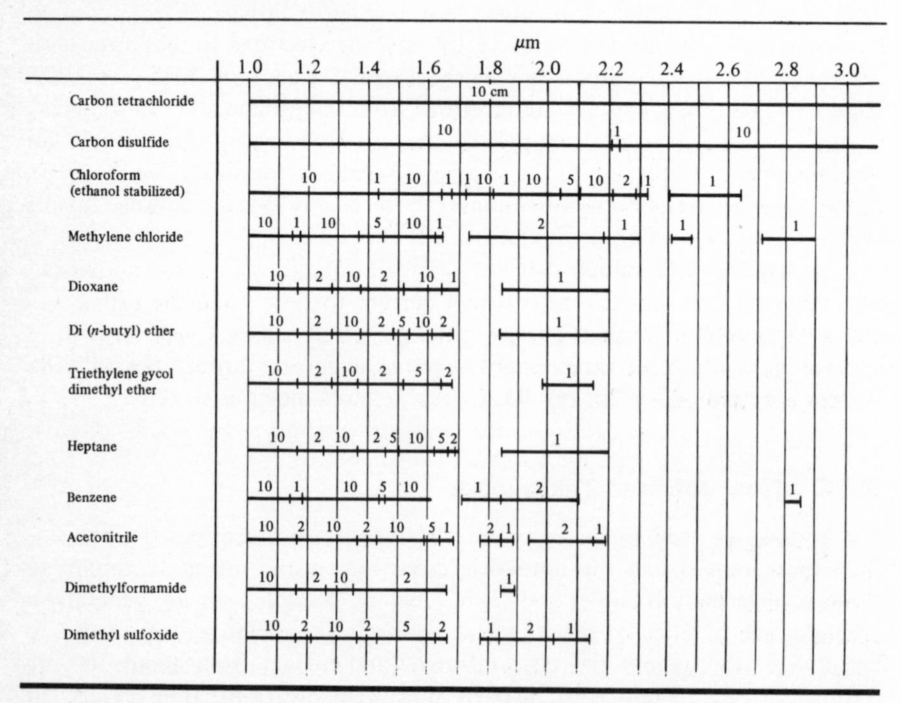

† From R. F. Goddu, p. 354 in C. N. Reilley (ed.), "Advances in Analytical Chemistry and Instrumentation," John Wiley & Sons, Inc., New York, 1960, by permission.

important to use as small a spectral slit as possible without introducing excessive instrument noise. Furthermore, especially for quantitative work, care must be taken to have a known and reproducible slit width.

Finally, it is strongly recommended that air-conditioned rooms be provided for near-infrared instruments, from the standpoint of both temperature and humidity control. The sensitivity of lead sulfide detectors is significantly temperature-dependent, being more sensitive at lower temperatures. At high humidity the water vapor in the light path will extract appreciable energy by absorption.

3-5D Near-Infrared Applications

The largest single use for near-infrared spectrophotometry, to date, has been for quantitative organic functional-group analyses. However, it has also found useful applications in qualitative analyses, hydrogen-bonding studies (and other solute-solvent interaction studies), process control (particularly for processes involving vinylation or epoxidation), and inorganic compound studies (with particular promise for the study of organometallic complexes).

Since quantitative organic functional-group analysis is the most important application, to date, of near-infrared spectrophotometry, a brief list of the more important functional groups that have been studied will be given.

C–H GROUPS

Terminal methylene and methyne groups give unique near-infrared spectra, and some compounds in which they have been studied with good results include polymers, olefins, epoxides, and cyclopropanes. As examples of the analytical sensitivities, 0.1 percent benzene in hydrocarbons and 0.04 percent olefin in hydrocarbons can be detected.

N–H GROUPS

In the quantitative determination of N—H groups, two regions are particularly important, the fundamental stretching region, at 2.8 to 3.0 μm and the first-overtone region, near 1.5 μm. The second-overtone region, near 1.0 μm, is also useful at times, though much less intense. Compounds in which the N—H group has profitably been studied include amines, amides, hydrazines, and miscellaneous compounds such as peptides, alkaloids, etc. As little as 0.04 percent amine in hydrocarbons can be detected.

O–H GROUPS

Compounds in which the fundamental O—H stretching band in the region of 2.7 to 3.0 μm has been studied include alcohols, phenols, hydroperoxides, carboxylic acids, oximes, and water. A few examples of the analytical sensitivities include the detection of 0.03 percent alcohol in hydrocarbons, 0.2 percent alcohol in acids, 0.005 percent acid in hydrocarbons, 0.5 percent acid in anhydrides, 0.004 percent water in hydrocarbons, 0.05 percent water in alcohols, and 0.04 percent water in carboxylic acids.

CARBONYL GROUPS

The first overtone of carbonyl groups appears at 2.8 to 3.0 μm. Compounds studied include esters, ketones, and acids.

OTHER GROUPS

Groups studied with some promise include S—H, P—H, nitriles, and various functional groups in organophosphorus compounds.

3-6 FAR-INFRARED SPECTROPHOTOMETRY

The far infrared may be considered to be the region from 25 μm to 1000 μm (1 mm), although, as with other spectral regions, the boundaries are not sharply defined. The region is defined by the boundaries of the adjacent spectral regions, i.e., the fundamental-infrared region on the short-wavelength end and the microwave region on the long-wavelength end.

The usefulness and current status of the far-infrared region have been dictated by the availability of commercial instruments. For example, it has been possible to extend the limits of prism spectrophotometers out to about 50 μm, and thus the region from 25 to 50 μm has come to be used almost routinely. With the development of inexpensive replica gratings of good quality, it has been possible to extend the range of conventional spectrophotometers out to about 300 μm. To go beyond 300 μm, special interferometric spectrophotometers are necessary, and instruments of this type have recently become commercially available. Thus, the far-infrared region is on the verge of full-scale exploitation for analytical purposes.

3-6A Information Obtainable in the Far Infrared

As might be expected, the far-infrared region furnishes information which is supplementary to that obtained in the fundamental-infrared region. For example, this region is useful for distinguishing between very similar compounds, such as isomers, whose spectra may appear qualitatively alike in the fundamental-infrared region. But its greatest value is probably in the special information it provides. For example, the lattice vibrations of crystalline materials generally fall in the far infrared. Similarly, the transition from a valence band to a conduction band of electrons in semiconductors very often requires energy in the far-infrared region. Of particular interest to chemists is the fact that fundamental-vibrational frequencies of many organometallic and inorganic substances fall in the far infrared, due to the heaviness of the atoms and the weakness of the bonds [see Eq. (3-4)].

3-6B Far-Infrared Instrumentation

There are three principal problems inherent in the experimental study of the far-infrared region. (1) There are very few sources that can be used, and only a small amount of energy is available in the far infrared. An associated problem is the large amount of interfering shortwave radiation emitted by all available sources. This must be removed by techniques which are almost unique to the far-infrared region. (2) There is a scarcity of suitable optical (window) materials for use in this region. (3) The widespread absorption due to the rotational bands of water vapor in the far infrared makes it desirable to remove all water vapor when working out to 50 μm and *essential* to do so *beyond* 50 μm. This can be done by flushing with dried air or nitrogen or by evacuation.

SOURCES

In the region out to 50 μm, the Nernst glower or Globar can be used. Beyond 50 μm, a medium- or high-pressure (1 to 100 atm) mercury arc confined in a quartz envelope is commonly used. High pressure has the advantage of giving a larger proportion of the emission at longer wavelengths, and the quartz envelope blocks radiation below about 50 μm. The mercury-arc lamp is an example of a gaseous discharge tube; mercury vapor, at high pressure, is excited by a high-voltage discharge between two oxide-coated electrodes, and the temperature rises to several thousand degrees; the excited mercury atoms emit radiant energy as their electrons fall back to the ground state. Even at high pressures, however, the preponderance of the emitted radiation lies below 50 μm, and a relatively low-intensity source must suffice.

WINDOW MATERIALS

Cesium iodide, the longest-wavelength-transmitting alkali halide, can be used in thin windows to about 50 μm. Crystal quartz is opaque from about 4.5 to 45 μm, but it transmits appreciably beyond 45 μm; thus thin (say 2-mm) quartz windows are useful beyond 50 μm while simultaneously filtering out shorter wavelengths. Most types of polyethylene are essentially free of bands beyond 25 μm, with occasional weak absorptions along about 50 and 150 μm. Diamond is the only known inorganic material transparent throughout the fundamental and far infrared, and it is very useful for detector windows as long as the windows do not exceed a diameter of about ¼ in.

MONOCHROMATORS

Out to 50 μm, the monochromator systems in use are much like the conventional systems used in the fundamental-infrared region, discussed in Sec. 3-2*A*. Beyond 50 μm, gratings can be used, at least out to about 300 μm, but a foreprism monochromator cannot be used to remove higher-order spectra, since no suitable prism material is known for this region. Instead, a series of transmission filters or reflection filters is used to reduce stray radiation and sort out overlapping grating orders. The Beckman IR 11 and Perkin-Elmer model 180 spectrophotometers are examples of commercially available instruments which use a series of gratings and transmission filters. The IR 11 covers the wavelength range of 12.5 to 300 μm, and incorporates a pneumatic air dryer to eliminate the water-vapor problem. The Perkin-Elmer model 180, discussed in Sec. 3-2*C* (see Figs. 3-20 and 3-21), covers the wavelength range of 2.5 to 320 μm and uses a dry-air or dry-nitrogen purge to eliminate water vapor.

INTERFEROMETRIC (FOURIER) SPECTROPHOTOMETERS

In 1927 Michelson pointed out the possible usefulness of interferometry for spectroscopic studies, but only recently has interferometry been applied to spectroscopy in the far-infrared region. The principles can best be described by using the Beckman RIIC model FS-720 Fourier spectrophotometer, diagramed in Fig. 3-43.

In Fig. 3-43, radiation from the water-cooled mercury lamp (*a*) is modulated with a chopper (*b*) rotating at 15 Hz and directed to a taut Mylar beam splitter (*c*). At this point, part of the radiation is reflected by the beam splitter toward the movable mirror (*d*), and part is transmitted toward the stationary mirror (*e*). The rays directed toward each mirror are reflected back to the beam splitter (*c*) and recombined. (At least half the radiation from each mir-

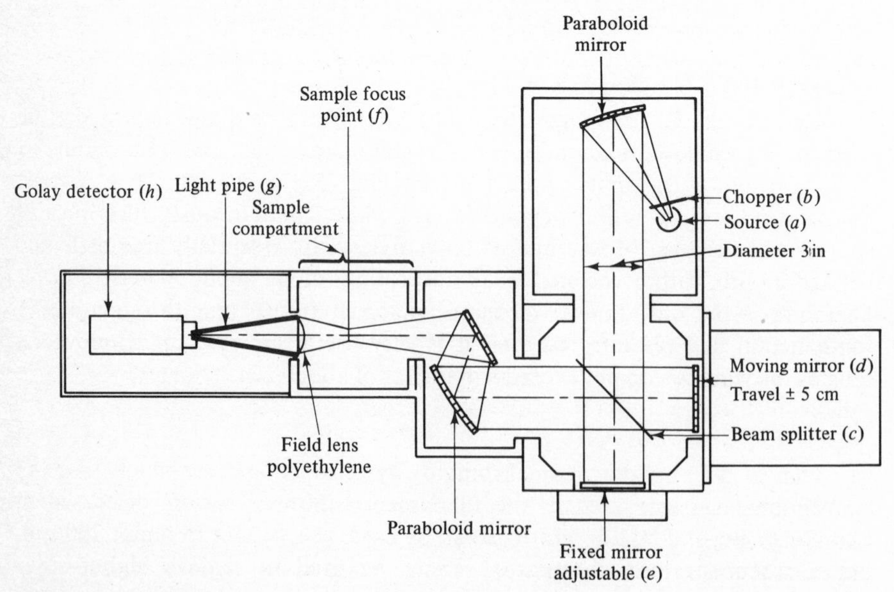

FIGURE 3-43 Schematic optical diatram of Beckman RIIC model FS-720 Fourier spectrophotometer. (*Beckman Instruments, Inc.*)

ror is permanently lost upon its return to the beam splitter, since at least half is directed back toward the source.) The recombined beam is then focused on the sample (*f*) in the sample compartment. The transmitted radiation is collected by a conical light pipe (*g*) and is funneled into the Golay detector (*h*).

The instrument contains no monochromator, and at all times the detector receives radiation of all optical frequencies radiated by the source and not absorbed by the sample or removed by optical filtering. This provides at least a theoretical advantage for interference spectrophotometry, since at any given time the detector is receiving information about all wavelengths simultaneously, and not just the narrow band of wavelengths received by the detector in a conventional spectrophotometer. This results in large signal-to-noise ratios.

The information reaching the detector is modified by interference effects taking place at the Mylar beam splitter. By moving the movable mirror at a known rate toward or away from the beam splitter, an *interferogram* can be plotted of radiation reaching the detector vs. the distance moved by the movable mirror. Though the interferogram does not by itself give explicit information about the spectrum from which it was produced, it *contains* the information as a Fourier transform, and a computer can be used to make the transform and give the conventional transmittance-vs.-wavelength (or wave number) spectrum.

To understand how this interference principle works, consider the effect of mirror position on wave interference shown in Fig. 3-44. To simplify the initial discussion, assume that the source emits a single monochromatic wavelength of radiation λ_1. At the beam splitter, radiation from the source will be partially reflected (ray 1 in Fig. 3-44*a*) and partially transmitted (ray 2 in Fig. 3-44*a*). Since in Fig. 3-44*a* the movable mirror and the stationary mirror are both at the same distance *d* from the beam splitter, radiation reflected by the two mirrors will arrive back at the beam splitter exactly in phase. Therefore, *constructive* interference will occur (see Sec. 1-4*D*).

If the movable mirror is now moved one-fourth wavelength further away from the beam splitter, as shown in Fig. 3-44*b*, ray 1 will arrive back at the beam splitter exactly out of phase with ray 2; that is, there will be one-half wavelength difference in their optical paths and *destructive* interference will occur. When the mirror is displaced exactly one-half wavelength by further movement of the movable mirror, *constructive* interference will occur, and so on.

If the radiation level of λ_1 falling on the detector is recorded as a function of *X*, the distance moved by the movable mirror, an interferogram of the monochromatic source is obtained, as shown in Fig. 3-45*a*. Note that destructive interference has occurred at points where the waveform crosses the median-energy line (dotted line) and the crests and troughs of the wave are points of constructive interference.

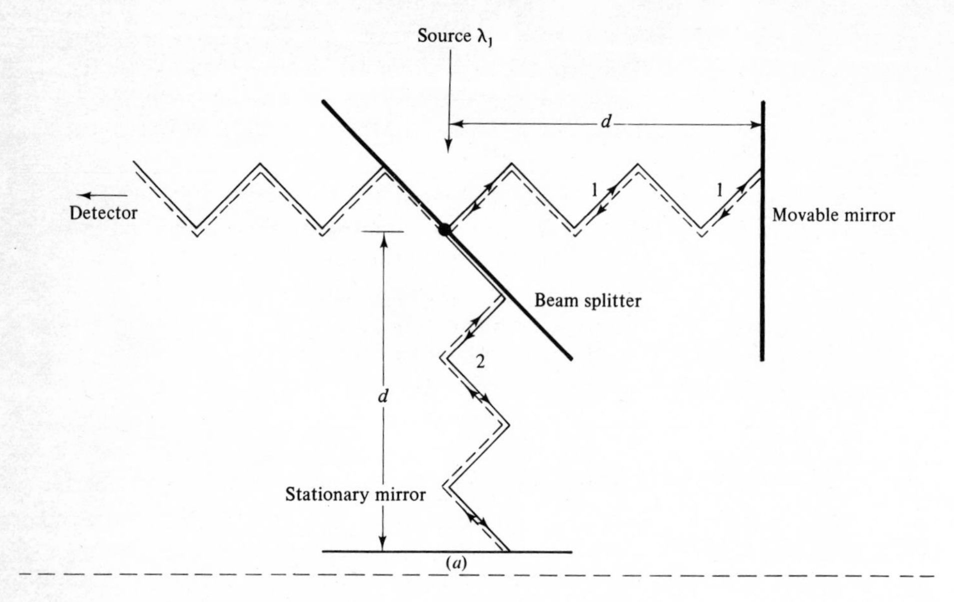

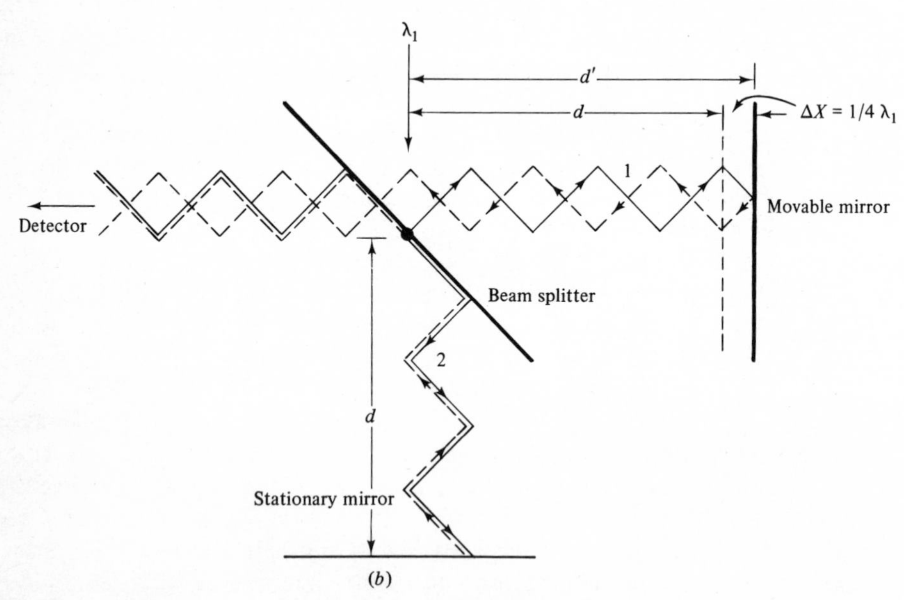

FIGURE 3-44 Effect of mirror position on wave interference: (*a*) constructive interference; (*b*) destructive interference. (*Beckman Instruments, Inc.*)

The waveform for monochromatic radiation of wavelength λ_2 is shown in Fig. 3-45*b*. If the source emitted both wavelengths λ_1 and λ_2, the resulting interferogram would be the *sum* of the two waves. If the source emitted polychromatic radiation consisting of many wavelengths $\Sigma\lambda_n$, the resulting interferogram would be the composite sum of all the individual waveforms and would have the general shape shown in Fig. 3-45*c*. The damped appearance of the waveform in Fig. 3-45*c* is fairly typical of all interferograms and should be understood. Interferograms will always show maximum energy at the point when the movable mirror and the stationary mirror are both at the same distance d from the beam splitter (the point $\Delta X = 0$ in Fig. 3-45*a* and *b*) since *all* wavelengths *constructively* interfere at this point. At other distances, the various wavelengths constructively or destructively interfere to various degrees.

In the operation of an interferometer, for mathematical reasons, two-sided interferograms are taken; i.e., the mirror is moved from some short distance (negative value of X), through zero path difference ($X = 0$), to some greater distance (positive value of X) to produce a symmetrical interferogram.

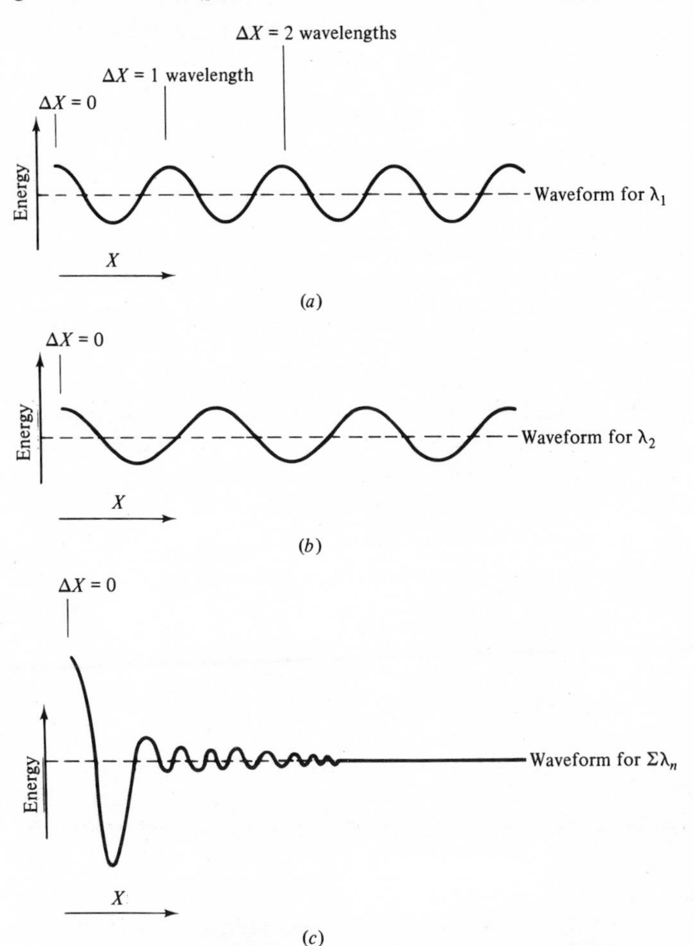

FIGURE 3-45 Interferograms of various wavelength sources: (*a*) monochromatic source of λ_1; (*b*) monochromatic source of λ_2; (*c*) polychromatic source consisting of many wavelengths λ_n. (*Beckman Instruments, Inc.*)

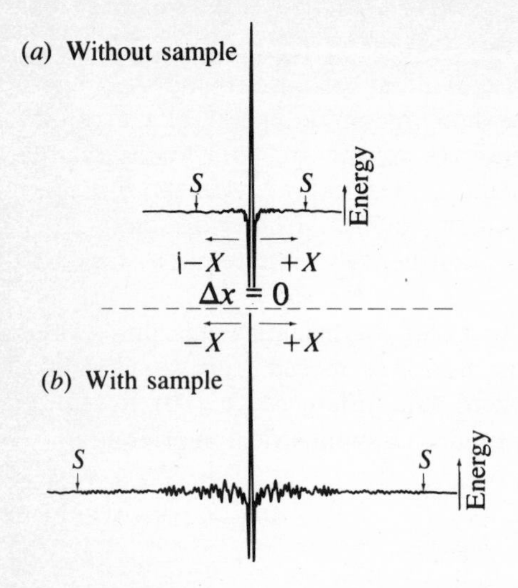

FIGURE 3-46 Typical two-sided interferograms (obtained with Beckman RIIC model FS-720). (*Beckman Instruments, Inc.*)

A typical two-sided interferogram from the Beckman RIIC model FS-720 is shown in Fig. 3-46. Because of the symmetry, the information contained in the right side of the interferogram (for positive displacements of the movable mirror) duplicates that contained in the left side (for negative values of X). In other words, the left- and right-hand portions are mirror images. (Points labeled S on the interferograms are points beyond which is it meaningless to move the mirror since modulation stops.)

So far nothing has been said about a sample; i.e., the interferogram shown in Fig. 3-46*a* was produced from an evacuated FS-720. If a *sample* is inserted, it may absorb *some of the wavelengths* of radiation, thus removing some of the wave *components* from the interferogram. Consequently, the shape of the resulting interferogram may be changed, as shown in Fig. 3-46*b*.

Though the interferogram obtained does not have the same appearance as the spectrum produced from a conventional spectrophotometer, the interferogram does *contain* the same kind of information. To obtain a meaningful spectrum of energy vs. wavelength or wave number, the interferogram must be analyzed to determine which wavelengths (or frequencies) of radiation are contributing to the total and in what amounts. This analysis can be performed with a large *digital computer* or with a smaller *analog computer.* With a digital computer the interferogram must be digitized and recorded on punched-paper tapes, cards, or magnetic tape. These data are then mathematically analyzed (taking the inverse Fourier transform) with the digital computer. Alternatively, the interferogram can be analyzed on the spot with a special-purpose analog computer, such as the Beckman RIIC model FTC-100 transform computer. With this on-line computer, the interferogram is digitized by an analog-to-digital converter (A/D) and the digital readings are stored in the ferrite memory core of the FTC-100. The interferogram is then reconstructed in analog form by a D/A converter and wave-analyzed to record the desired spectrum on an attached special *xy* recorder. Figure 3-47 shows the spectrum of benzimidazole from 200 to 20 cm^{-1} (50 to 500 μm) taken with the model FS-720 Fourier spectrophotometer and analyzed with the model FTC-100 computer. Thus, with an on-line computer, the experimenter can obtain the convenience and often the speed of conventional grating spectrometers. On the other hand, a small on-line computer generally has a limited storage capacity, which has the result of limiting the resolution of the final spectrum. Usually, however, greater resolution is required only for theoretical studies of gaseous samples giving very sharp and closely spaced absorption bands, and in such cases, the interferograms can be analyzed further with a large digital computer.

It is worth briefly comparing the advantages and disadvantages of far-infrared *interferometers* and *grating spectrophotometers.* Interferometers have the advantage of higher energy, especially at longer wavelengths, and go out to longer wavelengths. (The Beckman RIIC model FS-720 covers the wavelength range of 20 to 1000 μm, and the model FS-820 covers from 50 to 1000 μm; on the other hand, the wavelength limit for grating instruments is about 300 μm.) Furthermore, the interferometer is capable of higher resolution, especially when a digital computer is used. Cost of the instruments and convenience of operation are sometimes quite comparable; in fact it is claimed that the model FS-720 is actually simpler to operate than the model IR-11. If very high resolution and digital computation are required, the interferometer method involves time delays and may become quite costly. Furthermore, the interferometer is less foolproof, and instrumental difficulties may not be as apparent as with conventional spectrophotometers. Finally, band intensities

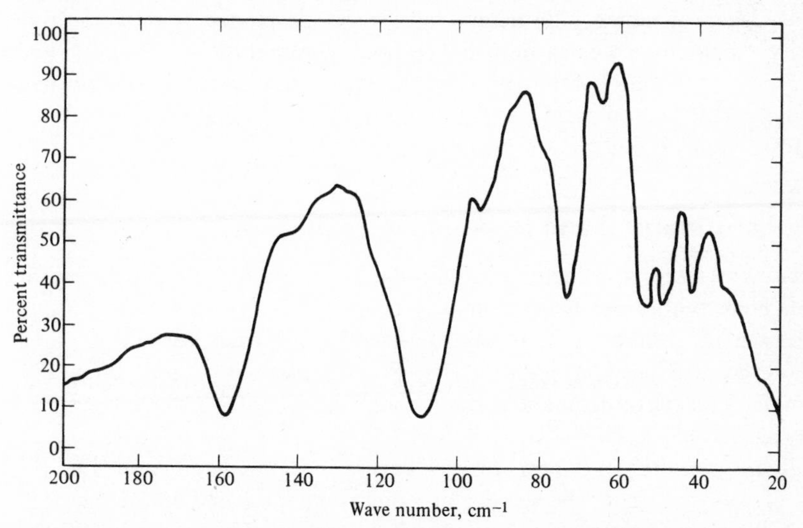

FIGURE 3-47 Far-infrared spectrum of benzimidazole obtained with Beckman RIIC model FS-720 interferometer and the FTC-100 transform computer. (*Beckman Instruments, Inc.*)

and the effect of noise are more uncertain with an interferometer, since the mathematical analysis is so complex. Thus, false bands, shoulders, and distortions may appear in the mathematically transformed spectrum, and one usually cannot tell a priori from a single run what is real information and what is noise. On the other hand, it is possible to improve the reliability of the data greatly by averaging several repeat runs.

DETECTORS

The Golay (pneumatic) detector, described in Sec. 3-2*A*, has been the detector of choice in the far infrared, though recently *pyroelectric crystals* have come into use for wavelengths out to about 300 μm. A good example of a pyroelectric-crystal detector is the triglycerine sulfate (TGS) detector used in the Perkin-Elmer model 180 (see Fig. 3-21). The TGS detector is considered a bolometer but functions differently from a conventional bolometer (see Sec. 3-2*A*). A TGS crystal is mounted between two parallel electrodes, and this unit responds to small temperature changes by changing *capacitance*. Thus, the TGS bolometer generates an output signal which is proportional to the change in capacitive impedance, whereas a conventional bolometer responds by changing *resistive* impedance (see Sec. 3-2*A*). The conventional bolometers described in Sec. 3-2*A* can at best be used only out to about 30 μm. The TGS detector is sufficiently sensitive to be used out to about 315 μm. The optimum response range is from about 40 to 300 μm (250 to 33 cm^{-1}), though it could be used from about 15 to 315 μm (667 to 32 cm^{-1}). The TGS detector used in the model 180 is sealed in a compartment with a polyethylene window and is operated at a temperature of about 38°C. The TGS crystal is hygroscopic, and absorbed water would cause it to lose sensitivity.

A number of other bolometers have been developed with very high sensitivity, including the carbon bolometer, germanium bolometer, and several superconducting bolometers. Unfortunately, however, all these devices require liquid-helium temperatures for effective operation and thus have found only limited application.

3-6C Far-Infrared Sample-Handling Techniques

Solid-state sampling is, in general, preferable to solution sampling in the far infrared. Here scattering, being inversely proportional to wavelength, is not nearly so serious a problem as in the conventional infrared. Very large particles, however, will produce poor results. Pressed plates to 1-mm thickness, Nujol mulls, and polyethylene and hydrocarbon wax matrices have been successfully used.

Solution sampling in the far-infrared region is mainly handicapped by the scarcity of good solvents. The solvent problem is compounded by the fact that relatively long path lengths (of the order of 1 mm) are often required for the more weakly absorbing bands common to the far infrared. A collection of 12 organic liquids for possible use as far-infrared solvents has been published.† Of these, hydrocarbons such as cyclohexane and benzene appear to be the most transparent; other solvents with fairly wide ranges of usefulness include CCl_4, CS_2, $CHCl_3$, and $CHBr_3$.

The most effective and economical *sample cells* for the far-infrared region appear to be those with polyethylene windows, though quartz can be used in the region of 50 to 330 μm. High-density polyethylene, with window thicknesses up to 1.5 mm, is commonly used.

3-6D Far-Infrared Applications

STUDIES OF THE PHYSICAL PROPERTIES OF SOLIDS

A few applications of particular value include the determination of lattice energies of crystalline materials, measurements of dielectric properties (including real and complex refractive indexes), and studies of the optical properties of semiconductors (including measurements of the transition energy of electrons in passing from valence to conduction bands). It should be noted that the influence of magnetic fields on optical properties of materials is of a magnitude to show up in the far infrared.

STUDIES OF INORGANIC AND ORGANOMETALLIC COMPOUNDS

The fundamental vibrational frequencies of many inorganic, organometallic, and even organic molecules containing heavy atoms all fall in the far infrared, due to the heaviness of the atoms and the weakness of the bonds [see Eq. (3-4)]. Thus, for these compounds, the far infrared should be particularly useful in elucidating molecular structure. A few types of compounds that have been fairly extensively studied include organophosphorus, organosulfur, and organometallic compounds. Of the latter, sandwich compounds have been of particular interest, and a fairly complete tabulation of their characteristic frequencies is available [12]. Since the far-infrared absorption frequencies of organometallic compounds are often sensitive to the particular metal ion present, this region should become extremely valuable in the study of coordination compounds. As an example of far-infrared spectra, Fig. 3-48 shows the spectrum of the copper chelate of 1-hydroxy-2-carbethoxycyclooctene taken with a Beckman model IR 11 spectrophotometer, using polyethylene plates to press a Nujol mull of the sample. The peaks marked *p* are due to absorption by the polyethylene plates, and the discontinuities in the frequency axis occur at points where semiautomatic filter and grating changes must be made. Figure 3-49 shows the far-infrared spectrum of $K_2S_2O_5$ taken with a Beckman RIIC model FS-720 Fourier spectrophotometer and an associated FTC-100

† H. R. Wyss, R. D. Werder, and H. H. Gunthard, *Spectrochim. Acta,* **20:** 573 (1964).

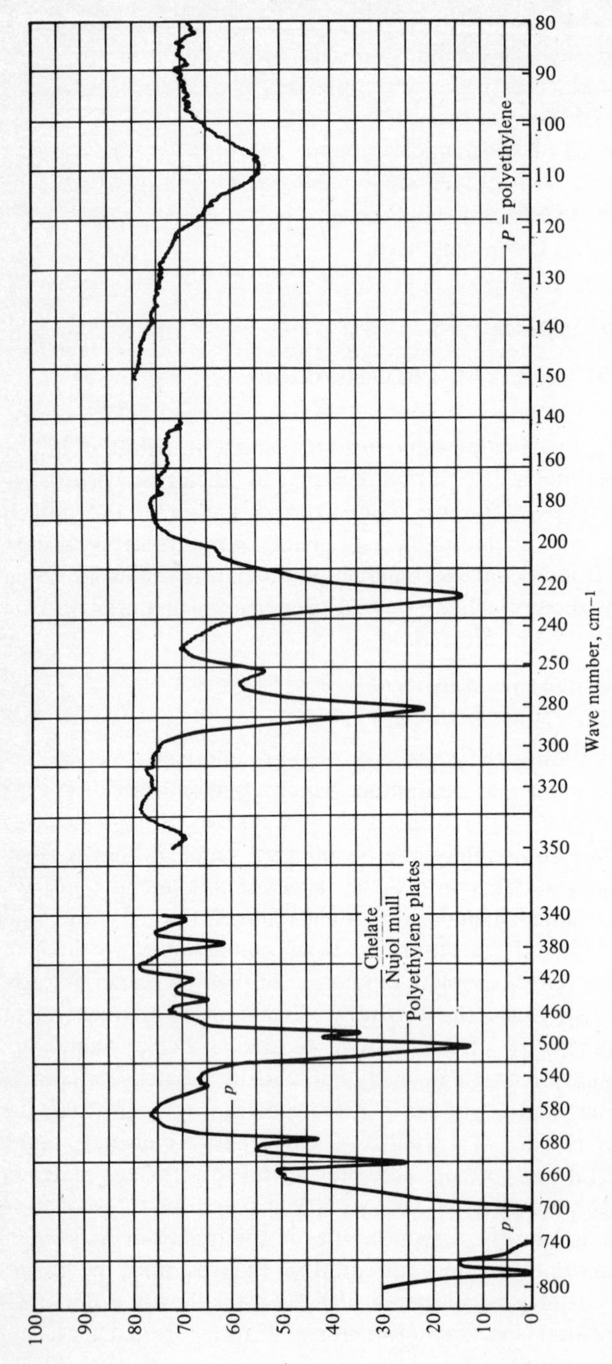

FIGURE 3-48 Far-infrared spectrum of a copper chelate (copper chelate of 1-hydroxy-2-carbethoxycyclooctene), taken with a Beckman model IR-11 spectrophotometer. (*Beckman Instruments, Inc.*)

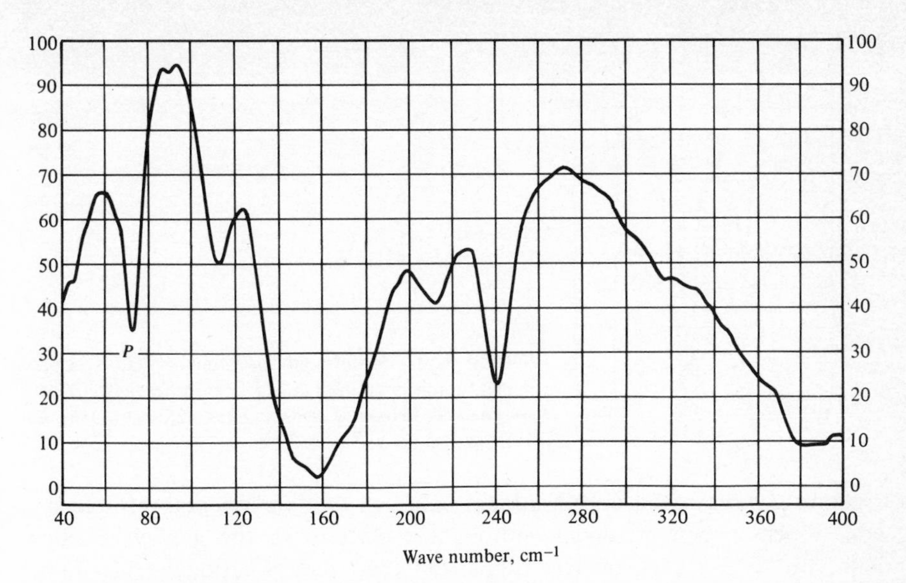

FIGURE 3-49 Far-infrared spectrum of $K_2S_2O_5$ obtained with Beckman RIIC FS-720 interferometer and FTC transform computer. (*Beckman Instruments, Inc.*)

transform computer. The sample was also a mull, and one peak in the spectrum is labeled *P* for polyethylene band.

QUALITATIVE ANALYSIS OF SIMILAR COMPOUNDS AND HIGH-MOLECULAR-WEIGHT COMPOUNDS

In general, in the low energy of the far-infrared region, it is to be expected that molecular vibrations would be particularly sensitive to changes in the overall structure of the molecule, to an even greater extent than in the fingerprint region of the fundamental infrared. Though this is true, characteristic frequencies also occur in the far infrared, and tables of correlations are available [12]. For qualitative-analysis purposes, the far infrared is particularly valuable for similar compounds, isomers, and polymers. For example, tetraphenylstannane, triphenylchlorostannane, and diphenyldichlorostannane have been shown to be qualitatively alike in the 2- to 15-μm region but to have marked differences in the far infrared. Similarly, many isomers appear to be identical in the fundamental infrared but differ in the far infrared due to differences in out-of-phase and in-phase frequencies. Finally, to illustrate the far-infrared spectrum of a polymer, Fig. 3-50 shows a portion of the spectrum of Teflon, $(CF_2)_n$, taken with a vacuum-grating spectrophotometer. The spectrum is analyzed in detail in the orginal reference.

ROTATIONAL SPECTRA OF GASES

The rotations of molecules are quantized and give rise to discrete systems of energy transitions. This, of course, causes the well-known rotational fine

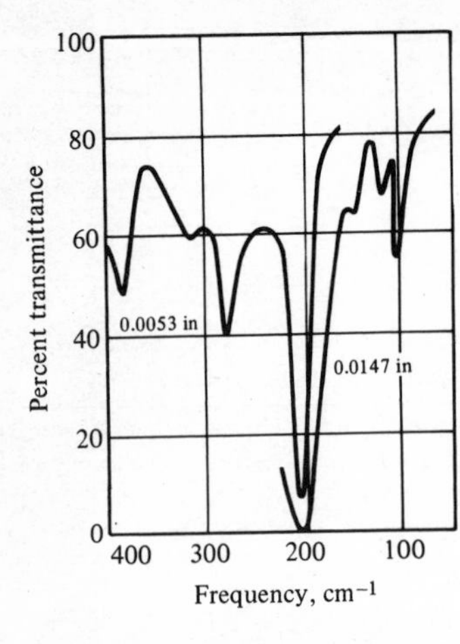

FIGURE 3-50 Far-infrared spectrum of Teflon, $(CF_2)_n$ polymer; numbers are film thickness in inches. [*After C. Y. Liang and S. Krimm, J. Chem. Phys.*, **25**:563 (1956), *by permission*.]

structure superimposed on vibrational bands in the fundamental-infrared region, which is best observed with light molecules in the gas phase. For molecules having a permanent dipole moment, pure rotational transitions, unaccompanied by vibrational transitions, can be used for infrared absorption. For light molecules the low energies involved fall in the far-infrared region. Molecules which have been studied in the far infrared include NH_3, methyl halides, H_2O, AsH_3, PH_3, NO_2, O_3, SO_2, CH_3CN, $CH_3{\equiv}CH$, HCF_3, $(CH_3)_3N$, HCl, and HBr. The most serious limitation on the use of the far-infrared region for such studies is that only the lightest molecules have transitions in this region. The rotational spectra of heavier molecules lie at *millimeter* wavelengths, which fall in the *microwave* region (see Chap. 7).

QUANTITATIVE ANALYSIS

Quantitative analysis in the far infrared has been limited by the relatively poor signal-to-noise ratio implicit in long-wavelength spectroscopy and by the relatively low intensity of the absorption bands obtained.

EXERCISES

3-1 Diagram the fundamental modes of vibration of (*a*) CO and (*b*) N_2. (*c*) The infrared spectrum of one of these gases shows a strong absorption band at 2143 cm^{-1} (4.67 μm) and a weak band at about 4260 cm^{-1} (2.35 μm). Which gas, CO or N_2, is responsible for the observed spectrum? Identify the vibration responsible for each of the two absorption bands.

Ans: (*a*) Symmetrical stretch, $\overleftarrow{C}=\overrightarrow{O}$ (infrared-active); (*b*) symmetrical stretch, $\overleftarrow{N}\equiv\overrightarrow{N}$ (infrared-inactive); (*c*) CO; the 4.67-μm band is the fundamental vibrational mode, and the 2.35-μm band is its first overtone.

3-2 How many fundamental vibrational frequencies would you expect to observe in the infrared absorption spectrum of CS_2?

Ans: CS_2 is linear, similar to CO_2, and of the four fundamental modes of vibration, an absorption band is found for the asymmetrical stretching mode and another for the doubly degenerate bending modes (the symmetrical stretching mode is infrared-inactive). (The infrared spectrum of CS_2 solution shows absorption bands at 4.63 and 6.65 μm.)

3-3 The infrared absorption spectrum of HCN shows three absorption bands at 3312 cm^{-1} (3.02 μm), 2089 cm^{-1} (4.79 μm), and 712 cm^{-1} (14.04 μm). Can you deduce from this whether HCN is linear or bent? Explain.

Ans: No; three absorption bands would be expected for both a linear and a bent configuration. In actuality, HCN is linear, and the three absorption bands are due to the CH stretch, the CN stretch, and the doubly degenerate CH bending modes.

3-4 The infrared spectrum of gaseous N_2O has three strong absorption bands, at 2224 cm^{-1} (4.50 μm), 1285 cm^{-1} (7.78 μm), and 579 cm^{-1} (17.27 μm). In addition, several weaker bands are observed, two of them being at 2563 cm^{-1} (3.90 μm) and 2798 cm^{-1} (3.57 μm). It is known that N_2O is linear, e.g., the absence of a Q branch in the first two absorption bands indicates linearity, but assume that it is not known whether the structural formula is N—N—O or N—O—N. Using the infrared data given, decide which of the two structures is correct. What vibrations can be assigned to the absorption bands mentioned?

Ans: N≡N=O gives three fundamental vibrational bands whereas N—O—N would only give two. The weak 2563-cm^{-1} band is the first overtone of the 1285-cm^{-1} fundamental, and the 2798-cm^{-1} band is a combination band composed of the 2224- and 579-cm^{-1} vibrations.

3-5 The bent triatomic molecule O—N—Cl has three strong infrared absorption bands, at 1799 cm^{-1} (5.56 μm), 592 cm^{-1} (16.89 μm), and 332 cm^{-1} (30.1 μm). Diagram the fundamental modes of vibration of O—N—Cl, and assign them to the absorption bands.

3-6 How many fundamental modes of vibration would you predict for (*a*) benzene, (*b*) toluene, (*c*) methane, (*d*) BF_3, (*e*) CH_2=CH_2, (*f*) ethene?

Ans: (*a*) 30, (*b*) 39, (*c*) 9

3-7 Estimate the frequency of absorption of the C≡N stretch in HCN.

Ans: 2006 cm^{-1} (observed 2089 cm^{-1})

3-8 Calculate the wavelength of absorption of the H—Cl stretching frequency. *Ans:* 3.50 μm

3-9 Many ultraviolet and visible spectrophotometers are single-beam instruments, but it is generally not practical to have single-beam infrared spectrophotometers. Explain why.

3-10 The overwhelming majority of infrared instruments have one chopper rather than two. In all these instruments, the chopper is located between the sample and the monochromator. What disadvantages are there to chopping *after* the sample? Why is it necessary, with one chopper, to chop after the sample rather than before?

3-11 In the two-chopper optical system shown in Fig. 3-21, sketch the intensity of light reaching the detector as a function of time over two cycles of chopper C, labeling the sample and reference portions of the wave, and explain why chopper D has *no effect* on the signal measured. *Hint:* On the same time axis as above, sketch the waveform (intensity vs. time) for detector signal that would result if only chopper D were used and continuous radiation were being passed through both the sample and reference, i.e., similar to the system used in Fig. 3-18; then compare the two waveforms, realizing that chopper C initiates the usable radiation.

3-12 The spectrum of a certain infrared sample cell, run empty, shows 14 peaks (or troughs) between 8 and 10 μm. Calculate the thickness of this cell.

3-13 A cell for measuring the infrared spectrum of gases is 10 cm long. Predict how many interference fringes should appear in the spectrum between 1200 and 1000 cm^{-1} if the cell were run empty. Would it be feasible to measure the cell length in this way? Explain.

3-14 A certain bromobenzene exhibits no absorption bands between 11 and 14.5 μm. What is the probable structure of this compound?

Ans: Hexabromobenzene

3-15 A chlorotoluene, C_7H_7Cl, has a single band at 12.50 μm. What is the correct structure of the compound?

3-16 Cite similarities and differences between the fundamental-infrared region and the near-infrared region, with respect to (*a*) instrumentation and (*b*) applications.

3-17 What instrumentation features do the near infrared, the fundamental infrared, and the far infrared have in common? What differences are there?

3-18 Give examples of chemical problems answerable only in the far-infrared region.

REFERENCES

INTRODUCTORY

1 CROSS, A. D.: "An Introduction to Practical Infrared Spectroscopy," Butterworths, London, 1960.
2 DYER, J. R.: "Applications of Absorption Spectroscopy of Organic Compounds," Prentice-Hall, Englewood Cliffs, N.J., 1965.

3 EWING, G. W.: "Instrumental Methods of Chemical Analysis," 3d ed., McGraw-Hill, New York, 1969.

4 SILVERSTEIN, R. M., and G. C. BASSLER: "Spectrometric Identification of Organic Compounds," 2d ed., Wiley, New York, 1967.

INTERMEDIATE

5 BRÜGEL, W.: "An Introduction to Infrared Spectroscopy," trans. by A. R. Katritzky and A. J. D. Katritzky, Methuen, London, 1962.

6 CONLEY, R. T.: "Infrared Spectroscopy," 2d ed., Allyn and Bacon, Boston, 1972.

7 COLTHUP, N. B., L. H. DALY, and S. E. WIBERLY: "Introduction to Infrared and Raman Spectroscopy," Academic, New York, 1964.

8 SMITH, L. A.: Infrared Spectrophotometry, in I. M. Kolthoff and P. J. Elving (eds.), "Treatise on Analytical Chemistry," pt. I, vol. 6, Wiley, New York, 1965.

9 RAO, C. N. R.: "Chemical Applications of Infrared Spectroscopy," Academic, New York, 1963.

10 EGLINTON, G.: Infrared and Raman Spectroscopy, chap. 3 in J. C. P. Schwarz (ed.), "Physical Methods in Organic Chemistry," Holden-Day, San Francisco, 1964.

11 WILLIAMS, D. H., and I. FLEMING: "Spectroscopic Methods in Organic Chemistry," McGraw-Hill, New York, 1966.

ADVANCED

12 FREEMAN, S. K.: "Interpretive Spectroscopy," Reinhold, New York, 1965. Good chapters on both far infrared and conventional infrared.

13 GODDU, R. F.: Near-Infrared Spectrophotometry, in C. N. Reilley (ed.), "Advances in Analytical Chemistry and Instrumentation," vol. I, Wiley-Interscience, New York, 1960.

14 ALPERT, N. L., W. E. KEISER, and H. A. SZYMANSKI: "Infrared: Theory and Practice of Infrared Spectroscopy," Plenum, New York, 1970.

15 EWING, G. W.: *J. Chem. Educ.,* **48**(9):A521 (1971). An excellent discussion of infrared detectors.

CATALOGS OF SPECTRA

16 AMERICAN PETROLEUM INSTITUTE: Catalogue of Infrared Spectrograms, *Texas A & M Univ. API Res. Proj.* 44, 1968.

17 BELLAMY, L. J.: "The Infrared Spectra of Complex Molecules," 2d ed., Wiley, New York, 1958.

18 Sadtler Standard Spectra, a continually updated subscription service, Sadtler Research Laboratory, Inc., Philadelphia.

4 Emission Spectroscopy

In emission spectroscopy a sample is excited by thermal or electric energy, and the radiation emitted by the excited sample is measured and used for qualitative and quantitative analysis. Whereas most spectroscopic techniques analyze for *molecules*, emission spectroscopy analyzes almost exclusively for *atoms*. The technique is therefore almost unsurpassed as a method of *elemental-metal* analysis. In principle, emission spectroscopy could be used for the identification and quantitative determination of *all* the elements in the periodic table. In practice, usually only the metals and metalloids (a total of about 70 different elements) are determined in this way. Samples can be in the form of solid, liquid, or gas, but the great majority of samples are solids (or evaporated solutions). Liquid samples are occasionally analyzed, but gaseous samples are only rarely determined. Generally little or no sample preparation is necessary.

Emission spectra are characteristic of *atoms* rather than *molecules* because the very large amounts of energy required to excite most chemical samples completely dissociate any chemical compounds to the free elements. Thus it is not feasible to use emission spectroscopy to determine the state of chemical combination of substances.

In emission spectroscopy it is the *valence* electrons of elements that are generally excited, and the emitted radiation usually consists of sharp, well-defined *lines*. All these lines fall in the ultraviolet, visible, or near-infrared region of the spectrum. Identification of the *wavelength* of these lines permits qualitative analysis of the elements present; measurement of their *intensities* permits quantitative analysis.

It may be useful to summarize the major advantages and disadvantages of emission spectroscopy. The outstanding *advantages* are the following:

1 It is an excellent method of elemental *trace* analysis, typically detecting elements at the parts-per-million level.

2 It can be used for *all* metals and metalloids and occasionally other elements as well.

3 It can be used for very *small* samples, typically a few milligrams, and occasionally even less than a milligram.

4 Samples can generally be analyzed *as received,* preliminary chemical separations usually being unnecessary.

5 The method is relatively *rapid,* particularly if a direct-reading instrument is available.

The *disadvantages* may be cited as follows:

1 The equipment is relatively *expensive,* especially the large instruments that may be required in analyzing heavy elements or very complex samples.

2 The precision and accuracy are limited to roughly 5 percent, which often makes it uncompetitive with other methods.

3 The sample (albeit a very *small* sample) is destroyed in the process of analysis.

4 The method is limited to the analysis of elements.

4-1 THEORY

The various types of emission spectra that a sample can emit, the excitation energy requirements of various substances, and the transitions responsible for line spectra will be described and common spectral terms will be defined.

4-1A Types of Emission Spectra

When thermal or electric energy is added to a sample, *continuous* spectra, *band* spectra, or *line* spectra may be emitted.

CONTINUOUS SPECTRA

A continuous spectrum is characterized by generally uninterrupted emission over a considerable wavelength region and by the absence of sharp lines or discrete bands. Continuous spectra are emitted by incandescent solids or solids like iron or carbon that are heated until they glow. These substances emit blackbody radiation, which depends principally on the temperature rather than the chemical composition. Therefore, continuous spectra are of no value for spectrochemical analysis and are avoided by volatilizing the sample before excitation.

BAND SPECTRA

A band spectrum is characterized by groups of lines so close together that under ordinary resolution conditions they appear to be smeared together into continuous bands. In emission spectroscopy bands are *caused* by excited molecules, and in general band spectra are avoided by operating at excitation conditions giving high enough energy for molecules to be broken up into individual *atoms*. Occasionally molecules resist decomposition, and emission bands may appear. One common example is cyanogen, CN, which is formed in electric arcs when carbon from electrodes combines with nitrogen from the air. Cyanogen bands can be a serious interference, particularly if weak lines are being sought in the cyanogen-band regions of 4216, 3883, and 3590 Å.†

LINE SPECTRA

A line spectrum consists of discrete, apparently irregularly spaced lines. This type of spectrum is caused by individual atoms or atomic ions which have been excited and are emitting their excess energy in the form of radiation (see Sec. 1-3*A*). Whereas *molecules* have vibrational and rotational energy levels which are superimposed on the electronic energy levels (thus giving a multitude of closely spaced lines which smear together and appear to be *bands*), *atoms* have only *electronic* energy levels, and thus the energy-level transitions which occur result in discrete *line* emissions. Line spectra are the predominant type in emission spectroscopy.

4-1B Excitation-Energy Requirements and Spectral Sensitivity

For an element in the zero oxidation state to emit a single spectral line, energy equivalent to the *excitation potential* of the element must be absorbed. (The excitation potential may be defined as the energy required to raise a valence electron from the ground state to the first excited state.) On the other hand, to produce the complete (multilined) emission spectrum of an element, energy equivalent to the *ionization potential* must be absorbed, where the ionization potential is the energy required to remove a valence electron completely and ionize the element.

Because a given emission source has a fixed amount of energy, ionization

† It is the practice in this book to use the wavelength units most commonly employed by specialists in each optical method. Thus, while nanometers (the older millimicrons) are most commonly used by workers in ultraviolet and visible spectrophotometry (Chap. 2), most emission spectroscopists use angstrom units (Å) for this same wavelength region.

potentials can serve as a guide to the relative sensitivities of various elements in emission spectroscopy. For example, the alkali metals and other elements with low excitation potentials can be determined with high sensitivity in emission spectroscopy, whereas nonmetals give relatively low sensitivities. With an energetic enough source, *all* the elements in the periodic table could be excited and their emission spectra produced. But the requirement of a very energetic source and the problem of low sensitivity are not the only drawbacks to the analysis of nonmetals by emission spectroscopy (in fact, sufficiently energetic sources *are* available). The other problem with elements having very high ionization potentials is that their emission lines are also very energetic and in fact usually fall in the vacuum-ultraviolet region (the wavelength region below 2000 Å), which is not readily accessible. For these reasons emission spectroscopy is commonly limited to some 70 elements, mainly the metals and metalloids.

The complexity of an emission spectrum depends on the electron configuration of the element involved. Thus, for example, the spectra of the alkali metals are very simple, consisting of only a dozen or so widely spaced lines ranging from the ultraviolet to the infrared. On the other hand, most of the transition elements give very complex spectra, and uranium, as a specific example, gives rise to thousands of lines. As a rule of thumb it may be noted that the sensitivity of emission spectroscopy is approximately inversely proportional to the complexity of the spectrum. Thus, for example, the familiar yellow lines of sodium will show up even if only traces are present. Anyone who has tried the qualitative-analysis flame (emission) test for sodium knows that contamination from ubiquitous traces of sodium is difficult to avoid. Conversely, in order to be able to observe any of the lines in the emission spectrum of uranium, a higher energy of excitation and larger amounts of uranium are required. In general, any element having a complex emission spectrum will give a lower sensitivity than an element with a simple spectrum because any given emission source has a fixed amount of energy, and if that energy is divided up among many different lines and excitation mechanisms, the sensitivity of any one line will be diminished.

4-1C Transitions Responsible for Line Spectra

When an atom is excited by thermal or electric energy, an outer electron in the atom will be raised to some higher electronic energy level. Almost immediately thereafter (roughly 10^{-8} s) it will return to its original level or to one of its intervening levels. In the process the atom will emit a single line for each transition that occurs. The energy E of an emitted line is given by

$$E = E_2 - E_1 = h\nu \qquad (4\text{-}1)$$

where E_2 represents the energy of the higher electron level and E_1 is the energy of the lower level. This energy difference may be equated to $h\nu$, where h is

Planck's constant and ν is the frequency of radiation emitted. The specific electronic transitions that can take place are governed by well-defined *selection rules*.

SPECTRUM OF HYDROGEN

The simplest atom is of course hydrogen, and it will serve as an excellent starting point for the discussion of emission spectra. Figure 4-1 shows the emission spectrum of hydrogen over a wide wavelength range. The spectrum is of the type most commonly observed in emission spectroscopy, namely, a line spectrum measured on a *photographic plate,* where the *intensity* of a given line is actually indicated by the *darkness* or *density* of a given line, although in Fig. 4-1 (and in subsequent spectra to be shown) the intensity is indicated schematically by the *thicknesses* of the lines.

In Fig. 4-1 the various lines are grouped or classified into five different *series* of lines located in various regions of the spectrum. The basis for this classification is the electronic energy level to which an excited electron *returns*. A hydrogen atom in the ground state will have an electron in the n = 1 ground state, but upon excitation the electron will be raised to any of an infinite number of definite energy levels farther from the nucleus. It will then fall back toward its original (n = 1) energy level. However, in a large collection of excited hydrogen atoms, various intermediate stops will be made at intervening energy levels. All transitions ending in the n = 1 energy level (E_2 - E_1, E_3 - E_1, E_4 - E_1, etc.) represent a series of lines discovered by *Lyman* in the far-ultraviolet region; all transitions ending in the n = 2 energy level (E_3 - E_2, E_4 - E_2, etc.) represent a series of lines discovered by *Balmer* in the visible region, etc. There is a large energy (and spatial) separation between the n = 1 levels and n = 2 levels, a smaller separation between the n = 2 and n = 3 levels, and successively smaller separations as the distance from the nucleus increases (see, for example, the energy-level diagram in Fig. 1-6). The fact that energy-level spacings become smaller as the distance from the nucleus

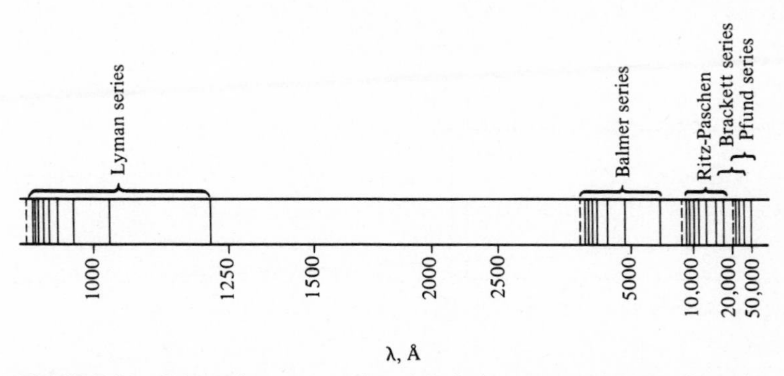

FIGURE 4-1 Schematic representation of the hydrogen-atom spectrum from the ultraviolet (on left) through the near-infrared regions. (*Adapted from G. M. Barrow, "Physical Chemistry," 2d ed., p.* 71, *McGraw-Hill Book Company, New York,* 1966, *by permission.*)

increases is the reason each series of lines shifts to lower energy (longer wavelength) as one proceeds from the Lyman series through the Pfund series.

Although the Bohr picture of a hydrogen atom, using only the principal quantum numbers n, can explain the most important features of the hydrogen emission spectrum, it cannot explain the fact that if the spectrum is examined with a spectrograph of very high resolving power, many of the lines turn out to be *two* very closely spaced lines, as illustrated in Fig. 4-2. Each pair of closely spaced lines is called a *doublet*. The splitting of the 6563- and 4861-Å lines is only 0.14 and 0.08 Å, respectively, which an ordinary emission spectrograph would not show. The splitting occurs because each of the principal energy levels is actually divided into subshells or sublevels, allowing transitions with only slightly differing energy to take place.

SPECTRA OF OTHER ELEMENTS

The spectra of elements heavier than hydrogen tend to become more complex (have more lines) because more outer electrons are involved, and a greater number of transitions can take place. However, the spectra of all the group I elements (H, Li, Na, K, etc.) are quite similar in their complexity because they all have just one electron in their outermost (valence) shell. In fact, the spectra of *ions* with one valence electron left, for example, He^{+}, Li^{2+}, Be^{3+}, are also similar in their complexity. But the actual spectra are *different* (the lines appear at different wavelengths, thereby characterizing the species involved) because the energy-level separations are different.

Spectrum of Lithium Figure 4-3 shows the complete spectrum of lithium and divides it into four different series of lines, called the *sharp, principal, diffuse,* and *fundamental* series. This classification system, in use for all elements except hydrogen, originally came from the *appearance* of the lines; later these terms were based more systematically on the type of excited-state subshell from which the electron transition originated. Thus, in the hydrogen spectrum, all lines in a given series involve electron transitions which *end up*

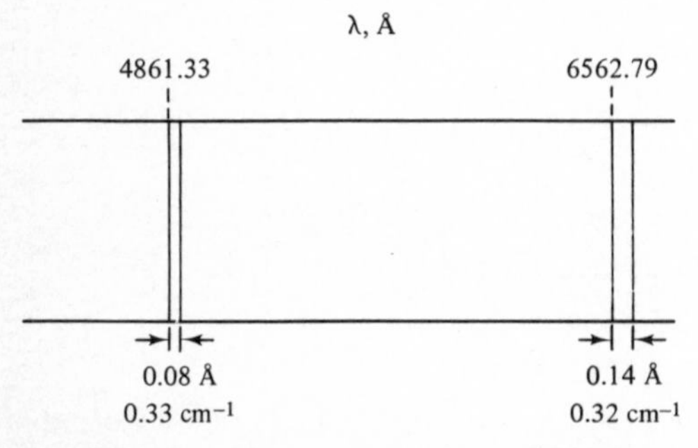

FIGURE 4-2 First two doublets of the Balmer series of hydrogen lines. (*From S. Walker and H. Straw, "Spectroscopy," vol.* 1, *p.* 13, *Chapman & Hall Ltd., London,* 1961, *by permission.*)

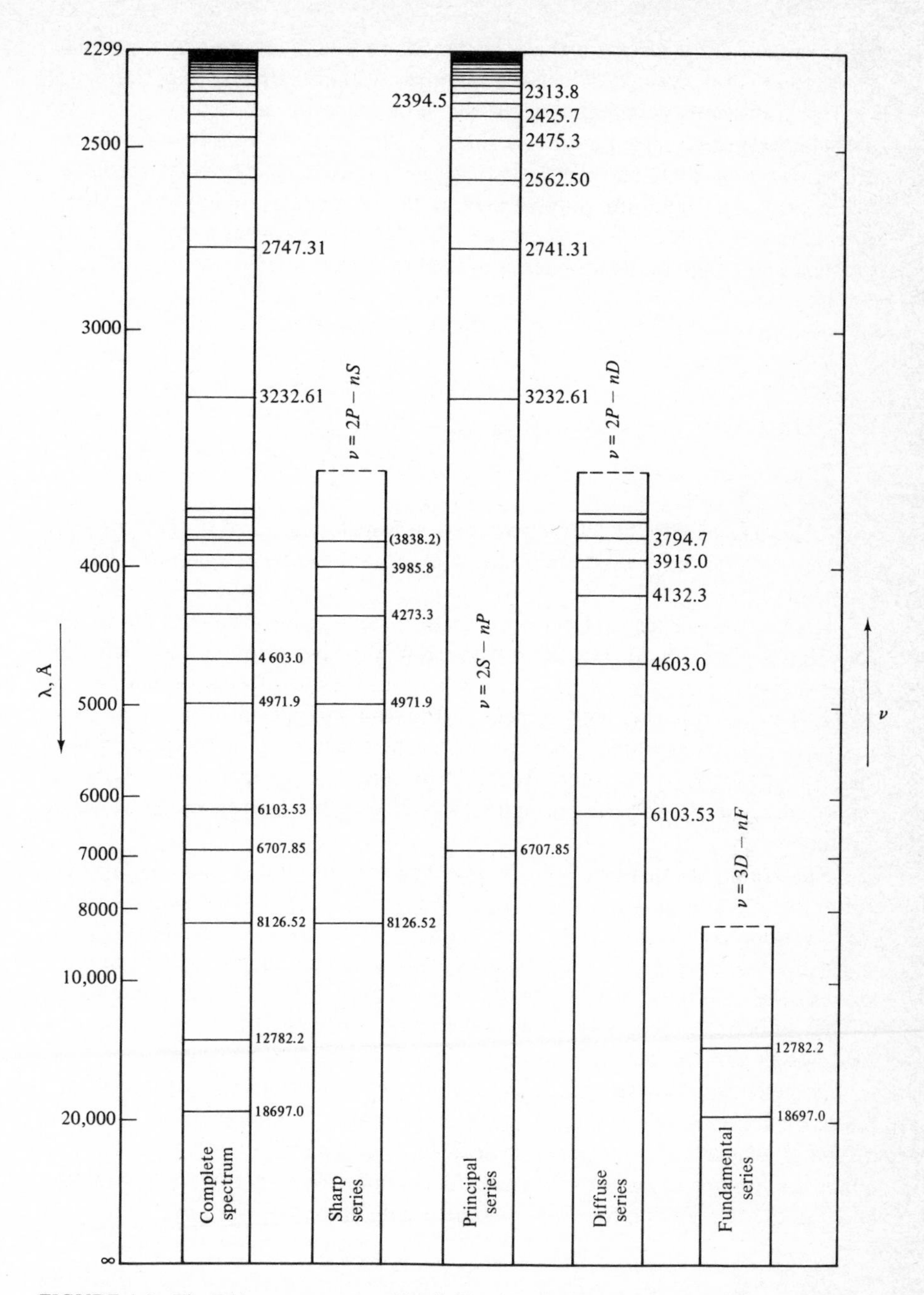

FIGURE 4-3 The lithium spectrum and its division into the sharp, principal, diffuse, and fundamental series. (*After W. Grotian, "Graphische Darstellung der Spectrum von Atom und Ionen mit ein, zwei, und drei Valenzelektronen," Springer-Verlag, Berlin,* 1928, *by permission.*)

in the same energy level; on the other hand, in the sharp, principal, diffuse, and fundamental system of classifying lines, all lines in a given series are electron transitions *originating from* the same type of subshell (*s*, *p*, *d*, or *f* subshells, respectively†). Figure 4-4 shows some of the transitions responsible for various emission lines of lithium. Classifying each energy level according to the principal quantum number n and the secondary quantum number l (where $l = 0, 1, 2, 3, \ldots$, represents an $s, p, d, f, \ldots$, subshell), the transitions responsible for each of the four series of the lithium atom are:

Principal series: $\nu = 2s - np = 2S - nP$

Sharp series: $\nu = 2p - ns = 2P - nS$

Diffuse series: $\nu = 2p - nd = 2P - nD$

Fundamental series: $\nu = 3d - nf = 3D - nF$

where the use of capital letters for each subshell is part of a spectroscopic convention to identify electronic states by means of *term symbols*, defined more fully later.

It should be noted that certain *selection rules* govern the electron transitions that are permitted. For these transitions the laws of quantum mechanics indicate that *any change in n* is permissible, but it must be *accompanied by* $\Delta l = \pm 1$. To illustrate this with respect to Fig. 4-4, transitions of the type $3s \rightarrow 2p$ or $4s \rightarrow 2p$ are *permitted*, but transitions between the same type of subshell (like $3s \rightarrow 2s$ or $4s \rightarrow 2s$, where Δl would be zero) are *forbidden*. Similarly, Δl cannot be more than 1; for example, $3d \rightarrow 2s$ or $4f$ to $2p$ transitions would be forbidden.

Spectrum of Sodium When the spectra of the alkali metals are examined in detail, it is found that each line in the *principal* and *sharp* series is really a doublet. For example, the sodium D "line" is a member of the principal series (due to a $3P \rightarrow 3S$ transition), and under moderate resolution it proves to be *two* lines, at wavelengths of 5896 and 5890 Å. The splitting is due to the influence of *electron spin*. A spinning electron produces a magnetic field, which can interact with and alter the total energy of a subshell.

An electron can spin in only one of two directions with respect to a reference field: *parallel*, where s (the spin quantum number) is $+\frac{1}{2}$, or *antiparallel*, where s is $-\frac{1}{2}$. Since the secondary quantum number l and the spin quantum number s are both measures of angular momentum, the interaction between the two gives a *resultant angular momentum j*, defined by

$$j = l + s \tag{4-2}$$

where s is the total spin and j can never be negative. Thus, for a sodium atom

† It was the emission spectroscopists' classification of lines as sharp, principal, diffuse, or fundamental that led to naming the electron subshells *s*, *p*, *d*, and *f*.

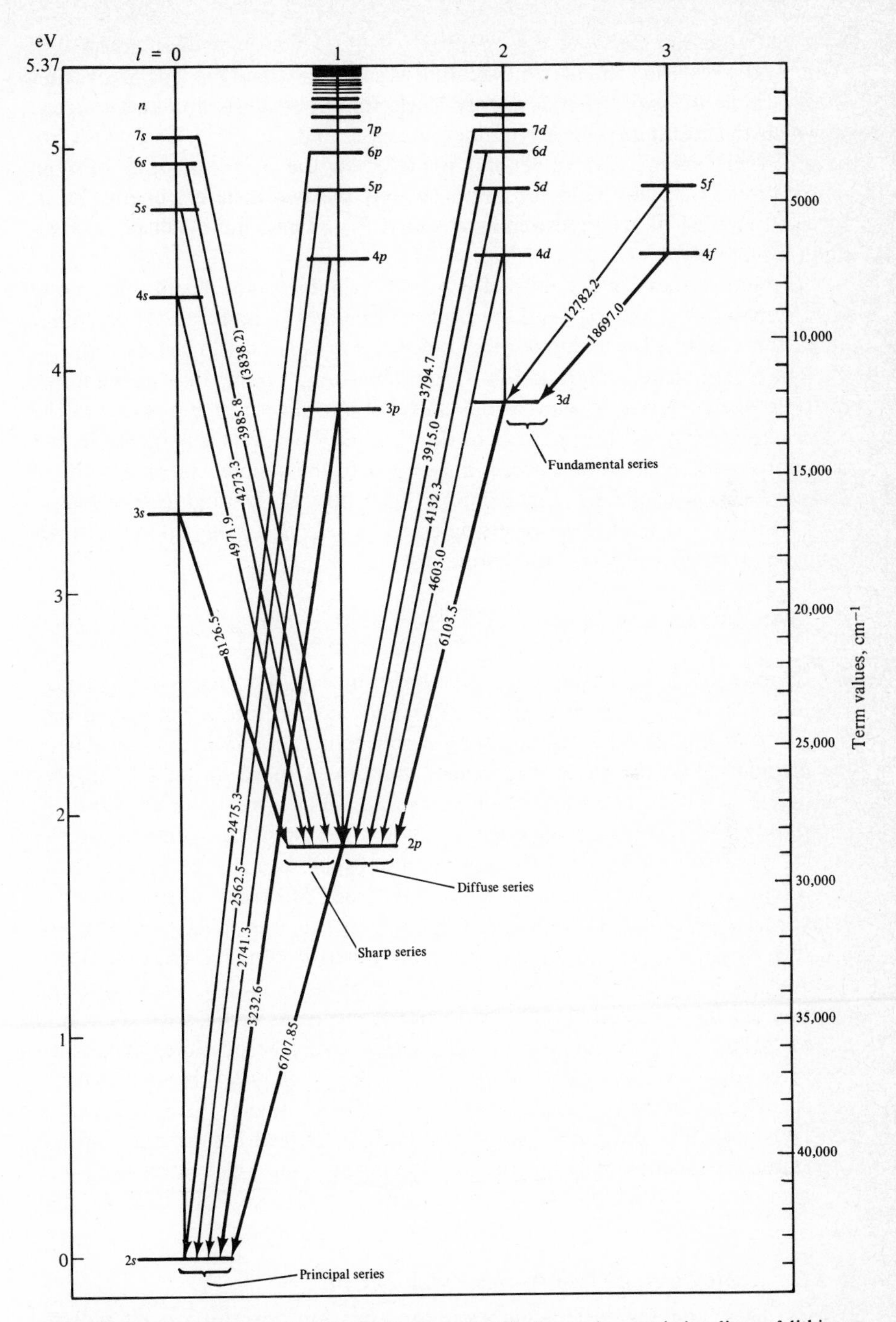

FIGURE 4-4 Some energy-level transitions responsible for various emission lines of lithium. (*Adapted from S. Walker and H. Straw, "Spectroscopy," vol.* 1, *p.* 24, *Chapman & Hall, Ltd., London,* 1961, *by permission.*)

with one unpaired electron, $s = \frac{1}{2}$ and $j = l + \frac{1}{2}$. Thus, for the *p* subshell of sodium, the resultant angular momentum is $j = 1 + \frac{1}{2}$ and $j = 1 - \frac{1}{2}$; in other words, there are two different energy levels for the *p* subshell of sodium (one in which the resultant angular momentum is $\frac{3}{2}$ and the other $\frac{1}{2}$). In spectroscopic *term symbols* the value of *j* is added to the *S, P, D,* or *F* subshell designation as a right-hand subscript, so that the two different energy levels for each *P* subshell are indicated as $P_{3/2}$ and $P_{1/2}$. Thus, the *P* subshells of sodium are doublets.

Similarly, the *d* and *f* subshells are each doublets, the *d* subshell having term symbols $D_{5/2}$ and $D_{3/2}$ and the *f* subshell being $F_{7/2}$ and $F_{5/2}$. On the other hand, the *s* subshell is a singlet level, since $j = \frac{1}{2}$ only [see Eq. (4-2) with $l = 0$]. The energy-level diagram for a sodium atom, taking into account the multiple energy levels of the various subshells, is shown in Fig. 4-5. On the scale used for Fig. 4-5 the splitting of the $P_{3/2}$ and $P_{1/2}$ levels is perceptible, but only barely, whereas the differences in energies of the doublet states in both the *D* and *F* levels would be imperceptible, and thus the *D* and *F* levels were drawn in Fig. 4-5 as if they were singlets. The *selection rules* taking into account the resultant angular momentum are

$$\Delta n = \text{any integral value} \qquad \Delta l = \pm 1 \qquad \Delta j = 0 \text{ or } \pm 1 \tag{4-3}$$

Numerous doublet transitions are shown on Fig. 4-5, but under ordinary resolution only the yellow lines at 5890 and 5896 Å are usually resolved. Notice from Fig. 4-5 that the splitting of subshells becomes less noticeable as the distance from the nucleus increases, making it more difficult to resolve multiplets involving subshells at higher values of *n*. On the other hand, in comparing the spectra of different elements, the splitting increases as the atomic number increases. This is illustrated in Fig. 4-6 for the alkali metals.

Elements with More than One Outer Valence Electron The more outer valence electrons there are in an atom the greater the multiplicity of splittings possible. Without going into detail, if the number of unpaired electrons is *N*, the multiplicity of energy levels that can be expected is $N + 1$. Thus, with the alkaline earths, for example, the two valence electrons could have their spin *paired,* making $N = 0$ and only *singlet* levels would be possible; or the two electrons could be *unpaired,* making $N = 2$, and *triplet* levels would be expected. The alkaline earths actually have a set of singlet *S, P, D,* and *F* subshells, but they also have a set of triplet *S, P, D,* and *F* subshells, and the spectra of the alkaline earths (especially the heavier ones) can be resolved into a singlet series and a triplet series. For further details, see Refs. 6, 8, 13, and 14.

4-1D Definition of Some Spectral Terms

Before discussing the instrumentation for emission spectroscopy it will be helpful to define a number of spectral terms used in the identification of lines and the interpretation of the results.

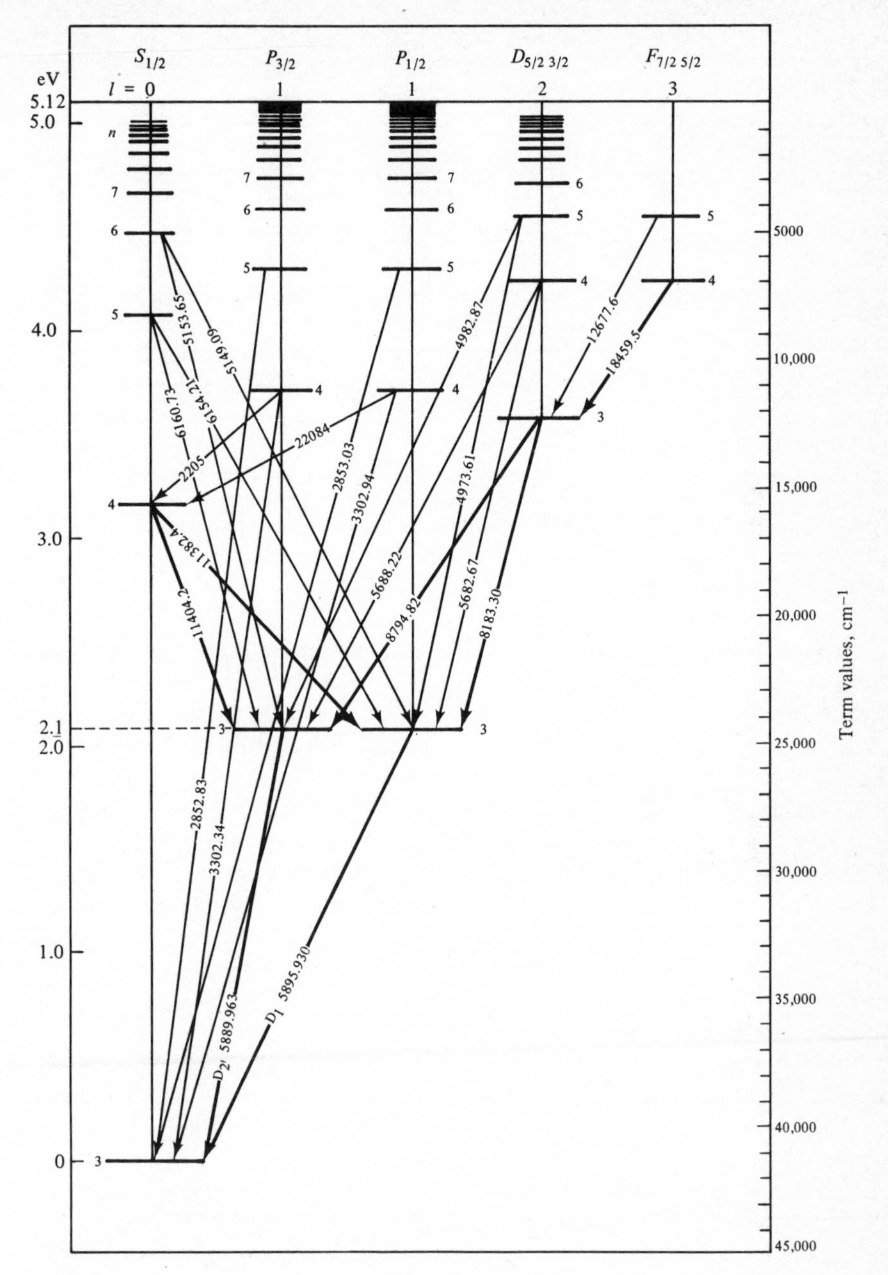

FIGURE 4-5 Some energy-level transitions responsible for various emission lines of sodium. (*After W. Grotian, "Graphische Darstellung der Spectrum von Atom und Ionen mit ein, zwei, und drei Valenzelektronen," Springer-Verlag, Berlin,* 1928, *by permission.*)

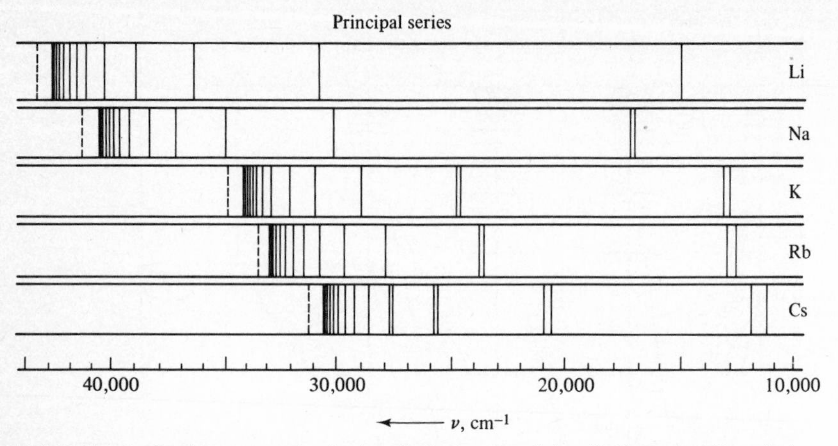

FIGURE 4-6 Doublets appearing in the principal series of the alkali metals. The splitting of the lines in the doublets of sodium and potassium has been exaggerated. (*From H. E. White, "Introduction to Atomic Spectra," McGraw-Hill Book Company, New York,* 1934, *by permission.*)

Raies Ultimes The raies ultimes (also called ultimate lines or persistent lines) are the most intense lines in a spectrum, and *the* raie ultime is the last line of an element to disappear as the quantity of sample is decreased to the vanishing point. Raies ultimes are useful reference lines to look for in beginning the interpretation of an emission spectrum.

Pressure Broadening The exact energy levels of an excited atom are affected by other atoms in its vicinity. It has already been mentioned that the energy levels of a *molecule* (and thus its spectrum) are considerably different from those of its constituent *atoms*. But atomic energy levels may be disturbed by neighboring particles even if no bond is formed. Collisions between atoms will affect energy levels, and thus if the pressure is increased, energy levels are perturbed and line widths are increased. Atomic gas at low pressure will emit the sharpest line spectra.

Self-Absorption Self-absorption is a common phenomenon in analysis for high concentrations of a given element. What happens is that radiation emitted by an atom can be reabsorbed by a similar atom in the ground state before that emission leaves the sampling area. This occurs because only a small fraction of volatilized atoms are excited and the unexcited atoms are ideal absorbers for the wavelength of radiation being emitted. Furthermore, most sources are hotter in the center than around the outer portion; thus emissions predominantly originate in the center, and emitted radiation is forced to pass through the cooler regions which contain mostly unexcited atoms. Hence, working curves often level off (tend to saturate) at high concentrations.

Self-Reversal Self-reversal is an extreme case of self-absorption, where the emitted intensity at the center of a line may actually be less (since it is efficiently self-absorbed) than the intensity at the fringes of the line. Self-reversal makes the interpretation of line intensity both difficult and inaccurate. It can generally be minimized by decreasing the amount of the sample.

4-2 INSTRUMENTATION

In emission spectroscopy the sample is excited and becomes the *source* of radiation, following which a *monochromator* is needed to select the desired wavelength and a *detector* is needed to measure the radiation. Table 4-1 lists for reference the kinds of excitation sources, monochromators, and detectors in use.

4-2A Excitation Sources

The function of the excitation source is to volatilize the sample and then excite valence electrons in the vaporized atoms to higher energy levels. The common sources of excitation energy in emission spectroscopy are the dc arc, the ac arc, and the ac spark. The flame is also a widely used (though lower-energy) emission source, but it will be discussed in Chap. 5. There are special advantages and applications of each type of source, and it is important to know which type of source is best for a given purpose.

DC ARC

Figure 4-7 shows the essential components of a circuit for a dc arc. The power supply is a dc voltage in the range of 50 to about 300 V, which could be supplied by batteries but usually is a rectified ac supply. Typically, a 220-V dc supply is used, and a current of 5 to 15 A is caused to flow across the arc gap in series with a variable resistor R (10 to 40 Ω) and an inductance coil L. The sample, in the form of a solid or a liquid, is placed on the lower electrode in the arc gap, the power is turned on, and the electrodes are momentarily brought together to start the current flow. Once the current begins to flow, the temperature in the gap rises rapidly, the resistance of the gap goes down, and

TABLE 4-1 Excitation Sources, Monochromators, and Detectors Used in Emission Spectroscopy

Excitation sources	Monochromators	Detectors (instrument name)
Dc arc (includes Stallwood Jet and Plasma Jet)	Prism	Eye (spectro*scope*)
Ac arc	Grating	Photographic plate (spectro*graph*)
Ac spark (includes laser microprobe)		Photomultiplier tube (spectro*meter*)
Flame (see Chap. 5)		

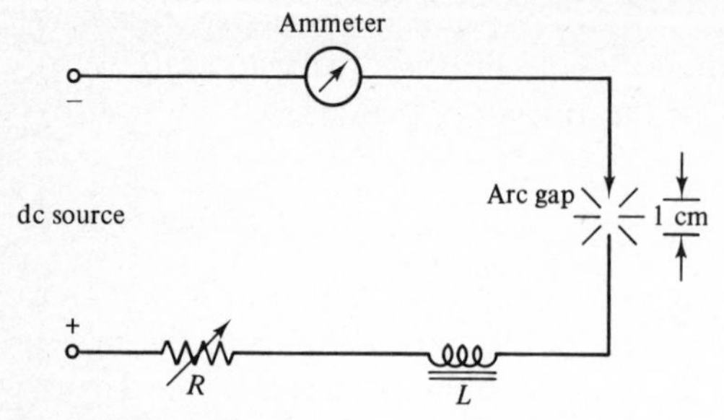

FIGURE 4-7 Circuit for a dc arc source (see text for description of circuit).

the electrodes can be separated by about a centimeter. The electric arc will jump across the gap. Typical arc temperatures range from about 4000 to 8000 K, and in general neutral *atoms* rather than ions are the emission species with this source.

The variable resistance in the circuit R can be used to adjust the current, and the inductance coil L helps *stabilize* the flow of current, even though the arc tends to wander from spot to spot on the surface of the lower electrode (the anode). The wandering is one of the prominent features of the dc arc, and thus unless the sample is very homogeneous or all the sample is volatilized, it is not a very reproducible source for quantitative analysis.

On the other hand, the very high temperatures generated in the dc arc makes it a very *sensitive* source, one that is very good for qualitative analysis. The excitation energy is mainly *thermal* rather than electric, and the energy is sufficient to excite all the metal elements with this source.

Stallwood Jet Modification An attachment for use with the dc arc, called the Stallwood Jet (Spex Industries), has been developed which corrects many of the faults traditionally associated with the dc arc. Figure 4-8 shows a diagram of a commercial version of the Stallwood Jet. A mixture of argon and oxygen gases (predominantly argon, generally in the ratio of 80:20 to 60:40) is passed through the gas inlet and swirls upward around the electrode as it is arced. A quartz enclosure helps to maintain a blanket of gas at a slight positive pressure around the arc column, thereby excluding ambient air, which has a high concentration of nitrogen. The gases swirling around the lower electrode cause the sample to burn more slowly and evenly, making greater precision and accuracy in quantitative analyses possible.

The Stallwood Jet increases sensitivity both by reducing background from cyanogen bands and by controlling the volatility and excitation energy. Whereas normal dc-arc excitation results in fractional volatilization of the most volatile components of the sample at the expense of the less volatile components, the Stallwood Jet minimizes fractional distillation and tends to give uniform volatilization of the entire sample. Furthermore, the ratio of argon to oxygen can be varied from element to element to optimize the excitation energy for the element of interest at the expense of the others. Table 4-2 gives the persistent lines and approximate sensitivities of about 70 elements and compares the detection limit of a Stallwood Jet to a regular open arc. As may be seen in Table 4-2, the Stallwood Jet often increases the sensitivity by a factor of 2 or more.

Plasma-Jet Modification Another modification for use with the dc arc is the *Plasma Jet* (sometimes called the plasma *arc*). This device has been designed exclusively for the analysis of liquids, and Fig. 4-9 shows the cross section of a Plasma Jet analyzer. In principle and in practice the Plasma Jet atomizer is very similar to a total-consumption flame-photometer atomizer (see Sec. 5-1*A* and Fig. 5-4*a*). A vertical-atomizer capillary dips into a beaker containing the sample solution, and the solution is drawn up the capillary by the venturi effect of argon or helium flowing through an outer concentric channel and past the upper end of the capillary. The liquid is aspirated into a chamber and swept into the dc arc. This causes a tremendous change to occur in the dc discharge temperature and the resulting spectrum. The discharge temperature jumps from about 6000°C to as high as 10,000°C, and the spectrum, instead of being characteristic of excited neutral atoms, becomes characteristic of excited *ions*.

The temperature in the Plasma Jet depends largely on the arc current and the gas flow rate. Increasing the gas flow increases the rate of sample consumption and the electric conductivity in the gap, and both the current and the temperature rise.

The advantages of the Plasma Jet include minimum sample preparation, high sensitivity, e.g., as little as 0.05 to 0.3 ppm of many elements are

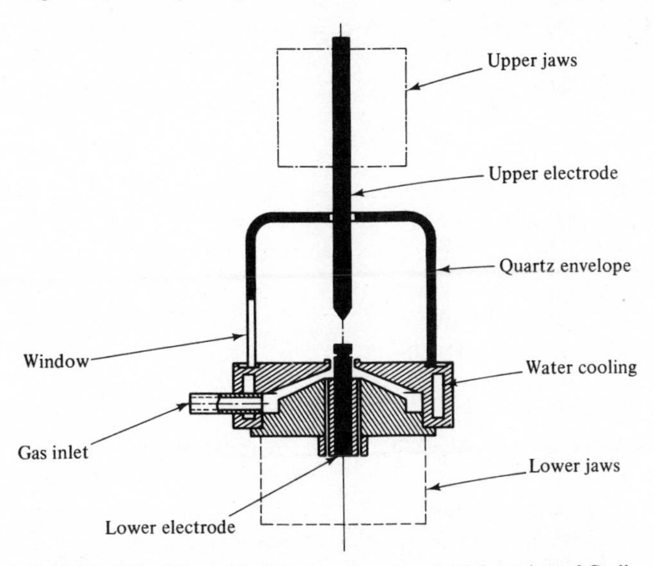

FIGURE 4-8 Schematic diagram of a commercial version of Stallwood Jet (see text for description). [*From G. H. Morrison (ed.), "Trace Analysis," p. 200, John Wiley & Sons, Inc., New York, 1965, by permission.*]

TABLE 4-2 Persistent Lines and Sensitivities of the Elements†,‡

Element	Line, Å	Limit of detection, ppm	
		Open arc	Stallwood Jet
Ag	3280.7	1	0.5
Al	3092.7	10	5
As	2349.8	100	50
Au	2428.0	<5	<5
B	2497.7	2	2
Ba	4554.0	5	2
Be	2348.6	0.1	0.5
	3130.4	0.5	0.02
Bi	3067.7	5	2
Ca	4226.7	1	0.2
Cd	3261.1	20	10
Ce	4186.6	1000	500
		10	5
Co	3453.5	20	5
Cr	2835.6	10	5
Cs	4555.4	500	500
Cu	3247.5	1	0.5
Dy	4046.0	5	2
Er	3692.6	5	2
Eu	4594.0	5	2
Fe	3020.6	5	2
Ga	4172.1	20	5
	2943.6	5	5
Gd	3422.5	10	2
Ge	2651.2	5	5
Hf	3194.2	50	10
Hg	2536.5	500	50
Ho	3891.0	5	2
In	4101.8	20	5
	3256.1	10	<5
Ir	3220.8	50	10
K	4044.1	>1000	1000
La	3949.1	5	2
Li	3232.6	500	500
Lu	2615.4	5	2
Mg	2795.5	0.2	0.2
Mn	2576.1	5	1
Na	3302.3	500	500
Nb	3094.2	10	5
Nd	4303.6	20	5
Ni	3414.8	20	5
P	2535.7	100	50
Pb	2833.0	20	5

TABLE 4-2 (Continued)

Element	Line, Å	Limit of detection, ppm	
		Open arc	Stallwood Jet
Pd	3242.7	10	5
Pr	4225.3	20	5
Pt	3064.7	<5	<5
Rb	4201.8	>1000	1000
Re	3460.5	50	10
Rh	3434.9	10	<5
Ru	3436.7	100	10
Sb	2598.0	50	50
Sc	4023.7	1	0.5
Se	2062.8	1000	500
Si	2881.6	1	0.5
Sm	4424.3	20	5
Sn	3175.0	20	5
Sr	4077.7	50	5
Ta	3311.2		
Tb	3509.2	50	10
Te	2385.8	500	100
Th	4019.1	500	500
Ti	3349.0	10	1
	3775.7	50	20
Tm	3462.2	5	2
U	4241.7	500	500
V	3093.1	20	5
W	4302.1	100	100
Y	3242.3	5	2
Yb	3289.4	1	0.5
Zn	3302.6	50	10
Zr	3496.2	50	10

† The data were obtained with a graphite matrix for all the elements save the rare earths, for which the matrix was lithium carbonate (Spex semiquantitative G and L standards). A 3.4-m spectrograph with a 600-lines-per-millimeter grating set for the first order was used.
‡ From G. H. Morrison (ed.), "Trace Analysis," p. 197, John Wiley & Sons, Inc., New York, 1965, by permission.

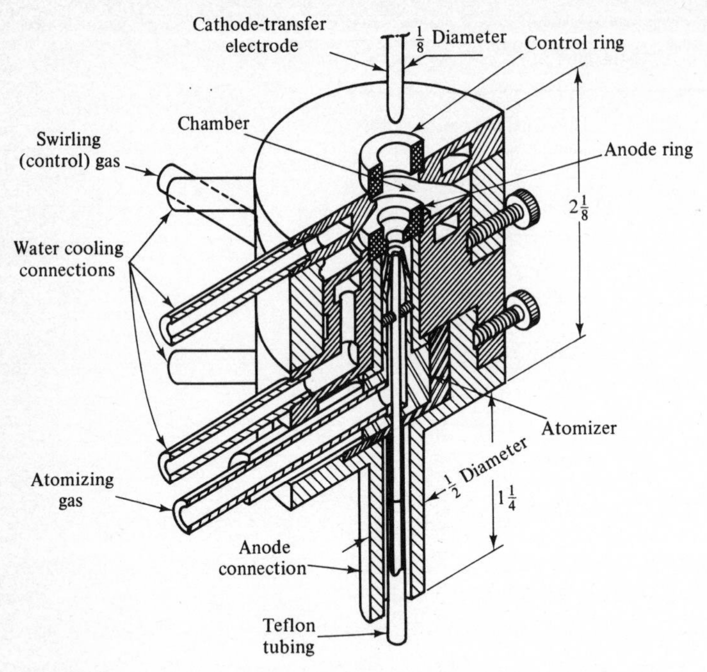

FIGURE 4-9 Cross section of a Plasma Jet liquid analyzer. Aspirated into a chamber, the sample is then mixed with argon or helium in a chamber between the anode and control rings. Forced upward through an orifice, the atomized mixture is arced at around 20 A dc. Instead of a conventional arc, a Plasma Jet develops. (*Spex Industries, Inc.*)

analyzable, high stability, adaptability to continuous analysis, and applicability to a wide variety of types of samples. A few examples of the types of samples that have been analyzed with the Plasma Jet include organic materials of low flashpoint, which cannot be analyzed with conventional emission sources; acids up to about 5 percent concentration; seawater (after dilution to prevent salt encrustation) for strontium; blood, for traces of aluminum; lubricating oils, for additives and wear metals; and high-temperature alloys, for major elements.

Some disadvantages of the Plasma Jet include occasional clogging of the atomizer capillary, critical aspiration rate (which in turn depends on the gas flow rate and other mechanical parameters), high background radiation, and high noise (sound) level. For many purposes the advantages outweigh the disadvantages, and the Plasma Jet should see increasing use in the years ahead.

AC ARC

Figure 4-10 shows the essential components of a circuit for an ac arc. A transformer steps the voltage up to 2000 to 5000 V ac, sufficient to allow the

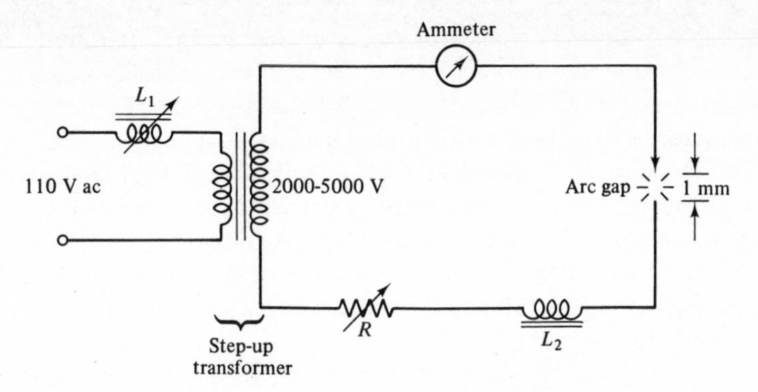

FIGURE 4-10 Circuit for an ac arc source (see text for description).

arc to jump the gap without having to close it first. The electrode area must be carefully shielded to protect the operator from the dangerously high voltages. The current is controlled with a variable inductance L_1 in the primary circuit, and currents in the range of 1 to 5 A are generally used.

Since the current is alternating at a frequency of 60 Hz, the arc extinguishes itself and reverses directions 120 times per second. This has the advantage of giving better reproducibility and more representative sampling than the dc arc, since the ac arc picks out a new surface area each cycle. In this way, the whole surface of the sample is arced and excited. Another consequence of the stop-and-start nature of this source is that the gap temperature is much lower than with the dc arc, with a concomitant lower sensitivity.

AC SPARK

Figure 4-11 shows the essential components of a circuit for an ac spark source (sometimes called a *Feussner circuit*). A step-up transformer boosts the voltage up to 10 to 40 kV, and the secondary circuit contains a capacitor C, an inductance coil L, a resistor R_2, a spark gap for the sample (analytical spark gap), and an auxiliary spark gap (synchronous spark gap). The auxiliary spark gap is operated by a motor whose rotation is synchronous with the al-

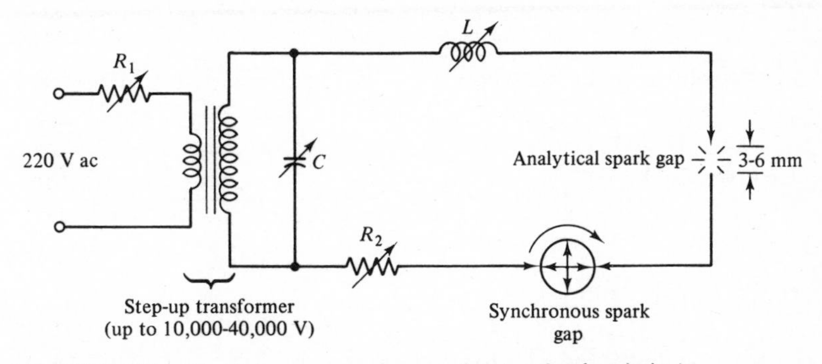

FIGURE 4-11 Circuit for an ac spark source (see text for description).

ternations of the line current. The capacitor stores up the charge for a half-cycle, and when it is charged to a maximum, the synchronous rotor allows the circuit to be completed and a spark jumps across the analytical gap. Each discharge is actually a damped oscillatory discharge, typically oscillating at a frequency of 10 MHz. Resistor R_1 is a current control, and resistor R_2, inductor L, and capacitor C control the frequency and current-voltage characteristics of the oscillatory discharge. The initial pulse of a discharge may give an extremely high current (1 to 2 kA) and temperature (up to 40,000 K), but the succeeding pulses are severely damped and the *average* current is usually only a few tenths of an ampere to a few amperes.

In contrast to the previously discussed sources, where the excitation in the gap is mainly thermal, the electrodes in the spark gap are relatively cold,† and excitation seems to be mainly *electrical* (bombardment with electrons). A consequence is that is is not a very sensitive source, since only minute amounts of sample are volatilized; therefore it is probably the least desirable source for qualitative analysis. On the other hand, the spark source is very stable and reproducible and is one of the best sources for quantitative analysis.

Laser Microprobe Figure 4-12 is a schematic diagram of a laser microprobe. A high-intensity pulsed laser beam is focused through a microscope onto a sample, immediately vaporizing the spot bombarded. Above the sample is a pair of high-purity graphite electrodes charged to about 1 to 2 kV. Though the electrodes are normally nonconducting, the laser pulse triggers a discharge between the electrodes by vaporizing a portion of the sample, and the radiation emitted from the sparklike discharge is photographed spectrographically.

† Whereas the *gap* temperature momentarily gets very hot with each discharge, the high-temperature flash lasts only about 1 μs and the *electrodes* remain relatively cool.

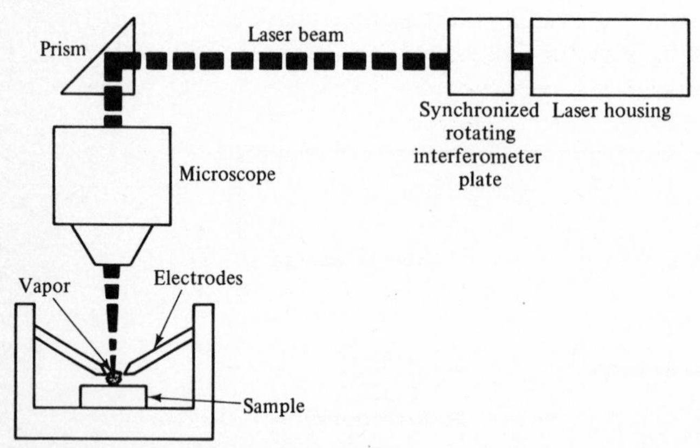

FIGURE 4-12 Schematic diagram of a laser microprobe. (*Jarrell-Ash Division, Fisher Scientific Company.*)

An advantage of the laser microprobe is its ability to be focused on a spot as small as 50 μm across, furnishing the biological researcher with a tool theoretically capable of examining the insides of individual cells. Whereas organic materials tend to ignite with most other types of sources, the laser localizes the vaporization and allows direct analysis. Even refractory nonconducting samples can be examined with this technique.

Attempts are being made to use narrower beams of laser radiation, which would increase the resolution still further. Metallurgists, for example, would be interested in a microprobe with a resolution of at least 5 μm, since at this level individual grains of impurities, precipitated intermetallics, and inclusions could be examined and identified. The major problem has been with the lenses used to focus the laser beam. When a laser beam is concentrated to a diameter below 50 μm, the immense energy in such a small area may melt the cement between lenses or shatter the lens itself. If this problem can be overcome, the laser microprobe will become an exceedingly powerful tool for structural studies.

ELECTRODES AND SAMPLE HANDLING

The electrodes in use in emission spectroscopy are basically of two types, *self electrodes* and *graphite electrodes*. Where the material to be investigated is a conductor and will withstand high temperatures, it may be desirable to use the material directly as the electrodes. This is often done, for example, when analyzing alloys or metal powders that can be pressed into a solid disk or cylinder.

Usually, however, the sample is placed in a small cavity of a graphite electrode, and graphite is used as the upper counterelectrode. Often the top electrode is brought to a point in a pencil sharpener. The bottom electrode usually has some type of cup or depression to hold the sample. Solid samples are usually ground and powdered before putting them in the cup, and liquids are often evaporated in the cup before arcing or sparking. Graphite is a fairly good conductor, and a common technique for increasing reproducibility (minimizing aimless wandering of the arc) is to mix a powdered sample with powdered graphite (called a *buffer*).

Many of the sample electrodes have center posts or are notched to give a narrow neck or both. The center post minimizes wandering of the arc source, improving the reproducibility, and the narrow neck decreases heat conductance away from the gap, improving the sensitivity.

4-2B Monochromators

Since prism and grating monochromators were discussed in detail in Secs. 1-4*F* and 2-2*B*, only their relative advantages and disadvantages for emission spectroscopic studies need be mentioned here.

PRISMS

Figure 4-13 is a schematic diagram for a typical prism spectrograph, the Littrow spectrograph. The name Littrow is given to any prism cut at angles of 30, 60, and 90° and having a reflective backing so that the light beam traverses the prism twice, thereby giving twice the amount of dispersion for a given amount of prism material. Note from Fig. 4-13 that a prism bends short wavelengths more than longer wavelengths; i.e., in this case blue is dispersed considerably, whereas red is reflected back with less dispersion. This is an illustration of the *nonlinear dispersion* characteristic of a prism. Nonlinear dispersion causes the resolution to decrease with increasing wavelength and creates something of a problem in identifying the wavelength of lines which lie along a nonlinear axis.

One of the advantages of using a photographic plate as a detector, as illustrated in Fig. 4-13, is that a large portion of the spectrum can be obtained on one photographic plate, with one prism setting and one burst of excitation energy. To view a region of the spectrum not falling on the photographic plate, it is merely necessary to rotate the prism setting a fixed increment and expose a second photographic plate. A typical large Littrow spectrograph with a quartz prism covers a wavelength range of from 2000 to 8000 Å in about three settings.

GRATINGS

In principle a grating could be substituted for the prism in Fig. 4-13, keeping the rest of the arrangement the same. In practice, several different kinds of mounts have some advantage over the simple arrangement of Fig. 4-13. Two of these, the Czerny-Turner mount and the Ebert mount, were

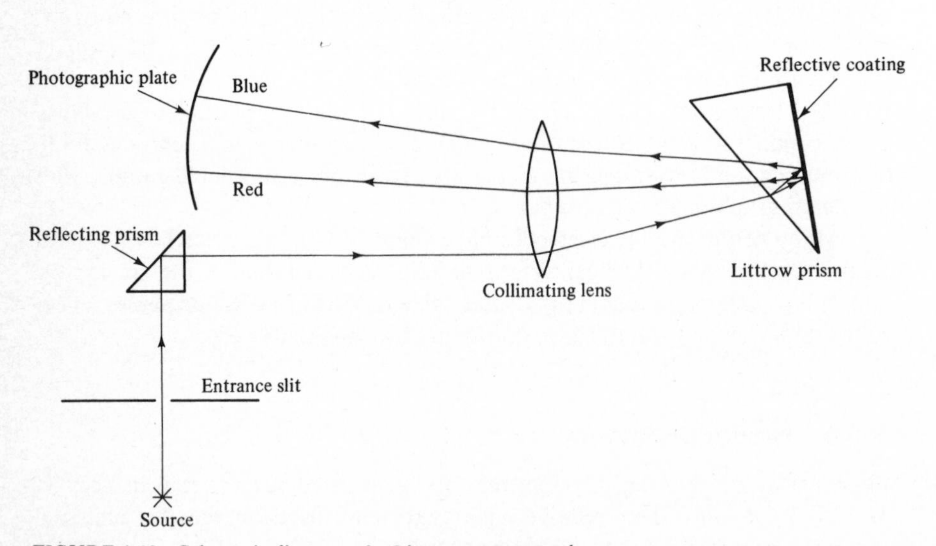

FIGURE 4-13 Schematic diagram of a Littrow spectrograph.

illustrated in Figs. 2-37 and 3-21, respectively. Reference 3 has a good discussion of various grating mounts.

For a number of reasons, gratings have now largely replaced prisms as the dispersing element. High on the list is the cost factor, since replica gratings of good quality are now being made inexpensively. More fundamental is the fact that gratings give *linear dispersion*; thus the resolution is constant and independent of wavelength, and the problem of identifying the wavelength of emission lines on a photographic plate is greatly simplified. Once a known reference line is identified (or preferably, two or more), measuring along the plate with a ruler will identify the wavelength of other lines. A *disadvantage* of gratings is that higher-order wavelengths overlap (see Sec. 1-4*E*), but overlap can usually be eliminated fairly easily with filters or by using a detector that is insensitive to the higher orders.

4-2C Detectors

The two main types of detection systems in use in emission spectroscopy are *photographic* detection and *photoelectric* (or *direct-reading*) detection. The direct-reading instruments are by far the fastest and most convenient instruments, but they are also the most expensive. Both types of instruments are widely used. Each type will be discussed very briefly; additional details are available in Refs. 3, 9, 10, and 13.

PHOTOGRAPHIC DETECTION

The identification of the wavelength of lines on a photographic plate is not difficult, particularly if a grating instrument is used (see Sec. 4-2*B*), but quantitatively evaluating the *intensity* of lines is both an art and a science and requires the knowledge and careful control of numerous factors. Before outlining them it may be useful to summarize the advantages and disadvantages of photographic detection.

There are five principal *advantages* to using photographic plates for detection in emission spectroscopy:

1 A large number of spectral lines are recorded simultaneously (see Sec. 4-2*B*).

2 The photographic plate gives a permanent record of the spectrum.

3 A photographic emulsion can integrate the emission intensity over a period of time.

4 Photographic emulsions have a high sensitivity throughout the ultraviolet and most of the visible region, which is the region of most interest in emission spectroscopy.

5 The cost of an emission spectrograph (with *photographic* detection) is

significantly less than the cost of an emission spectrometer of comparable quality with *photoelectric* detection, which gives direct readout.

The two main *disadvantages* of using photographic plates for detection are the time and effort necessary to process the plates (both to develop them and to interpret them) and the necessity for stringent control of conditions (temperature, development time, etc.). Many of these advantages and disadvantages will become apparent in the discussion of the key properties of photographic emulsions and the steps involved in carrying out a quantitative analysis, which follows.

Two features of photographic emulsions should be understood if a photographic plate is to be used for quantitative analysis. First, photographic emulsions obey the *reciprocity law,* which states that the density of an exposed line depends only on the *total* energy absorbed by the emulsion, regardless of the intensity and duration of irradiation. This can be stated mathematically as

$$D = k \int_{0}^{t=\tau} \epsilon \, dt \approx k\epsilon\tau \qquad (4\text{-}4)$$

where D = density of exposed line on emulsion
k = proportionality constant which depends on type of film and wavelength used
ϵ = average energy (or average intensity) of radiation absorbed per unit time
t = time
τ = total exposure time

Equation (4-4) essentially says that a photographic emulsion is capable of intergrating the response over a period of time, so that even if the source intensity *fluctuates* (as all sources will), as long as it fluctuates around some *average* intensity (or, as long as we irradiate long enough to *establish* a reproducible average intensity), the density of the exposed line will be the same as if the source intensity were absolutely constant.

A second important property of photographic emulsions which affects the interpretation of line density is the *contrast* or response characteristics of the film used. This property of photographic film may be quantified through the use of a characteristic curve like that in Fig. 4-14. Density, plotted on the y axis, is measured by taking the logarithm of the ratio of the intensity of a beam of light passing through a clear portion of the film to the intensity of the same beam after passing through an exposed portion of the film. Exposure, plotted on the x axis, is the relative amount of energy to which the film has been exposed; e.g., it could be various amounts of exposure *time* while irradiating at constant intensity. A characteristics curve like Fig. 4-14 is called an H and D curve after Hurter and Driffield, who discovered the relationship. A large portion of the curve is linear, and the slope γ of this portion is a quantitative measure of the film's *contrast*. A high value of γ indicates that the

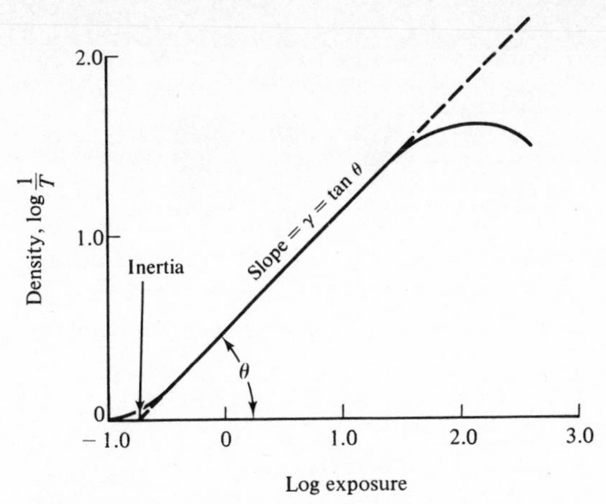

FIGURE 4-14 Typical photographic emulsion characteristic (H and D) curve.

film has strong contrast, whereas a low value of γ indicates low contrast. In general, a high contrast is desirable. The value of γ depends on the development time, which must therefore be carefully controlled for reproducible results. Another characteristic of the particular photographic film is its inertia, identified in Fig. 4-14. Inertia is the x-axis intercept of the linear portion of the curve. The inertia is a measure of the minimum energy necessary to activate the film.

It will be useful to outline the steps involved in interpreting an exposed photographic plate in terms of the concentration of the element giving rise to a given emission line. For details of each step, consult Refs. 6, 10, and 13.

Step 1: Determine the H and D Curve The most difficult step in this process is to determine an H and D curve for the particular emission source, wavelength, and photographic film type to be used. One of the best ways of evaluating exposure is to measure an emission line while rotating a logarithmic step-sector disk in front of the entrance slit (between the source and the entrance slit; see Fig. 4-13). Figure 4-15 shows a logarithmic step-sector disk and its relationship to the entrance slit. The sector shown is a disk with four circular steps cut away. The length of the arc of each step is twice the previous step; thus, as this sector rotates in front of the slit while excitation and photographic recording are occurring, step 1 on the photographic plate will get maximum exposure, step 2 will only get half the exposure, step 3 will get half the exposure of step 2, etc. Thus, each line appearing on the photographic plate will be made up of segments which will vary in density by stepwise amounts from one end of the line to the other. Figure 4-15 illustrates how four lines of different intensities might show up on a photographic plate after only a short exposure time. In Fig. 4-15 the line density is indicated schematically by relative line widths. Of the four lines shown, λ_4 represents the most intense line, λ_2 the second most intense, λ_3 the third most intense,

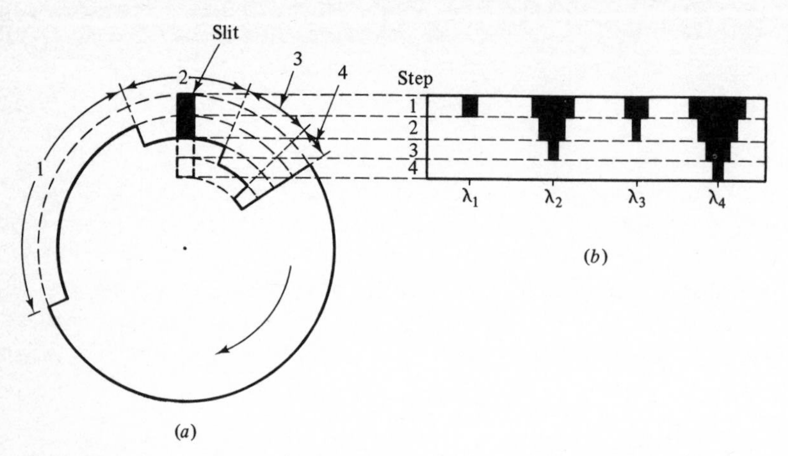

FIGURE 4-15 (*a*) Logarithmic step-sector disk in front of entrance slit. (*b*) Schematic illustration of a four-line spectrum resulting from use of logarithmic step-sector disk (see text for explanation).

and λ_1 the weakest. A line that is not very intense (like λ_1) would have time to blacken the plate only in the small area that is exposed almost constantly (step 1), whereas intense lines (like λ_4) will blacken the plate to almost the entire height of the slit. Only the line at λ_4 would be sufficiently intense to prepare a four-point H and D curve, plotting the density of steps 1 to 4 vs. 1 through 4 arbitrary units of exposure. Other types of sectors are in use (see, for example, Ref. 6).

To measure the density of a given line (or the density of a *step* of a given line), a *densitometer* should be used. The intensity I_0 through a clear portion of a film is first measured, after which the emission line of interest is moved into the light beam and the light intensity I transmitted through the emission line is measured. The density is given by

$$D = \log \frac{I_0}{I} \qquad (4\text{-}5)$$

The photographic plate can also be moved vertically as well as horizontally, so that the density of a single step can be measured (see Fig. 4-15).

Step 2: Prepare a Working Curve Once the H and D curve is prepared, the remaining step is typical of most instrumental methods of analysis: a working curve relating the measured parameter to concentration is prepared. In this case, exposure is plotted vs. concentration. A series of standards of known concentration is run, the density of the lines determined with a densitometer, and the exposure interpolated from the H and D curve.

An additional refinement, necessary for high accuracy, is to include an *internal-standard* element in both the unknown sample and the calibration standards. An internal standard is an element *other* than the one being analyzed for, which is added to sample and standards in a known concentration. By plotting the *ratio* of the intensities (or exposures) of the element sought to the internal-standard element vs. concentration of the element sought, any fluctuations in excitation conditions (or development conditions) should be compensated for. This is because any fluctuations in conditions should presumably affect the intensity of the internal-standard element in the same way as the element sought. In choosing an internal-standard element, it is important that the two elements to be compared have similar rates of volatilization and preferably have lines which are close together in wavelength so that the same H and D curve can be used for both elements. Two lines which satisfy these two criteria are known as a *homologous pair,* and tables listing such pairs are available [9].

PHOTOELECTRIC DETECTION

A number of companies manufacture *direct-reading* emission spectrometers, in which a series of photomultiplier tubes is used for detection instead of a photographic plate. Such instruments are less versatile than spectrographs using photographic plates, since the number of emission lines simultaneously being detected depends on the number of photomultiplier tubes the instrument is equipped with, and this number is necessarily limited by spacing and other problems. Furthermore, direct-reading spectrometers are more expensive than spectrographs. However, direct-reading instruments are very fast, convenient, and more precise than spectrographs. Whereas quantitative error in photographic detection is at least 1 to 2 percent under favorable conditions, photoelectric detection is capable of precision as good as 0.5 percent or better under favorable conditions.

Almost all direct readers are custom-built to analyze for a specific number of elements, although a modest amount of flexibility may be provided for. Typically, a direct reader is a grating instrument with an opaque barrier following the focal curve of the instrument and pierced by 12 or more slits located at wavelengths appropriate to the elements to be analyzed. Behind each slit is mounted a photomultiplier tube, and the output of each tube is recorded automatically. One of the elements observed is usually an internal standard, with which the rest are compared. In some instruments the slits and detectors are movable to allow different elements to be analyzed for, but in most instruments such modifications are difficult and impractical.

Almost all direct readers attempt to compensate for fluctuations in excitation conditions by integrating the detector signal over a period of time. Integration is usually accomplished electronically with a suitable capacitor-resistor circuit in the output of each photomultiplier tube. The voltage across the capacitor at the end of the exposure is a function of the accumulated detector current, and therefore the capacitor voltage is the quantity read out. A typical integration time is about 25 to 40 s.

In summary, photographic detection is to be preferred for nonroutine qualitative analysis, since all emission lines are recorded. On the other hand, where a fixed number of elements must routinely be analyzed for, and where speed and high quantitative precision are needed, a direct-reading spectrometer may more than compensate for its extra cost.

4-3 APPLICATIONS

For a comprehensive summary of applications of emission spectroscopy, the biannual reviews in *Analytical Chemistry* should be consulted [15]. The following brief list will give a sampling of the great variety of materials and elements being analyzed for by emission spectroscopy.

1 Ceramics, for both trace and major constituents

2 High-purity acids and other analytical reagents, for a variety of trace metal impurities

3 Metals, for trace metal constituents as well as oxygen and nitrogen

4 Aluminum, for traces of cobalt

5 Graphite, for traces of cobalt, nickel, molybdenum, and vanadium

6 Spent nuclear fuels, for the rare earths

7 Vitreous substances, e.g., glass and slags; refractories, e.g., clays; and various inorganic and organic substances, all analyzed for trace metals

8 Blood, for trace amounts of calcium, copper, and zinc

9 Mild, for trace amounts of zinc, manganese, and molybdenum

10 Pancreas tissue, for zinc (using the laser-microprobe technique)

Qualitative analysis of unknown samples is one of the most widespread applications of emission spectroscopy. It frequently forms the basis for planning a complete chemical analysis, and the fact that only a minute amount (10 mg or less) of sample material must be expended for this preliminary information is often a particular advantage. Examples include the analysis of thin films on the inner surface of lamp bulbs and numerous cases in forensic chemistry.

Qualitative analysis is extremely reliable by emission spectroscopy, since each element gives a variety of emission lines. Generally, wavelengths are established by a direct comparison with one or more spectra. Most laboratories build up their own reference library of emission spectra, so that comparisons can be made line by line with plates obtained with the same spectrograph and preparative conditions. Alternatively, wavelength tables may be consulted in any of a number of references, e.g., Refs. 10, 11, and 13. Since one photographic plate usually has room for several spectra, it is good practice to photograph the emission spectrum of a known element alongside the spectrum of the unknown sample on the same photographic plate. Iron is often used as a reference element, since its spectrum has a great many prominent lines.

Quantitative analysis was discussed in detail in Sec. 4-2*C*. For routine analyses emission spectroscopy is excellent, but for occasional quantitative analyses there is often a more convenient or precise instrumental technique that would be preferable, e.g., atomic absorption spectroscopy (see Chap. 5).

EXERCISES

4-1 Explain why emission spectroscopy is poorly suited for the determination of the halogens.

4-2 Why do excited molecules emit band spectra whereas excited atoms emit line spectra?

4-3 Using Fig. 4-1, estimate the longest wavelength at which hydrogen atoms would *absorb*. *Hint:* What is lowest energy transition possible for a hydrogen atom in the ground state?

Ans: 1216 Å, the first line in the Lyman series

4-4 Compare the dc arc, the ac arc, and the ac spark as excitation sources for emission spectroscopy. Cite instances where each type would be the preferred source, and give reasons for your choice.

4-5 What are the relative advantages and disadvantages of prisms and gratings for emission spectroscopy?

4-6 Why are direct-reading emission spectrometers considered less versatile than emission spectrographs?

4-7 Using a simple schematic diagram, design a single-channel scanning emission spectrometer, i.e., a spectrometer that has only one photomultiplier detector but which will continuously scan the entire wavelength region and automatically record the emission intensity as a function of wavelength.

4-8 Speculate why single-channel emission spectrometers (see Exercise 4-7) are not available.

4-9 A series of experiments using increasing exposure time is carried out with an emission spectrograph, and it is found that the most intense line in the spectrum turns into a doublet at all exposures exceeding a certain value. Explain this phenomenon.

REFERENCES

INTRODUCTORY

1 CLARK, G. L. (ed.): "The Encyclopedia of Spectroscopy," pp. 99–330, Reinhold, New York, 1960. Various sections written by a variety of authors on most aspects of emission spectroscopy. Lacks continuity, but good as a reference for specific topics.

2 EWING, G. W.: "Instrumental Methods of Chemical Analysis," pp. 164–175, 3d ed., McGraw-Hill, New York, 1969. A brief but clear introduction to the field.

3 STROBEL, H. A.: "Chemical Instrumentation," pp. 110–139, Addison-Wesley, Reading, Mass., 1960. A concise account of the essential aspects of emission spectroscopy.

INTERMEDIATE

4 SCRIBNER, B. F., and M. MARGOSHES: Emission Spectroscopy, chap. 64 in I. M. Kolthoff and P. J. Elving (eds.), "Treatise on Analytical Chemistry," pt. I, vol. 6, Wiley-Interscience, New York, 1965. Useful and reasonably detailed.

5 MITTELDORF, A. J.: Emission Spectrochemical Methods, chap. 6 in G. H. Morrison (ed.), "Trace Analysis," Wiley-Interscience, New York, 1965. An excellent discussion of some of the newer developments in emission spectroscopy.

6 NACHTRIEB, N. H.: "Principles and Practice of Spectrochemical Analysis," McGraw-Hill, New York, 1950. Though not up to date, it contains a useful coverage of most of the fundamental aspects of emission spectroscopy.

7 SAWYER, R. A.: "Experimental Spectroscopy," 3d ed., Dover, New York, 1963. Includes a chapter on the vacuum-ultraviolet region.

8 WALKER, S., and H. STRAW: "Spectroscopy," vol. 1, Chapman and Hall, London, 1961. An excellent discussion of the theory of spectroscopy with a historical perspective.

9 WILLARD, H. H., L. L. MERRITT, JR., and J. A. DEAN: "Instrumental Methods of Analysis," 4th ed., pp. 280–308, Van Nostrand, New York, 1965. A good treatment, stressing instrumentation.

ADVANCED

10 AHRENS, L. H., and S. R. TAYLOR: "Spectrochemical Analysis," 2d ed., Addison-Wesley, Reading, Mass., 1961. An excellent overall treatment, with many data and working guidelines.

11 BRODE, W. R. : "Chemical Spectroscopy," 2d ed., Wiley, New York, 1943. Though not up to date, it contains detailed tables of spectral lines and useful operating principles.

12 CALDER, A. B.: "Photometric Methods of Analysis," pp. 1–87, Hilger, London, 1969. Brief but useful.

13 HARRISON, G. R., R. C. LORD, and J. R. LOOFBOUROW: "Practical

Spectroscopy," Prentice-Hall, Englewood Cliffs, N.J., 1948. A detailed treatment including tables of spectral lines arranged according to element and according to wavelength.

14 HERZBERG, G.: "Atomic Spectra and Atomic Structure," Dover, New York, 1944. A classic; a detailed theoretical treatment, clearly written.

15 BARNES, R. M.: *Anal. Chem.*, **46**(5):150R (1974). A thorough review of emission spectroscopy; these reviews appear every two years.

5

Flame Photometry, Atomic Absorption Spectroscopy, and Atomic Fluorescence Spectroscopy

Flame photometry is an emission-spectroscopic method in which a flame is used as the excitation source and an electronic photodetector is used as the measuring device. Flame emission (FE) photometry is mainly a method of *quantitative* analysis and is one of the most sensitive and precise methods available for the alkali metals, most of the alkaline-earth metals, and at least half a dozen other metal elements. By making a wavelength scan of the emission spectrum (flame *spectro*photometry†), it is also possible to do *qualitative* analysis, but this application is severely limited, compared with ordinary emission spectroscopy, since the energy of the flame is sufficient to excite only about 30 to 50 elements, the number depending on the type of flame used.

Atomic absorption (AA) spectroscopy is closely related to flame photometry in that a flame is used to atomize the sample solution, putting the elements to be analyzed in the form of atomic vapor, but in atomic absorption a separate source of monochromatic light, specific for the element to be analyzed for, is passed through the atomic vapor and the amount of

† To be consistent with the nomenclature adopted throughout this book the name *photometer* should be reserved for *filter* instruments, and *spectrophotometer* should be used with grating or prism instruments capable of scanning the spectrum. Out of deference to commonly accepted practice, flame photometry is used in this chapter as a generic term, regardless of whether the instrument uses a monochromator or a filter.

absorption is measured. Figure 5-1 schematically compares the experimental arrangement of an FE photometer and an AA spectrometer. In flame photometry the sensitivity is proportional to the number of atoms that have been *excited,* whereas in atomic absorption the sensitivity depends on the number of atoms in the *ground state.* Often only a small percentage of the atoms are excited in the flame. Therefore, atomic absorption often provides much greater sensitivity than flame photometry for a number of elements, for example, Mg, Zn, Cd, Pb, and about 12 more. Furthermore, atomic absorption tends to be freer of interferences and procedurally simpler than flame photometry, accounting for the spectacular rise in popularity of atomic absorption in recent years, It should be emphasized that AA spectroscopy, in spite of many claims, has not made flame photometry obsolete. The two methods should be regarded as complementary, each being superior in sensitivity for certain elements.

Atomic fluorescence (AF) spectroscopy, the newest of the three techniques of this chapter to be developed, essentially goes one step further.

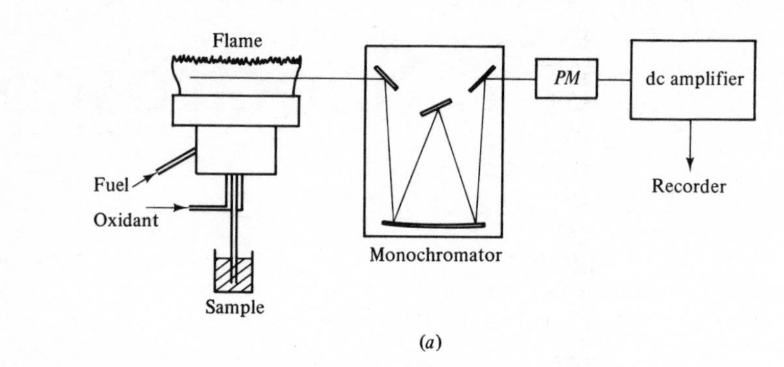

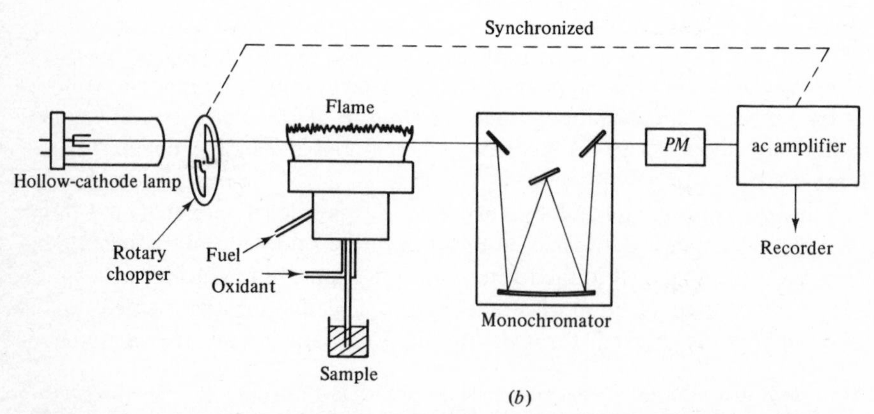

FIGURE 5-1 Instrument arrangements for (*a*) flame emission and (*b*) atomic absorption spectrometers. *PM* = photomultiplier tube. (*From G. W. Ewing, "Instrumental Methods of Chemical Analysis," 3d ed., p.* 177, *McGraw-Hill Book Company, New York,* 1969, *by permission.*)

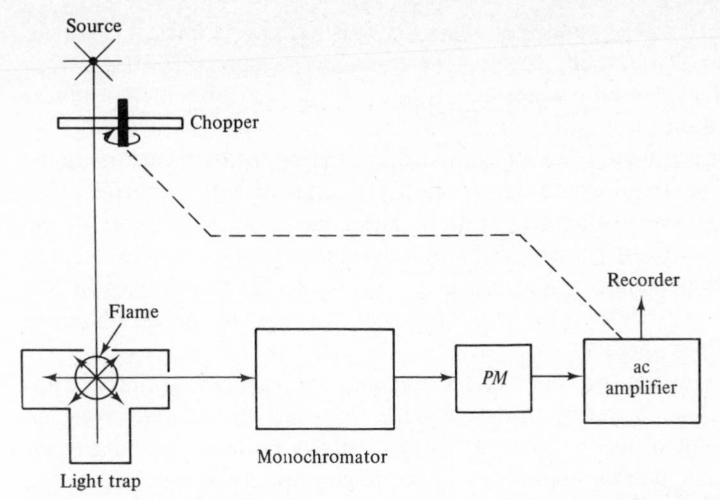

FIGURE 5-2 Schematic diagram of an atomic fluorescence spectrometer.

Whereas in AA spectroscopy the *absorption* of radiation from a hollow-cathode lamp is measured, in AF spectroscopy one measures the resonance *emission* which occurs after the absorption step, the emission being measured at right angles to the beam of exciting radiation. Figure 5-2 illustrates the experimental arrangement for an AF spectrometer. Atomic fluorescence competes with atomic absorption more than with flame photometry. Its biggest potential advantages over atomic absorption appear to be that it is more readily adaptable to the use of a continuous source than atomic absorption and also its sensitivity is directly proportional to the source intensity used, whereas the sensitivity of atomic absorption is relatively unaffected by the source intensity. Thus, the big hope for AF spectroscopy is the development of very intense continuous sources. In certain instances, AF spectroscopy already appears to be the method of choice, e.g., in the determination of zinc and cadmium and possibly mercury.

5-1 THEORY

Much of the theory of emission spectroscopy discussed in Chap. 4, particularly regarding the types of emission spectra, energy requirements, and transitions responsible for line spectra (Secs. 4-1*A* to 4-1*C*), is directly applicable here, and the reader is referred to those sections. The use of a *flame* excitation source instead of an *electrical* excitation source, as is used in the usual arc and spark emission spectroscopy (Chap. 4), results in some important advantages and disadvantages.

The two main *advantages* of *flame* emission and absorption methods are that the spectra produced are very *simple* and quantitative results tend to be more *reproducible*. The simplicity of the spectra results from the lower excitation energy of the flame, and thus there are fewer emission lines. This is advantageous because it greatly lessens the problems of spectral interferences from overlapping lines and bands of other elements and also means that a high-resolution monochromator is not required.

The greater reproducibility of quantitative analyses using flame methods, compared with electrical excitation, is simply a result of better control of all the variables involved in flame excitation. The analyst must be aware of the many variables that are present in flame photometry if they are to be controlled to give high accuracy and precision. There are fewer variables in AA and AF spectroscopy than in flame photometry, but most of the variables are the same in all three methods.

The two main *disadvantages* of FE methods are that the available excitation *energy is too low* for many elements and that the *sample must be dissolved.* As pointed out in Chap. 4, the dc arc, for example, can excite *all* the metal elements; on the other hand, in flame photometry, depending on the type of flame used, only about 30 to 45 elements can be excited. This lower energy is not as much of a disadvantage in AA and AF spectroscopy as it is in FE photometry, since the only function of the flame in AA and AF spectroscopy is to atomize the sample and put it in the form of atomic vapor, i.e., it need not also excite the sample, and thus AA (and theoretically AF) spectroscopy is applicable to many more elements than flame photometry. About 70 elements have been determined by AA spectroscopy.

The requirement that the sample be dissolved for flame methods can be a real disadvantage compared with ordinary emission spectroscopy, where an alloy, for example, can be excited directly as a solid. On the other hand, many samples are already in solution or readily put in solution, and for such samples flame methods are actually more convenient than the ordinary emission methods (for a comparison with ordinary emission methods, see the end of Sec. 4-2*A*).

5-1A Characteristics of Flames

The reactions occurring in flames are exceedingly complex, and while much is known about flames, much remains to be learned [7, 8, 10]. For the analyst using any of the three flame methods described in this chapter a detailed knowledge of all aspects of flames is not essential, but one should understand enough about the characteristics of flames and their function to be able to make a wise choice in such important parameters as burner type, gases, relative flow rates, and height above the burner tip for observation of the emission or absorption process. There are no simple answers to any of these questions, but certain principles and guidelines will help in making a wise choice.

FUNCTION AND REQUIREMENTS OF FLAMES

The flame serves three basic functions: (1) it is used to transform the sample to be analyzed from the liquid (or suspended solid) state into the gaseous state; (2) it is used to decompose the molecular compounds of the element of interest into individual atoms (preferably) or at least into simpler molecules; and (3) in flame photometry, the individual atoms (or simple molecules) are excited by the energy of the flame.

The two main requirements of a satisfactory flame are that it have the proper temperature and gaseous environment to perform these functions and that the flame background not interfere with the observation to be made.

TYPES OF FLAMES AND THEIR TEMPERATURES

A typical premixed flame consists of an inner cone, outer cone, and interconal zone. The *inner cone* is generally a region of *partial combustion,* i.e., without thermal equilibrium, which is being heated by conduction and radiation from the hotter region just above it (see Fig. 5-3). In this region intermediate oxidation products are being formed. Light emission (from the fuel gases, not the sample), ionization, and the concentration of free radicals in this zone are extraordinarily high, and this region is rarely used for any type of analytical work.

Immediately above the inner-cone region is the *interconal zone*. This is the so-called hot part of the flame, corresponding to essentially complete com-

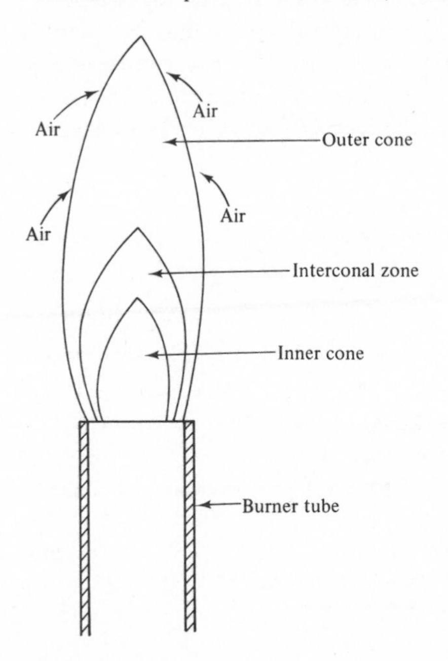

FIGURE 5-3 Schematic diagram of the structure of a flame.

bustion and nearly complete thermodynamic equilibrium. This part of the flame is almost always used in flame photometric analysis and usually in AA and AF spectroscopy as well. The height of this zone above the burner varies considerably with the type of burner, type of gases being burned, and their rate of flow.

The *outer-cone* region (see Fig. 5-3) is a secondary combustion zone, where partially oxidized products like carbon monoxide may complete their combustion. This region is cooled by the surrounding air and is generally a less useful region of the flame.

One of the most important properties of a flame is its *temperature,* since heat energy is the primary form of energy responsible for decomposing samples into individual atoms and for exciting the atoms formed. Since the interconal zone enjoys nearly complete thermodynamic equilibrium, the temperature of this zone is used to characterize a flame.

The maximum temperatures of various types of flames are given in Table 5-1. The temperature must be regarded as only approximate due to the difficulty of accurately measuring the temperature and the fact that the temperature depends on the ratio of gases used and the region of the flame where the temperature is measured.

Notice in Table 5-1 that *oxygen* gives temperatures roughly 1000°C hotter than air. This is because combustion is more rapid in oxygen than in air and, more important, when air is used, heat is wasted in heating up the large amounts of nitrogen present (roughly 80 volume percent nitrogen).

Nitrous oxide, N_2O, decomposes in a flame to give a mixture of two parts of nitrogen to one part of oxygen, and therefore nitrous oxide gives flame temperatures roughly between those of air and oxygen. Flames consisting of acetylene and nitrous oxide are becoming extremely important in flame photometry and AA spectroscopy. Both acetylene-oxygen and acetylene–nitrous oxide flames are capable of forming atomic vapor (and various amounts of excitation) with about 60 elements.

TABLE 5-1 Maximum Temperatures of Various Flames, °C†

	Support gas (or oxidant)		
Fuel	Air	Oxygen	Nitrous oxide
Propane	1925	2800	
Hydrogen	2100	2780	
Acetylene	2200	3050	2955‡
Cyanogen	2330	4550	

† Data from Ref. 5, p. 319, except as noted.
‡ From J. B. Willis, *Nature,* **207:** 715 (1965).

As Table 5-1 shows, a *cyanogen*-oxygen mixture gives temperatures approaching those of a dc arc (see Sec. 4-2*A*), and thus the cyanogen-oxygen flame is a potential rival of the dc arc in excitation sensitivity.† A number of workers, notably Vallee and coworkers [14], have studied the analytical aspects of a cyanogen-oxygen flame, and while its superiority over conventional flames for the detection and determination of a number of elements has been demonstrated, acceptance of the flame has been very slow, probably because of the explosive nature of cyanogen-oxygen mixtures and the high toxicity of cyanogen [15].

Propane gas is fairly widely used for flame photometry, in spite of its relatively low flame temperatures (see Table 5-1). The low temperatures of propane-air mixtures are usually quite adequate for most alkali-metal analyses, as in clinical work. In many hospitals, for example, fire codes prohibit the use of more flammable gases, like acetylene, unless the bottles are isolated outside and the gas piped in. In instances like these, propane or even natural gas is usually used. *Natural gas* (containing mainly methane, propane, and butane) gives a flame temperature equivalent to propane, but it is much less desirable than tank propane because of the lower and more variable pressure of natural city gas.

The *hydrogen* flame is a very good choice for flame-photometric analysis of alkali metals because it gives an extremely low emission background. A hydrogen-air flame is practically nonluminous in the visible region of the spectrum. The only significant radiation it emits is from OH bands below 350 nm. The major disadvantage of hydrogen is its relatively rapid consumption (roughly 12 liters/min). Also, it is more explosive than propane or natural gas.

For most flame analyses, hot flames are advantageous, mainly because chemical interferences are fewer the hotter the flame. Since interferences are discussed in detail in Sec. 5-1*D*, it suffices to mention here that various types of compounds form in flames, interfering with all three methods of flame analysis. If the flame temperature is high enough, most of these compounds are decomposed into free atoms, thereby eliminating or minimizing this source of interference. Similarly, hot flames are generally advantageous in flame photometry because they provide more energy for excitation of metal atoms. On the other hand, if the temperature is too high, a high proportion of the neutral atoms may become ionized, according to the equation

$$M^0 + \text{heat} \rightarrow M^{n+} + ne^- \qquad (5\text{-}1)$$

Since this reduces the population of neutral atoms M^0, the sensitivity of all three methods of flame analysis is decreased if the flame temperature is too

† The comparison on the basis of temperature is not completely valid, since in use the cyanogen-oxygen flame would be appreciably cooled by aspiration of liquid samples into the flame whereas the dc arc almost always is used to excite solid samples and is not comparably cooled.

high. While these arguments point to an optimum flame temperature for each metal being analyzed for, there is often an effective way of suppressing ionization in very hot flames. This is accomplished by adding an excess of an easily ionized element, called an *ionization suppressant,* which has the effect of creating a large density of electrons in the flame and thus repressing the ionization of the element sought, as typified by Eq. (5-1). This is also called the *radiation-buffer technique.* Generally the ionization suppressant is added in a fairly high concentration, say 100 to 1000 ppm (μg/ml). Thus, except where it is not feasible to add an ionization suppressant, it is usually desirable to use a hotter rather than a cooler flame.

TYPES OF BURNERS

There are two basic types of burners in use, the main differences hinging on whether or not the fuel and oxidant gases are mixed before entering the flame area. The first type is called a *total-consumption* or *turbulent-flow burner* (also, but less commonly, a *diffusion burner*), and the second type is called *premix* or *laminar-flow burner.* Figure 5-4 shows representative burners of both types.

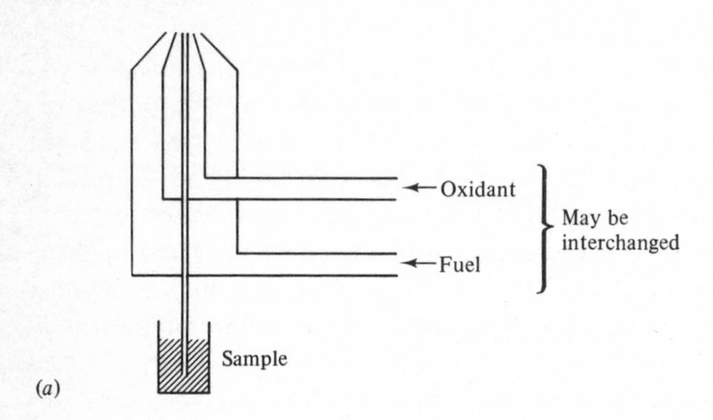

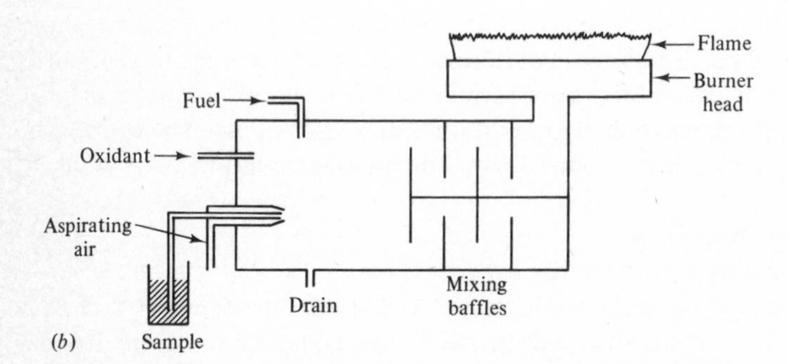

FIGURE 5-4 Two main types of burner designs: (*a*) total-consumption (or turbulent-flow) type; (*b*) premix (or laminar-flow) type. (*From G. W. Ewing, "Instrumental Methods of Chemical Analysis,"* 3d *ed., p.* 180, *McGraw-Hill Book Company, New York,* 1969, *by permission.*)

The *total-consumption* burner aspirates the sample directly into the flame by the venturi effect of fuel and oxidant rushing past the tip of a palladium capillary tube, the other end of which is dipping into the sample solution. *All* the sample traveling up the capillary tube reaches the flame, and hence the name *total consumption.* The high speed with which the oxidant emerges from the very narrow annulus means that a *turbulent* mixing of fuel, oxidant, and sample liquid occurs at the burner tip; hence the name turbulent-flow burner. This turbulence is effective in mixing fuel and oxidant, promoting complete combustion, but it also causes the burner to be noisy, like a blowtorch, the turbulence and hissing noise being accentuated when the spray enters the flame. For this reason, total-consumption burners are generally enclosed in housings with acoustical insulation. The major problem with total-consumption burners is that they produce droplets in the flame of widely differing sizes, and many of the large droplets travel entirely through the flame without totally evaporating. Even for medium-sized droplets that are totally evaporated in the flame, many chemical compounds may not be completely decomposed because the flame has been significantly cooled in evaporating the heavy load of liquid or the remaining residence time in the flame may be too short to complete the decomposition. Thus, chemical *interferences* are more severe in total-consumption burners. Furthermore, the suction capillary tube of this type of burner is easily clogged, especially if the sample solution contains any particulate matter. Finally, the alignment of the capillary with respect to the fuel and oxidant openings is extremely critical and subject to distortions from salt incrustation around the burner tip. The total-consumption burner has been widely used for flame emission and to a certain extent, for AF studies, but it is almost never used for atomic absorption; it is beginning to lose ground to the premixed, or laminar-flow, burner for many FE and AF analyses.

The *premix*, or *laminar-flow*, burner was developed after the total-consumption burner, which it will probably supersede for all but a few specialized applications. In this type of burner (see Fig. 5-4*b*) aspirated sample, fuel, and oxidant are thoroughly mixed before reaching the burner opening and thence the flame. The orifice consists of a long narrow slit, and the gases emerge in a smooth, nonturbulent fashion, i.e., in *laminar* flow. The flame is likewise nonturbulent, noiseless, and stable. The long flame path is a great advantage for atomic absorption and appears also to be an advantage for FE analyses.

Another advantage of the premix burner is that only a uniform fine spray of sample droplets reaches the flame. Larger drops from the aspirator impinge on the walls of the mixing chamber and baffles and are drained off. Depending on the flow rate of gases used and the design of the premixing chamber (or *spray chamber,* as it is often called), as much as 90 percent or more of the sample may go down the drain and be wasted, thereby contributing to a loss of sensitivity. However, this loss must be balanced against the loss of large droplets that pass directly through a total-consumption burner without being evaporated.

Typically, premix burners will handle solutions up to several percent total solids without clogging. The clogging at higher solids content results from precipitation of the material at the hot slot from which the flame gases emerge. If the slot is widened, the clogging is reduced but the possibility of *flashback* increases. The problem of flashback is widely misunderstood and deserves discussion.

Flashback, sometimes called strike-back, occurs when the flame strikes back into the burner tube causing a noisy but generally harmless explosion. Flashback can occur if the velocity of the gases is lower than the burning velocity. Oxygen and acetylene burn with such a tremendous velocity (about 1130 cm/s) they cannot be used together in a premix burner. Air and acetylene burn with a greatly reduced velocity (160 cm/s), making it quite easy to avoid flashback but at the cost of a much lower flame temperature (see Table 5-1). Fortunately, nitrous oxide and acetylene combine the advantages of low burning velocity (180 cm/s) with a flame temperature which is close to that of oxygen-acetylene (see Table 5-1). Therefore, the nitrous oxide–acetylene flame is relatively safe with respect to flashback and is becoming one of the most widely used flames for both FE and AA studies. (For AF analyses, it is desirable to avoid hydrocarbon flames altogether, since CO and CO_2 in the flame are serious quenchers of fluorescence; here, oxyhydrogen flames are perhaps the most widely used [29].) Proper design of the burner also minimizes the chances of flashback. A thick metal head with a relatively narrow slot allows rapid heat conduction, minimizing the chance for the flame to be maintained if it begins striking back through the slot.

As Fig. 5-4*b* shows, the unused liquid sample is removed from the spray chamber through a drain. It is important that the drain tube not be open to the air, or much of the oxidant gas will escape through the drain opening, causing (with acetylene) a luminous and very sooty flame.

5-1B Sequence of Events in a Flame

The sequence of events occurring in rapid succession in the flame can be summarized as follows:

1 The water or other solvent is evaporated, leaving minute particles of dry salt.

2 The dry salt is vaporized, or converted into the gaseous state.

3 A part or all of the gaseous molecules are progressively dissociated to give free neutral atoms or radicals. These neutral atoms are the species which absorb in AA spectroscopy and are the potentially emitting species in FE and AF spectroscopy.

4 A part of the neutral atoms may be thermally excited or even ionized. The fraction that is thermally excited is important in FE analyses, since return of the excited electron to the ground state is responsible for the emission of light which is measured.

5 A portion of the neutral atoms or radicals in the flame may combine to form new gaseous compounds. Since the formation of these compounds reduces the population of neutral atoms in the flames, they constitute chemical interferences in all three methods of flame analysis, as discussed in Sec. 5-1*D*.

FORMATION OF FREE ATOMS

The efficiency with which the flame produces neutral atoms is of equal importance in FE, AA, and AF spectroscopy. Table 5-2 shows the fraction of free atoms present in an air-acetylene flame (at roughly 2200°C) compared with a nitrous oxide–acetylene flame (at roughly 3000°C) for a few selected elements. It may be seen that the hotter nitrous oxide–acetylene flame appears to be more effective in creating free neutral atoms than the cooler air-acetylene flame. The alkali metals are an exception, probably because they become appreciably ionized in the hotter flame. The overall conclusion from these data is that these two flames (particularly the nitrous oxide–acetylene flame) are very efficient at forming neutral atoms. If these data are reasonably

TABLE 5-2 Fraction of Free Atoms Formed in Flames†

For a Given Element, the Fraction is the Ratio of the Number of Free Atoms to the Total Number of Atoms in all States Present in the Flame at Any Instant

	Conditions	
Element	Air-C_2H_2 flame	N_2O-C_2H_2 flame
Li	(0.20)‡	
Na	0.68	0.31
Mg	1.09	1.00
Ca	0.069	0.50
Sr	0.087	0.98
Ba	0.0019	0.11
Zn	0.62	0.54

† Data from Ref. 9 except as noted.
‡ From L. de Galan and J. D. Winefordner, *J. Quant. Spectrosc. Radiat. Transfer*, 7: 251 (1967).

accurate, there is probably little to be gained (in terms of fraction of free atoms) by using flames other than these.

FORMATION OF EXCITED ATOMS

The fraction of free atoms that are thermally excited is governed by a Boltzmann distribution, as follows:

$$\frac{N^*}{N^0} = Ae^{-\Delta E/kT} \tag{5-2}$$

where N^* = number of excited atoms
N^0 = number of atoms remaining in ground state
A = constant for particular system
ΔE = difference in energies of the two levels
k = Boltzmann's constant
T = flame temperature, K

The fraction of atoms N^*/N^0 excited to the first excited energy level at various temperatures is given in Table 5-3 for a few elements. The striking feature about these data is the relatively small fraction of ground-state atoms that are excited at ordinary flame temperatures. Only if the flame temperature is very high does the fraction of atoms excited become appreciable. That the fraction of atoms excited depends critically on the temperature of the flame emphasizes the importance of controlling flame temperature carefully in FE analysis. Conversely, since the fraction of atoms in the ground state is very large, small fluctuations in flame temperature are not as important in AA analysis, and this is a decided advantage of AA spectroscopy. One should not carry this argument too far, however. For example, the smallness of the fraction of atoms excited is often advanced as evidence that atomic absorption is inherently capable of higher sensitivity than flame emission. This argument is worth examining briefly.

TABLE 5-3 Fraction of Atoms in First Excited State at Various Temperatures†

Element	Emission line, nm	Temperature, K		
		2000	3000	4000
Cs	852.1	4×10^{-4}	7×10^{-3}	3×10^{-2}
Na	589.0	1×10^{-5}	6×10^{-4}	4×10^{-3}
Ca	422.7	1×10^{-7}	4×10^{-5}	6×10^{-4}
Zn	213.9	7×10^{-15}	6×10^{-10}	2×10^{-7}

† Reprinted with permission from A. Walsh, *Spectrochim. Acta*, **7**: 108 (1955), Pergamon Press.

If the intensity of the emission signal in FE spectrometry is designated as I_{FE}, it follows that this signal will be directly proportional to the number of excited atoms N^*, according to the equation

$$I_{FE} = k_{FE}N^* \tag{5-3}$$

where k_{FE} is a proportionality constant which depends on the efficiency of the emission process. On the other hand, in AA spectroscopy the signal which is important is really ΔI_{AA}, the *difference* in the intensity of the source when the sample is in the flame from when the sample is absent. This signal difference depends on the number of atoms in the ground state N^0, given by

$$\Delta I_{AA} = k_{AA}N^0 \tag{5-4}$$

where k_{AA} is a different proportionality constant, depending on the intensity of the source and the efficiency of the absorption process. If k_{FE} were equal to k_{AA}, then obviously the fraction of excited atoms (N^*/N^0) would be the sole factor determining the relative sensitivity of flame emission and atomic absorption. Because of spectral radiance factors and relative line widths in the two processes, however, k_{AA} is often smaller than k_{FE}, making flame emission more sensitive than atomic absorption for a number of elements like aluminum, calcium, strontium, gallium, indium, vanadium, and the alkali metals [29]. Further details may be found in Refs. 6 and 9.

FORMATION OF IONIZED ATOMS

The ionization that occurs in a flame rarely goes beyond the loss of one electron and can be represented by the equilibrium equation

$$A \rightleftarrows A^+ + e^- \tag{5-5}$$

where A = neutral atom
A^+ = its positive ion
e^- = free electron

This dissociation process will be concentration- or pressure-dependent, since one species is dissociating into two. Assuming the partial pressure of metal atoms in the flame to be 10^{-6} atm, the percent ionization of the alkali and alkaline-earth metals at three different flame temperatures is given in Table 5-4. As the partial pressure of metal atoms in the flames increases, the percentage ionization decreases, as would be expected from consideration of the Le Châtelier effect on Eq. (5-5). Table 5-4 shows that easily ionized elements like rubidium, cesium, and barium are highly ionized even at relatively low flame temperatures; at the temperature of acetylene-oxygen almost all the elements are significantly ionized. As mentioned earlier, the degree of ionization of the element to be analyzed for can be suppressed by the addition of a fairly high

TABLE 5-4 Percent Ionization of Alkali and Alkaline-Earth Metals in Various Flames†

Element	Ionization potential, eV	Air-propane, 2200 K	Hydrogen-oxygen, 2450 K	Acetylene-oxygen, 2800 K
Li	5.37	<0.01	0.9	16.1
Na	5.12	0.3	5.0	26.4
K	4.32	2.5	31.9	82.1
Rb	4.16	13.5	44.4	89.6
Cs	3.87	28.3	69.6	96.4
Ca	6.11	<0.01	1.0	7.3
Sr	5.69	<0.1	2.7	17.2
Ba	5.21	1.0	8.6	42.8

† From J. A. Dean, "Flame Photometry," p. 42, McGraw-Hill Book Company, New York, 1960, by permission.

concentration of another element which is more easily ionized (a radiation buffer or ionization suppressant). Generally, suppressing ionization in this way is preferable to using a lower-temperature flame, since chemical interferences are more serious at lower temperatures (see Sec. 5-1*D*).

5-1C Background

Background must be interpreted as a signal that is present when the element sought is absent, and as such, the sources of background are different in each of the three methods of flame analysis. Of the three methods, background contributions are least in AF spectroscopy, particularly when a line source is used, since the only reasonable source of background would be trace amounts of the element sought in the reagents used, and this is easily corrected for by running a blank. With a continuous source, some background contribution from *scattering* is also possible, but this can usually be detected and corrected for by scanning over the wavelength region of interest.

In AA spectroscopy, many workers are unaware of potential background errors, and the fact that they are difficult to measure with ordinary AA equipment makes the threat of undetected background errors a serious problem. Background in atomic absorption can be caused by molecular absorption and by scattering from particles in the flame. With the usual monochromatic line source (a hollow-cathode tube) there is no good way of evaluating this background quantitatively. Sometimes a superficial background check can be made by using a second emission line from the hollow-cathode lamp, preferably one very close in wavelength to the analytical line of interest. A more efficient way of evaluating the background and correcting for it is to use an auxiliary continuous source, e.g., a hydrogen or deuterium lamp, in addition to the hollow-cathode lamp and either scan the wavelength region in the immediate vicinity of the line of interest (single-beam instrument) or automatically and electronically compensate for it (double-beam instrument). With a single-beam instrument the background can be estimated by means of a line drawn across the base of the absorption line, a technique called the *base-line* method. A number of AA instrument manufacturers have accessories available, e.g., Perkin-Elmer's deuterium background corrector, which allow these background corrections to be conveniently made.

The background contributions in flame photometry are by far the most substantial. Background is contributed from two sources, the *flame* emission background and emission from the *sample matrix*. There is no precise, characteristic spectrum that can be associated with the flame of a given fuel, since the spectrum is very dependent on such flame conditions as the fuel-to-oxidant ratio and the temperature. However, it is instructive to consider the emission spectrum of a typical nitrous oxide–acetylene flame, as shown in Fig. 5-5. At first sight, the brightness of the flame and the complexity of the spectrum might discourage anyone from trying FE photometry for traces of metals in the presence of this background. However, with high-resolution (preferably grating) spectrometers, a very narrow wavelength region away from the major background lines and bands can usually be selected and the background thus kept to within tolerable limits. It should be clear that it is essential to correct the emission readings for this flame background, but this is usually readily accomplished by taking a reading on a solvent blank. Organic solvents used to enhance the emission intensity may contribute greatly

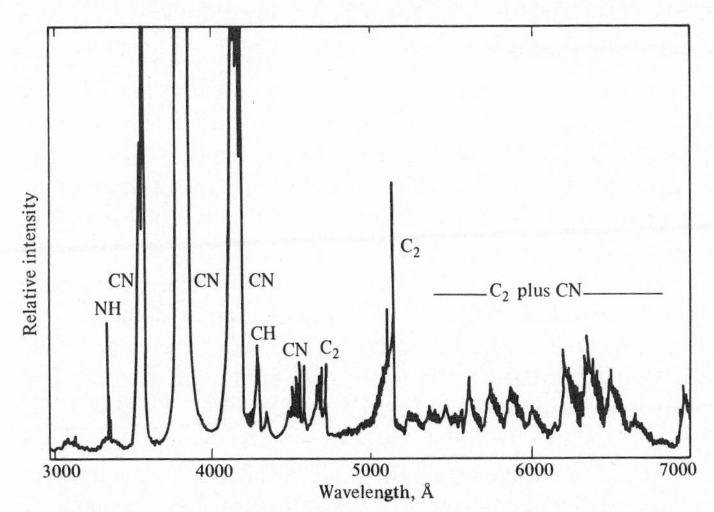

FIGURE 5-5 Emission spectrum of a nitrous oxide–acetylene flame. [*From E. E. Pickett and S. R. Koirtyohann, Anal. Chem.*, **41**(14):37A (1969). *Copyright by the American Chemical Society*.]

to the flame background, and this is a major disadvantage of such solvents. It has already been mentioned that the hydrogen flame gives a very low flame background, and thus the all-important signal-to-background ratio is often most favorable with this flame.

Actually, however, radiation from the flame itself is often of less concern than the additional background contribution from the *sample matrix*. The best way of evaluating and correcting for this contribution is to slowly scan over the wavelength region of the line of interest and use the *base-line* method to make the correction. Alternatively, where a manual monochromator is used or where it is necessary to conserve sample, fixed readings at 1½ to 2 times the bandpass of the instrument can be made on either side of the nominal wavelength. However, readings at fixed wavelength settings can be in error if unresolved fine structure or background lines are present, and a continuous wavelength scan is the only reliable way of observing these problems. It is desirable, though not always practical, to make such scans on samples from which the element being sought has been removed, in order to rule out the possibility of an interfering line which is unresolved from the analysis line.

5-1D Interferences

The classification of interferences, the seriousness of interferences, and the means of dealing with interferences have been treated in a variety of ways, often with conflicting conclusions. This section will classify interferences in a very simple way and attempt to put in perspective the relative seriousness of the various kinds of interferences in FE, AA, and AF spectroscopy.

SPECTRAL INTERFERENCES

Spectral interference may be defined as radiation reaching the detector from a source other than the element being sought. In other words, spectral interferences are caused by the overlapping of a line or band of an impurity element at the wavelength of the element to be measured. For example, in the flame photometric analysis of sodium at 590 nm, the presence of calcium in the sample may give a spectral interference because calcium can form the thermally stable compound CaOH in the flame and CaOH emits an orange band of radiation which overlaps 590 nm. As another example, depending on the resolution of the instrument used, the presence of manganese in a sample may interfere with the flame photometric analysis of potassium at 404.4 nm, since manganese gives an intense emission line at 403.3 nm. In both these examples, the interference can be minimized (or in the second case even eliminated) by the use of a monochromator having sufficient resolution to ensure that only the sharp emission line of interest is measured. In general, when the spectral interference is an overlapping *band*, as with CaOH, high resolution cannot theoretically *eliminate* the interference, but the emission intensity of most bands is so low compared to the intensity of an emission line that high resolution usually reduces the relative signal strength from a band to very low proportions. If the interference cannot be prevented by increased resolution, it may be possible to select other spectral lines where mutual interference does not occur, or it may be necessary to perform a separation, perhaps by ion exchange or selective solvent extraction.

In AA spectroscopy direct spectral interferences are far less common than in flame photometry, since in atomic absorption the source line is extremely narrow and specific (typical line half-widths from hollow-cathode tubes are of the order of 0.01 to 0.05 Å). However, it is not true, as is often implied in the literature, that there are no spectral interferences in atomic absorption. Table 5-5 lists six pairs of lines that give mutual interference in AA spectroscopy. Thus, for example, in the analysis of copper the line at 3247.5 Å is usually the logical choice, but if europium is present, another, less sensitive line would have to be used in order to avoid interference. It should be pointed out, however, that these are the only six pairs of lines that have been found thus far to give mutual interference in atomic absorption, and of these, only the copper line at 3247.5 Å is commonly chosen as the first choice of wavelengths to be used for analysis. In AF spectroscopy with the usual line source, the number of spectral interferences is even less than in atomic absorption.

Flame background, already discussed, can also be considered a type of spectral interference.

CHEMICAL INTERFERENCES

Chemical interferences are those in which some kind of chemical compound is present or is formed in the flame which decreases the population of free metal atoms. Chemical interferences affect FE photometry, AA spectroscopy, and AF spectroscopy similarly.

TABLE 5-5 Pairs of Elements Giving Mutual Spectral Interference in AA Spectroscopy†

Pair number	Element and line, Å			
1	Cu	3247.540	Eu	3247.530
2	Si	2506.899	V	2506.905
3	Fe	2719.025	Pt	2719.038
4	Al	3082.155	V	3082.111
5	Hg	2536.52	Co	2536.49
6	Mn	4033.073	Ga	4032.882

† Data from Ref. 9.

Many elements form oxides or hydroxides in the flame. Typical components of a burned gas mixture are atomic or molecular hydrogen or oxygen and neutral or negatively charged hydroxyl radicals. These reactive species, particularly oxygen atoms and hydroxyl radicals, may induce the formation of oxides or hydroxides, as follows:

$$M^0 + O = MO \tag{5-6}$$

$$M^0 + OH = MOH \tag{5-7}$$

Among the alkali metals, lithium, rubidium, and cesium are known to form hydroxides in the flame, and most other metals have tendencies to form both oxides and hydroxides. The best way to minimize these interferences is to use as hot a flame as possible, since the degree of dissociation of all these compounds increases as the flame temperature increases.

Other kinds of chemical interferences may occur when components such as aluminum, boron, chromium, iron, phosphate, and sulfate are present in a sample. All these substances have in common their ability to exist as anions (aluminate, borate, etc). As such, these anions can crystallize with the metal ion being sought as the droplets of spray are evaporating, forming compounds of the general type MX, which must then be vaporized and dissociated according to the equilibria

$$MX(s) \rightleftarrows MX(g) \tag{5-8}$$

$$MX(g) \rightleftarrows M^0(g) + X^0(g) \tag{5-9}$$

Either of the two steps can be slow. For example, salts of high-melting compounds are slow to be converted into the gaseous state [Eq. (5-8)], and this will limit the population of neutral metal atoms (M^0). Other compounds may have high dissociation energies, causing Eq. (5-9) to be the limiting step.

A *premix* burner will give smaller chemical interferences of this type than a total-consumption burner, thanks to a more effective vaporization step [Eq. (5-8)]. Similarly, the use of hotter flames will speed up and shift both Eqs. (5-8) and (5-9) to the right. However, a more effective way of circumventing this type of interference is to use *chemical additives,* either releasing agents or protective chelating agents. A *releasing agent* is a metal ion which also forms compounds with the interfering anion. When it is added to the sample and standards in fairly high concentration, it preferentially ties up the interfering anion and releases the element being sought. For example, addition of a few hundred parts per million of lanthanum, strontium, or magnesium salts is effective in overcoming interferences due to phosphate. The determination of calcium is a specific case where phosphate interference can be severe.

In *protective chelation,* a chelating agent is added to complex the metal ion being sought, to prevent it from combining with the interferent in the solution phase. To be effective, the chelating agent must readily be decomposed in the flame. Ethylenediaminetetraacetic acid (EDTA) is one of the most effective protective chelating agents. Various polyhydroxy alcohols have also been used.

The use of various miscible organic solvents, notably methanol, acetone, and methyl isobutyl ketone, has been found in a number of cases to give very significant enhancements, particularly in flame photometry [5,13]. There is some evidence that a contributing factor in the FE enhancement is the increase in *chemiluminescence* in the flame; radicals and radical ions in the flame emit ultraviolet and visible radiation, which in turn serves as an excitation source in addition to the thermal energy of the flame. The fact that this effect decreases the population of ground-state atoms could be disadvantageous in AA and AF spectroscopy. Another problem reported with premix burners is an increase in chemical interferences when organic solvents are used [9]. For example, in the AA analysis of calcium with an air-acetylene flame, the interference from phosphate is more serious if acetone or alcohol is present than when aqueous solutions are used. The increased chemical interference is attributed to a need to reduce the flow of acetylene to avoid excessive flame richness when organic solvents are present, and this, in turn, reduces the flame temperature since the solvent vapor has a lower heat of combustion than the acetylene it replaces [9].

PHYSICAL INTERFERENCES

Alteration of the physical properties of the solution (viscosity, surface tension, density, vapor pressure) by *salts, acids,* or *organic substances* can cause a transport interference, as well as changes in flame temperature, rate of evaporation of solvent and solute, and form of the flame. Variations in these properties usually cause a bigger problem with premix burners than with total-consumption burners, but changes in surface tension or viscosity will obviously effect the rate of atomization with a total-consumption burner to an extent that cannot be overlooked. The use of an organic solvent may increase the rate of transport and give a greater proportion of small droplets.

With premix burners, the vapor pressure and surface tension will influence droplet size. Added salts and acids hinder the evaporation of solvent, and the larger droplets are more likely to condense on the walls of the spray chamber and be lost. The solution of these problems is to be sure that the physical properties of the sample solutions and standard solutions are as similar and as constant as possible. Sometimes dilution of these solutions will achieve this purpose, and other times it may be necessary to add a high concentration of some salt or acid to both the test solution and comparison standards. Lithium chloride is often added for this purpose.

5-2 INSTRUMENTATION

5-2A Basic Instrument Arrangements

As already noted, the three techniques discussed in this chapter have much in common, and their differences arise mainly from the physical differences involved in their respective modes of signal measurement.

It is apparent from Fig. 5-1 that an AA spectrometer is basically an FE spectrometer to which a separate radiation source (generally a hollow-cathode lamp) has been added. In order to reject the FE signal and accept only the AA signal, the hollow-cathode source is modulated (usually with a mechanical chopper, as shown in Fig. 5-1*b*), and an ac amplifier (instead of the dc amplifier usually used in a flame photometer) is tuned to the frequency of the source modulation.

With an AF spectrometer (Fig. 5-2) the arrangement is almost identical to that of atomic absorption, except that the radiation from the flame is viewed at right angles to the light path from the source. Also, a baffled light trap is usually used around the flame to minimize the effect of scattered radiation.

It is feasible to use an unmodulated source and dc amplifier in AF spectroscopy if measurements are made in the ultraviolet region below 3000 Å, where flame background is low.

5-2B Flame Photometers

There is a large variety of commercial FE photometers in widespread use, ranging all the way from very simple single-beam filter photometers to multichannel spectrometers with automatic background correction. In addition to instruments designed specifically as flame photometers, many AA instruments now becoming available are designed to perform flame emission also. For a fairly complete listing of the large variety of commercially available flame photometers, see Refs. 3, 11, and 15.

EXAMPLES OF FLAME PHOTOMETERS

Figure 5-6 is a schematic diagram of a typical prism flame spectrometer. For the actual monochromator details, which would be a little more sophisticated than in Fig. 5-6, the reader can consult Sec. 2-2*B*. In principle the prism of Fig. 5-6 could be replaced with a *filter* or a *grating*, thereby simulating all the basic types of flame photometers. In practice the *grating* flame spectrometer is better illustrated with the Ebert mount shown in Fig. 5-1*a*. Another very popular type of grating mount is the Czerny-Turner mount, which is similar to the Ebert mount shown except that two separate concave mirrors take the place of the one large collimating mirror at the bottom of the monochromator in Fig. 5-1*a*; see, for example, Figs. 2-37 and 3-23.

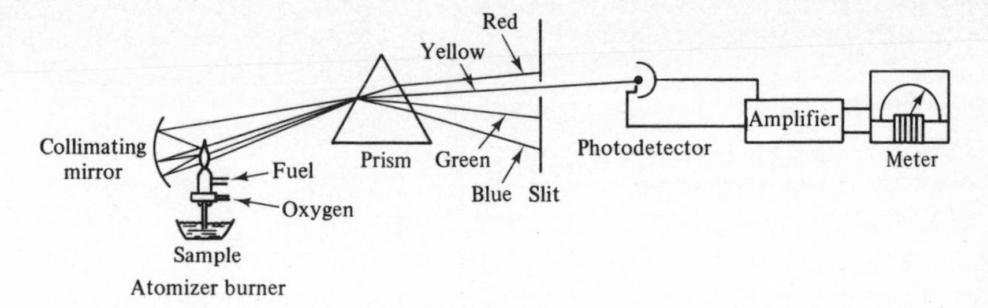

FIGURE 5-6 Schematic diagram of a prism flame spectrometer. (*From J. A. Dean, "Flame Photometry," p.* 67, *McGraw-Hill Book Company, New York*, 1960, *by permission.*)

The burner shown in Fig. 5-6 is a total-consumption type, but a premix burner could also be used. Note that a collimating mirror is used behind the flame to increase the emission intensity. Such backing mirrors are often omitted with premix burners having a long flame path (see Fig. 5-1*a*), especially with instruments used for both flame emission and atomic absorption. Similarly, it is sometimes necessary to omit the backing mirror in multiple-beam instruments, where there is insufficient room for them.

Figure 5-7 schematically illustrates the principles of an internal-standard flame photometer. Typically, lithium is the internal-standard element used, and an equal concentration is added to the standard and sample solutions. By measuring the ratio of the intensity of the element sought to that of lithium, many potential errors are minimized. In the schematic diagram of Fig. 5-7 a grating is used, and the lithium photodetector is shown in a fixed position at the 671-nm lithium line, whereas the detector for the element sought is movable, depending on which element is being determined. Actually, the most common arrangement for an internal-standard instrument is to use *filters*, one of which will pass only the lithium line and the other will pass the line of interest. The internal-standard method is widely used in flame photometry, but since it often is not completely understood, it will be discussed in some detail.

THE INTERNAL-STANDARD METHOD

The internal-standard method can be employed on any ordinary flame photometer by simply taking successive readings on the internal-standard element (usually lithium) and then on the element sought, from which the ratio of the two intensities is calculated. This ratio is then plotted vs. the concentration of the element sought to form a calibration curve (the concentration of the internal-standard element being held constant). While this procedure eliminates certain errors, it nevertheless is not as convenient or as accurate as using an instrument like that in Fig. 5-7, which gives a direct and simultaneous reading of the ratio of intensities. Obviously, when successive readings with a single-beam instrument are taken, the effects of mo-

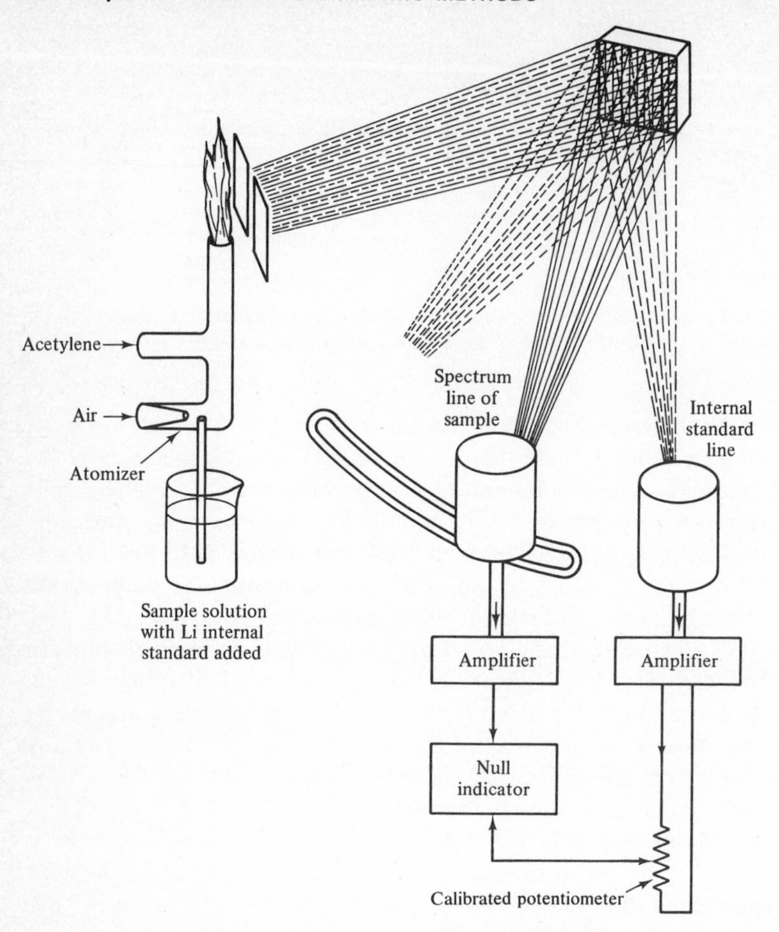

FIGURE 5-7 Schematic diagram of an internal-standard flame photometer. (*From E. J. Bair, "Introduction to Chemical Instrumentation," p. 23, McGraw-Hill Book Company, New York,* 1962, *by permission.*)

mentary fluctuations in flame characteristics, caused, for example, by fluctuations in fuel or oxidant pressures, or electronic drift will not be eliminated, as they would be with the direct-reading instrument of Fig. 5-7. However, in neither case can the internal-standard method solve all these problems, and often the method is abused to the point of giving less accurate results than could be obtained without an internal standard.

There are two general requirements for the element being chosen as the internal standard: (1) it should not be present appreciably in the original sample and (2) it should give an emission line that reacts to interferences similarly to the line of the element sought. One or both of these requirements is often violated, sometimes with disastrous results. Table 5-6 gives some line pairs that react very similarly to most interferences. Some misgivings have lately arisen regarding the use of lithium, because its emission is more sensitive to location in the flame than that of other elements and because changes in gas composition and OH content of the flame have a greater effect on lithium than on sodium or potassium [15]. Calcium is not a very good internal-standard element because it is a common impurity.

In using the internal-standard method for a new application, the suitability of a given internal-standard element should be tested by determining the direction and magnitude of the interferences affecting the two emissions separately and in conjunction. It is hoped that the interference effects on the two lines will be of the same *magnitude*, but in any event they must at least be of the same *sign*, so that the interference effects will be largely canceled when the ratio is taken. Otherwise, it would be far better not to use an internal standard.

The requirement that the internal-standard element not be present appreciably in the original sample is sometimes violated unexpectedly, and it is a good idea to test this possibility with new samples before adding the internal-standard element. An alternative approach is to use an element already present in the sample as the internal standard, provided its concentration can be duplicated in the calibration solutions. Usually this is feasible only if the element is a major constituent of the sample. Still another approach is to use a suitable part of the flame background adjacent to the analytical line as the internal standard [15].

Even a poor choice of internal-standard element will usually allow

TABLE 5-6 Line Pairs Useful for Internal-Standard Analysis†

Analytical line, nm	Internal-standard line, nm
Na 589	Li 671
K 767	Li 671
Ca 554	Li 671
Ca 630	Li 671
Ba 554	Li 671
Ca 554	Na 589
Li 671	K 767
Li 671	Co 352.9
Cu 324.8	Ag 328.0
Fe 386.0	Co 387.1

† From R. Herrmann and C. T. J. Alkemade, "Chemical Analysis by Flame Photometry," p. 166, John Wiley & Sons, New York, 1963, by permission.

minimization of irregularities of atomization and variable condensation of the liquid in the spray chamber of a premix burner. Similarly, systematic errors due to differences in viscosity and surface tension are considerably reduced. The effect of flame flicker is likewise reduced, but, as already mentioned, the effect of radiation interference may or may not be reduced, depending on the similarity of the effects on the two elements. Possible pipetting errors and the additional labor that the use of an internal standard entails should also be considered.

PRESSURE REGULATION FOR THE BURNER-ATOMIZER

The two main types of burner-atomizers, the *total-consumption* type and the *premix* type, were described in Sec. 5-1*A*. One of the most critical features of flame photometry is the regulation of fuel-gas and oxidant-gas flow, affecting both the stability of the flame and the rate of sample atomization. Generally the gases to be used are contained in steel cylinders under pressure, although air from a compressor is sometimes used, and occasionally city gas is used with a booster pump. In this case it is necessary to have high-quality regulating valves for careful adjustment of the pressure and flow rate, plus gauge and/or flowmeters to allow accurate and continuous monitoring of the gas flow. Ordinary needle valves, by themselves, cannot achieve constant flow and pressure, independent of changes in pressure in storage tanks or other gas lines. The most satisfactory type of regulating valve for this purpose is the *diaphragm* pressure-regulating valve.

Figure 5-8 shows the construction of a typical single-stage pressure regu-

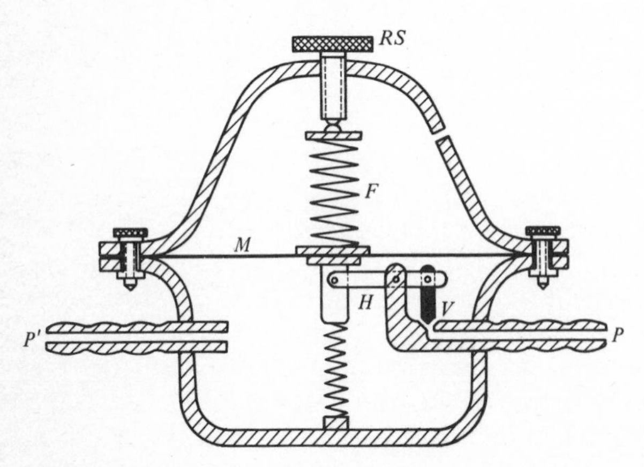

FIGURE 5-8 Schematic representation of the construction of a single-stage pressure regulator (diaphragm type). M = membrane, H = lever, V = valve, F = spring, RS = regulating screw, P' = high-pressure (source) side, P = low-pressure (burner) side. (*From R. Herrmann and C. T. J. Alkemade, "Chemical Analysis by Flame Photometry," p.* 91, *John Wiley & Sons, Inc., New York,* 1963, *by permission.*)

lator. To close off and *stop* the gas flow, the regulating screw (*RS* in Fig. 5-8) is loosened (turned counterclockwise), which removes the tension on the coil spring (*F*) and allows the valve (*V*) to close. To *start* the gas flow, the regulating screw is tightened (turned clockwise) applying tension on the membrane (*M*) through the coil spring and allowing the valve to open and take a position of stable equilibrium. The rate of gas flow is proportional to the tension put on the spring *F*, which in turn is determined by the position of the regulating screw *RS*. Even if the source pressure (P') changes, the flexible diaphragm (or membrane, *M*) responds to keep the burner pressure relatively constant. Thus, if P' decreases, the pressure differential causes the membrane to move down, further opening the valve *V* to maintain pressure *P*.

A single-stage regulator with a large membrane may be adequate for a good air-compressor source if the pressure reduction is not great (from say 40 to 20 lb/in^2), but if it is necessary to reduce the pressure from a high-pressure steel cylinder, e.g., at 2000 lb/in^2, a single-stage regulator will be inadequate. For this purpose a two-stage pressure regulator is needed. These regulators are essentially two single-stage units mounted in series, the first reducing the pressure from, say, 2000 to 75 lb/in^2, and the second unit being used to bring this pressure down to a value slightly higher than that needed. A two-stage regulator outwardly looks the same as a single-stage regulator. A single regulating screw provides the pressure adjustment needed. Both single-stage and two-stage regulators generally have two pressure gauges attached to them, one on the high-pressure side monitoring the source (tank) pressure and one on the exit (burner) side. Many flame photometers have their own two-stage regulators built into them, for final and exacting adjustment of the pressure flow. Usually about a 2:1 pressure ratio is desirable between the gas pressure coming from the two-stage regulator on the tank (source) and the final pressure supplied the burner by means of the controls on the flame photometer.

TYPES OF DETECTORS

Because of its high sensitivity and its ability to use a small slit width for high resolution, a *photomultiplier* tube is by far the best detector. The RCA 1P28 or EM-1 is used over the range of approximately 200 to 750 nm, giving a high quantum efficiency and a low dark current. For longer wavelengths, no completely satisfactory photomultiplier tube is currently available, and a red-sensitive *phototube* must be used. Moderately priced flame photometers use phototube detectors throughout the wavelength range covered, and many simple filter flame photometers use only a *photovoltaic* cell. Further details on these detectors can be found in Sec. 2-2*B*.

5-2C Atomic Absorption Spectrometers

Commercially available AA spectrometers have been developed at a rapid rate over the last few years, and there are now well over 20 commercial models available. Models range from very simple single-beam instruments to highly automated complex designs. Most instruments are also designed to perform FE measurements, and some are capable of AF measurements as well. Descriptions of commercial instruments can be found in Refs. 20, 21, and 23.

SINGLE- AND DOUBLE-BEAM INSTRUMENTS

Most instruments use a single-beam optical system, of the type shown in Fig. 5-1*b*. Figure 5-9 shows a block diagram of a typical double-beam AA spectrophotometer. With the double-beam system the light from the source is split, usually by means of a rotating sector mirror, into two beams, one of which passes through the flame, while the second (the reference beam) is deviated around the flame. At a point beyond the flame the two beams are recombined, and their ratio is electronically measured.

The advantages and disadvantages of a double-beam system have been debated [23, 25, 27]. Clearly, the double-beam system cannot overcome instability and noise from the burner, since the burner is in only one of the two beams. What the double beam can do is to overcome the effects of source fluctuations and drift and changes in detector sensitivity with time. Detector stability is not a serious problem, and therefore the question is whether the flame or the hollow cathode is the limiting factor in the stability of the instrument. Recent burner designs give a relatively stable flame, but fluctuations in flame background are still a potential problem. One definite advantage of a double-beam system is that warm-up time of the hollow-cathode lamp is greatly reduced, but preheating units for hollow-cathode tubes tend to reduce this advantage. Over a prolonged run, e.g., with sequential analysis of a large number of samples, the freedom from electronic drift can be a decided

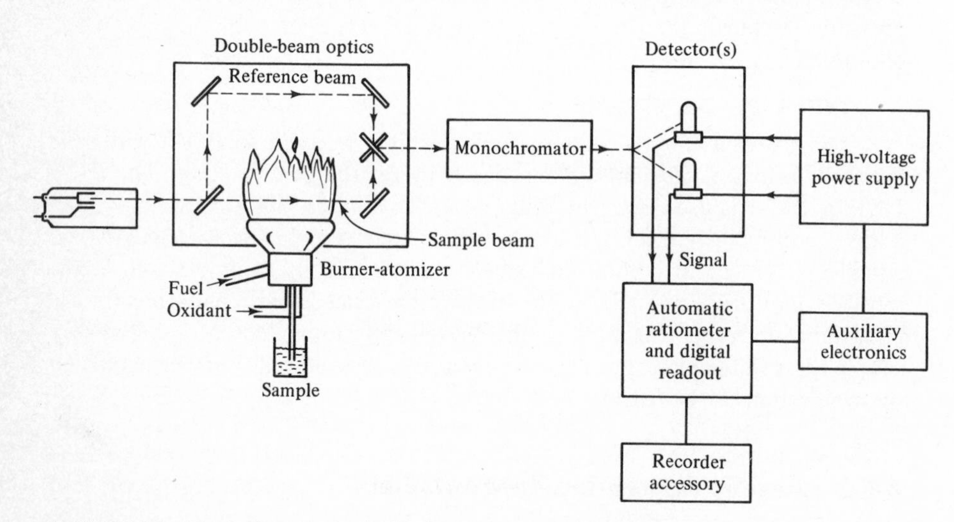

FIGURE 5-9 Block diagram of a double-beam AA spectrophotometer.

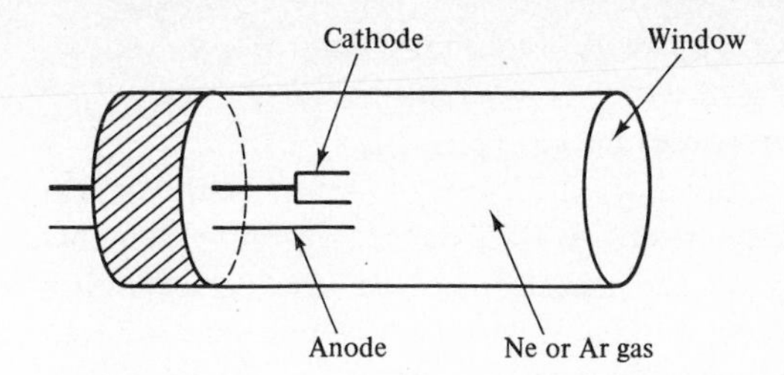

FIGURE 5-10 Schematic diagram of a hollow-cathode lamp.

advantage. Another advantage of the double-beam system is that incorporation of a background compensation system† is possible. Balanced against these advantages are the increased cost and complexity of a double-beam system. For the majority of analyses for which atomic absorption is used, a single-beam system may be adequate.

SOURCES

The most critical component of an AA spectrometer is the source. Because the width of AA lines is so narrow (on the order of 0.01 Å), it is very difficult to measure the absorption accurately against the background of a continuous source. Attempts are still being made to develop very intense continuous sources that would be usable, but so far it has been feasible only to adopt such lamps as auxiliary sources, mainly for the purpose of making background corrections [9]. Work is also being carried out on electrodeless discharge tubes, powered by microwave generators [25], but so far none of these sources measure up to hollow cathodes for overall reliability, speed, and convenience.

A schematic diagram of a *hollow-cathode lamp* is shown in Fig. 5-10. The cathode is cup-shaped and made of the metal which is to be analyzed for. The anode material is not critical. Neon or argon at a few millimeters pressure is used as a filler gas. The front window of the lamp is usually quartz, although glass may be used for certain elements with intense lines in the visible region. After a short warm-up period, application of 100 to 200 V between the two electrodes produces a glow discharge with most of the emission coming from within the hollow cathode. The excitation process is thought to be at least a two-step process, involving first a sputtering, or vaporization, of metal atoms along with some ionization of the filler gas, followed by excitation of the neutral metal atoms by collision with the neon or argon ions.

With certain elements, notably nickel, it has been possible to increase the intensity of the source by inserting two auxiliary electrodes, with their own power supply, which are designed to pass a low-energy stream of electrons and help to excite the sputtered metal. A limited number of such high-brightness

† H. L. Kahn, *At. Absorpt. Newsl.*, 7:40 (1968).

lamps are now available, but most are costly, require an extra power supply, and have relatively short lifetimes.

Though continuous-spectrum sources do not yet appear feasible, a step in this direction has been taken in the form of *multielement lamps*. With varying degrees of success, lamps containing from two to about six different elements have been developed. Depending on the metals involved, the cathodes are made from alloys, intermetallic compounds, or mixtures of powders sintered together. Many presently available combinations can be used without disadvantage compared with single-element lamps and can usually be obtained at less cost. But other combinations may lack the intensity or introduce the possibility of spectral interferences. A few multielement lamps that appear to work well are a combination of chromium, manganese, and copper; a combination of calcium, magnesium, and aluminum; and a combination of calcium and zinc.

BURNER-ATOMIZERS

Almost all AA spectrometers employ premix burners, though notable exceptions are the Jarrel-Ash instruments, which use total consumption. A number of *nonflame atomizers* for atomic absorption have become commercially available, including the Perkin-Elmer model HGA-70 (more recently the HGA-2000) heated-graphite furnace, the Varian Techtron model 61 and 63 carbon-rod atomizer, and the Instrumentation Laboratory model 355 flameless sampler (tantalum-ribbon furnace). These atomizers are proving capable of much higher sensitivities than flame cells and can handle much smaller samples. The heated-graphite furnace will be used to illustrate this development.

Figure 5-11 shows a cross section of the Perkin-Elmer model HGA-2000 heated-graphite furnace. It consists of a hollow graphite cylinder a few centimeters long and a few millimeters in cross section placed along the sample beam in place of the conventional burner-atomizer. Electrodes at the end of the cylinder are connected to a low-voltage high-current supply, which is capable of delivering up to 500 A at up to 10 V. The sample solution, ranging in volume from only 1 to 50 μl, is inserted through a small hole in the center of the cylinder with the aid of a microliter syringe. By using a small current and raising the cylinder walls to a temperature of about 100°C, the sample is first dried by evaporation. The temperature of the cylinder can then be elevated long enough to ash any residue. Finally, full power is applied, and the elements in the sample are atomized rapidly and efficiently. The element of interest produces an absorption signal which lasts from a few milliseconds to a few seconds—long enough to record the peak absorption.

To prevent oxidation of the sample and the graphite tube, a continuous purge of the furnace enclosure is made with an inert gas like argon or nitrogen. To protect the operator from burn hazards and to make it possible

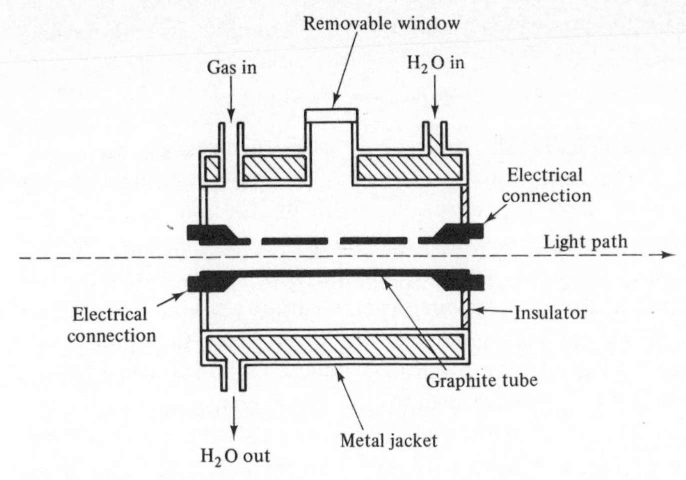

FIGURE 5-11 Cross section of the Perkin-Elmer model HGA-2000 heated-graphite furnace. (*Perkin-Elmer Corp.*)

to drop the temperature to ambient rapidly, the metal jacket around the cylinder is water-cooled before analysis of the next sample. Samples can be run as fast as two per minute. Preliminary results indicate that the heated-graphite analyzer can be used for most of the elements which can be determined by conventional atomic absorption, and with improved sensitivity. The minute amount of sample solution that is necessary puts the absolute detection limit on the submicrogram level for most elements, competing with neutron-activation analysis in sensitivity. The model HGA-2000 may be used with almost any Perkin-Elmer AA spectrophotometer and probably could be adapted to many other instruments as well.

5-2D Atomic Fluorescence Spectrometers

Atomic fluorescence spectroscopy is not yet a widely accepted method of analysis. In principle, the instrumental arrangement of an AF spectrometer is similar to that of an AA spectrometer (compare Figs. 5-2 and 5-1*b*). In fact, most of the AF studies to date have been made with modified AA spectrometers [30]. Technicon Corp. has announced an instrument designed explicitly for AF analysis, an automated AF spectrometer for multielement analysis. The instrument uses pulsed hollow-cathode lamps, a rotating interference-filter wheel, a flame cell, and logic circuitry to measure the fluorescence of each metal when the proper interference filter is in place. This instrument can analyze six elements per sample and 100 samples per hour [29].

Although there are many instrumentation similarities between AA and

AF, there also are some differences, mainly in the *sources* and the *burner-atomizers*.

SOURCES

To be useful analytically the sources in AF spectroscopy must be extremely intense over the center of the absorption line. Although narrow-line sources are preferred in most instances, the width of the line profile is not nearly so critical as in AA spectroscopy, and there is hope that very intense continuous sources can be developed that will become widely useful in AF spectroscopy. In general, the conventional hollow-cathode sources now being used in AA spectroscopy are inadequate for AF work, since they lack sufficient intensity. In certain cases high-brightness hollow cathodes, described in Sec. 5-2*C*, are usable, but they are not available for many elements.

The most successful and widely used sources for AF studies have been *electrodeless discharge tubes*. Figure 5-12 gives a schematic diagram of an electrodeless discharge lamp. The lamp itself consists of a sealed quartz tube containing a small quantity of the required metal element, usually as the io-

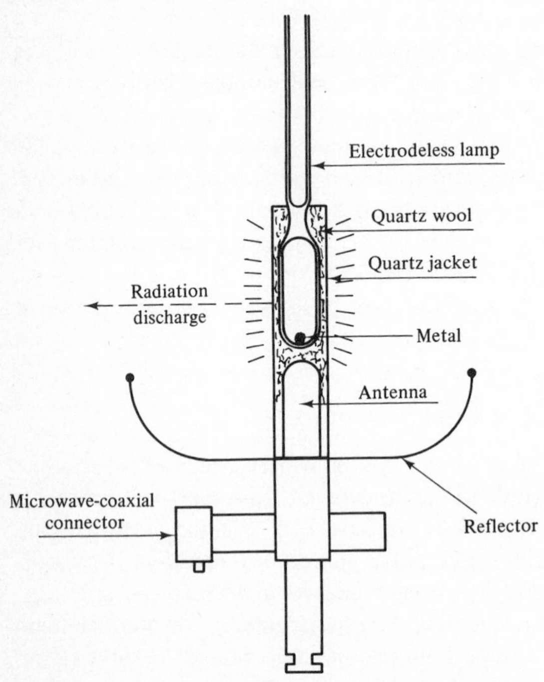

FIGURE 5-12 Schematic diagram of an electrodeless discharge lamp with microwave excitation. [*Reprinted with permission from J. M. Mansfield, Jr., M. P. Bratzel, Jr., H. O. Norgordon, D. O. Knapp, K. E. Zacha, and J. D. Winefordner, Spectrochim. Acta,*23*B*:393 (1968), *Pergamon Press.*]

dide, under a pressure of between 1 and 4 mm of argon. The lamp is placed in a high-frequency microwave field, which causes excitation of the atomic vapor in the lamp. Optimum performance and lifetime are a function of the power of the microwave field, the lamp temperature, and the uniformity of the microwave radiation around the lamp.

There are a number of advantages and disadvantages of the currently available electrodeless discharge lamps. The lamps themselves are low in price and have exceedingly long lifetimes. Furthermore, they emit very sharp spectral lines of much higher intensity than can be obtained from hollow-cathode lamps. However, the microwave generator required is quite costly, and, more seriously, the tubes vary widely in characteristics and require critical tuning by moving the lamp within the field or careful adjustment of the field itself. Warm-up time is at least 30 min and in extreme cases over 2 h. Not all elements give emissions which are intense enough for practical AF applications.

BURNER-ATOMIZERS

A variety of flame and nonflame atomizers have been used for AF spectroscopy. Total-consumption burners have been most widely used despite their relatively poor aspiration efficiencies and droplet dispersion distribution. They have the advantage of being relatively inexpensive, giving high-temperature flames, being easily adaptable to any solvent type, and producing large concentrations of atomic absorbers in the flame gases. However, such burners produce turbulent flames with a significant noise component due to scattering of exciting light from unevaporated solvent particles in the flame gases. Premix burners have been used in a number of studies in order to minimize light scattering from large droplets and particles in the flame and to produce a larger flame area.

One of the problems unique to AF spectroscopy is that *radiation-quenching* processes must be minimized. The major quenchers in flames are CO, CO_2, and N_2. Thus, usually nonhydrocarbon flames such as hydrogen-oxygen (total-consumption burners) or hydrogen-air (premix burners) are used to avoid CO and CO_2. A mixture of argon, oxygen, and hydrogen has been found to be a useful flame with premix burners. The argon-oxygen mixture often gives much higher sensitivity with hydrogen than air, apparently because of the low quenching efficiency of argon but possibly also because of a higher flame temperature. Acetylene-air and natural gas–air have also been used with premix burners.

A number of nonflame atomizers have also been used in AF spectroscopy, including a graphite furnace, a heated graphite filament, a metal-tube furnace, and a heated metal loop. All nonflame cells are electrically heated. These nonflame atomizers are described in Refs. 29 and 30.

5-3 APPLICATIONS

5-3A Comparison of the Detection Limits of Flame Emission, Atomic Absorption, and Atomic Fluorescence

Perhaps the characteristic of single greatest interest to the analyst in comparing FE photometry, AA spectroscopy, and AF spectroscopy is the detection limit for each element. Unfortunately, there is much disparity between published detection limits and the claims of various instrument manufacturers. Furthermore, detection limits change rapidly with new developments. Nevertheless, it is useful to compare detection limits based on what it is hoped are the best and most recent data available. Table 5-7 compares the detection limits by the three methods for 37 elements. For details of the conditions used for each element, the original publications should be consulted [9, 29].

Table 5-7 shows that each of the three flame methods is clearly superior to the other two for certain elements. For example, flame emission is best for the alkali metals, the alkaline earths (except for magnesium), and aluminum. On the other hand, atomic absorption appears to be superior for elements like beryllium, magnesium, molybdenum, rhodium, and silicon, and atomic fluorescence seems superior for silver, bismuth, cadmium, copper, mercury, tellurium, and zinc. The detection limits for a number of elements approach the spectacular, particularly for calcium, lithium, and strontium by flame emission, magnesium by atomic absorption, and silver, cadmium, mercury, and zinc by atomic fluorescence. The detection limits of these elements by flame methods are superior to almost any other method of chemical analysis. In terms of convenience and relative freedom from interferences, the various flame methods compete favorably with most other methods for a large number of the elements lists in Table 5-7. Thus it is clear that flame methods are coming to the forefront of methods available for the analysis of metals.

5-3B Flame Photometric Applications

QUALITATIVE ANALYSIS

Flame photometry is used mainly for quantitative analysis, but, unlike the other two flame methods, it can also be used for qualitative analysis of a complex mixture by scanning the wavelength spectrum and recording emission intensity. One of the inherent problems in qualitative analysis is caused by the flame background, which may mask a substance or be mistaken for a component in the sample. Furthermore, the fact that flame photometers are single-beam instruments means that the response of the instrument will vary much like the response curve of the detector. A scan of the solvent blank is very helpful in evaluating the magnitude of these two problems. While useful for qualitative analysis in certain instances, flame photometry ranks well below ordinary emission spectroscopy for this purpose.

TABLE 5-7 Comparison of Flame Emission, Atomic Absorption, and Atomic Fluorescence†

Element	Wavelength, Å	Detection limit, μg/ml		
		Flame emission	Atomic absorption	Atomic fluorescence
Ag	3281	0.02	0.0005	0.0001‡
Al	3962	0.005‡	0.04	0.1
As	2350, 1937, 1937	50	0.1	0.1
Au	2676, 2428, 2676	4	0.01	0.005
Ba	5536	0.001‡	0.05	
Be	2349	0.1	0.002‡	0.01
Bi	2231	2	0.05	0.005‡
Ca	4227	0.0001‡	0.0005	0.02
Cd	3261, 2288, 2288	2	0.00006	0.000001‡
Co	3454, 2407, 2407	0.05	0.005	0.005
Cr	4254, 3579, 3579	0.005	0.005	0.05
Cu	3274, 3247, 3247	0.01	0.003	0.001‡
Fe	3720, 2483, 2483	0.05	0.005	0.008
Ga	4172, 2874, 4172	0.01	0.07	0.01
Ge	2652	0.5	0.1	0.1
Hg	2537	40	0.2	0.0002‡
In	4511, 3039, 4511	0.005	0.05	0.1
K	7765	0.0005‡	0.005	
Li	6708	0.00003‡	0.005	
Mg	2852	0.005	0.0003‡	0.001
Mn	4031, 2795, 2795	0.005	0.002	0.006
Mo	3903, 3133, 3133	0.1	0.03‡	0.05
Na	5890	0.0005‡	0.002	
Ni	3415, 2320, 2320	0.6	0.005	0.003
Pb	4058, 2833, 4058	0.2	0.01	0.01
Pd	3635, 2746, 3405	0.05	0.02	0.04
Rb	7800	0.001‡	0.005	
Rh	3692, 3435, 3692	0.3	0.03‡	3
Sb	2598, 2175, 2311	20	0.07	0.05
Si	2516, 2516, 2040	5	0.1‡	0.6
Se	1960	ND§	0.1	0.04
Sn	2480, 2246, 3034	0.3	0.03	0.05
Sr	4607	0.0002‡	0.004	0.03
Te	2383, 2143, 2143	200	0.1	0.005‡
Tl	3776, 2768, 3776	0.02	0.02	0.008
V	4379, 3184, 3184	0.01	0.02	0.07
Zn	2138	50	0.002	0.00002‡

† Reprinted by permission from J. D. Winefordner and R. C. Elser, *Anal. Chem.*, **43**(4): 24A (1971) and (alkali metals and barium) Ref. 9, p. 29A. Copyright by the American Chemical Society.

‡ This is the best detection limit of the three methods if the detection limit is clearly superior by a factor of 3 or more. A factor of 3 is used as the criterion of superiority in an effort to avoid the problem caused by different methods of defining and measuring detection limits used by various authors.

§ Not detectable.

QUANTITATIVE ANALYSIS

For quantitative analysis in flame photometry it is necessary to prepare a working curve or calibration curve of emission intensity (in arbitrary units, usually 0 to 100) vs. concentration of the element being sought. Generally this curve will be linear only over a small concentration range. Characteristically, the curve will almost always bend toward the abscissa at high concentrations due to *self-absorption*, and at low concentrations it may or may not bend upward due to *ionization* of the element in the hot flame.

In *self-absorption* some of the light emitted by a given element is *reabsorbed* by other unexcited atoms of that same element before the light can emerge from the flame. Take sodium as an example. The calibration curve for sodium is generally linear over a concentration range of 0 to 1 ppm. However, above 1 ppm self-absorption begins to become important because the population of unexcited sodium atoms in the flame begins to be significant. The reason for self-absorption can be understood better by reference to Fig. 5-3. Whereas excitation and emission take place mainly in the *interconal zone*, the light produced must pass through the cooler *outer-cone region*, where there is a relatively high population of vaporized (but not excited) atoms of the same element. Since the emitted radiation is of precisely the correct wavelength to be absorbed by ground-state atoms in the outer cone, self-absorption will occur, causing the radiation intensity to be less than expected. Self-absorption occurs with all elements in the flame, but the concentration at which it becomes serious varies from element to element.

SPECIFIC APPLICATIONS

Flame photometry has been applied in almost every analytical field, including biology, agriculture, and all types of industries. Since many reviews are available [11, 13, 15], only a brief listing of a few applications will be given here.

Flame photometry is extremely important in the analysis of biological fluids and tissues. Routinely analyzed are elements like sodium, potassium, calcium, and iron, and other elements are determined occasionally [13, 15]. In soil analysis elements like sodium, potassium, aluminum, calcium, cobalt, and iron are frequently determined [13, 15]. Natural and industrial waters, glass, cement, petroleum products, and metallurgical products are but a few of the materials routinely analyzed by flame photometry [13, 15].

5-3C Atomic Absorption Applications

Atomic absorption is an excellent method of quantitative trace-metal analysis. Since the method is based on the *absorption* of electromagnetic radiation, Beer's law applies (see Sec. 1-5*B*), and working curves of *absorbance* vs. concentration are prepared. An idea of the wide applicability of atomic absorption and some specific procedural directions can be found in

Ref. 24. The scope of AA analyses is easily as broad as that of flame photometry, and complete reviews of applications can be found in Refs. 11, 19, 21, and 25 to 27.

5-3D Atomic Fluorescence Applications

Relatively few applications of AF spectroscopy to real samples have emerged to date. Since an emission is being measured, the calibration curves necessary for quantitative analysis consist of a plot of fluorescence intensity (in arbitrary units, say 0 to 100) vs. concentration of metal ion. It is often possible to obtain a relatively linear curve over a concentration range of 10^3 to 10^5, which is usually a slightly larger range than is possible in FE photometry and significantly greater than the range possible in AA spectroscopy.

A few specific examples of the application of atomic fluorescence to real samples include the determination of trace wear metals in jet-engine lubricating oils; the determination of nickel in gas, oils, and petroleum distillates; the determination of copper, iron, and lead in hydrocarbon fuels; and the determination of lead in blood and urine [29]. A complete résumé of analytical applications can be found in the biannual reviews of *Analytical Chemistry* [11].

EXERCISES

5-1 When nitrogen molecules, N_2, are present in a flame because air or nitrous oxide support gas is used, the temperature is appreciably lower than when pure oxygen is the support gas (see Table 5-1). Postulate a mechanism by which nitrogen molecules extract energy from the flame.

5-2 One of the potential problems of premix or laminar-flow burners is that the flame may strike back, particularly if oxygen support gas is used or if the flow rate of the gases falls below the burning velocity. Why do turbulent-flow burners *never* strike back?

5-3 Summarize the possible advantages and disadvantages of adding a miscible organic solvent to a solution to be analyzed by flame methods. List the ways in which the organic solvent can alter the measured results.

5-4 Why are spectral interferences less severe in AA and AF spectroscopy than in FE photometry?

5-5 Fluctuations in flame temperature can be considered a physical interference. Discuss the relative seriousness of flame-temperature fluctuations in FE photometry and AA spectroscopy.†

† A paper of interest on this topic is L. De Galan and J. D. Winefordner, *Anal. Chem.*, **42**:1412 (1966).

5-6 What are the major problems or disadvantages of AF spectroscopy? What are the advantages?

5-7 (*a*) For AF spectroscopy, what are the disadvantages of acetylene-air or acetylene-oxygen flames? (*b*) Why are hydrogen and oxygen not used with a premix burner? (*c*) Why can a mixture of hydrogen, oxygen, and argon be used with a premix burner, and what advantages does it have over hydrogen-air for AF studies?

5-8 When long-path slot burners first became available, it was believed they would not be suitable for FE photometry because of *self-absorption*. This does not appear to be a problem. Explain why.

REFERENCES

GENERAL

Elementary

1 EWING, G. W.: "Instrumental Methods of Chemical Analysis," 3d ed., McGraw-Hill, New York, 1969. Chapter 8 on Flame Spectroscopy gives a brief but clear overview of flame photometry, atomic absorption, and atomic fluorescence.

2 SIGGIA, S.: "Survey of Analytical Chemistry," McGraw-Hill, New York, 1968. A very brief but practical introduction to flame photometry and atomic absorption.

Intermediate

3 MAVRODINEANU, R., and H. BOITEUX: "Flame Spectroscopy," Wiley, New York, 1965. A beautifully illustrated book containing information on all aspects of flames, flame regulation, and spectra. Both theoretical and practical.

4 WEBERLING, R. P., and J. F. COSGROVE: Flame Emission and Absorption Methods, chap. 7 in G. H. Morrison (ed.), "Trace Analysis," Wiley-Interscience, New York, 1965. A brief but useful summary of pertinent aspects of flame photometry and atomic absorption.

5 WILLARD, H. H., L. L. MERRITT, JR., and J. A. DEAN: "Instrumental Methods of Analysis," 4th ed., Van Nostrand, New York, 1965. Chapter 11 on Flame Photometry is a practical and useful treatment. Touches on atomic absorption also.

Advanced

6 ALKEMADE, C. T. J.: *Appl. Opt.*, **7**:1261 (1968). Discusses science vs. fiction in atomic absorption.

7 FRISTROM, R. M., and A. A. WESTENBERG: "Flame Structure," McGraw-Hill, New York, 1965. A detailed treatment of flame structures.

8 GAYDON, A. G., and H. G. WOLFHARD: "Flames," 3d ed., Chapman & Hall, London, 1970. A clear and detailed treatment of flames.

9 PICKETT, E. E., and S. R. KOIRTYOHANN: *Anal. Chem.*, **41**(14):28A (1969). A critical discussion of the relative merits of flame photometry and atomic absorption.

10 PUNGOR, E.: "Flame Photometry Theory," Van Nostrand, New York, 1967. A valuable treatment of flames, with a brief tie-in with flame photometry and atomic absorption.

11 WINEFORDNER, J. D., and T. J. VICKERS: *Anal. Chem.*, **46**(5):192R (1974). One of a series of complete reviews of all aspects of flame spectrometry, appearing every 2 years.

FLAME PHOTOMETRY

12 BERRY, J. W., D. G. CHAPPELL, and R. B. BARNES: *Ind. Eng. Chem., Anal. Ed.*, **18**:19 (1946). A description of the first instrument employing the internal-standard method for flame photometry, with a good discussion.

13 DEAN, J. A.: "Flame Photometry," McGraw-Hill, New York, 1960. A thorough, though slightly outdated, treatment of all aspects of flame photometry.

14 FUWA, K., R. E. THIERS, and B. L. VALLEE: *Anal. Chem.*, **31**:1419 (1959). Describes a well-designed burner for cyanogen flame spectroscopy and gives references to previous analytical work with cyanogen flames.

15 HERRMANN, R., and C. T. J. ALKEMADE: "Chemical Analysis by Flame Photometry," 2d ed., Wiley-Interscience, New York, 1963. An excellent treatment of all aspects of flame photometry.

16 POLUÉKTOV, N. S.: "Techniques in Flame Photometric Analysis," Consultants Bureau, New York, 1961. A good, practical treatment of flame photometry.

17 RAMÍREZ-MUÑOZ, J.: Flame Emission Spectrometry, chap. 2 in J. D. Winefordner (ed.), "Spectrochemical Methods of Analysis," Wiley-Interscience, New York, 1971. A fairly good, up-to-date treatment of flame photometry.

18 VALLEE, B. L., and R. E. THIERS: Flame Photometry, chap. 65 in I. M. Kolthoff and P. J. Elving (eds.), "Treatise on Analytical Chemistry," pt. I, vol. 6, Wiley-Interscience, New York, 1965. A good, brief treatment of the important aspects of flame photometry.

ATOMIC ABSORPTION

19 AMERICAN SOCIETY FOR TESTING AND MATERIALS: Atomic Absorption Spectroscopy, *ASTM Spec. Tech. Pub.* 443, Philadelphia, 1969. A good treatment of the physical and chemical aspects of atomic absorption, plus a series of special applications.

20 CHRISTIAN, G. D.: *Anal. Chem.*, **41**(1):24A (1969). An interesting account of the role of atomic absorption in medicine.

21 ELWELL, W. T., and J. A. F. GIDLEY: "Atomic Absorption Spectrophotometry," Pergamon, New York, 1966. A very useful monograph, with procedures for each of the elements.

22 FUWA, K.: Flame Absorption Spectrometry, chap. 3 in J. D. Winefordner (ed.), "Spectrochemical Methods of Analysis," Wiley-Interscience, New York, 1971. A good, up-to-date treatment of atomic absorption.

23 KAHN, H. L.: *J. Chem. Educ.*, **43**:A7, A103 (1966). A good discussion of atomic absorption instrumentation, in two parts.

24 PERKIN-ELMER CORP.: "Analytical Methods for Atomic Absorption Spectrophotometry," Perkin-Elmer Corp., Norwalk, Conn., 1968. A loose-leaf procedural manual, originally published in 1964, and updated approximately every 2 years with supplements.

25 REYNOLDS, R. J., and K. ALDOUS: "Atomic Absorption Spectroscopy," Barnes & Noble, New York, 1970. A practical guide to atomic absorption, very well written.

26 ROBINSON, J. W.: "Atomic Absorption Spectroscopy," Dekker, New York, 1966. A good treatment of atomic absorption, with elemental procedures.

27 SLAVIN, W.: "Atomic Absorption Spectroscopy," Wiley-Interscience, New York, 1968. A thorough treatment, with procedures and example applications.

ATOMIC FLUORESCENCE

28 SMITH, R.: Flame Fluorescence Spectrometry, chap. 4 in J. D. Winefordner (ed.), "Spectrochemical Methods of Analysis," Wiley-Interscience. New York, 1971. A good, thorough, and up-to-date treatment of atomic fluorescence.

29 WINEFORDNER, J. D., and R. C. ELSER: *Anal. Chem.*, **43**(4):24A (1971). A review of the theoretical and practical aspects of atomic fluorescence, with a comparison of detection limits with flame photometry and atomic absorption.

30 WINEFORDNER, J. D., and J. M. MANSFIELD: Atomic Fluorescence Spectrometry, chap. 13 in G. G. Guilbault (ed.), "Fluorescence," Dekker, New York, 1967. One of the best treatments of atomic fluorescence, but now slightly out of date.

6

Raman Spectroscopy

Raman and infrared spectra give basically the same kind of molecular information, and either method can be used to supplement or complement the other. The Raman effect was discovered in 1928 by C. V. Raman, an Indian physicist, and during the 1930s Raman measurements were more widely used than the infrared, mainly because Raman measurements could be directly recorded photographically while infrared measurements had to be recorded manually. About 1945, automatic-recording infrared instruments became commercially available, and an overwhelming shift in popularity to infrared spectrophotometry took place, leaving Raman spectroscopy relatively little used, except in Russia, where it continued to be used more extensively than infrared. With the development of efficient Raman spectrophotometers using electronic instead of photographic detection, Raman spectroscopy experienced some return to popularity, and the incorporation (about 1965) of the laser as an excitation source has greatly increased the popularity of this important tool.

Some of the advantages of Raman spectroscopy, particularly with respect to infrared spectrophotometry, are the following:

1 Water is an excellent Raman solvent, whereas water cannot be used in infrared studies.

2 Glass cells can be used in Raman spectroscopy, whereas salt cells, which are difficult to handle, must be used in the infrared.

3 Raman spectra are usually simpler than the corresponding infrared spectra, primarily because overtone and combination effects are relatively small compared with the principal Raman frequencies. Thus, overlapping bands are much less common in Raman spectroscopy.

4 Totally symmetric modes of vibration can be studied by the Raman effect, whereas they are not observed in infrared spectroscopy.

5 The polarization of Raman spectra adds a valuable extra dimension to the information. This extra information in vibrational assignments or structure determinations often makes it possible to rule out otherwise acceptable assignments or structures.

6 Raman studies provide information for the calculation of chemical equilibrium constants and other thermodynamic properties.

7 By the nature of the Raman effect, one instrument and a single continuous scan can be used to cover the entire range of molecular vibration frequencies, from the group-frequency region through the lattice-vibration region (equivalent to an infrared range from about 2 to 200 μm). At least two infrared instruments and a number of range changes would ordinarily be required to cover this range in the infrared.

8 The intensity of a Raman line is directly proportional to concentration, whereas there is a logarithmic (Beer's law) relationship for conventional spectrophotometry. Thus, quantitative analysis is often more convenient in Raman spectroscopy and sometimes more accurate.

Before the development of the laser as an excitation source, Raman spectroscopy suffered from the following disadvantages:

1 Samples had to be restricted to clear, colorless, nonfluorescent liquids.

2 The low intensity of the Raman effect required relatively concentrated solutions.

3 Much larger volumes of sample solution were needed than for infrared spectrophotometry.

These three disadvantages were major factors in limiting the use of Raman spectroscopy, but the helium-neon laser source now available has gone a long way toward overcoming them. Difficulties due to colored or fluorescent liquids are greatly reduced, solids can be handled, intensities are greater, and samples as small as 1 μl can now be used in Raman work.

6-1 THEORY

Raman found that when molecules are irradiated with monochromatic light, a portion of the light is scattered; most of this scattered radiation (about 99 percent) has the original frequency (Rayleigh scattering), but a small portion

(<1 percent) is found at other frequencies. The *difference* in frequency between these new frequencies (now called *Raman lines* or *bands*) and the original frequency is characteristic of the molecule irradiated and numercially identical with certain of the vibrational and rotational frequencies of that molecule.

Rayleigh scattering and Raman scattering can be explained as follows. In Rayleigh scattering, the oscillating electric and magnetic forces constituting the incident light induce a dipole moment in molecules; they in turn radiate light of the same frequency (*elastic* scattering) as the incident or exciting radiation but in all directions. Rayleigh scattering always accompanies Raman scattering.

In Raman scattering, some of the energy of the incident light may be used up in exciting a molecule to a higher vibrational or rotational energy level, and the radiation emitted (scattered) by the molecule (called a *Stokes' line*) will be of correspondingly lower energy and frequency. Alternatively, since some of the molecules encountered by the incident radiation will already be in higher vibrational or rotational energy states, the molecule may *contribute* this extra energy to the scattered photon, resulting in emitted radiation (called an *anti-Stokes' line*) with energy higher than that of the incident radiation by an amount corresponding to the vibrational or rotational transition energy of the molecule. (Since Raman scattering involves a change in energy, the collision between the molecule and the incident photon is *inelastic*.)

These phenomena can be portrayed more accurately in the energy-level diagram shown in Fig. 6-1, where $h\nu_0$ can represent almost *any* energy of an incident photon. When the incident photon interacts with a molecule in any of its stable states, say E_0 or E_1, the energy of the molecule is momentarily raised by the amount $h\nu_0$. If the resulting energy state ($E_0 + h\nu_0$ or $E_1 + h\nu_0$) is not a stable allowed energy level of the molecule, the molecule immediately returns to a lower energy state and the corresponding amount of energy is emitted (scattered).† If the molecule returns to its *original* state, the frequency emitted is identical to the incident frequency and *Rayleigh scattering* results, as shown in Fig. 6-1; but if the molecule goes to an energy state other than the original one, *Raman scattering* results. In practice, the anti-Stokes' lines are much weaker than the Stokes' lines and are generally ignored since they furnish the same information as the Stokes' lines.

For a particular mode of vibration to appear in the Raman spectrum, i.e., to be *Raman-active*, the molecule's *polarizability* must change during the course of this vibration.‡ The polarizability of a molecule is the ability of the molecule to be polarized under the action of an electric field such as the alternating field of a light wave, and it can be defined in terms of the dipole

† If the energy state $E_0 + h\nu_0$ or $E_1 + h\nu_0$ is an allowed energy level, the incident light is *absorbed*; any radiation which is subsequently reemitted when the molecule returns to a lower energy state is called *fluorescent radiation*.

‡ If a *rotation* were studied, it too would have to be accompanied by a change in polarizability in order to be Raman-active; however, rotational transitions are of little analytical importance in Raman spectra.

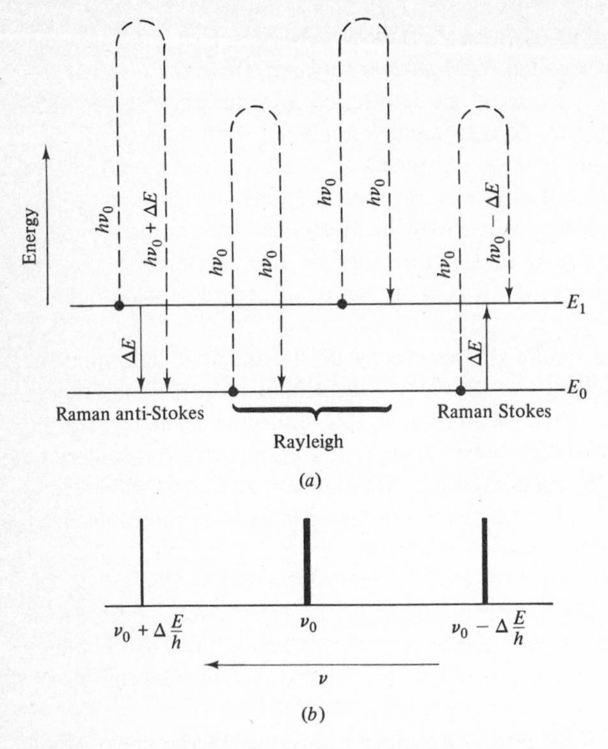

FIGURE 6-1 Rayleigh and Raman scattering: (*a*) energy diagram; dotted arrows show the entire transition, and solid arrows represent the net transition; (*b*) spectrum which results using a photographic plate; width of lines represent relative scattering intensities.

moment D produced by the electric field E:

$$D = \alpha E \tag{6-1}$$

where α is the polarizability. The polarizability is thus a measure of the efficiency with which a varying electric field will induce a dipole moment in a molecule. In measuring the Raman effect, visible light is almost always used as the incident radiation. Usually the 435.8-nm line from a mercury arc or the 632.8-nm line from a helium-neon laser is used (see Sec. 6-2*A*); ultraviolet light could be used, but it is not as widely applicable, not only because it is absorbed by many substances but also because it may cause molecular dissociation and fluorescence. Under the influence of visible (or ultraviolet) incident light, only electrons (not nuclei) oscillate, since nuclei cannot follow the rapid oscillations. Thus, polarizability measures the ease of displacement of electrons by the electric field, and the polarizability will change, even if only slightly, whenever the interatomic distance changes.†

† Polarizability can be expected to depend on molecular *rotation* also, since the ease of displacing electrons will depend on the orientation of the molecule to the field.

6-1A Correlation between Infrared and Raman Spectra

Infrared and Raman spectra tend to be complementary because of their differing requirements for activity. Whereas a change in polarizability must occur in order for a vibrational mode to be Raman-active, it will be recalled from Chap. 3 that a change in dipole moment is necessary for a vibrational mode to be infrared-active. The practical manifestations of these differing requirements can be summarized with one absolute rule and two additional generalizations.

RULE 1: MUTUAL EXCLUSION

This absolute rule, with no known exceptions, says that *for all molecules with a center of symmetry, transitions that are allowed in the infrared are forbidden in the Raman spectrum;* and conversely, *transitions that are allowed in the Raman spectrum are forbidden in the infrared.* As a simple example, an O_2 molecule has only one fundamental mode of vibration ($3N - 5 = 1$; see Sec. 3-1*B*), which is a symmetrical stretching vibration, and this will be *infrared-inactive* since no change in dipole moment occurs during the vibration. On the other hand, this vibration will be *Raman-active,* since there will be a change in polarizability during the vibration; i.e., there will be a change in the ease of displacing electrons during the stretch since the bond distance changes during the stretch, and therefore the binding energy of the bonding electrons will change. Other examples of this type will be given later.

It is particularly useful to apply the rule of mutual exclusion to *functional-group*† absorptions. For example, the C═C stretching vibration in olefins is usually absent or very weak in infrared spectra, but the corresponding Raman line is relatively strong. This is illustrated in Fig. 6-2 with the Raman and infrared spectra of pentene-2, aligned with a common wavenumber scale. In the Raman spectrum, note the strong band at 1675 cm^{-1}, resulting from the stretching vibration of the carbon-carbon double bond; in the infrared spectrum this absorption band is absent.

GENERALIZATION 2: MUTUALLY ALLOWED TRANSITIONS

It is generally true that *for all molecules that do not have a center of symmetry the transitions will be both infrared- and Raman-active.* Since most molecules (and most functional groups) have no center of symmetry, there is often some correlation between Raman and infrared spectra. For example, in the infrared and Raman spectra of pentene-2, shown in Fig. 6-2, the characteristic C—H stretching and bending vibrations, found in the regions of 3000 and 1400 cm^{-1}, respectively, show up in both types of spectra.

Whereas the rule of mutual exclusion has no exceptions, the statement above is presented as a generalization because group theory predicts that there

† It will be recalled from Sec. 3-1*D* that the region from about 1300 to 4000 cm^{-1} is known as the fundamental group-frequency region in the infrared, since in this region absorption is more or less dependent only on the functional group and not on the complete molecular structure.

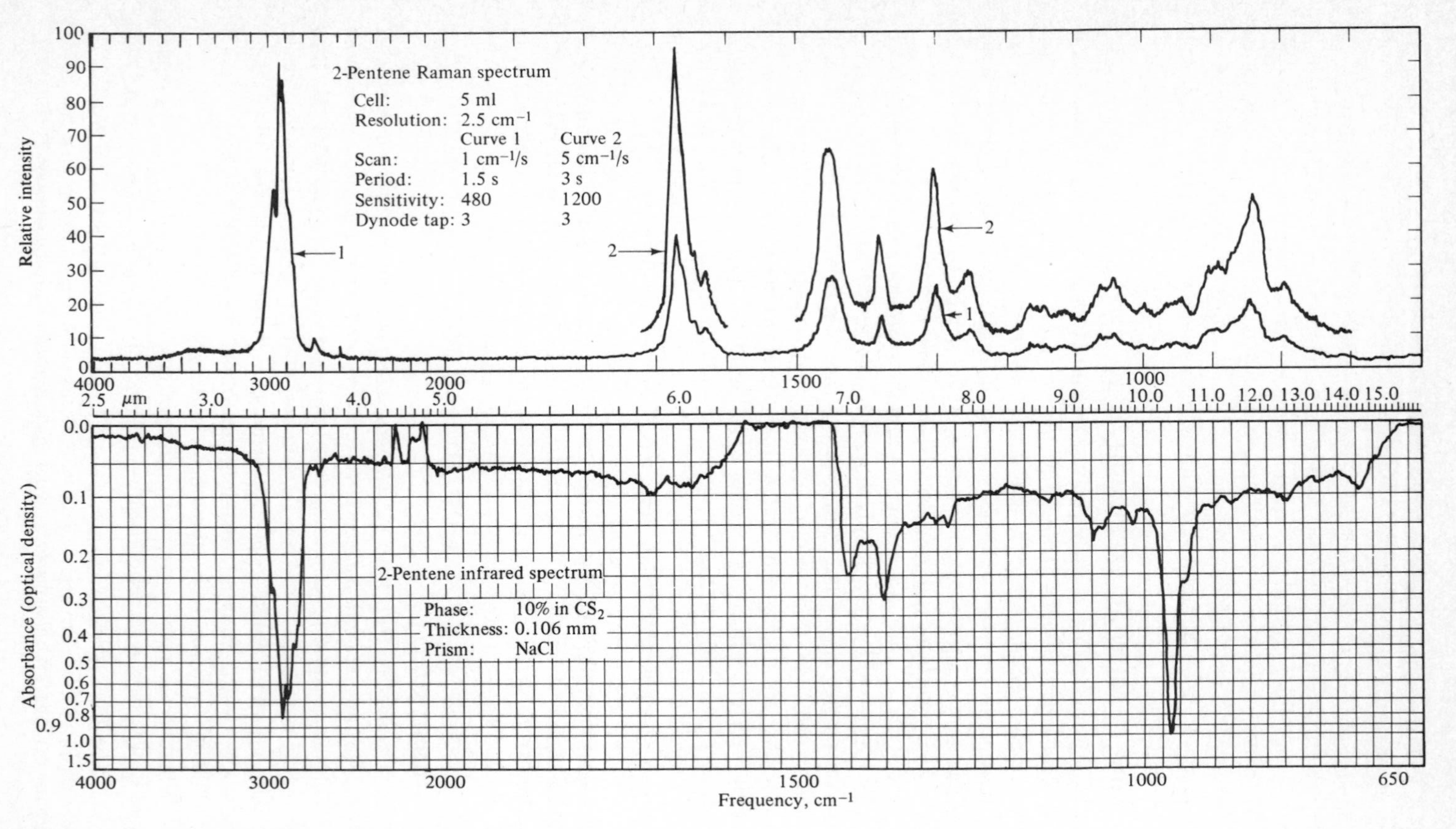

FIGURE 6-2 Raman and infrared spectra of pentene-2. (*Varian Associates.*)

could exist an unusual molecule that does *not* have a center of symmetry yet follows the rule of mutual exclusion, i.e., has an allowed infrared transition that is forbidden in the Raman spectrum, or vice versa. For example, if cyclopentane existed as a perfectly flat pentagon instead of in its actual slightly puckered configuration, a symmetrical stretching vibration (ring-breathing mode) would be *infrared-inactive* and *Raman-active*, even though it would have no center of symmetry.

GENERALIZATION 3: MUTUALLY FORBIDDEN TRANSITIONS

Although the previous two statements cover the great majority of molecules and transitions, *there are a few cases where transitions are forbidden in both the Raman effect and the infrared.* A good example is the twisting mode of vibration of the planar ethylene molecule, shown in Fig. 6-3. Since ethylene is planar and symmetrical, it has no permanent dipole moment and no *change* in dipole moment occurs with the twisting mode shown in Fig. 6-3, making this mode *infrared-inactive.* Furthermore, no change in polarizability occurs during this twisting mode because for small amplitudes of vibration, at least, no change in the ease of displacing electrons will occur, and thus, this mode is also *Raman-inactive.*†

The following examples demonstrate correlations that can be made between infrared and Raman spectra and further illustrate the rule and generalizations.

Example 6-1 Diagram the fundamental modes of vibration of CO_2, and predict which of the modes will be infrared-active and which will be Raman-active.

Answer CO_2 being linear, will have $3N - 5 = 4$ fundamental modes of vibration (see Sec. 3-1*B*), which can be diagramed as

$$\overleftarrow{O}=C=\overrightarrow{O} \qquad \nu_1 \quad \text{symmetrical stretch}$$

$$\overrightarrow{O}=\overleftarrow{C}=\overrightarrow{O} \qquad \nu_2 \quad \text{asymmetrical stretch}$$

$$\downarrow O=\overset{\uparrow}{C}=O \downarrow \qquad \nu_3 \quad \text{in-plane bending}$$

$$\underset{+}{O}=\underset{-}{C}=\underset{+}{O} \qquad \nu_4 \quad \text{out-of-plane bending}$$

Of these four modes of vibration, ν_1 involves no change in dipole moment

† This example does not violate rule 1, the rule of mutual exclusion; rule 1 does not imply that all transitions that are *forbidden* in one *must* occur in the other but rather that transitions *allowed* in one are forbidden in the other.

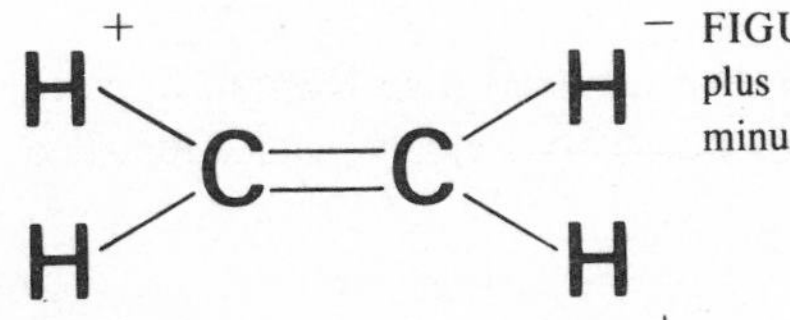

FIGURE 6-3 Twisting mode of vibration of ethylene. A plus sign signifies movement toward the reader and a minus sign movement away; signs reverse every half-cycle.

and will thus be *infrared-inactive*; on the other hand, ν_1 involves a change in polarizability (since the ease of displacing bonding electrons will change during the vibration), making this frequency *Raman-active.* In practice, the Raman band for ν_1 appears at 1388 cm^{-1}.

Frequency ν_2 will be *infrared-active*, since the dipole moment changes during the vibration, but the mode will be *Raman-inactive,* since the polarizability change produced at each atom will be exactly nullified by the antisymmetrical displacement of the atom across the symmetry center from it and the *net* effect on the polarizability is zero. In practice, the infrared band for ν_2 appears at 2349 cm^{-1}.

Frequencies ν_3 and ν_4 are degenerate, or occur at the same frequency, and these modes will be *infrared-active* and *Raman-inactive.* In practice, the ν_3, ν_4 band appears in the infrared at 667 cm^{-1}.

Note that the Raman activity could also have been predicted simply from a knowledge of rule 1 and the infrared activity.

Example 6-2 Diagram the fundamental modes of vibration of acetylene, C_2H_2, and predict which of the modes will be infrared-active and which will be Raman-active.

Answer Acetylene, being linear, has $3N - 5 = 7$ fundamental modes of vibration, which may be diagramed as follows:

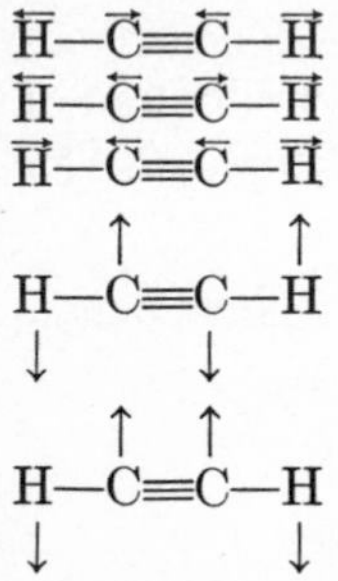

$$\overleftarrow{H}—\overrightarrow{C}\equiv\overleftarrow{C}—\overrightarrow{H} \qquad \nu_1 \qquad \text{C—H symmetrical stretch}$$

$$\overleftarrow{H}—\overleftarrow{C}\equiv\overrightarrow{C}—\overrightarrow{H} \qquad \nu_2 \qquad \text{C}\equiv\text{C stretch}$$

$$\overrightarrow{H}—\overleftarrow{C}\equiv\overleftarrow{C}—\overrightarrow{H} \qquad \nu_3 \qquad \text{C—H asymmetrical stretch}$$

$$\begin{array}{cccc} & \uparrow & & \uparrow \\ H— & C\equiv & C— & H \\ \downarrow & & \downarrow & \end{array} \qquad \nu_4 \qquad \text{trans C—H bending (doubly degenerate)}$$

$$\begin{array}{cccc} & \uparrow & \uparrow & \\ H— & C\equiv & C— & H \\ \downarrow & & & \downarrow \end{array} \qquad \nu_5 \qquad \text{cis C—H bending (doubly degenerate)}$$

Note that both ν_4 and ν_5 are doubly degenerate, i.e., there is a corresponding out-of-plane bending for each of the in-plane bending modes shown, thus accounting for the seven fundamental modes of vibration.

It can be predicted that only ν_3 and ν_5 will be infrared-active, and since acetylene has a center of symmetry, rule 1 predicts that only ν_1, ν_2, and ν_4 can be Raman-active.

In practice, ν_3 and ν_5 appear at 3287 and 729 cm^{-1}, respectively, in the infrared, and ν_1, ν_2, and ν_4 appear at 3374, 1974, and 612 cm^{-1}, respectively, in the Raman.

The mutual exclusion of bands in the infrared and Raman spectra can be used to prove that C_2H_2 has a center of symmetry, and a detailed analysis of the number of combination and overtone bands found in the combined infrared and Raman spectra can be used to prove that aceylene is linear.

Example 6-3 Diagram the fundamental modes of vibration for H_2O, and predict which modes of vibration will be infrared-active and which will be Raman-active.

Answer Nonlinear H_2O will have $3N - 6 = 3$ fundamental modes of vibration, as follows:

Diagram	Mode	Description
↑ O ↙/ \↘ H H	ν_1	symmetrical stretch
O ↖/ \↗ H ↓ H	ν_2	bending
↙ O ↗/ \↘ H H	ν_3	asymmetrical stretch

All three modes of vibration are infrared-active, and all three modes are Raman-active as well, although only ν_1, the symmetrical stretch, gives a strong Raman band. That these transitions are allowed in both the infrared and Raman spectra is in agreement with generalization 2, and proves that water does not have a center of symmetry.

Example 6-4 Diagram the twisting mode of vibration of N_2O_4 (planar) and predict whether this mode will be infrared- or Raman-active.

Answer

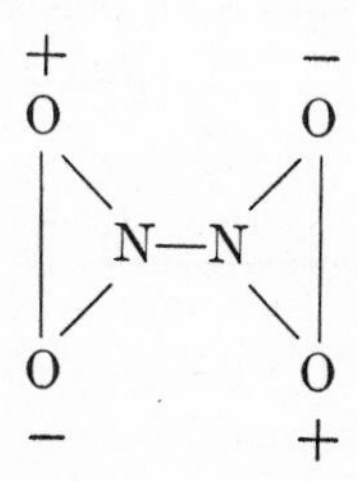

This mode will be neither infrared-active nor Raman-active, since neither a change in dipole moment nor a change in polarizability occurs. Thus, this is another example of one of the few cases where the transitions are mutually forbidden (generalization 3).

6-1B Polarization Measurements

Whereas most spectroscopic lines are characterized by two basic quantities, the *frequency* and the *intensity*, Raman lines are further characterized by a third basic quantity, the *degree of polarization*, or *depolarization factor*. In other words, Raman lines possess the interesting and valuable property of being *polarized*,† which means that Raman emissions have a greater intensity in one plane of vibration than in another. Measurement of the polarization of a Raman line gives useful extra information, since the degree of polarization is directly related to the symmetry of the molecular vibration from which the line originates.

In defining the depolarization factor, two cases must be considered, depending upon whether *natural* (unpolarized) incident light (e.g., radiation from a mercury arc) or *plane-polarized* incident light (e.g., a helium-neon laser source) is being used. Figure 6-4 depicts natural incident radiation moving along the y axis and being observed along the x axis after interacting with the sample at the origin. The depolarization factor ρ_n is defined as the ratio of the intensity of scattered light polarized perpendicular to the xz plane $I_\perp$ to that po-

† Although the words "polarization" and "polarizability" are similar, they have different meanings. Polarization refers to a property of a beam of radiation, whereas polarizability is a molecular property and determines whether or not a particular molecular vibration is Raman-active.

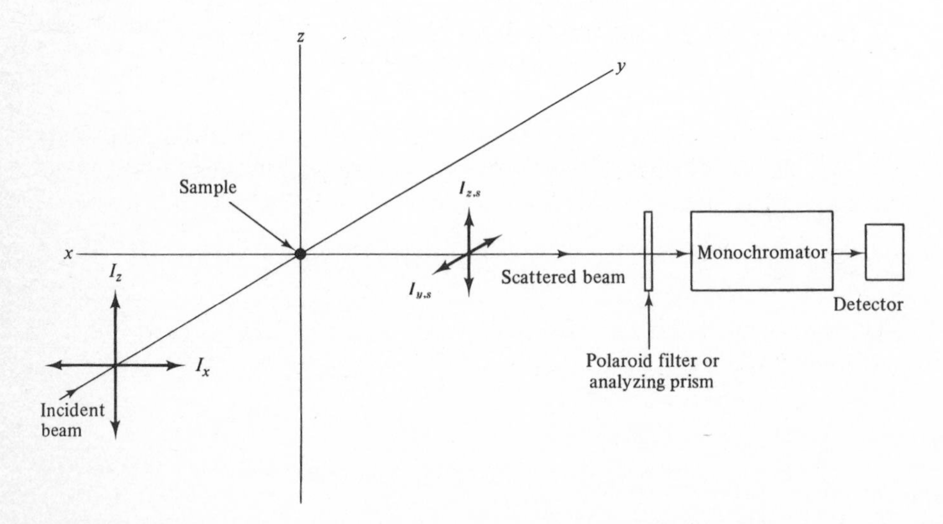

FIGURE 6-4 Measurement of depolarization factor using natural incident light.

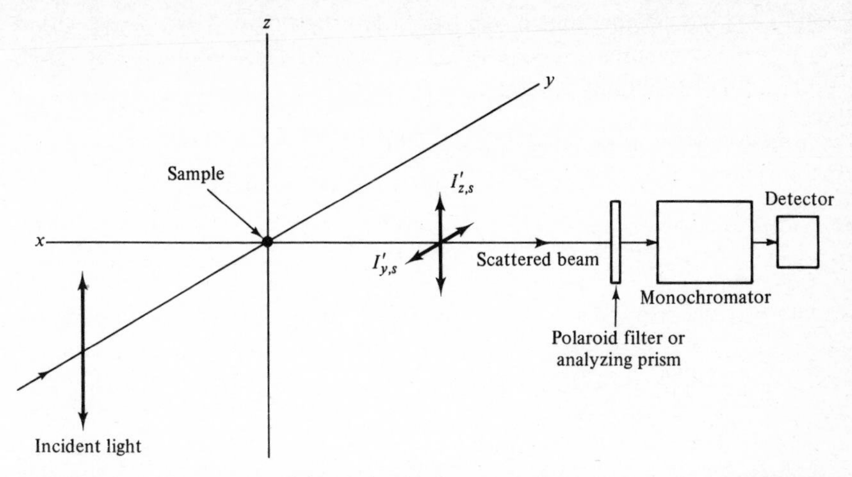

FIGURE 6-5 Measurement of depolarization factor using plane-polarized incident light.

larized parallel to the xz plane $I_{\parallel}$. Thus, from Fig. 6-4,

$$\rho_n = \frac{I_{\perp}}{I_{\parallel}} = \frac{I_{y,s}}{I_{z,s}} \tag{6-2}$$

where the subscript n denotes *natural* incident light and the subscript s denotes *scattered* light measured 90° to the incident light. It has been shown by Born† that the theoretical maximum value of ρ_n is $6/7$ and occurs for all nonsymmetrical vibrations giving rise to Raman lines. Thus, a nonsymmetrical vibration gives rise to relatively little polarization and is said to be *depolarized* (relatively nonpolarized). On the other hand, totally symmetrical vibrations yield *polarized* Raman lines, with values of ρ_n less than $6/7$ and frequently approaching zero. Thus, the depolarization factor gives useful additional information in assigning frequencies to particular modes of vibration within the molecule.

Plane-polarized incident radiation, e.g., laser radiation, is diagramed in Fig. 6-5. The depolarization factor is now denoted by ρ_p (where the subscript p stands for plane-polarized incident light), and is defined from Fig. 6-5 as

$$\rho_p = \frac{I'_{\perp}}{I'_{\parallel}} = \frac{I'_{y,s}}{I'_{z,s}} \tag{6-3}$$

The theoretical maximum value for ρ_p is $3/4$, which again occurs for all nonsymmetrical vibrations giving rise to Raman lines. The relationship between ρ_n and ρ_p is given by

$$\rho_n = \frac{2\rho_p}{1 + \rho_p} \tag{6-4}$$

† M. Born, "Optik," Edwards, Ann Arbor, Mich., 1943.

Example 6-5 Show that a depolarization factor of $6/7$, measured with a mercury-arc source, is equivalent to a depolarization factor of $3/4$ measured with a laser source.

Answer Solving Eq. (6-4) for ρ_p yields

$$\rho_p = \frac{\rho_n}{2 - \rho_n}$$

Therefore, if $\rho_n = 6/7$,

$$\rho_p = \frac{6/7}{2 - 6/7} = \frac{3}{4}$$

Figure 6-6 illustrates the Raman polarization spectrum of CCl_4. The Raman band at 459 cm^{-1} is strongly polarized, as indicated by a depolarization factor of only 0.007, whereas the bands at 314 and 218 cm^{-1} are considered depolarized and have depolarization factors of about 0.75. Unequivocally, the band at 459 cm^{-1} must correspond to the totally symmetrical stretching mode of vibration of CCl_4, whereas the bands at 314 and 218 cm^{-1} correspond to nonsymmetrical vibrations.

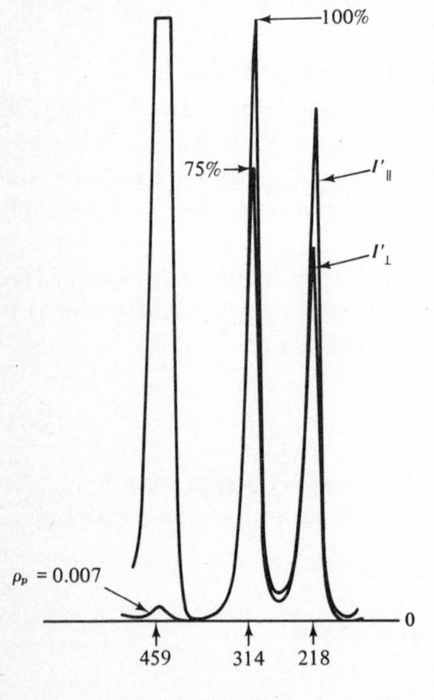

FIGURE 6-6 Raman polarization spectrum of CCl_4, obtained with a Perkin-Elmer model LR-1 laser Raman spectrometer; spectral slit width, 10 cm^{-1}; time constant, 3 s; sample 100% CCl_4, liquid. (*Perkin-Elmer Corp.*)

Example 6-6 Figure 6-7 depicts the Raman polarization spectrum of *p*-dioxane. Using Fig. 6-7, identify the Raman bands corresponding to totally symmetric vibrations and those corresponding to nonsymmetrical vibrations.

Answer All bands marked with a *P* for polarized in Fig. 6-7 correspond to totally symmetrical vibrations, whereas those marked with a *D* for depolarized correspond to nonsymmetrical vibrations.

6-2 INSTRUMENTATION

6-2A Radiation Sources

Before the advent of the laser source, the low-pressure mercury arc was the most commonly used Raman source. Today continuous gas lasers are largely replacing the mercury arc as a Raman source for the following reasons:

1 Laser radiation is highly monochromatic. Whereas a mercury arc emits a number of emission lines (at 2537, 3650, 4047, 4358, 5461, and 5770 to 5790 Å), laser radiation can be concentrated at a single wavelength, for example, 6328 Å in the helium-neon gas laser and 4880 Å in the argon-ion gas laser. The mercury arc must be carefully filtered to allow only one wavelength (generally 4358 Å) to irradiate the sample, whereas no filtering is needed with the laser source. Furthermore, the bandwidth of the laser line is smaller than the bandwidth of any other light source. The narrow bandwith permits Raman frequencies of very low energy to be measured, comparable to energies measured in far-infrared spectroscopy.

2 Laser beams are very intense. Although the present power level of continuous gas lasers is relatively low (about 50 mW, or less, compared with about 50 W for the 4358-Å line of a typical mercury arc), all the laser energy is concentrated in a very narrow, parallel beam having a relatively constant cross-sectional area over great distances, thereby achieving *specific* intensities (intensity per area) comparable to conventional sources.

3 Laser beams, being highly focused, permit unusually small sample sizes to be analyzed. Traditionally, Raman spectroscopy required large samples, on the order of 50 ml, although recent refinements such as multireflection cells allow as little as 0.2 ml to be examined in some cases. With the introduction of laser excitation, samples as small as 1 μl or less can be used.

4 Laser beams, being very well collimated, simplify the optical geometry of associated monochromators and permit more precise corrections for reflection losses.

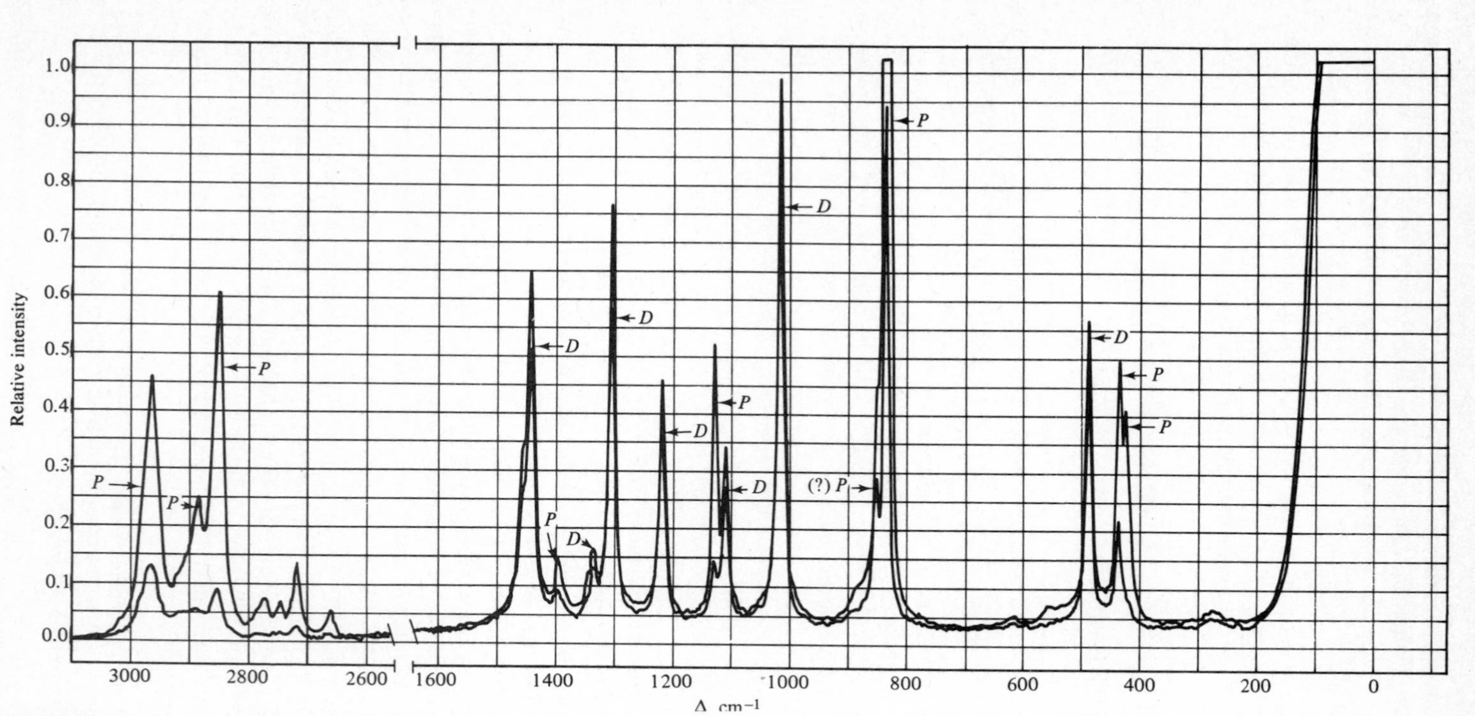

FIGURE 6-7 Raman polarization spectrum of *p*-dioxane, obtained with a Cary model 81 laser Raman spectrophotometer; spectral slit width, 5 cm^{-1}; time constant, 4 s. (*Varian Associates.*)

5 Laser beams, having well-defined polarization, allow more accurate depolarization measurements to be made.

Two different laser sources and the mercury arc will be discussed.

THE HELIUM-NEON LASER

This laser has an emission wavelength of 6328 Å, which is almost ideal for analytical work. In general, shorter-wavelength sources present greatly increased fluorescence difficulties, and longer-wavelength sources lead to severe losses in detector efficiencies, as well as complicating the demands on the monochromator for spectral purity and wavelength range. Furthermore, the orange-red color of the 6328-Å line allows a greater range of colored compounds to be studied than is possible with the blue 4358-Å line of the mercury arc. The 4358-Å line is almost useless for any compound that is even slightly colored, since even a trace of yellow will cause absorption and render Raman spectroscopy virtually impossible. On the other hand, with the 6328-Å laser line, yellow, brown, and red compounds can be studied without difficulty, and even blue and green compounds can be studied if they are strong Raman scatterers. An example of the latter is iodine in chloroform.

The first helium-neon laser was built in 1961, though all the necessary principles were known much earlier. The word *laser* is an acronym for *light amplification* by *stimulated emission* of *radiation.* In order to understand how a laser works, three principles of physics must be clearly understood: (1) *stimulated emission,* (2) *population inversion,* and (3) *optical resonance.* Stimulated emission, postulated by Einstein in 1917, is the basis of laser operation and occurs when a photon strikes an already excited atom (an atom that has a photon "in storage"), thereby causing that atom to emit its stored photon. However, stimulated emission will occur only when the energy of the impinging photon is exactly equal to the energy of the stored photon. Figure 6-8 shows the relationship of stimulated emission to the more ordinary processes of absorption and spontaneous emission. When a photon is *absorbed* by an atom (Fig. 6-8*a*), the energy of the photon is converted into internal energy of the atom; i.e., an electron, shown by a dot in Fig. 6-8, is raised to a higher energy level. The excited atom will sooner or later return to the ground state, spontaneously emitting a photon, as shown in Fig. 6-8*b*. The typical lifetime for an excited state is 10^{-8} s, though it may be as long as 10^{-4} s, or longer. During the period when the atom is still excited, it can be *stimulated* to emit a photon if it is struck by an outside photon having precisely the energy of the stored photon (see Fig. 6-8*c*). Thus, the incoming photon is now joined by a second photon from the excited atom, resulting in a gain or amplification of photons. Moreover, the second photon wave will be precisely in phase with the photon wave that triggered its release, giving a perfectly *coherent* (in-phase) beam of radiation.

The second essential principle is population inversion. In order for the stimulated emission process to continue and result in a net gain in the number

(*a*) Absorption

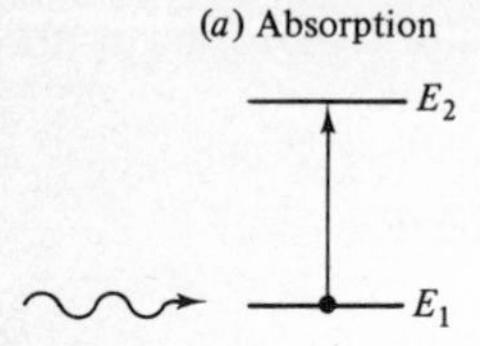

(*b*) Spontaneous emission

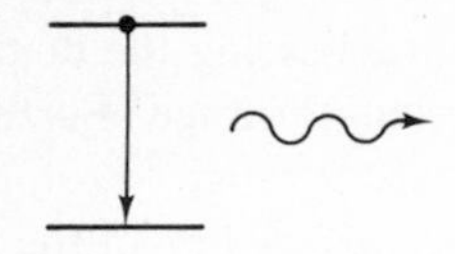

(*c*) Stimulated emission

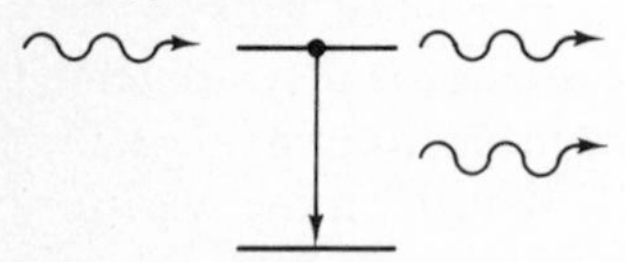

FIGURE 6-8 Various absorption and emission processes: (*a*) photon absorption; (*b*) spontaneous emission; and (*c*) stimulated emission.

of photons, the active medium must contain more excited atoms than unexcited ones. This condition does not ordinarily occur in nature, since, according to the Boltzmann distribution, there will be an increasingly larger population of atoms at decreasingly lower energies. Thus, it is necessary to cause a population inversion to achieve suitable conditions for laser operation. In the helium-neon laser, population inversion is achieved by making use of an unusual property of helium. An excited helium atom, with an electron in a 2*s* level, cannot lose this energy by radiation and can lose it only by collision with another atom which has a comparable energy level available. The neon 3*s* levels are comparable in energy to the 2*s* (singlet) level in helium, as shown in Fig. 6-9, and thus when an excited helium atom collides with a ground-state neon atom, a transfer of energy from helium to neon can take place. The excited 3*s* levels of neon have a relatively long lifetime before spontaneously decaying to the 2*p* ground state with the emission of a 6328-Å photon, and so it is possible to build up a larger population of neon atoms in the 3*s* state than in the 2*p* state (population inversion). The spontaneous emission of a single photon at 6328 Å can trigger a whole cascade of similar photons by the process of stimulated emission. In practice, population inversion is achieved by mixing a large proportion of helium with a small proportion of neon (the optimum ratio is about 7:1), and the mixture is exposed to ionizing radio-frequency or dc voltage. Helium atoms are ionized by electron bombardment, but they rapidly return to the excited 2*s* state, where they remain until collision with neon. The buildup of a population of atoms in an excited state is sometimes called *optical pumping.*

The third principle essential to the operation of a laser is optical resonance, whereby the amplification process is made to build up or strengthen until a very intense beam is formed. In practice, the laser action is staged in a tube, closed off with mirrors to form a *resonant cavity.* Making the spacing between the mirrors an integral multiple of the desired wavelength means that there will be a buildup of energy at the desired wavelength (resonant frequency) and a loss of energy at other wavelengths. Figure 6-10 illustrates how the buildup of laser intensity occurs. The dots represent a random population of excited neon atoms, and in Fig. 6-10*a* they are shown emitting photons in all directions. Photons that happen to get started along the tube axis are reflected by end mirrors, whereas photons starting in other directions are lost. As photons pass through the tube, stimulated emission causes a rapid growth in photon intensity (Fig. 6-10*b*), and the repeated reflections cause an enormous buildup in intensity. If one mirror is semitransparent, a portion of the traveling beam will escape through it (Fig. 6-10*c*).

A schematic illustration of a complete helium-neon laser in given in Fig. 6-11. The tube is filled with a 7:1 mixture of helium and neon gas for optimum output of the 6328-Å laser line. Although radio-frequency excitation was used in the early models, high-voltage (2- to 3-kV) dc excitation is preferred today. To start the laser, 5 to 10 kV dc is used, but thereafter the beam

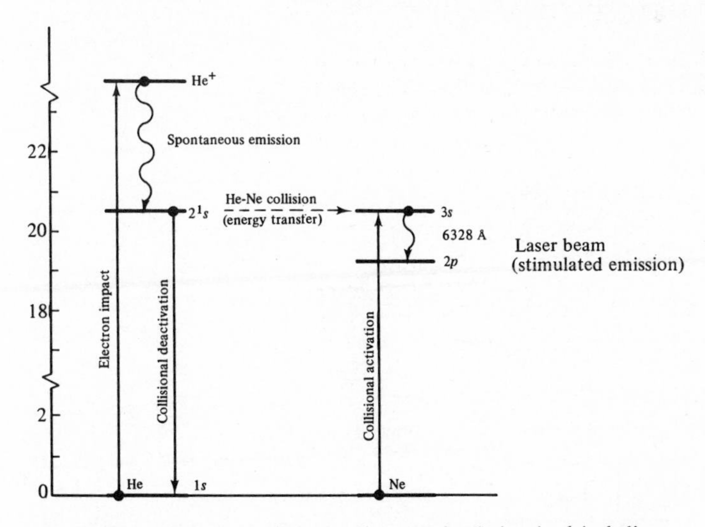

FIGURE 6-9 Schematic diagram of energy levels involved in helium-neon laser. [*From A. D. White and J. D. Rigden, Proc. I.R.E.*, **50**:1697 (1962), *by permission.*]

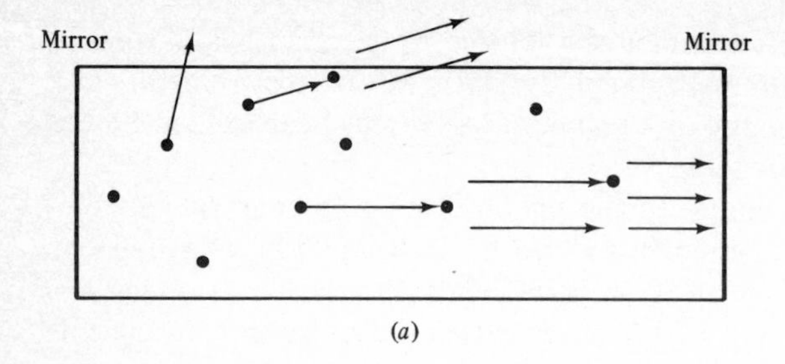

(a)

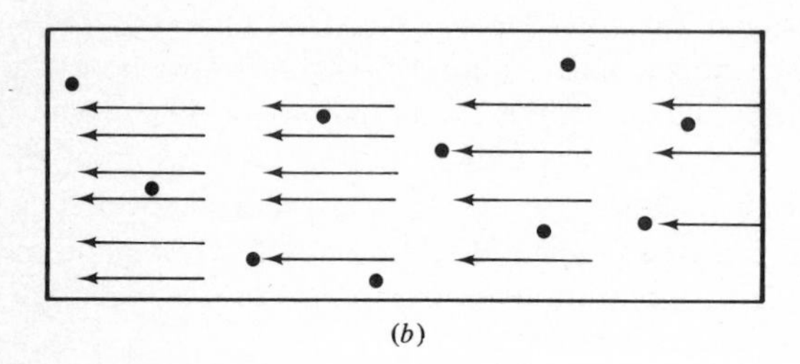

(b)

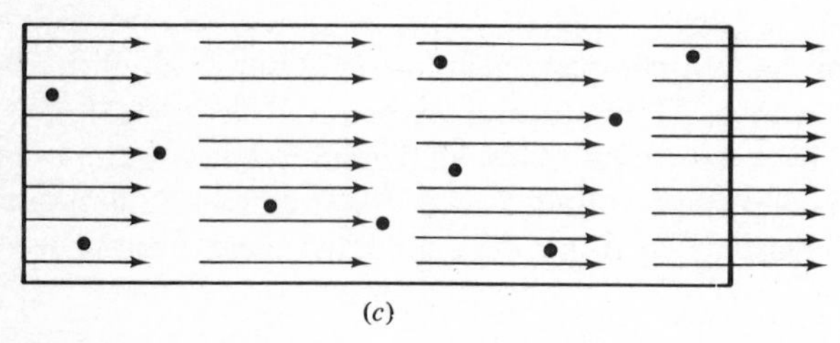

(c)

FIGURE 6-10 Buildup of laser action. Dots represent gaseous atoms; see text for explanation. (*From A. L. Schawlow, Optical Masers. Copyright © 1961 by Scientific American, Inc. All rights reserved.*)

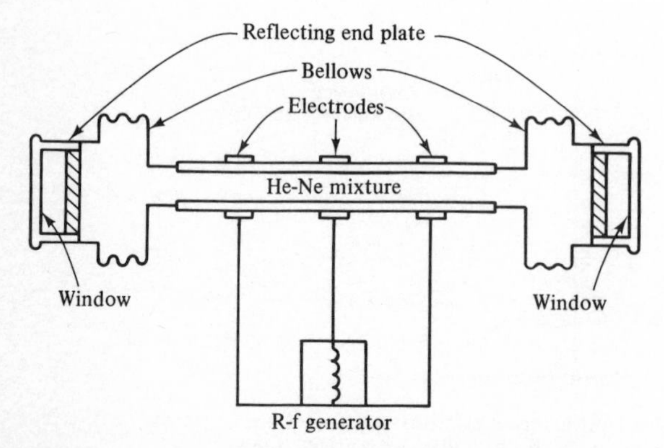

FIGURE 6-11 Simplified schematic diagram of a helium-neon laser. (*From B. A. Lengyel, "Introduction to Laser Physics," p. 55, John Wiley & Sons, Inc., New York, 1966, by permission.*)

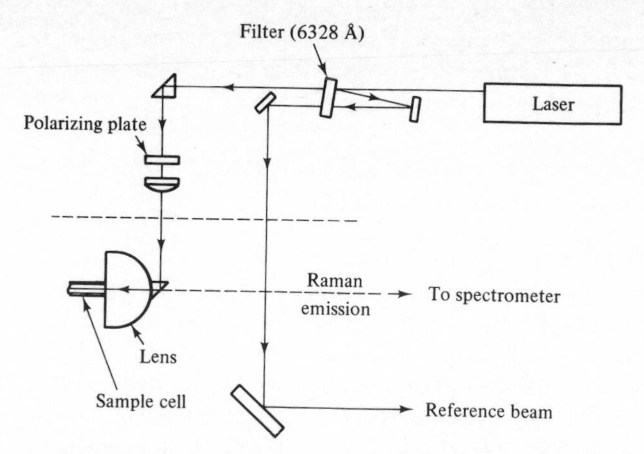

FIGURE 6-12 Arrangement of laser tube and sample cell in Cary model 81 Raman spectrometer. (*From H. A. Szymanski, "Raman Spectroscopy," p.* 90, *Plenum Publishing Corporation, New York,* 1967 *by permission.*)

can be maintained on the lower operating voltage. In another recent change, the flat mirrors of Fig. 6-11 have largely been replaced by one or two spherically curved mirrors, which are much less critical to adjust and give a better-defined laser beam. One of the mirrors should have a reflection coefficient as close to unity as possible, with the other mirrors transmitting about 4 percent at 6328 Å. The laser tubes now in use are generally between 0.5 and 1.5 m long and yield powers from 0.5 to 100 mW at the 6328-Å line.

In laser Raman spectrometers, the sample cell is placed outside the laser cavity. Figure 6-12 shows the laser-tube–sample-cell arrangement used in the

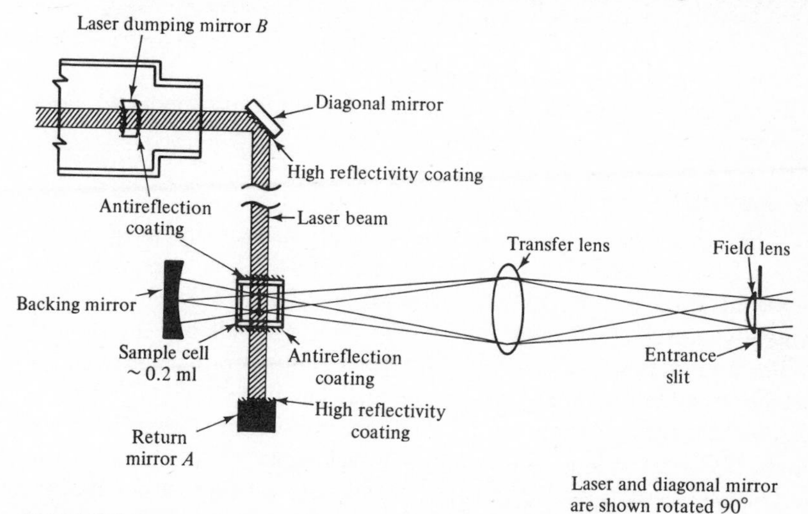

FIGURE 6-13 Arrangement of laser tube and sample cell in Perkin-Elmer model LR-1 Raman spectrometer. (*Perkin-Elmer Corp.*)

Cary 81 Raman spectrometer, and Fig. 6-13 shows the arrangement used in the Perkin-Elmer model LR-1 Raman spectrometer. The return mirror in Fig. 6-13 is used to reflect the laser beam back through the cell to the end window of the laser, where it is again reflected, allowing multiple passes through the sample cell. Thus, the backing mirror gathers Raman emissions that would otherwise be lost and diverts them into the spectrometer.

THE ARGON-ION LASER

Although there are numerous other types of lasers being developed, the argon-ion laser seems to hold the most promise an another Raman source, for two main reasons: (1) It yields very intense lines at 4880, 4965, and 5145 Å, and these blue to green colors should thus complement the red helium-neon laser, especially for blue and green samples. (2) The power delivered in these argon-ion lines can be as high as 500 mW to 1 W or more, which is about an order of magnitude higher than the best achieved with the helium-neon laser. Thus, more intense Raman spectra should be obtainable. Unfortunately, the lifetime of the argon-ion laser is still rather short, making it too costly for general use at present. However, the potential of this source merits a brief discussion of how it works.

There is considerable evidence that the excitation process in an argon laser involves two steps: (1) an inelastic collision with an electron takes an argon atom (with an outer-electron configuration of $3s^2 3p^6$) to the ground state of Ar^+ ($3s^2 3p^5$); (2) another electron collision excites an outer electron of the argon ion to one of several $4p$ states, and a subsequent transition to a $4s$ state results in an emission line. Thus, transitions between various levels in the $4p$ and $4s$ states result in the three closely spaced laser lines at 4880, 4965, and 5145 Å. The 4880-Å line is the most intense and is the only emission if the amount of argon is kept small. Again, spontaneous emission triggers the reaction, and stimulated emission of the otherwise delayed $4p$ to $4s$ transition permits the amplification necessary for laser action.

An argon-ion laser accessory is available for the Cary model 81 laser Raman spectrometer.

THE MERCURY ARC

Until the development of the laser, the mercury arc was the most widely used Raman source. Mercury has strong emission lines at approximately 2537, 3650, 4057, 4358, 5461, 5770, and 5790 Å, that at 4358 Å being the most generally useful for Raman studies. To eliminate all but the desired emission line and minimize a small but significant continuous background, filters must be used. For example, to absorb lines below 4358 Å, a saturated aqueous sodium nitrite solution can be used, and a rhodamine dye (Du Pont rhodamine 5 GDN extra) will remove higher wavelengths.

The Toronto mercury-arc lamp is one of the most widely used for Raman studies. Mercury electrodes are located at each end of a glass helix coil, and the electrodes are cooled with circulating thermostated water to keep the mercury vapor pressure low and thus reduce the background continuum. The arc emission takes place through the helix coil, and air blowers are used to cool the helix. The sample cell, in the form of a tube, is placed in the center of the helix coil, and a filter jacket goes between the sample tube and the helix source. A magnesium oxide reflector is sometimes used on the outside of the helix.

6-2B Monochromators

A Raman spectrometer uses a monochromator to scan the Raman emission lines emerging from the sample beam at right angles to the source beam. To illustrate the operation of Raman spectrometers, two commercial instruments, the Perkin-Elmer model LR-1 and the Cary model 81, will be briefly described.

A block diagram of the Perkin-Elmer model LR-1 is shown in Fig. 6-14, and a photograph of the complete instrument is shown in Fig. 6-15. Light from the helium-neon laser beam (at 6328 Å) enters the sample compartment horizontally, from the rear of the instrument, and Raman scattering from the sample cell is focused on the monochromator entrance slit. When depolarization measurements are to be made, the Raman emission is first passed through an analyzer prism before entering the monochromator (see Fig. 6-14). The monochromator is a double-pass, Littrow-mounted grating type, a schematic of which is shown in Fig. 6-16. A 13-Hz chopper is used between the first and second passes, and the detector is made to respond only to this alternating signal, thereby decreasing interference from stray radiation. The wavelength is scanned automatically, scanning speeds ranging from 4 to 440 cm^{-1}/min.

The Cary model 81 Raman spectrometer uses a double monochromator.

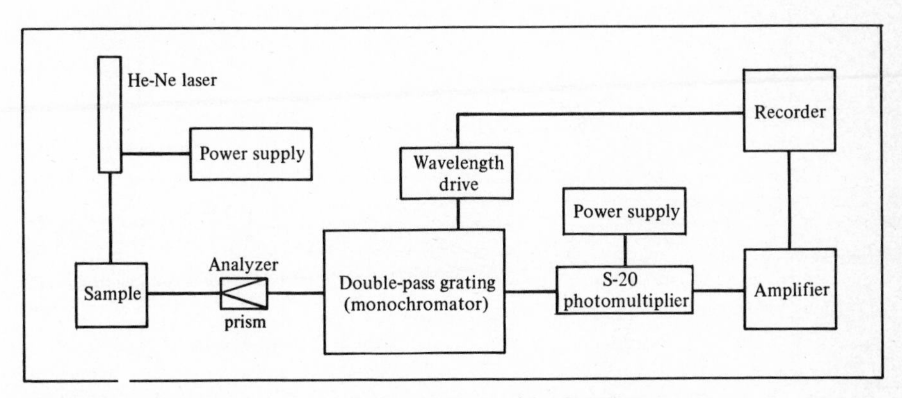

FIGURE -14 Block diagram of Perkin-Elmer model LR-1 Raman spectrometer. (*Perkin-Elmer Corp.*)

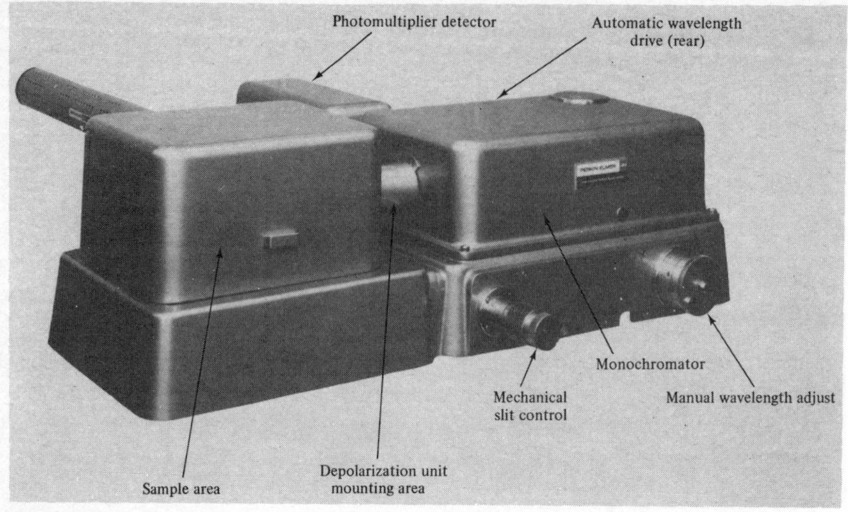

FIGURE 6-15 Perkin-Elmer model LR-1 Raman spectrometer. (*Perkin Elmer Corp.*)

The instrument is shown in Fig. 6-17, and an optical diagram is given in Fig. 6-18. Light from laser source *A* is filtered by isolation filter *B*, which passes energy of 6328 Å. This energy passes through prism *C*, half-wave plate *D* (which rotates the plane of polarization), and focusing lens *E* into the very small prism *F*, which directs it into the sample. Raman light from the sample is focused by lens *G* onto the cell exit lens *H*, which directs it through polarizer *I* and collimating lens *J* to the first image slicer *K*. Image slicers improve the light-gathering power of the instrument by taking all the rays emerging from the end of the sample cell and concentrating them to the exact size of the

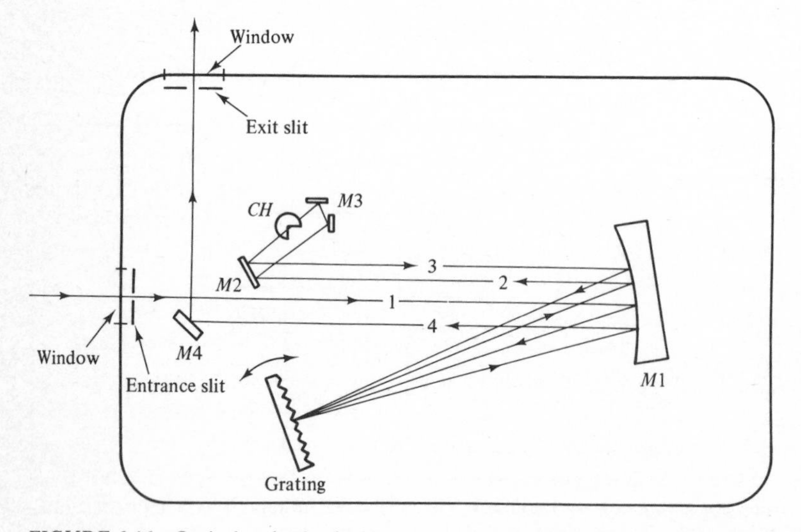

FIGURE 6-16 Optical path of a double-pass grating monochromator. (*Perkin-Elmer Corp.*)

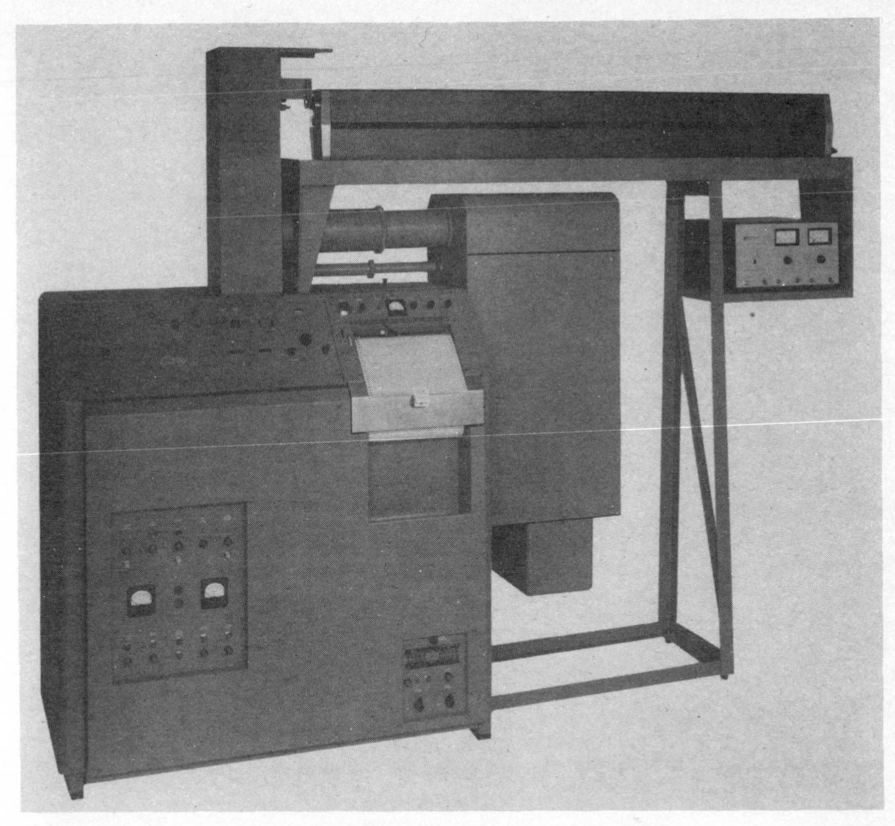

FIGURE 6-17 Cary model 81 Raman spectrometer. (*Varian Associates.*)

double entrance slit. Image slicer K divides the beam into 20 images, which are directed onto the 10 sections of the second image slicer L, each section receiving two superimposed images. Image slicer L, in turn, focuses the 10 compound images onto two narrow strips of light in the plane of double entrance slit S_1. Prisms M and P serve only to change the direction of the light. From collimating mirror R_1, the beam is reflected to grating T_1, then reflected by mirror U through double intermediate slit S_2 to the second monochromator. From S_2 it is reflected by mirror V and the second collimator R_2 to the second grating T_2, and then directed through double exit slit S_3. Lenses W and X direct the beam to rotating mirror Y, where it is chopped and directed to photomultiplier Z_1 by lenses h, i, and j. The signal from photomultiplier Z_1 is compared to the reference signal developed by photomultiplier Z_2. Concurrently, part of the laser radiation is reflected by filter B to mirror b_1, which directs it back through filter B into an auxiliary optical train (elements b_2, b_3, c, d, e, f, g) to rotating mirror Y. It is then directed through lens l to reference photomultiplier Z_2 for comparison with the Raman signal from photomultiplier Z_1.

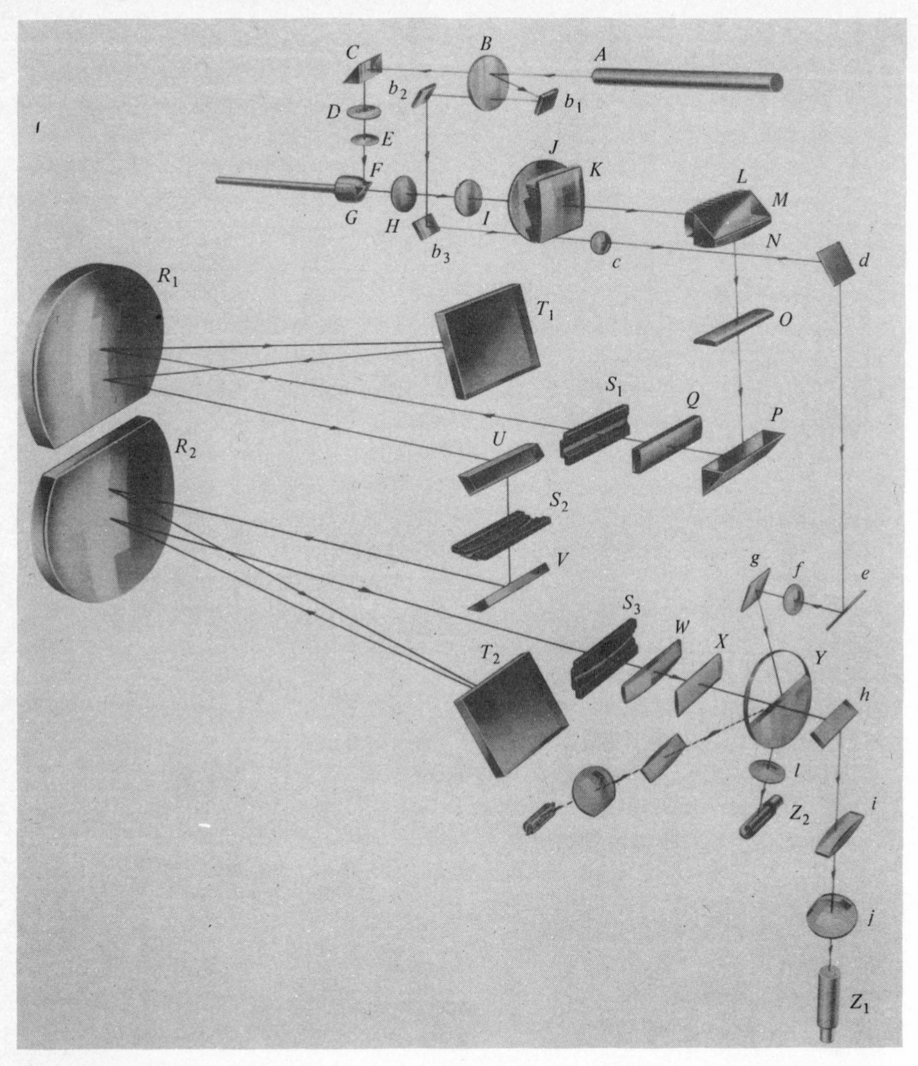

FIGURE 6-18 Optical diagram of Cary model 81 Raman spectrometer with laser source. (*Varian Associates.*)

Cary-Varian has made two new laser Raman instruments available; model 82 is a research instrument with a triple monochromator, and model 83 is a lower-cost instrument for routine analytical applications. Another commercially available instrument is the Jeol, Ltd., model JRS-S1 laser Raman spectrophotometer.

6-2C Detectors

Until about 1946, photographic plates were the universal method of detecting Raman emissions. The low intensity of Raman lines usually necessitated long exposure times and limited the interpretation of spectra to qualitative or at best semiquantitative analysis. Today, spectrometers with photomultiplier detectors have almost completely replaced the photographic recording spectrographs. Both the Perkin-Elmer LR-1 and the Cary 81 Raman spectrometers use photomultipliers with an S-20 response, which means that the photocathode is coated with a material giving a spectral response like that shown in Fig. 6-19. Typical of such photomulipliers in use is the 14-stage RCA 7265 photomultiplier, which has a photocathode coated with a semitransparent layer of antimony, potassium, sodium, and cesium.

The S-20 type spectral response is one of the best so far developed for the red region, but the falloff at longer wavelengths is the major factor in limiting the range of most Raman spectrometers to about 0 to 4000 cm^{-1} when the 6328-Å helium-neon laser line is used as a source and to about 0 to 5000 cm^{-1} when the 4358-Å line from a mercury arc is used (see Exercise 6-1).

Example 6-7 Show that a Raman spectrometer equipped with a helium-neon laser source and a photomultiplier detector with an S-20 response will be limited to a wave-number range of 0 to 4000 cm^{-1}.

Answer The helium-neon line at 6328 Å corresponds to about 15,800 cm^{-1}. Since the most important Raman lines are the Stokes' lines, falling

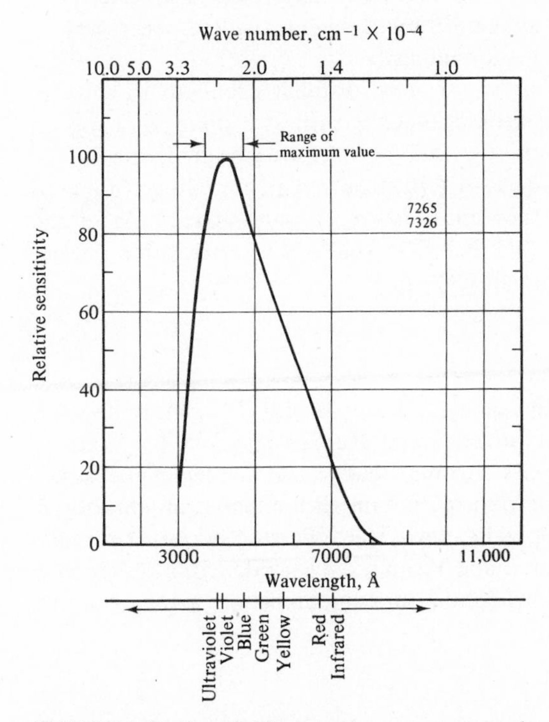

FIGURE 6-19 Type S-20 spectral-response curve used in detectors for Raman spectrometers. (*RCA Corp.*)

at lower energy than the source line, a Raman shift of 4000 cm^{-1} would thus fall at 15,800 − 4000 = 11,800 cm^{-1}. This corresponds to 847 nm, which approximates the long-wavelength limit of a S-20 type response (see Fig. 6-19).

6-3 SAMPLE HANDLING

Gases, liquids, and solids can be studied by Raman spectroscopy. The techniques and sample sizes vary somewhat according as a mercury arc or a laser source is being used. In general, liquids are the easiest to handle; volumes ranging from 0.2 to 65 ml are analyzable with a mercury-arc source and volumes from 1 μl to 2.5 ml with a laser source. Since Raman emissions are relatively weak compared with the intensities dealt with in absorption spectrophotometry, liquid samples are usually not diluted before analysis. For dissolving solids or when dilution proves desirable, solvents are selected on the basic of their Raman spectra, as well as on their solvating power and possible chemical interaction with the solute. Water, which absorbs so strongly in the infrared region, is an ideal solvent for polar compounds.

Gaseous samples have been handled with mercury-arc sources by going to relatively large volumes, high pressures, and multipass cell compartments. Very little has thus far been done with respect to analyzing gases with laser sources, but undoubtedly multiple-reflection techniques will be necessary in order to get sufficient Raman emission intensity.

Solid sampling has been one of the most difficult problems to solve in Raman spectroscopy, in spite of the fact that the earliest Raman work was on solids. However, with the laser source, it is now feasible to analyze solids in the form of single crystals, powders, or KBr pellets, which are like those used in infrared work except that they must be more concentrated. With the mercury-arc source, single crystals and KBr pellets give reasonable results, while powders give very poor Raman scattering.

6-4 APPLICATIONS

This chapter has emphasized the many similarities and differences between the information obtained from infrared and Raman spectra. To attack a particular problem, it is sometimes advantageous to use one technique rather than the other, whereas to obtain information on all the modes of vibration it is usually necessary to use both techniques. The following major factors affecting the choice between infrared and Raman data (aside from the obvious and overriding consideration of availability of equipment) are arranged in the order of decreasing importance.

SAMPLE STATE

If the sample is *gaseous*, infrared is generally the better technique, for two reasons. (1) The efficiency of Raman scattering is inherently much less than that of infrared absorption, and since the intensity of gaseous spectra is often a problem even in the infrared, it becomes formidable with Raman spectroscopy. (2) Whereas large multipass gaseous cells have been developed for Raman spectrometers using a mercury-arc source, very little has been done so far with the otherwise more advantageous laser sources, presumably because a laser source has such a small and well-defined beam that the intensity problems with gaseous samples are accentuated.

With *liquid* samples, either technique can be used with about equal ease. However, when appreciable water is present, Raman spectroscopy is the exclusive choice.

With *solid* samples, Raman spectroscopy with a laser source probably has a slight edge over the infrared, since it is now feasible to obtain Raman spectra directly on single crystals and powders, as well as on KBr pellets. However, the large variety of techniques that have been developed for examining solids with an infrared spectrophotometer, including oil mulls, melts, pellets, and attenuated total reflection, prevents the choice between Raman and infrared from being decisive.

CONCENTRATION OF SAMPLE

The inherent differences in intensity between Raman and infrared spectra make infrared the method of choice for dilute solutions, whereas concentrated solutions are better suited for Raman spectroscopy.

WAVELENGTH RANGE

Until recently, very few infrared instruments were available that went out into the far infrared (beyond 25 μm, or 400 cm^{-1}). Today specialized infrared spectrophotometers go out to about 50 μm (200 cm^{-1}), and Fourier spectrophotometers go out to 1000 μm (10 cm^{-1}); nonetheless, few laboratories possess these instruments. A single Raman spectrometer, on the other hand, routinely covers the range 100 to 4000 cm^{-1} (or 2.5 to 100 μm), with the width of the source line (Rayleigh line) setting the lower limit of the detectable frequency.

OVERTONES

A well-established experimental rule of thumb says that the intensity of overtones and combination tones relative to that of fundamentals in the Raman effect is very small (0.01 or less) whereas in the infrared the ratio is about 0.1. This difference gives the Raman spectra two big advantages over infrared spectra: (1) it simplifies the Raman spectrum and makes possible an easy differentiation between overtones and fundamentals; (2), it sometimes permits a clear analytical delineation between two substances in a mixture where infrared methods fail because the overtones of one substance overlap the fundamentals of the other.

The following specific examples of useful Raman applications are grouped according to common objectives and listed in the order of decreasing use.

6-4A Structure Determination

The introductory comments about structure determination made in the chapter on infrared spectrophotometry (Sec. 3-4*A*) are entirely applicable here. Raman spectroscopy strongly complements the information obtained from infrared spectrophotometry, and in fact the two must be used together in order to obtain full information about bond angles, bond stiffness, and other molecular parameters.

The procedure for determining the molecular structure of a not too complicated molecule, using infrared and Raman spectra, can be outlined as follows:

1 Observe both spectra.

2 Assume a molecular model and predict the number of Raman lines and infrared absorption bands it should have [a simple approach, with examples, was outlined in the theory section (Sec. 6-1); a more rigorous approach using group theory is explained in Ref. 10].

3 Compare the observed spectra with the predictions for the model.

4 Accept or reject the model.

More often than not this procedure cannot be applied because the molecule is too complex (involves too many degrees of freedom) or yields two models with similar spectral predictions, or because certain frequencies fail to appear in one spectrum or the other. In the last two cases, the difficulties may be resolvable with the help of isotopic derivatives, e.g., using deuterium.

With complex molecules it is usually possible to reach conclusions of value using vibrational frequencies and intensities to deduce the presence or absence of certain functional groups in a molecule, using methods completely analogous to those used in qualitative analysis by infrared spectrophotometry (see Secs. 3-4*A* and 3-4*B*). A frequency-assignment chart for Raman spectra is given in Fig. 6-20.† It should be no surprise that the group correlations for Raman spectra agree in most cases with those given for infrared spectra (compare Fig. 6-20 with Fig. 3-36). For example, for both Raman and infrared spectra O—H stretches occur in the vicinity of 3600 cm^{-1}, N—H stretches at about 3400 cm^{-1}, C—H stretches at about 3000 cm^{-1}, C≡C stretches at about 2200 cm^{-1}, and C=C stretches at about 1600 cm^{-1}. On the other hand, separate correlation charts are desirable since in some cases a correlation is unique for either Raman or infrared spectra. For example, internal double bonds in *trans*-olefinic hydrocarbons, represented by the formula

† Further details may be found in H. J. Sloane, The Use of Group Frequencies for Structural Analysis in the Raman Compared to the Infrared, pp. 15–36 in C. D. Craver (ed.), "Polymer Characterization," Plenum, New York, 1971.

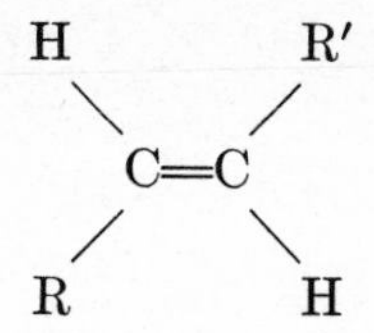

have a strong Raman line near 1675 cm^{-1}, whereas the corresponding infrared band is absent or very weak. On the other hand, a useful infrared band at 720 cm^{-1}, which indicates the presence of four or more methylene groups in a chain, has no counterpart in the Raman spectrum.

In the Raman correlation chart shown in Fig. 6-20, a given functional group is shown to appear over a *range* of frequencies, rather than at a fixed frequency, since the neighboring atoms and indeed the molecule as a whole exerts an influence on the fundamental vibration. The shift in the Raman frequency of a given functional group can yield valuable information about the molecule. For example, as Fig. 6-20 shows, the ethylenic group, in general, may be found anywhere from about 1600 to 1680 cm^{-1}. For specific compounds, it may be noted that the Raman line for ethylene is at 1620 cm^{-1}, but for propylene and other derivatives of the type $CH_2{=}CHR$ (where R is an aliphatic radical) it is about 1647 cm^{-1}. If the compound is $CH_2{=}CHCl$, the ethylene line occurs at 1608 cm^{-1}. If the compound is $CH_2{=}CHCHO$, the line occurs at 1618 cm^{-1}. It is interesting to note that if a CH_2 or R group is introduced on one side of the double bond, e.g., to give $CH_2{=}CHCH_2Cl$, the shift is nearly normal, i.e., about 1640 cm^{-1}. On the other hand, if the CH_2 or R groups are added on both sides of the double bond, e.g., giving $CH_2CH{=}CHR$, cis-trans isomerism becomes possible, and the cis form is al-

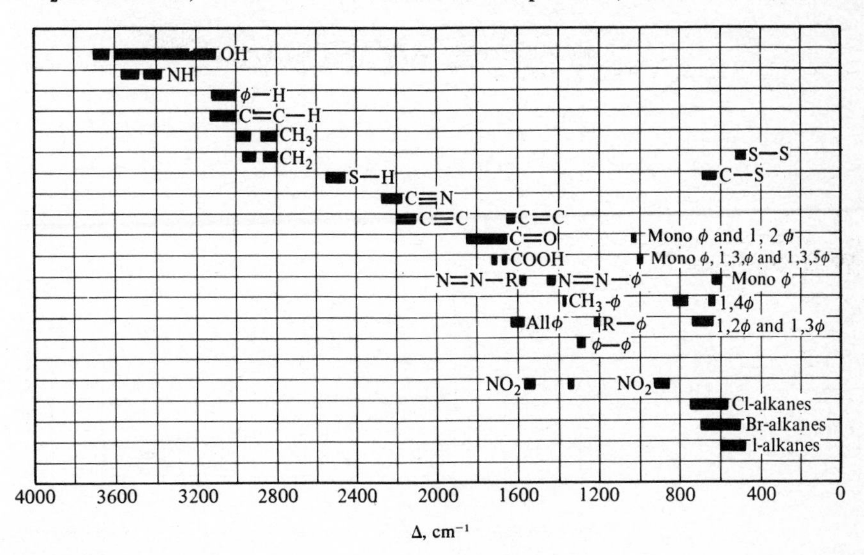

FIGURE 6-20 Frequency-assignment chart for Raman spectra. (ϕ denotes C_6H_5—.) (*Varian Associates.*)

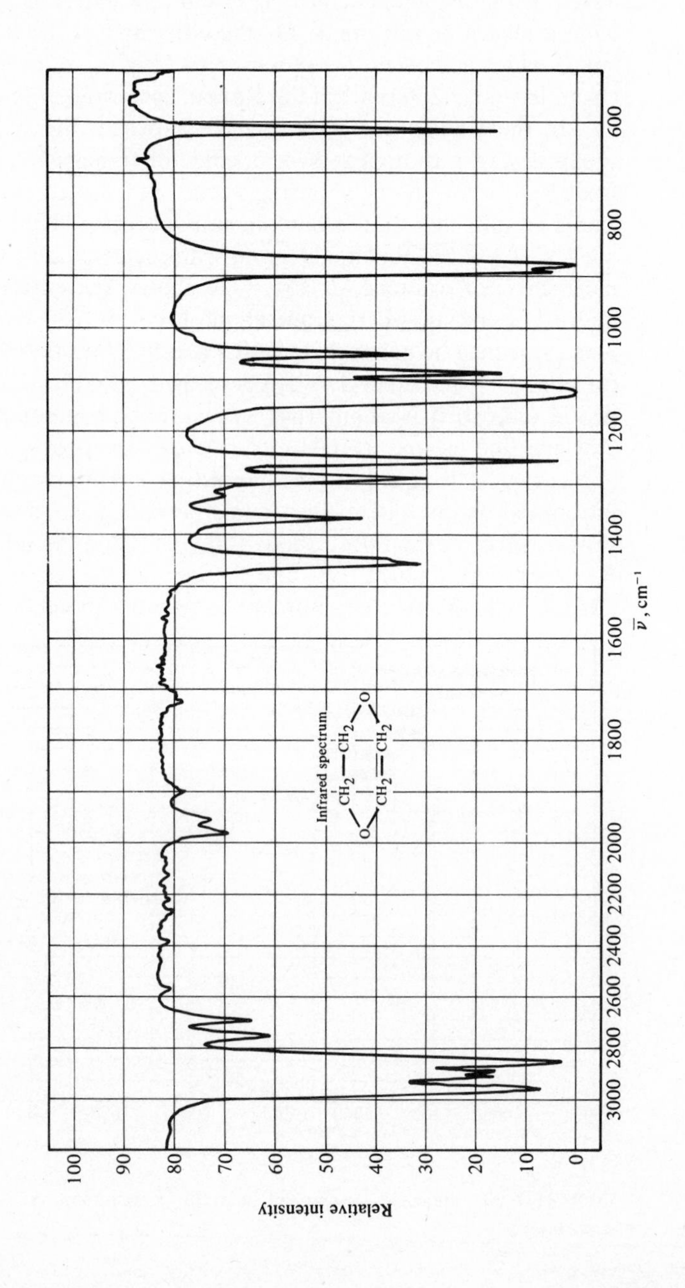
Infrared spectrum
$O\langle CH_2—CH_2 \rangle O$, $CH_2—CH_2$
Relative intensity
100
90
80
70
60
50
40
30
20
10
0
3000
2800
2600
2400
2200
2000
1800
1600
1400
1200
1000
800
600
$\bar{\nu}$, cm^{-1}

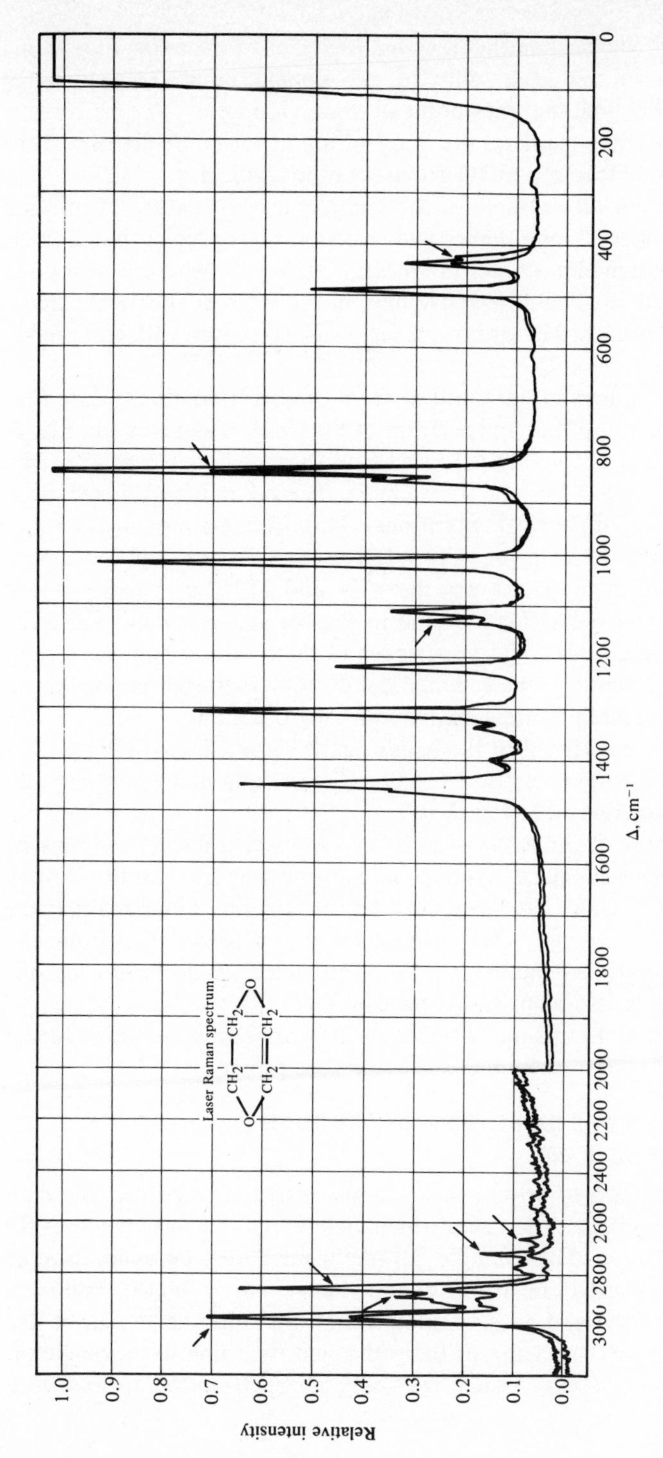

FIGURE 6-21 Infrared and Raman spectra of *p*-dioxane. (*Varian Associates.*)

ways about 20 cm^{-1} lower than the trans but higher than in compounds with a CH_2 or R group on one side only of the double bond. In substituted hydrocarbons of this type the lines occur at about 1658 cm^{-1} for cis isomers and 1674 cm^{-1} for trans isomers, the lines falling at lower frequencies if the substituent group is chlorine, a CHO group, or other radical.

Figure 6-21 gives an example of the complementary nature of infrared and Raman techniques. Group theory and selection rules predict that dioxane should have 36 fundamental modes of vibration ($3N - 6$, where $N = 14$; see Sec. 3-1*B*), of which 18 should be active only in the Raman and 18 should be active only in the infrared. The spectra in Fig. 6-21 agree well with this prediction.

The additional information obtained from polarization studies (see Sec. 6-1*B*) is also shown in the Raman spectrum of Fig. 6-21. The arrows in Fig. 6-21 point to bands that are highly polarized (specifically, bands near 425, 840, 1130, 2660, 2720, 2850, 2890, and 2955 cm^{-1}). These bands can therefore be assigned to totally symmetrical vibrations. This information allows other interesting deductions to be made with relative ease. For example, the most intense peaks in the two spectra are the 875- and 1125-cm^{-1} bands in the infrared spectrum and the 840-cm^{-1} band in the Raman spectrum. Since the 840-cm^{-1} band is polarized, it can be assigned to the totally symmetrical ring-breathing mode analogous to the symmetrical C—O—C stretch in open-chain ethers. The corresponding antisymmetrical ring-breathing modes in the infrared spectrum are assigned to the bands near 875 and 1125 cm^{-1}, the latter analogous to the antisymmetrical C—O—C stretching mode normally observed in open-chain ethers (see Fig. 3-36).

Another example of the value of polarization measurements can be seen by examining the closely spaced doublet at 425 and 440 cm^{-1} in the Raman spectrum (Fig. 6-21). Since one peak (425 cm^{-1}) is highly polarized and the other is not, it is immediately obvious that these two peaks arise from distinctly different vibrational modes, a conclusion which would have been difficult to make without this polarization measurement.

To conclude this discussion of structure determination, several specific examples of the usefulness of Raman spectra will be given.

DETERMINING THE LOCATION OF GROUPS ON A BENZENE RING

The Raman spectra of ortho-, meta-, and para-substituted benzene derivatives have characteristic differences which enable one to decide between the positions of substitution. For example, all meta-substituted benzenes have an intense, strongly polarized line at 995 cm^{-1}, which is not present in ortho and para compounds. Ortho and para compounds, on the other hand, can be distinguished by the richer spectrum of the ortho and by a line in the neighborhood of 625 cm^{-1} in the para, which is usually absent from the ortho derivatives.

DISTINGUISHING BETWEEN CIS AND TRANS ISOMERS

In general, Raman spectra are very useful for distinguishing between cis and trans isomers, particularly in conjunction with infrared spectra. The basis for this is that many isomers of the trans structure have either a genuine center of symmetry or a sufficient approximation thereto to make operative the selection rule that excludes from the Raman effect all vibrations antisymmetrical to the center. This rule does not apply to cis isomers, which do not have a center of symmetry. One can therefore determine which of a pair of isomers is the cis and which is the trans because the trans isomer will have a smaller number of strong Raman lines. The differentiation is more positive in conjunction with infrared spectra, since in the infrared spectrum of the trans isomer, absorption bands corresponding to strong Raman lines should be weak or missing, whereas strong infrared bands should be weak or missing in the Raman spectrum. This procedure is not applicable when the trans isomer departs markedly from the centrosymmetric, since the differences between the Raman spectra of cis and trans isomers become small.

DETERMINING THE STRUCTURE OF SUBSTANCES IN AQUEOUS SOLUTION

Since infrared spectrophotometry cannot be used for studying molecules dissolved in water, Raman spectroscopy promises to be extremely valuable for

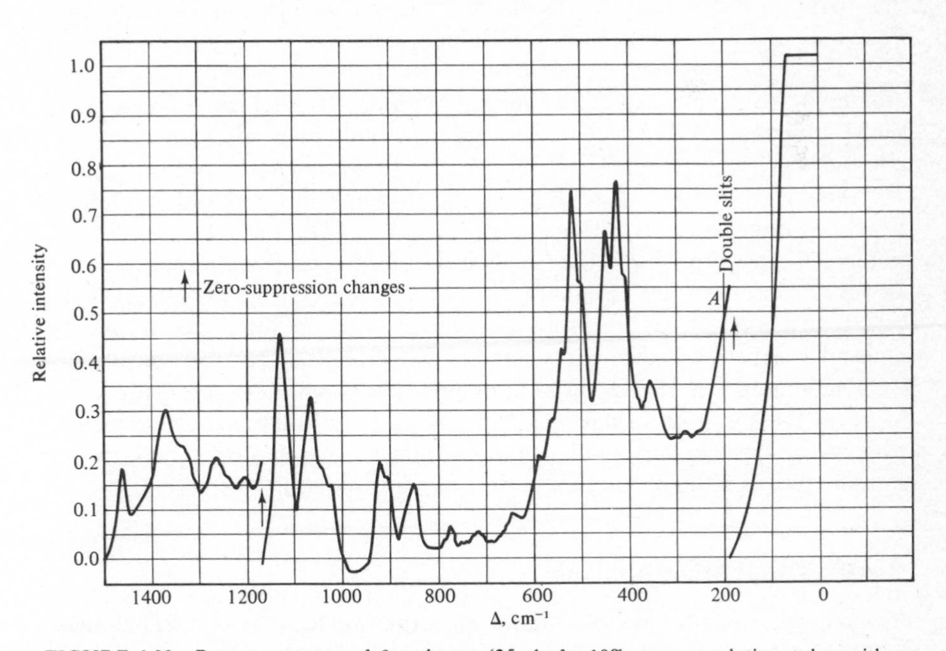

FIGURE 6-22 Raman spectrum of β-D-glucose (25 μl of a 10% aqueous solution, taken with a Cary model 81 laser Raman spectrometer). (*Varian Associates.*)

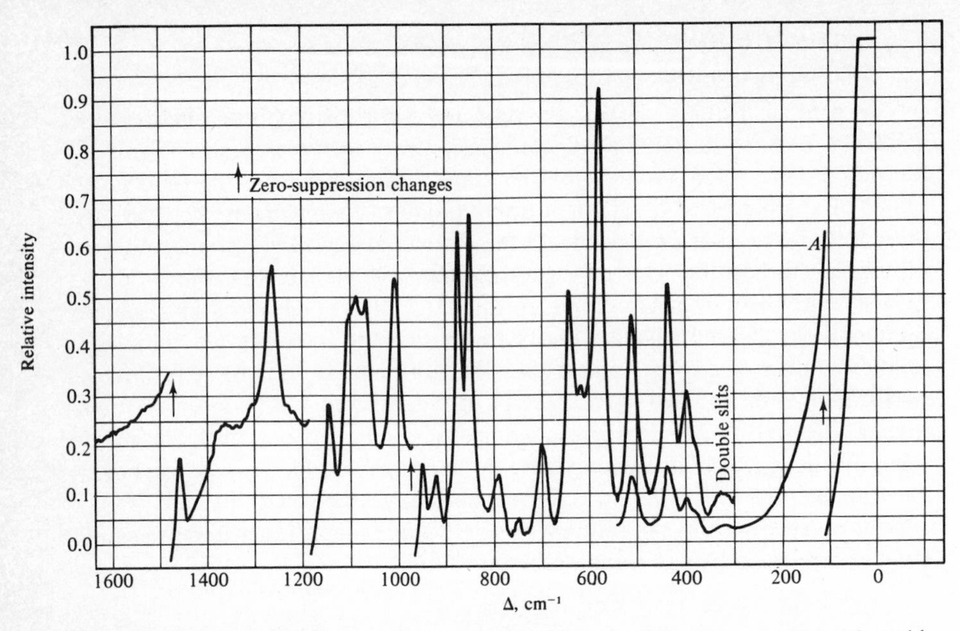

FIGURE 6-23 Raman spectrum of D(−)-arabinose (25 μl of a 10% aqueous solution, taken with a Cary model 81 laser Raman spectrometer). (*Varian Associates.*)

this purpose, particularly for biochemists. One example may be seen from Figs. 6-22 and 6-23, which show the Raman spectra of 10% aqueous solutions of β-D-glucose and D(−)-arabinose, respectively. Although glucose and arabinose are structurally quite similar, a comparison of Figs. 6-22 and 6-23 shows a striking difference in their spectra, both in band frequencies and intensities.

Another example where Raman spectroscopy has revealed interesting structural information is with gases such as CO_2, SO_2, and NH_3 dissolved in water. Perhaps surprising is the fact that the vibrational Raman spectra of the dissolved substances are essentially the same as those of the pure gases, indicating that the process of solution has resulted in little change in the molecular structure and in particular has not produced spectroscopically detectable amounts of H_2CO_3, H_2SO_3, or NH_4OH. Even more surprising is the fact that the spectrum of aqueous NH_3 shows some rotational fine structure with energy spacings very similar to those shown by gaseous NH_3, indicating a considerable freedom of rotation of the ammonia molecule in solution.

6-4B Qualitative Analysis

Raman spectroscopy uses essentially the same procedures as infrared spectrophotometry for qualitative analysis (Sec. 3-4*B*). Some of the practical factors on which a choice between infrared or Raman may be made were discussed at the beginning of Sec. 6-4, and the use of a Raman frequency-assignment chart was discussed in Sec. 6-4*A*. Like the infrared spectrum, the Raman spectrum of a compound is unique, though it must be realized that differences between the spectra of homologs may be small if the molecules are large. The Raman spectrum of a mixture of unreacting components is the superposition of the Raman spectra of the individual components. With an unknown sample an identification can be made by a simple matching process if the spectra of the possible components of the sample are available. This process becomes less reliable as the number of components in the sample increases, and in general it is seldom possible to make reliable analyses with more than six components. Auxiliary information like Raman polarization measurements or infrared spectra can greatly improve the reliability of the analysis.

6-4C Quantitative Analysis

Raman spectroscopy offers great possibilities for rapid, easy, and accurate analysis of mixtures that are troublesome with any other method. Raman spectroscopy has two big advantages over infrared spectrophotometry when it comes to quantitative analysis. First, the height of Raman peaks (emission intensity) varies linearly with concentration, whereas with infrared spectrophotometry (and all absorption techniques) there is a logarithmic relationship between concentration and the transmitted light (Beer's law). The principal consequence of this is that analytical accuracy is inherently greater for Raman spectroscopy than for infrared spectrophotometry, i.e., for the same relative photometric error dT/T the relative error in concentration dC/C is greater for an absorption method of analysis than for an emission method (see Exercise 6-13).

A second advantage of Raman spectroscopy is the relative simplicity of the spectra compared with the infrared. In the Raman effect overtones and combinations are almost completely absent, and emission bands tend to be narrow and have less overlap than infrared absorption bands where rotational fine structure is superimposed on the vibrational transitions. Thus, if a sample does not contain too many components, it is usually possible in a Raman spectrum to find a characteristic band that is free of spectral interferences. A standard sample of each component is used to determine the proportionality between peak height and concentration. If only one component is of interest, it is not necessary to determine the others. If band overlaps do occur, they can be resolved by setting up and solving simultaneous linear equations under the assumption that the measured intensity at a particular spectral position is made up of contributions from the overlapping components which are proportional to their concentrations.

As an example of what can be done with a very complex mixture, Nicholson† has analyzed an eight-component mixture containing benzene,

† D. E. Nicholson, *Anal. Chem.*, **32**:1634 (1960).

isopropylbenzene, 1,2-, 1,3-, and 1,4-diisopropylbenzenes, 1,3,5- and 1,2,4-triisopropylbenzenes, and 1,2,4,5-tetraisopropylbenzene. The average error for each component was about 1 percent.

Although it is difficult to make general statements about the accuracy of quantitative analysis by Raman spectroscopy, major components can usually be determined to better than 1 percent, and an accuracy of 0.2 percent can be obtained with a good instrument under optimum conditions. Again it should be emphasized that Raman spectroscopy is not sensitive enough for low-concentration measurements, where absorption techniques have a decided advantage.

6-4D Determination of Thermodynamic Properties

By means of statistical mechanics it is possible to calculate heat contents, free energies, entropies, and heat capacities from fundamental vibrational frequencies, bond distances, and bond angles, all obtained from Raman and infrared spectra. These thermodynamic data are often more reliable than the few direct calorimetric values found in the literature. As an example of the usefulness of such data, one could calculate the equilibrium constant for a reaction that has never been carried out in the laboratory.

6-4E Miscellaneous Applications

In this section some miscellaneous features of Raman spectroscopy will be illustrated, with particular emphasis on applications of laser instruments that would not have been feasible with earlier, mercury-arc Raman instruments.

Figure 6-24 shows the Raman spectrum of benzenechromiumtricarbonyl, $(C_6H_6)Cr(CO)_3$, obtained with a laser Raman spectrometer. This spectrum would have been almost impossible to obtain with a mercury-arc source, for four reasons.

1 This organometallic compound is yellow and like most colored compounds would strongly absorb the blue light from the mercury arc.

2 This compound undergoes rapid photodecomposition when exposed to the 4358-Å Hg line, whereas the red laser line at 6328 Å does not cause decomposition.

3 This compound in solution is strongly fluorescent when exposed to the 5461-Å line from the mercury source. In contrast, the 6328-Å laser radiation is too low in energy to excite the electronic transition responsible for the fluorescence interference.

4 This sample was run as a powdered solid, which ordinarily gives a

high background with mercury sources due to Tyndall scattering, but with the laser source such scattering is small (see Fig. 6-24).

Figure 6-25 compares the spectrum of a solution of naphthalene in carbon tetrachloride with the spectrum of a solid pellet of naphthalene. Notice that the solid sample gives a greater scattering of the exciting line, as evidenced by the high background even out beyond 200 cm^{-1}. The three strong peaks at 218, 314, and 459 cm^{-1} present in the solution spectrum and absent in the solid spectrum of Fig. 6-25 are due to the CCl_4 solvent, as may be realized from looking at the CCl_4 spectrum in Fig. 6-26.

Figure 6-26 shows a normal spectrum of CCl_4 compared with a high-resolution scan of the 459-cm^{-1} peak obtained with a Cary model 81 laser Raman spectrometer. It should be noted that high resolution reveals a splitting of the 459-cm^{-1} peak into a triplet, and the splitting is due to the three most abundant isotopes of chlorine. Note that only about 3 cm^{-1} separates the bands. It is also worth noting that the spectra were obtained using less than 5 μl of sample, which is a far smaller sample than can be used with the mercury-arc source.

To avoid giving the impression that the laser source has rendered the mercury-arc source obsolete, Fig. 6-27 shows the Raman spectrum of oxygen using the Cary model 81 Raman spectrometer with a multipass gas cell and a mercury-arc source. The regularly spaced rotational bonds of oxygen are

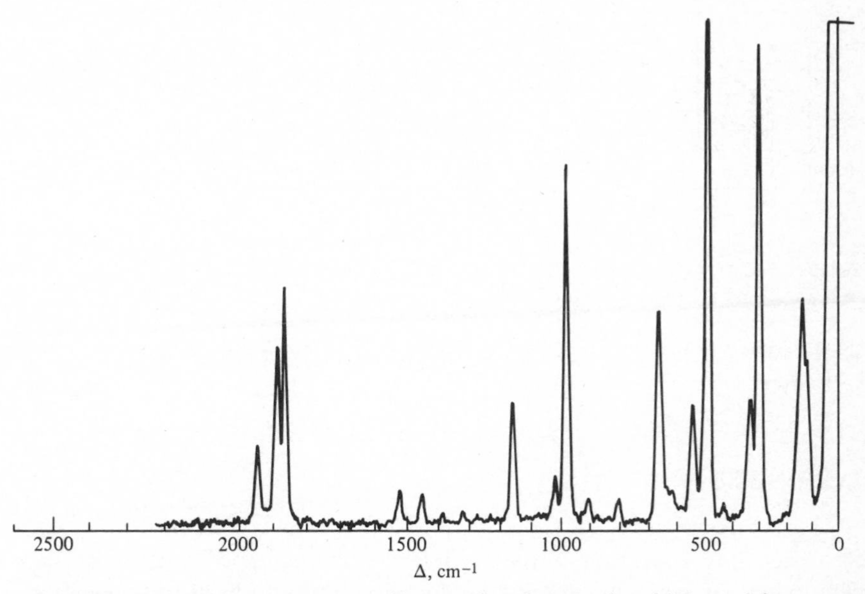

FIGURE 6-24 Laser Raman spectrum of benzenechromiumtricarbonyl (32 mg of the pure compound as a powdered solid, taken with a Perkin-Elmer model LR-1 laser Raman spectrometer). (*Perkin-Elmer Corp.*)

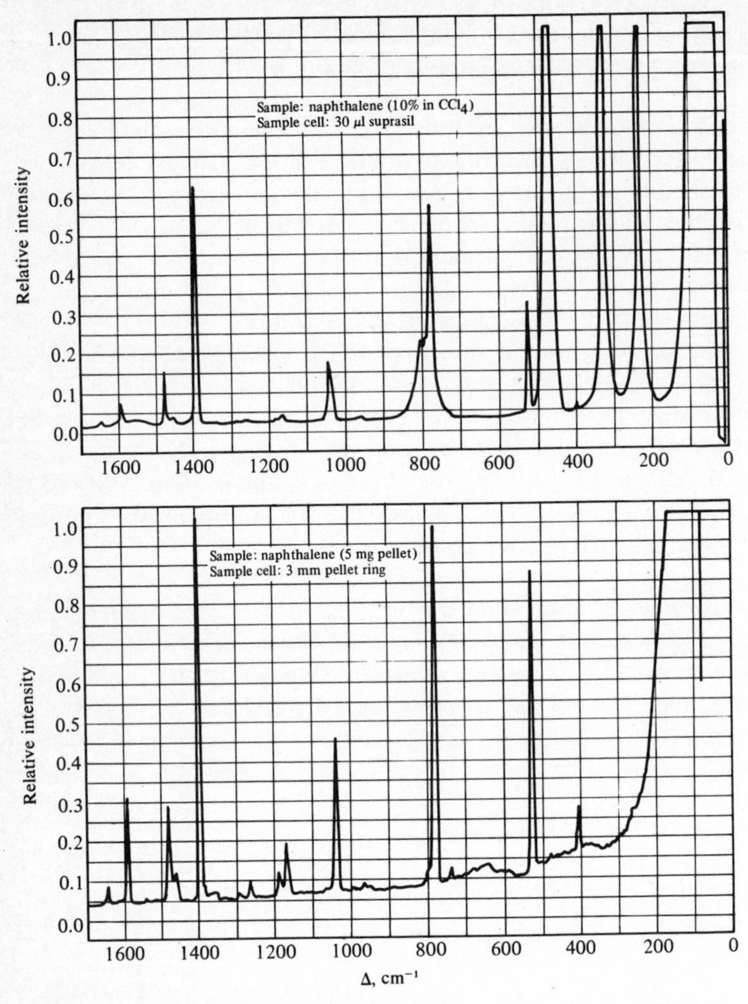

FIGURE 6-25 Comparison of the laser Raman spectrum of naphthalene run as a solid and as a solution. (*Varian Associates.*)

clearly seen. At present it is not feasible to examine gaseous samples with a laser source.

EXERCISES

6-1 The Cary model 81 Raman spectrometer, equipped with a Toronto mercury-arc source (4358 Å) and photomultipliers with S-20 response, has a wave-number range of 0 to 5000 cm^{-1}. (*a*) Calculate the absolute wavelength, in nanometers, corresponding to a Raman shift of 5000 cm^{-1}. (*b*) Calculate

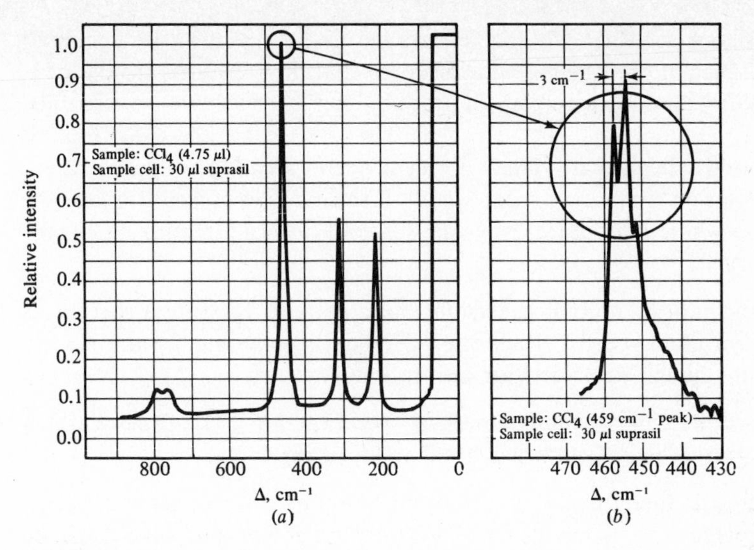

FIGURE 6-26 (*a*) Normal and (*b*) high-resolution spectrum of CCl_4. (*Varian Associates.*)

the maximum wave-number range the above instrument would have if the response of the S-20 detector were the only factor governing the wave-number range. (*c*) Speculate why the above instrument is limited to a range of 0 to 5000 cm^{-1} when it would seem that the detector would be sensitive to a wider range of wave numbers. (Note that when a laser source is used, the full response range of the detector is used.)

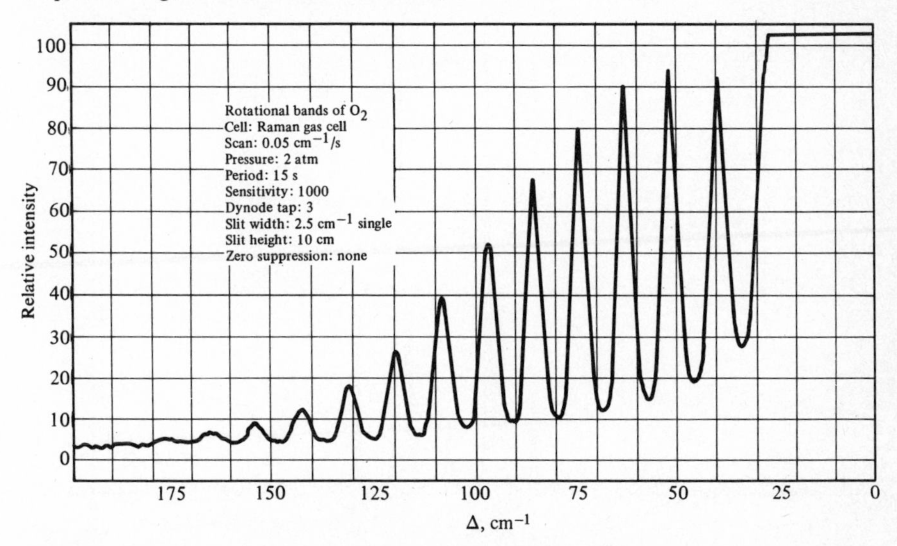

FIGURE 6-27 Raman spectrum of gaseous oxygen, obtained with Cary model 81 Raman spectrometer equipped with a mercury-arc source and a multipass gas cell. (*Varian Associates.*)

6-2 Suppose a weak Rayleigh line appears but no Raman lines are apparent. Is the appearance of the Rayleigh line *evidence* that the substance irradiated is Raman-active? Could it mean that the Raman lines are simply too weak to be observed?

Ans: No, Rayleigh scattering is not evidence of a *change* in polarization; weak Rayleigh scattering may occur simply if the average polarization for the molecule as a whole is nonzero, whereas Raman scattering requires a *change* in polarization.

6-3 Does the rule of mutual exclusion apply to pure rotational spectra as well as to vibrational spectra? Illustrate by indicating whether O_2 and H_2O in the vapor state give infrared or Raman rotational spectra. *Ans:* yes

6-4 Diagram the fundamental modes of vibration for CO_2. Predict which modes are infrared-active and which are Raman-active.

6-5 Diagram the fundamental modes of vibration of ethylene (planar C_2H_4), and predict which of the modes will be infrared-active and which will be Raman-active.

6-6 Diagram the fundamental models of vibration of cyanogen (linear N≡C—C≡N), and predict which of the modes will be infrared-active and which will be Raman-active.

6-7 Alkenes show a Raman frequency of 1642 cm^{-1}, which is due to the C=C stretching motion. Calculate the corresponding *wavelengths* in the

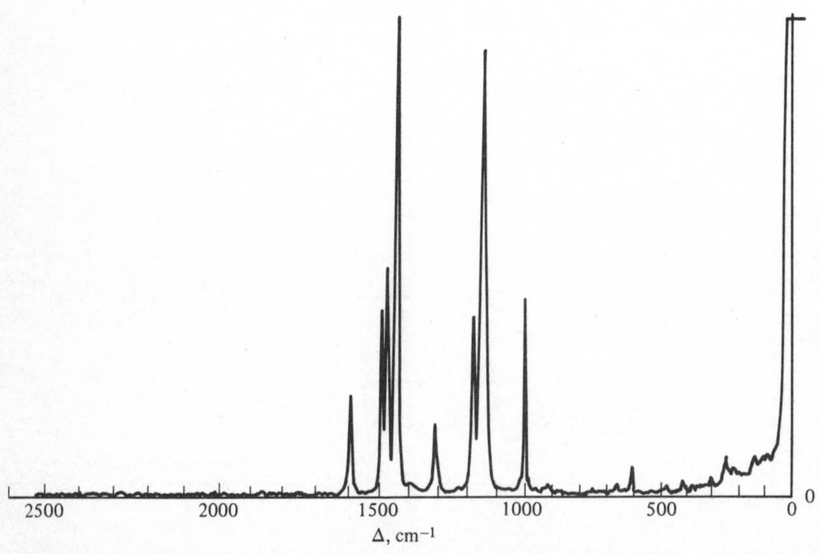

FIGURE 6-28 Spectrum of an unknown (azobenzene, benzene, or cyclohexane). (*Perkin-Elmer Corp.*)

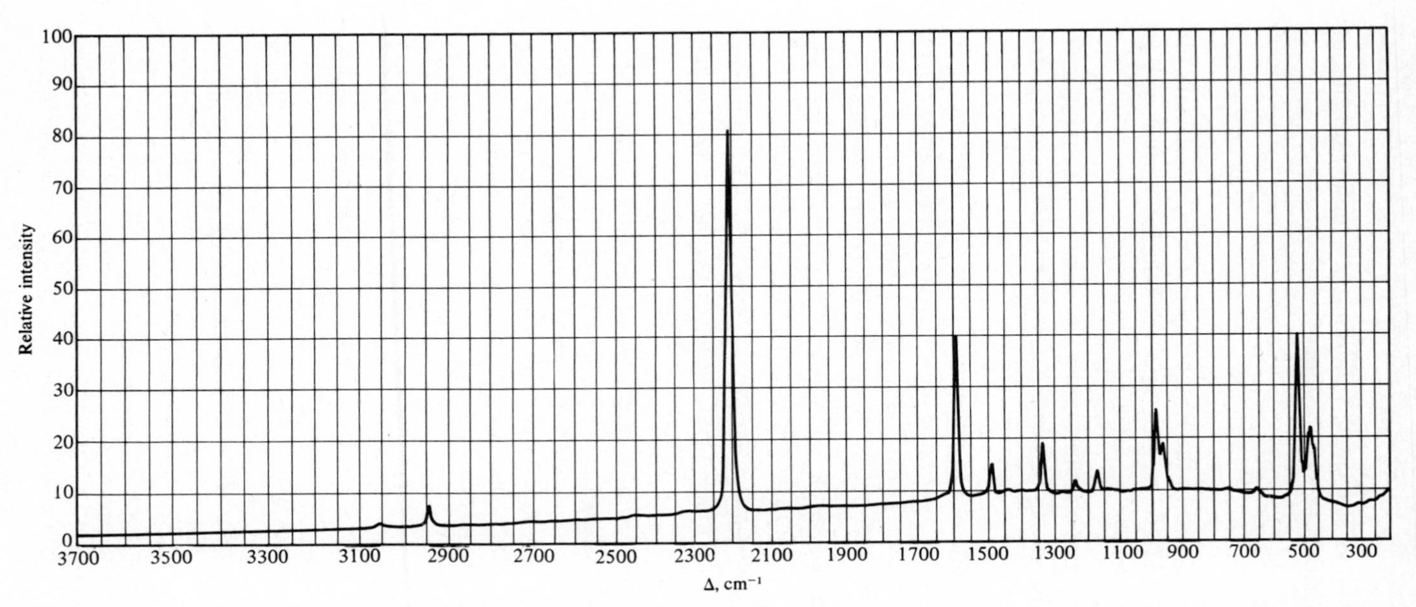

FIGURE 6-29 Spectrum of an unknown (isobutylcyclopentane, 2-methyl-1-butene, diphenylbutadiene, or octanol). (*Courtesy of Dr. Ronald O. Kagel, Dow Chemical Co.*)

Raman spectrum (*a*) if the 4358-Å exciting line is used; (*b*) if the 6328-Å exciting line is used.

6-8 Acetylene has two C—H stretching vibrations, a symmetrical one at 3374 cm^{-1} and an unsymmetrical one at 3287 cm^{-1}:

$$\overleftarrow{H—}C{\equiv}C\overrightarrow{—H} \qquad \overleftarrow{H—}C{\equiv}\overleftarrow{C—}H$$

(*a*) Which vibration will be active in the Raman spectrum? (*b*) Calculate the Raman wavelengths for a 6328-Å exciting line. (*c*) Which vibration will be active in the infrared? (*d*) Calculate the infrared wavelength.

6-9 Raman lines due to rotational transitions are usually spaced about 75 cm^{-1} or less. Calculate the energy of such a transition in kilocalories per mole.

6-10 The largest Raman shifts due to vibrational transitions are about 3500 cm^{-1}. (*a*) How many kilocalories per mole and electronvolts per mole are required to make such a transition? (*b*) If the 5461-Å line was the radiation source, could the resulting Raman line be detected in the visible region?

6-11 The Raman spectrum shown in Fig. 6-28 could be for azobenzene, benzene, or cyclohexane. Use the data given in this chapter to identify it.

6-12 The Raman spectrum shown in Fig. 6-29 may be of isobutylcyclopentane, 2-methyl-1-butene, diphenylbutadiene, or octanol. Use the data given in this chapter to identify it.

6-13 In general, for the same relative photoelectric error dT/T the relative error in concentration dC/C is greater for an *absorption* method of analysis than for an *emission* method. An exception is found when percent $T = 36.8$, when both methods give the same relative error. Show that this is so. *Hint:* With emission methods, $T = kC$, whereas with absorption methods $\log T = -k'C$. Derive equations relating dT/T to dC/C.

REFERENCES

INTRODUCTORY

1 ROSENBAUM, E. J., J. E. STEWART, and F. F. CLEVELAND: Raman Spectroscopy, pp. 675–684 in G. L. Clark (ed.), "The Encyclopedia of Spectroscopy," Reinhold, New York, 1960.

2 HIBBEN, J. H.: Raman Spectra, pp. 389–410 in W. G. Berl (ed.), "Physical Methods in Chemical Analysis," 2d ed., vol. 1, Academic, New York, 1960.

3 MELOAN, C. E., and R. W. KISER: "Problems and Experiments in Instrumental Analysis," Merrill, Columbus, Ohio, 1963. Contains useful exercises.

4 HARRISON, G. R., R. C. LORD, and J. R. LOOFBOUROW: "Practical Spectroscopy," pp. 506–530, Prentice-Hall, Englewood Cliffs, N.J., 1948. Outdated, but still useful.

INTERMEDIATE AND ADVANCED

5 BEATTIE, I. R.: He/Ne Laser Raman Spectroscopy, *Chem. Br.*, **3**:347–352 (1967). Good comparison of arc and laser sources and survey of materials analyzable.

6 HAWES, R. C., K. P. GEORGE, D. C. NELSON, and R. BECKWITH: Laser Excitation of Raman Spectra, *Anal. Chem.*, **38**:1842–1847 (1966).

7 SZYMANSKI, H. A. (ed.): "Raman Spectroscopy," Plenum, New York, 1967.

8 COLTHUP, N. B., L. H. DALY, and S. E. WIBERLEY: "Introduction to Infrared and Raman Spectroscopy," Academic, New York, 1964. Emphasis is on infrared, but gives an excellent treatment of group theory and the classification of molecules.

9 WEISSBERGER, A. (ed.): "Technique of Organic Chemistry," vol. 9, Wiley-Interscience, New York, 1966. Particularly useful for group theory and structural correlations.

10 HERZBERG, G.: "Molecular Spectra and Molecular Structure," vol. 2, "Infrared and Raman Spectra of Polyatomic Molecules," Van Nostrand, Princeton, N.J., 1945. Excellent for theory and data.

11 CLEVELAND, F. F.: Raman Spectra, chap. 6 in E. A. Braude and F. C. Nachod (eds.), "Determination of Organic Structures by Physical Methods," Academic, New York, 1955. Particularly good for the interpretation of spectra and applications to structure determinations.

12 WILSON, M. K.: Infrared and Raman Spectroscopy, chap. 3 in F. C. Nachod and W. D. Phillips (eds.), "Determination of Organic Structures by Physical Methods," vol. 2, Academic, New York, 1962. A fairly rigorous treatment of theory and structural correlations, with emphasis on infrared.

13 GROSSMAN, W. E. L.: Raman Spectrometry, *Anal. Chem.*, **46**,(5):345R (1974). One of a series of reviews appearing every 2 years, which should be consulted for latest developments.

LASERS

14 BROTHERTON, M.: "Masers and Lasers," McGraw-Hill, New York, 1964. An elementary treatment.

15 SCHAWLOW, A. L.: Optical Masers, *Sci. Am.*, **204**(6):52 (1961). A lucid description of lasers.

16 GARRETT, C. G. B.: "Gas Lasers," McGraw-Hill, New York, 1967. An intermediate-level treatment.

17 WHITE, A. D., and J. D. RIGDEN: Continuous Gas Maser Operation in the Visible, *Proc. IRE*, **50**:1697 (1962).

18 LENGYEL, B. A.: "Introduction to Laser Physics," Wiley, New York, 1966. An elementary treatment.

7

Microwave Spectroscopy

The microwave region of the electromagnetic spectrum lies between the far-infrared region and the radio-frequency region and corresponds to a wavelength of about 1 mm to 10 cm,† or a frequency of 300 to 3 GHz (1 GHz = 10^9 Hz). The most widely used portion to the microwave region is 10 to 50 GHz. As with most regions of the spectrum, these boundaries are not fixed. On the long-wavelength end, the boundary between microwave spectroscopy and electron spin resonance (esr) spectroscopy is arbitrary, since esr instruments also use microwave hardware (waveguides, klystrons, etc.). In practice, however, the delineation between microwave and esr spectroscopy is clear-cut, since microwave spectroscopy is based on the absorption of microwave energy by *freely rotating molecules,* causing a change in their rotational energy. On the other hand, in esr spectroscopy, microwave radiation is used to cause transitions between *electron spin* energy levels with the sample placed in an external magnetic field.

On the short-wavelength end of the microwave region, the boundary between the microwave region and the far-infrared region is even more arbitrary since experimental difficulties have thus far prevented the two regions from even meeting. Currently the far infrared extends out to about 0.1 mm (100 μm), and the microwave region starts about 1 mm, leaving the region in between virtually inaccessible.

† In wave-number units this is 10 to 0.1 cm^{-1}.

Microwave spectroscopy is mainly limited to gases which have a permanent dipole moment. Furthermore, the gases must be at a low pressure so as not to inhibit free rotation. A few liquids and solutions have been studied in the microwave region, but in general they give broad and uninteresting absorption peaks of limited usefulness. Therefore the discussion in this chapter will be largely confined to gaseous samples.

For gaseous samples with a permanent electric dipole moment, microwave spectroscopy provides sharp spectral lines and a wealth of detailed information on the structure of the molecules. The multitude of very sharp absorption lines that appear provide a highly distinctive fingerprint of the absorbing material, and analysis of the line widths, shapes, and spacing often allows accurate evaluation of such properties as bond angles, interatomic distances, molecular dipole moments, and even atomic masses.

Most of these applications will be clear after a brief consideration of the theory of microwave spectroscopy.

7-1 THEORY

In the discussion of infrared spectroscopy in Chap. 3 it was shown that rotational transitions of molecules contribute to infrared absorption in two ways: (1) in the fundamental-infrared region (2.5 to 50 μm) rotational transitions *accompany* vibrational transitions and contribute fine structure to the observed spectra (see Sec. 3-1*F*); (2) in the far-infrared region (50 to 1000 μm), pure rotational transitions are responsible for the spectra observed, as in the microwave region, but the relatively high energy of the far-infrared region is suitable for studying only the lightest of gases (see Sec. 3-6*D*). To study the rotation of heavier gas molecules, microwave radiation is required.

In general, microwave spectra are due to transitions between different rotational energy levels *within* the same vibrational and electronic state, and thus the rotational properties of a molecule can be described quite well by picturing the molecule as a rigid body which can rotate freely in space. It will be shown shortly that since the rotational behavior of a rigid body is determined by its moments of inertia, which in turn are functions of the geometry and mass distribution, the microwave spectrum is determined primarily by the geometric structure of the molecule. However, occasionally slight perturbations or interactions modify and enrich the observed microwave spectrum, and these subtleties often give valuable information about vibrational energies and even electronic structure of a molecule. Examples of such rigid-body perturbations are centrifugal distortion of bonds at high rotation energies, nuclear-quadrupole spin interactions, and electron spin interactions.

7-1A Interaction of Radiation with Rotating Molecules

The first requirement for the interaction of microwave radiation with a rotating molecule is that the molecule have a permanent electric dipole

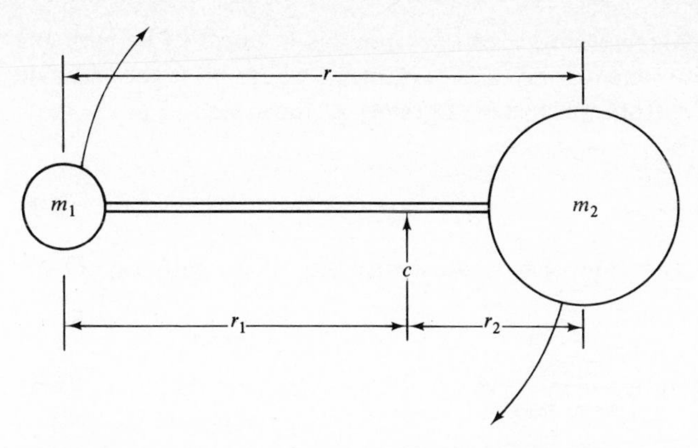

FIGURE 7-1 Rotation of a rigid diatomic molecule about its center of gravity. [c = center of gravity; m_1, m_2 = mass of atoms 1 and 2; r = the molecular bond length (equal to $r_1 + r_2$).]

moment (see Sec. 3-1*A*). Thus, very symmetrical molecules such as N_2, CO_2, and CH_4 have no microwave spectrum. [An important exception to this rule is molecular oxygen, which has no permanent electric dipole moment but nonetheless absorbs in the microwave region at about 60 GHz; the reason for this apparent anomaly is associated with the paramagnetic nature of oxygen (see Sec. 13-5*A*), which allows a *magnetic* dipole interaction to occur]. Furthermore, since the intensity of a microwave absorption line is proportional to the square of the permanent dipole moment, highly polar molecules will in general have strong microwave spectra. Other selection rules which limit the *number* of allowed transitions will be given in the next section.

7-1B Linear Molecules

Most microwave spectra, and especially those of linear molecules, are unique for the regularity of their line spacings and the ease and accuracy with which the spectra can be used to calculate such molecular properties as *moments of inertia, reduced masses,* and *bond lengths.* To understand why, consider a simple diatomic molecule like HBr or CO which can be pictured as in Fig. 7-1 as a rigid dumbbell capable of rotating end over end about point c, its center of gravity. From the definition of a center of gravity in physics it follows that at mechanical equilibrium

$$m_1 r_1 = m_2 r_2 \tag{7-1}$$

From Fig. 7-1,

$$r = r_1 + r_2 \tag{7-2}$$

Combining Eqs. (7-1) and (7-2) gives

$$r_1 = \frac{m_2}{m_1 + m_2} r \qquad \text{and} \qquad r_2 = \frac{m_1}{m_1 + m_2} r \tag{7-3}$$

A very useful way of characterizing a rotating body is in terms of its *moment of inertia* I, defined in a general way as $I = \Sigma m_i r_i^2$, where m_i is the mass of particle i at a distance r_i from the center of gravity of the system. For the two-particle system of Fig. 7-1,

$$I = m_1 r_1^2 + m_2 r_2^2 \tag{7-4}$$

To replace the r_1 and r_2 terms with r, substitute Eq. (7-3) into Eq. (7-4), giving

$$I = \frac{m_1 m_2^2}{(m_1 + m_2)^2} r^2 + \frac{m_1^2 m_2}{(m_1 + m_2)^2} r^2 \tag{7-5}$$

which simplifies to

$$I = \frac{m_1 m_2}{m_1 + m_2} r^2 = \mu r^2 \tag{7-6}$$

where μ is the *reduced* mass, defined as

$$\mu = \frac{m_1 m_2}{m_1 + m_2} \tag{7-7}$$

Equation (7-6) conveniently defines the moment of inertia in terms of the atomic masses and the bond length.

For a direct interpretation of microwave spectra an expression for the *energy* of rotation is needed. The kinetic energy (KE) of an object of mass m is

$$\text{KE} = \tfrac{1}{2} m v^2 \tag{7-8}$$

where v is a linear velocity. However, for rotary motion it is preferable to introduce the angular velocity ω, given by

$$\omega = \frac{v}{r} \tag{7-9}$$

where ω is in radians per second. Substituting Eq. (7-9) into Eq. (7-8) gives

$$\text{KE} = \tfrac{1}{2} m r^2 \omega^2 \tag{7-10}$$

Since the moment of inertia for an object of mass m is

$$I = m r^2 \tag{7-11}$$

we can rewrite Eq. (7-10) as

$$\text{KE} = \tfrac{1}{2} I \omega^2 \tag{7-12}$$

Thus, for this model of a diatomic molecule, Eq. (7-12) indicates that the rotational energy is given simply by the angular velocity ω and the moment of inertia I. Classically, any angular velocity and energy should be possible. For a system of molecular dimensions, however, the Schrödinger equation should be applied, and the solution of this equation reveals that only *certain* rotational energy states are allowed. In terms of angular momentum $I\omega$, the only rotational states allowed are those which satisfy the expression

$$I\omega = \frac{h}{2\pi}\sqrt{J(J+1)} \qquad J = 0, 1, 2, \ldots \tag{7-13}$$

where J is called the *rotational quantum number* and h is Planck's constant. In Eq. (7-13) the restriction to integral values is a consequence of the Schrödinger equation. When Eq. (7-13) is substituted into Eq. (7-12), the allowed rotational energy levels can be deduced as

$$E_J = \tfrac{1}{2}I\omega^2 = \frac{1}{2}\frac{(I\omega)^2}{I} = \frac{h^2}{8\pi^2 I}J(J+1) \qquad J = 0, 1, 2, \ldots \tag{7-14}$$

where E_J is used to denote the energy (in ergs) of a given rotational energy level. Equation (7-14) is the most important equation in microwave spectroscopy and shows that the allowed rotational energies depend only on the moment of inertia I and on the integer quantum number J.

When Eq. (7-14) is used, the allowed energy levels can be plotted as in Fig. 7-2. Clearly, if $J = 0$, the $E_J = 0$ and the molecule is not rotating at all. As J increases, the rotational energy increases, and in principle there is no limit to the rotational energy the molecule may have. In practice, there will be a point at which the centrifugal force of a rapidly rotating diatomic molecule is greater than the strength of the bond and the molecule is disrupted, but this point is usually not reached at normal temperatures.

For spectral purposes the *differences* between levels is more important than the absolute values of the energy levels. A rather sophisticated solution of the Schrödinger wave equation shows that the only rotational transitions which can be induced are those which satisfy the *selection rule* $\Delta J = \pm 1$. All other transitions are spectroscopically *forbidden*. Since microwave spectra are almost always studied by observing the *absorption* rather than the *emission* of radiation, the only part of the selection rule that is of interest is $\Delta J = +1$. For a molecule in the $J = 0$ state (the *ground rotational state,* in which no rotation occurs), the energy of incident radiation required to raise it to the $J = 1$ state can be calculated from Eq. (7-14) or deduced from Fig. 7-2 as

$$E_{J=0\rightarrow 1} = \frac{h^2}{4\pi^2 I} \tag{7-15}$$

Similarly, if a molecule is raised from the $J = 1$ to the $J = 2$ level, the energy

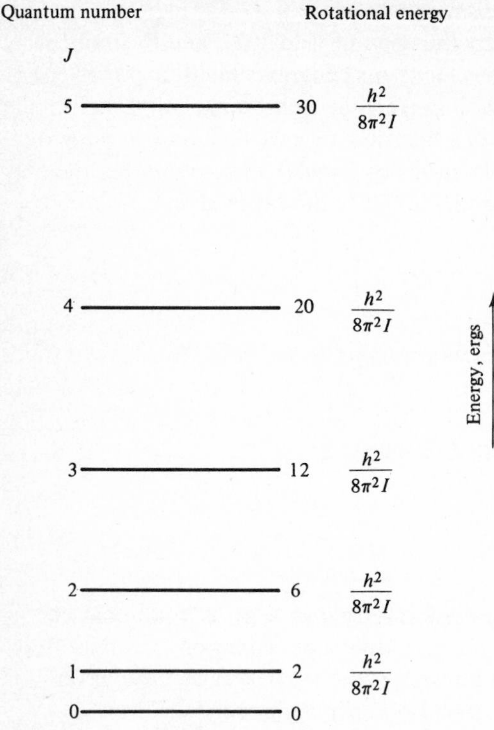

FIGURE 7-2 The allowed rotational energies of a diatomic molecule [calculated from Eq. (7-14)]. (*After G. M. Barrow, "Physical Chemistry," 2d ed., p. 336, McGraw-Hill Book Company, New York, 1966, by permission.*)

of incident radiation required is

$$E_{J=1\to 2} = 2\frac{h^2}{4\pi^2 I} \tag{7-16}$$

For subsequent transitions of $J = 2$ to $J = 3$, $J = 3$ to $J = 4$, etc., the energy required is $3(h^2/4\pi^2 I)$, $4(h^2/4\pi^2 I)$, etc. In other words, each successive transition requires an amount of energy equivalent to $h^2/4\pi^2 I$ more than the preceding transition, as illustrated by Fig. 7-3*a*. Thus the absorption spectrum which results, as shown in Fig. 7-3*b*, is a series of lines with a constant separation of $h^2/4\pi^2 I$.

Many linear molecules have been studied in the microwave region of the spectrum and show absorptions which can be correlated with the previous equations. As a simple example consider the absorption spectrum of CO, given in Fig. 7-4. The frequency separation between lines is 115 GHz, and from this information the J levels in each transition can be assigned. Further-

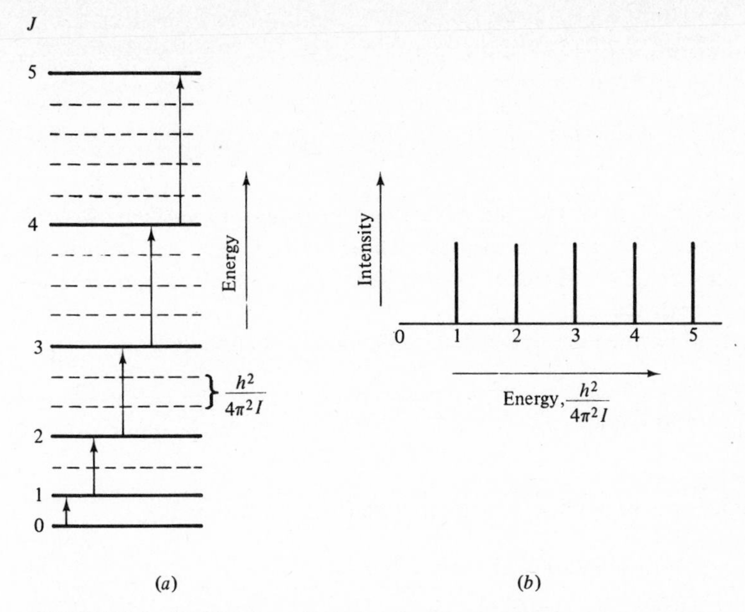

FIGURE 7-3 (*a*) The allowed rotational energy transitions and (*b*) resulting spectrum for a rigid diatomic molecule. (*After C. N. Banwell, "Fundamentals of Modern Spectroscopy," p.* 30, *McGraw-Hill Book Company, New York,* 1966, *by permission.*)

more the moment of inertia and bond length can be calculated, as the following example shows.

Example 7-1 Using the microwave spectrum of CO (Fig. 7-4), (*a*) verify the *J*-level assignment for each absorption peak; (*b*) calculate the moment of inertia for CO; and (*c*) calculate the bond length for CO.

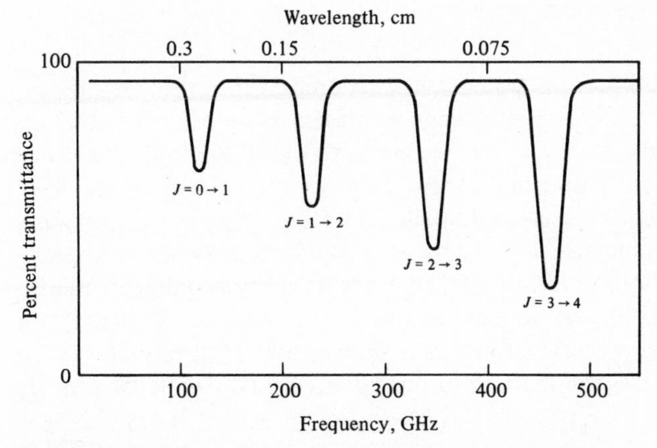

FIGURE 7-4 Schematic representation of the microwave absorption spectrum of CO.

Answer (*a*) From Fig. 7-4 the frequency *separation* between lines is 115 GHz, and from Eq. (7-15) and the fact that $\Delta E = h\nu$, it follows that

$$\nu = \frac{h}{4\pi^2 I} = 115 \text{ GHz}$$

As indicated by Eq. (7-15), the absorption peak at 115 GHz must correspond to the $J = 0 \rightarrow 1$ transition, whereas the $J = 1 \rightarrow 2$ transition should occur at 230 GHz [Eq. (7-16)], etc. Thus, the J assignments of Fig. 7-4 are vertified.

(*b*) From part (*a*) the line-spacing assignment was made as

$$\frac{h}{4\pi^2 I} = 115 \text{ GHz}$$

and from this the moment of inertia is readily calculated as

$$I = 1.45 \times 10^{-39} \text{ g-cm}^2$$

(*c*) To calculate the bond length r we now need only calculate the reduced mass from Eq. (7-7) and invoke Eq. (7-6):

$$\mu = \frac{m_1 m_2}{m_1 + m_2} = \frac{(12)(15.99)}{27.99} \frac{1}{6.02 \times 10^{23}} \text{ g/molecule}$$

$$= 1.14 \times 10^{-23} \text{ g}$$

$$r^2 = \frac{I}{\mu} = \frac{1.45 \times 10^{-39} \text{ g-cm}^2}{1.14 \times 10^{-23} \text{ g}} = 1.27 \times 10^{-16} \text{ cm}^2$$

$$r = 1.13 \times 10^{-8} \text{ cm} = 1.13 \text{ Å}$$

In the above calculation, values have been rounded off to three significant figures. Actually, microwave spectra can be measured with extremely high precision, and in fact the $J = 0 \rightarrow 1$ transition in CO has been measured to be 115.2706 Hz.† Thus, the precise calculation of the CO bond length is 1.128277 Å. The fact that a molecule is continually vibrating during the rotational measurements means that the bond length is really some effective or average value, and it is probably not meaningful to carry as many significant figures. Nonetheless, the reader should be impressed with the potential precision with which microwave spectroscopy allows the dimensions of molecules to be measured. Table 7-1 gives the moment of inertia and bond length for a series

† O. R. Gilliam, C. M. Johnson, and W. Gordy, *Phys. Rev.*, **78**:140 (1950).

TABLE 7-1 Moments of Inertia and Bond Lengths of Some Diatomic Molecules from Studies of Rotational Energy-Level Spacings†

Molecule	Moment of inertia, g-cm^2	Bond length, Å
HF	1.34×10^{-40}	0.917
HCl	2.65	1.275
HBr	3.31	1.414
HI	4.29	1.604
CO	14.5	1.128
NO	16.5	1.151
NaCl	129	2.361
CsCl	393	2.904

† From Gordon M. Barrow, "The Structure of Molecules," copyright © 1963, W. A. Benjamin, Inc., Menlo Park, California.

of diatomic molecules as measured from rotational energy-level spacings. Again note the precision of the measured bond lengths.

Polyatomic linear molecules have two or more internuclear distances, which obviously cannot be directly evaluated from one datum, the moment of inertia. In such cases the measurement of the spectra of different isotopically substituted molecules, whose distances can be assumed to be unaltered by isotopic substitution, provides additional data which often make the calculation of individual bond lengths possible. For example, consider carbonyl sulfide, OCS, which is linear in the same way as CO_2 or CS_2 (these two have zero dipole moments, preventing their study by microwave spectroscopy). From the microwave spectrum of ordinary OCS, $^{16}O^{12}C^{32}S$, the moment of inertia can be calculated as 138.0×10^{-40} g-cm^2, but obviously this one datum does not permit the r_{CO} and r_{CS} bond lengths to be calculated. However, by isotopically substituting ^{34}S for ^{32}S in the molecule, the new rotational spectrum now leads to a moment of inertia of 141.4×10^{-40} g-cm^2. By deriving a new equation for the moment of inertia in terms of r_{CO} and r_{CS} [3, pp. 46–48], two equations in two unknowns can be set up and solved, giving

$$r_{CO} = 1.165 \text{ Å} \qquad \text{and} \qquad r_{CS} = 1.558 \text{ Å}$$

7-1C Nonlinear Molecules

Whereas any linear molecule has only one moment of inertia, other molecules have two or three different principal moments of inertia, and analysis of the microwave spectra of such molecules can sometimes lead to values for them.

It is convenient to classify molecules as *symmetric tops* if two of their three principal moments of inertia are equal and *asymmetric tops* if all three moments of inertia are different. For example, methyl fluoride

```
H
 \
H—C—F
 /
H
```

is a symmetric-top molecule, since the end-over-end rotation in the plane of the paper has the same moment of inertia as the end-over-end rotation perpendicular to the plane of the paper, whereas rotation about the C—F bond axis gives a different moment of inertia. This last rotation can be imagined to give the appearance of a spinning top, and hence the name of the class. Other examples of symmetric tops are benzene and ethane. Examples of asymmetric tops, with all three moments of inertia different, are water, chloroethane, and chlorobenzene.

The energy-level and absorption-spectrum patterns for the rotation of nonlinear molecules are similar to those of linear molecules but more complicated, as illustrated in Fig. 7-5. Because of the symmetry of a *symmetric-top molecule* the spectrum is still relatively simple, and it is usually possible to assign transitions and calculate moments of inertia. Again, isotopic molecules are used to measure the various bond lengths and angles. *Slightly asymmetric-top* molecules become more complex to analyze, but it is often possible to ex-

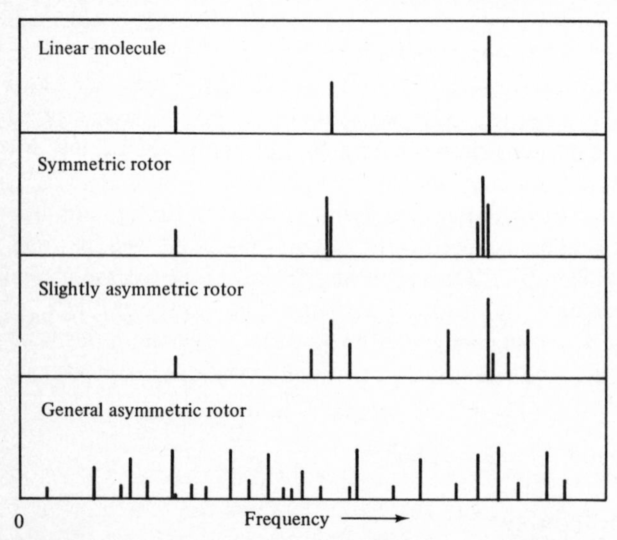

FIGURE 7-5 Typical microwave spectral patterns for various classes of molecules. (*From D. R. Lide, Jr., "Advances in Analytical Chemistry and Instrumentation," vol.* 5, *p.* 243, *John Wiley & Sons, Inc., New York,* 1966, *by permission.*)

tract accurate molecular data from their spectra. *General asymmetric-top* molecules give spectra that are very complex, as illustrated in Fig. 7-5, and it is not possible to derive general energy equations describing the allowed transitions. However, by treating each molecule as a separate case, structural parameters have been accurately determined in a number of cases [3, p. 53].

7-1D Extensions and Subtleties

Before leaving the theoretical discussion, mention should be made of three kinds of complications affecting microwave spectra: (1) the effect of nonrigidity of molecules, (2) nuclear and electron spin interactions, and (3) the effect of external fields. Each of these complications will be briefly described.

EFFECT OF NONRIGIDITY OF MOLECULES

The equations developed so far have been based on a model which is rigid, the so-called *rigid-rotor approximation.* Actually, as any real molecule rotates faster and faster, centrifugal distortion sets in and the bonds, like springs, become slightly stretched. This centrifugal distortion leads to progressively larger effective moments of inertia and therefore smaller energy-level spacings [see, for example, Eq. (7-15)]. In a linear molecule this means that the spacing between successive spectral lines is not really constant but decreases slightly with increasing J, that is, with increasing speed of rotation. With a symmetric-top molecule the splitting of the spectral lines tends to become wider with increasing J. With asymmetric-top molecules the effect of centrifugal distortion is much more complicated.

NUCLEAR AND ELECTRON SPIN INTERACTION

Although microwave spectroscopy is fundamentally concerned only with the way *molecules* rotate, in certain cases *nuclei* and *electrons* rotating *within* molecules can interact with and affect molecular rotation, thereby altering microwave spectra. Of the nuclear spin interactions, only when a nucleus has a spin greater than $\frac{1}{2}$ is the interaction important. Thus, many common nuclei such as ^{12}C, ^{16}O, ^{34}S, etc. with zero nuclear spin have *no* effect on the spectrum, and others such as ^{1}H, ^{17}F, and ^{31}P, with spins of $\frac{1}{2}$, usually have a negligible effect. But molecules containing, for example, nitrogen (^{14}N has a spin of 1) or the heavy halogens (^{35}Cl, ^{37}Cl, ^{79}Br, and ^{81}Br, all of which have a spin of $\frac{3}{2}$, and ^{127}I with a spin of $\frac{5}{2}$) are said to exert a *nuclear quadrupole* interaction, splitting the spectral lines, usually into doublets, triplets, or quartets.

Electron spin interactions can occur if the electronic angular momentum is nonzero, and this occurs if the molecule contains an odd number of electrons, as is the case with NO, NO_2, and ClO_2 or free radicals. Electron spin interactions are infrequent in microwave spectroscopy, but, like nuclear spin interactions. they have the effect of splitting spectral lines and in general complicating the spectrum.

EFFECT OF EXTERNAL FIELDS

Microwave spectra may be affected by imposing an external *magnetic* field (Zeeman effect) or an external *electric* field (Stark effect). The *Zeeman effect* is of slight importance in microwave spectroscopy because except with paramagnetic molecules such as NO, NO_2, and free radicals, extremely high magnetic fields are required to produce significant splittings. On the other hand, the *Stark effect* is extremely important, and the discussion would not be complete without a brief consideration of it. A more detailed discussion can be found in Ref. 8.

Experimentally the Stark effect requires placing an electric field across the sample perpendicular to the direction of the radiation beam. Since a molecule exhibiting a rotational spectrum must have an *electric dipole moment,* it follows that the external electric field will interact with the molecule and alter its rotational energy levels. Thus, the absorption lines of the spectrum will be *shifted,* and by an amount which depends on the extent of the interaction. This interaction, in turn, depends on the magnitude of the applied field E and the dipole moment μ of the molecule. For a *linear* molecule the shift is

$$\Delta\nu = k(\mu E)^2 \qquad \text{linear molecule} \qquad (7\text{-}17)$$

whereas for a symmetric-top molecule the shift is

$$\Delta\nu = k'\mu E \qquad \text{symmetric top} \qquad (7\text{-}18)$$

where k and k' are proportionality constants. With asymmetric-top molecules the relationship is a little more complex, since three different components of the electric dipole must be considered, but Eq. (7-17) holds approximately for such molecules. As a result of Eqs. (7-17) and (7-18), the Stark shift allows an accurate measurement of dipole moments, and this is an important consequence of the Stark effect. Moreover, since the measurements are made on very dilute gas samples, the dipole moment observed is intrinsically accurate, unhampered by solvent effects, molecular interactions, etc.

Another valuable application of the Stark effect is in the assignment of observed spectral lines to particular J values. Without the Stark effect the assignment of J values is not always obvious. For example, the line of lowest frequency which is observed *may* happen to correspond with $J = 0$, or it may be that it is the first *observable* line of a series, either because the lines are intrinsically very weak or the instrument used is limiting. In the presence of an electric field, however, a line will be split into $2J + 1$ components; i.e., in the absence of an external field, each line is said to be $(2J + 1)$-degenerate because rotations can occur in $2J + 1$ orientations in space at the same energy. Thus, for example, a line which is a singleton in the absence of a field but which is split into five components in the presence of a field is unambiguous evidence of a $J = 2$ ground state; i.e., set $2J + 1 = 5$ and solve for J.

Finally, it turns out that if the applied electric field is *oscillating* (generally in the form of a square wave), this *Stark modulation* leads to greatly improved instrument sensitivity (signal-to-noise ratio). The use of Stark modulation will be discussed further in Sec. 7-2*B*.

7-1E The Shape and Intensity of Spectral Lines

The shape and intensity of a microwave absorption line depend on experimental conditions of temperature and pressure. The pressure of the sample affects mainly the shape of a line, whereas the temperature of the sample affects mainly the relative line heights.

At relatively high *pressures* (even 1 torr may be considered relatively high pressure in microwave spectroscopy),† absorption lines appear relatively broad, typically of the order of 25 MHz at half intensity. At these pressures line width is controlled by molecular collisions, which in turn disturb the state of the molecule and cause a wide spread of kinetic energies. As the pressure is gradually lowered, the lines at first become narrower, without significantly changing in peak height or frequency. Eventually, however, a point is reached where the line width and shape do not vary but the peak height decreases proportionately with pressure (and *concentration*). The narrowing of the lines as the pressure is reduced is brought about by a decrease in molecular collisions. As the pressure is decreased still further, a limiting line shape is reached, with the intrinsic width determined by the doppler effect and collisions with the cell walls. Doppler broadening results from the fact that gas molecules traveling at different velocities emit or absorb radiation at slightly different frequencies, producing a gaussian line shape. Collisions with the walls of the cell produce a broadening of absorption lines because each collision disturbs the energy state of the molecule, just as in collisions between molecules. The particular line width obtained depends on the molecular weight of the gas, the frequency of the radiation, and the temperature.

To understand how *temperature* affects the relative line intensities one must first understand that all transitions which satisfy the selection rule $\Delta J = \pm 1$ have an equal probability of occurring. In other words, the intrinsic probability that a single molecule in the $J = 0$ state will move to the $J = 1$ state is the same as that of a single molecule moving from the $J = 1$ to the $J = 2$ state. Since at a given temperature and in a large collection of molecules there will be a Boltzmann distribution of molecules in each of a large number of different levels, the spectral lines will be of differing intensities because the intensity of each line is directly proportional to the number of molecules in each level.

† Typical sample gas pressures are of the order of 10^{-4} torr in microwave spectroscopy.

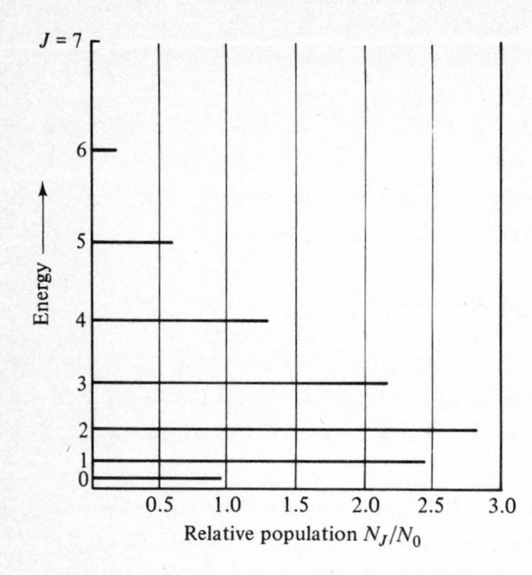

FIGURE 7-6 The relative populations of the lower rotational energy levels of HCl at 25°C. (*From G. M. Barrow, "Physical Chemistry," 2d ed., p. 340, McGraw-Hill Book Company, New York,* 1966, *by permission.*)

Even at ordinary temperatures the components of a population of gas molecules are in a large number of different rotational energy levels, as can be illustrated with HCl, which has one of the smallest moments of inertia (only H_2, which of course is spectroscopically inactive, has an appreciably smaller moment). The small moment of inertia of HCl means that it will have a more widely spaced rotational energy pattern than other molecules [see Eq. (7-14)]. For HCl the energy-level spacing factor [$h^2/4\pi^2 I$; see Eq. (7-15)] is 8.1×10^{-15} erg, and this value is small compared with the room-temperature value of kT of 4.14×10^{-14} erg. Thus, it is clear that most molecules have many rotational energy levels in the energy range of zero to kT.

The distribution of molecules in the various rotational energy levels at a given temperature can be calculated from Boltzmann's distribution equation

$$N_J = g_J N_0 e^{-E_J/kT}$$

where g_J is a statistical factor equal to the degeneracy $2J + 1$ for a linear molecule. Figure 7-6 shows the relative population N_J/N_0 of the seven lowest energy levels for HCl at 25°C. At higher temperatures the population of the higher energy levels increases, and thus it is to be expected that the relative intensities of spectral lines would change markedly with temperature.

7-2 INSTRUMENTATION

Although numerous companies manufacture the various components used in microwave spectrometers, and many research instruments made from these components are in use, few companies produce complete spectrometers at this time. Therefore, this discussion will be directed largely toward acquainting the potential user with the simplest available microwave components and with their physical assemblage into functioning spectrometers. References 6 and 8 contain more details.

7-2A A Basic Microwave Spectrometer

Figure 7-7 gives the block diagram of a simple microwave spectrometer. The basic components of even the simplest spectrometer are (1) a *source* of radiation, generally a klystron tube but sometimes a backward-wave oscillator (BWO); (2) a *wavemeter,* for accurately measuring the wavelength or frequency of the radiation; (3) a *sample cell* through which the radiation passes; (4) a *detector,* which is usually a silicon or germanium crystal detector; and (5) an *oscilloscope* or *recorder* for observing the spectrum. Although microwave radiation could be handled with mirrors and lenses, it is far more efficient to transmit the radiation through a *waveguide,* a rectangular copper, brass, silver, or gold-plated brass pipe. Since the dimensions of the waveguide should be comparable to the wavelength of radiation, waveguides of several sizes are required to cover the full microwave spectrum.

The absorption cell itself is constructed from a section of waveguide, with thin (0.001- to 0.005-in) mica windows cemented to the ends. The cell may vary in length from a few feet to perhaps 100 ft, 10 ft being a typical length for a waveguide cell in the 1-cm wavelength region (20 to 40 GHz). In general, the higher the frequency the shorter the cell that can be used.

RADIATION SOURCES

These are normally vacuum-tube oscillators; klystron oscillators are the most widely used and BWOs come next in importance. The output of such tubes is highly monochromatic, and the frequency can be varied over a limited range, obviating the need for a prism or grating to single out successive frequencies. Certain solid-state devices, such as the Gunn diode, are beginning to

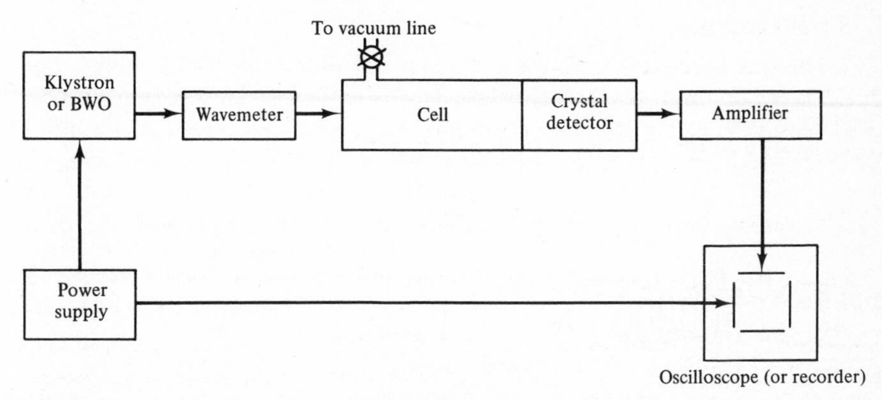

FIGURE 7-7 Block diagram of a simple microwave spectrometer. (*After J. H. Goldstein in I. M. Kolthoff et al. (eds.), "Treatise on Analytical Chemistry," pt. I, vol.* 5, *p.* 3238, *John Wiley & Sons, Inc., New York,* 1964, *by permission.*)

replace klystrons and BWOs. They are simpler and can be tuned over a wider range. The output of a klystron may be scanned continuously by turning a screw which mechanically deforms the tube, or it may be swept electronically by applying a sawtooth voltage to a suitable electrode. In a BWO the frequency control is entirely electronic. Further details on klystron oscillators are given in Sec. 13-5*B*. A brief description of the operation of a BWO will be given here.

Figure 7-8 is a schematic diagram of a BWO. A beam of electrons travels through a slow-wave coil (so named because the electromagnetic radiation generated travels down the tube more slowly than the electrons in the helix coil), and electromagnetic waves are generated within the tube. As the electron beam travels down the tube, it synchronizes with the backward wave generated in the helix and the electrons tend to be bunched. These electron bunches can effectively transfer kinetic energy to the field of the wave, and as a result of this energy transfer the traveling wave in the tube increases in amplitude as it progresses. Most of the backward wave being generated is removed through the output port, but a portion of the backward wave is fed back into the traveling-wave tube, setting up oscillations in the tube.

WAVEMETERS

These are usually of the cavity type, which operate by absorbing a small amount of energy when the length of the cavity is properly related to the wavelength of the radiation. It usually consists of a cylindrical cavity of suitable dimensions in which a movable plunger is mounted. The position of the plunger can be rather precisely adjusted by a micrometer screw movement. In practice the plunger knob is turned until the resonant frequency is achieved, as indicated by an absorption signal on the recorder or oscilloscope. The calibrated reading on the micrometer screw gives the frequency, accurate to within a few megahertz.

DETECTORS

The most convenient detector is a crystal diode (the 1N23, 1N26, and 1N53 types are most commonly used). This is mounted in a waveguide holder attached to the end of the absorption cell. The crystal diode generates a direct current which is a function of the intensity of the microwave radiation

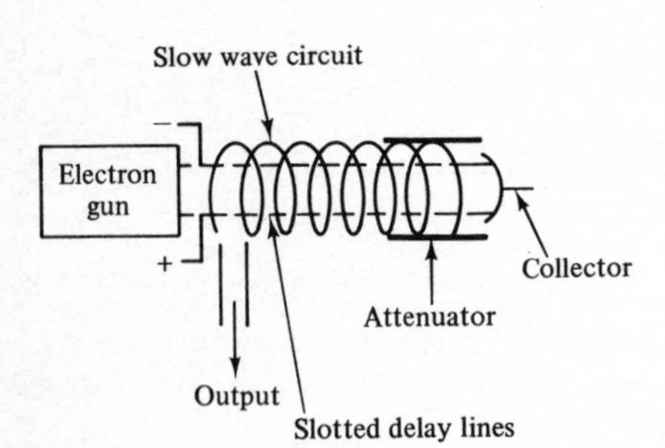

FIGURE 7-8 Schematic diagram of a BWO. (*After G. D. Sims and I. M. Stevenson, "Microwave Tubes and Semiconductor Devices," p.* 186, *John Wiley & Sons, Inc., New York,* 1963, *by permission.*)

reaching it. Since the output current is rather small, it must be amplified considerably before being fed to the oscilloscope or recorder.

7-2B A Stark-modulated Spectrometer

The simple spectrometer diagramed in Fig. 7-7 is suitable for the accurate measurement of intense absorption lines and for scanning a narrow absorption range but lacks sensitivity and versatility for two reasons. (1) The efficiency of the absorption process in the microwave region is relatively low. It is rare to have the net power absorption as high as 1 percent, and often it is as low as 1 ppm. Such small changes in the transmitted microwave power are difficult to detect by direct measurement, but great improvements can be achieved by using Stark modulation. (2) The relative instability of a direct measuring system makes it impractical to scan a large frequency range without regular recalibration and retuning.

Figure 7-9 gives a block diagram of a simple Stark-modulation microwave spectrometer. The main modification from the spectrometer shown in Fig. 7-7 is that the Stark absorption cell contains a central electrode running the entire length of the cell and insulated from it (see Fig. 7-9*b*). To this central electrode is applied a high-frequency (5- to 100-kHz but usually 100-kHz) square-wave voltage, which is zero-biased. In other words, for one half-cycle there is no voltage difference between the electrode and the waveguide, and absorption of the microwave radiation by the gas occurs at the normal frequency. For the other half-cycle there is a large voltage which shifts (and slightly broadens) the absorption peak as described in Sec. 7-1*D* (Stark effect), so that microwave power is no longer absorbed. Thus, if the source frequency is set for the absorption peak, a 100-kHz oscillating signal will appear at the detector crystal; this ac signal is amenable to high amplification. When the source frequency is off the absorption peak, no signal will reach the detector. In this way, the sensitivity can be greatly increased over that of direct measurements. Furthermore, variations in the incident power level are automatically removed from the spectral display (in effect, the derivative of the transmitted radiation is being measured), and thus noise is minimized and stability improved. By using a recorder along with a motor to drive the klystron or BWO frequency, a large frequency range can be scanned automatically and conveniently. A single waveguide cell can be used throughout most of the microwave region, excluding only the very low and very high frequencies. But the oscillator, detector, and other waveguide components must be changed about every 10 to 20 GHz. The Hewlett-Packard Company markets a commercial spectrometer which covers the 8.2- to 12.4-GHz region, and Tracerlab (Waltham, Mass.) had models covering the 18- to 26.5- and 26.5- to 40-GHz ranges. Cell compartments are generally thermally insulated to allow measurements above or below room temperature. Temperatures used range from that of dry ice to as high as 300°C.

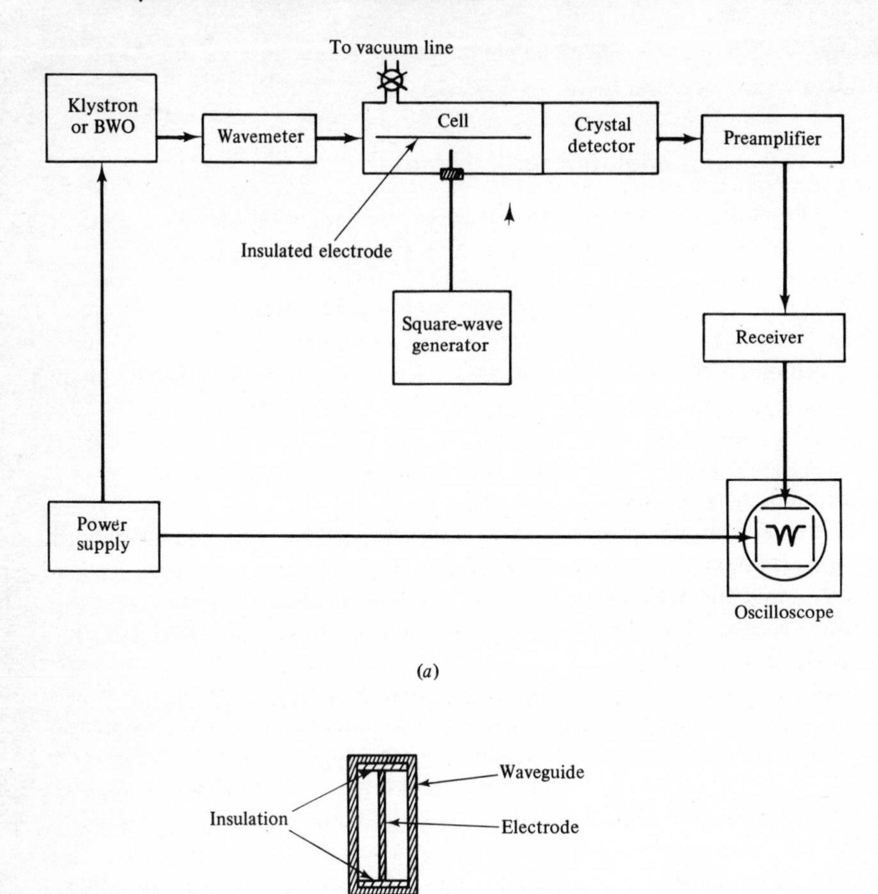

FIGURE 7-9 (*a*) Block diagram of a simple Stark-modulation microwave spectrometer, and (*b*) cross-sectional view of Stark cell. [*After J. H. Goldstein, in I. M. Kolthoff et al. (eds.), "Treatise on Analytical Chemistry," pt. I, vol. 5, p.* 3272, *John Wiley & Sons, Inc., New York,* 1964, *by permission.*]

7-2C Sample Handling

Since oxygen absorbs in the microwave region, it is standard practice to freeze the sample with dry ice or liquid nitrogen first and then remove air by pumping. The sample is finally warmed up until it vaporizes into the manifold at a suitable pressure. It is then admitted to a previously evacuated waveguide cell. At the usual operating pressures (10^{-1} to 10^{-4} torr), microgram quantities suffice to fill the cell, but allowance should be made for surface adsorption and manifold volume. The use of a compact vacuum system will minimize the latter. Sample contamination is to be avoided since the contaminants participate in pressure broadening or may introduce other spectral lines. Repeated flushing of the waveguide cell with fresh portions of the sample can be carried out if impurities from the previous sample are suspected. Ammonia is a notorious contaminant because of its strong microwave absorption and its tendency to be adsorbed on the cell walls. If ammonia is used for frequency calibration, it is recommended that a separate short cell be reserved for this purpose. In extreme contamination cases it may be necessary to heat the cell under vacuum to facilitate desorption.

7-3 APPLICATIONS

Before outlining the applications of microwave spectroscopy, it will be useful to review the conditions that must be satisfied in order to observe and characterize microwave spectra.

The first requirement is that the substance to be analyzed have a permanent electric dipole moment. The lack of a dipole moment rules out many compounds, and many others that are weakly polar with small moments give only weak microwave absorption peaks. As instruments become available with higher sensitivity (higher signal-to-noise ratio) this limitation will diminish. Such weakly polar hydrocarbons as propylene, isobutane, and propane have already been studied in considerable detail [4].

Second, the sample should be in the form of a gas, and 50- to 100-μm vapor pressure is typically required. Sometimes compounds of low volatility can be analyzed using heated cells. Cells have been constructed for use at temperatures as high as 1000°C, but at present they are very inconvenient to use. As high-temperature cells become commercially available, the applications of microwave spectroscopy should greatly expand.

Despite these requirements and limitations there are a large number of molecules whose microwave spectra can be conveniently measured. The applications of such spectra are mainly for qualitative and quantitative analysis and structure elucidation.

7-3A Qualitative Analysis

Microwave spectroscopy is an excellent technique for qualitative identification of gaseous compounds. The microwave spectrum is a property of the molecule as a whole, giving a unique fingerprint for the molecule. Absorption lines are very sharp and readily resolved with the instruments available. The information-handling capacity of the microwave region is extremely high. If we consider the readily accessible range of, say, 10 to 50 GHz, and take a typical line width of 0.1 MHz (10^{-4} GHz), there is space in the spectrum for 400,000 lines. This means that it should be possible, in principle, to identify the constituents of a very complex mixture with little risk of interference from the overlapping lines of different compounds.

The highly specific nature of the rotational spectrum can be illustrated by the fact that not only will different isotopic species always have distinctly different spectra, but even altering the position occupied by the isotope will alter the spectrum. For example, propane-1-^{13}C and propane-2-^{13}C have quite different spectra. Rotational isomers, even those which interconvert too rapidly for chemical separation, can be distinguished very easily. In short, the slightest change in the geometry or mass of a molecule is almost certain to alter the spectrum drastically. It is doubtful whether any other experimental technique allows the molecule to be characterized so uniquely.

By the same token, however, it must be pointed out that microwave spectroscopy is fundamentally unsuited for characterizing new compounds. Whereas techniques such as infrared spectroscopy and nuclear magnetic resonance are powerful tools for characterizing new compounds, providing valuable clues to the presence of particular functional groups and the environment of particular atoms, the microwave spectrum is a property of the molecule as a whole, and it is not possible to relate certain parts of the spectrum to certain parts of the molecule. Through detailed and complex analysis of a spectrum there may be occasional exceptions to these statements, but, in general, the use of microwave spectra for qualitative analysis requires tables of previously observed microwave frequencies. In contrast to infrared spectrophotometry, comparison spectra are unnecessary, since the detailed frequency information defines the spectrum sufficiently well for purposes of qualitative analysis. Tables of prominent microwave lines are available [5, and the references cited therein].

7-3B Quantitative Analysis

The use of microwave spectroscopy for quantitative analysis is not widespread and presents special problems, but at the same time it has some unique and specific advantages. The biggest problem stems from the fact that the peak height and especially the line width usually depend on the composition of the sample, due to the collision-broadening process. In other words, it is not sufficient to assume that the peak height will be proportional to the pressure (or concentration) of the absorbing gas molecule unless it has been shown that the other constituents of the gaseous mixture behave ideally in the collision-interaction process. There are two main solutions to this problem: (1) If the substance being measured absorbs very strongly, it may be possible to dilute both the standards and the sample with a large amount of a noninterfering gas such as helium or nitrogen. Then most collisions will occur with the diluent molecules, and collision broadening can be assumed to be nearly constant even though the composition of the sample changes. (2) A more widely applicable approach is to use comparison standard mixtures whose overall compositions are near that of the unknown. In some cases it may be possible to use a standard-addition technique, whereby portions of the unknown are used as a sample matrix to which are added known concentrations of the constituent being sought, and in this way a series of reliable comparison standards is prepared. The constituent being sought must be a minor constituent of the sample for this technique to be successful.

A good example of the potential of quantitative microwave analysis was reported by Southern et al. [9], who in determining isotopic abundances were able to determine ^{15}N in the range of 0.38 to 4.5 percent within ± 3 percent and ^{13}C in the range of 1.1 to 10.0 percent within less than 2 percent. Only 0.00015 mol of gas was required for a determination, and the procedure took only 10 to 15 min per sample. Some other examples are the determination of the decomposition products of nitrocellulose† and a study of the kinetics of vapor-phase reactions.‡ As microwave spectrometers become more available and analysts become more familiar with the techniques, the analytical possibilities of microwave spectroscopy should expand greatly.

7-3C Structure Elucidation

The use of microwave spectroscopy to measure structural features of molecules, as outlined in Sec. 7-1, should be mentioned again. The analysis of a microwave spectrum can provide very accurate values of interatomic distances, angles, dipole moments, and bond properties, as well as considerable information on the internal dynamics of the molecule. This is an area in which microwave spectroscopy has much to contribute.

EXERCISES

7-1 In determining bond lengths for polyatomic linear molecules by microwave spectroscopy, the assumption is made that bond length is not altered by isotopic substitution. Explain how this assumption can be checked.

Ans: Repeat measurments with still a third isotopically labeled isotope, for example, $^{16}O^{12}C^{32}S$, $^{16}O^{12}C^{34}S$, $^{18}O^{12}C^{32}S$. Bond lengths must agree.

7-2 Using the data in Table 7-1, predict the frequency at which the $J = 2 \rightarrow 3$ absorption peak for HBr would appear.

7-3 Calculate the reduced mass of HF. *Ans:* 15.9×10^{-24} g

7-4 Using the data in Table 7-1, (*a*) predict the frequency at which the $J = 2 \rightarrow 3$ absorption peak for HCl would occur. (*b*) Is this transition in the microwave region? (*c*) Compare this absorption frequency with the frequency for a comparable transition in HBr (Exercise 7-2) and explain the differences.

Ans: (*a*) 62.1 cm^{-1}; (*b*) no

† B. T. Hicks, T. Turner, W. Kendrick, and V. C. Fiora, *Ballist. Res. Lab. Memo. Rep.* 703, 1953.

‡ A. N. Brown, Ph.D. thesis, Texas A & M College, Texas Station, 1954.

7-5 In microwave spectroscopy, rotation about the bond axis of a linear molecule cannot be responsible for any absorption peaks. Explain.

7-6 How can a Stark-modulated microwave spectrometer be used to observe the Stark effect? Explain.

7-7 Show that Eq. (7-3) can be obtained by combining Eqs. (7-1) and (7-2).

REFERENCES

INTRODUCTORY

1 BARROW, GORDON M.: "The Structure of Molecules," Benjamin, New York, 1963.
2 BARROW, GORDON M.: "Physical Chemistry," 2d ed., McGraw-Hill, New York, 1966.
3 BANWELL, C. N.: "Fundamentals of Molecular Spectroscopy," McGraw-Hill, New York. 1966.

INTERMEDIATE

4 GOLDSTEIN, J. H.: Microwave Spectrophotometry, chap. 62 in I. M. Kolthoff, P. J. Elving, and E. B., Sandell (eds.), "Treatise on Analytical Chemistry," pt. I, vol. 5, Wiley-Interscience, New York, 1964.
5 LIDE, DAVID R., JR.: Analytical Applications of Microwave Spectroscopy, in C. N. Reilley and F. W. McLafferty (eds.), "Advances in Analytical Chemistry and Instrumentation," vol. 5, pp. 235–277, Wiley-Interscience, New York, 1966.

ADVANCED

6 GORDY, W., W. V. SMITH, and R. F. TRAMBARULO: "Microwave Spectroscopy," Wiley, New York, 1953. One of the first texts in the field; very readable and informative.
7 INGRAM, D. J. E.: "Spectroscopy at Microwave and Radiofrequencies," 2d ed., Butterworths, London, 1967. A concise and useful compilation, particularly on experimental aspects.
8 TOWNES, C. H., and A. L. SCHAWLOW: "Microwave Spectroscopy," McGraw-Hill, New York, 1955. A detailed and authoritative coverage of the field.
9 SOUTHERN, A. L., H. W. MORGAN, G. K. KEILHOLTZ, and W. V. SMITH: *Anal. Chem.*, **23**:1000 (1951). A good example of a quantitative analysis for ^{15}N and ^{13}C.

8 Fluorometry and Phosphorimetry

Most substances that absorb ultraviolet or visible light energy dissipate excess energy as heat, through collisions with neighboring atoms or molecules. However, a number of important substances lose only *part* of this excess energy as heat and emit the remaining energy as electromagnetic radiation of a longer wavelength than that absorbed. This process of emitting radiation is collectively termed *luminescence,* further classified as either *fluorescence* or *phosphorescence*.† The distinction between the two is based on the mechanism or path by which the substance returns to the ground state, although a practical distinction is based on how soon after absorption the emission occurs.

In general, the absorption process raises a molecule from the ground electronic and vibrational state to an excited electronic and vibrational state. When a molecule has an excited electronic state that is more stable than usual, the excess vibrational energy will be dissipated by collisions while the molecule remains in the excited electronic state; later the molecule will either return *directly* to the ground electronic state by the emission of radiation (*fluorescence*) or, less commonly, it may shift to a metastable triplet level before emitting radiation (*phosphorescence*). A typical lifetime for an excited

† *Chemiluminescence* is the same as fluorescence except that the energy of excitation is produced in a chemical reaction; when chemiluminescence occurs in a living system, e.g. a firefly, it is called *bioluminescence*. These phenomena will not be discussed here.

electronic state is about 10^{-8} s, and materials exhibiting fluorescence generally reemit excess radiation within 10^{-8} to 10^{-4} s of absorption. The lifetime of phosphorescence is much longer than fluorescence, generally ranging from about 10^{-4} to 20 s or longer.

Although comparatively few substances exhibit fluorescence or phosphorescence, a number of very important compounds of biological, pharmacological, and organic interest do, and where applicable, fluorometry and phosphorimetry offer unique analytical advantages of high sensitivity and selectivity. Most substances exhibiting fluorescence or phosphorescence are large, rigid, multicyclic organic molecules. Rigidity is sometimes the result of complex formation with a metal ion, and in such instances the luminescence can be used as a means of analysis for the metal. Examples of substances that have been analyzed fluorometrically include steroids, estrogens, and tranquilizing alkaloids. Examples of molecules that can be made to phosphoresce include benzene, naphthalene, and tryptophan.

In the sections that follow, the theory of fluorescence and phosphorescence will be briefly described, after which sections will deal with instrumentation and applications.

8-1 THEORY

In order to understand and apply luminescence to analyses intelligently, it is essential to become familiar with the nature of excited states and excited-state processes in molecules.

8-1A Excited-State Processes in Molecules

Before luminescence can occur, a substance must absorb radiation of the proper wavelength. When a quantum of light of the proper wavelength impinges on a molecule, it will be absorbed in about 10^{-15} s (the time required for a molecule to go from one electronic state to another). An excited molecule can exist as such for only about 10^{-7} or 10^{-8} s before it will eliminate some or all of the excess energy. One of three things will happen to the excited molecule: (1) it will reemit a photon of the same frequency as it absorbed (resonance emission); (2) it will emit an infrared photon, thus losing vibrational energy and ending up at the lowest vibrational level of the excited electronic state; or (3) it will undergo *radiationless* loss of vibrational energy through collisions, ending up in the lowest vibrational level of the excited electronic state. The first two alternatives are extremely rare and in practice occur only with gases at very low pressures. With typical systems, e.g., gases at ordinary pressure, solutions, or solids, thermal relaxation of a vibrationally excited molecule is more likely to occur long before the molecule has had time to reemit a photon.

Once a molecule arrives at the lowest vibrational level of an excited state, it again has three possible modes by which it may lose the remaining excess energy: (1) it may undergo *radiationless* loss of electronic energy through collisions or other interactions; (2) it may emit an ultraviolet or visible (*fluorescent*) light photon; or (3) it may undergo a transition to a metastable *triplet* state and some time thereafter return to the ground state, usually by emission of an ultraviolet or visible (*phosphorescent*) light photon. In general, the three modes of electronic deactivation are listed in the order of decreasing probability. In other words, most molecules that absorb ultraviolet or visible light dissipate all this excess energy as heat. Relatively few molecules exhibit fluorescence, and still fewer exhibit phosphorescence. Nonetheless, in a number of vitally important compounds, e.g., aromatic hydrocarbons, radiationless loss of electronic energy is of low probability, and fluorescence and phosphorescence assume great importance. Before proceeding further, it is necessary to define the nature of the excited states, particularly the triplet state which is an essential part of the mechanism of phosphorescence.

SINGLET AND TRIPLET STATES

The names *singlet* and *triplet* arise from multiplicity considerations of atomic spectroscopy and simply define the number of *unpaired* electrons in the absence of a magnetic field. In general, if n is the number of unpaired electrons, there will be $(n + 1)$-fold degeneracy (equal energy states) associated with the electron spin, regardless of the molecular orbital occupied. Thus, if there are *no* unpaired electrons ($n = 0$), there is only $n + 1$ or *one* spin state, and this is called a *singlet* state. Similarly, the names doublet, triplet, quartet, quintet, etc., refer to systems having 1, 2, 3, 4, . . . unpaired electrons, respectively.

Fluorescence and phosphorescence involve *molecules,* and most molecules in their ground state, i.e., lowest energy state, have no unpaired electrons (singlet state). When ultraviolet or visible radiation of the proper frequency is absorbed by such a molecule, one or more of the paired electrons (usually a π electron; see Sec. 2-1*B*) is raised to an *excited* singlet state, in which the spin of the promoted electron has not changed and the net spin is still zero. Figure 8-1 schematically illustrates the difference between a molecule in the ground singlet state (Fig. 8-1*a*) and the excited singlet state (Fig. 8-1*b*). Figure 8-1*c* represents a molecule in an excited *triplet* state: one set of electron spins have become unpaired, and the two unpaired electrons make it a triplet state. A direct transition from the singlet ground state (Fig. 8-1*a*) to the triplet excited state (Fig. 8-1*c*) is *forbidden* (has extremely small probability), and thus, as will now be described, the only feasible mechanism for population of the triplet excited state is by conversion of an excited singlet state to an excited triplet state.

SUMMARY OF LUMINESCENCE PROCESSES

Figure 8-2 is a schematic energy-level diagram which summarizes the energy changes involved in absorption, fluorescence, and phosphorescence. In

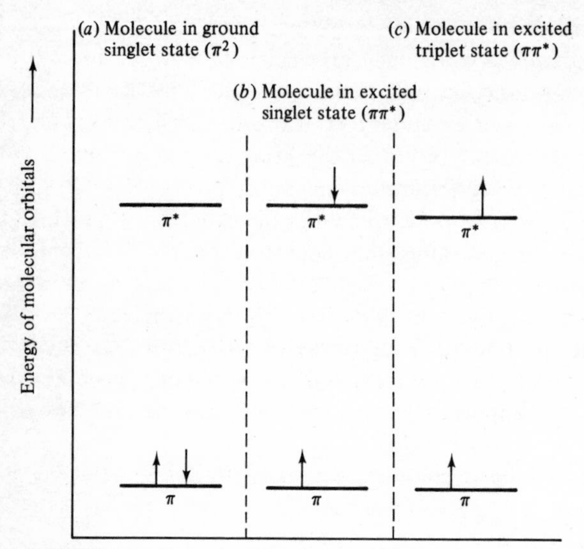

FIGURE 8-1 Examples of molecules with π electrons in various spin configurations.

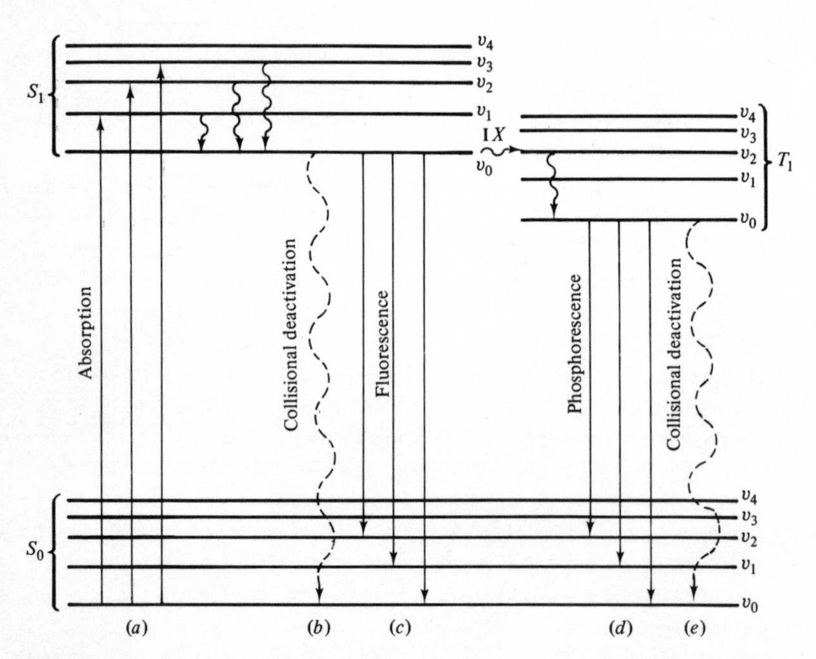

FIGURE 8-2 Schematic energy-level diagram illustrating the energy changes involved in absorption, fluorescence, and phosphorescence. (Straight arrows represent radiation photons; wavy arrows represent radiationless collisional deactivation; v = vibrational levels, S_0 = singlet ground state, S_1 = singlet first excited state, T_1 = triplet first excited state.)

absorption (Fig. 8-2*a*) the molecule is raised from the ground electronic and vibrational state to an excited electronic state and any of a number of possible excited vibrational states. The molecule will then undergo vibrational deactivation by collisions to the lowest vibrational level of the excited singlet state. From this lowest excited singlet state, one of three things will probably occur, depending on the molecule involved and the conditions.

If this excited singlet state is relatively unstable, the molecule will return to the ground state by a *radiationless* collisional deactivation (Fig. 8-2*b*). On the other hand, molecules with relatively stable excited singlet states will return to the ground electronic state by *fluorescent* photon emission (Fig. 8-2*c*). The excited singlet state can be stabilized and the efficiency of fluorescent emission improved by working at low concentrations of solute, at low temperatures, and with high-viscosity solvents, e.g., glycerol, all of which minimize collisional deactivation.

For most molecules the vibrational energy-level spacings are similar in the ground and excited electronic states, so that the fluorescence emission spectrum is often an approximate mirror image of the absorption spectrum, with the fluorescence spectrum shifted to longer wavelengths; this is illustrated in Fig. 8-3 with the absorption and fluorescence spectra of anthracene.

Finally, for molecules with relatively stable excited singlet states capable of undergoing a transition to a *triplet* state, *phosphorescence* emission (Fig. 8-2*d*) may occur. The process of crossing from a singlet (no unpaired electron) state to a triplet (two unpaired electrons) state is called an *intersystem crossing,* labeled IX on Fig. 8-2. The mechanism for intersystem crossing in-

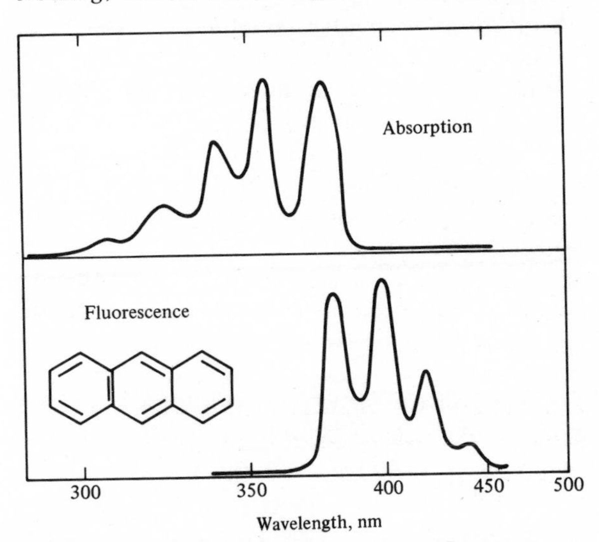

FIGURE 8-3 Absorption and fluorescence spectra of anthracene, showing approximate mirror symmetry. (*From D. H. Whiffen, "Spectroscopy," p.* 164, *John Wiley & Sons, Inc., New York,* 1966, *by permission.*)

volves *vibrational coupling* between the excited singlet state and an excited triplet state. Since the excited triplet state in general is at a lower energy than the comparable excited singlet state, the molecule after intersystem crossing will once again have excess vibrational energy and will undergo radiationless vibrational deactivation to the zeroth vibrational level of the triplet state. The lowest triplet state has a very long lifetime, relatively speaking (of the order of 10^{-4} to many seconds), and therefore, once the molecule is in the lowest triplet state, whether phosphorescence emission (Fig. 8-2*d*) or collisional deactivation (Fig. 8-2*e*) will occur depends largely on the environmental conditions.

It is important to select the proper environmental conditions to observe or accentuate phosphorescence. Phosphorescence is almost never observed at room temperature (biacetyl is one exception), since the very long lifetime of the triplet state usually enhances collisional loss of energy. On the other hand, if the molecule is placed in a *rigid* medium so that collisional processes are minimized, phosphorescence emission will be observed. The most common technique of immobilizing the system is to freeze the phosphorescent molecules in special glass, one of the most popular of which is EPA, a mixture of ethyl ether, isopentane, and ethyl alcohol in a ratio of 5:5:2 which is frozen at liquid-nitrogen temperatures (77 K).

8-1B Molecular-Structure Factors in Luminescence

The relationship between structure and luminescence will be briefly outlined here. A more detailed treatment can be found in Refs. 4, 5, and 13.

The first requirement for fluorescence or phosphorescence is a molecular structure which will absorb ultraviolet or visible radiation. In general, the greater the absorbancy of a molecule the more intense its luminescence. This requirement virtually eliminates aliphatic and saturated cyclic organic compounds. Molecules containing conjugated double bonds, especially those with a high degree of resonance stability, are particularly fruitful for study. Aromatic hydrocarbons, particularly rigid, planar, multicyclic structures, are ideal. Aromatic rings containing heteroatoms often exhibit better phosphorescence than fluorescence.

Substituents often have a marked effect on the luminescence of molecules. These effects usually correlate with their effects on the absorption spectra. Though no rigid rules can be given, a few generalities may be helpful:

1 Electron-donating groups often enhance fluorescence because they tend to increase the transition probability between the lowest excited singlet state and the ground state. Particularly useful groups are $—NH_2$ and —OH.

2 Substituent groups which have only a small interaction with π-electron systems, such as $—SO_3H$, $—NH_3^+$, and alkyl groups, tend to have little effect on luminescence.

3 The introduction of a high-atomic-number atom into a π-electron system usually enhances phosphorescence and decreases fluorescence.

4 Electron-withdrawing groups, such as —COOH, —NO_2, —N=N—, and halides, diminish or even destroy fluorescence.

The fact that rigid, planar structures promote fluorescence can be illustrated by comparing the structure of fluorescein with phenolphthalein:

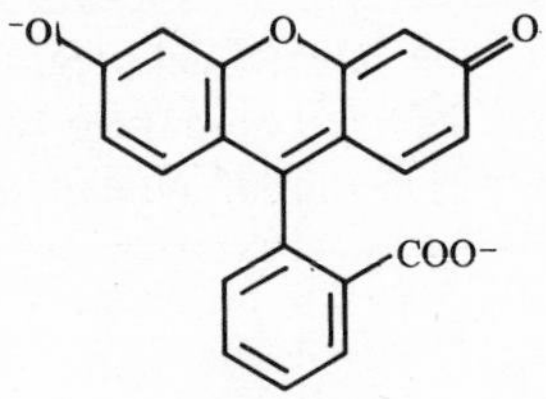

Fluorescein

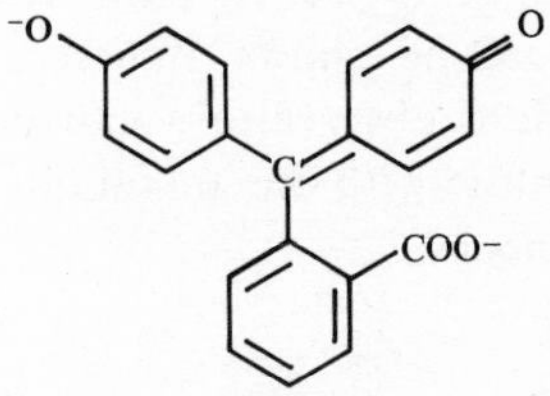

Phenolphthalein

Fluorescein exhibits very intense fluorescence in liquid solution, but phenolphthalein does not, despite the structural similarity. The principal effect of increasing molecular rigidity is to decrease vibrations; this minimizes intersystem crossing to a triplet state and collisional heat degradation.

In general, the formation of chelates with metal ions also promotes fluorescence by promoting rigidity and minimizing internal vibrations. This can be illustrated by comparing the structure of pontachrome blue black R (sodium 2,2′-dihydroxy-1,1′-azonaphthalene-3-sulfonic acid, designated pontachrome BBR), with its aluminum chelate:

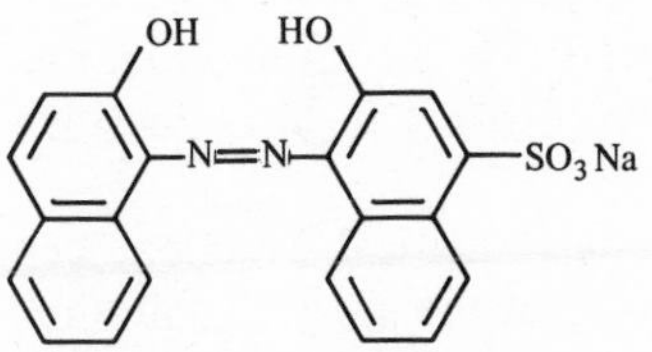

Pontachrome BBR

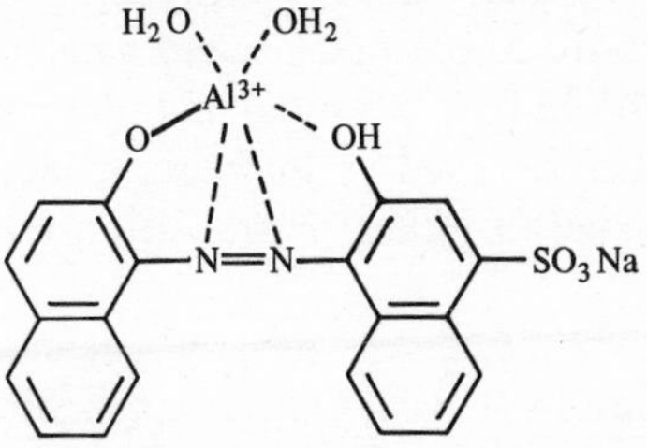

Al^{3+} chelate of pontachrome BBR

Pontachrome BBR is a dye which does not fluoresce in the absence of aluminum, but in the presence of aluminum at a pH of about 4.5 it shows a red fluorescence.† In the absence of aluminum the naphthalene groups of pontachrome BBR are free to rotate around the azo group, whereas when aluminum is chelated, the molecule is held in a rigid planar position. In a similar manner, a number of substances can be used as fluorescent indicators in che-

† A. Weissler and C. E. White, *Ind. Eng. Chem. Anal. Ed.*, **18**:530 (1946).

lometric titrations. For example, calcein (a substituted derivative of fluorescein [8, p. 108]) is an excellent indicator for calcium and copper. With free metal ion available for chelation calcein has an intense yellow-green fluorescence, but when all the metal ion is complexed, e.g., with EDTA, calcein will not fluoresce.

The pH of the solution often has a marked effect on the fluorescence of compounds, mainly by altering the charge and resonance forms of the chromophore. For example, in neutral or alkaline solution aniline exhibits a fluorescence in the visible region, but the visible fluorescence disappears if aniline is acidified (aniline shows fluorescence in the ultraviolet region regardless of pH). These observations can be explained by comparing the resonance forms of aniline with the anilinium ion, in which the amine cannot contribute to ring resonance:

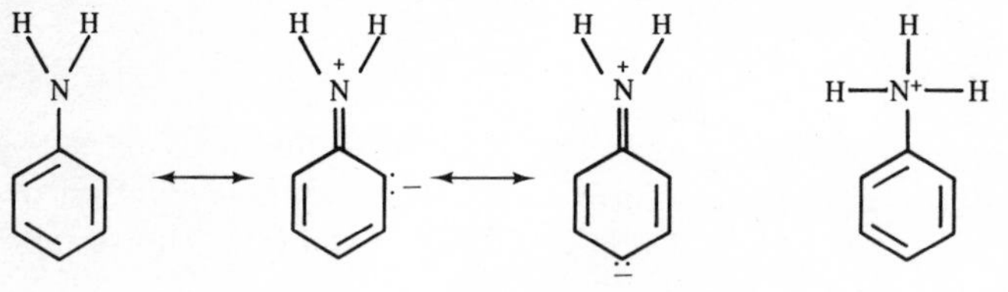

Some resonance forms of aniline

Anilinium ion

In acid solution a positive charge is fixed at the nitrogen atom, and thus anilinium ion has only the same resonance forms as benzene, which also exhibits fluorescence only in the ultraviolet region. On the other hand, in neutral or basic solution aniline has three additional resonance structures, resulting in a more stable first excited singlet state and a longer wavelength of fluorescent radiation. In a similar manner some substances are so sensitive to pH that they can be used as fluorescent indicators in pH titrations. Fluorescein, which shows an indicator transition range from pH 4 to 6, is one such example.

Finally, two processes, oxygen quenching and chemical reaction, can *compete* with luminescence. Oxygen quenching is one of the most troublesome aspects of fluorometry and phosphorimetry and requires that all solutions be deaerated before analysis. In some cases, oxygen decreases the fluorescence intensity of an organic compound by oxidizing it; usually, however, oxygen quenches excited singlets and triplets by promoting intersystem crossings through collisions with the excited species and forming a transitory charge-transfer complex. The ground state of an oxygen molecule is a triplet, and this paramagnetic nature of oxygen apparently increases spin-orbit coupling with electronically excited molecules. Other paramagnetic gases, for example, NO, behave likewise, but oxygen is the only common one.

A second process which sometimes competes with luminescence is a chemical reaction by a molecule in the excited state. Since it may have a completely different electron distribution from the ground state, it can thus react differently. Consider 9,10-anthraquinone, for example. In the ground state, anthraquinone dissolved in alcohol is quite stable and shows no tendency to react. However, when anthraquinone absorbs radiation, it goes first to the excited singlet and then to the triplet state. Though this excited species is capable of phosphorescence, it is also now very reactive and rapidly abstracts a hydrogen atom from the alcohol, being reduced to 9,10-dihydroxyanthracene and at the same time oxidizing the alcohol to its corresponding aldehyde. Actually, the 9,10-dihydroxyanthracene produced is *fluorescent,* and thus in this case the chemical reaction quenches phosphorescence and induces fluorescence.

8-1C Relation between Luminescence Intensity and Concentration

The intensity of fluorescent radiation I_f or phosphorescent radiation I_p will be directly proportional to the concentration of *fluorophors* or *phosphors* present in the sample. Although the fundamental relationships between the measured emission intensity and the concentration of luminescent material are essentially the same in fluorometry and phosphorimetry, the concentration range over which the two methods of analysis are applicable are somewhat different and will be treated separately.

FLUOROMETRY

The intensity of fluorescent radiation I_f will be directly proportional to the concentration C of the absorbing species, but *only* at very low concentrations, which may be shown as follows.

If Beer's law holds (see Sec. 1-5B), the fraction of transmitted light T is given by

$$T = \frac{I}{I_0} = e^{-abC} \tag{8-1}$$

where I_0 = incident intensity
I = transmitted intensity
a = absorptivity
b = length of cell path
C = concentration of absorbing substance

The corresponding fraction of light absorbed is then

$$1 - \frac{I}{I_0} = 1 - e^{-abC} \tag{8-2}$$

When Eq. (8-2) is rearranged, the *amount* of light absorbed is

$$I_0 - I = I_0(1 - e^{-abC}) \tag{8-3}$$

The intensity of fluorescent radiation which is measured I_f will be related to the amount of light absorbed:

$$I_f = k\phi_f(I_0 - I) \tag{8-4}$$

where k is a proportionality constant that depends on the instrument (the efficiency of the detector system and the geometry) and ϕ_f is the quantum efficiency of fluorescence (the ratio of quantum emitted to the quantum absorbed in a unit of time). The quantity in parentheses in Eq. (8-3) can be expanded by an exponential series as

$$1 - e^{-abC} = 2.3abC - \frac{(2.3abC)^2}{2!} + \frac{(2.3abC)^3}{3!} - \cdots \tag{8-5}$$

At low concentrations, where the term abC is less than about 0.05, all but the first term of Eq. (8-5) can be dropped. Therefore, when Eqs. (8-3) to (8-5) are combined, the fluorescent intensity at low concentrations will be given by

$$I_f = k\phi_f I_0(2.3abC) \tag{8-6}$$

Thus, for very dilute solutions (of the order of a few parts per million or less) the fluorescent intensity will be directly proportional to concentration. However, at higher concentrations deviations will result because of the higher-order terms in Eq. (8-5). Figure 8-4 gives a typical calibration curve showing how the fluorescent intensity depends on concentration and temperature for naphthalene. Note that at the optimum temperature, −196°C, the calibration curve merely levels off (saturates) at high concentration. However, at higher temperatures the curve may actually bend downward at high concentration, a behavior believed to be due to self-absorption. In other words, at high concentrations the fluorescent emission may be reabsorbed by other sample molecules before emerging from the solution.

Effect of Temperature on Fluorescence Intensity In concluding the discussion on the relationship between concentration and fluorescence intensity, special mention should be made of temperature problems in conducting a quantitative fluorescence analysis. Since most fluorescence instruments use sources of very high intensity, heating of the sample or sample chamber can occur. An increasing temperature usually *decreases* fluorescence intensity since the increased random motion of molecules increases the probability of collisional deactivation of the excited molecule. Temperature coefficients are typically about 1.0 to 1.2 relative fluorescence units per Celsius degree. Most

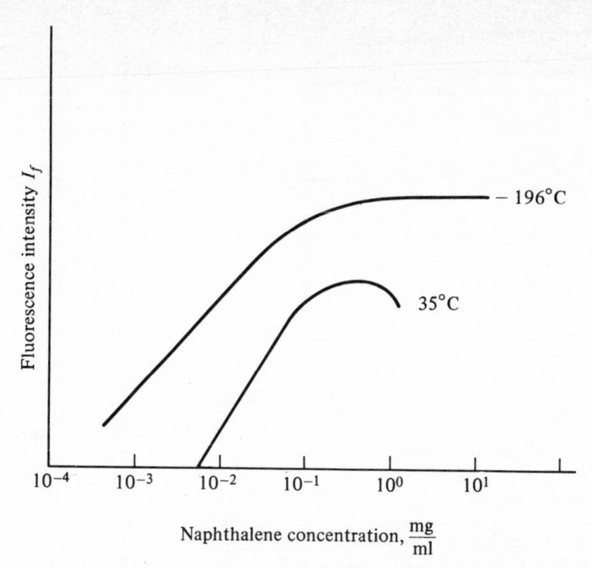

FIGURE 8-4 Dependence of fluorescent intensity upon concentration and temperature for naphthalene.

instruments minimize temperature effects by providing shutter mechanisms to minimize exposure time and thermally insulating the sample holder. Nonetheless, some heating still occurs, and for very precise quantitative measurements temperature regulation of the sample holder is required.

The same drawbacks would hold for phosphorescence measurements, except that most phosphorescence measurements are made at constant temperature in an insulated Dewar flask at liquid-nitrogen temperatures.

PHOSPHORIMETRY

Exactly the same considerations as discussed for fluorometry are important in relating phosphorescent intensity I_p to concentration, and thus the final equation is

$$I_p = k\phi_p I_0 (2.3abC) \tag{8-7}$$

where k depends on the instrument used and ϕ_p is the quantum efficiency of phosphorescence. Again this equation is valid only when the concentration is low (abC is less than 0.05, or preferably less than 0.01) but it turns out that the *range* of concentration over which the calibration curve is linear is generally somewhat larger in phosphorimetry than in fluorometry. Figure 8-5 gives a typical phosphorimetry calibration curve, obtained for metycaine hydrochloride at 77 K. Note that the curve is linear over a concentration range of about 10^3, whereas a typical fluorometry calibration curve is linear only over a range of about 10^2 (see Fig. 8-4).

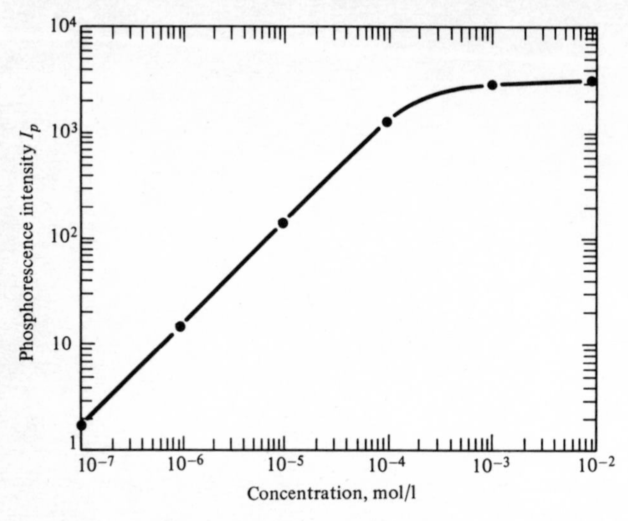

FIGURE 8-5 Typical phosphorimetry calibration curve (metycaine hydrochloride at 77 K). (*From D. M Hercules, "Fluorescence and Phosphorescence Analysis," p.* 174, *John Wiley & Sons, Inc., New York,* 1966, *by permission.*)

8-2 INSTRUMENTATION

Although the principal components of any luminescence instrument (source, monochromator, sample cell, and detector) differ little from those used in ultraviolet and visible absorption spectrophotometry (Chap. 2), the nature of the phenomena being measured causes significant differences in instrument design. For example, whereas absorption measurements are nearly always made after passing a beam of radiation straight through the sample cell, fluorescence and phosphorescence measurements are always made at an angle (usually 90°) to the incident radiation beam. As another example of the inherent differences, recall that in making ultraviolet and visible absorption measurements a filter photometer is usually inferior to a spectrophotometer in both selectivity and sensitivity (see Sec. 2-2*B*). However, in fluorescence measurements a filter instrument may show comparable or even superior sensitivity to a more complicated monochromator instrument, though a filter instrument will lack the versatility and some of the selectivity of a monochromator instrument. Finally, recall that the finest ultraviolet and visible absorption spectrophotometers use double-beam operation (see Sec. 2-2*C*), whereas true double-beam operation is not practical on a luminescence spectrometer. Therefore the problems of single-beam operation need to be considered.

These and other differences in instrumentation make it necessary to discuss the principles of luminescence measurements in some detail. The emphasis will be on fluorescence instruments, since they are much more widely used than phosphorescence instruments.

8-2A Some Terms Defined

The word fluoro*o*meter is often used interchangeably with fluor*i*meter, just as phosphoro*o*meter is used in place of phosphor*i*meter. Following the lead of most American instrument manufacturers this book uses fluor*o*meter and phosphor*i*meter.

FLUOROMETERS VS. SPECTROFLUOROMETERS

The distinction between *fluorometers* and *spectrofluorometers* (as well as between *phosphorimeters* and *spectrophosphorimeters*) is exactly analogous to the distinction between filter photometers and spectrophotometers in ultraviolet and visible absorption spectrophotometry (see Sec. 2-2*A*). In other words, a fluorometer uses *filters* to isolate the wavelength of excitation or emission, whereas a spectrofluorometer utilizes one or two (usually two) monochromators to isolate the wavelength desired. Thus, a spectrofluorometer is capable of scanning the wavelength spectrum continuously (and usually automatically), whereas a fluorometer is always a manual instrument and is best suited for measurements at one or two wavelengths, a change in filters being required each time the wavelength is changed.

EXCITATION SPECTRA VS. EMISSION SPECTRA

In dealing with luminescence spectra (plots of intensity vs. wavelength), it will soon become apparent to the reader that since *two* monochromators are generally involved, two different kinds of spectra can be recorded. Specifically, luminescence spectrometers will have a monochromator between the source and the sample (*excitation* monochromator) and another between the sample and the detector (*emission* monochromator). If the emission monochromator is set at a fixed wavelength (usually the wavelength of maximum fluorescence or phosphorescence emission) and the excitation monochromator is allowed to vary, an *excitation* spectrum results. On the other hand, if the excitation monochromator is set at a fixed wavelength (almost always the wavelength of maximum absorption) and the emission monochromator is allowed to vary, an *emission* spectrum results. For analytical applications in fluorometry and phosphorimetry the *emission* spectrum is used, but it is standard practice with a spectrofluorometer or spectrophosphorimeter to first run an excitation spectrum in order to confirm the identity of the substance and to select the optimum excitation wavelength. Figure 8-6 shows a typical excitation and superimposed emission spectrum for the same compound measured with a spectrofluorometer. The excitation spectrum is identical, in theory at least, with the absorption spectrum of the compound, and thus is useful for confirmation purposes. The emission spectrum, of course, falls at longer wavelengths than the excitation spectrum and is an approximate mirror image of the excitation spectrum. The fact that

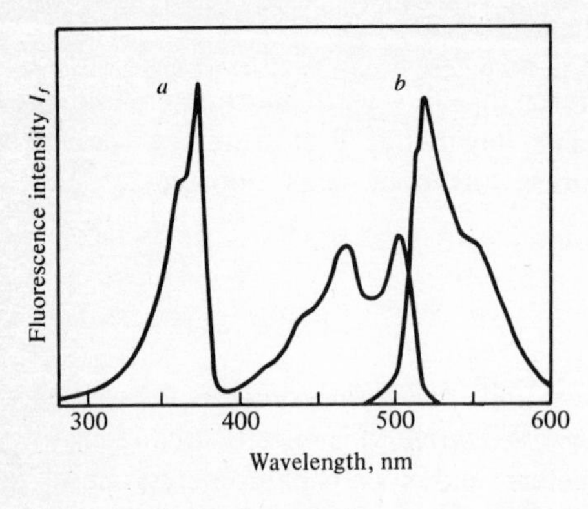

FIGURE 8-6 (*a*) Excitation and (*b*) emission spectra of 2,1,3-naphtho(2,3-c)-selenadiazole measured with a spectrofluorometer. [*From P. Cukor, J. Walzyck, and P. F. Lott, Anal. Chim. Acta,* **30**:473 (1964), *by permission.*]

the emission spectrum is not an exact mirror image of the excitation spectrum is due not only to molecular energy-level considerations (see Sec. 8-1*A*) but also to various instrumentation difficulties, which will be discussed next.

TRUE (OR CORRECTED) SPECTRA VS. APPARENT (OR UNCORRECTED) SPECTRA

One of the inherent problems of luminescence spectrometers is that there is no good way of getting true double-beam operation. Since all luminescence spectrometers are basically single-beam instruments, all the problems of single-beam operation (instability of light sources, detectors, etc.) are potentially present. However, two particularly troublesome problems are that the intensity of any light source used will vary with wavelength and the sensitivity of any detector used will vary with wavelength. Because very few instruments correct for these two problems, the exact shape of the excitation or emission spectra produced by such instruments will depend greatly on precise characteristics of the instrument used to make the measurements; they should be termed *apparent* (or *uncorrected*) spectra. On the other hand, it is possible to correct the apparent spectra for the source and detector characteristics, and the resulting spectra, which will then be independent of the particular instrument used to make the measurement, are termed *true* (or *corrected*) spectra. Although it is possible to correct an apparent spectrum manually when the characteristics of the source and detector used are known, this is at best a crude and time-consuming process. Certain commercially available instruments have devices, available either as accessories or already built in, which will automatically make these corrections. The principles involved in the design of such automatic correctors will be described in Sec. 8-2*C*.

8-2B General Principles of Luminescence Instrumentation

BLOCK DIAGRAM OF A FLUOROMETER

A block diagram of the essential components of a typical fluorometer or spectrofluorometer is shown in Fig. 8-7. In a fluorometer filters are used (the excitation filter is often called the *primary* filter, and the emission filter is often called the *secondary* filter); in a spectrofluorometer grating or prism monochromators are used.

Since fluorescence radiation is emitted in all directions by the sample, some variation in the angle of detection with respect to the excitation beam is possible. However, the right-angle arrangement shown in Fig. 8-7*a*, which minimizes detection of excitation beam and stray light, is most advantageous for dilute solutions or gases. For solid samples or ones that are very opaque, the arrangement shown in Fig. 8-7*b* is the best choice.

MODIFICATIONS NECESSARY TO MEASURE PHOSPHORESCENCE

To measure phosphorescence instead of fluorescence, there must be a delay between the time when the sample is irradiated and the luminescence is observed. There are several ways of obtaining this delay, one of which is illustrated in Fig. 8-8. The rotating shutter is simply a hollow cylinder with one or more slits cut in the circumference. By rotating it a delay is introduced

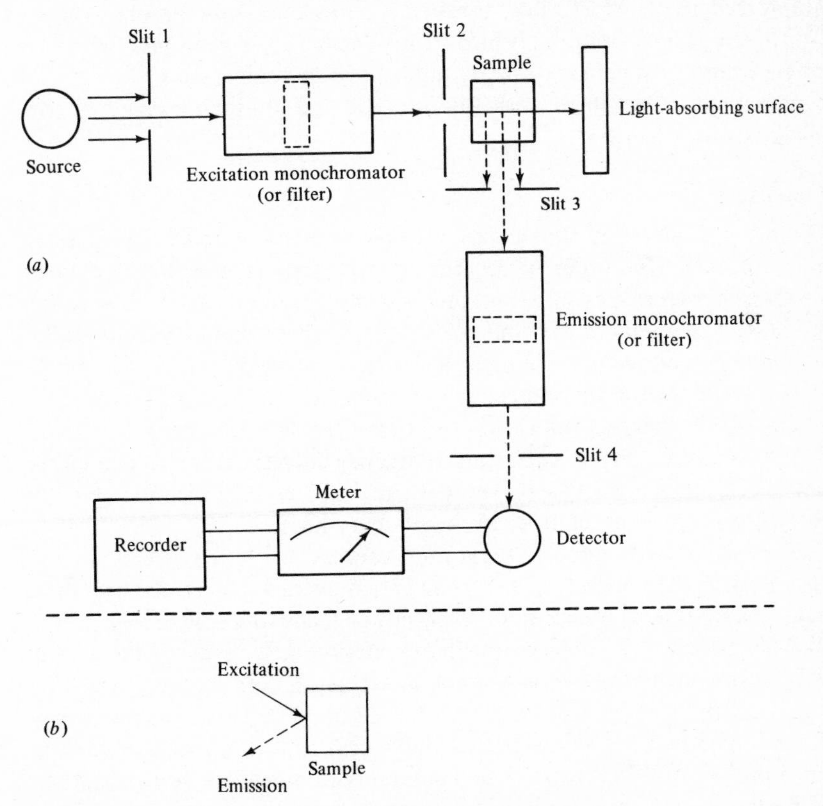

FIGURE 8-7 Block diagram of a typical fluorometer or spectrofluorometer. Detection at (*a*) right angles (used with most samples) and (*b*) at acute angles (for samples that are solid or very opaque).

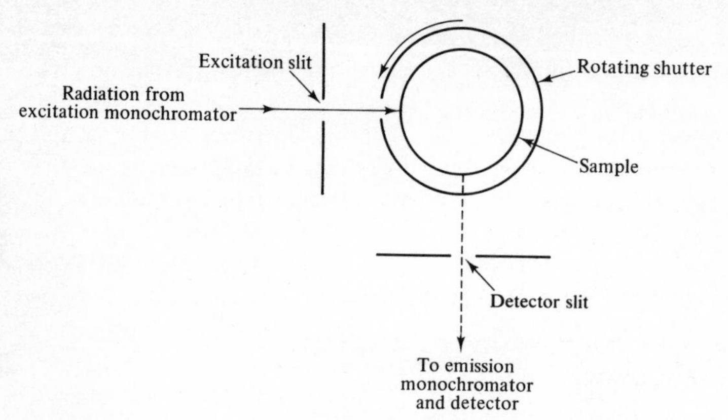

FIGURE 8-8 Use of a rotating shutter to permit phosphorescence measurements.

between the time of irradiation and observation of phosphorescence. The only other significant difference between a phosphorimeter and a fluorometer is in the sample cell itself. Most phosphorescence emissions must be observed at liquid-nitrogen temperatures, which introduces complications, whereas fluorescence measurements are usually made at room temperature.

Components common to both fluorescence and phosphorescence instruments will now be described.

SOURCES

The most widely used sources are dc *xenon, xenon-mercury,* or *mercury* arc lamps, listed in the order of decreasing popularity. Figure 8-9 compares the spectral characteristics of xenon-arc and mercury-arc lamps. The xenon arc produces essentially a continuous spectrum throughout the ultraviolet region, whereas the mercury arc gives a discrete-line spectrum. A continuous source is preferable for luminescence spectrometers and is applicable to a larger number of analyses than a line source. On the other hand, if a compound to be analyzed has an excitation frequency corresponding to one of the mercury lines, the use of this source can lead to significantly greater sensitivity due to the high intensity of the mercury lines.

The most critical factors regarding sources are their *stability* and *intensity*. If lamps are operated at their specified average power, the arc tends to wander. Operation at higher than rated power tends to stabilize the arc and give higher intensity but reduces the lamp lifetime and leads to decreasing source intensity with time. It is helpful in terms of both stability and longevity to cool the sources with water or forced air. Most commercial xenon lamps are cooled by convection, but mercury arcs cooled in this way tend to wander. Voltage stabilizers should be standard equipment on power supplies for the source.

Changes in lamp intensity with age as well as changes in monochromator-detector response usually make it necessary to calibrate the fluorescence emission of the instrument before each analysis. For this purpose a fluorescent reference compound is used. Commercial uranium glass or an acidic quinine solution are two standards in use. The frequency with which this calibration need be made is minimized in instruments equipped with automatic spectra correctors, discussed in Sec. 8-2*C*.

MONOCHROMATORS

Fluorometers using *filters* are the most common luminescence instruments. Filters that have sharp cutoff regions and low-fluorescence properties can be obtained from a number of optical supply companies. Interference filters that give particularly narrow bandwidths and good rejection of stray light are also available.

Most commercial spectrofluorometers and spectrophosphorimeters use grating monochromators. Gratings have the advantage of giving linear dispersion, better dispersion in the visible region, and less light loss than a prism system. Quartz *prisms* give better dispersion in the ultraviolet region and do not have second- and third-order spectra to contend with.

Selection of the proper slit width is important in spectroluminescence instruments. A compromise between sensitivity and resolution is necessary.

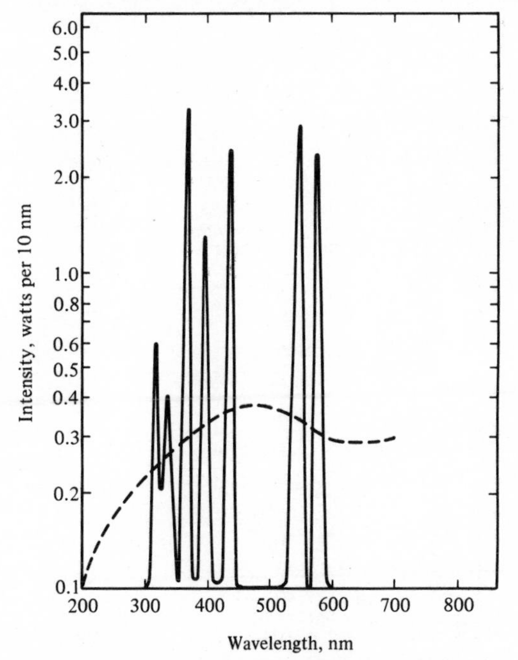

FIGURE 8-9 Spectral characteristics of 150-W xenon-arc (*dashed curve*) and 100-W mercury-arc (*solid curve*) lamps. [*From P. F. Lott, J. Chem. Educ.*, **41** (6):A422 (1964), *by permission.*]

Large slit widths give maximum sensitivity but minimum resolution and increase the probability of stray and scattered light. The best compromise is often obtained by adjusting the slits on the excitation monochromator separately from the emission monochromator. When recording an *excitation* spectrum it is usually advantageous to keep the excitation slits small (for good resolution) while opening the emission slits wider (for good sensitivity). Conversely, when recording an *emission* (fluorescence *or* phosphorescence) spectrum, it is avantageous to widen the excitation slits and narrow the emission slits.

DETECTORS

It is vital that very sensitive detectors be used. Most instruments use high-gain photomultiplier tubes, such as the RCA 1P28 and 1P21 tubes. The spectral-response range of these tubes is from 210 to 670 and 310 to 670 nm, respectively. Both tubes have a high gain and a low dark current, with the 1P21 slightly superior in both respects.

8-2C Fluorometers and Spectrofluorometers

The general principles of instrument design will be illustrated with a few typical commercially available instruments. For a complete review and discussion of various commercially available instruments, see Refs. 9 and 10.

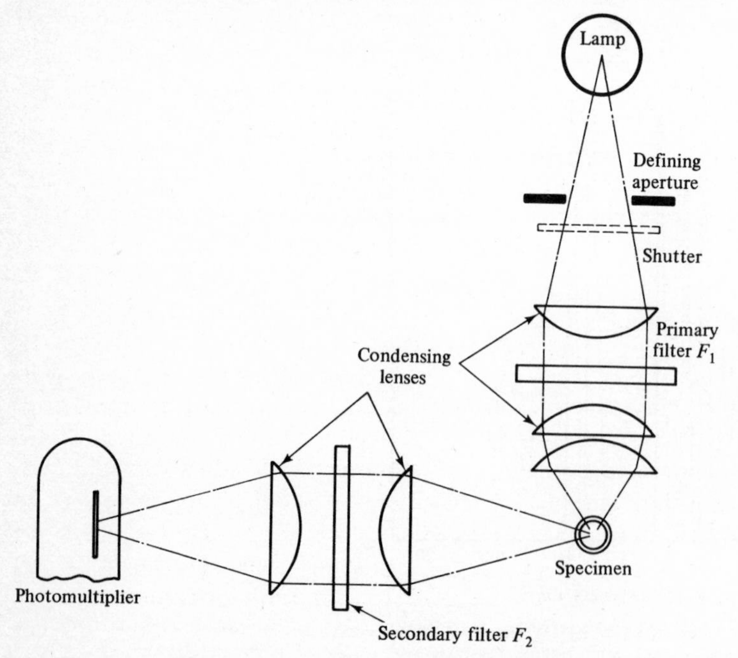

FIGURE 8-10 Optical system of a Farrand model A-4 fluorometer. (*Farrand Optical Co., Inc.*)

FLUOROMETERS

Figure 8-10 illustrates the basic arrangement for a typical single-beam filter fluorometer. It employs an 1P21 photomultiplier, a xenon-arc source, stabilized power supplies for both lamp and photomultiplier, quartz optics, and interference filters. Ordinarily the primary filter F_1 is selected to transmit ultraviolet but not visible radiation, and the secondary filter F_2 is selected to transmit visible and absorb ultraviolet radiation. An accessory is available for calibrating the instrument.

A wide range of fluorometers is commercially available, starting from units even simpler than the one illustrated in Fig. 8-10, all the way to double-beam, ratio-recording fluorometers [9]. A ratio-recording fluorometer incorporates a reference solution and alternately irradiates the sample and the reference solution, while the meter indicates the ratio of the two luminescence signals. This system minimizes the effect of temperature changes and variations in source and detector.

SPECTROFLUOROMETERS

Figure 8-11 gives a schematic diagram of a typical spectrofluorometer, the Aminco-Bowman spectrofluorometer. The *source* for the Aminco-Bowman instrument is normally a high-pressure xenon arc, fed by a stable dc power supply. The output of the lamp passes through a slit and onto the *excitation monochromator.* Light reflected from the excitation monochromator passes through a series of slits into the *sample.* Fluorescence radiation from the sample leaves the cell compartment at right angles to the excitation path and strikes the *emission monochromator.* The fluorescence leaving the emission monochromator is directed through a slit to the *photomultiplier detector.* The output of the photomultiplier is further amplified and then displayed on a meter, an oscilloscope, or a recorder (or all three).

Details of the excitation and emission monochromators are not shown in Fig. 8-11, but their arrangement is of the Czerny-Turner type (see, for example, Figs. 2-37 and 3-23). The wavelength setting on each of the monochromators can be adjusted manually, or an automatic scan can be obtained by a constant-speed motor operating through a cam connected to the monochromator. An xy recorder is used, and the position of the x axis on the recorder or oscilloscope is determined by a potentiometer coupled to the wavelength cam. The y axis on the oscilloscope or recorder is a measure of the fluorescence intensity. The amplifier has the usual sensitivity-adjustment switches and a dark-current control to cancel out the dark current of the photomultiplier.

A number of *accessories* for the Aminco-Bowman instrument include fixed or adjustable slits, different photomultiplier tubes for various regions of the spectrum, an interchangeable grating for extending the range to 1200 nm, a device for cooling the photomultiplier in order to lower the dark current, a xenon-mercury arc source, temperature baths, a polarization accessory, a

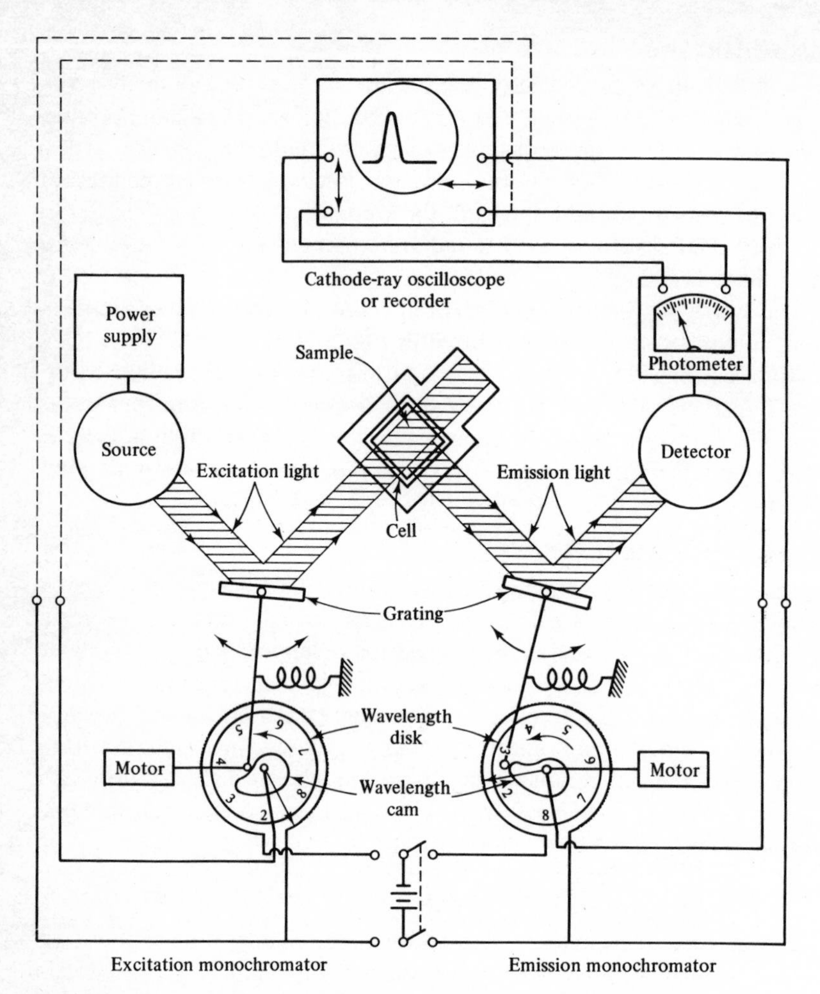

FIGURE 8-11 Schematic diagram of an Aminco-Bowman spectrofluorometer. See text for explanation (*American Instrument Co.*).

rotary turrent capable of holding three samples, an automatic titration assembly, a corrected-spectra attachment, and a phosphorescence accessory which includes the necessary shutter mechanism and liquid-nitrogen Dewar.

Automatic Spectrum Correctors In Sec. 8-2*A* the differences between true (or corrected) spectra and apparent (or uncorrected) spectra were defined and discussed. The great majority of spectrofluorometers give uncorrected spectra. The instrumental principles involved in correcting spectra automatically will now be presented.

There are essentially two separate corrections to be made. First, it is necessary to keep the *source energy* essentially constant as the wavelength is scanned. Probably the best way to accomplish this is to split off a portion of the source radiation and feed it to a *thermocouple* in a control loop (see Fig. 8-12). The output of a thermocouple depends only on the *energy* of the light signal and is independent of the wavelength. The output of the thermocouple is maintained at a constant level by means of a servo attached to the slits of the excitation monochromator. Since the servo keeps the thermocouple signal constant, it simultaneously keeps the energy which is emerging from the excitation monochromator constant, regardless of changes in source intensity.

The second correction, correcting for the variation in the spectral response of the *detector*, is more difficult. Probably the best system employs what might be called an "electric cam," which is nothing more than a potentiometer which generates a *predetermined* correction program (voltage vs. wavelength). This predetermined voltage output can be used in a number of ways to correct the energy of the detector. In Fig. 8-12, it is used to operate a second slit servo, thereby controlling the amount of light energy emerging from the emission monochromator. In other instruments, e.g., in the American Instrument Company corrected-spectra attachment, the predetermined voltage is fed directly to a multiplier-divider circuit in the amplifier following the photomultiplier, and the multiplier-divider circuit makes the appropriate correction in the amplitude of the signal fed to the readout device. In still other instruments the predetermined voltage activates an optical-density wedge, which adjusts the amount of energy reaching the photomultiplier.†

To calibrate an electric cam to compensate for the spectral characteristics of a particular photomultiplier tube, a neutral scattering material, e.g., a block of magnesium oxide, is substituted for the sample and the wavelength is then scanned, using the excitation system as a source of constant energy. The recorded graph which results will be a plot of relative photomultiplier output

† One other system, used, for example, in the G. K. Turner Associates model 210 Spectro, avoids the predetermined correction program. It employs a reference lamp, and the ratio of the sample fluorescence signal to the reference signal is used to correct for nonlinearity of the photomultiplier. Details of this system can be found in Refs. 4 and 9.

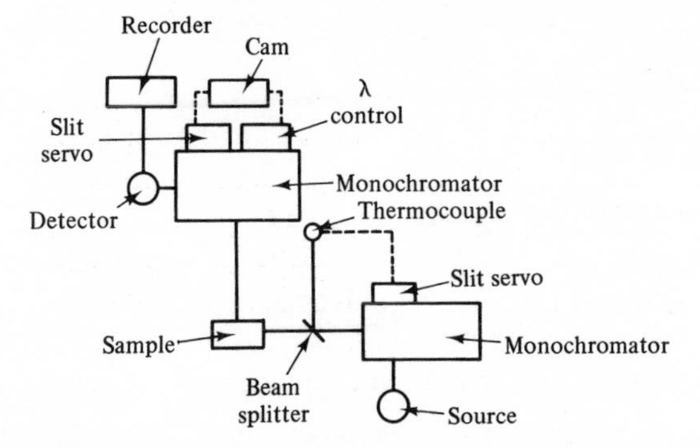

FIGURE 8-12 Block diagram of an instrument which provides automatic spectra corrections (Perkin-Elmer model 195 spectrofluorometer). (*Perkin-Elmer Corp.*)

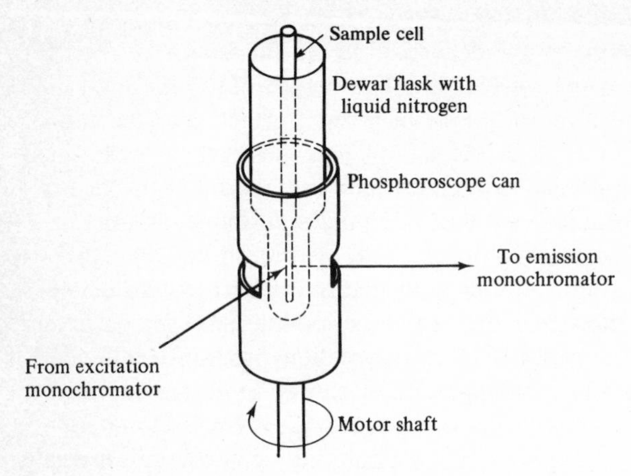

FIGURE 8-13 Schematic diagram of a rotating-can phosphoroscope. (*Reprinted from Ref.* 5, *p.* 376, *by courtesy of Marcel Dekker, Inc.*)

vs. wavelength, and this graph can then be used as the basis for adjusting each of the potentiometers connected to the tapped potentiometer.

8-2D Spectrophosphorimeters†

A spectrophosphorimeter is identical to a spectrofluorometer except that a phosphorescence instrument must be equipped with (1) a rotating-shutter device (commonly called a *phosphoroscope*) and (2) a sample system which allows the sample to be monitored at liquid-nitrogen temperatures.

PHOSPHOROSCOPES

A rotating shutter, or phosphoroscope, was schematically illustrated in Fig. 8-8. There are two major types in use. The rotating-can phosphoroscope (Fig. 8-13), used in the Aminco spectrophosphorimeter consists of a hollow cylinder with one or more slits equally spaced in the circumference. As the can is turned by a variable-speed motor, the sample is first illuminated and then darkened; while it is dark, its phosphorescence is allowed to pass onto the emission monochromator and be measured. The *resolution time* is the length of time between the cutoff of each excitation pulse and the time when the phosphorescence enters the monochromator, and this, in turn, is a function of the motor speed, the size and spacing of the cuts, and the relative radial positions of the cuts to each other.

The other major type of phosphoroscope, called the *Becquerel* or *rotating-disk* type, is illustrated in Fig. 8-14. This type is used in several luminescence instruments giving corrected spectra because it is more versatile

† Filter phosphorimeters are commercially available only as accessory attachments for filter fluorometers.

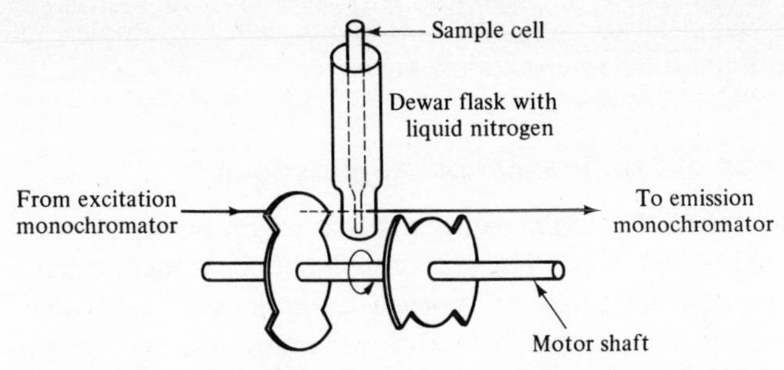

FIGURE 8-14 Schematic diagram of a Becquerel rotating-disk phosphoroscope. (*Reprinted from Ref.* 5, *p.* 378, *by courtesy of Marcel Dekker, Inc.*)

than the rotating-can type. It consists of two disks with notches cut at equal intervals and mounted on a common axis turned by a motor. Once again the excitation is made to occur out of phase with the emission measurement. A greater number of openings can be cut in the disks than in the can, and therefore shorter decay times can be measured with the rotating-disk phosphoroscope. Typically, the shortest decay time that can be measured with a rotating can is about 1 ms, whereas appreciably shorter decay times can be measured with the rotating-disk phosphoroscope.

SAMPLING SYSTEMS

In most phosphorimetric studies, samples are placed in small quartz tubes (1 mm ID by 3 mm OD), which are then placed in liquid nitrogen (77 K at 1 atm) and held in a quartz Dewar flask.

Of all possible coolants, e.g., liquid nitrogen, air, oxygen, rare gases, nitrous oxide, and other liquids or solids, only liquid nitrogen is sufficiently pure to be transparent and nonluminescent at all wavelengths between 200 to 800 nm, as well as being safe, convenient, and inexpensive.

The most common solvent in use is EPA, which freezes to give a clear, rigid glass.† However, even with this solvent, some "snow" formation is inevitable, and a *rotating* sample cell has been developed which minimizes this problem [6]. Rotation also significantly reduces sample-positioning errors, one of the major contributing factors to poor precision in phosphorimetry. This rotating sample cell increases the precision in phosphorimetric measurements by a factor of 10 or more.

8-3 APPLICATIONS

Some applications of fluorometry and phosphorimetry have already been mentioned. Fluorometric analysis and, to a lesser extent, phosphorimetric analysis have now come of age, as evidenced by the fact that hundreds of publications in these fields appear each year. This section outlines the general

† EPA is a 5:5:2 mixture of ethyl ether, isopentane, and ethanol.

scope of luminescence applications and then presents select examples of specific fluorometric and phosphorimetric applications.

8-3A General Scope of Luminescence Applications

The types of compounds to which luminescence analysis might be expected to be applied were discussed in Sec. 8-1*B*. The critical points to be discussed here include a comparison of the kind of *sensitivity* and *selectivity* that luminescence methods offer in comparison to absorptiometric methods and a *comparison* of relative advantages and disadvantages of fluorometry and phosphorimetry.

SENSITIVITY

Luminescence techniques in many cases are considerably more sensitive than absorptiometric methods, mainly because in absorptiometry the detector must distinguish a small difference between two large signals, each of which has appreciable detector noise superimposed. On the other hand, in fluorometry and phosphorimetry, the detector need only detect a small positive signal against a black background or, more properly, against a very small background noise. The sensitivity of luminescence methods is often 10 to 10^3 times greater than the sensitivity of absorptiometric methods.

SELECTIVITY

In general, luminescence methods are much more specific and selective than absorptiometric methods. On a probability basis it is clear why this is true. There are many molecules that will *absorb* but not *reemit* radiation. Thus, there are inherently fewer interferences when a luminescence technique is used.

In addition, another parameter can be used to enhance selectivity. In luminescence measurements both the wavelength of excitation and the wavelength of emission can be selected to minimize interferences, whereas in absorptiometry only the wavelength for absorption is selectable. Therefore, if two luminescent materials differ in *either* their excitation spectra or their emission spectra, it may be feasible to analyze for one selectively in the presence of the other (see Exercise 8-12).

COMPARISON OF FLUOROMETRY AND PHOSPHORIMETRY

Since phosphorimetry is somewhat more complicated experimentally and the equipment has been commercially available a shorter time, far fewer applications have been worked out for phosphorimetry than for fluorometry. Nonetheless, the potential of phosphorimetry appears to be great, and it is clear that the two methods are complementary.

A number of comparative analytical studies have been carried out, including a series of phenyl-substituted silanes, the analysis of anticoagulants and sulfa drugs in blood, and aspirin and salicylic acid in blood serum and plasma. In each of these cases some of the compounds are better analyzed by phosphorimetry and others by fluorometry [5, 6].

When samples are complex and interferences are likely, phosphorimetry often proves more selective than fluorometry because phosphorescence decay times offer an additional parameter for differentiation. In other words, after cutoff of the exciting radiation the phosphorescence will decay exponentially with time, and there are at least two ways of differentiating two molecules with different decay times. One way is to vary the phosphoroscope (shutter) speed. If a sufficient difference in decay times exists, it is possible to slow the shutter to the point where only the slower-decaying species is emitting, whereas a very fast shutter speed will measure the emission of both species. This method is known as *phosphoroscopic resolution*. A second way of differentiating two different phosphorescent species is to stop the phosphoroscope after the sample has been excited and measure the emission signal as a function of time. A plot of the logarithm of the phosphorescence signal vs. time after termination of excitation will be a composite first-order decay curve analogous to a decay curve for a mixture of radioisotopes. If a sufficient difference in decay times exists, the curve can be resolved into its component parts. This method is called *time-resolved phosphorimetry* [5]. In both methods it is desirable for the decay times to differ by a factor of approximately 10, but decay times differing by less than a factor of 5 have been resolved [5].

Finally, the sensitivity of phosphorescence methods is often appreciably greater than for fluorescence methods, for two main reasons. First, wider slit widths and bandwidths are possible in phosphorimetry than in fluorometry because scattering problems are essentially nonexistent in phosphorimetry but relatively severe in fluorometry. Here scattering includes the light scattered from optical surfaces or dust in the instrument as well as from small particles that may be present in the sample solution.

Second, quantum efficiencies are often much larger for phosphorescence than for fluorescence. Much of this improvement is of course due to the fact that phosphorimetry is conventionally carried out at 77 K whereas fluorometry is conventionally carried out at room temperature (300 K). Furthermore, the low temperatures and rigid media in phosphorimetry make quenching less of a problem than in fluorometry.

In summary, whenever fluorometry and phosphorimetry offer comparable sensitivity, fluorometry is probably the method of choice since it is experimentally less complicated. On the other hand, phosphorimetry may be the method of choice when fluorometry is insufficiently sensitive or selective.

8-3B Some Fluorometric Applications

Since fluorometry is such a well-established analytical tool, a review article, e.g., Ref. 10, should be consulted. This section will list a few specific examples, classified according to the type of analysis.

ORGANIC ANALYSIS

It is feasible to conduct a fluorometric analysis for a great many aromatic compounds (see Sec. 8-1*B*), and advantage of this fact has been taken in the analysis of aromatic air pollutants. Samples include cigarette smoke, air-pollutant concentrates, and automobile exhausts. As a specific example, *benzopyrene* has been analyzed in the nanogram range, a sensitivity which is comparable to gas-liquid chromatography with electron-capture detectors, while surpassing it in giving definitive information for qualitative identification. Whereas gas chromatography has only a retention time on which to base an identification, fluorometry offers both an emission and an excitation spectrum for identification purposes [7].

INORGANIC ANALYSIS

In general, inorganic ions do not exhibit fluorescence.† However, a great many inorganic ions are capable of forming with aromatic organic molecules chelate complexes which may be highly fluorescent, even when the organic molecule itself does not exhibit fluorescence (see Sec. 8-1*B*). This has provided the basis for very sensitive analyses of many elements including most of the transition elements. In addition, some anions such as fluoride or cyanide have been analyzed based on their ability to quench the fluorescence of a chelate [7]. Often fluorometric titrations can be carried out with extremely simple equipment, such as a mercury lamp equipped with a black-glass envelope and a darkened room or a dark box [2, 8].

BIOCHEMICAL AND BIOMEDICAL ANALYSIS

Fluorometry is widely used in biochemical analysis, in biomedical research, and in routine clinical analysis. The main reasons for this are the relatively high sensitivity and, more important, the relatively high specificity that fluorometry has for many compounds of biological interest. Thus, even though biological fluids are extremely complex, many substances can be analyzed without tedious isolation procedures. In addition, fluorometry provides a unique tool for the investigation of cellular processes at the molecular level, e.g., in studies on the interaction of biologically active materials with nucleic acids and in studies on the structure and function of proteins.

Numerous reviews of this field have been published [7, 10, 11], and a chapter in Ref. 8 gives some detailed clinical fluorometric procedures. Only a few examples will be mentioned here.

Amino acids and proteins can be studied fluorometrically, although the only free amino acids capable of luminescence are the aromatic amino acids tryptophan, tyrosine, and phenylalanine. Two other possibilities for luminescence analysis exist, however. Some proteins contain a fluorescent coenzyme, such as reduced nicotinamide adenine dinucleotide (NAD) or py-

† Uranium salts are an exception.

ridoxal phosphate, and these chromophores are useful in the fluorescence analysis of many proteins. The other possibility is to label a protein synthetically with a fluorescent chromophore, and this has been done in various research studies of protein structure.

Pyrimidines, purines, and nucleic acids are another class of biological compound studied fluorometrically, though most have rather weak fluorescence unless the pH and temperature are optimum. Acridine dyes and carcinogens bind to nucleic acids and often introduce significant fluorescence, which has been studied.

Fluorometry has become a valuable tool for the qualitative and quantitative analysis of enzymes and for studing enzyme kinetics and mechanisms. In many of these studies the high sensitivity and selectivity of fluorometry makes it possible to carry out in situ in tissues without separation or isolation.

PHARMACEUTICAL CHEMICAL ANALYSIS

There is a great need for analytical procedures for determining low concentrations of various drugs, and in a number of cases fluorometry has proved very effective. For example, the tranquilizer reserpine exhibits fluorescence, although it is usually oxidized first to give an even more highly fluorescent product. Similarly, lysergic acid diethylamide (LSD) exhibits intense fluorescence, and as little as 0.05 μg can be detected as a fluorescent spot on thin-layer chromatograms, and as little as 0.003 μg can be extracted from biological materials and identified fluorometrically [7].

AGRICULTURAL CHEMICAL ANALYSIS

Among the agricultural chemicals successfully analyzed by fluorometry are organothiophosphorous pesticides (using thin-layer chromatography), insecticidal carbamates, and methylenedioxyphenyl synergists, which are used to enhance the activity of insecticides. Trace amounts of amprolium and thiobendazole in feeds have likewise been analyzed by fluorometry.

8-3C Some Phosphorimetric Applications

As in fluorometry, the greatest number of applications in phosphorimetry are undoubtedly in biology and medicine, but phosphorimetric methods are only beginning to be accepted for routine analysis.

Of the constituents normally found in blood, only tryptophan shows significant phosphorescence. This is probably fortunate, because it makes possible the measurement of trace concentrations of a variety of drugs in blood against a very low phosphorescence background. Some drugs and analgesics analyzed phosphorimetrically in blood with great sensitivity include sulfa drugs, aspirin, phenacetin, and caffeine. By extraction procedures low concentrations of procaine, cocaine, phenobarbital, and chlorpromazine in blood serum and cocaine and atropine in urine have been determined [4]. The analysis of carcinogenic hydrocarbons and a variety of antimetabolites are other examples of phosphorimetric procedures appearing to have great promise [6].

EXERCISES

8-1 (*a*) Distinguish fluorescence from phosphorescence in terms of the mechanism by which an excited molecule is deactivated. (*b*) How would you distinguish the two experimentally?

8-2 (*a*) What is the ground-state degeneracy of a free-radical molecule with one unpaired electron? (*b*) What is the next higher spin state of a free-radical molecule; i.e., what is the degeneracy if another electron becomes unpaired?

Ans: (*a*) Twofold degeneracy, or doublet state; (*b*) fourfold degeneracy, or quartet state.

8-3 Summarize (or, where necessary, speculate) how each of the following environmental conditions affects the intensity or wavelength of luminescence: (*a*) temperature; (*b*) pH; (*c*) presence of oxygen; (*d*) presence of metal ions; (*e*) solvent polarity.

Ans: All are spelled out at various places in the chapter except (*e*), a hint on which can be found in discussion of solvent effects in Sec. 2-1*B*; or see Ref. 5, p. 76.

8-4 Speculate whether each of the following compounds will exhibit luminescence, and (where applicable), indicate approximately in what wavelength region it might be observed and/or what special conditions might be necessary: (*a*) acetone, (*b*) benzene, (*c*) glyceraldehyde, (*d*) naphthalene, (*e*) phenolphthalein, (*f*) anthracene, (*g*) phenol, (*h*) aniline, (*i*) aniline hydrochloride, (*j*) pyridine, and (*k*) dihydroxybenzene.

Ans: (*a*),(*c*),(*e*),(*i*), and (*j*) not expected to be luminescent; (*k*) would require a metal ion for chelation to become fluorescent; the others would exhibit luminescence starting from the ultraviolet and working through the visible in the order of (*b*),(*g*),(*h*),(*d*), and (*f*).

8-5 It was stated in the text that the proportionality constant k in Eq. (8-4) depends on the efficiency of the detector system and the geometry of the instrumental arrangement. Elaborate on these two factors and explain what specific steps could be taken to increase the value of k.

8-6 Compare the relative sensitivity of fluorometry and phosphorimetry. If you use Figs. 8-4 and 8-5 as a basis for comparison, be sure that the concentration units are made the same. (Is it valid to answer this question using Figs. 8-4 and 8-5? Why or why not?)

8-7 What causes a fluorescence emission spectrum to be the approximate mirror image of an excitation spectrum? Cite reasons, both molecular and instrumental, why the fluorescence emission spectrum is often different from the mirror image of the excitation spectrum.

8-8 (*a*) Is it possible to do a quantitative analysis using phosphorescence emission with a simple fluorometer? (*b*) What, if any, advantages would there be to causing a time delay between excitation and observation of luminescence?

Ans: (*a*) Theoretically, yes; however, few samples will phosphoresce unless frozen. (*b*) Light scattering by turbid samples is eliminated, and in certain cases the phosphorescence measurement will be more selective (fewer interferences).

8-9 Why is true double-beam operation (alternately observing sample and reference) feasible for a fluorometer but not for a spectrofluorometer?

8-10 Suppose that a double-beam spectrophotometer is converted into a fluorescence spectrophotometer with suitable attachments. What would be wrong or what would be the *value* of using a solvent blank in the reference beam in determining a fluorescence spectrum, analogous to the way an absorption spectrum is measured with a double-beam instrument?

Ans: There would be little or no value, assuming of course that the solvent is chosen to be nonfluorescent (a prime requirement of a good solvent). A nonfluorescent reference would give zero reference signal, and therefore would be of little, if any, value.

8-11 Many luminescence methods of analysis are much more sensitive than comparable absorptiometric methods. Considering that the quantum efficiency for luminescence emission can be no higher, and in fact is usually lower, than the quantum efficiency for absorption, explain how luminescence methods can be more sensitive.

8-12 Suppose that two luminescent materials fluoresce at essentially the same wavelength whereas their excitation spectra show absorption bands with maxima that are partially separated (the bands significantly overlap). Suggest a spectrofluorometric procedure that might be capable of quantitatively analyzing for each of the two materials without performing a chemical separation.

REFERENCES

INTRODUCTORY

1 EWING, G. W.: "Instrumental Methods of Chemical Analysis," 3d ed., chap. 4, McGraw-Hill, New York, 1969. A clear, brief introduction.

2 WILLARD, H. H., L. L. MERRITT, JR., and J. A. DEAN: "Instrumental Methods of Analysis," 4th ed., chap. 13, Van Nostrand, New York, 1965. Moderately detailed.

3 CONRAD, A. L.: Fluorometry, chap. 59 in I. M. Kolthoff and P. J. Elving (eds.), "Treatise on Analytical Chemistry," pt. I, vol. 5, Wiley-Interscience, New York, 1964. A capsule summary of fluorometry only.

INTERMEDIATE

4 HERCULES, D. M. (ed.); "Fluorescence and Phosphorescence Analysis," Wiley-Interscience, New York, 1966. A lucid monograph on the most important aspects of fluorometry and phosphorimetry.

5 GUILBAULT, G. G. (ed.): "Fluorescence," Dekker, New York, 1967. Covers most aspects of fluorometry and phosphorimetry.

6 MCCARTHY, W. J.: Phosphorescence Spectrometry, chap. 8 in J. D. Winefordner (ed.), "Spectrochemical Methods of Analysis," Wiley-Interscience, New York, 1971. A well-written summary of recent advances in phosphorimetry.

7 VAN DUUREN, B. L., and T. L. CHAN: Fluorescence Spectrometry, chap. 7 in J. D. Winefordner (ed.), "Spectrochemical Methods of Analysis," Wiley-Interscience, New York, 1971. Intermediate-level treatment of fluorometry.

8 WHITE, C. E., and R. J. ARGAUER: "Fluroescence Analysis," Dekker, New York, 1970. Covers practical aspects.

9 LOTT, P. F.: Instrumentation for Fluorometry, *J. Chem. Educ.*, **41**(5):A327 (1964); **41**(6):A421 (1964). Covers commercially available fluorometers and spectrofluorometers.

ADVANCED

10 WHITE, C. E., and A. WEISSLER: Fluorometric Analysis, *Anal. Chem.*, **44**(5):182R (1972). A thorough review of fluorometry and phosphorimetry; one of a series published biannually.

11 KONEV, S. V.: "Fluorescence and Phosphorescence of Proteins and Nucleic Acids," trans. from Russian by S. Udenfriend, Plenum, New York, 1967. Excellent.

12 PRINGSHEIM, P.: "Fluorescence and Phosphorescence," Wiley-Interscience, New York, 1965. Very detailed.

13 BECKER, R. S.: Fluorescence and Phosphorescence Spectroscopy, chap. 4 in W. West (ed.), "Chemical Applications of Spectroscopy," Wiley-Interscience, New York, 1968. Molecular and quantum aspects.

9
Refractometry and Interferometry

One of the simplest optical methods of analysis is based on measurement of the refractive index n. In *refractometry* a direct measurement of the refractive index of a sample is made, whereas in *interferometry* the difference in refractive index between a standard and a sample is measured. Interferometry is a little more complicated than refractometry, but since it is a differential method, it is capable of much higher precision and can be used to monitor extremely small differences in refractive index between standard and sample.

The refractive index can be used to characterize a substance in the same way that melting points and boiling points are used. Although the refractive index by itself is insufficient to characterize a substance, few substances have identical refractive indexes at a given temperature and wavelength, and thus in conjunction with other data it can be extremely useful for confirming identity and purity. Other applications include analyzing process streams and mixtures; estimating properties of polymers such as molecular weight, size, and shape; and calculating physical properties such as reflectivity and optical dispersion. Measurements of refractive index are nondestructive, require only small samples, and can be made simply and rapidly.

Whereas refractometry is used widely for liquids and occasionally for solids, interferometry is restricted to gases and occasionally liquids. The emphasis in this chapter will be on refractometry, with only a brief treatment of interferometry.

9-1 THEORY

In this section refractometry will be dealt with first, followed by interferometry. The discussions of reflection, refraction, diffraction, dispersion, and interference in Chap. 1 form the basis for the discussion to follow.

9-1A Refractometry

Whenever a ray of light passes obliquely from one medium into another of different density, its *direction* is changed; this is called *refraction*. If the second medium is optically denser than the first, the ray will become more nearly perpendicular to the surface, as shown in Fig. 9-1. The primary cause for the change in direction of the light ray on entering a new medium is a change in *velocity* v of the light in the two media. Experiments have shown, for example, that yellow light from a sodium lamp slows down from 3.0×10^{10} cm/s in a vacuum to 2.25×10^{10} cm/s upon entering water; in othe words, it is traveling only about 75 percent as fast in water. The greater the density of a substance, the lower the speed of light in that substance.

The *index of refraction n* of a substance is defined as the ratio of the sines of the angles of *incidence* and *refraction*; the sines of these angles in turn are directly proportional to the velocities of the light in the two media, as given by

$$n = \frac{\sin i}{\sin r} = \frac{v_m}{v_M} \tag{9-1}$$

The chief variables affecting the measuring of n are wavelength and temperature, which should therefore be specified. If the incident ray passes from a dense to a less dense medium, n will be less than 1. More commonly, however, the refractive index is measured as light passes from air to a denser medium, and thus n is usually a number greater than 1. For example, the *refractive* index of water (air-water path) is about 1.33; for crown glass, $n \approx 1.50$, and for carbon disulfide, $n \approx 1.63$, all measured with the D lines of sodium at room temperature. In general the values of n for organic liquids range from about 1.3 to 1.7. Some other values of n are 1.75 for dense flint glass, 2.42 for diamond, and 2.62 to 2.90 for titanium dioxide or synthetic rutile. The refractive index of some other common substances is given in Table 1-2. The refractive index based on a vacuum as a reference medium would be only about 0.03 percent higher than that based on air.

The slowing down of a light beam upon entering a denser medium is caused by an interaction of the electromagnetic radiation with the electron clouds present in the sample and consequently is a function of the number of electrons present and their state of binding. Therefore it is not surprising that there is a close parallel between refractive index, dielectric constant, and density. In a homologous series of nonpolar organic compounds, for example, the refractive index increases fairly regularly with increase in length of the

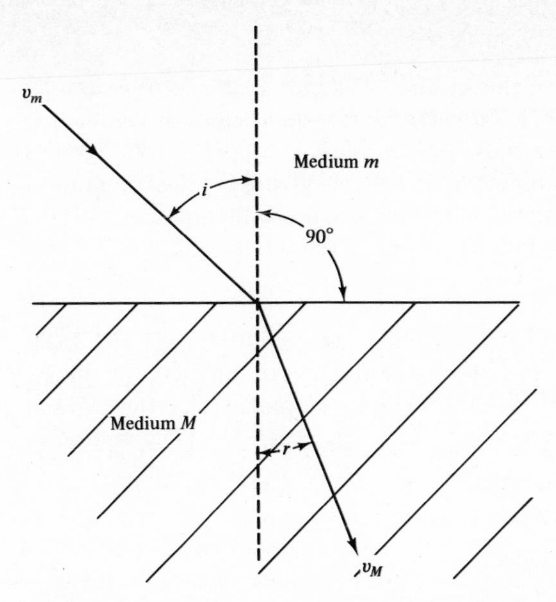

FIGURE 9-1 Refraction of a light ray as it passes from one medium m to a denser medium M. The angle of incidence i, the angle of refraction r, and the velocity v are indicated.

carbon chain, and this trend is paralleled by the increases in density and dielectric constant [9].

If a light beam enters a gaseous sample, it encounters mostly empty space, and therefore its velocity is but slightly affected. If the same sample is condensed into the liquid state, the velocity of the light beam is slowed down in direct proportion to the increased density of the medium. If the observed refractive index were adjusted to take into account the density of the medium, the result would be a measure of the electronic properties of the molecules themselves and relatively independent of temperature, pressure, and state of aggregation. One of the best ways of taking the density of a substance into account is by the Lorentz-Lorenz *specific-refraction* equation

$$r = \frac{n^2 - 1}{n^2 + 2}\frac{1}{d} \tag{9-2}$$

where r = specific refraction
n = observed refractive index
d = density of medium

Since the refractive index is a dimensionless quantity, the specific refraction has units of reciprocal density. The specific refraction is a valuable means of identifying a substance and can serve as a criterion of its purity.

The particular atoms and molecules that make up the medium can be taken into account through the *molar refraction* R, given by

$$R = rM \tag{9-3}$$

where M is the molecular weight of the substance and r is the specific refraction as defined in Eq. (9-2). Note that R has dimensions of volume per mole. The molar refraction is a more or less additive property of the groups or elements comprising the compound, and tables giving the contribution of various atoms to the molar refraction are available in the literature, e.g., Refs. 8 and 9. An abridged set of values of "atomic" refractions is given in Table 9-1.

Example 9-1 Acetic acid has a refractive index at 20°C (n_D^{20}) of 1.3698. From this calculate (*a*) the specific refraction r and (*b*) the molar refraction R. (*c*) Use Table 9-1 to calculate the molar refraction R, and compare it with the observed value calculated in part (*b*). (The molecular weight of acetic acid is 60.0, and the density at 20° is 1.049 g/cm³.)

Answer (*a*) From Eq. (9-2),

$$r = \frac{n^2 - 1}{n^2 + 2}\frac{1}{d} = \frac{1.3698^2 - 1}{1.3698^2 + 2}\frac{1}{1.049} = 0.2154 \text{ cm}^3/\text{g}$$

(*b*) From Eq. (9-3),

$$R = rM = (0.2154)(60.0) = 12.93 \text{ cm}^3/\text{mol}$$

(c) From the formula of acetic acid,

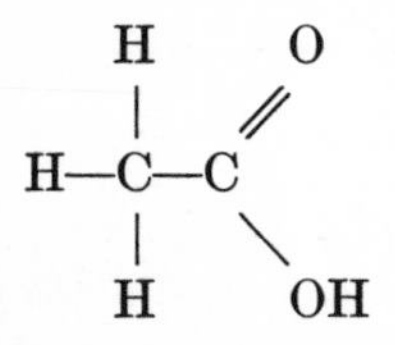

we can tabulate the following molar-refraction contributions (see Table 9-1):

4 hydrogens	4 × 1.100 =	4.400
2 carbons	2 × 2.418 =	4.836
1 carbonyl oxygen	=	2.211
1 hydroxyl oxygen	=	1.525
Total		12.972

Note that the molar refraction estimated from the atomic contributions (12.97) agrees very well with the observed molar refraction (12.93).

In many cases, even small differences between the observed molar refraction and the calculated value (assuming additivity of the various atomic contributions) allow important conclusions to be drawn concerning in-

TABLE 9-1 The Contribution of Various Atoms to Molar Refraction at Room Temperature†

Atom	Group	R, cm³/mol (sodium D line)
H		1.100
C	Single bond	2.418
	Double bond	1.733
	Triple bond	2.398
O	Hydroxyl	1.525
	Ether, ester	1.643
	Carbonyl	2.211
F		1.2
Cl		5.967
Br		8.748
I		13.900
N	Primary amine	2.322
	Secondary amine	2.502
	Tertiary amine	2.840

† Data from Ref. 6, p. 397, and Ref. 9, p. 1171.

termolecular and intramolecular forces and the electronic structure of molecules. A number of examples may be found in Ref. 9.

Example 9-2 Methyl formate, $H{-}\overset{\overset{O}{\|}}{C}{-}OCH_3$, has the same empirical formula as acetic acid. Using Table 9-1, estimate the molar refraction of methyl formate, and compare it with the molar refraction of acetic acid calculated in Example 9-1.

Answer

4 hydrogens	4 × 1.100 =	4.400
2 carbons	2 × 2.418 =	4.836
1 carbonyl oxygen	=	2.211
1 ester oxygen	=	1.643
Total		13.090

From Example 9-1 $R_{HAc} = 12.97$, and therefore,

$$\% \text{ difference} = \frac{13.09 - 12.97}{12.97} \times 100 = 0.93\%$$

Even though the percentage difference is only about 1 percent, the precision of ordinary refractive-index measurements makes it very easy to observe such differences in molar refraction.

9-1B Interferometry

In interferometry, the difference in refractive index between a sample and a standard is measured using the phenomenon of *interference* (see Sec. 1-4*D*).

The basic principles of an interferometer can be explained with Fig. 9-2. Parallel monochromatic light passes through two small slits, S_1 and S_2, and *diffraction* occurs at each opening (see Sec. 1-4*E*). In the *absence* of the sample cell (C), there will be a point (O) where the two beams will arrive in phase (path S_1O is exactly equal to path S_2O) and a bright line will result. In addition to the bright line at point O, there will be bright lines at points Q_1 and Q_2, where the optical paths differ by exactly one wavelength, e.g., distances $S_2Q_1 - S_1Q_1 = 1\lambda$. Thus, *constructive interference* will take place at points Q_1, O, and Q_2, as well as other points further from O where the optical paths differ by other integral values of wavelength (2λ, 3λ, etc.). In between those points, *destructive interference* will take place, and dark zones will result. P_1 and P_2 represent points where the two beams differ in length by exactly half a wavelength, and dark *lines* appear because of complete cancellation.

Now suppose that a substance with a slightly greater refractive index than air is placed at C in Fig. 9-2. Since the velocity of light through C will be decreased, this has the effect of increasing the optical length of the beam S_2O

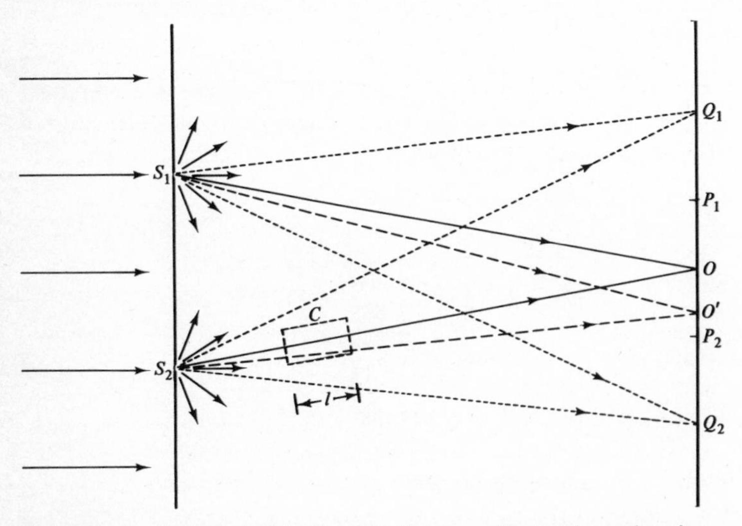

FIGURE 9-2 Optical principle of an interferometer. See text for explanation.

by an amount Δl, defined as

$$\Delta l = l(n - n_0) \tag{9-4}$$

where l = length of cell C
n = refractive index of substance C
n_0 = refractive index of the original medium (air)
Thus, the two beams will no longer arrive in phase at point O but at some other point O' which is now optically equidistant from S_1 and S_2. If we let N be the number of fringes (each made up of a dark and a light band) that appear between points O and O' at a given wavelength of light λ, then N is given by

$$N = \frac{\Delta l}{\lambda} \tag{9-5}$$

or

$$N = \frac{l(n - n_0)}{\lambda} \tag{9-6}$$

With monochromatic light it is difficult to determine how many interference fringes are involved in the shift, since all bands look alike. To avoid this difficulty, white light is used instead of monochromatic light. Now the central band is the only one which is pure white. Bands on either side of the central band are fringed with blue toward the center of the system and red toward the outer edge. The next adjacent bands are even more highly fringed, and after six or seven bands, the diffusion is so great that the entire field is uniformly white. Thus, with substance C in the path of one of the beams, it is a simple matter to count the number of bands N by which the central band has been shifted. If the cell length l, the wavelength λ, and the refractive index of the reference media n_0 are known, the refractive index of the sample n can be calculated from Eq. (9-6). In practice, however, an interferometer is rarely used for making absolute measurements of the refractive index n. Instead, it is more often used for measuring changes or *differences* in refractive indexes, which in turn may be related to the concentration of gases and solutions. In commercial interferometers, to be discussed in the next section, matched cells are put in each of two beams, one cell being the reference and the other being the sample. To avoid counting bands or fringes, a fixed band pattern is set up below the desired pattern, and a variable glass compensator is pivoted in the sample path until the two patterns are in alignment. The value of N is obtained from a reading of the compensator micrometer screw. Further details are given in the next section.

9-2 INSTRUMENTATION

In this section some general principles of instrument design will be presented and illustrated with commercial instruments. A variety of different instruments are reviewed elsewhere [5, 9].

9-2A Importance of Wavelength and Temperature Control

In refractometric measurements the precision achieved depends on the degree with which the wavelength and temperature are controlled. In interferometric measurements these variables are less critical but still important.

WAVELENGTH

The refractive index of a substance changes with wavelength, and usually the variations are nonlinear (see Sec. 1-4*A*). Because of this, monochromatic radiation must be used for precise measurements of refractive index.

Refractive-index measurements are usually made in the visible region, both because visible sources are more readily available and because most substances are colorless and do not absorb in the visible region. If a substance absorbed strongly, not only would it be more difficult to detect the resulting signal but *anomalous dispersion* would occur (see Sec. 1-4*F*). The most widely used source is the sodium discharge lamp, which has an intense yellow doublet (the D line) at 589.0 and 589.6 nm. The sodium lamp is simple to use since no filters are required for most work. Mercury, mercury-cadmium, and hydrogen lamps are sometimes used, all of which require filters.

In certain instruments white-light sources are used. In interferometers, white light is an advantage, since the colored fringes are easier to identify, as described in Sec. 9-1*B*. In certain simple refractometers, e.g., the Jelley-Fisher refractometer, white light is tolerated since refractive indexes are estimated only to three decimal places. In many other fairly precise refractometers, e.g., Abbe, Pulfrich, and immersion refractometers, white-light sources are used, but the white light can be tolerated because these refractometers are equipped with *compensating,* or *Amici prisms,* which in effect "sort out" yellow light and get rid of other wavelengths. Amici prisms will be discussed in the next section. In the most precise refractometers, e.g., the *precision Abbe refractometer*, monochromatic light must be used.

TEMPERATURE

One of the disadvantages of refractometric analysis is the need to control the temperature carefully. The majority of liquids show a decrease in refractive index of about 0.00045 unit per Celsius degree rise. Water is somewhat less sensitive, decreasing about 0.0001 unit per degree rise. Solids are still less sensitive, changing roughly 0.00001 unit per degree. In general, the decrease in refractive index with increasing temperature is due both to a decrease in density and a decrease in dielectric constant of the medium.

For a refractive-index measurement to be reliable in the fourth decimal place, which is about the precision normally sought, liquid samples must be thermostated to at least ±0.2°C. To secure a precision of ±0.000001, the temperature must be regulated to ±0.01°C. Thermostating is ordinarily accomplished by circulating fluid from a constant-temperature bath through a jacket surrounding the sample.

One of the advantages of interferometry is that temperature regulation is much less critical than for refractometry, as long as the two cells used for the differential measurement are at the same temperature. By using very simple temperature control, e.g., an air jacket or pumping water thermostated to about ±0.5°C through a jacket containing both cells, it is possible to determine the difference in refractive index of two solutions to about 0.0000001.

9-2B Refractometers

The principles of the two main types of refractometers, *critical-angle refractometers* and *image-displacement refractometers*, will be described.

CRITICAL-ANGLE REFRACTOMETERS

With an instrument of this class the observer sees a field through a telescopic eyepiece which is divided into two portions, a light field and a dark field. The sharp boundary between the light and dark fields is the *critical ray*. The concept of a critical ray can be understood by reference to Fig. 9-3, which shows the prism arrangement in a refractometer. Light passes through a thin (about 0.1-mm) layer of sample and then enters the *refracting* prism P_2. (P_1 is an *illuminating* prism, the upper surface of which is rough-ground; the rough surface acts as the source of an infinite number of rays which pass through the liquid sample layer in all directions.) Radiation which just *grazes* the surface of prism P_2 is barely able to enter prism P_2, and upon entering the prism makes an angle ϕ_c with a line drawn perpendicular to the surface of the prism. ϕ_c is called the *critical angle* for this critical ray, and its numerical value depends on the wavelength used and on the refractive indexes of the sample n and refracting prism n_p. The important point is that no ray can make an angle greater than the critical angle ϕ_c, since the source of any such rays cannot enter the prism (the critical ray is the most tangential ray that can enter the prism). Thus, all rays entering prism P_2 more nearly perpendicular to the surface than the critical ray (see the dotted rays on Fig. 9-3) will be refracted to the *right* of the critical ray in Fig. 9-3 and will illuminate the right half of the telescopic eyepiece. The left half of the eyepiece remains dark since no rays are being refracted at an angle greater than ϕ_c.

Since the critical angle ϕ_c depends on the refractive index of the sample

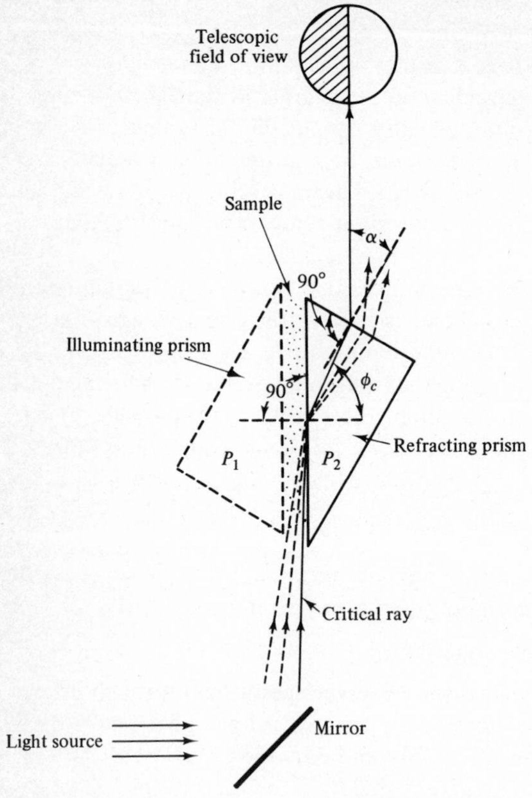

FIGURE 9-3 Approximate prism arrangement and field of view produced in an Abbe refractometer. The sample thickness is greatly exaggerated. See text for explanation.

for a given refracting prism and wavelength of light, the measurement of refractive index with a critical-angle refractometer, like an Abbe, amounts to measuring the critical angle ϕ_c. What is actually measured with an Abbe refractometer is angle α, the angle of emergence of the critical ray, but angle α is the complement of angle ϕ_c (see Exercise 9-5).

Refractometers based on measuring the critical angle far outnumber all other types and include the Abbe, the precision Abbe, the Pulfrich, and the immersion or dipping refractometers. A few more details about the Abbe and precision Abbe refractometers will be given; complete descriptions of the others can be found in Refs. 5, 7, and 9.

The Abbe Refractometer The schematic diagram and picture of a typical Abbe refractometer in Fig. 9-4 give a more accurate representation of the role played by the *illuminating* prism P_1 than that in Fig. 9-3. As can be seen in Fig. 9-4, the light enters the illuminating prism from below, being reflected from the source by a mirror. One to three drops of the liquid sample are placed between prisms P_1 and P_2, which are hinged together at their lower edges for access to the sample. The prisms are partially hollowed to permit circulation of thermostated fluid. The prisms are mounted on an axis at their center so that they can pivot in scanning the upper edge of refracting prism P_2 in the process of measuring the angle α. A movable arm attached to the prisms contains a graduated scale magnifier at the end open to the light. The function of the movable arm and its scale magnifier is to project a graduated scale on the upper edge of refracting prism P_2, and this scale, in turn, is viewed through the permanently mounted telescopic eyepiece (see Fig. 9-4). The standard Abbe scale is graduated directly in n_D units to the nearest 0.001 unit, and the magnifier allows the next decimal place to be estimated to ±0.0001. Indexes are measurable over the range of 1.30 to 1.70. Only a few substances have refractive indexes outside these limits.

Before discussing other types of refractometers, it is in order to discuss the function of the *compensator* prisms shown in Fig. 9-4. This is a combination of prisms, often called an *Amici prism,* which allows white light to be used as the source for the Abbe in routine work. A more detailed view of an Amici compensator is given in Fig. 9-5. The different varieties of glass chosen for an Amici compensator disperse all wavelengths except yellow light in the vicinity of the sodium D line, which is allowed to pass. Thus, these prisms compensate for the use of white light and act as a miniature monochromator.

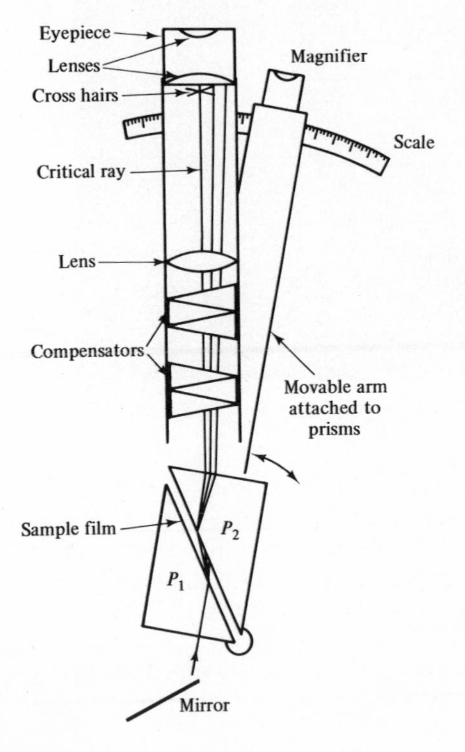

FIGURE 9-4*a* Schematic diagram of an Abbe refractometer. See next page for picture.

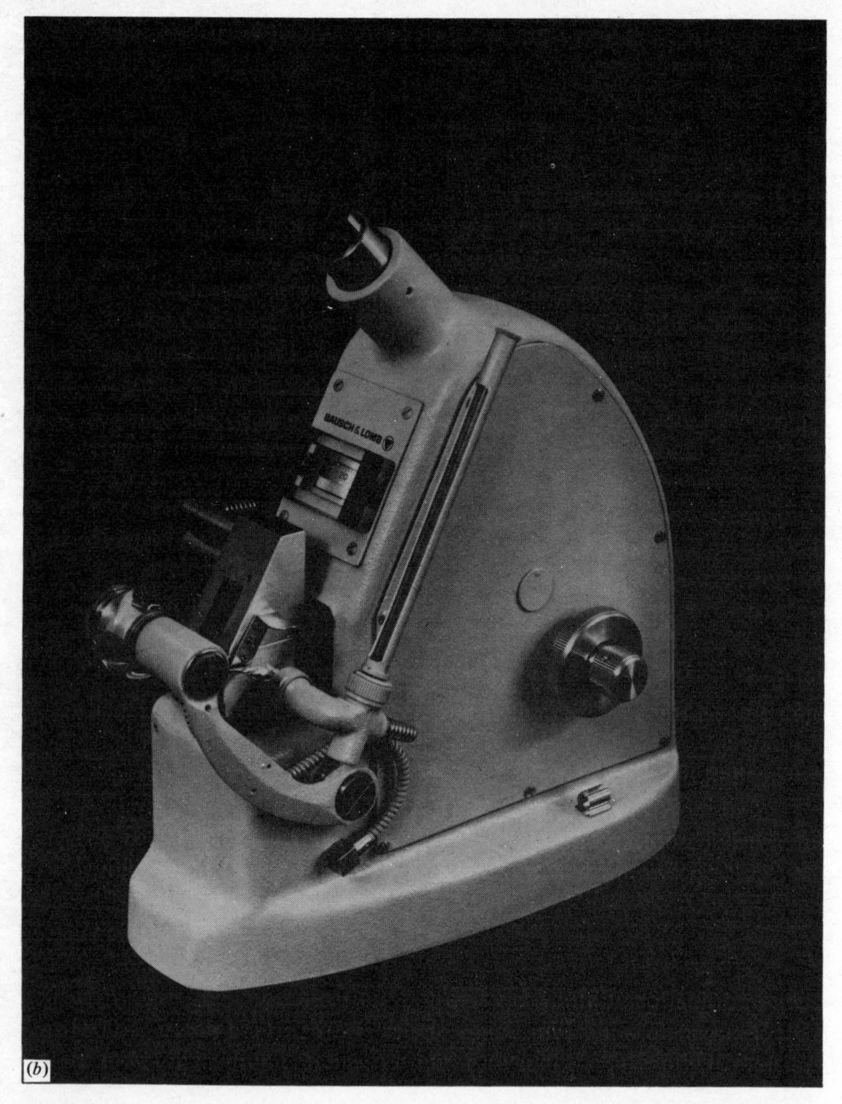

FIGURE 9-4*b* Picture of an Abbe refractometer. (*Analytical Systems Division, Bausch and Lomb.*)

However, the resolution is far from perfect, and the best accuracy possible with white light and an Amici compensator is about ± 0.001 refractive-index unit. Many immersion or dipping refractometers also use Amici prisms.

The Precision Abbe In the precision Abbe the reliability and accuracy have been improved by a factor of about 3 by increasing the size and quality of the refracting prism and using a diffusing prism with a smooth face. In this way a greater proportion of the incident radiation enters at a grazing angle

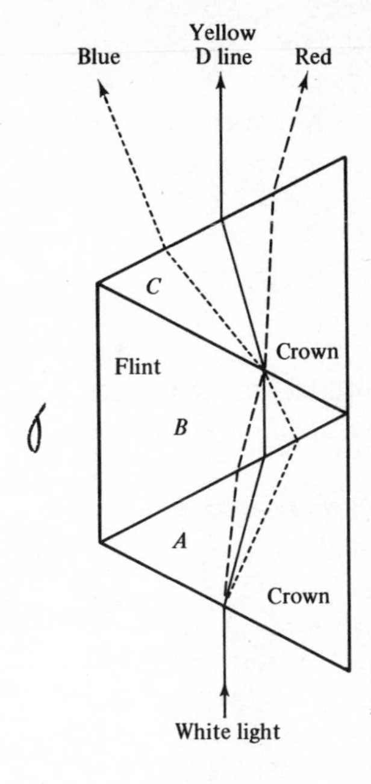

FIGURE 9-5 The dispersion of white light with an Amici compensator. (*After H. A. Strobel, "Chemical Instrumentation," 2d ed., p.* 597, *Addison-Wesley Publishing Company, Reading, Mass.*, 1973, *by permission.*)

giving a sharper boundary line. In addition, the direct-reading scale of the regular Abbe is replaced with a finely graduated scale with a vernier, and a table is supplied to convert the readings into refractive indexes.

IMAGE-DISPLACEMENT REFRACTOMETERS

In an image-displacement refractometer, instead of measuring the displacement of the light-dark boundary due to the critical ray, the actual displacement of the refracted ray is measured with respect to the incident ray. This gives a more direct measurement of the refractive index. In this method, the sample is shaped into the form of a prism, and the angular displacement of light on passing through the sample prism is a measure of the index of refraction. Liquids are poured into a hollow prism-shaped vessel. With the sample acting as the prism in a simple spectroscope, the angular displacement of the slit image gives the refractive index.

The optimum precision and accuracy of an image-displacement spectrometer are about two orders of magnitude better than a regular Abbe refractometer, or about $\pm 1 \times 10^{-6}$ refractive-index unit. Furthermore, it is capable of handling a limitless range of refractive indexes and a much greater range of wavelengths. Since any angle of refraction can be measured, there is no limit on the value of n. With quartz optics, even ultraviolet and near-infrared observations are possible. However, extremely precise temperature control (± 0.002°C for liquids) is required to achieve optimum accuracy, and thus the method is not well suited for routine work. Examples of image-displacement instruments include the simple Jelley-Fisher refractometer (accurate to about ±0.002 refractive-index unit), the Eykman hollow-prism refractometer, and differential refractometers using two hollow prisms set in opposition to each other. These and other instruments are described in detail elsewhere [5, 7, 9].

9-2C Interferometers

The theory of interferometers was discussed in Sec. 9-1*B*, and the importance of wavelength and temperature control was discussed in Sec. 9-2*A*. In this section a typical commercial interferometer, the Zeiss portable gas and water interferometer, will be described briefly. Other commercial interferometers are described in Refs. 7 and 9.

A schematic diagram of the *Zeiss portable gas and water interferometer* is given in Fig. 9-6. The source S is a small 4-V tungsten lamp, furnishing white light. A lens L renders the light parallel and directs it through two glass plates (P_1 and P_2) and then on through two cell compartments, C_1 and C_2. (The function of the glass plates, P_1 and P_2, will be described shortly.) After passing through the cells and rectangular apertures R_1 and R_2, the two light beams strike a mirror M and are reflected back upon themselves, again passing through the cells. Finally, by means of the lens L, the two beams are reunited at O, forming a series of interference fringes.

In addition to these two interfering beams of light, another pair proceed through the instrument identically except that they pass *below* and not through the sample cells and glass plates. These two beams form a second system of interference fringes at O, which serve as reference lines.

Glass plates P_1 and P_2 are used to bring the sample fringes into alignment with the reference fringes, and in the process of making this adjustment a

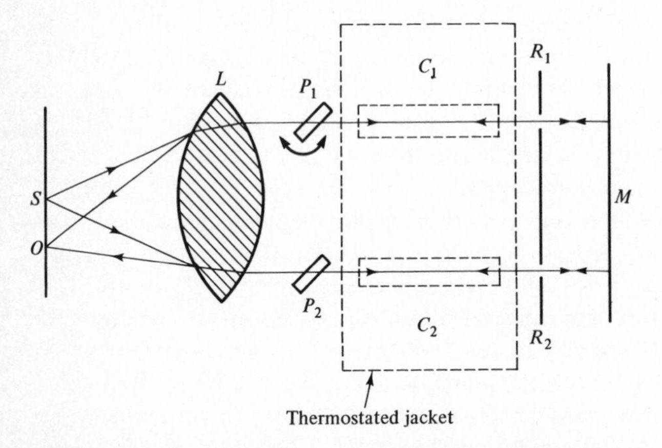

FIGURE 9-6 Schematic diagram of the Zeiss portable gas and water interferometer. See text for explanation. (*Zeiss Optical Measuring Tools and Instruments, Scherr Tumico Company.*)

quantitative measure of the interferometric shift is obtained. This is accomplished in the following way. P_1 and P_2 are nominally arranged in the two beams at an angle of about 45° to the beam. The position of P_2 is fixed, whereas P_1 is attached to a micrometer screw and can be rotated, thus increasing or decreasing its effective thickness. With solvent in cell C_1 and a dilute sample solution in cell C_2, plate P_1 is rotated by means of a micrometer drum until the upper (sample) and lower (reference) sets of interference fringes coincide. The reading on the micrometer drum is recorded and is directly proportional to the difference in refractive indexes between the solution and the pure solvent. These arbitrary readings are usually converted directly into concentration units by means of a calibration curve and are rarely converted into refractive-index units. Gases as well as liquids can be handled in this way.

Cells C_1 and C_2 are enclosed in a jacket through which thermostated air or water can be circulated. As long as the two cells are at exactly the same temperature, the interferometric readings will be precise to $\pm 1 \times 10^{-7}$ refractive-index unit. To use the same calibration curve the temperature should be within ±0.5°C of the temperature for which it was prepared.

9-3 APPLICATIONS

Refractometry and interferometry will be considered together, so that their relative advantages and disadvantages will be clear. Since refractometry gives absolute measurements of refractive indexes whereas interferometry measures small differences in refractive indexes, there will generally be a decided advantage in choosing one method over the other for a specific case.

9-3A Qualitative Analysis

The index of refraction is widely used as a physical constant in the identification of a substance and as a criterion of its purity. Only a few drops of material are needed. Periodic calibrations are desirable and are essential for reliability out to the fourth decimal place and beyond. Three good calibration liquids are water (n_D^{25} = 1.33250), chloroform (n_D^{25} = 1.44293), and benzene (n_D^{25} = 1.49790). Refractive-index data on other substances can be found in a chemistry handbook.

In identifying *liquids* from literature data, only refractive-index differences greater than 0.005 are significant. The lack of accuracy, particularly in older literature, was caused in many cases by insufficient purity or faulty technique, e.g., inadequate temperature control during measurement or calibration errors in the instruments used.

A very convenient microscale method of determining whether an unknown liquid is a *single component* consists of observing the field in an Abbe refractometer as the sample is allowed to evaporate. If the boundary between dark and light fields moves significantly during evaporation, this is evidence that the liquid is a *mixture* with two or more components.

It is generally less convenient to measure the refractive index of *solids* than liquids. Large crystals can be measured with an Abbe or Pulfrich refractometer by swinging the illuminating prism (P_1 in Figs. 9-3 and 9-4) out of the way and placing the sample directly on the refracting prism (P_2 in Figs. 9-3 and 9-4). However, the surface of the sample in contact with the prism must be carefully polished to make good contact.

Alternately, and more commonly, *immersion* methods are employed for solids. One method, extensively used by mineralogists, is to find a liquid having the same index as that of the sample (making the sample appear invisible). The refractive index of the liquid is then taken as the refractive index of the solid. Less tedious techniques involving a series of bracketing liquids have been developed [8], but the accuracy limit is about ± 0.002 unit. Perfectly formed or crushed specimens can be measured.

The *dispersion* of the refractive index is another useful physical characteristic. It is simply the difference in refractive index at two different wavelengths. Mercury-vapor and sodium lamps have been used to measure a term identified as $n_G - n_D$. Besides yielding another piece of confirmatory data, dispersion values have the additional advantage of being about 30 percent more sensitive to changes in composition than a refractive index alone. The petroleum industry has fully exploited this technique, and correlations of dispersion with composition are available.†

While the above applications are all handled most conveniently by refractometry, many of them could be carried out with more precision using interferometric methods. If a liquid standard with a refractive index within about 0.0006 of the sample can be found, it is possible to determine differences in refractive index between the two liquids with an error only in the eighth decimal place. An obvious application of this is in determining the purity of ultrapure liquids.

9-3B Quantitative Analysis

When the identity of a solution is known, refractive-index measurements often offer a simple, fast, and reliable means of quantitative analysis. For many solutions containing a *single* solute the index is found to be a linear function of concentration; more precisely, the following equation holds:

$$n - n_0 = kC \tag{9-7}$$

where n = refractive index of solution
n_0 = refractive index of solvent
C = concentration, generally, g per 100 ml
k = proportionality constant

In practice a calibration curve should be prepared, plotting $n - n_0$ vs. C.

† R. C. Wockher, Quantitative Determination of Aromatic Hydrocarbons by New Method, *Ind. Eng. Chem., Anal. Ed.*, **11**:614 (1939).

If a *binary* system, e.g., two liquids, is to be analyzed, it is sufficient to plot the refractive index of the solution vs. the percent composition (volume percent is most convenient and generally gives better linearity than weight percent). Linearity is seldom observed over a large range of composition (say 0 to 100 percent, or even 0 to 50 percent), but over a small range (say 20 percent) it usually holds. Linearity is most likely to be obtained if the constituents are chemically similar. Some systems show maxima or minima. For example, an ethanol-water mixture goes through a refractive-index maximum at about 79.3 weight percent of ethanol.

Occasionally, even *ternary* systems can be analyzed if refractive-index measurements are made at two different temperatures and if the temperature coefficients of at least two of the components differ appreciably. For example, *n*-butanol–acetone–water mixtures have been analyzed, with accuracies in the order of 2 to 3 percent for the major components.

To analyze *gaseous* mixtures, very high precision is required, since the refractive index varies very little with composition. It is here that interferometric analysis excels.

A few specific examples where quantitative analysis is routinely carried out by refractometry include determining the water content of milk; determining the protein content of aqueous solutions and the albumin content of blood serum; determining the total sugar content of water, food, liquors, beer, sap, and other fluids; determining the sulfur content of rubber; determining the unsaturated-oil content of butter, fats, vegetable oils, and seeds; determining the total aromatic content of petroleum hydrocarbons; determining salinity; and numerous applications in determining the concentration of acids, bases, salts, and organic substances in dilute aqueous solutions. Further details of these and other applications are given in Refs. 7 to 9.

Refractometry (and to some extent interferometry) is finding extensive use in *quality-control applications*, both qualitative and quantitative analysis. Often the measurements are made on-stream, monitoring the composition of process streams continuously. Often the refractive-index measurement is made photoelectrically and continuously recorded, and usually a *differential* refractometer is used to minimize the need for precise temperature control. Details may be found in Refs. 4 and 7.

9-3C Determination of Physical Properties

Refractometry is sometimes used to evaluate certain physical properties that are difficult to measure in any other way. Examples include measuring *diffusion coefficients*, wherever diffusion gives rise to refractive-index gradients in solution; determining the *degree of crystallinity* of polymers (microvoids in a sample preclude the use of density measurements); and calculating *dipole moment* from dielectric constants (a dipole moment depends on the refractive index as well as the dielectric constant, and both must be accurately known at a given temperature [9]).

9-3D Structure Determinations

The molar refractivity R of a compound [defined by Eqs. (9-2) and (9-3); see Sec. 9-1] can be calculated from literature values of atomic refraction. Differences between the calculated and experimental values of R can be interpreted in terms of internal structural complexities and help unravel questions of molecular structure. In conjunction with spectra and other physical measurements, refractive-index data can yield corroborative evidence of the configuration, molecular structure, and electronic binding of a substance.

EXERCISES

9-1 In your own words, so that you are sure of what they mean, define (*a*) diffraction, (*b*) reflection, (*c*) refraction, (*d*) critical angle, (*e*) interference pattern.

9-2 At 20°C the refractive index of carbon tetrachloride is 1.4573, and the density is 1.595 g/cm^3. Calculate the molar refraction. *Ans:* 26.51

9-3 At 20°C the refractive index of benzene is 1.4979, and the density is 0.879 g/cm^3. Calculate the specific refraction and the molar refraction.

9-4 Using the contribution of the various atoms (Table 9-1), calculate for nitrobenzene (*a*) the molar refraction, (*b*) the specific refraction, and (*c*) the refractive index. At 20°C the density is 1.210 g/cm^3, and the experimental value of the refractive index is 1.5524.

9-5 Using Fig. 9-3, show that the angle α (the angle actually measured with an Abbe refractometer) is the complement of angle ϕ_c, the critical angle.

9-6 With an Abbe refractometer, must the sample have a refractive index which is less than that of the refracting prism. Why?

9-7 Why is it necessary to correct the Abbe refractive-index scale when wavelengths other than the sodium D line are used?

9-8 At 25°C, a Pyrex test tube is found to be invisible in a mixture of benzene and carbon tetrachloride that is 0.345 mol fraction benzene. Assuming that the refractive index of a benzene–carbon tetrachloride mixture varies linearly with mole fraction, what is the refractive index of the Pyrex glass? At 25°C the refractive index of carbon tetrachloride is 1.4576, and that of benzene is 1.4979. *Ans:* 1.472

9-9 At 20°C the refractive index of benzene is 1.4979, and that of nitrobenzene is 1.5524. Calculate the refractive index of a mixture consisting of 10 ml benzene and 40 ml nitrobenzene. What assumption must be made?

REFERENCES

INTRODUCTORY

1 EFRON, A.: "Light," Rider, New York, 1958. A clear elementary treatment of all properties of light, including refraction, reflection, dispersion, and interference.

2 JENKINS, F. A., and H. E. WHITE: "Fundamentals of Optics," 3d ed., McGraw-Hill, New York, 1957. Includes a lucid and detailed description of refraction and interference.

3 HARDY, A. C., and F. H. PERRIN: "The Principles of Optics," McGraw-Hill, New York, 1932. A clear description of the principles of an Abbe and Pulfrich refractometer.

4 SIGGIA, S.: "Survey of Analytical Chemistry," McGraw-Hill, New York, 1968. A brief description of the most important aspects of refractometry and interferometry.

INTERMEDIATE

5 STROBEL, H. A.: "Chemical Instrumentation," Addison-Wesley, Reading, Mass., 1960. An excellent overview of refractometry, with a brief treatment of interferometry. Table 9-2, p. 241, summarizes the important characteristics of all types of refractometers and interferometers.

6 WILLARD, H. H., L. L. MERRIT, JR., and J. A. DEAN: "Instrumental Methods of Analysis," 4th ed., Van Nostrand, New York, 1965. A good intermediate treatment of both refractometry and interferometry.

ADVANCED

7 TILTON, L. W., and J. K. TAYLOR: Refractive Index Measurement, pp. 411–462 in W. G. Berl (ed.), "Physical Methods in Chemical Analysis," 2d ed., vol. 1, Academic, New York, 1960. A clear, definitive treatment of refractometry and interferometry, with emphasis on instrumentation and measurement details.

8 LEWIN, S. Z., and N. BAUER: Refractometry and Dispersometry, chap. 70 in I. M. Kolthoff, P. J. Elving, and E. B. Sandell (eds.), "Treatise on Analytical Chemistry" pt. I, vol. 6, Wiley-Interscience, New York, 1965. The principles of refractometry are described in detail. Does not cover interferometry or instrumentation.

9 BAUER, N., K. FAJANS, and S. Z. LEWIN: Refractometry, chap. 18 in A. Weissberger (ed.), "Physical Methods of Organic Chemistry," 3d ed., vol. I, pt. II, Wiley-Interscience, New York, 1960. A thorough, comprehensive treatment of theory, practice, and instrumentation of refractometry and interferometry. Table XVIII, p. 1278, is a useful guide to choosing the best type of instrument for a given purpose. This chapter is indispensable for the active practitioner.

10

Spectropolarimetry and Circular Dichroism Spectrometry

A great many substances characteristically rotate the plane of polarized radiation. These substances, which are said to be *optically active* or to possess *optical rotatory power,* are characterized by a lack of symmetry in their molecular or crystalline structure. The extent to which the plane of polarized radiation is rotated varies widely from one active compound to another. For any given compound, the extent of rotation depends on the number of molecules in the path of the radiation or, in the case of solutions, upon the concentration and the length of the containing vessel. It is also dependent upon the wavelength of the radiation and the temperature. In *polarimetry,* the extent to which the plane is rotated is measured at a single wavelength (generally at 589 nm, the D line of a sodium-vapor lamp). In *spectropolarimetry,* the optical activity is measured as a function of wavelength, and the resulting spectra yield much more structural information than polarimetry. The change of optical rotation with wavelength is also known as *optical rotatory dispersion* (ORD).

Circular dichroism (CD) is closely related to rotatory dispersion. In principle it provides the same structural information as ORD, but often it is more valuable. Circular dichroism is an *absorption* phenomenon, whereas ORD is a *dispersion* phenomenon. As pointed out in Chap. 1, plane-polarized radiation can be resolved into two beams which are said to be *circularly polarized* in opposite senses. If in a given medium the *indexes of refraction* for the left- and right-hand circular components are not the same, the plane of polarization will be rotated; this is *optical rotation*, the basis of ORD. On the other hand, if there is a difference in *absorption* of the left- and right-hand circular components of plane-polarized light upon passing through a given medium, *circular dichroism* results.

10-1 THEORY

Although optical rotation was used in some of the earliest structural studies, the fundamental theory is still not thoroughly understood; e.g., it is not possible to calculate precisely the optical activity to be expected for a simple asymmetric molecule. There are complex molecular theories that do quite well in explaining the observed facts, and there are numerous empirical rules that are extremely useful in predicting and interpreting ORD and CD spectra, but a comprehensive and fundamental theory has yet to be devised.

10-1A Comparison of Absorption, Dispersion, Circular Dichroism, and Rotatory Dispersion

ABSORPTION AND DISPERSION

In Chap. 1 *absorption* and *dispersion* were defined, and the relationship between them noted (Sec. 1-4*F*). The fact that the two phenomena are related may be seen by reference to Fig. 10-1. Figure 10-1*a* is called a *dispersion* curve because it indicates how light of various wavelengths would be dispersed by a prism made of the material being studied. Figure 10-1*b* is obviously an *absorption* spectrum, and it should be noted that a marked change in refractive index occurs at whatever wavelength a substance absorbs. Except for the wavelength region where there is an absorption band, the value of the refractive index *increases* with *decreasing* wavelength; this is called *normal dispersion*. On the other hand, in the wavelength region of an absorption band, the value of the refractive index *decreases* with *decreasing* wavelength; this is called *anomalous dispersion*. At approximately the wavelength of maximum absorption, the refractive index becomes unity, which essentially means that the absorption process is making no net contribution to the refractive index at that wavelength (see definition of refractive index in Sec. 1-4*A*).

Equations have been developed that quantitatively relate the absorptivity a to the refractive index n as a function of wavelength for a particular substance [9]. In principle, absorption and dispersion reveal equivalent information about molecular structure. Experimentally it is generally much easier and more accurate to make absorption measurements than to make

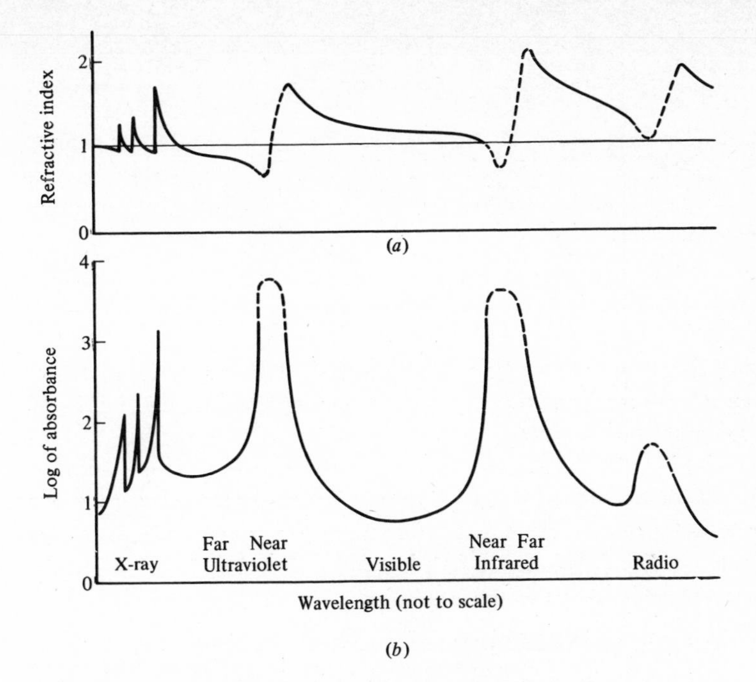

FIGURE 10-1 (*a*) Refractive index and (*b*) absorbance as functions of wavelength over the whole electromagnetic spectrum. Schematic for a hypothetical substance. (*From G. W. Ewing, "Instrumental Methods of Analysis,"* 3*d ed., p.* 15, *McGraw-Hill Book Company, New York*, 1969, *by permission.*)

refractive-index measurements. But whereas accurate absorption measurements can be made *only* in the immediate region of an absorption band, accurate refractive-index measurements can be made at almost all wavelengths *except* at an absorption band. This complementary feature can be used to advantage. As an example, a completely saturated compound that cannot be analyzed spectrophotometrically either in the visible or ultraviolet regions will give definitive refractive-index measurements in both regions.

ROTATORY DISPERSION AND CIRCULAR DICHROISM

In the preceding discussion of absorption and dispersion it was not important to specify whether the measurements were being made with polarized or unpolarized light, and in fact they are almost always made with unpolarized light. To make CD or ORD measurements, however, polarized light must be used.

Plane and Circularly Polarized Light Plane-polarized light (Sec. 1-4*G*) can be resolved into two *circular* components, as shown by Fig. 10-2. If we let **E** be the magnitude of an electric field vector oscillating sinusoidally along a determined direction in space (plane-polarized light), then $\mathbf{E}_R$ and $\mathbf{E}_L$ are two

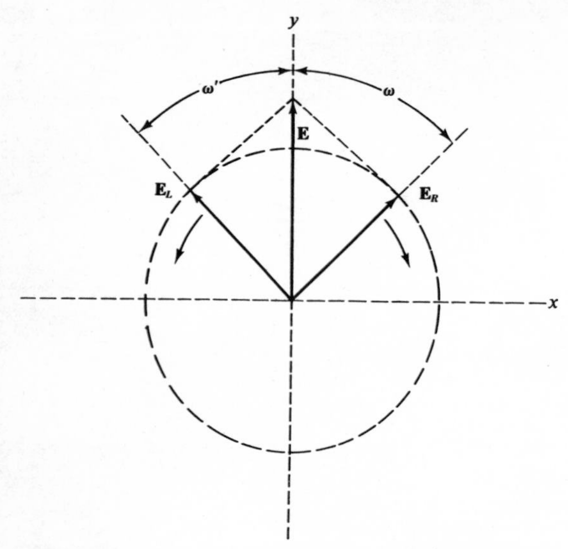

FIGURE 10-2 End view of the electric field vector **E** as the resultant of two rotating vectors, $\mathbf{E}_R$ and $\mathbf{E}_L$ (plane-polarized light).

rotating vectors into which **E** can be resolved. In other words, plane-polarized light is equivalent to the sum of two vector components of equal magnitude, $\mathbf{E}_R$ and $\mathbf{E}_L$, one rotating clockwise and one rotating counterclockwise with the same frequency, so that at any moment the two vectors make equal angles ($\omega = \omega'$) with their resultant **E**.

Rotatory Dispersion When plane-polarized light travels through an optically active material, vectors $\mathbf{E}_R$ and $\mathbf{E}_L$ will rotate at *different speeds* and at a given moment (or at a given distance through the material) the angles with respect to the original plane will be different ($\omega \neq \omega'$). Figure 10-3 illustrates this for the case where the right circularly polarized ray $\mathbf{E}_R$ travels faster than the left circularly polarized ray $\mathbf{E}_L$ ($\omega > \omega'$). Thus, the resultant **E** is still plane-polarized, but its plane of polarization is rotated through an angle α. Materials which allow the right circularly polarized ray $\mathbf{E}_R$ to travel faster than the left $\mathbf{E}_L$ are *dextrorotatory*, and α is assigned a positive (+) sign. Conversely, the medium is said to be *levorotatory* when the plane-polarized wave makes a *negative* angle with the *y* plane ($\omega < \omega'$).

Since the *speed* of a light wave traveling through a material medium is a function of its index of refraction *n* [see Sec. 1-4*A* or Eq. (1-14)], it becomes apparent that the criterion for a material which is said to be optically active is that the material exhibit different indexes of refraction for left and right circularly polarized light (these indexes are designated n_L and n_R, respectively). Thus, it can be shown that the angle of rotation α is given by

$$\alpha = \frac{\pi}{\lambda(n_L - n_R)} \tag{10-1}$$

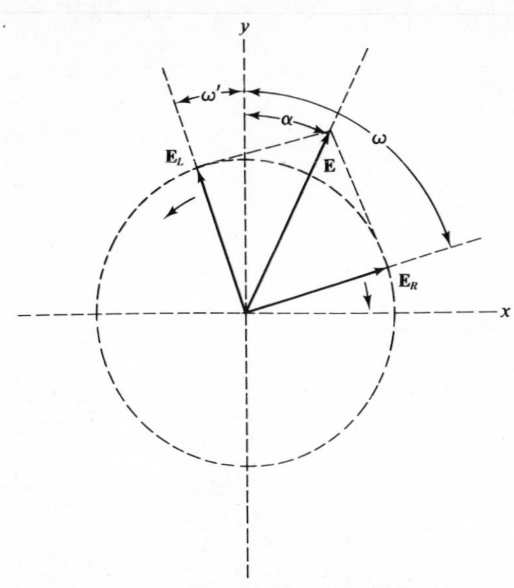

FIGURE 10-3 End view of electric field vectors on passage of plane-polarized light through an optically active material (α is the observed rotation).

where α = rotation, rad/unit length (the length having the same units as λ)
λ = wavelength of incident light
n_L = index of refraction for left circularly polarized light
n_R = index of refraction for right circularly polarized light

The rotation α can be converted into *specific rotation* $[\alpha]$ (having units of degrees per decimeter per unit concentration) as follows:

$$[\alpha] = \frac{\alpha}{C}\frac{1800}{\pi} \tag{10-2}$$

where C is the concentration in grams per milliliter of solution and the factor $1800/\pi$ is the conversion factor to put α into units of degrees per decimeter. In ORD measurements it is common to modify the optical rotation units still further, introducing the term molecular rotation $[\Phi]$ defined by

$$[\Phi] = \frac{[\alpha]M}{100} \tag{10-3}$$

where M is the molecular weight of the optically active substance and the factor 100 is arbitrarily introduced to give $[\Phi]$ a more convenient magnitude.† The *molecular rotation* is the most suitable experimental quantity for comparing rotations of different substances, since comparison is then made on a mole-for-mole basis.

Returning to the fundamental equation of optical activity, Eq. (10-1), we

† In some publications, particularly in carbohydrate chemistry, the factor of 100 is omitted.

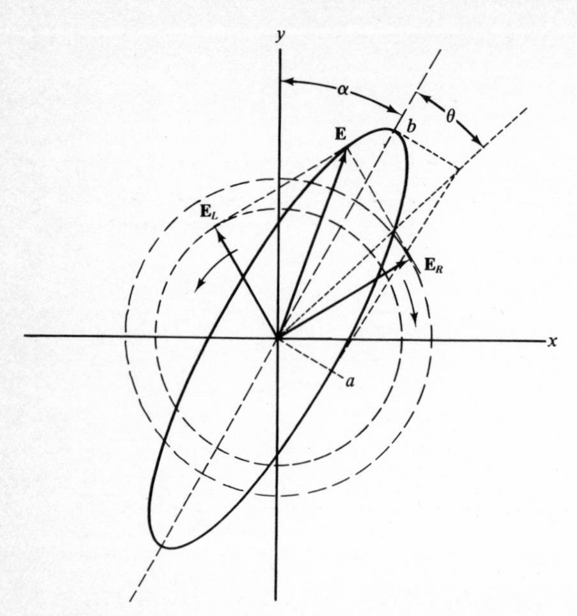

FIGURE 10-4 End view of electric field vectors on passage of plane-polarized light through an optically active material (the wavelength of light is near an absorption band).

should point out that quantity $n_L - n_R$ is called *circular birefringence* and all optically active materials must have nonzero values for it. Furthermore, the indexes of refraction $n_L - n_R$ will vary *differently* with wavelength, and it is this variation of optical activity with wavelength which gives an ORD curve.

Circular Dichroism When plane-polarized light of the proper wavelength† travels through an optically active material, not only will the vectors $\mathbf{E}_R$ and $\mathbf{E}_L$ rotate at different *speeds*, i.e., exhibit circular birefringence, but they will be differentially *absorbed*. Thus, in the spectral region in which optically active absorption bands are present, the *length* of the vector $\mathbf{E}_R$ will no longer be equal to $\mathbf{E}_L$, and their resultant **E** no longer oscillates along the circumference of a circle. Figure 10-4 illustrates this for a dextrorotatory substance which absorbs the left circularly polarized ray $\mathbf{E}_L$ more strongly than the right $\mathbf{E}_R$. In Fig. 10-4 the head of the resultant vector **E** now traces out an ellipse, the resulting light is said to be *elliptically polarized,* and the material is said to exhibit *circular dichroism.* Not only does the plane of polarization rotate at an angle α, but the resulting ellipse can be characterized with the angle θ, which is the tangent of the angle of ellipticity.‡ By analogy with Eq. (10-1), it

† As inferred from the earlier discussion and Fig. 10-1, the wavelength of light used must be at or near a wavelength where the substance *absorbs* in order to be able to observe significant circular dichroism.

‡ In Fig. 10-4 tangents are drawn at both the minor axis a and the major axis b, and θ is the angle (from the major axis) made by a vector drawn through the intersection of the two tangents. It follows that $\tan \theta = a/b$.

could be shown that the angle of ellipticity θ is given by

$$\theta = \frac{\pi}{\lambda}(a_L - a_R) \tag{10-4}$$

where a_L and a_R are the absorptivities (sometimes called extinction coefficients or absorption coefficients) for the left and right circularly polarized rays, respectively. Again, by analogy with the specific rotation $[\alpha]$ in Eq. (10-2), the *specific ellipticity* $[\theta]$ is defined by

$$[\theta] = \frac{\theta}{lC} \tag{10-5}$$

where θ = angle, deg
l = path length, dm
C = concentration, g/ml of solution

Similarly, the molecular ellipticity $[\Psi]$ is defined by

$$[\Psi] = \frac{[\theta]M}{100} \tag{10-6}$$

Returning to the fundamental equation of circular dichroism, Eq. (10-4), we should emphasize again that it is the quantity $a_L - a_R$ which is responsible for the circular dichroism of a material. Since a_L and a_R vary with wavelength, measurements of this difference (or θ) as a function of wavelength result in CD curves, examples of which will be given in Sec. 10-1*B*. Whereas the angle of rotation α can be measured *directly* with a polarimeter or spectropolarimeter, the angle of ellipticity θ can be measured only *indirectly,* essentially by measuring the absorption first with left-hand and then with right-hand circularly polarized light and taking the *difference* in absorption as specified by Eq. (10-4). In practice, CD instruments use a device known as a *quarter-wave plate* to resolve plane-polarized light into *either* left or right circularly polarized light. If, for example, the left circularly polarized vector E_L were removed with a quarter-wave plate (see Fig. 10-3), the resulting right circularly polarized light could be compared to a right-hand helix. Thus, the head of vector E_R (from Fig. 10-3) is rotating continuously around the axis of propagation. Most commercial CD instruments use an alternating voltage to transform plane-polarized light first into left and then into right circularly polarized light in the course of one cycle of alternating voltage. The difference in absorption during one cycle is electronically detected and read out, giving what appears to be a *direct* CD measurement. Further details will be given in Sec. 10-2.

10-1B Classification of ORD and CD Curves

ROTATORY DISPERSION CURVES

ORD curves are classified into two main types, *plain* curves and *Cotton effect* curves. *Plain curves* result when the compound being studied has no absorption band within the wavelength region being studied, and Fig. 10-5 gives examples of some ORD plain curves. Plain curves are further classified as *positive* or *negative,* according as the rotation of the compound becomes more positive or more negative with decreasing wavelength (see Fig. 10-5).

Cotton effect curves† result when the ORD measurements are made in a wavelength region where the substance absorbs, and Fig. 10-6 shows a typical curve. To avoid confusion with absorption curves, the maximum and minimum of a Cotton effect curve are called the *peak* and *trough,* respectively, and collectively they are referred to as the *extrema* of the curve. When the peak occurs at a higher wavelength than the trough, as in Fig. 10-6, the curve is a *positive* Cotton effect curve. Conversely, when the trough occurs at a higher wavelength, it is a *negative* Cotton effect curve. The vertical distance between the peak and the trough is called the *amplitude a,* and it is conventionally expressed in hundreds of degrees (see Fig. 10-6). The horizontal distance between the peak and trough is called the *breadth b* of the curve (see

† The Cotton effect was discovered by the French physicist A. Cotton. See A. Cotton, *Compt. rend.*, **120**:989 (1895).

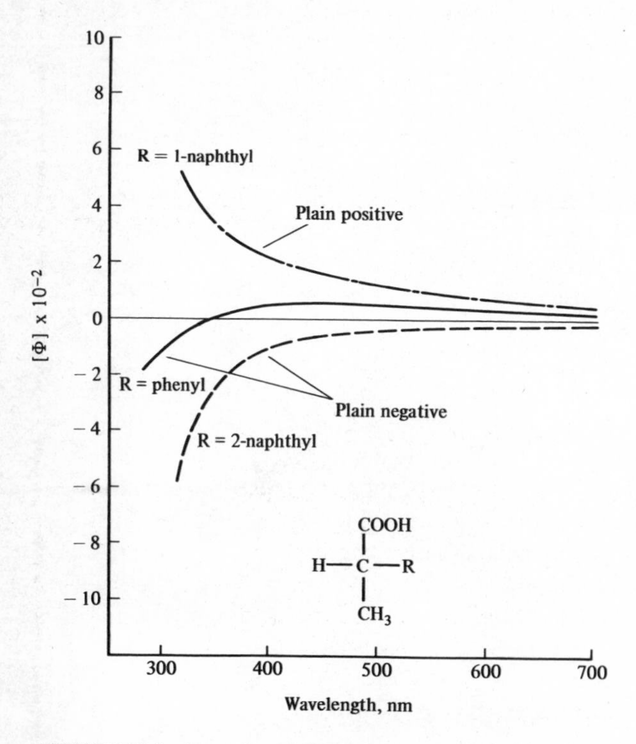

FIGURE 10-5 Examples of ORD plain curves (some substituted propionic acids in 0.1 *N* sodium hydroxide). [*Reprinted with permission from B. Sjöberg, Acta Chem. Scand.*, **14**:287 (1960).]

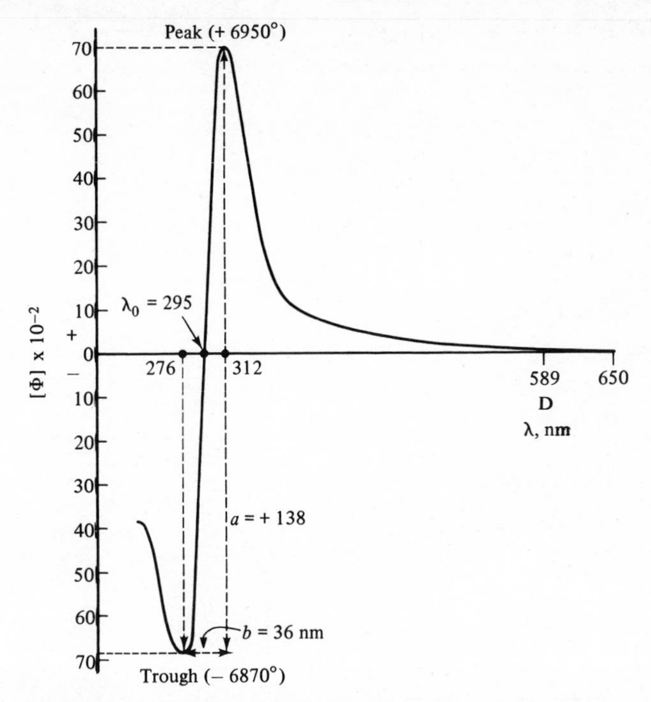

FIGURE 10-6 Typical ORD Cotton effect curve (3β-hydroxy-5α-androsten-17-one, showing a positive Cotton effect). (*From P. Crabbé, "Optical Rotary Dispersion and Circular Dichroism in Organic Chemistry," p.* 17, *Holden-Day, Inc., San Francisco,* 1965, *by permission.*)

Fig. 10-6). The breadth of the curve is not widely quoted, but the amplitude is an important characteristic of the compound being studied.

The point λ_0 on the ORD curve where the molecular rotation is zero corresponds roughly to the wavelength of maximum absorption in the ultraviolet or visible absorption spectrum. In Fig. 10-6 this point is 295 nm.

Whereas a Cotton effect curve with one geometrical maximum and one geometrical minimum is called a *single* Cotton effect curve (for example, Fig. 10-6), it is possible to have two or more peaks with a corresponding number of troughs, giving *multiple* Cotton effect curves. Figure 10-7 shows the ORD curve of testosterone, which is a double negative Cotton effect curve.

CIRCULAR DICHROISM CURVES

Circular dichroism curves are all measured in a wavelength region where the substance absorbs since CD is inherently an absorption phenomenon. Figure 10-8 shows the CD curve (and for comparison, the ultraviolet absorption curve) for the compound for which the ORD curve was given in Fig. 10-6. In general, the peak of a CD curve will coincide fairly well with the maximum of the ordinary absorption curve, but the shapes of the two curves

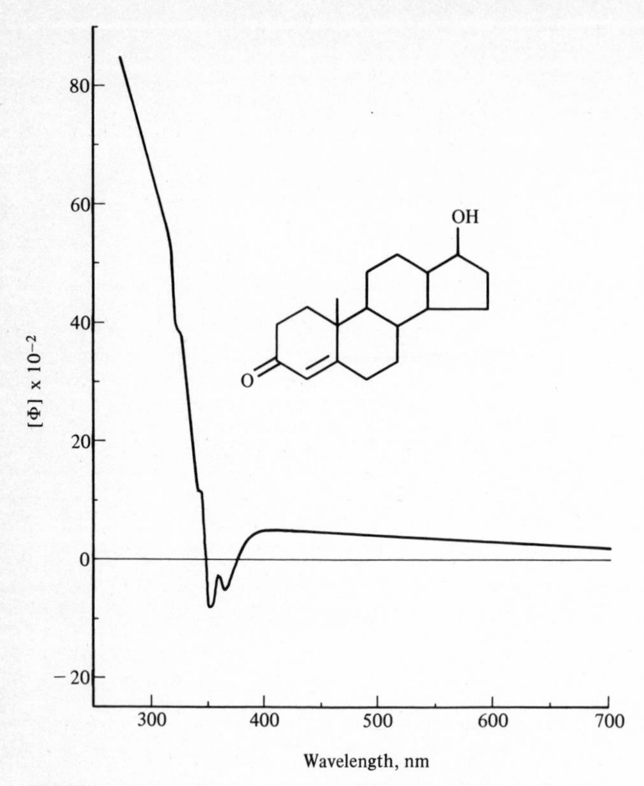

FIGURE 10-7 ORD curve of testosterone (a double negative Cotton effect curve). (*After P. Crabbé, "Optical Rotary Dispersion and Circular Dichroism in Organic Chemistry," p.* 18, *Holden-Day, Inc., San Francisco,* 1965, *by permission.*)

are not completely alike. Just as ORD curves are classified as being positive or negative, so too CD curves are said to have either a *positive CD* (a maximum, like the curve in Fig. 10-8), or a *negative CD* (a minimum). For the same compound, the sign of the CD curve will be the same as the sign of the ORD Cotton effect curve (see Figs. 10-6 and 10-8).

10-1C Molecular Origin of Optical Activity

The concept of *chromophores* is a familar one in absorption spectroscopy, but it is much less accepted in polarimetric analysis. It might be assumed from various publications on polarimetry that the rotation of light has its origin at the site of one or more asymmetric carbon atoms in the molecule, independent of any absorption. This interpretation is inaccurate, as has been learned from ORD and CD studies. The concept of the optically active chromophore acquired importance with ORD curves, since it is in the region of an absorption band that dispersion is most significant. In CD studies the significance of optically active chromophores is even greater, since measurable values of CD

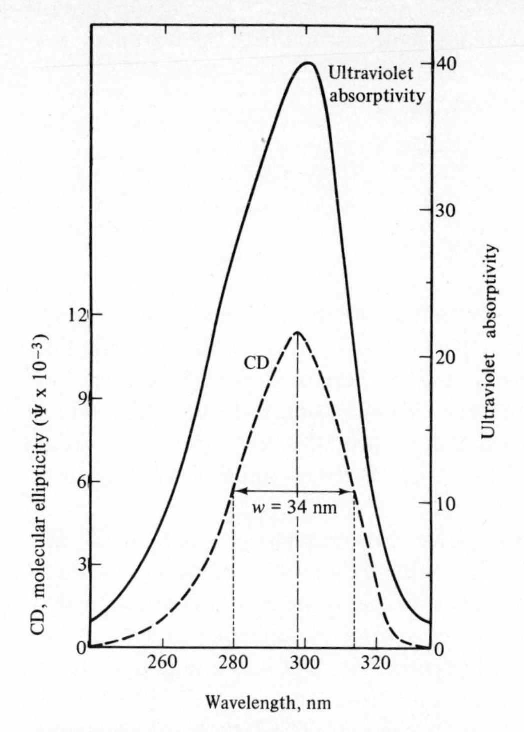

FIGURE 10-8 Circular dichroism and ultraviolet absorption curves of 3β-hydroxy-5α-androsten-17-one. (*From P. Crabbé, "Optical Rotary Dispersion and Circular Dichroism in Organic Chemistry," p. 21, Holden-Day, Inc., San Francisco,* 1965, *by permission.*)

absorption are observed only in the immediate wavelength region of the optically active chromophore.

Chromophores are generally considered to be only those groups possessing mobile electrons, allowing absorption in the visible or near ultraviolet regions. Examples of chromophoric groups include double bonds, hydroxyl groups, and conjugated or nonconjugated carbonyl groups (see Sec. 2-1*B*). Such groups will be referred to here as *active* chromophores. If it were not for experimental difficulties, *all* bonds could be considered chromophores, since in the far ultraviolet *all* bonds absorb (see Sec. 2-5). Although these "shortwave" absorbing groups have no influence on CD measurements in the near ultraviolet or visible regions, they mildly influence ORD and polarimetry measurements. In general, the closer the wavelength of polarimetric analysis to the absorption maximum, the greater the influence.

Optically active chromophores can be classified into two types: (1) those which are optically active because they are inherently asymmetric and (2) those which are inherently symmetric but which are asymmetrically perturbed by the neighboring environment. The first type is rare in organic molecules.

One example is hexahelicene, an aromatic hydrocarbon with the formula

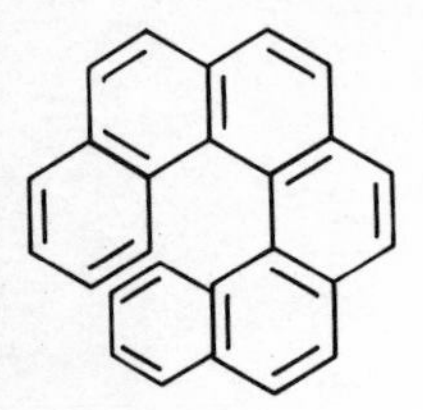

The spatial model of the molecule shows that the two terminal rings overlap, so that the molecule is not planar but helicoidal. As a result, the molecule has no center or plane of symmetry and is therefore optically active. The molecule may be thought of as one large chromophore, a chromophore which is inherently dissymmetric. Twisted biphenyls are other examples of this class of optically active chromophore. Often there is no asymmetric carbon atom present with this type.

The second type of optically active chromophores, those which are inherently symmetric but which become optically active by induction, is much more common and important. A typical example is the carbonyl group, which is a chromophore having the elements of symmetry indicated in Fig. 10-9*a*. Thus, the carbonyl group is inherently symmetric, and in a compound which possesses a plane of symmetry (like cyclohexanone, in Fig. 10-9*b*), the chromophore will be optically *inactive.* On the other hand, the carbonyl group in 2-methylcyclohexanone (Fig. 10-9*b*) is optically *active*, since the methyl group in position 2 induces asymmetry in the chromophore. Since this latter type of optical activity is *induced* by the electron environment around the chromophore rather than being *inherently* optically active, the magnitude of the Cotton effect is often considerably smaller than in the first type of chromophore. However, the magnitude of the optical activity of this second class of chromophores is not proportional to the absorptivity of light, and often substances which absorb only weakly give pronounced optical activity.

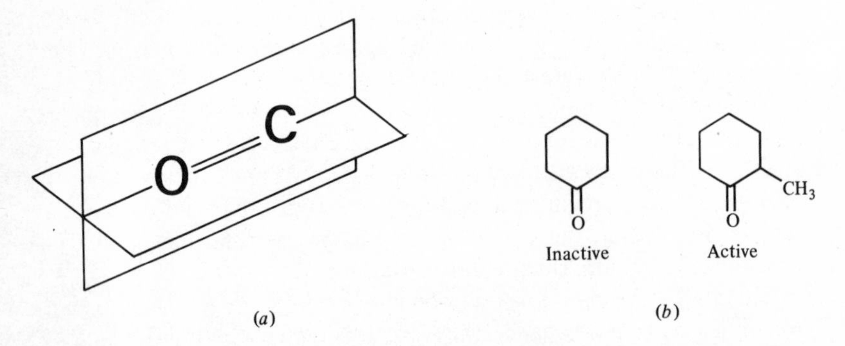

FIGURE 10-9 Elements of symmetry of the carbonyl group: (*a*) carbonyl group alone; (*b*) carbonyl group in two types of compounds.

10-2 INSTRUMENTATION

10-2A Spectropolarimeters

Spectropolarimeters are best studied after first considering the essential components of a simple polarimeter, shown in Fig. 10-10. Light from a monochromatic source passes through a *polarizer*† and becomes *plane-polarized*; i.e., its vibrations are restricted to one plane, the orientation of which is determined by the orientation of the polarizer. After traversing the sample, the light reaches *the analyzer,* a prism similar to the polarizer. With a simple polarimeter, the detector is the human eye or a phototube.

The intensity of the light reaching the detector varies as the analyzer is rotated and is a minimum (essentially zero) when the plane of transmission of the analyzer is at 90° to the plane of polarization of the light incident upon it. In principle, the zero setting of the polarimeter is first determined in the absence of an optically active sample (best determined by filling the sample tube with solvent). Next the sample tube is filled with a solution of the material to be studied, and the angle through which the analyzer must be rotated to again extinguish the light gives the optical rotation of the solution.

These same principles can be incorporated into an automatic recording spectropolarimeter, but a fundamental problem in the simple polarimeter just described also causes difficulty in automation: the detector, whether it is the eye or a phototube, is required to identify the position of the analyzer when the transmitted radiation is zero, and this cannot be done with precision (with the eye, the human mind must "remember" very small differences in intensity as the analyzer is rotated; with an electronic detector, the instrument goes dead as the null is approached). To avoid this problem, various methods have been adopted for stopping at a *reference point* which is other than zero intensity. With a simple polarimeter, this is usually accomplished with a *half-shade* device, whereby a small polarizing prism is added to block one-half of

† The polarizer is usually a compound prism made of a double-refracting material such as quartz or calcite; see Sec. 1-4*G*.

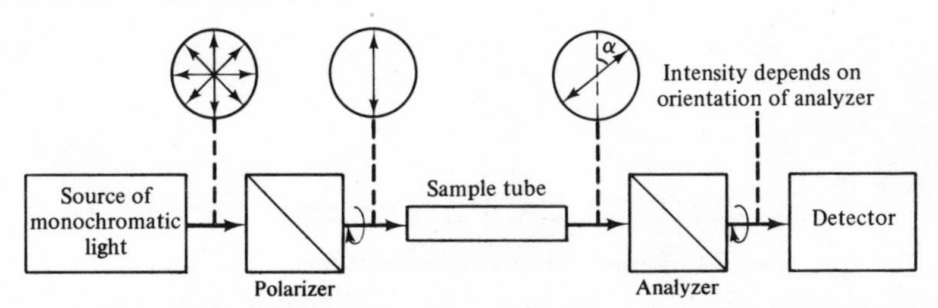

FIGURE 10-10 Essential components of a simple polarimeter. The circles represent cross sections of the light beam as viewed from the analyzer; the lines indicate the planes of polarization (sample shown is dextrorotatory). (*After J. C. P. Schwarz, "Physical Methods in Organic Chemistry," p.* 215, *Oliver & Boyd, Edinburgh,* 1964, *by permission.*)

the light beam. The half-shade prism is permanently oriented with its polarizing axis at an angle of a few degrees from that of the polarizer. There is then a particular position of the analyzer at which the rays passed in the two halves of the beam are just equal in power. This is a more satisfactory reference point than complete extinction, since the visual observation consists of exactly matching the powers of two half-beams at some intermediate level, for which the eye is well suited.

In principle the same approach can be taken in automated instruments. For example, if the light beam were to be modulated in such a way that first one half the beam and then the other half reached the detector, the signal from the detector would be an *alternating* current if the light intensities were unequal but a *direct* current if they were equal.

Commercial spectropolarimeters are null-balance instruments which achieve the null reading in one of two ways. Some instruments use a *mechanical* null-balance device, in which the analyzer or polarizer is turned by a servomotor. Other instruments use an *electrical Faraday effect* null balance, in which the polarizer and analyzer positions can be fixed and the rotation of the sample is compensated electrically using the Faraday effect (the production of optical activity in normally inactive materials by a magnetic field).

MECHANICAL NULL BALANCE

Figure 10-11 is a schematic diagram of an instrument which uses a mechanical null-balance system. As in all spectropolarimeters, a *monochromator* is used to produce light of variable wavelength, generally over the region from 200 to 700 nm but in some instruments down to 185 or 190 nm with nitrogen flushing. In the Durrum-Jasco spectropolarimeter shown in Fig. 10-11 a modulating motor operating at 12 Hz is used to rock the polarizer back and forth through an angle of $\pm 1°$. In this way the beam passing through the sample is modulated with respect to its state of polarization. The servomotor responds only to the 12-Hz frequency and causes the servomotor to adjust the analyzer continuously to the point where the 12-Hz signal will be

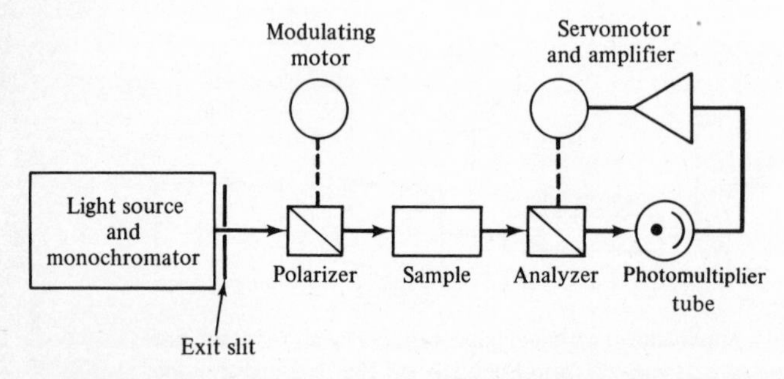

FIGURE 10-11 Schematic diagram of a spectropolarimeter using a mechanical null-balance system. (*Jasco Incorporated.*)

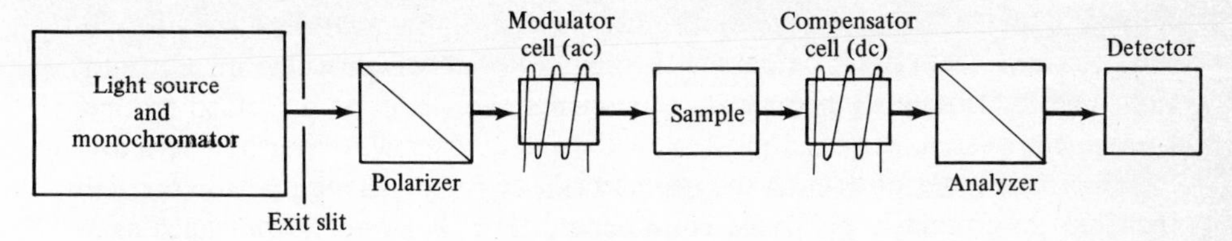

FIGURE 10-12 Schematic diagram of a spectropolarimeter using an electrical Faraday effect null-balance system.

symmetrically displaced about the null point. The servomotor simultaneously positions the recording pen. As the wavelength is scanned, a constant optical rotation will give a steady pen reading, but as the optical rotation changes, the error signal to the servomotor increases and a pen deflection results.

ELECTRICAL NULL BALANCE

A *Faraday cell* is used in instruments which obtain a null balance electrically. The cell is simply a cylindrical rod of glass or other material (or preferably a cylindrical cell usually containing water) placed inside a coil. If this coil is connected to a dc source, the plane of polarization of polarized light passing through the cell is rotated. If the coil is connected to an alternating source, the direction of the plane of polarization oscillates at the same frequency as the alternating current.

A schematic diagram of an instrument which uses an electrical Faraday effect null-balance system is given in Fig. 10-12. In the system illustrated (used, for example, in Bendix recording spectropolarimeters) neither the polarizer nor the analyzer moves. The polarized beam is rotationally *modulated* and passed through the sample cell. If the sample rotates the light and causes a departure from a previous null point, this is sensed electrically by the photomultiplier detector. The null is restored electronically by passing enough current through the *compensator cell* to introduce a counterrotation equal and opposite to that of the sample. The amount of current fed to the compensator is recorded.

An instrument which uses both mechanical null balance and the Faraday effect is the Cary model 60 spectropolarimeter. This instrument uses a vitreous-silica rod as a Faraday cell to modulate the plane of polarization at a 60-Hz frequency and a servo system which operates the polarizer to find the null balance. The analyzer is immovable.

10-2B Circular-Dichroism Spectrometers

The measurement of the circular dichroism of an optically active substance involves measuring the difference between the absorption coefficients of the substance in right and left circularly polarized light. To obtain circularly polarized light the beam of radiation must first be plane-polarized, after which the polarized beam must be passed through a device which will resolve it into right and left circularly polarized components. This is done by retarding one component relative to the other by exactly one-quarter wavelength. There are basically two types of devices in commercial use for resolving plane-polarized light into its circularly polarized components. The first uses *prisms*, such as a *Fresnel rhomb*. The second uses an electro-optic retardation modulator, known as the *Pockels cell*.

PRISM MODULATORS

The Cary CD accessory for the model 14 spectrophotometer is an example of a system using a prism for a quarter-wave retarder. Figure 10-13 shows the optical assembly which is placed in the sample beam for CD measurements. An equivalent unit with the Fresnel rhomb oriented to pass the opposite circularly polarized component is placed in the reference beam. By placing sample in *both* beams of the spectrophotometer the difference in absorbance due to right and left circularly polarized light will automatically be measured.

ELECTRO-OPTIC MODULATORS

The Pockels cell consists of a Z-cut plate of ammonium dihydrogen phosphate, a tetragonal crystal which has only one optical axis in the absence of an electric field (the z axis, along which the light beam normally passes) and two optical axes when an electric field is applied; i.e., it exhibits the Pockels effect. By applying a relatively high alternating voltage (in the kilovolt range) across the z axis, the electro-optically induced axes will alternately be $+45°$ and $-45°$ to the original plane of polarization. The net effect of this is to transform the plane-polarized light first into right and then into left circularly polarized light in the course of one cycle of the applied alternating voltage. If the sample is circularly dichroic, i.e., if it absorbs right and left circularly polarized radiation differently, then the net light beam monitored by the detector will be *elliptically polarized*. This type of device thus permits a direct (rather than a

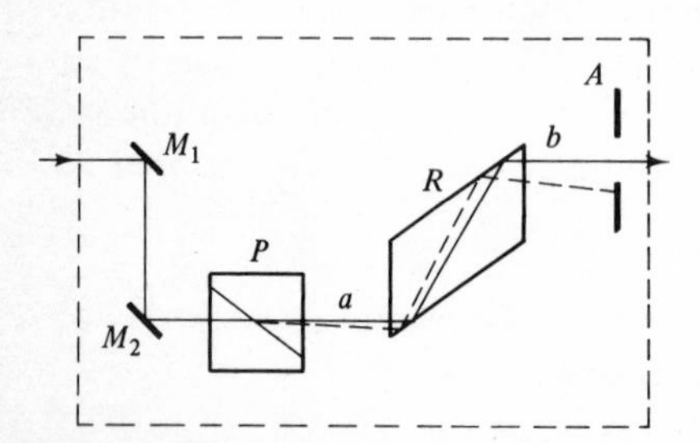

FIGURE 10-13 Optical assembly of prism-type quarter-wave retarder for CD measurements. Radiation enters at the left, is displaced downward by mirrors M_1 and M_2, plane-polarized by the compound prism P, and passed through the Fresnel rhomb R; here it undergoes two internal reflections which introduce a phase retardation of one-fourth wavelength, thus producing circular polarization. The mask A eliminates the extraordinary ray, while permitting the ordinary ray to pass. The entire unit fits into the sample chamber of a standard spectrophotometer. A second unit, oppositely orientated, is required for the reference pattern. The sample is placed at b for CD measurements and at a for ORD measurements. (*Varian Associates.*)

differential) measurement of circular dichroism.† Examples of instruments using the Pockels cell principle are the Dichrograph (manufactured by the Société Jouan, available through the Farrand Optical Co., Inc.) and the CD adaptor accessory for the Cary model 60 spectropolarimeter.

10-3 APPLICATIONS

10-3A Comparison of Information Obtainable from CD and ORD

It was emphasized in the theory section that CD and ORD phenomena are both manifestations of the same property of molecules (their optical activity), and in principle it is possible to transform either kind of data to the other. However, there are practical considerations which often make it advantageous to measure one or the other.

The basis for deciding between the two techniques hinges largely upon whether or not the molecule to be studied contains an optically active chromophore which absorbs at a wavelength accessible to the available instrument. If the molecule does absorb in the accessible-wavelength region, CD measurements are probably to be preferred over ORD measurements for the following reasons. First, CD measurements tend to be specific for a given chromophore, whereas ORD measurements are affected by all chromophores, even those absorbing only in the far ultraviolet. This is not disturbing to ORD measurements if the continuous background from distantly absorbing chromophores is weak in comparison with the Cotton effect being studied. However, it sometimes happens that the characteristic contribution of the chromophore can no longer be seen. A typical example is given in Fig. 10-14. On the ORD curve of

† In actuality, the output of the photomultiplier will contain an ac component superimposed on an average component, the alternating current being related to difference in transmission of the sample for right and left circularly polarized light. This complex signal is converted by electronic means to give direct recorder readout of the circular dichroism of the sample [9].

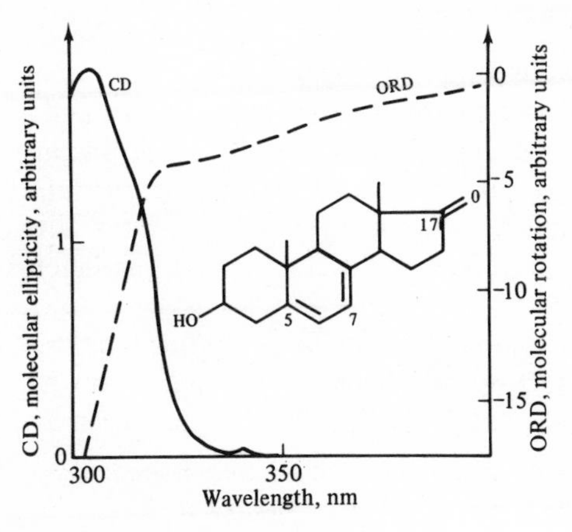

FIGURE 10-14 ORD and CD curves for the 17-ketone group of a steroid. (*After L. Velluz, M. Legrand, and M. Grosjean, "Optical Circular Dichroism,"* p. 19, *Academic Press Inc., New York,* 1965, *by permission.*)

Fig. 10-14 the Cotton effect is completely concealed, even though the optically active carbonyl group at the 17 position absorbs strongly near 300 nm. The Cotton effect is obscured because of the presence of the conjugated double bonds, also optically active, which produce a negative continuous background of high intensity. On the other hand, the CD curve clearly shows the Cotton effect from the carbonyl group (maximum near 300 nm). Thus, the CD measurements are more specific and informative in this case.

A second advantage of CD measurements is that CD curves are more easily analyzed than ORD curves, especially when multiple transitions are present. For example, when an ORD curve shows several Cotton effects in a given wavelength region (multiple maxima and minima), it is very difficult to separate the contributions from specific chromophores. On the other hand, CD curves inherently show greater resolution, and it is often possible to assign chromophores to specific CD peaks.

A third advantage of CD curves over ORD curves comes in the quantitative analysis of optically active mixtures. The greater specificity and resolution of CD curves is an obvious advantage, since ORD curves will have background contributions which are difficult to evaluate.

If the molecule to be studied does not absorb in the accessible wavelength region, CD is no longer applicable, whereas ORD will still provide useful information (plain curves). Examples of important optically active compounds which often fall in this category are amino acids, proteins, and polypeptides. Trends of plain curves with pH and other reaction conditions have led to valuable information about their structure and configuration. On the other hand, in some cases it may be possible to prepare derivatives of the compound which will have suitable absorption bands, thereby introducing a Cotton effect that can be studied by CD measurements. To be most valuable, the chromophore should be chemically added in the immediate vicinity of the asymmetric center whose configuration is to be studied.

10-3B Summary of Applications

ORD and CD measurements are most useful in solving structural, conformational, and chemical problems involving optically active compounds. However, a partial listing of additional applications would include *quantitative analysis* (particularly for sugars and complex mixtures), *purity* and *identity* determinations, studies of *reaction rates, equilibria studies,*† and *preparative studies* where changes in rotation are useful criteria for determining the progress of a reaction.

The structure of an optically active compound can be determined in three different ways: (1) by comparing the CD or ORD curve of the unknown with curves of analogous compounds whose absolute configurations have been es-

† The inversion of sucrose, a hydrolysis, and the mutarotation of glucose, an epimerization, are two reactions which owe their name to the rotational changes accompanying them.

tablished; (2) from theoretical considerations, e.g., using the so-called octant rule or axial haloketone rule [6–9]; or (3) by preparing a derivative of a functional group located on the asymmetric center. It should be emphasized, however, that the structure or identity of a substance is difficult to arrive at using only CD or ORD data. Only in conjunction with other physical and spectroscopic data can structural decisions be conclusive.

EXERCISES

10-1 Is the presence of an asymmetric carbon atom required for a compound to be optically active?

10-2 Explain in detail how the mechanical null-balance system of a spectropolarimeter works.

10-3 The two principal means of measuring the circular dichroism of a sample might be termed direct and indirect. Explain in detail the principles of each method.

10-4 When 2 g of an organic substance is dissolved in 150.0 ml of water, the solution in a 20-cm-long tube reads $+2.787°$ in a polarimeter with sodium D light; distilled water reads $+0.018°$. Calculate the specific rotation of the substance in units of milliliter-degrees per gram-decimeter.

10-5 If the optical rotation of the compound in Exercise 10-4 is measured with a spectropolarimeter, speculate whether the rotation would increase or decrease as the scan goes toward shorter wavelengths. Explain.

10-6 Explain why chromophores in the far ultraviolet region influence ORD and polarimetry measurements but have no influence on CD measurements.

REFERENCES

POLARIMETRY

1 WILLARD, H. H., L. L. MERRITT, JR., and J. A. DEAN: "Instrumental Methods of Analysis," 4th ed., Van Nostrand, New York, 1965. Chapter 15 gives an excellent summary of polarimetry.

2 STROBEL, H. A.: "Chemical Instrumentation," Addison-Wesley, Reading, Mass., 1960. Chapter 10 is a useful summary of polarimetry.

3 EWING, G. W.: "Instrumental Methods of Analysis," 3d ed., McGraw-Hill, New York, 1969. Chapter 10 summarizes polarimetry and leads into ORD and CD.

SPECTROPOLARIMETRY

4 SCHWARZ, J. C. P. (ed.): "Physical Methods in Organic Chemistry," Oliver & Boyd, Edinburgh and London, 1964. Chapter 6 gives an excellent introduction and is especially useful for understanding the correlations between structure and optical rotation.

5 STRUCK, W. A., and E. C. OLSON: Optical Rotation-Polarimetry, chap. 71 in I. M. Kolthoff and P. J. Elving (eds.), "Treatise on Analytical Chemistry," pt. I, vol. 6, Wiley-Interscience, New York, 1965. Although the emphasis is on polarimetry, the section on ORD is concise and useful.

6 DJERASSI, C.: "Optical Rotatory Dispersion," McGraw-Hill, New York, 1960. A thorough and detailed treatment.

7 JIRGENSON, B.: "Optical Rotatory Dispersion of Proteins and Other Macromolecules," Springer-Verlag, New York, 1969. Although the title identifies the scope of this book, the introduction will be valuable to all users.

CIRCULAR DICHROISM

8 VELLUZ, L., M. LEGRAND, and M. GROSJEAN: "Optical Circular Dichroism," Academic, New York, 1965. An excellent summary of all aspects of CD, including comparisons with ORD.

9 ABU-SHUMAYS, A., and J. J. DUFFIELD: Circular Dichroism: Theory and Instrumentation, *Anal. Chem.*, **38**(7):29A (1966). Particularly good for instrumentation principles.

10 BEYCHOKS, S.: Circular Dichroism of Biological Macromolecules, *Science*, **154**:1288 (1966). Good summary of relationship between ORD and CD and the application of CD to biological compounds.

COMBINED ORD AND CD

11 FOSS, J. G.: Absorption, Dispersion, Circular Dichroism, and Rotary Dispersion, *J. Chem. Educ.*, **40**:592 (1963). Excellent unified treatment of the observable relationships between these phenomena.

12 CRABBÉ, P.: "Optical Rotatáry Dispersion and Circular Dichroism in Organic Chemistry," Holden-Day, San Francisco, 1965. A lucid summary of all aspects of the two fields, with extensive applications.

11 Turbidimetry and Nephelometry

Turbidimetry and nephelometry are techniques of analysis based on the scattering of light by particles suspended in solution. The two techniques differ only in the manner of measuring the scattered radiation. In *turbidimetry* the source radiation is passed directly through the sample solution and the decrease in intensity is measured. In *nephelometry* the beam of radiation is measured at an angle (usually 90°) to the incident beam.

Because of the difference in the angle of measurement, turbidimetry is best suited for determining relatively high concentrations of suspended particles, whereas nephelometry is most suited for determining very low concentrations.† Thus, if a given suspension does not scatter strongly, e.g., the transmittance is greater than about 95 to 98 percent, turbidimetry should not be used since a comparison would have to be made of two large quantities of nearly equal values. In this instance nephelometry would be much more sensitive and precise, since the small amount of scattered light would be measured against a black background. On the other hand, for denser suspensions turbidimetry is the method of choice since it can accurately measure small changes in transmitted intensity when the absolute intensity is relatively low, whereas nephelometric measurements under these conditions suffer from interference effects that tend to saturate the response curve.

† The prefix *nephel-* comes from the Greek word for cloud.

Turbidimetry and nephelometry are used on gaseous, liquid, or even transparent solid samples in greatly varying applications. Whenever precipitates form which are difficult to filter, e.g., due to small particle size or a gelatinous nature, they usually make ideal suspensions to be measured by light-scattering techniques, replacing gravimetric operations. Such applications are common in analytical laboratories, clinical laboratories, and in process plants. A particularly valuable area of application is in air- and water-pollution studies, where the two techniques are used in determining the clarity and in controlling the treatment of potable water, water-plant effluents, and other types of environmental waters. Similarly, light-scattering measurements are used to ascertain the concentration of smog, fog, smoke, aerosols, etc. Further details on these and other applications will be given in Sec. 11-3.

Although the terms turbidimetry and nephelometry are largely restricted to applications whereby the *concentration* of particulate matter in suspension is being measured, mention should be made of various other applications of light-scattering measurements. These include the determination of particle size and shape and the determination of molecular weights (especially for polymers). These applications use many of the same principles as turbidimetry and nephelometry, but the measurements are considerably more complex and are largely outside the scope of this chapter. For a detailed discussion of these applications, see Refs. 3 and 5 to 7.

11-1 THEORY

The concepts of *reflection* and *scattering* were discussed in Secs. 1-4*B* and 1-4*C*. In this section it will be shown how these principles apply to the practice of turbidimetry and nephelometry.

11-1A General Principles

REFLECTION VS. SCATTERING

Reflection and scattering play an important role in turbidimetry and nephelometry. The criterion for deciding whether reflection or scattering is responsible for the deviation of light from its original path is based on the size of the suspended particles compared with the wavelength of light used. Specifically, if the suspended particles have dimensions about the same order of magnitude or *smaller* than the incident wavelength, the light will be *scattered,* whereas if the particles are larger than the wavelength of light, *reflection* will occur. The distinction is important because it affects the sensitivity of the measurement as well as how the measurement is made. For *nephelometric* measurements it is desirable that the suspended particles be small with respect to the wavelength used, so that scattering rather than reflection predominates. This is because scattering gives a symmetrical pattern of secondary rays in space, and this optimizes the intensity observed 90° to the primary beam. Larger particles cause a smaller fraction of the light to be deviated at right angles to the primary beam, since appreciable destructive interference occurs at angles in the region perpendicular to the primary beam. Actually, particles which are equal to or larger than the wavelength of light cause a preponderance of the scattered or reflected light to be deviated in a *forward* direction, rather than at right angles or backward (see Sec. 1-4*C*). Therefore, with large particles, there are advantages to measuring the scattered or reflected radiation at angles less than 90° from the primary beam, say in the region of 5 to 20°, or even 45° [8]. Most instruments currently available allow measurements to be made only at 90°. One example of a commercial instrument which allows angles other than 90° to be used will be described in Sec. 11-2.

Just as the suspended particles should not be too large for optimum scattering efficiency, neither should they be too small, or the scattering efficiency likewise falls off. For wavelengths in the ultraviolet and visible regions of the spectrum the optimum particle size is in the range of about 0.1 to 1 μm (colloidal size†).

For *turbidimetric* measurements particles larger than the wavelength of light are somewhat less of a problem, since the measurement is based on the total radiation removed from the primary beam, regardless of the mechanism by which it is removed or the angle through which it is deviated. However, the relationship (similar to Beer's law) between "absorbance" (see Sec. 11-1*B*) and concentration becomes nonlinear with larger particles because an increasing fraction of the deviated light is propagated in the forward direction and still reaches the detector. (Some of this is due to multiple reflections between large particles, but most of it is forward-angle scattering.)

VARIABLES AFFECTING MEASUREMENTS

The amount of radiation removed or deviated from the primary radiation beam depends on (1) the *concentration* of the particles, (2) the ratio of *refractive indexes* of the particle and its surrounding medium, (3) the *size* and *shape* of the particles, and (4) the *wavelength* of incident light. *Concentration* is the measurement objective in turbidimetry and nephelometry, and its relationship to the measured light intensity will be discussed in Secs. 11-1*B* and 11-1*C*. There must be a *refractive-index* difference between the particle and its surrounding medium if either reflection or scattering is to occur (see Sec. 1-4*C*). Suffice it to say that it is sometimes advantageous to change solvents in order to increase the refractive-index differences. The other two variables, *particle geometry* and *wavelength,* are particularly important variables for the analyst to control and merit a more detailed discussion.

Particle Geometry The most critical variable to the success of turbidimetric or nephelometric analysis is the control of particle size and shape. In turbidimetry, for example, the transmittance does not depend only on the mass of suspended material per unit volume, i.e., the concentration, but rather upon the number of light-blocking particles per unit volume and their cross-sectional area, as the following example shows.

† Colloids range in size from about 10 to 10,000 Å, or about 0.001 to 1 μm.

Example 11-1 Calculate the cross-sectional blocking area of 1.00 g of a solid with a density of 1.50 g/cm^3 assuming the solid is distributed as a monolayer of spherical particles across a light beam and that the particles have radii of (*a*) 0.1 μm and (*b*) 10 μm.

Answer (*a*) Since 1 μm = 10^{-6} m = 10^{-4} cm, the volume of a single particle with a radius of 10^{-5} cm is

$$\frac{4}{3}\pi(10^{-5})^3 = 4.19 \times 10^{-15}\ \text{cm}^3$$

The number of particles in 1.00 g is

$$1.00\ \text{g blocking material} \times \frac{1}{1.50}\ \text{cm}^3 \frac{1}{4.19 \times 10^{-15}} = 1.59 \times 10^{14}$$

The blocking area of 1.00 g is

$$(\text{Area of one particle})(\text{number of particles}) = [\pi(10^{-5})^2](1.59 \times 10^{14}) = 4.99 \times 10^4\ \text{cm}^2$$

(*b*) Similarly, if each particle has a radius of 10^{-3} cm, the blocking area of 1.00 g is 4.99×10^2 cm^2. Thus, the more finely divided material in part (*a*) blocks the light more effectively.

Particles in suspension are rarely, if ever, of uniform size. Therefore it is extremely important that all samples and calibration solutions with which they are compared should have the same distribution of small, medium, and large particles. This, in turn, requires that samples and standards be prepared under identical conditions, which is often no easy task. In producing particles by precipitation the size of the resulting particles may vary widely with changes in concentration of reactants, temperature, agitation, pH, presence of nonreactive materials, the order of mixing of reactants, and the time allowed for particle growth. Hence it is important to control all these conditions with the utmost care. Variations in particle growth are the most common cause of error in turbidimetry and nephelometry.

Often surface-active agents are added to precipitant solutions to stabilize the colloidal state and prevent the growth of large particles through agglomeration. Glycerol, gelatin, dextrin, and gum arabic are the materials most often used in turbidimetry and nephelometry. These function as protective colloids, being adsorbed on the particle surface, thus interrupting particle growth and stabilizing particle size. More work needs to be done on the use of modern surface-active agents such as polyethylene glycol, polyvinyl alcohol, sulfonated naphthalenes, and water-soluble cellulose ethers.

Wavelength of Incident Light The wavelength of light is generally more critical in turbidimetry than it is in nephelometry because in turbidimetry it is imperative to prevent or minimize *absorption,* which is not so important in nephelometry. Obviously, in measuring the transmittance of a solution, there is no way of distinguishing between radiation that has been removed by absorption and radiation that has been removed by scattering. Thus, in turbidimetry, it is important to choose a wavelength where the sample solution does not absorb strongly. If the sample solution is colored, often it suffices to use light of that same color (see Sec. 2-1*C*). For clear solutions with dark particles, light in the red or even near infrared may show minimum absorption. In nephelometry absorption is much less of a problem, and white light is usually used as a convenience. In both turbidimetry and nephelometry the calibration standards should be colored similarly to the sample solutions to minimize absorbance effects.

Small particles give *Rayleigh scattering,* and the intensity of scattering is inversely proportional to the fourth power of the wavelength (see Sec. 1-4*C*). Thus, blue light is scattered more efficiently than red light, which accounts for the blue color of the sky. An understanding of Rayleigh scattering allows a simple test for optimum *particle size* to be made in nephelometry. Thus, if *small* particles are illuminated with white light and viewed at 90°, a bluish color will be noted (Rayleigh scattering). On the other hand, if the particles are too large, white light will be scattered, largely because of reflection. This test is sufficiently sensitive to be carried out with the eye when the particles are very uniform in size but may require a wavelength scan with a spectrophotometer when a large range of particle sizes is involved. If by this test the particles turn out to be too large for Rayleigh scattering, it may be possible to improve the sensitivity by measuring the scattering at less than 90° (small-angle scattering), or it may be necessary to reform (reprecipitate) the particles.

11-1B Turbidimetry

In turbidimetry the transmittance of a primary beam of radiation is measured, where

$$T = \text{transmittance} = \frac{I}{I_0} \tag{11-1}$$

I_0 is the intensity of incident light, preferably measured after passing through a comparison cell containing solvent, and I is the intensity of light after passing through the cell containing the sample. The transmitted radiation is related to the concentration C of suspended material by an equation completely analogous to Beer's law [Eqs. (1-38) and (1-39)]:

$$S = \log \frac{I_0}{I} = kbC \tag{11-2}$$

where S may be called the "turbidence" due to scattering (analogous to the term absorbance), k is a proportionality constant, sometimes called the *turbidity coefficient,* and b is the path length. Another proportionality constant called the *turbidity* τ, equal to $2.303k$, is sometimes used. The value of the proportionality constant depends on the particle size and shape, wavelength, and refractive indexes of the suspended and suspending media. Strictly speaking, Eq. (11-2) is valid only for small particles, where Rayleigh scattering is the only mechanism of attenuation, and for low concentrations of suspensions, where multiple scattering is unlikely. However, the suspension must not be too dilute or the transmitted intensity I will be too similar to the incident intensity I_0 for accurate measurement. In real cases there are appreciable departures from Eq. (11-2), just as there are departures from Beer's law. In practice, analyses are made by preparing a working curve and plotting S vs. known concentrations of scattering material.

11-1C Nephelometry

In turbidimetry it is possible to derive a theoretical equation relating the transmittance to the concentration of suspended particles [(Eq. (11-2)], but in nephelometry it is not practical to relate the scattered intensity to the concentration by any simple theoretical equation. This is because the scattered intensity depends in a complicated way on the properties of the scattering suspension and the angle and geometry of the measuring instrument. The best that can be done is to relate the scattered intensity I_s empirically to the concentration of suspended particles C by the approximate equation

$$I_s = k_s I_0 C \tag{11-3}$$

where k_s is an empirical constant for the system, I_0 is the incident intensity, and all measurements are made under identical conditions.

To carry out a quantitative analysis in nephelometry, a working curve of I_s/I_0 is plotted vs. the concentration of suspended particles under carefully controlled conditions. Actually, since Eq. (11-3) is only an approximate relationship, log (I_s/I_0) is often plotted vs. C, not because of any theoretical significance, but to conform with the more usual spectrophotometric and turbidimetric practices.

11-2 INSTRUMENTATION

Much of the instrumentation used in turbidimetry and nephelometry is very similar to the spectrophotometric devices described in Chap. 2. Only special features will be described here.

11-2A Instrument Components

SOURCES

Although white light may be used with nephelometers, it is advantageous to use monochromatic radiation, and it is necessary to use monochromatic radiation with turbidimeters to minimize absorption. In either case the source should be of high intensity, and wherever possible (where absorption and fluorescence do not preclude it) short wavelengths should be used in order to increase the efficiency of Rayleigh scattering. The mercury arc, with filters to isolate one of its lines, is one common source. Another is a tungsten lamp with a monochromator or filters.

DETECTORS

Ordinary detectors such as phototubes may be used with turbidimeters, but photomultiplier tubes should be used with nephelometers since the intensity of scattered radiation is usually quite small. Most nephelometers have the detector fixed at an angle 90° to the primary beam, but for maximum versatility and sensitivity it would be desirable to be able to vary the detector angle, particularly at an angle closer to the primary beam. In some research instruments the detector is mounted on a circular disk that turns to allow the detector to be used at various angles.

CELLS

Although cylindrical cells are often used, it is far preferable to use cells with flat faces. This is to minimize reflections and multiple scattering from the cell walls. A semioctagonal cell designed for use with the Brice-Phoenix light-scattering photometer is shown in Fig. 11-1. The "octagonal" faces allow measurements to be made at 0, 45, 90, or 135° to the primary beam. If a cell is to be routinely used for a fixed angle of measurement, walls through which light is not to pass should be painted a dull black to absorb unwanted radiation and minimize stray radiation.

11-2B Turbidimeters

Most turbidity measurements are carried out with ordinary colorimeters or spectrophotometers. Simple visual instruments like the Parr turbidimeter or the Duboscq colorimeter are also used.

An unusual and interesting turbidimeter is the Du Pont model 430, shown schematically in Fig. 11-2. This turbidimeter is much more sensitive to

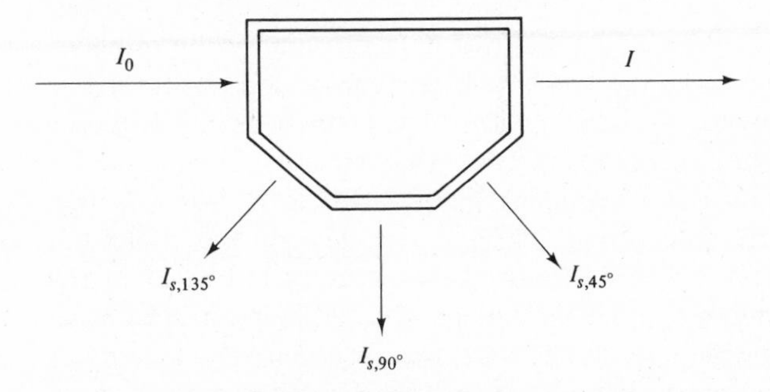

FIGURE 11-1 A semioctagonal cell designed for light-scattering measurements. (*After G. W. Ewing, "Instrumental Methods of Chemical Analysis," 3d ed., p.* 152, *McGraw-Hill Book Company, New York,* 1969, *by permission.*)

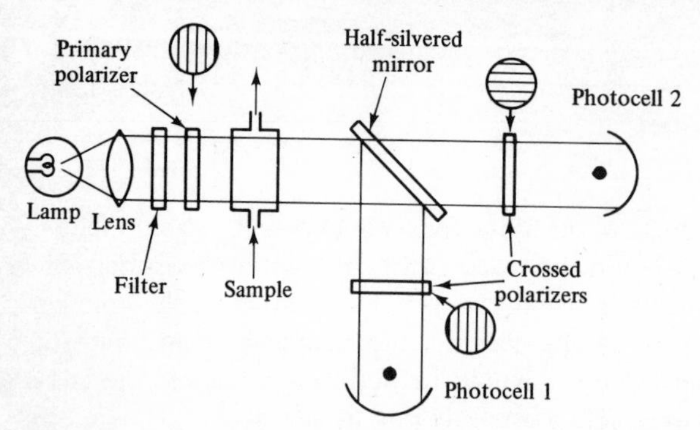

FIGURE 11-2 Schematic diagram of a Du Pont model 430 turbidimeter. (*Du Pont Instruments.*)

low concentrations of suspended particles than an ordinary turbidimeter, based on the fact that scattering causes the plane of polarization to be changed. Thus, the incident beam is plane-polarized with the primary polarizer (see Fig. 11-2). After passing through the sample, the beam is split into two parts with a half-silvered mirror and detected with two separate photocells. With no suspended particles in the sample solution, photocell 1 will give a maximum response and photocell 2 will give minimum or zero response; the ratio of signal 2 to signal 1 is taken as a measure of the concentration of suspended particles. As the concentration of suspended particles in the sample increases, the response of photocell 2 increases while that of 1 decreases, and the ratio of the two signals is thus a sensitive measure of the turbidity. This double-beam arrangement also minimizes the problem of absorption by the particles or the solution, but it cannot be used with solutions that contain optically active substances. The instrument can be used as an online monitor for flowing streams, as well as for individual samples.

11-2C Nephelometers

Nephelometric measurements are often carried out on ordinary fluorometers (see Sec. 8-2). In addition, some spectrophotometers can be adapted for use as nephelometers, and even visual-matching instruments such as the Duboscq colorimeter can be modified for nephelometric measurements.

An example of a moderately complex light-scattering photometer that can be used for the determination of particle size, shape, and molecular weight, in addition to nephelometric measurements, is given in Fig. 11-3. The photomultiplier tube is mounted on a turntable and could be positioned at almost any angle, but for nephelometric measurements it is generally positioned 45 or 90° to the primary beam. The undeviated primary beam passes into a black tube called a light trap (T in Fig. 11-3).

Other commercially available nephelometers, ranging from the very simple to more elegant, versatile, and expensive designs are available from

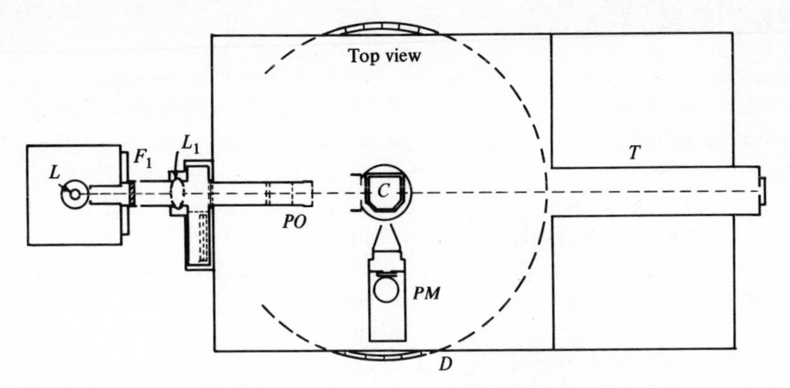

FIGURE 11-3 Simplified schematic diagram of a Brice-Phoenix light-scattering photometer. C = semioctagonal cell, D = graduated disk to which deflector is attached, F_1 = monochromatic filter, L = mercury lamp, L_1 = achromatic lens, PM = photomultiplier tube, PO = demountable polarizer, T = light-trap tube. (*Phoenix Precision Instrument Co.*)

Hach Chemical Co., Fisher Scientific Co., Coleman Instruments, Inc. (a division of Perkin-Elmer Corp.), American Instrument Co., and Leeds and Northrup Co.

11-3 APPLICATIONS

11-3A Inorganic Analysis

The most widespread use of light-scattering measurements is to determine the clarity of all kinds of water and beverage products. Other examples include the determination of sulfate (as $BaSO_4$), carbonate (as $BaCO_3$), chloride (as AgCl), fluoride (as CaF_2), cyanide (as AgCN), calcium (as oxalate or oleate), and zinc (as ferrocyanide). The sulfate determination is particularly widely used and serves for the routine determination of total sulfur in coal, coke, oils, rubber, and other organic materials. A useful carbon dioxide determination is to bubble the gas through an alkaline solution of a barium salt and then analyze for the barium carbonate suspension. Many other examples are given in Refs. 4 and 5.

11-3B Organic Analysis

Applications to the analysis of food and beverages form the biggest group here. Included are the analysis of turbidity in sugar products such as maple syrup and molasses, the clarity of citrus juices, and the clarity of various alcoholic beverages. Other examples include determining the benzene content of alcohol by dilution with water (and causing the benzene to become an immiscible suspension) and the determination of alkaloids by precipitation as phosphomolybdates.

11-3C Biochemical Analysis

A very widespread, routine application of turbidimetry is to measure the amount of growth of a test bacterium in a liquid nutrient medium. In addition, turbidimetry is used to measure amino acids, vitamins, and antibiotics. Nephelometry has been used for the estimation of protein and the determination of yeast, glycogen, and of beta and gamma globulins in blood serum and plasma. Nicotine can be analyzed by precipitation with tungstosilicic acid, or it can be determined in the presence of nornicotine by precipitation with potassium iodomercurate. Other examples include the determination of ribonuclease, sulfate in urine, and the thrombin fibrinogen (clot capacity) in plasma. References 4 and 5 should be consulted for further details.

11-3D Titration Analysis

Numerous workers have used the measurement of turbidity as a way of measuring the end point in a titration. A few examples include the titration of 0.01 M fluoride with calcium, 10^{-4} to 10^{-6} M bromide with silver, sulfate with barium, and ketones in alcohol-ketone mixtures by titration with water. Reference 5 lists over 50 different precipitation reactions that have been monitored by turbidimetry or nephelometry.

11-3E Continuous On-Stream Monitoring

Turbidimetry and nephelometry are becoming extremely important tools in the continuous monitoring of air and water pollution. Dust and smoke are monitored in air, and any form of turbidity is monitored in various types of natural waters. Automation of turbidimetric measurements is widely used in the petroleum industry, and specific automatic analyzers for substances such as sulfate and hydrogen sulfate are becoming available [4].

11-3F Particle-Size Classification

The application of turbidimetry to the classification of particle size is beyond the scope of this chapter, but it is sufficiently important to be mentioned. Usually sedimentation techniques are used, based on Stokes' law, which says that particles in suspension fall at rates in proportion to the square of their radii. Thus, turbidimetry is used as a monitor to determine the rate of settling of various sized particles [4].

EXERCISES

11-1 What is the main criterion for deciding whether turbidimetry or nephelometry should be used in the analysis of a medium containing suspended particles?

11-2 For very small particles (diameters no more than 5 or 10 percent of the wavelength of incident radiation), Rayleigh scattering occurs, in which the intensity of scattered light is proportional to λ^{-4} [see Eq. (1-17)]. Speculate on the magnitude of the wavelength exponent for *real* chemical systems in which particles cover a range of sizes and are usually larger than Rayleigh particles. *Hint:* Recall that reflection of white light often occurs with large particles.

Ans: λ^{-3} and λ^{-2} are typical values.

11-3 (*a*) From your knowledge of the general nature of scattered radiation, give an approximate equation relating the scattered intensity I_s to the difference in intensity between the incident and transmitted intensities $I_0 - I$. (*b*) What physical and chemical factors affect the relationship between I_s and $I_0 - I$? (*c*) Can I_s, as measured at some fixed angle, ever be greater than $I_0 - I$? (*d*) Why is it that I_s can be measured with much greater precision and accuracy at low concentrations of suspended particles than the transmitted intensity I/I_0?

11-4 In the turbidimetric analysis of sulfate using a Beckman DU spectrophotometer at a wavelength of 355 nm, a certain sample in a 1.00-cm cell is found to have a turbidance S of 0.121. If the turbidity coefficient k of sulfate is 1.73×10^3 l/(mole)(cm) at this wavelength and in this concentration region, what is the concentration of sulfate? *Ans:* 6.9×10^{-5} mol/l

11-5 It is important to measure the absorbance of a certain dissolved solute, but there is present in the solution a cloudy precipitate of some other solute. Simple means of removing the cloudy solute (filtration, centrifugation) fail. Devise a means of correcting the absorbance for the turbidity. What problems are involved, and how accurate do you think the correction will be?

11-6 In a spectrophotometric study in the visible region a certain colored solution gives an absorbance A of 0.15 in a 1-cm cell. When the solution is removed from the spectrophotometer, the solution appears cloudy in addition to being colored. By a series of measurements (see Exercise 11-5), it is determined that the turbidity τ is 0.012 at the wavelength used for the absorbance measurement. What is the true absorbance of the solution?

Ans: 0.145

REFERENCES

INTRODUCTORY

1 BLAEDEL, W. J., and V. W. MELOCHE: "Elementary Quantitative Analysis," 2d ed., pp. 882–885, Harper & Row, New York, 1963. Brief, but clear summary.

2 EWING, G. W.: "Instrumental Methods of Chemical Analysis," 3d ed., pp.151–159, McGraw-Hill, New York, 1969. A capsule summary of attributes and problems.

INTERMEDIATE

3 STROBEL, H. A.: "Chemical Instrumentation," pp. 71–80 and 217–227, Addison-Wesley, Reading, Mass., 1960.

4 HOCHGESANG, F. P.: Nephelometry and Turbidimetry, chap. 63 in I. M. Kolthoff, P. J. Elving, and E. B. Sandell (eds.), "Treatise on Analytical Chemistry," pt. I, vol. 5, Wiley-Interscience, New York, 1964.

ADVANCED

5 KRATOHVIL, J. P.: Light Scattering, *Anal. Chem.*, **38**(5):517R (1966). A thorough review.

6 BILLMEYER, F. W., JR.: Principles of Light Scattering, chap. 56 in I. M. Kolthoff, P. J. Elving, and E. B. Sandell (eds), "Treatise on Analytical Chemistry," pt. I, vol. 5, Wiley-Interscience, New York, 1964. Emphasis on particle-size and molecular-weight measurements.

7 OSTER, G.: Light Scattering, chap. 32 in A. Weissberger (ed.), "Physical Methods of Organic Chemistry," 3d ed., vol. I, pt. III, Wiley-Interscience, New York, 1960. A detailed treatment, with emphasis on measuring size and molecular weight of polymers.

8 SLOAN, C. K.: Angular Dependence Light Scattering Studies of the Aging of Precipitates, *J. Phys. Chem.*, **59**:834 (1955). The effect of aging of precipitates and the sensitivity dependence of various scattering angles.

12

X-Ray Methods: Absorption, Emission, and Diffraction

Since a variety of x-ray techniques and methods are in use, for our purpose we shall classify all methods as falling in one of three main categories: absorption, emission, and diffraction methods.

X-ray absorption methods are analogous to absorption methods in other regions of the electromagnetic spectrum. A beam of x-rays is passed through the sample, and the attenuation, or fraction of x-ray photons absorbed, is taken as a measure of the concentration of the absorbing substance. Absorption techniques are undoubtedly the least used of the various x-ray methods but are helpful in certain cases for elemental analyses and thickness measurements.

X-ray emission methods are those in which x-rays are generated from within the sample, and by measuring the wavelength and intensity of the generated x-rays qualitative and quantitative elemental analysis is possible. X-ray emission methods can be further classified according to the method used to excite x-rays in the sample. In *direct* x-ray excitation the sample is bombarded with a beam of electrons, and the x-rays generated in the sample are measured. In practice, this technique is relatively little used, since the sample generally has to be mounted inside an evacuable x-ray tube. However, a more convenient modification of this approach, *electron-probe microanalysis*, has been developed (see Sec. 12-3*C*).

The alternate (*indirect*) approach to generating x-rays in a sample is to use other x-rays (primary x-rays) to excite the sample. The primary x-rays must be more energetic than the minimum absorption energy of the sample. The emitted x-rays (secondary x-rays) will be of less energy and characteristic of the element being excited. This method is usually called *x-ray fluorescence* spectroscopy, and as an analytical tool for qualitative and quantitative analysis it is probably more important than all other x-ray methods combined.

Finally, *x-ray diffraction* methods are based on the scattering of x-rays by crystals, the scattering being caused by the electron atmosphere of the atom. The wavelength of a scattered x-ray remains unchanged. A diffraction pattern is produced by such scattering only when certain geometrical conditions as expressed by the Bragg law or Laue conditions are satisfied. Diffraction patterns may be used to identify molecules and to determine the atomic and molecular structure. Though x-ray diffraction methods are extremely important for determining atomic and molecular structures, a thorough treatment will not be given in this chapter since the development of any kind of expertise in x-ray diffraction requires specialized training which is beyond the scope of this book. Only the general principles and applications of x-ray diffraction will be discussed.

12-1 GENERAL THEORY

12-1A Origin of X-Rays

The x-ray region of the electromagnetic spectrum is bounded on the short-wavelength end by the gamma-ray region and on the long-wavelength end by the vacuum-ultraviolet region and consists of wavelengths in the range of about 0.1 to 100 Å. As with most other regions of the electromagnetic spectrum, the precise boundaries are not clearly defined. The range of 0.7 to 2.0 Å is the region most useful for analytical purposes.

The process of producing x-rays may be visualized in terms of the Bohr atom picture given in Fig. 12-1. An inner-shell electron (usually a K- or L-shell electron) can be knocked out of the target atom by a high-energy electron or a high-energy (primary) x-ray. Following the loss of the inner-shell electron one of the outer electrons will fall into the vacated orbital, with the simultaneous emission of an x-ray photon. The energy of the emitted x-ray will correspond exactly to the difference in energy between the two levels involved. For example, if a K-shell electron is ejected and an L-shell electron takes its place, the resulting K_α x-ray (see Fig. 12-1) will have an energy given by

$$E_{K\alpha} = E_L - E_K \tag{12-1}$$

where E_L and E_K represent the energies of the L- and K-shell electrons, respectively. The customary notation used to identify the emitted x-ray is illustrated in Fig. 12-1. The *K series* of lines is obtained when an electron in the innermost K shell is dislodged; the *L series* of lines is obtained when an

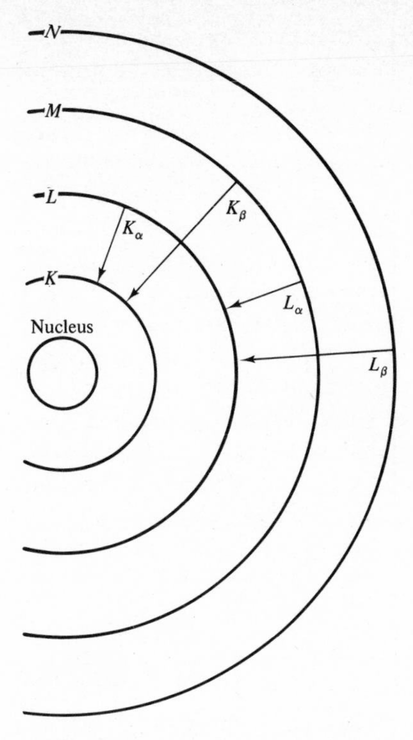

FIGURE 12-1 Bohr atom representation of some transitions responsible for x-rays.

electron in the L shell is ejected; etc. The subscripts, α, β, γ, etc. are used to identify the origin of the electron which filled the vacated spot, with a transition from the closest shell being labeled an α transition, a transition from the next nearest neighbor being a β transition, etc. Additional numerical subscripts (K_{α_1}, K_{β_2}, etc) are used to identify the particular subshell or spin state within a main shell from which the transition originated.

In practice the spectrum of x-rays emitted from a given target element may be either a series of sharp and characteristic lines, as implied by the preceding discussion, or may be a much more complicated continuum of x-rays with fairly sharp spectral lines superimposed on the continuum. The difference depends on whether primary x-rays (in the former case) or high-energy electrons (in the latter) are used to bombard the target element and eject an inner-shell electron. Figure 12-2 shows a typical spectrum emitted by a molybdenum target which has been bombarded by electrons accelerated at 35 kV. If primary x-rays had been used to excite the target element, the continuum would be absent and only the sharp emission lines would be present.

12-1B Production of X-Rays

Since all x-ray *tubes* depend on electrons for excitation, the x-ray continuum is important to understand. It is due very simply to the rapid deceleration of the bombarding electrons because of multiple interactions with the electrons

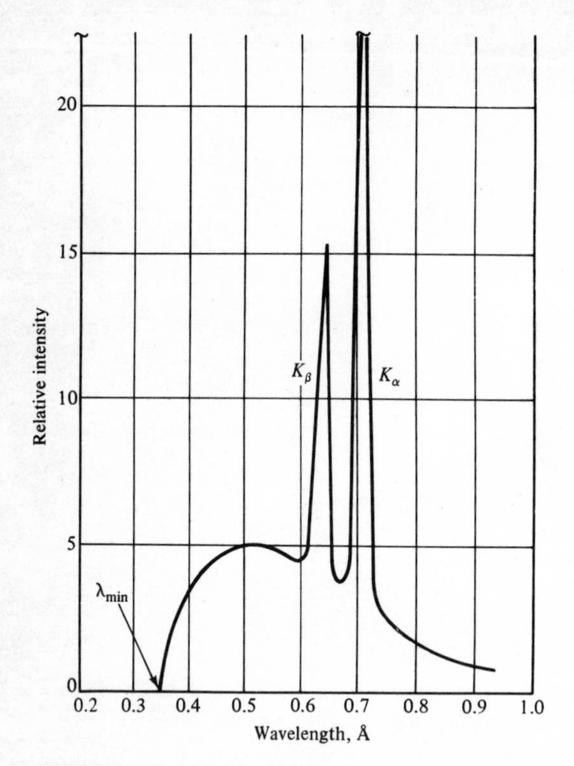

FIGURE 12-2 Intensity distribution from a molybdenum-anode x-ray tube operated at 35 kV. (*After G. H. Brown and E. M. Sallee, "Quantitative Chemistry," p.* 537, *Prentice-Hall, Inc., Englewood Cliffs, N.J.*, 1963, *by permission.*)

in the target as the bombarding electrons pass through the target material. The energy lost in slowing down will be converted into a continuum of x-radiation (sometimes called bremsstrahlung), and there will be a sharp minimum wavelength λ_{min} (maximum frequency) corresponding to the maximum energy of the electrons (see Fig. 12-2). This cutoff wavelength is inversely proportional to applied voltage, as given by

$$\lambda_{min} = \frac{hc}{Ve} = \frac{12{,}400}{V} \tag{12-2}$$

where h = Planck's constant
c = speed of light in vacuum
e = charge on electron
V = accelerating voltage applied across x-ray tube, V

By substitution of the appropriate values for the constants a factor of 12,400 results when λ_{min} is in angstroms. The wavelength of *maximum intensity* of the continuum can be estimated to occur at about 1.5 times the short-wavelength minimum.

Example 12-1 Calculate the minimum wavelength emitted by an x-ray tube operated at 40 kV. Also estimate the wavelength of maximum intensity.

Answer

$$\lambda_{\min} = \frac{12{,}400}{40{,}000} = 0.310\ \text{Å}$$

$$\lambda \text{ of max intensity} \approx 1.5\,(0.31\ \text{Å}) = 0.46\ \text{Å}$$

The x-ray continuum is a very useful feature of an x-ray tube, providing a versatile source of x-rays for irradiating various elements. The *intensity*, or total power, of the emission spectrum of an x-ray tube, which amounts to the integrated intensity of the output spectrum I_{int}, depends on the *electric power* applied to the tube iV, in watts, where i is the current in amperes and V is the voltage across the tube in volts, and the *efficiency* of x-ray production. In practice the efficiency can only be empirically evaluated, but a useful equation for estimating the integrated output intensity is

$$I_{\text{int}} = 1.4 \times 10^{-9} iZV^2 \tag{12-3}$$

where Z is the atomic number of the target element used in the x-ray tube. Equation (12-3) becomes less reliable as V decreases, but nonetheless it emphasizes two important things about the production of x-rays by electron bombardment. First, it is obvious that this means of producing x-rays is inefficient. In typical cases, less than 1 percent of the electron energy appears in the x-ray beam. Almost all the rest of the energy is degraded to heat, and thus special provisions for cooling the target often are necessary. Obviously the higher the electric power applied to the tube the more severe the heating problem becomes.

Second, it should be noted from Eq. (12-3) that the output intensity depends on the atomic number Z of the target anode used in the x-ray tube. In practice, the spectral distribution in the continuum is relatively constant for various target elements [and thus, for example, the cutoff wavelength $\lambda_{\min}$ given by Eq. (12-2) is independent of Z], but the absolute intensity is greater for higher atomic numbers. Tungsten is the most frequently used target element in x-ray tubes because of its high atomic number and because its high melting point allows a higher tube current to be used. When this tube is operated at 50 kV, the emitted x-rays are capable of exciting K radiation from elements in the periodic table up to about cerium ($Z = 58$). In heavier elements the energy of x-rays required to knock out K-shell electrons is too great for the tungsten continuous-spectrum source, although this tungsten radiation is quite efficient in exciting L spectra or M spectra. Tungsten x-ray tubes capable of operating at 100 kV have been developed, allowing K spectra of elements up to gold ($Z = 79$) to be excited. This is a distinct advantage, since K spectra, in general, are simpler and more intense then L or M spectra.

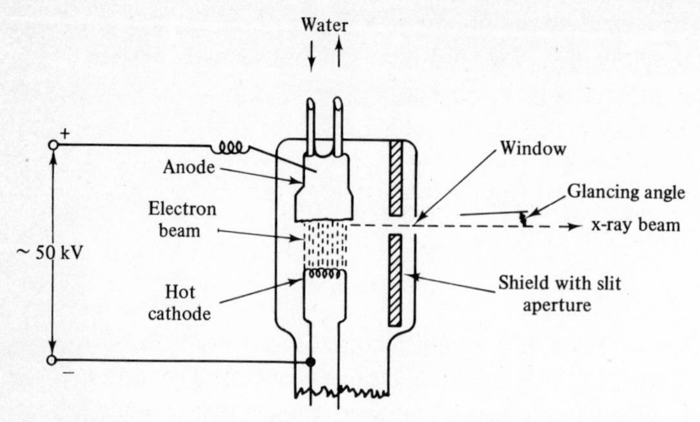

FIGURE 12-3 Schematic diagram of filament and target of an x-ray tube.

Sometimes other factors become important in the choice of target materials for x-ray tubes. For example, if tungsten were to be analyzed for by x-ray fluorescence spectroscopy, it would be better to use a tube with another target element, such as molybdenum, since some of the primary radiation from the tungsten tube could be scattered by the sample and reach the detector, interfering with the measurement of the tungsten fluorescent radiation. Another factor in the choice of target elements is their variable heat effects, which in turn determine the thickness of window to be used in the x-ray tube. In general, thinner windows can be used when lower-atomic-number anodes are used, and thin windows can be advantageous when using long-wavelength x-rays, as in the analysis of light elements.

A schematic diagram of the filament and target of a typical x-ray tube is given in Fig. 12-3. The cathode is heated by means of a filament current, and the electrons that are emitted are accelerated through a high voltage to the target anode. Electrons striking the anode produce x-rays and heat energy and a significant portion of the x-rays pass through the window. The large amount of heat produced at the target anode must be dissipated; circulating water is usually used for this purpose.

12-1C Interaction of X-Rays with Matter

X-rays can interact with matter in three important ways: *absorption, scattering,* and *diffraction.*

ABSORPTION

In contrast to most absorption processes at longer wavelengths, the absorption of x-rays depends entirely on *atomic* rather than molecular properties.† In other words, the extent of absorption by a given element de-

† Atomic absorption spectroscopy also depends on atomic properties (see Chap. 5).

pends only on the number of atoms of that element in the path of the x-rays and is *independent* of the physical or chemical state of that element. The amount of absorption by a given number of bromine atoms, for example, will be the same regardless of whether the bromine is in the form of a monatomic gas, a diatomic gas, a liquid, or a solid or is present as a compound such as potassium bromide or bromobenzene. This simplicity is a unique feature of x-ray absorptiometry.

The absorption of x-rays follows Beer's law, which may be written

$$I = I_0 e^{-\mu l \rho} \tag{12-4}$$

where I_0 = incident intensity of x-rays
I = intensity after passing through absorbing sample
l = path length, cm
μ = mass-absorption coefficient, cm^2/g
ρ = density of absorbing material, g/cm^3

The mass-absorption coefficient μ is a function only of the wavelength of x-ray used and the atomic number of the absorbing element. Tables of mass-absorption coefficients as a function of atomic number and wavelength have been published [6].

If a plot of mass-absorption coefficient vs. wavelength for a given element is prepared, sharp discontinuities called *absorption edges* are found. Figure 12-4 shows such a plot for three different elements, lead, copper, and aluminum, over a large range of wavelengths (note that it is a log-log plot). The mass-absorption coefficient may be regarded as a measure of the probability of absorption of an incoming x-ray. As Fig. 12-4 shows, the probability of absorption increases as the wavelength increases, until an absorption edge is reached, when a sharp dropoff occurs. The dropoff occurs at a wavelength corresponding to the binding energy of the electron involved. At x-ray energies equal to or slightly greater than the absorption edge, there is a high probability that the x-ray will interact and eject the electron in question, whereas when the x-ray energy is less (wavelength longer) than the absorption edge, the probability for ejection of that electron falls sharply. The mass-absorption coefficient does not fall to zero at the absorption edge, for two reasons. First, even though the electron in question no longer can be removed, other, more weakly held electrons may now interact with the x-rays. Furthermore, there is a second mechanism, besides ejection of electrons, which contributes to the observed absorbance. This is *scattering* of the x-rays by the electrons. Scattering increases in importance at shorter wavelengths (decreasing λ) and lower atomic numbers (lower Z), but except when dealing with light elements (like carbon, nitrogen, and oxygen) scattering contributes very little to the overall mass-absorption coefficient.

Between absorption edges, the mass-absorption coefficient can be related to the wavelength λ and atomic number Z by the approximate empirical equa-

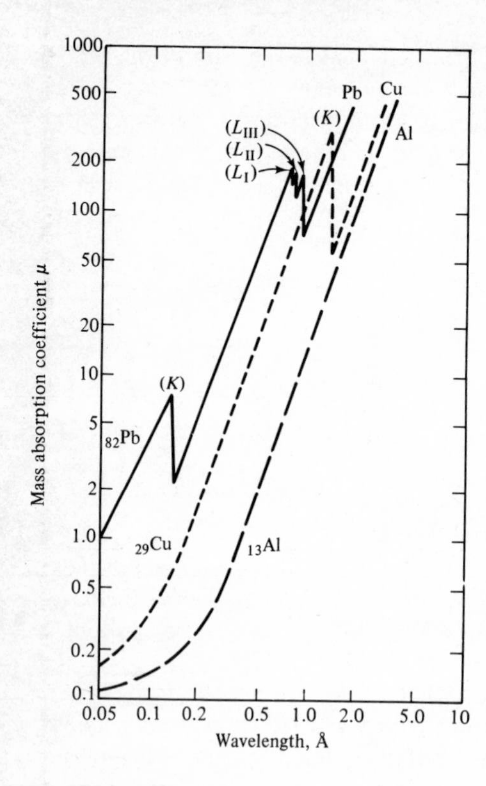

FIGURE 12-4 Mass-absorption coefficients for lead, copper, and aluminum as a function of wavelength. (*After H. A. Liebhafsky, H. G. Pfeiffer, and F. H. Winslow, in I. M. Kolthoff and P. J. Elving (eds.), "Treatise on Analytical Chemistry," pt.* I, *vol.* 5, *p.* 3088, *John Wiley & Sons, Inc., New York,* 1964, *by permission.*)

tion

$$\mu \approx \frac{CN}{A} Z^4 \lambda^3 \tag{12-5}$$

where C = proportionality constant
N = Avogadro's number
A = atomic weight of absorbing element

This equation assumes that photoelectric absorption is the only absorption process of importance, and thus the equation becomes more accurate as Z and λ increase (and scattering becomes less important).

The important features of Fig. 12-4 can now be better understood. First, lead is seen to have a K edge and three L edges, corresponding to the ejection of K electrons and L electrons. The L electrons have less binding energy and are ejected at longer wavelengths than the K electrons. Furthermore, the L electrons have three slightly different energy levels, giving three closely spaced edges. Over the wavelength range shown in Fig. 12-4 copper has only the K edge, and aluminum has no absorption edges, the binding energy of their electrons obviously being less than the binding energy of the higher-atomic-number element, lead. With copper and particularly aluminum, the slope starts to fatten out at short wavelengths, due to an increase in scattering and corresponding decrease in photoelectric absorption. As the wavelength increases, the slopes between absorption edges approach the value of 3 for the log-log plot, as would be expected from Eq. (12-5). Finally, it should be noted from Fig. 12-4 that the value of the absorption coefficient increases as Z increases, as Eq. (12-5) predicts.

There are various ways in which these principles can be utilized for analytical purposes. Three of these ways may be termed x-ray *absorptiometry* because they are based on Beer's law of absorption, Eq. (12-4). A fourth method is to measure the *emission* of longer-wavelength (fluorescent) x-rays after irradiation with primary x-rays. These four methods can be explained by referring to the lead mass-absorption coefficient curve in Fig. 12-4.

The first absorption method is completely analogous to spectrophotometry in the ultraviolet, visible, or infrared region of the spectrum. Specifically, *monochromatic* x-rays, preferably at a wavelength just on the short-wavelength side of an absorption edge, are used to irradiate the sample, and the extent of absorption is then related to the number of atoms of the element being sought that are present. Unfortunately, this method is rather unspecific, since almost all elements that are present will absorb to varying degrees, and furthermore the sensitivity is also relatively low. In short, this method is usually impractical and rarely used.

A second method of carrying out x-ray absorptiometry is to irradiate the sample sequentially with several monochromatic wavelengths above and below a given element's absorption edge; by extrapolation the absorption edge is located and used to identify the element. The decrease in the measured absorption coefficient at the edge gives the amount of the element present. This method is specific for the element in question, and although it is relatively insensitive and requires some manipulation of the data to obtain the desired result, it nonetheless has found a fair amount of usefulness in routine analyses, especially in determining chlorine, sulfur, and bromine in hydrocarbons.†

The third method of carrying out x-ray absorptiometry is analogous to colorimetry with white light, in that polychromatic beams of x-rays are used to irradiate the sample. The biggest advantage of this method comes from the high intensity of the x-ray beams, making it possible to construct relatively simple and rugged photometers. Thus, while the sensitivity may be relatively high, the method is very unspecific.

Finally, the fourth way in which these absorption principles can be used for analytical purposes is to measure the x-rays emitted from the sample, rather than to measure the absorption of the primary x-rays used to irradiate the sample. Here again a polychromatic beam of primary x-rays is used to irradiate the sample; x-rays that are shorter in wavelength than the absorption

† H. A. Liebhafsky, *Anal. Chem.*, **25**:688 (1953).

edge of interest will cause photoelectric ejection of an electron from the target atom, and the subsequent rearrangement of electrons in the target atom results in emission by the sample of characteristic secondary (fluorescent) x-rays, which are measured. This method, called *x-ray fluorescence spectroscopy,* is very specific and can be used for qualitative elemental analysis. It is fairly sensitive, although the specificity is its biggest advantage (see Sec. 12-3).

SCATTERING AND DIFFRACTION

Since the phenomenon of *scattering* is the basis of *diffraction,* they are considered together. Scattering occurs when electrons in the path of the electromagnetic x-radiation are forced to oscillate at the same frequency as the primary radiation, thereby acting as small oscillators and emitting electromagnetic radiation in all directions at the same frequency as the primary radiation. The sum of all the scattered waves coming from an orderly array of electrons in a crystal lattice will result in reinforced waves traveling in certain directions and out-of-phase, or diminished, wavefronts in other directions. The reinforced waves are said to be *diffracted* by the crystal planes. Every crystalline substance scatters the x-rays in its own unique diffraction pattern, producing a fingerprint of its atomic and molecular structure.

The conditions for diffraction are governed by Bragg's law, which can be understood by reference to Fig. 12-5. Reinforcement of reflected rays emerging from two different planes will occur if the difference in the path lengths of the two rays is equal to a whole number of wavelengths. In Fig. 12-5 the path difference between two parallel x-rays, one striking the top crystal plane at A and the other striking the second crystal plane at B, is equal to $CB + BD$; and since

$$CB = BD = l$$

$n\lambda$ must equal $2l$ for reinforcement, where n is an integer. However, from the

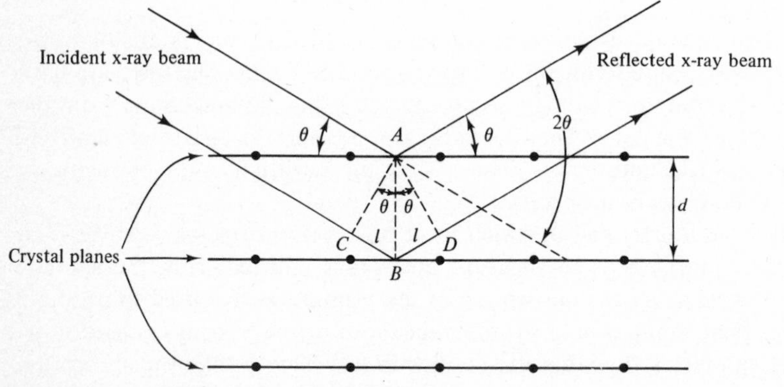

FIGURE 12-5 Diffraction of x-rays from a set of crystal planes.

geometry of Fig. 12-5 it can be seen that

$$l = d \sin \theta$$

where d is the lattice spacing. Hence the ultimate reinforcement condition is

$$n\lambda = 2d \sin \theta \tag{12-6}$$

which is the familiar Bragg's law.

In Eq. (12-6), the integer n is called the *order* of the reflection, and it should be noted that a second-order beam of wavelength $\lambda/2$ will be reflected at the same angle as a first-order beam of wavelength λ. Hence a crystal cannot be used to distinguish between an x-ray of wavelength λ, $\lambda/2$, $\lambda/3$, etc. However, from Eq. (12-5) it should be noted that the mass-absorption coefficient changes greatly with wavelength, and if it is necessary to distinguish between various orders of reflection, it is possible to insert a suitable absorber between the crystal and detector. In this way the relative intensities of the various orders can be estimated.

X-rays can be used for diffraction studies of crystals because the wavelengths of x-rays are comparable in magnitude to the spacings in crystals (see Exercise 12-5). Bragg's law is obeyed so well that it is possible to use Eq. (12-6) for highly precise determinations of the lattice spacing d. Alternatively, crystals of known spacing can be used like a grating for precise resolution and measurement of the wavelength λ of x-rays. In the latter application the detector is placed at an angle 2θ from the incident x-ray beam (see Fig. 12-5), and the wavelength or radiation being measured is accurately calculated from Eq. (12-6).

Scattering is important not only because it is responsible for x-ray diffraction but because it usually contributes to the background encountered when the intensities of analytical lines are being measured. Whereas most of the scattering processes discussed correspond to coherent, or *Rayleigh,* scattering, the background problem is due largely to incoherent, or *Compton,* scattering. Rayleigh scattering may be thought of as arising when an x-ray photon interacts directly with a tightly bound electron, causing the electron to oscillate and emit energy at the same wavelength as the original x-ray. On the other hand, if the electron is only loosely bound, the x-ray may impart only a fraction of its energy to the electron and the emitted (scattered) radiation is at a longer wavelength (decreased energy) than the original x-ray. This results in a *continuum* of scattered x-rays at longer wavelengths than the primary x-ray, and this usually represents an undesirable background. Compton scattering becomes more severe with shorter wavelengths of the x-ray and lower atomic number of the target element.

12-1D Detection of X-Rays

In order to detect x-rays it is necessary to convert the x-ray energy into another form of energy which is more readily measured. There are four different devices for doing this: (1) photographic plates; (2) gas-filled detectors, such as Geiger counters and proportional counters; (3) solid-state scintillation counters; and (4) solid-state semiconductors. All four depend on the ability of x-rays to ionize matter and differ only in the subsequent fate of the electrons produced by the ionization process.

Photographic plates form the simplest type of detector. Here the x-rays induce a photochemical reaction which reduces silver halides to free silver via an ionization process. The intensity of the x-ray beam will determine the number of silver atoms produced, and this in turn can be estimated from the blackening of the film. This type of detector is widely used in diffraction studies, since it clearly reveals the entire diffraction pattern on a single film, but it is rarely used for the other x-ray methods since a quantitative measure of the intensity from the film blackening requires a densitometer, the operation of which can be both time-consuming and subject to considerable error.

Gas-filled detectors are perhaps the most widely used in x-ray fluorescent spectroscopy and are being increasingly used in modern x-ray diffraction spectrometers, although the other types of detectors have certain advantages for some applications. The two most important types of gas-filled detectors are the Geiger counter and the proportional counter.

A schematic drawing of a *Geiger tube* (often called a Geiger-Müller tube) is given in Fig. 12-6. The Geiger tube is filled with an inert gas like argon, and a positive potential of 800 to 2500 V is applied to a central wire anode. When ionizing radiation such as an x-ray enters the active volume of the Geiger tube, collision with the filling gas produces an ion pair; the electron produced migrates toward the center wire, while the positive ion migrates toward the outer electrode. The electron is accelerated by the potential gradient and causes the ionization of numerous other argon atoms, with the result that the initial ionizing event gives rise to an avalanche of electrons traveling toward the central anode. By this internal amplification process a single ionizing

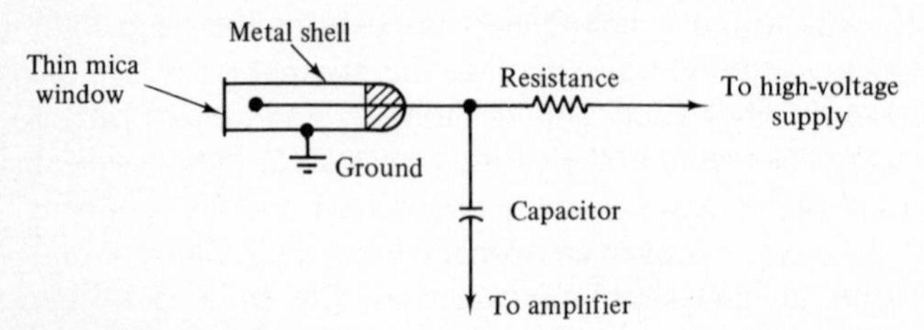

FIGURE 12-6 Schematic drawing of a Geiger tube. (*After G. H. Brown and E. M. Sallee, "Quantitative Chemistry," p.* 552, *Prentice-Hall, Inc., Englewood Cliffs, N.J.*, 1963, *by permission.*)

radiation may result in an output pulse of 1 to 10 V, which is easily measured with relatively simple electronic circuitry. Thus, the biggest advantage of the Geiger over other counters is the simple associated electronic circuitry, making it an inexpensive and relatively trouble-free detector. However, there are a number of disadvantages which should be considered. First, the tube takes a relatively long time to recover from each counting event (a typical dead time is about 250 μs), and thus its use should be restricted to relatively low counting rates (500 counts per second is about the highest count rate a Geiger tube can tolerate without large counting errors). Furthermore, the size of the voltage pulse coming from a Geiger tube will be independent of the energy of the x-ray which caused the ionization, and thus a Geiger tube cannot be used to monitor the energy of the ionizing radiation, something that is possible with proportional counters, scintillation counters, and semiconductor counters. And finally, the efficiency of a Geiger tube falls off rapidly at wavelengths below about 1 Å or so, since high-energy radiation will usually pass through the active volume of the Geiger tube without interacting with the counter gas.

A *proportional counter* is similar in construction to a Geiger counter. Xenon or krypton generally replaces argon as the filling gas, since heavier rare gases are more easily ionized. The major difference between a Geiger counter and a proportional counter lies in the fact that a proportional counter is operated at a voltage below the *Geiger plateau.* In this lower-voltage region the voltage pulses produced by ionizing radiation are directly *proportional* to the engery of the radiation. With the proper electronic circuitry, it is possible to count x-rays of a particular energy selectively. Another advantage of a proportional counter is that the dead time is relatively short (approximately 0.2 μs), and thus a proportional counter can be used for relatively high count rates without significant error (coincidence losses). The sensitivity and efficiency for counting short wavelengths of x-rays is comparable to that of a Geiger counter. To count long-wavelength (10 Å or more) x-rays, proportional counters are often made in the form of *flow counters,* permitting extremely thin windows to be used and thus allowing low-energy x-rays to pass through without being significantly absorbed. Extremely thin, e.g., quarter-mil, or 0.00025-in, windows (such as Mylar or polypropylene) allow appreciable gas permeation, and thus the filling gas must be constantly replenished with a flow of fresh counter gas. The major disadvantage of the proportional counter, compared with a Geiger counter, is that the associated electronic circuitry must be more complex and expensive because the output signal from the proportional counter is relatively low, requiring linear amplification; the stability of the applied voltage to the proportional tube is more critical; and pulse-height discrimination requires relatively sophisticated circuitry. For specific applications, however, e.g., counting x-rays from light elements (long wavelength), and for work involving high count rates, the advantages of the proportional counter outweigh the extra cost.

The third type of detector, the *scintillation counter*, has many of the advantages of proportional counters, but the principles of detection are different. The usual scintillation crystal for x-ray analysis consists of a large sodium iodide crystal activated with a small amount of thallium. The absorption of an x-ray photon in such a crystal causes pulses of visible light to be emitted by the crystal, and this light is detected by a photomultiplier tube. Figure 12-7 shows the construction of a scintillation counter schematically. The light pulses generated in the scintillation crystal strike the light-sensitive surface of the photomultiplier and cause electrons to be ejected. The photomultiplier tube then acts to multiply the number of electrons by means of a series of accelerating grids, or *dynodes,* each of which (from left to right in Fig. 12-7) is charged at successively higher positive voltages. Thus, electrons produced at the photocathode are accelerated to the first dynode, and at the surface of this dynode each original electron ejects several more electrons; these, in turn, are accelerated on to the second and subsequent dynodes. By the time the final collector anode is reached, amplification factors of about 10^6 are typically achieved. By keeping the amplification of the photomultiplier constant, i.e., by applying constant and stable voltages to the dynodes, the output signal from the photomultiplier will be directly proportional to the *energy* of x-ray absorbed in the scintillation crystal. This is true because the number of light photons produced in the scintillation crystal is directly proportional to the energy of the incident x-ray, and in turn the number of electrons ejected by the photocathode is directly proportional to the number of light photons. With a constant photomultiplier amplification factor, the final output of the photomultiplier will be a large number of electrons (or a voltage pulse) which is directly proportional to the energy of the incident x-ray. By feeding the output of the photomultiplier into a pulse-height analyzer, the detection system can be made to count only x-rays of a desired energy.

The principal advantage of the scintillation counter over Geiger and proportional counters is its efficiency in detecting high-energy x-rays. Geiger and

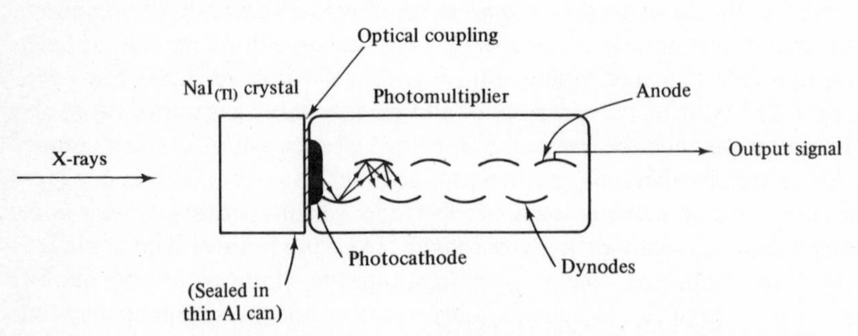

FIGURE 12-7 Schematic drawing of a scintillation-crystal photomultiplier detector. (*After L. S. Birks, "X-Ray Spectrochemical Analysis," p.* 51, *John Wiley & Sons, Inc., New York,* 1969, *by permission.*)

proportional counters are very inefficient at counting x-rays with energies of about 15 keV or more (about 1 Å or less) because such high-energy radiation passes through the active volume of the counter without being absorbed. On the other hand, a scintillation counter will completely absorb x-rays of the highest energy used in x-ray analysis, i.e., up to about 50 keV, corresponding to a wavelength of roughly 0.2 Å. In general, the efficiency of detecting x-rays with a scintillation counter is superior to that of any gas-filled counter for all energies above about 6 keV (roughly 2 Å or less); at x-ray energies below 6 keV the efficiency of scintillation counters falls off, since at these energies x-rays have difficulty getting into the scintillation crystal (they are absorbed by the protective aluminum covering surrounding the crystal or are reflected at the surface of the crystal). An additional problem inherent in trying to measure low-energy radiation with a scintillation crystal is a low-energy noise, present in all commercial photomultipliers and caused by a small number of thermal electrons randomly emitted by the electrodes. The dead time of a scintillation counter is short (approximately 0.2 μs, comparable to that of a proportional counter), and this allows for counting at high rates. Scintillation counters are discussed in more detail in Sec. 14-1*B*.

The fourth type of detector used in x-ray studies is the *solid-state semiconductor*. In a semiconductor detector the electrons produced by x-ray bombardment are promoted into conduction bands, and the current which flows is directly proportional to the incident x-ray energy. A fundamental advantage of the semiconductor detector is that the energy-conversion process is extremely efficient. Where it takes about 30 eV to produce an ion pair in an inert gas and about 50 eV to produce a light photon in a thallium-activated sodium iodide phosphor, a typical semiconductor requires only about 3 eV to produce a positive hole (the semiconductor equivalent of an ion pair). Thus a given x-ray photon will produce a much larger number of events in a semiconductor than in any other kind of detector, with the inherent advantage that the statistical spread of the output pulses from a given x-ray energy will be much smaller than for other detectors. This means that the theoretical resolution of two different x-rays of only slightly different energies will be much better with a semiconductor detector than with any other detector. A practical disadvantage of these detectors, however, is that the semiconductor must be kept at very low (usually cryogenic) temperatures, to minimize the noise and prevent deterioration in the detector characteristics. Germanium and silicon semiconductors find considerable application in gamma-ray spectroscopy, but the application of these devices for x-ray detection is still in the early stages.

12-2 ABSORPTION METHODS

Absorption methods are the least used of the various x-ray methods, but three types of applications can be cited.

12-2A Elemental Analysis with a Polychromatic Beam of X-Rays

This method is analogous to colorimetric analysis with white light and has many of the same general advantages and disadvantages. Most important are simplicity of equipment and nonspecificity of absorption. The simplest system readily analyzed by x-ray absorption is one in which the element to be determined is the sole heavy component in a material of low atomic weight. A number of important determinations fall in this category, including lead (as the tetraethyl compound) in gasoline and chlorine or other heavy halides in organic compounds. In all these cases the heavy elements absorb x-rays much more strongly than carbon or hydrogen, and thus simple and routine x-ray determinations are possible.

12-2B Elemental Analysis with a Monochromatic Beam of X-Rays

For this method to be satisfactory it is desirable that the elements sought be the only heavy elements present. However, monochromatic x-rays allow a reasonable degree of specificity to be achieved through the use of the absorption-edge technique. Here, absorption measurements are made as a function of wavelength, and by extrapolating to the absorption edge, both qualitative and quantitative analyses can be carried out. The method has been used to analyze for heavy elements in glass, silicates, and other minerals and salts.

12-2C Determination of Film Thickness

This method is a widely used industrial application of x-ray absorptiometry [4, 5]. Applications range from the continuous monitoring and control of the thickness of steel sheets emerging from a steel mill to determining the thickness of a film plated on a substrate of different composition, e.g., the thickness of tin plate on steel.

12-3 EMISSION METHODS

The most important x-ray emission method is x-ray *fluorescence spectroscopy*. A beam of sufficiently short-wavelength x-radiation is used to bombard a sample, which in turn emits characteristic x-rays of longer wavelength. Measuring the wavelength of the emitted lines makes qualitative elemental analysis possible, and measuring the intensities of the emitted lines makes quantitative elemental analysis possible. Although the method is fairly sensitive, often being able to detect minor constituents down to parts per ten thousand with accuracy on the 1 to 5 percent error level, the biggest advantages of the method are its *specificity* and its *simplicity*. The high specificity arises from the fact that emitted x-rays are uniquely characteristic of the element doing the emitting. Furthermore, x-ray spectra contain relatively few lines, with the result that there are few cases where lines from different elements overlap one another. The simplicity of the method arises mainly from the fact that the results are relatively independent of whether the sample is a liquid or a solid (or even a gas, if the concentration is sufficient), and thus no special sample preparation is required. Furthermore, the method is nondestructive, so that the sample is not harmed in the process of analysis.

The biggest limitation of the method is the number of elements that can be determined. The method works best for elements of atomic number 22 (titanium) through 55 (cesium), but with special provisions and/or lower sensitivity, elements as light as magnesium (atomic number 12) and as heavy as uranium (atomic number 92) can be determined. The major problem in dealing with light elements is that they emit x-rays of relatively long wavelength, which in turn are strongly absorbed by any matter in their path, including air in the instrument, windows on the detector, and covers on the sample. Instruments are now available that can be evacuated, and the method has been extended with fair results to as light an element as boron (atomic number 5). With elements heavier than atomic number 55, the main problem is the lack of practical x-ray tubes to generate x-rays of sufficiently short wavelength (high energy) to remove the K electrons. Elements heavier than atomic number 55 can be routinely analyzed by x-ray fluorescence spectroscopy, but only by using the L series of emission lines, with a significant (roughly tenfold) decrease in intensity over that of the K series of lines.

12-3A Instrumentation

A typical arrangement for an x-ray fluorescence spectrometer is shown schematically in Fig. 12-8. The sample is often rotated to improve uniformity of exposure. As with all types of spectrometer the three principal components are a *source,* a *monochromator,* and a *detector.* In this case, as shown in Fig.12-8, the source is an x-ray tube, the monochromator is a pair of collimators and an analyzer crystal, and the detector is usually a Geiger or proportional tube but may be a scintillation counter or solid-state semiconductor. X-ray tubes were discussed in Sec. 12-1*B*, and detectors were discussed in Sec. 12-1*D*. For the analyzer crystal a material like NaCl or LiF is usually used because the lattice spacings are of the right order of magnitude (about 1 Å) to act as a diffraction grating for x-rays, based on Bragg's law (see Exercise 12-5). To scan the wavelength spectrum the analyzing crystal is rotated slowly while the detector is rotated twice as fast (see Fig. 12-8). These rotations are automatically handled by the goniometer assembly. When the lattice spacing d of the analyzer crystal and the angle 2θ that the detector makes with respect to the beam incident to the analyzer crystal are known, Bragg's law [Eq. (12-6)] can be used to calculate the wavelength of x-ray being detected.

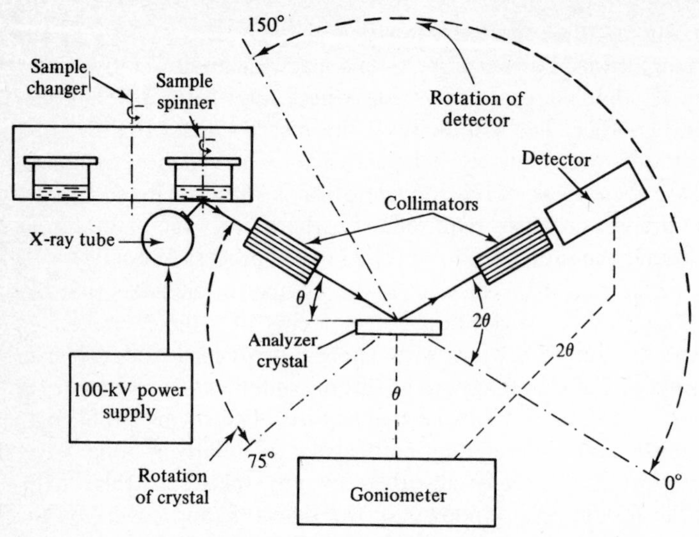

FIGURE 12-8 Typical arrangement of components in an x-ray fluorescence spectrometer.

A typical spectrum of a certain high-temperature alloy is shown in Fig. 12-9. Notice that the abscissa is usually presented in units of 2θ, though the conversion to wavelength units is readily made with the Bragg equation. Note from Fig. 12-9 that for every K_α peak there is a K_β peak appearing at a shorter wavelength (lower 2θ angle). For a given element the K_β peak is always of lower intensity than the K_α peak, since the transition of an electron from an M to a K shell is much less probable than the L to K transition.

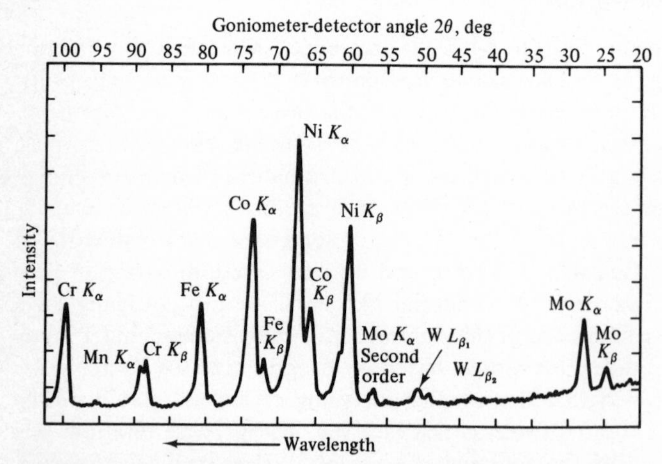

FIGURE 12-9 A typical x-ray fluorescence spectrum of a high-temperature alloy. [*Reprinted with permission from R. M. Brissey, Anal. Chem.,* **25**:190 (1953). *Copyright by the American Chemical Society.*]

In order to carry out a *qualitative analysis* using a spectrum like Fig. 12-9, the wavelength corresponding to the various peaks is calculated from Eq. (12-6), and reference is made to a table of characteristic x-ray wavelengths, e.g., Ref. 5. For *quantitative analysis* a peak of maximum intensity (and minimum background) is chosen, and a count of x-ray pulses is made over a sufficiently long period of time to give a satisfactory counting precision. Standard samples should be run to prepare a working curve.

12-3B Sample Handling

Samples are best handled as *liquids*. Liquid samples should exceed a certain minimum depth so that they will appear infinitely thick to the primary x-ray beam, but this is usually only about 5 mm. Any solvent used should not contain heavy atoms, since they absorb a higher percentage of the radiation than would be absorbed by a lighter element. For example, it is preferable to use HNO_3 and water as solvents rather than H_2SO_4 or HCl.

Solid samples introduce an additional complication due to scattering. Particle size and shape are important and determine the degree to which the incident beam is absorbed or scattered. It is best to grind both standards and samples to the same mesh size, preferably finer than 200 mesh. An internal standard can be added to the sample to correct for errors due to differences in the packing density of the samples. Before analyzing it is best to press powdered samples into a wafer, using a metallurgical specimen press, or convert the sample into a solid solution by fusion with borax.

12-3C Applications

The general scope of x-ray fluorescence spectroscopy was discussed in the introduction to Sec. 12-3. In the analysis of major constituents x-ray fluorescence can rival the accuracy of wet chemical methods, provided that standards virtually identical in composition with the unknown are available for direct comparison. For trace analysis it is usually difficult to detect an element present in less than 1 part in 10,000, but in favorable cases, e.g., the determination of nickel in an aluminum alloy, the limit of detectability may be as low as a few parts per million. The method is especially attractive for elements which lack reliable wet chemical methods, e.g., niobium, tantalum, and the rare earths. X-ray fluorescence spectroscopy is useful for the analysis of nonmetallic specimens, especially since conventional emission spectroscopy requires that a sample be an electrical conductor.

Electron-probe microanalysis deserves brief mention. This technique was developed in 1951 as a method for the analysis of isolated surface areas only 1 μm in diameter. A beam of electrons is collimated into a fine pencil of 1-μm cross section and directed at the sample surface exactly on the spot to be analyzed. Electron bombardment causes characteristic x-rays to be generated from the elements at the point source, and the intensities generated are considerably higher than with ordinary fluorescent excitation. The limit of detectability is about 10^{-14} g. The relative accuracy is generally 1 to 2 percent if the concentration is greater than a few percent and if adequate standards are available [1, 4].

12-4 DIFFRACTION METHODS

In x-ray diffraction, a monochromatic beam of x-rays is used to irradiate a crystalline solid, and each atom in the crystal is capable of scattering the x-ray beam. The sum of all the scattered waves results in a diffraction pattern which is unique for each crystalline substance. There are two experimental approaches to x-ray diffraction studies, the *single-crystal* method and the *powder* method.

In the single-crystal method it is necessary to have a relatively large crystal (optimum dimensions are of the order of 0.05 to 0.5 mm, which is large enough to be readily seen by the eye), and the crystal is rotated about one axis while it is irradiated with monochromatic x-rays. The photographic pattern which results is a vast array of dots or spots, corresponding to reflection of radiation from various atoms in the crystal lattice. Only one mounting is necessary if the crystal lattice is symmetrical, like a cubic lattice, but two or more separate mountings may be necessary to completely characterize a crystal with low symmetry. The single-crystal method is the most powerful and widely used method, but it is also the most exacting in sample preparation and the more complicated to interpret. The single-crystal method is used for most organic compounds and all but the simplest inorganic compounds.

In the powder method the sample is a crystalline material which is ground to a fine powder. The various crystals in the sample will present all possible orientations to the x-ray beam, and instead of individual spots, continuous cones of diffracted rays are produced. Figure 12-10 shows the experimental arrangement used in the powder method and illustrates the type of pattern that results. The diffraction pattern obtained is just like that which results from mounting a single crystal and turning it through all possible angles. For each crystal plane there will be some one angle at which the Bragg law will be satisfied, and by an analysis of the spacing pattern on the exposed film the type of crystal and spacings between planes can be calculated. Unfortunately, for fairly complex crystal systems, particularly those with low symmetry, many planes in the crystal happen to have equal or nearly equal spacing, and even if these planes have different directions in the crystal, the powder method will superimpose the reflection from these planes. Thus, for complex systems, the powder method loses much useful information, and the more precise but tedious single-crystal technique must be used.

It is appropriate to conclude this section with a brief comparison of x-ray diffraction with two other newer methods, *neutron diffraction* and *electron dif-*

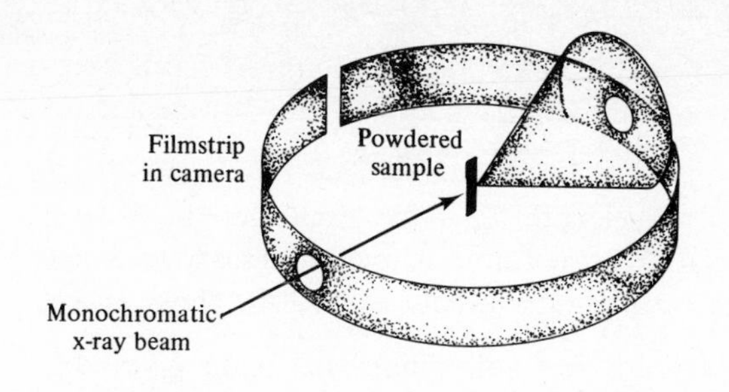

FIGURE 12-10 The experimental arrangement in the x-ray powder diffraction method. (*From G. M. Barrow, "Physical Chemistry," 2d ed., p.* 402, *McGraw-Hill Book Company, New York,* 1966, *by permission.*)

fraction. In x-ray diffraction scattering is caused by the electron atmosphere of each atom, and since hydrogen atoms have only one electron each, they are very poor scattering centers. Hence it is difficult or even impossible to locate hydrogen atoms in an x-ray diffraction pattern. On the other hand, in *neutron* diffraction a hydrogen atom is an ideal scattering center; this makes neutron diffraction a useful adjunct to x-ray diffraction. Examples of compounds which have been successfully studied by the neutron method include α-resorcinal, sodium sesquicarbonate, and potassium hydrogen bisphenylacetate, all of which involve hydrogen bonding.

Electron diffraction is of much less importance in the study of solids than either x-ray or neutron diffraction, since electrons have very poor penetrating power through solids. However, electron diffraction has been widely used for gases. Since in gases the molecules are far apart and randomly orientated, the only *reinforced interference* (which constitutes scattering or diffraction) to occur arises solely from scattering within different parts of the *same molecule.* The amount of information contained in the resulting diffraction pattern is somewhat limited, and the method is therefore restricted to fairly simple molecules. Nevertheless, it has proved valuable in the determination of bond lengths and molecular conformations for such compounds as 1,2-dichloroethane, butadiene, *cis-* and *trans*-decalin, and various glyoxals.

EXERCISES

12-1 Identify the transitions responsible for each of the following lines: (*a*) K_{α_1}, (*b*) K_{β_2}, (*c*) L_{β_1}.

12-2 A certain x-ray tube is operated at 50 kV. (*a*) Calculate the short-wavelength limit for the x-rays being emitted. (*b*) Estimate the wavelength of maximum intensity. *Ans:* (*a*) 0.248 Å; (*b*) 0.37 Å

12-3 For zirconium the cutoff wavelength λ_{min} for excitation of *K*-level electrons is about 0.70 Å. Calculate the minimum voltage necessary to excite *K* electrons in an x-ray tube with a zirconium target. *Ans:* about 18 kV

12-4 A powder diffraction spectrum for ammonium benzoate gives diffraction rings with sin θ values of 0.41198, 0.39758, 0.35806, and 0.34006. Assuming that an x-ray tube emitting radiation at 1.54 Å is used, calculate the *d* spacings of ammonium benzoate at these diffraction angles using the Bragg equation.

12-5 Assume that a diffractometer can be used over a 2θ range of 8° to 150° (a θ range of 4 to 75°; see Fig. 12-5). If the measurements are restricted to first-order reflections, what is the range of wavelengths that can be used to measure the diffraction pattern of LiF, having a lattice spacing of 2.014 Å?
Ans: 0.28 to 3.9 Å

12-6 Calculate the goniometer setting 2θ for the emission lines of strontium and chromium when lithium fluoride is the analyzing crystal.
Ans: for Sr K_{α_1}, 25.2°

12-7 An x-ray fluorescence spectrometer equipped with a tungsten target x-ray tube and a lithium fluoride analyzer crystal was used to irradiate an unknown metal with x-rays. The strongest emission line was found at a 2θ setting of 62.4°. Using data from a handbook or other x-ray reference, (*a*) calculate the wavelength of the emission line, and (*b*) identify the metal.

REFERENCES

INTRODUCTORY

1 WILLARD, H. H., L. L. MERRITT, JR., and J. A. DEAN: "Instrumental Methods of Analysis," 4th ed., Van Nostrand, New York, 1965.
2 BARROW, G. M.: "Physical Chemistry," 2d ed., McGraw-Hill, New York, 1966. Good treatment of x-ray diffraction.
3 SIM, G. A.: Diffraction Methods, chap. 8 in J. C. P. Schwartz (ed.), "Physical Methods in Organic Chemistry," Holden-Day, San Francisco, 1964.

INTERMEDIATE

4 LIEBHAFSKY, H. A., H. G. PFEIFFER, and E. H. WINSLOW, X-Ray Methods, chap. 60 in I. M. Kolthoff and P. J. Elving (eds.), "Treatise on Analytical Chemistry," pt. I, vol. 5, Wiley-Interscience, New York, 1964.

5 LIEBHAFSKY, H. A., H. G. PFEIFFER, E. H. WINSLOW, and P. D. ZEMANY: "X-Ray Absorption and Emission in Analytical Chemistry," Wiley, New York, 1966.
6 JENKINS, R., and J. L. DEVRIES: "Practical X-ray Spectrometry," Springer-Verlag, New York, 1967. Excellent practical book.
7 BIRKS, L. S.: "X-Ray Spectrochemical Analysis," 2d ed., Wiley-Interscience, New York, 1969. Recommended for x-ray fluorescence spectroscopy.

ADVANCED

8 AZAROFF, L. V.: "Elements of X-Ray Crystallography," McGraw-Hill, New York, 1968.
9 BERTIN, E. P.: "Principles and Practice of X-Ray Spectrometric Analysis," Plenum, New York, 1970. Emphasizes x-ray emission (fluorescence) spectroscopy.

13

Nuclear Magnetic Resonance and Electron Spin Resonance Spectroscopy

Nuclear magnetic resonance (nmr) spectroscopy and electron spin resonance† (esr) spectroscopy are both absorption techniques, but in contrast to the other absorption techniques discussed in this book, the sample to be studied must be placed in a powerful magnetic field before electromagnetic radiation will be absorbed by *nuclei* (in nmr) or *electrons* (in esr). Without such a magnetic field the *spin states* of nuclei and electrons are *degenerate*, i.e., possess the same energy, and no energy-level transition is possible. In the presence of a powerful magnetic field, the energy of certain nuclei and electrons is resolved into separate levels, and electromagnetic radiation of the proper frequency can cause transitions between spin states. Nuclear magnetic resonance uses radio-frequency radiation, which is the lowest-energy electromagnetic radiation of any optical method of analysis. At present, magnetic fields of the order of 5 to 23 kG are used in nmr spectrometers, and depending on the nuclei being studied, radio-frequency radiation in the range of 5 to 100 MHz (corresponding to a wavelength range of 60 to 3 m) is used. In esr studies, weaker magnets of about 3 kG are used, and a much higher energy of electromagnetic radiation, about 9 GHz (corresponding to a wavelength of about 3 cm, which is

† Also known as electron paramagnetic resonance (epr).

in the *microwave* region) is used. (See Fig. 1-4 for the various regions of the electromagnetic spectrum.)

Electron spin resonance and, especially, nuclear magnetic resonance are powerful tools for investigating molecular structure. Of the two, esr is more limited in applications, since this kind of resonance occurs only if *unpaired* electrons are present. In organic chemistry it is largely limited to free radicals and certain organometallic compounds, and in inorganic chemistry it is essentially limited to transition-metal ions. Nonetheless, within this sphere of application it is an extremely useful tool for obtaining electron-density information and performing structural analysis. On the other hand, nmr is probably the single most powerful tool for structural elucidation at the disposal of the organic chemist. It should be emphasized, however, that nmr cannot replace older methods such as ultraviolet or infrared spectroscopy but is complementary to them. The newness and relative importance of nmr can be appreciated from the fact that the first observations of the nmr signals were made by Purcell at Harvard and Bloch at Stanford in 1945, and the first application to organic chemistry (spectrum of ethyl alcohol) was made in 1951. In 1952 Purcell and Bloch were awarded the Nobel prize in physics for their discovery.

13-1 THEORY OF MAGNETIC RESONANCE

Magnetic resonance spectra are more complex than other types of absorption spectra, requiring greater care and insight in their measurement and interpretation. In this section a physical picture of the magnetic resonance process will be presented. Derivations of equations and a more rigorous treatment are available in the references [5, 14, 15, 22–25].

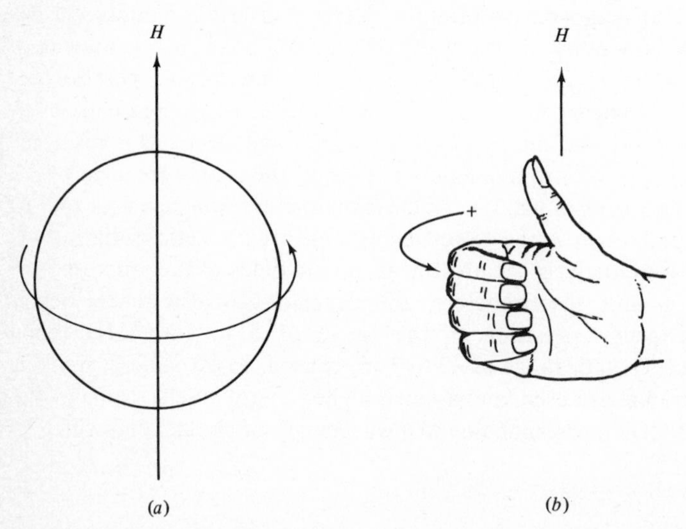

FIGURE 13-1 (*a*) The spin of a proton and (*b*) the direction of the magnetic field generated.

13-1A Magnetic Properties of Electrons and Atomic Nuclei

All electrons and the nuclei of many isotopes possess an intrinsic mechanical spin. The spin of a proton is depicted in Fig. 13-1*a*. Because of this spin and the associated electric charge, a magnetic field is generated. For a positively charged particle the direction of the magnetic field can be predicted from the right-hand rule, illustrated by Fig. 13-1*b*. (Spinning electrons, being negatively charged, generate a magnetic field that can be predicted from the left-hand rule.)

Any bar magnet has a *magnetic moment* μ, which is the product of the pole strength and the length of the magnet. Any spinning, charged particle behaves like a bar magnet and therefore has a magnetic moment. Without knowing the precise shape of the charged particle (little is known of the shape of electrons or nuclei), it is impossible to calculate the magnetic moment. However, the magnetic moment can be experimentally measured and used to characterize a given nucleus or electron. Table 13-1 gives the magnetic moments of a number of isotopes and an electron. If a spinning charged particle like a proton is placed in a uniform magnetic field of strength H, as shown in Fig. 13-2, the nuclear spin axis will align itself at some angle θ with respect to the direction of the field and the particle will *precess* about the axis of the field like a toy gyroscope. The magnetic moment μ is a vector quantity, and its direction is illustrated in Fig. 13-2. The frequency of the precession of the spinning particle is of special interest and is directly proportional to the strength

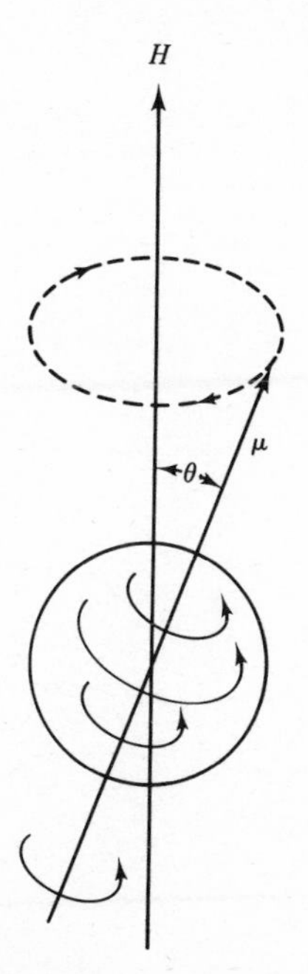

FIGURE 13-2 Precession of a spinning proton about the direction of a steady applied magnetic field H.

TABLE 13-1 Some Magnetic Properties of Various Nuclei and an Electron

Isotope	Natural abundance, %	Nmr frequency, MHz, for 14,092-G field	Relative sensitivity for equal numbers of nuclei at constant frequency	Magnetic moment, units of nuclear magneton, 5.05×10^{-24} erg/G	Spin
^{1}H	99.98	60.000	1.000	2.793	½
^{2}H	0.0156	9.211	0.409	0.857	1
^{11}B	81.17	19.250	1.60	2.688	3/2
^{13}C	1.108	15.085	0.251	0.702	½
^{14}N	99.64	4.335	0.193	0.404	1
^{17}O	0.037	8.136	1.58	−1.893	5/2
^{19}F	100	56.446	0.941	2.627	½
^{23}Na	100	16.000	1.32	2.216	3/2
^{29}Si	4.70	11.900	0.199	−0.555	½
^{31}P	100	24.288	0.405	1.130	½
^{35}Cl	75.4	5.879	0.490	0.821	3/2
Free electron		39.405	658	−1836	½

of the applied magnetic field, as given by the Larmor equation

$$\omega_0 = 2\pi\nu = \gamma H \tag{13-1}$$

where ω_0 = angular velocity
ν = frequency of precession
γ = proportionality constant

The proportionality constant is given the name *magnetogyric ratio* (sometimes, but less properly, *gyromagnetic ratio*). The magnetogyric ratio is another fundamental property of the specific particle in question, and its relationship to the magnetic moment μ will be given later. An important feature of Eq. (13-1) is that it shows that the precessional frequency is independent of the angle of inclination θ of the particle axis to the field direction.

Quantum mechanics correctly predicts that only certain orientations in the magnetic field are allowed, the number of orientations being determined by the *spin number,* symbolized by S for the electron and I for atomic nuclei. A knowledge of spin numbers is important to predicting the number of transitions and the magnitude of energy involved in nmr experiments.

SPIN NUMBER

For *electrons* the only spin number S that is needed is $\frac{1}{2}$. For atomic *nuclei,* however, the spin number I may assume values of 0, $\frac{1}{2}$, 1, $\frac{3}{2}$, . . . up to at least $\frac{13}{2}$ for different nuclei; $I = 0$ denotes no spin. The spin number of a nucleus is the resultant of the spins of protons and neutrons which constitute the nucleus. Each proton and neutron has a spin of $\frac{1}{2}$, but the resulting spin number of a nucleus cannot be decided without knowing whether the spins are aligned *parallel* or *antiparallel,* and to date this information can only be obtained experimentally. For example, a deuterium nucleus, which contains one proton and one neutron, could be predicted to have a spin of either 1 or 0, according as the spins of the neutron are aligned parallel or antiparallel. Experimentally, the deuteron is found to have a spin of 1 (see Table 13-1), indicating parallel alignment. A few simple rules serve as a helpful guide toward classifying nuclei:

1 If the mass number of the nucleus is *even*, the spin will have zero or integral values.

2 If *both* the mass number and the atomic number are even, the spin will be zero.

3 If the mass number is odd, the spin will be half-integral.

Example 13-1 Classify each of the following isotopes according to whether it will have integral, half-integral, or zero spin: ${}^{1}_{1}H$, ${}^{3}_{1}H$, ${}^{6}_{3}Li$, ${}^{7}_{3}Li$, ${}^{10}_{5}B$, ${}^{11}_{5}B$, ${}^{12}_{6}C$, ${}^{13}_{6}C$, ${}^{14}_{6}C$, ${}^{14}_{7}N$, ${}^{15}_{7}N$, ${}^{16}_{8}O$, ${}^{17}_{8}O$, ${}^{19}_{9}F$, ${}^{31}_{15}P$, ${}^{32}_{15}P$.

Answer All isotopes with odd mass numbers (namely ^1H, ^3H, ^7Li, ^{11}B, ^{13}C, ^{15}N, ^{17}O, ^{19}F, and ^{31}P) will have half-integral spins, like $\frac{1}{2}$, $\frac{3}{2}$, $\frac{5}{2}$, etc. Experimentally, it is found that ^7Li and ^{11}B have $I = \frac{3}{2}$; ^{17}O has $I = \frac{5}{2}$; and all the rest of these isotopes have $I = \frac{1}{2}$.

All isotopes with both an even mass number and even atomic number (namely $^{12}_{6}C$, $^{14}_{6}C$, and $^{16}_{8}O$) will have $I = 0$. This agrees with experimental observations.

All the rest of these isotopes, having even mass numbers but odd atomic numbers (namely $^{6}_{3}Li$, $^{10}_{5}B$, $^{14}_{7}N$, and $^{32}_{15}P$), will have integral spin numbers, like 1, 2, 3, ... Experimentally, it is found that ^{10}B has $I = 3$, whereas the other three isotopes all have $I = 1$.

ALLOWED ORIENTATIONS

In a uniform magnetic field a spinning particle can have only $2S + 1$ or $2I + 1$ orientations. Thus, an electron ($S = \frac{1}{2}$) can have only two possible orientations, and the same is true for such isotopes as $^{1}_{1}H$, $^{3}_{1}H$, $^{13}_{6}C$, $^{19}_{9}F$, etc., all of which have $I = \frac{1}{2}$. On the other hand, an isotope like $^{6}_{3}Li$ ($I = 1$) can assume any of three possible orientations, and $^{7}_{3}Li$ ($I = \frac{3}{2}$) can have four possible orientations. Figure 13-3 illustrates the allowed precessional orientations for nuclei having a spin of $\frac{1}{2}$, 1, and $\frac{3}{2}$, respectively.

13-1B Nuclear Resonance

A proton in a static external magnetic field H can have only $2I + 1 = 2$ orientations. The *low-energy* orientation corresponds to that state in which the nuclear magnetic moment is aligned *parallel* to the external magnetic field, illustrated in Fig. 13-2. If an alternating radio-frequency field is imposed at right angles to the fixed magnetic field (along the x axis of Fig. 13-2), and if the radio-frequency field has precisely the same frequency as the precessional frequency of the proton, the radio-frequency energy will be absorbed by the

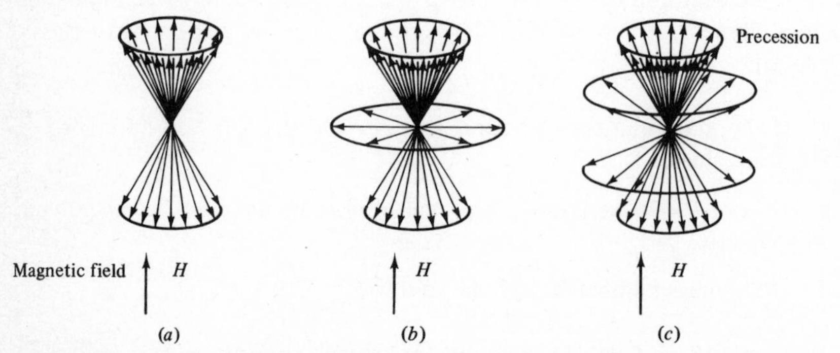

FIGURE 13-3 Allowed orientations of angular-momentum vectors for three different nuclei: (*a*) nucleus with $I = \frac{1}{2}$, for example, a proton; (*b*) nucleus with $I = 1$, for example, ^6Li; (*c*) nucleus with $I = \frac{3}{2}$, for example, ^7Li.

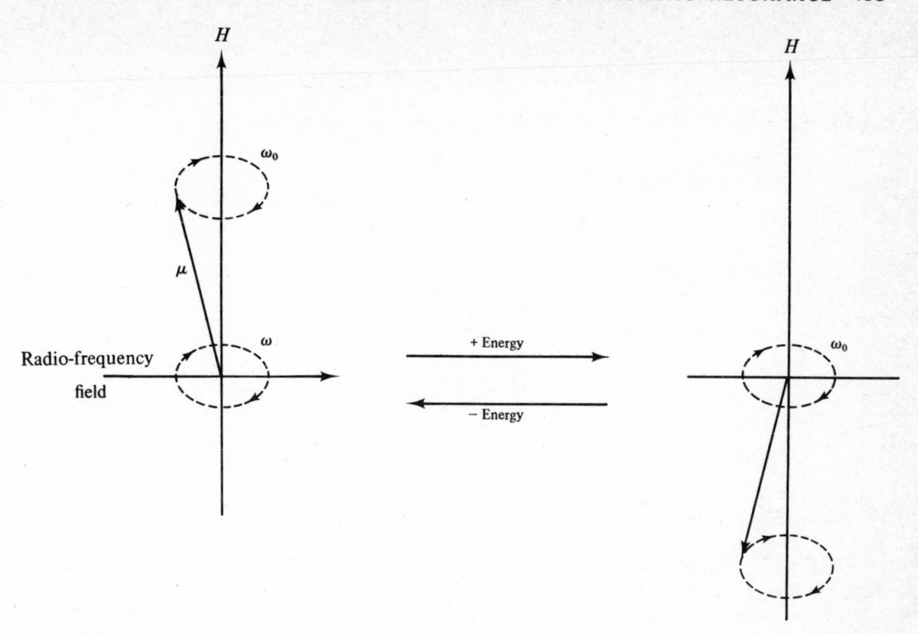

FIGURE 13-4 Interaction of a radio-frequency field with a proton precessing in an applied magnetic field H.

spinning proton and the nucleus will flip to its *high-energy* orientation, corresponding to that state in which the nuclear magnetic moment is aligned *antiparallel* (opposed) to the external field (see Fig. 13-4). The radio-frequency field has a rotating magnetic field of angular velocity ω, which will undergo *resonance interaction* with the nucleus if $\omega = \omega_0$ where ω_0 is the angular velocity of the precessing nucleus. The frequency ν of the radio-frequency radiation can be calculated from

$$\Delta E = h\nu = \frac{\mu H}{I} \tag{13-2}$$

It should be noted from Eq. (13-2) that the energy-level splitting ΔE depends on the strength of the magnetic field H. Figure 13-5 illustrates how the spacing between energy levels increases as the applied field strength increases. Note that in the *absence* of a magnetic field the two spin states of a proton are *equal* in energy; i.e., the system is doubly degenerate.

Example 13-2 Calculate the frequency of radiation needed for resonance absorption by a proton in a magnetic field of 14,092 G.

Answer From Table 13-1 the magnetic moment of a proton μ_p can be calculated to be

$$\mu_p = (2.793)\,(5.05 \times 10^{-24}\ \text{erg/G}) = 1.41 \times 10^{-23}\ \text{erg/G}$$

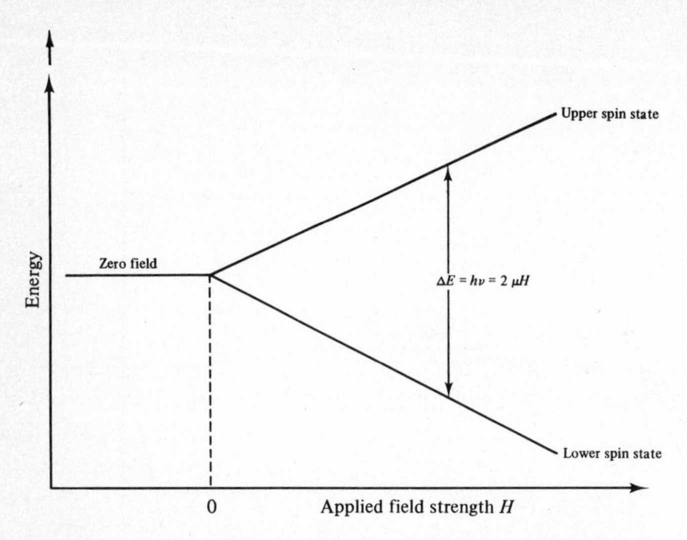

FIGURE 13-5 The splitting of nuclear spin energy levels in a uniform magnetic field *H*. Levels shown are for a proton or other particle with $I = \frac{1}{2}$.

From Eq. (13-2),

$$\nu = \frac{\mu H}{hI} = \frac{(1.41 \times 10^{-23}\ \text{erg/G})\ (14{,}092\ \text{G})}{(6.63 \times 10^{-27}\ \text{erg-s})\ (\frac{1}{2})}$$

$$= 60 \times 10^{6}\ \text{s}^{-1} = 60\ \text{MHz}$$

Example 13-3 Calculate the magnetic field strength required to give resonance absorption by protons at 100 MHz.

Answer From Eq. (13-2),

$$H = \frac{Ih\nu}{\mu} = \frac{(\frac{1}{2})\ (6.63 \times 10^{-27}\ \text{erg-s})\ (100 \times 10^{6}\ \text{s}^{-1})}{1.41 \times 10^{-23}\ \text{erg/G}} = 23.5 \times 10^{3}\ \text{G}$$

The magnetic moment μ is related to the spin number I through the magnetogyric ratio γ:

$$\mu = \frac{\gamma h I}{2\pi} \tag{13-3}$$

All three properties of nuclei (magnetic moment, spin number, and magnetogyric ratio) are presently evaluated by experiment, rather than from theory. As Eq. (13-3) shows, however, any two of the three parameters completely characterize the nucleus.

Example 13-4 Calculate the magnetogyric ratio for a proton.

Answer From Eq. (13-3),

$$\gamma = \frac{2\pi\mu_p}{hI} = \frac{(2)(3.14)(1.41 \times 10^{-23}\ \text{erg/G})}{(6.63 \times 10^{-27}\ \text{erg-s})(\frac{1}{2})}$$

$$= 2.67 \times 10^{4}\ \text{G-s}^{-1}$$

POPULATION OF NUCLEAR MAGNETIC ENERGY LEVELS AND RELAXATION

In ordinary absorption spectroscopy there is little concern about the relative population of various energy levels in atoms or molecules, and it is taken for granted that an absorption signal will be constant with time, rather than decreasing with time because it uses up the ground-state population. In nmr spectroscopy the relative population of various energy levels and the *change* in population during irradiation have a direct bearing on the experimental conditions chosen and the type of spectra obtained.

There are two principal reasons why these matters are of little concern in ordinary spectroscopy and of great concern in nmr spectroscopy. First, in any absorption process the intensity of the transition depends on the number of atoms or molecules in the state corresponding to the starting point of the transition. (The intensity also depends on the intrinsic *probability* that the transition will occur, but this has no bearing on the problem of relative population.) We can calculate the relative distribution of molecules in various energy levels with the Boltzmann distribution law, which may be expressed as

$$\frac{n_i}{n_j} = e^{-\Delta E/kT} \tag{13-4}$$

where n_i = number of molecules in upper state i
n_j = number of molecules in lower state j
ΔE = energy difference between the two states

When ΔE is much greater than kT, n_i is negligible with respect to n_j, and n_i becomes appreciable only when ΔE is of the same magnitude as kT. In ordinary spectroscopy ΔE is usually large with respect to kT, at least for high-energy states, and a preponderance of molecules exists in a lower energy state. In nmr spectroscopy there are few allowed energy levels (only two for a proton), and there is only a small excess of nuclei in the low-energy spin state.

Example 13-5 Calculate n_2/n_1, the ratio of hydrogen nuclei in the high- and low-energy states, respectively, in a magnetic field of 10,000 G at 27°C.

Answer From Eq. (13-2), $\Delta E = h\nu = \mu H/I$; and since $I = \frac{1}{2}$ for a proton, $\Delta E = 2\mu H$. Therefore,

$$\frac{\Delta E}{kT} = \frac{2\mu H}{kT}$$

$$\frac{\Delta E}{kT} = \frac{(2)\,(2.793)\,(5.05 \times 10^{-24}\ \text{erg/G})\,(10{,}000\ \text{G})}{(1.3805 \times 10^{-16}\ \text{erg/K})\,(300\ \text{K})} = 6.80 \times 10^{-6}$$

$$\frac{n_2}{n_1} = e^{-(6.8\times10^{-6})} = 0.999993$$

Thus, at room temperature for every 1,000,000 nuclei in the upper level there will be 1,000,007 in the lower level, which is an extremely small difference in population of the two energy levels.

In ordinary absorption spectroscopy there is a large population difference, even at high temperatures (see Exercise 13-10). The population of the lower state is so large that the relative distribution is not significantly altered during absorption. In magnetic resonance absorption, however, absorption may lead to *saturation* if a sufficiently strong radiation field is used. In other words, as the collection of nuclei continually absorbs radio-frequency radiation, the excess of nuclei in the lower state may diminish, with an accompanying decrease or even disappearance of the absorption signal. The significance of saturation will be discussed at the end of this section.

A second difference between nmr spectroscopy and ordinary spectroscopy is in the mechanism by which molecules in an excited energy state return to the lower energy state after irradiation. Electronic, vibrational, or rotational states lose energy through collisions between molecules. (It is these interactions, incidentally, which tend to destroy fine structure in spectra by limiting the lifetimes of the various states.) However, *nuclei* are isolated from each other by the electron clouds around each nucleus. Nonetheless, this electronic shielding does not completely preclude a small magnetic interaction between neighboring nuclei, and this feeble interaction is responsible for keeping the ground-state population larger than that of the upper states. The various ways in which a nucleus in the upper spin state returns to a lower spin state without emitting radiation are called *relaxation processes*.

Relaxation processes are not only responsible for the establishment and maintenance of the absorption condition but also control the width of the spectral lines by controlling the lifetime of a given state. From the uncertainty principle it is known that as the lifetime of an excited state increases, the "natural" or true line width decreases. With ordinary spectroscopic techniques the lifetime of the excited state is usually not of great importance, because the resolving power of the optical instrument usually limits the spectral line width and the true line width is seldom observed. In nmr spectroscopy, however, the instrumental resolution available is far superior to that available in ordinary spectroscopy, by virtue of the precision inherent in the generation and detection of signals electronically. In nmr and esr it is usually possible to observe the true line width.

Relaxation Processes There are two kinds of relaxation processes, *spin-spin* relaxation and *spin-lattice* relaxation. In *spin-spin* relaxation a nucleus in its upper spin state transfers its energy to a *neighboring* nucleus of the same isotope by a *mutual exchange of spins.* This relaxation process (sometimes called transverse relaxation) does not change the relative spin-state populations and therefore does nothing to maintain the absorption condition, but it does shorten the lifetime of a given nucleus in the higher state and therefore contributes to a widening of line widths. In *spin-lattice*† relaxation (sometimes called longitudinal relaxation) the energy of the nuclear spin system is converted into *thermal energy,* and it is this process that is directly responsible for maintaining the unequal distribution of spin states. For a spin-lattice interaction to occur, a given nucleus in a high-energy spin state must be *properly oriented* with respect to the lattice molecules, in which case energy is transferred to the lattice molecules, giving them extra translational or rotational energy. The total energy of the system remains unchanged, and a Boltzmann distribution of nuclei is obtained, with an excess of nuclei in the lower states.

Effect of Relaxation Processes on Line Width Either spin-spin or spin-lattice relaxation processes, or both, may control the natural line width. In solid and very viscous liquids molecular motions are greatly restricted, so that few nuclei will have the proper orientation for spin-lattice relaxation during a given period of time. As a result, most solids and viscous liquids have very *long* spin-lattice relaxation times. On the other hand, this restricted motion leads to a high probability of *spin-spin* interactions, and so spin-spin relaxation times in solids and viscous liquids are very *short* (typically of the order 1 ms), and this dominates the relaxation process, giving rise to very *broad* resonance lines. In liquids of lower viscosity, on the other hand, molecules undergo random motion (brownian movement), and spin-spin interactions are greatly decreased, along with some increase in the frequency of spin-lattice interactions. In general, as the viscosity decreases, the overall relaxation time tends to increase (a typical relaxation time for a nonviscous organic liquid is of the order of 1 s), giving sharper resonance lines (typically about 1 Hz in width), and spin-lattice relaxation times become more important in determining line widths.

Other Factors Affecting Line Width Two additional factors influence the width of a spectral line, both of which may be thought of as special types of spin-lattice interactions. The first of these is termed *paramagnetic*

† Although *lattice* generally connotes a *solid* framework, here it refers to the general framework of sample and solvent molecules, regardless of whether the molecules are solid, liquid, or gaseous.

broadening and results from the presence of paramagnetic molecules or ions in the sample. Paramagnetic components, e.g., dissolved oxygen, possess *electron* magnetic moments more than 1000 times larger than *nuclear* magnetic moments; this leads to strong spin-lattice interactions, greatly shortened spin-lattice relaxation times, and greatly broadened nmr lines. Thus a sample should be deoxygenated before a spectrum is determined.

The second special type of spin-lattice interaction involves nuclei which possess an *electric quadrupole moment.* Resonance signals for protons attached to an element, like nitrogen, that has an electric quadrupole moment will frequently be broadened. Nuclei with spin numbers greater than ½ (for example, the nitrogen nucleus has $I = 1$) have electric quadrupole moments because they have a nonspherical electric charge distribution, and the unsymmetrical local electrostatic fields that are generated provide an additional electrical mode of interaction with the lattice. These electrical interactions are effective over a much longer range than ordinary magnetic interactions. Thus nuclei with quadrupole moments usually have very short spin-lattice relaxation times, and the resonance signals of these nuclei (or of protons attached to them) are correspondingly broadened.

SATURATION

If enough absorption occurs for the population of the high spin state to exceed that of the low spin state, saturation occurs and absorption ceases. Adequate spin-lattice relaxation is a necessary condition for the continued observation of radio-frequency absorption. The degree of saturation depends on the power of the radio-frequency signal and the spin-lattice relaxation time. When the spin relaxation times are very long (e.g., a viscous liquid) even a moderately powerful radio-frequency signal may cause saturation. The phenomenon of saturation determines the minimum concentration of nuclei that can be studied in nmr since the strength of the resonance signal depends on the concentration and the power of the applied radio-frequency field. Were it not for saturation, the signal strength could be maintained at a measurable level for low concentrations simply by raising the power of the radio-frequency field.

The intentional saturation of interfering resonances while measuring a resonance being sought is a technique known as *spin decoupling*; it is discussed in Sec. 13-1*F*.

13-1C Magnetic Shielding of Atomic Nuclei (The Chemical Shift)

If the phenomenon of nmr depended solely on *nuclear* properties, it would be of little interest to chemists. Fortunately, however, the electron environment around the nucleus has a small but measurable effect on the resonance frequency, and this permits a wealth of chemical information to be obtained from nmr. Extranuclear electrons affect the measured resonance frequency by magnetically screening the nucleus so that the magnetic field felt by a given nucleus is not quite the same as the applied field. Thus, the nature of the neighboring groups and the types of chemical bonds will affect the electron density around a given nucleus and hence the degree of magnetic shielding. For example, an electron-withdrawing group can be predicted to *decrease* the shielding of the nucleus to which it is attached.

The effect of shielding at a given nucleus is sometimes represented in terms of a shielding constant σ, defined by

$$H = H_0(1 - \sigma) \tag{13-5}$$

where H is the local field *at the nucleus* and H_0 is the applied magnetic field. The significance of the shielding constant (or screening constant) can be understood by viewing it as a fraction which can vary from 0 (no screening) to 1 (total shielding). (In actual practice there is no known way of evaluating the true shielding constant or true field H that a molecule experiences; all that can be measured is the *relative* shielding constants of various nuclei.) Thus, as will be seen later, shielding constants in use are simply arbitrary numbers based on a reference. Since most nmr spectrometers operate at constant frequency, it is important to realize that the degree of shielding determines the amount by which the applied field H_0 must be increased for the field at the nucleus H to satisfy the resonance condition

$$\nu = \frac{\gamma H}{2\pi} \tag{13-1}$$

where γ is the gyromagnetic ratio of the nucleus, defined earlier.

Figure 13-6 illustrates the theoretical line spectrum we might expect for *p*-xylene. There are two ways of identifying the resonance lines that appear, i.e., determining their cause. First, the relative intensity of the lines will be in

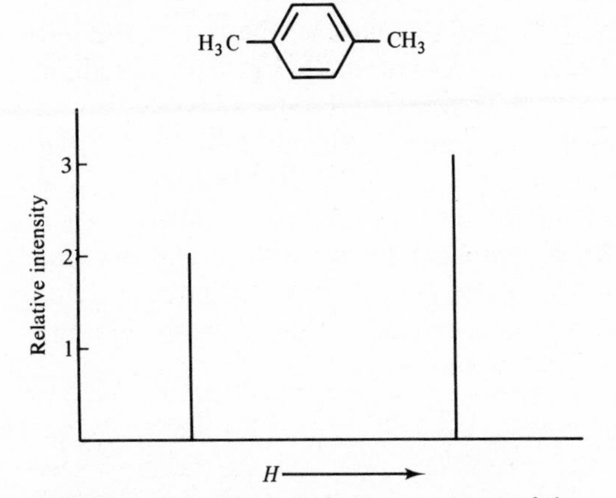

FIGURE 13-6 Theoretical nmr spectrum of hydrogen resonance lines for *p*-xylene.

the same ratio as the number of protons responsible for each line. Thus, in *p*-xylene one line will result from the four chemically equivalent protons on the ring, and another line will be due to the six protons of the methyl groups, accounting for the 2:3 ratio of relative intensities. (With real spectra, which show *bands* instead of lines, the *area* under each peak must be integrated to evaluate the relative intensities.) This proportionality is a unique advantage that other spectroscopic techniques lack and occurs because there is only one possible spin transition for protons. Thus the transition probability is the same for all protons.

The second way that resonance lines can be identified is by their location on the x axis (their field-strength location). By convention, spectra are represented with low field on the left end and high field on the right. Thus, absorption lines from less shielded or deshielded nuclei appear on the left (or down field), while more highly shielded nuclei have resonance lines at high field. With *p*-xylene we might predict that since the methyl protons are in a region of higher electron density than the aromatic-ring protons, the methyl resonance line should be the one at higher strength, in agreement with Fig. 13-6.

Differences in the position of resonance lines are called *chemical shifts* because they depend on structural features within the molecule. To put chemical shifts on a meaningful quantitative basis a standard reference compound is chosen which gives a virtually invariant line position, from which chemical shifts can be measured. A convenient reference compound in proton spectroscopy is tetramethylsilane, $Si(CH_3)_4$ (TMS). It can be added in small amounts (<1 percent) directly to the solution to be examined, giving a single, sharp absorption signal at a higher field than almost all other protons.† TMS has other advantages: it is chemically inert (it will not associate with itself or with other molecules); it is volatile (its boiling point is 27°C, which means that precious samples can be easily recovered afterward, if needed); and it is miscible with most common organic solvents. Because TMS is insoluble in water and deuterium oxide, it cannot be used in aqueous solutions. A suitable reference for aqueous solutions appears to be the methyl groups of sodium 2,2-dimethyl-2-silapentane-5-sulfonate, $(CH_3)_3SiCH_2CH_2SO_3Na$ (DSS).

Another problem in expressing meaningful chemical shifts is that electronic shielding σH_0 depends on the strength of the applied field H_0, and therefore chemical shifts differ for instruments operating at a different field strength or oscillator frequency. Most commercially available instruments for proton resonance operate at 60 or 100 MHz, but some are available at 30 or 40 MHz. To express the chemical shift independently of the field strength, the chemical-shift parameter δ is often used, defined as

$$\delta = \frac{H - H_{\text{ref}}}{H_{\text{ref}}} \tag{13-6}$$

† Since silicon is less electronegative than carbon, the electron density around the methyl groups in TMS is higher than that around the methyl groups in, for example, tetramethylmethane.

where H and H_{ref} are the respective field strengths corresponding to resonance for a particular nucleus in the sample H and reference H_{ref}. Since resonance frequency ν and field strength H are directly related by Eq. (13-1), spectra are often calibrated in frequency units (hertz), and thus Eq. (13-6) can be rewritten as

$$\delta = \frac{\Delta\nu \times 10^6}{\text{oscillator frequency}} \tag{13-7}$$

where $\Delta\nu$ is the difference in absorption frequencies of the sample and reference. In either case, δ is dimensionless and is expressed as parts per million. Since the chemical-shift parameter δ has the disadvantage of increasing in magnitude down field, another system for assigning resonance lines is based on arbitrarily assigning the TMS line a value of 10.00 ppm. The chemical shifts are then described in terms of τ (called *tau values*) defined as

$$\tau = 10.00 - \delta \tag{13-8}$$

The τ-value convention is extensively used by organic chemists and will be emphasized in this book.

The chemical shift is one of the most important quantities measured in nmr spectroscopy, being directly analogous to measuring the frequency of absorption in, say, ultraviolet or infrared spectroscopy. Chemical shifts are small, particularly for protons, since a proton involved in a covalent bond has only two electrons circulating about it. Shifts for the majority of hydrogen-containing groupings are only about 10 ppm. Fortunately, radio-frequency techniques make it possible to measure *differences* in line position with great accuracy, to about 0.1 Hz.

Before examining some of the specific mechanisms responsible for chemical shifts in various compounds, it will be helpful to consider a few specific examples of chemical shifts. Figure 13-6 presented the theoretical two-line spectrum of *p*-xylene. If the spectrum had been obtained in deuterochloroform, $CDCl_3$, one of the best nmr solvents, with a small amount (say 0.2 percent) of TMS present, the large peak from the protons of the methyl group would be found at $\tau 7.7$ and the smaller peak would be found at $\tau 2.95$. In addition to the small TMS reference peak assigned the value of $\tau 10.00$, spectra obtained in deuterochloroform often have a small peak at $\tau 2.7$, usually due to small amounts of chloroform which almost invariably contaminate the deuterochloroform employed in nmr work.

The line spectrum of toluene in deuterochloroform is shown in Fig. 13-7. As with *p*-xylene, the toluene spectrum has two main peaks, but the spectra are clearly discernible since the line-intensity ratio for toluene is 5:3 compared with 2:3 for *p*-xylene. The line at $\tau 2.80$ is from the five protons on the benzene ring, and the line at higher field is from the three equivalent protons of the methyl substituent. The small peak at $\tau 2.70$ in Fig. 13-7 may be assigned to chloroform.

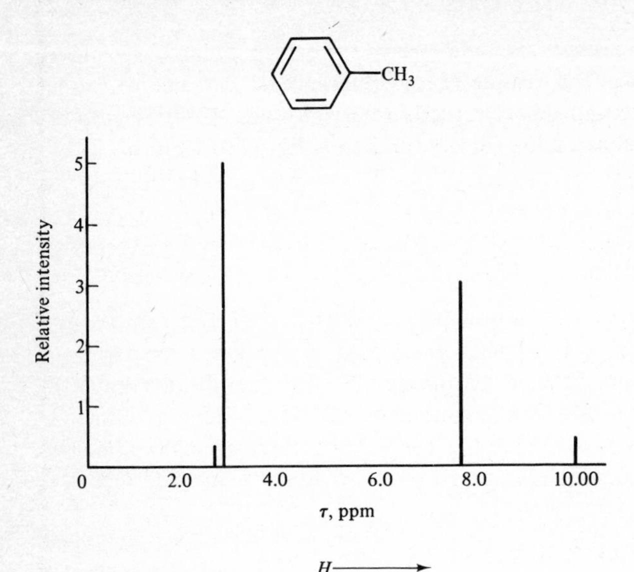

FIGURE 13-7 Proton magnetic resonance line spectrum of toluene in deuterochloroform.

Chemical intuition is useful, up to a point, in predicting the relative chemical shifts of various functional groups, as the following examples illustrate.

Example 13-6 Predict the relative chemical shifts, from high field to low field, of the methyl protons in the following types of compounds:

$$CH_3\text{—}\overset{|}{N}\text{—};\ CH_3\text{—}O\text{—};\ CH_3\text{—}F;\ CH_3\text{—}\overset{|}{\underset{|}{C}}\text{—}$$

Answer Considering the relative electronegativity of nitrogen, oxygen, fluorine, and carbon (increasing in the order C, N, O, and F), we would expect $CH_3\text{—}\overset{|}{\underset{|}{C}}\text{—}$ to absorb at the highest field and $CH_3\text{—}F$ at the lowest field. The actual assignments found experimentally are approximately

$CH_3\text{—}C\text{—}$	$\tau 9.1$
$CH_3\text{—}\overset{|}{N}\text{—}$	$\tau 7.7$
$CH_3\text{—}O\text{—}$	$\tau 6.7$
$CH_3\text{—}F$	$\tau 5.7$

Example 13-7 Predict the relative field location, i.e., high field or low field, for the resonance of an acidic proton, e.g., in RCOOH.

Answer An acidic proton would be very unshielded and would therefore be expected to absorb at very low fields. In practice the acidic protons in compounds of the type RCOOH are found to absorb at about -0.8τ.

Example 13-8 Predict the relative chemical shifts of protons in ethylene compared with acetylene.

Answer Of the two compounds, we would expect the ethylene protons to be more shielded than the protons in acetylene; that is, τ for ethylene is greater than τ for acetylene. In practice, however, it is the other way around, with the protons of acetylene found at $\tau 7.65$ and ethylene protons at $\tau 4.67$. Clearly, chemical intuition has led us astray here, and the reasons will be explained next.

SHIELDING MECHANISMS

Electronic shielding arises from the circulation of electrons induced by the applied magnetic field. In the absence of an external magnetic field the electrons in most organic molecules will not shield the nuclei magnetically because the molecules have no net electron spin. Free radicals are an exception. The magnetic field created by the induced circulation *opposes* the applied magnetic field and is called *diamagnetic* shielding. For protons it is useful to recognize three main kinds of diamagnetic shielding, arising from three kinds of induced electron circulation: (1) *local* diamagnetic shielding, (2) *neighboring* diamagnetic shielding by anisotropic groups, and (3) interatomic diamagnetic circulation *in aromatic rings*.

Local Diamagnetic Shielding The origin of local diamagnetic shielding can be understood by referring to Fig. 13-8. The applied magnetic field H_0 induces the circulation of electrons in the direction shown in Fig. 13-8, and the secondary magnetic field generated by the circulating electrons *opposes* the applied magnetic field.† Local diamagnetic shielding always *reduces* the apparent magnetic field at the proton and consequently is a source of *positive* shielding (the applied magnetic field must be *increased* to overcome the diamagnetic shielding and bring the proton into resonance). The *degree* of shielding clearly depends on the electron density around the proton, the shielding increasing with increasing electron density around the proton. When the relative electronegativity of substituent groups (inductive effect) is used, it is possible to predict the effect of various substituent groups on the degree of shielding.

Example 13-9 Arrange the methyl halides (CH_3Br, CH_3Cl, CH_3I, and CH_3F) in the order of increasing field strength.

Answer Fluoride can be expected to give the *least* shielding and iodide the most. Therefore, in the order of increasing field strength we would predict CH_3F, CH_3Cl, CH_3Br, CH_3I. This is confirmed in practice, with resonance peaks occurring at 5.74, 6.95, 7.32, and 7.84 τ, respectively.

† The direction of the secondary magnetic field can be predicted from the direction of electron circulation using the left-hand rule of physics. Lenz's law predicts that the secondary magnetic field must oppose the applied field; if the secondary field were to *add* to the applied field, all matter would serve to amplify magnetic fields, in violation of the law of conservation of energy.

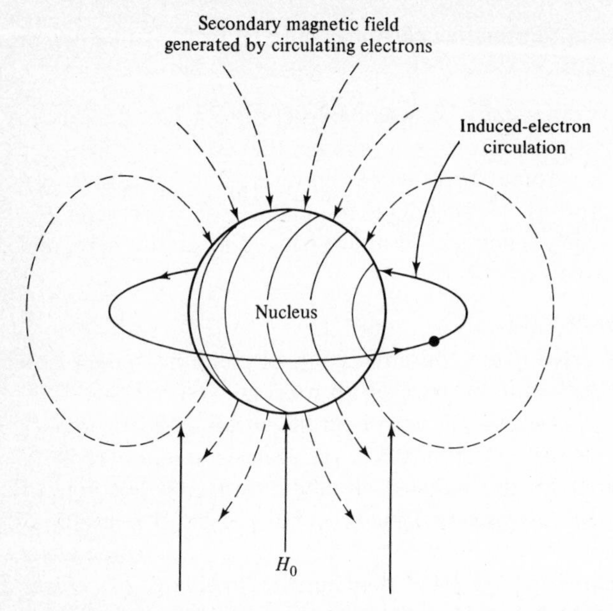

FIGURE 13-8 The origin of local diamagnetic shielding in an atom.

Example 13-10 Arrange methane and its mono-, di-, and trisubstituted phenyl derivatives in the order of increasing field strength.

Answer The phenyl group is more electron-withdrawing than a hydrogen atom, and thus triphenylmethane would be the least shielded and methane the most. This order is borne out experimentally, with the following absorption sequence:

$(C_6H_5)_3CH$ 4.37τ $(C_6H_5)_2CH_2$ 6.08τ

$C_6H_5CH_3$ 7.66τ CH_4 9.77τ

Neighboring Diamagnetic Shielding by Anisotropic Groups Shielding due to the circulation of electrons about *neighboring* atoms will be effective at nearby protons only if the secondary magnetic fields are anisotropic. This is illustrated in Fig. 13-9 with acetylene. If the axis of the acetylene molecule is lined up parallel to the applied field, as in Fig. 13-9*a*, the electrons within the annular π molecular orbitals are readily induced into circulation and the magnetic field generated serves to *shield* the acetylenic protons.† When the axis of the acetylene molecule is lined up perpendicular to the applied field, as in Fig. 13-9*b*, the π-electron circulation is severely restricted. Because of this

† A consideration of the *inductive* effect of the acetylene group would lead to the prediction that the acetylenic protons would be relatively *deshielded,* certainly with respect to, say, ethylene protons that absorb at 4.67τ. However, acetylenic protons absorb at 7.65τ, and this high degree of shielding is due to the secondary magnetic fields shown in Fig. 13-9*a*.

pronounced anisotropy, the shielding which the acetylenic protons experience will vary considerably with orientation as the molecule rotates randomly in solution. Thus, the shielding experienced by the acetylenic protons will not average to zero, as it would if an isotropic group were bound to a proton.

The shielding of aldehydic protons is another example of neighboring diamagnetic shielding due to anisotropy. Figure 13-10 illustrates the secondary magnetic field generated by the induced circulation of π electrons in a carbonyl group. Whereas a triple bond has axial symmetry and permits the electrons to circulate cylindrically *around* the axis of the bond, as illustrated in Fig. 13-9 for acetylene, a double bond lacks axial symmetry and the induced electron circulation must take place *along* the bond axis, in the π orbitals above and below the bond axis, as illustrated in Fig. 13-10. Just as the triple bond was shown to be anisotropic, so also is the double bond, giving a much greater electron circulation in the orientation shown in Fig. 13-10 than when the axis of the carbonyl group is parallel to the applied field. As can be seen from Fig. 13-10, the secondary magnetic field generated causes *shielding*

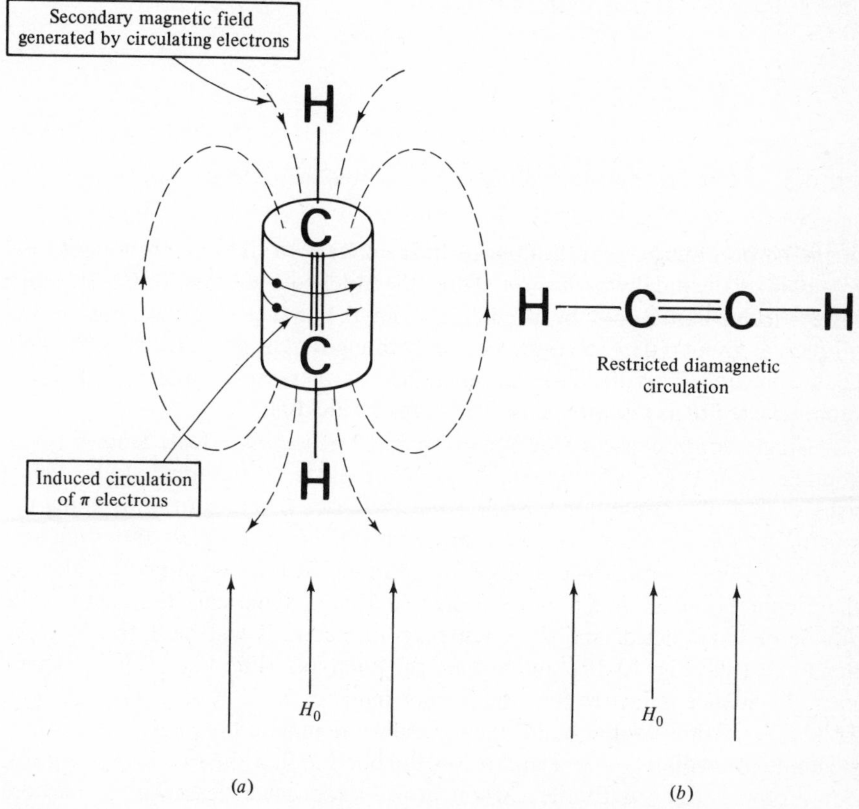

FIGURE 13-9 The origin of diamagnetic shielding of acetylenic protons due to the anisotropy of acetylene. Axis of molecule is (*a*) parallel and (*b*) perpendicular to applied field.

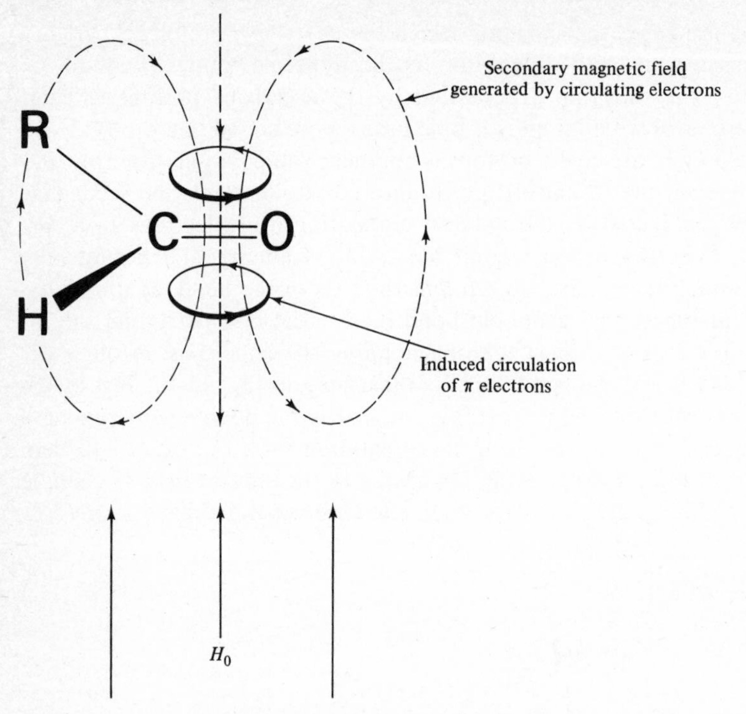

FIGURE 13-10 Deshielding of an aldehydic proton due to diamagnetic anisotropic effects.

at the carbon and oxygen atoms (of little interest to us) but pronounced *deshielding* at the aldehydic *proton*. Thus, the aldehydic proton absorbs at much lower fields than would be predicted from the inductive effect and in fact absorbs at about 0.03 τ, a region normally thought of as being restricted to very acidic protons. This *decrease* in shielding (or increase in magnetic field) is often referred to as *paramagnetic shielding* (deshielding).

Single bonds cause a small but observable shielding effect. Since a single bond has axial symmetry, one might predict that the shielding would be analogous to that of a triple bond. Such is not the case. Whereas protons along the axis of a triple bond experience pronounced *shielding*, protons along the axis of a single bond experience a slight *deshielding*. This is explained by viewing the sigma bond as much more isotropic (more spherical) than either the double or triple bond, and the circulation of electrons will be both *along* the bond axis (like Fig. 13-10), and *around* the bond axis (like Fig. 13-9*a*). Experimental evidence indicates that the former mode of circulation is favored (Fig. 13-11). As with a double bond, the secondary magnetic field generated causes shielding immediately above and below the bond and at the two carbon atoms, but protons attached to the carbon atoms experience deshielding. The deshielding effect is smaller than that for double bonds.

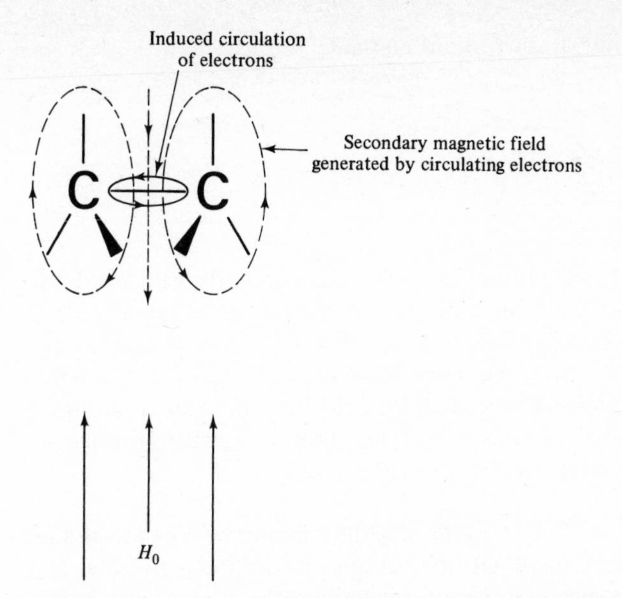

FIGURE 13-11 Diamagnetic shielding by the carbon-carbon single bond.

Example 13-11 Arrange the following saturated hydrocarbons in the order of increasing field strength at which their protons will absorb: CH_4, RCH_3, R_2CH_2, and R_3CH.

Answer The effect of successive alkyl substituents on a proton attached to carbon might be expected to cause *increased* shielding if the shift were controlled by the inductive effect, since successive substitutions of protons by alkyl groups would increase the electron density around the remaining proton. However, if the anisotropic effect (effect of nonspherical electron distribution) predominates, as it generally does, adding alkyl groups will cause *deshielding* and absorption will occur in the order R_3CH, R_2CH_2, RCH_3, and CH_4 (arranged from low to high field).

The fact that the anisotropic effect does predominate is borne out by the absorption of methine, methylene, methyl, and methane protons in $(CH_3)_3CH$, $(CH_3)_2CH_2$, CH_3CH_3, and CH_4 occurring at 8.5, 8.75, 9.12, and 9.77 τ, respectively.

Example 13-12 Predict whether there will be a difference in chemical shift for equatorial vs. axial protons in cyclohexane. If there is a difference, predict which will absorb at higher field.

Answer It is convenient to think of the carbon-carbon single bonds as having a deshielding cone along the bond axis. The problem can be simplified by looking only at the protons on the C_1 atom:

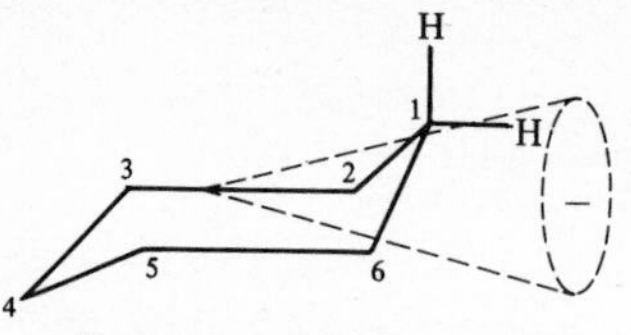

Cone represents a deshielding (−) zone

The axial and equatorial protons are orientated similarly with respect to the C_1—C_2 and C_1—C_6 bond axes, and therefore those bonds do not cause any differences in chemical shift. However, only the equatorial proton is within the deshielding cone of the C_2—C_3 (or C_5—C_6) bond. Thus, equatorial protons should be down field from the axial protons. In practice, the equatorial protons on various rigid six-membered rings are found 0.1 to 0.7 ppm down field from axial protons.

Finally, a special type of neighboring shielding should be mentioned, that of *steric congestion* or *van der Waals deshielding*. It is an experimental fact that if protons or other atoms in different parts of a molecule are brought sufficiently close for van der Waals repulsion to operate, mutual *deshielding* will occur. For example, in the conformationally rigid cyclohexanone chair system I (made rigid by being embedded in a steroid skeleton, for example), the proton H* will resonate at a lower frequency when R is CH_3 than when R is H because of electron repulsion between the electron-charge clouds. This type of steric interaction always causes *deshielding*, but the effects are usually small, generally of the order of 1 ppm or less.

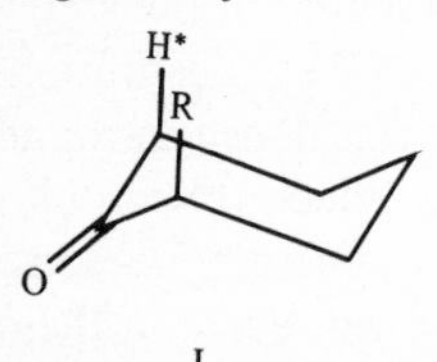

I

Interatomic Diamagnetic Circulation in Aromatic Rings This third type of diamagnetic shielding is associated with aromatic and pseudo-aromatic rings, having cyclically delocalized π electrons which are readily induced into circulation by an applied magnetic field. Large ring currents are thus induced, and correspondingly large shielding effects result. Such systems are far from being spherically symmetrical (in fact they are markedly anisotropic), and thus the effects of the induced field on a rigidly attached proton in the molecule do not average to zero for all possible orientations of the ring with respect to the applied field. Aromatic rings therefore provide a strong source of long-range shielding and deshielding, and such demonstrations of ring currents are probably the best evidence available for aromaticity.†

† Long-range shielding applies to shielding that operates through space rather than directly through chemical bonds; the latter is short-range shielding. Thus, of the three main kinds of diamagnetic shielding discussed, the first is short range and the last two are long range.

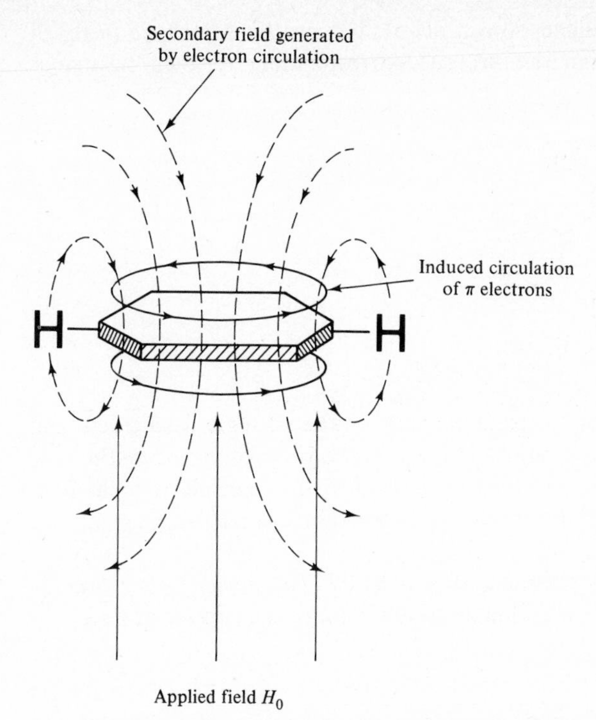

FIGURE 13-12 The magnetic field generated by the induced circulation of π electrons in benzene.

Figure 13-12 illustrates the magnetic field generated by the induced circulation of π electrons in benzene. As the figure shows, the secondary magnetic field generated by the induced ring current causes pronounced shielding at the center of the ring but *deshielding* outside, in the plane of the ring where the protons lie. Thus, the protons of benzene have a chemical shift of 2.73 τ, which is a lower applied field than would be necessary if it were not for this deshielding.

Example 13-13 The methylene protons in a normal aliphatic chain have a typical chemical shift of 8.75 τ. Predict whether the central methylene protons in 1,4-decamethylenebenzene, shown below, will resonate higher or lower than 8.75 τ.

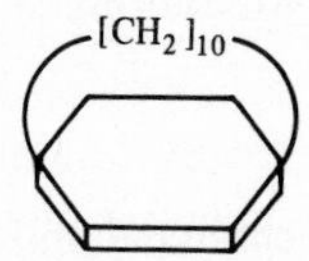

Answer The central methylene protons are held over the center of the benzene ring and thus will be shielded and resonate at a higher field than 8.75 τ. In practice, the absorption peak is located at about 9.20 τ.

Example 13-14 Annulene, shown below, has proton resonance peaks at $1.1\,\tau$ and $11.8\,\tau$. Assign the protons corresponding to each resonance peak.

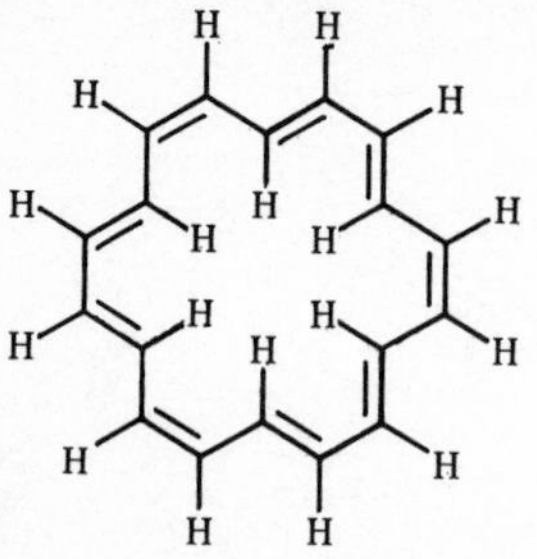

Answer The 12 protons *outside* the ring will be strongly deshielded, and therefore must be absorbing at $1.1\,\tau$, while the six protons *inside* the ring will be strongly shielded, and account for the unusually high field absorbance at $11.8\,\tau$.

Example 13-15 Acetophenone, shown below, has absorption peaks at 7.4, 2.6, and $2.1\,\tau$. Assign each absorption peak to the correct protons.

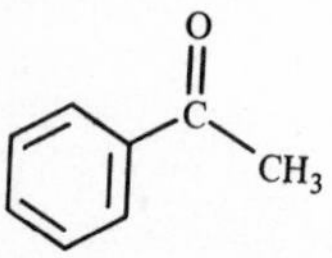

Answer First, it is to be expected that protons on the benzene ring will be strongly deshielded compared with the methyl protons. Therefore, the $7.4\,\tau$ peak will be assigned to the methyl protons, with the other two peaks assigned to ring protons. Of the ring protons, the two ortho protons can be expected to experience a different field (due to the proximity of the acetyl group) than is experienced by the meta and para protons. More specifically, the circulating π electrons in the carbonyl bond will cause additional deshielding of the ortho protons, and thus the $2.1\,\tau$ peak may be assigned to the ortho protons and the $2.6\,\tau$ peak to the meta and para protons.

Methyl protons are among the most highly shielded protons in organic molecules and generally absorb in the vicinity of $9.0\,\tau$ (this is the case, for example, in saturated hydrocarbons). In acetophenone, the deshielding effect of the carbonyl group accounts for the much lower field absorption ($7.4\,\tau$).

CHEMICAL-SHIFT CORRELATIONS

Figure 13-13 gives the general range of proton chemical shifts for some common functional groups. The ranges indicated in Fig. 13-13 are usually too wide to be of much use for precise structural assignments but serve to classify

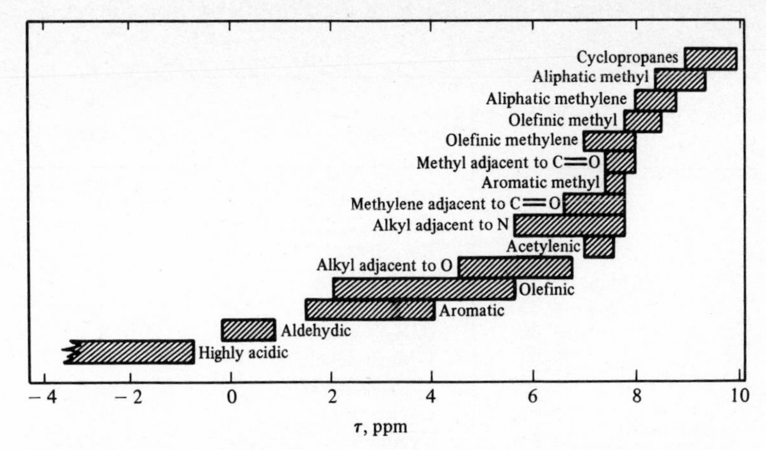

FIGURE 13-13 Approximate chemical-shift ranges for protons in common types of organic molecules. (*From J. C. P. Schwarz, "Physical Methods in Organic Chemistry," p.* 174, *Oliver & Boyd, Edinburgh,* 1964, *by permission.*)

bonds in a preliminary way. References 11, 14 to 16, and 18 give detailed correlation charts and compilations of spectra. Table 13-2 gives some examples of chemical shifts observed for protons in specific compounds.

13-1D Spin-Spin Interactions

Three main features of nmr spectra are used in their interpretation: (1) the *positions* of spectral bands (i.e., the chemical shifts), which are important because they give a clue to the structural environment of the protons involved; (2) the *intensities* of the bands, which give a measure of the relative number of protons involved; (3) and the *multiplicities* of the bands (or splitting of peaks into multiple bands instead of single peaks), which give information about the number of protons on adjacent carbon atoms, thereby conveying structural and often stereochemical information. This third aspect of nmr spectra, due to spin-spin interactions, will now be discussed.

Figure 13-14 shows the nmr spectrum and formula of 1,1,2-trichloroethane. The small peak at 2.72 τ is due to traces of chloroform in the $CDCl_3$ solvent, whereas the small peak at 6.32 τ is associated with a trace impurity. The triplet centered at 4.23 τ can be attributed to the methine proton (*b*), which is adjacent to two chlorine atoms. The doublet centered at 6.05 τ can be attributed to the two methylene protons (*a*), adjacent to one chlorine atom. The inductive effect caused by the electronegativity of the chlorine atoms can be used to explain the relative chemical shifts of the two types of protons, but this does not explain why the peaks occur as a triplet and a doublet, respectively.

The explanation for the peak splitting lies in the spin-spin coupling between the protons on adjacent carbon atoms. If we first consider the local field experienced by the two equivalent methylene protons (*a*), it is evident

TABLE 13-2 Typical Chemical Shifts Observed for Protons in Specific Compounds†

Compound	Shift τ, ppm
Methyl protons:	
$(CH_3)_4Si$	10.000
$CH_3(CH_2)_3CH_3$	9.15
$(CH_3)_4C$	9.08
CH_3CH_2OH	8.83
CH_3CH_2Cl	8.60
CH_3CH_2Br	8.34
$(CH_3)_2C{=}CH_2$	8.299
CH_3CN	8.026
$(CH_3)_2S$	7.942
CH_3COCH_3	7.915
CH_3I	7.84
$CH_3C_6H_5$	7.66
CH_3Br	7.38
CH_3Cl	7.00
$(CH_3)_2O$	6.73
CH_3OH	6.62
$CH_3OC_6H_5$	6.27
CH_3F	5.70
CH_3NO_2	5.72
Methylene protons:	
Cyclopropane	9.78
$CH_3(CH_2)_4CH_3$	8.75
Cyclohexane	8.56
Cyclopentane	8.49
Bicyclo[2.2.1]hepta-2,5-diene	8.05
Cyclopentanone	7.98
1,3-Cyclohexadiene	7.74
$(CH_3CH_2)_2CO$	7.61
$(CH_3CH_2)_2N$	7.58
$CH_3COCH_2COOCH_3$	6.52
$(CH_3CH_2)_2O$	6.64
$CH_3COCH_2COCH_3$	6.45
CH_3CH_2OH	6.41
Tetrahydrofuran	6.37
$HC{=}CCH_2Cl$	5.91
$HC{\equiv}CCH_2OH$	5.82
$C_6H_5CH_2OH$	5.61
Methine protons:	
Chlorocyclopropane	7.05
$(CH_3)_2CHNH_2$	7.05
Chlorocyclohexane	6.08
$(CH_3)_2CHOH$	6.05
$(CH_3)_2CHCl$	5.88
$(CH_3)_2CHBr$	5.83
$(CH_3)CCHCH_2Br$	8.20
Bicyclo[2.2.1]heptane	7.81
Bicyclo[2.2.1]hepta-2,5-diene	6.53
Olefinic protons:	
$(CH_3)_2C{=}CH_2$	5.4
$(CH_3)_2C{=}CHCH_3$	4.79
1-Methylcyclohexene	4.70
$(C_6H_5)_2C{=}CH_2$	4.60
Cyclohexene	4.43
Cyclohexa-1,3-diene	4.22
$CH_3CH{=}CHCHO$	3.95
Cyclopentadiene	3.58
$Cl_2C{=}CHCl$	3.55
Bicyclo[2.2.1]hepta-2,5-diene	3.35
cis-Stilbene	3.51
trans-Stilbene	3.01
Acetylenic protons:	
$HOCH_2C{\equiv}CH$	7.67
$ClCH_2C{\equiv}CH$	7.60
$C_6H_5C{\equiv}CH$	7.07
$CH_3COC{\equiv}CH$	6.83
Aromatic protons:	
Pyrrole (β hydrogen)	3.93
Furan (β hydrogen)	3.72
Pyrrole (α hydrogen)	3.47
Mesitylene	3.36
Thiophene (β hydrogen)	2.94
Toluene	2.91
Thiophene (α hydrogen)	2.83
Benzene	2.73
Furan (α hydrogen)	2.64
C_6H_5CN	2.46
Naphthalene	2.27
Pyridine (γ hydrogen)	2.64
Pyridine (β hydrogen)	3.015
Pyridine (α hydrogen)	1.50
Aldehydic protons:	
$(CH_3)_2N{-}CHO$	2.16
CH_3OCHO	1.97
$(CH_3)_2CHCHO$	0.44
$C_6H_5CH{=}CHCHO$	0.37
CH_3CHO	0.284
p-$CH_3OC_6H_4CHO$	0.199
C_6H_5CHO	0.035

† From J. C. Davis, Jr.: "Advanced Physical Chemistry: Molecules, Structures, and Spectra." Copyright © 1965, The Ronald Press Company, New York.

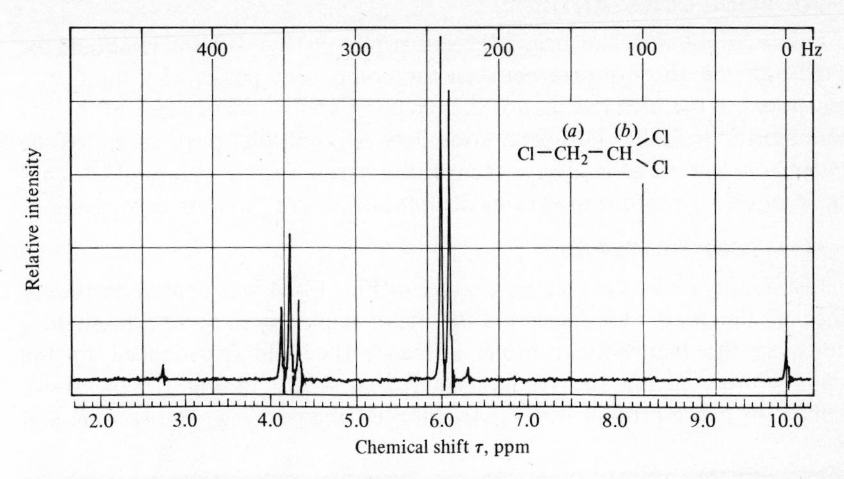

FIGURE 13-14 Nuclear magnetic resonance spectrum of 1,1,2-trichloroethane obtained with Varian A-60 nmr spectrometer using $CDCl_3$ solvent and TMS internal reference. (*From Varian Associates, "High Resolution NMR Spectra Catalogue," spectrum* 2, *National Press, Palo Alto, Calif.*, 1962, 1963, *by permission.*)

that the precise value of the field will depend on the orientations of the nuclear magnet of the methine proton (b). Thus, the adjacent methine proton will sometimes have its spin lined up *parallel* to the applied field, and at other times it will be antiparallel. Since these two spin states are equally probable, half the time the field from the methine proton will *augment* the applied field, and half the time it will *diminish* the applied field experienced by the two equivalent methylene protons (a). Figure 13-15 illustrates how the resonance energy of the methylene protons is affected (perturbed) by the two spin states of the methine proton. If the resonance energy of the methylene protons were not perturbed by spin-spin coupling, the spectrum would show a singlet line at a position midway between the components of the doublet (in practice, this would be observed at low resolution or if the applied field were relatively inhomogeneous in strength). Because of the spin-spin interaction, the doublet (two peaks of approximately equal intensity) is observed with high-resolution instruments (having applied fields that are highly homogeneous).

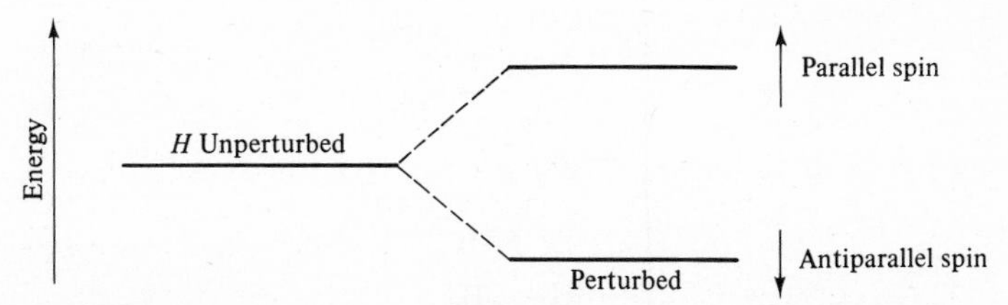

FIGURE 13-15 Schematic energy diagram showing how the field experienced by a given proton is perturbed by the two spin states of an adjacent proton.

COUPLING CONSTANTS

Before explaining the origin of the triplet in Fig. 13-14, it should be pointed out that the distance between the component peaks of a doublet is proportional to the effectiveness of the coupling and is independent of the applied magnetic field H_0. This separation between multiplet peaks is called the *coupling constant J* and is always recorded in hertz. Typical values of the coupling constant for protons in various configurations are given in Table 13-3.

SPLITTING PATTERNS

Just as the doublet centered at 6.05τ in Fig. 13-14 was caused by the influence of the methine proton on the field experienced by the methylene protons, so the methylene protons influence the field experienced by the methine proton, giving the triplet centered at 4.23τ in Fig. 13-14. If we designate the two *a* protons of 1,1,2-trichloroethane (see Fig. 13-14) as *a′* and

TABLE 13-3 Typical Proton Spin-Spin Coupling Constants

Type of structure	*J*, Hz	Type of structure	*J*, Hz
$H_2C<$ (geminal H–C–H)	12–15	>C=C< with CH on one carbon and H on the other (allylic CH–C=C–H)	4–10
>C=CH₂ (both H on same carbon)	0–3	>C=CH—CH=C<	10–13
H–C=C–H	Cis 6–14 Trans 11–18	>CH—C≡CH	2–3
>CH—CH< (free rotation)	5–8	>CH—OH (no exchange)	5
Ring H:		>CH—CHO	1–3
Ortho	7–10	—CH(CH₃)₂	5–7
Meta	2–3	—CH₂—CH₃	7
Para	0–1		

a'', then at a given moment there are four equally probable spin arrangements with respect to the applied field. The spins of a' and a'' are either (1) both parallel, (2) both antiparallel, (3) a' parallel and a'' antiparallel, or (4) a' antiparallel and a'' parallel with respect to the applied field. The effects of these four arrangements on the resonance energy of the methine proton are schematically indicated in Fig. 13-16. Obviously when both proton spins are parallel, the field experienced by the methine proton is augmented, whereas when both spins are antiparallel, the effective field is diminished. However, there is twice as much probability that one proton will be parallel and the other antiparallel, canceling any net effect on the field, and thus it can be predicted that the center peak will be unshifted and twice as intense as the two side peaks.

Another useful way of predicting this splitting pattern is to look at the splitting caused by one proton at a time. In other words, in the previous example, we would first predict that the two possible orientations of the a' proton would split the methine resonance into a doublet. Then, the two possible orientations of the a'' proton would further split each component of the original doublet into a second doublet. This process is represented diagramatically in Fig. 13-17. Since the coupling constant J will be identical for a' and a'' (a' and a'' are chemically equivalent), the doublets overlap in the center to give a triplet of relative intensity 1:2:1. The coupling constant in 1,1,2-trichloroethane (applicable to both the triplet and the doublet) is 6 Hz.

NATURE OF COUPLING

Before proceeding with other examples, it should be made clear that spin-spin coupling occurs through intervening bonds rather than directly through space. The transmission of spin-state information between coupled nuclei can be pictured as follows. First, the spin of a given proton will tend to polarize the spin of the nearest bonding electron in such a way that the spins will be paired, for example H↑ e↓. The spin of that bonding electron will in turn cause the spin of its companion bonding electron to be paired with it, H↑ e↓ e↑, and so on through to the next proton. Thus, the spin direction at one

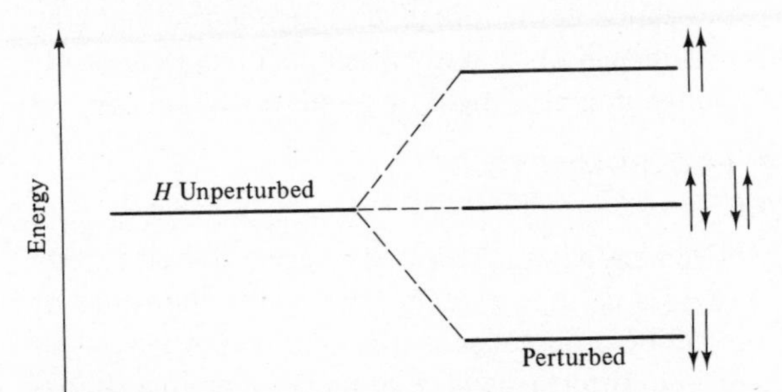

FIGURE 13-16 Schematic energy diagram showing how the various spin states of two methylene protons can perturb the field experienced by a proton at a nearby site.

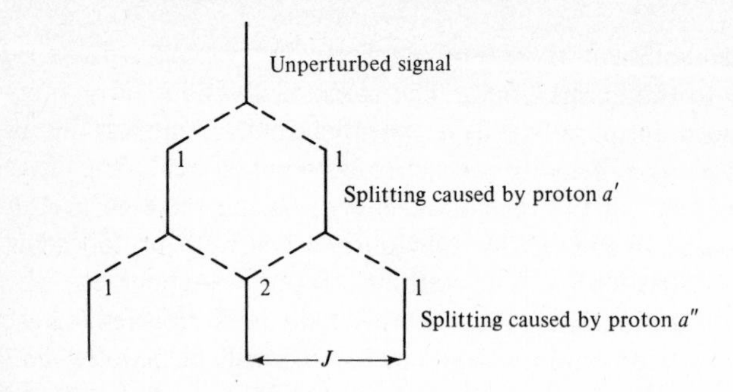

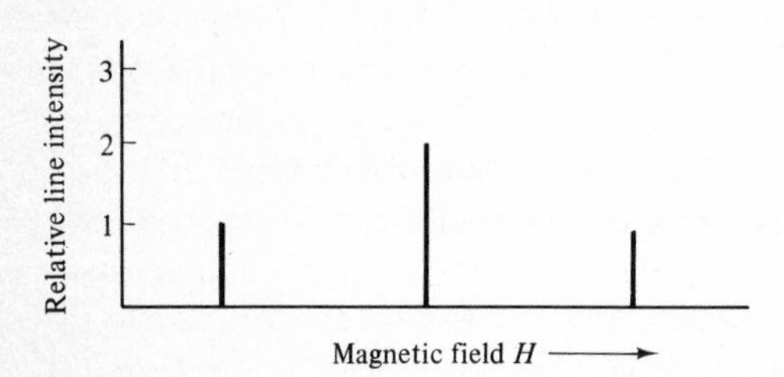

FIGURE 13-17 (*a*) Construction of a triplet splitting diagram and (*b*) resulting line spectrum. The relative intensity of each line in (*a*) is indicated by the numbers given; note that when lines converge, the intensity is additive.

proton nucleus will be transmitted through the intervening bonding electrons, causing a small (perturbing) *electron* spin field at a neighboring proton nucleus. This field will either augment or diminish the applied field, depending on the direction of the spin of the originating proton. Coupling is ordinarily not important beyond three bonds (and in fact is usually small even between adjacent atoms) unless there is electron delocalization, as in aromatic or unsaturated systems, or ring strain, as in small rings or bridged systems.

This simple picture of spin-spin interaction in no way depends on the strength of the applied field, and indeed, as stated earlier, coupling constants (in hertz) are independent of applied magnetic field or oscillator frequency.

FIRST-ORDER RULES FOR PREDICTING BAND MULTIPLICITIES

The interpretation of spin-spin coupling patterns is of great importance in structural analysis, since it provides definite evidence regarding the order or sequential arrangement of proton-bearing carbon atoms in a molecule. Unfortunately, a large number of nmr spectra involve spin-spin coupling that is too complex to interpret without a detailed mathematical (quantum-mechanical) analysis. On the other hand, for all compounds which satisfy some simplifying conditions, there is a very simple set of rules, known as first-order rules, for interpreting coupling patterns. Two conditions must be met: (1) the chemical shift $\Delta\nu$ separating two interacting groups must be large compared with the magnitude of the splitting (as a rule of thumb, $\Delta\nu/J$ should be 6 or more); (2) each nucleus in one group must interact equally with every nucleus in the second group; i.e., there must be just one value for the coupling constant J for the mutually interacting protons. As will be seen, the first condition is necessary in order to prevent a complex scrambling (partial overlap) of the absorption peaks of the two groups, often resulting in the appearance of many more lines than would be predicted from first-order rules. The second condition is also necessary to prevent the appearance of additional lines, as well as other changes in the spin patterns, although spectral changes due to violations of this condition can often be accounted for with reasonable ease, as will be shown.

It is difficult to generalize about the kinds of compounds or chemical groups that will satisfy the conditions for first-order spin splitting, but certainly first-order behavior is almost wholly restricted to *aliphatic* rather than aromatic systems. The second condition applies largely to aliphatic systems having free rotation about intervening bonds, so that coupling between mutually interacting protons will be equivalent. Since coupling constants between protons on adjacent carbon atoms are typically about 7 Hz in aliphatic systems (see Table 13-3), condition 1 requires that the difference in chemical shifts between the interacting groups of protons be about 42 Hz. This is less than 1 ppm† with any ordinary nmr spectrometer, which is easily achieved as long as two groups of protons are in a reasonably different chemical environment (see Table 13-2). For example, protons in such systems as $—OCH_2CH_3$ and $RCOOCH_2CH_2NR_2$ satisfy the first-order conditions.

Example 13-16 Predict whether or not the ring protons on the following disubstituted phenyl compound will follow first-order rules. Explain.

H_B H_A

CH_3O— —NO_2

$H_{B'}$ $H_{A'}$

Answer First note that the two protons ortho to the nitro group (A and A') are chemically equivalent to one another and should absorb at one frequency, whereas the two protons ortho to the methoxy group (B and B') should absorb at another frequency. Without even knowing whether the two groups of protons satisfy the *first* condition, it can be predicted that they will not satisfy the *second* condition, since protons A and A' are not *equally* coupled to protons B and B'. For example, proton A

† Most nmr spectrometers operate at 60 MHz, and thus 42 Hz is 42 Hz/(60 $\times$ 10^6 Hz) $\times$ $10^6 \approx$ 0.7 ppm.

should be *strongly* coupled to proton B, but proton A should be only *weakly* coupled to proton B'. (Table 13-3 confirms this difference in coupling; ortho protons have $J \approx 8$ Hz, whereas para protons have $J = 0$ to 1.) Therefore it can be predicted that first-order rules would not hold for these ring protons.

When the two conditions are met, the following three rules govern the number of peaks in a band, their relative intensities, and their arrangement:

1 The number of lines in a multiplet is $N + 1$, where N is the total number of neighboring protons. This rule is applicable only to *proton* spectra or other nuclei which have a spin of ½; a more general expression, applicable to nuclei of any spin number I is $2^n I + 1$, where n is the total number of adjacent nuclei of spin I.

2 The relative intensities of the components of a multiplet are given, ideally, by the coefficients in the expansion of $(1 + X)^N$, where N (used in rule 1) is again the number of neighboring protons. (Note that only the *coefficients* are of importance here, and therefore X is simply a variable that need not be explicitly defined.)

3 Peaks are symmetrically arranged about the chemical shifts of the groups (making group chemical shifts easy to measure) and are separated from each other by the coupling constant (making the coupling constant easy to measure).

Example 13-17 Predict the number of lines and the relative intensities of each line for a band split into multiplets for five successive cases starting with one neighboring proton and working through five neighboring protons.

Answer When $N = 1$, a doublet with 1:1 peak intensities is observed. When $N = 2$, a triplet with 1:2:1 intensities is observed; i.e., since $(1 + X)^2 = 1 + 2X + X^2$, the coefficients are 1, 2, and 1, respectively. When $N = 3$, a quartet with relative intensities of 1:3:3:1 is observed; that is, $(1 + X)^3 = (1 + X)(1 + 2X + X^2) = 1 + 3X + 3X^2 + X^3$. When $N = 4$, five peaks of relative intensities 1:4:6:4:1 are observed. When $N = 5$, six peaks of relative intensities 1:5:10:10:5:1 are observed.

Example 13-18 Predict the proton magnetic resonance line spectrum for 2-chloro-5-nitrothiophen

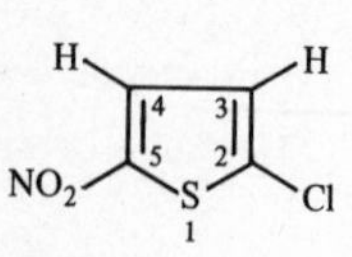

H H
4 3
5 2
NO_2 S Cl
1

Answer The protons in the 3 and 4 positions are not chemically equivalent, and they will have different chemical shifts. Since the nitro group is more electron-withdrawing than the chloro group, the proton in the 4 position may be expected at a lower field than the proton in the 3 position (see Table 13-2 for some typical chemical shifts). By the first-order rules each band can be expected to be split into a 1:1 doublet, as follows:

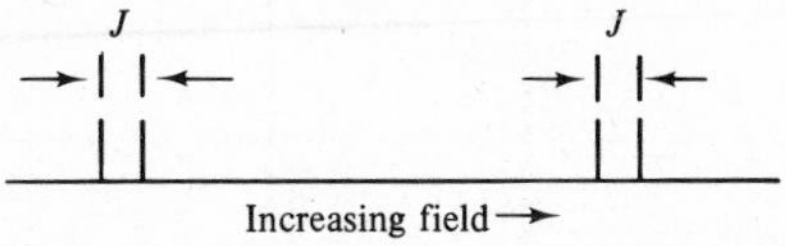

J J

Increasing field →

Example 13-19 Predict the proton magnetic resonance line spectrum for 2,5-dichlorothiophen.

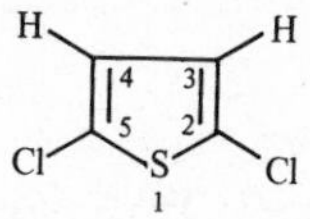

H H 4 3 5 2 Cl S Cl 1

Answer In this compound the protons in the 3 and 4 positions are chemically equivalent, and thus only a single peak will be observed. Although the two protons are actually spin-coupled to each other through the intervening bonds, no splitting is ever *observed* between chemically equivalent protons, since quantum-mechanial selection rules prohibit such interactions.

Example 13-20 The 60-MHz proton magnetic resonance spectrum of diethyl succinate is shown in Fig. 13-18. Assign each of the bands based on the relative chemical shifts and spin-spin splittings that you would expect from the chemical formula. Determine the coupling constants and decide whether this compound follows first-order splitting rules.

Answer Since diethyl succinate has a center of symmetry, the two ethyl groups on each end are identical and all four protons in the center ethylene, $—CH_2—CH_2—$, groups are identical. We would expect to find the methyl, $—CH_3$, groups at highest field ($8.75\,\tau$), and split into a 1:2:1 triplet through coupling with the methylene, $—CH_2—$, protons. The adjacent methylene groups would be at much lower field due to electron-withdrawing oxygen ($5.83\,\tau$) and should be split into a 1:3:3:1 quartet through coupling with the methyl protons. Since all four protons in the center ethylene group are chemically equivalent, a single peak should be observed at a field intermediate between the methyl and methylene protons ($7.4\,\tau$).

It can be deduced from Fig. 13-18 that the coupling constant between the methyl and methylene protons, $J_{CH_3-CH_2}$, is about 7 Hz ($\Delta\tau$ =

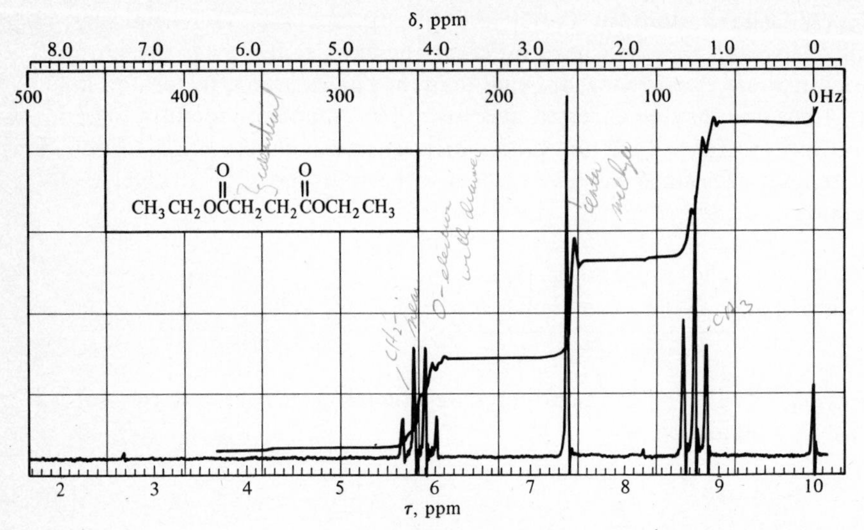

FIGURE 13-18 Proton nmr spectrum of diethyl succinate, obtained with a Varian A-60 spectrometer, using TMS internal standard and $CDCl_3$ solvent. (*From R. H. Bible, Jr., "Interpretation of NMR Spectra," p.* 44, *Plenum Publishing Corporation, New York,* 1965, *by permission.*)

0.12), and it has this same value in both the triplet and the quartet, a necessary condition for first-order splitting. The other condition for first-order splitting, that $\Delta\nu/J \geqq 6$, likewise is satisfied, since the chemial shift between the methylene protons and methyl protons is 250 − 75 Hz = 175 Hz, and therefore $\delta\nu/J$ is $^{175}/_{7} = 25$, a more than sufficient chemiccal shift.

The other small bands in Fig. 13-18 are easily assigned. First, the small band at 2.7 τ is due to chloroform impurity in deuterochloroform; the small peak at 8.2 τ must be an impurity in the compound; and the signal at 10 τ is the TMS reference band.

Finally, a check on the assignment of the protons in diethyl succinate can be made with the integral signal shown in Fig. 13-18. Since there are six methyl protons, four methylene protons, and four ethylene protons, the total band areas for these three groups should be in the ratio of 6:4:4, and examination of the integral signal in Fig. 13-18 confirms this, in agreement with the assignments made.

Example 13-21 The proton magnetic resonance spectrum of 1-nitropropane is shown in Fig. 13-19. Assign each of the bands and draw a splitting diagram to account for the splitting patterns and relative intensities observed.

Answer First, the peak at 10 τ is assigned to the TMS reference. Of the various protons in nitropropane, the methyl, $—CH_3$, protons would be the most shielded (8.97 τ) and would be split into a triplet by the adjacent methylene, —CH_2—, protons. The methylene group adjacent to the nitro group would be the most deshielded (5.65 τ) and would also be split into a triplet by the middle methylene protons. It should be noted that the triplet at 5.65 τ is smaller in area than the triplet at 9.87 τ, consistent with the relative number of protons in the methylene and methyl groups, further confirming the assignments.

Notice also that the inner peak of each triplet is slightly increased in intensity. This type of distortion is very common and is helpful to notice because the rising slant is *toward* the absorption of the group of protons responsible for the observed multiplicity. Thus, in this case, the two triplets indicate that the protons responsible for their splitting should absorb *between* the triplets, which is seen to be the case in Fig. 13-19.

Finally, and perhaps of greatest interest, is the multiplet centered at 7.93 τ, which is assigned to the middle methylene group. Since the coupling constant (about 7 Hz) between various protons in the CH_3CH_2— group is very nearly equal to the coupling constant between the various —CH_2—CH_2—NO_2 protons, the middle methylene proton is coupled by a total of five adjacent neighbors, resulting in six peaks having the approximate ratio of intensities of 1:5:10:10:5:1. (It is fortuitous in this case that the various coupling constants are nearly identical, since the methyl protons and end methylene protons are chemically quite different, but the fact that they are not *precisely* identical accounts for the slight broadening of the sextet peaks in Fig. 13-19; when neighboring protons have significantly different coupling constants, the simple $N + 1$ rule no

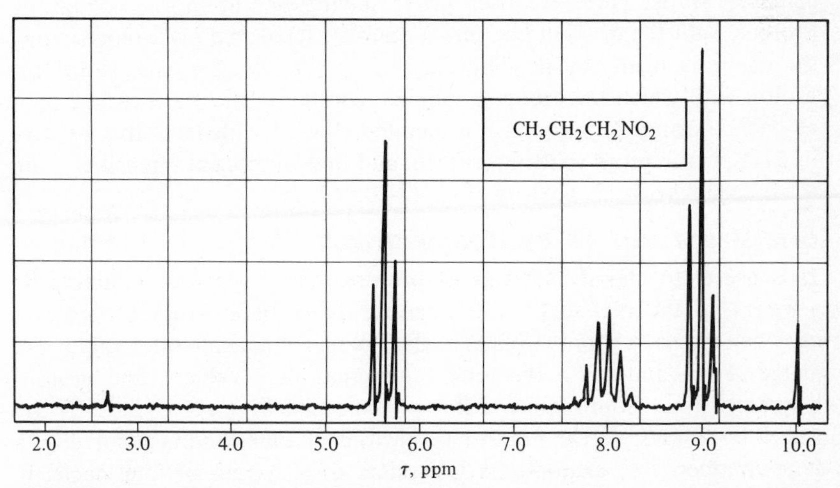

FIGURE 13-19 Proton nmr spectrum of 1-nitropropane obtained with a Varian A-60 spectrometer using TMS internal standard and $CDCl_3$ solvent. (*From Varian Associates, "High Resolution NMR Spectra Catalogue," spectrum* 42, *National Press, Palo Alto, Calif.*, 1962, 1963, *by permission.*)

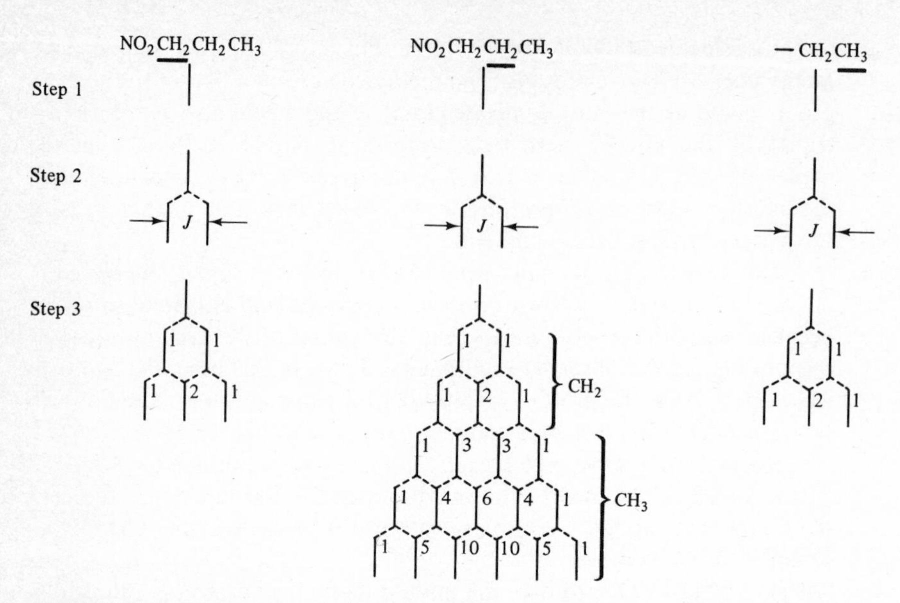

FIGURE 13-20 Construction of a first-order splitting diagram for 1-nitropropane. The numbers beside each line indicate the relative intensity of the line.

longer holds, as will be discussed in the section on higher-order spin patterns.)

The construction of a splitting diagram for nitropropane is outlined in Fig. 13-20. The first step is to draw a line for each group of chemically equivalent protons, locating the lines in positions relative to their chemical shifts. Next, for each group having one or more neighboring protons, split the original line into a doublet, separated by J, representing the coupling from *one* neighboring proton. Finally, for each additional proton with which the group is coupled, continue splitting each line until all neighboring protons are accounted for. To deduce the relative intensities of a given splitting pattern, add the component intensities from the preceding row, as shown in Fig. 13-20.

CLASSIFICATION OF PROTON SYSTEMS

It is useful to classify systems of protons giving spin-spin coupling by using letters of the alphabet to designate characteristic kinds of protons. Protons which have approximately the same chemical-shift value are designated A, B, and C (in the order of increasing τ values), and protons which absorb at positions quite different from the first set (A, B, C) are designated by X and Y. The number of protons of each kind is denoted by a subscript number. For example, AB_2X refers to a system of four nuclei in which one (A) is coupled to a group of two equivalent nuclei (B_2) with a coupling constant of the same order of magnitude as the difference in chemical shift between A and B_2, and one or more of these three nuclei are in turn coupled to the fourth nucleus (X) which has a very different chemical shift. If the signals of three different sets of coupled protons are far apart from each other in the spectrum, the third group of protons is designated as M. For example 1-nitropropane, $CH_3CH_2CH_2NO_2$, considered previously in Example 13-21, would be called an $A_3M_2X_2$ system.

Example 13-22 Classify the proton systems of 2-chloro-5-nitrothiophen and 2,5-dichlorothiophen, the spin patterns of which were considered in Examples 13-18 and 13-19.

Answer 2-Chloro-5-nitrothiophen would be an AX system, and 2,5-dichlorothiophen would be an A_2 system.

Example 13-23 Classify the protons of methyl ethyl ether, $CH_3OCH_2CH_3$, using the ABX convention.

Answer There is very little coupling between the methyl group and the ethyl group, and therefore it is best to classify each group separately. The methyl group would constitute an A_3 system, while the ethyl group would be an A_3X_2 system.

HIGHER-ORDER SPIN PATTERNS

Deviations from first-order patterns come either from having $\Delta\nu < 6J$ or from having unequal coupling constants between protons. In general, three changes in spin patterns may be observed when the two first-order conditions are not met: (1) the simple ratios of peak intensities are distorted; (2) extra peaks occur; and (3) the spacings between peaks may become unequal. Unfortunately, there are no simple rules for interpreting higher-order spin patterns analogous to the first-order rules, but it will be useful to point out some of the trends so that higher-order spectra can be easily recognized. (For a more detailed analysis of higher-order spin patterns the references should be consulted.)

It is possible to differentiate first-order systems from higher-order systems by means of the ABX classification system. The first condition for higher-order systems, that $\Delta\nu/J$ must be equal to or greater than 6, means that only systems of the general type A_aX_x or $A_aM_mX_x$ will be first order, whereas all systems of the AB type will be higher order. Furthermore, an A_aX_x or $A_aM_mX_x$ system may show first-order deviations through unequal coupling constants (deviations from the second condition), and it is convenient to designate this type of deviation by priming each letter of the system. For example, an A_2X_2 system would be first order, but an $A_2'X_2'$ system would be higher order due to deviations from the second condition. Likewise, an A_2B_2 system would be higher order due to deviations from the first condition, but an $A_2'B_2'$ system would be higher order due to deviations from *both* conditions.

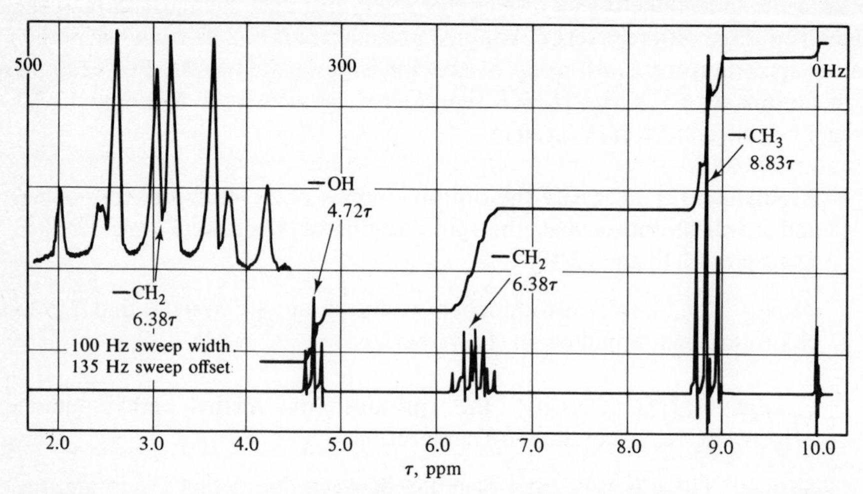

FIGURE 13-21 Nuclear magnetic resonance spectrum of very pure ethanol. The sample was neat; i.e., no solvent was used. Obtained with a Varian A-60 spectrometer using a 500-Hz sweep width, except for upper tracing, which represents a fivefold scale expansion of the sweep axis in the region of 6.4τ. Coupling constants measured from the spectrum are J_{AM} = 5.0 Hz and J_{MX} = 7.2 Hz. (*After J. R. Dyer, "Applications of Absorption Spectroscopy of Organic Compounds," p. 70, Prentice-Hall, Inc., Englewood Cliffs, N.J., 1965, by permission.*)

Example 13-24 The nmr spectrum of pure ethyl alcohol is shown in Fig. 13-21. Classify ethanol according to the ABX system.

Answer The three distinct types of protons in ethanol (hydroxyl, —OH; methylene, —CH_2—; and methyl, —CH_3) characterize ethanol as an AM_2X_3 system. First-order splitting rules predict that the hydroxyl and methyl peaks should each appear as a triplet, in agreement with Fig. 13-21, since each peak is split by the middle methylene group. However, first-order splitting rules predict that the methylene multiplet should be a quintet, provided that all four neighboring protons are equally coupled to the methylene protons; that is, $J_{AM} = J_{MX}$. The fact that the multiplet at 6.38τ contains many more than five peaks is evidence that higher-order splitting (due to unequal coupling constants) has occurred. Thus ethanol is an $A'M'_2X'_3$ system.

13-1E Effect of Chemical Exchange on Spin-Spin Interactions

In the previous section the spectrum of a very pure sample of ethanol was discussed (Fig. 13-21). This spectrum is somewhat idealistic because of the unusually high purity of the ethanol. If only moderately pure ethanol is examined by nmr, a spectrum like Fig. 13-22 results. (Actually, the spectrum of Fig. 13-22 is for reagent-grade ethanol to which a trace of hydrochloric acid has been added; the spectrum may be considered as typical of ordinary grades of anhydrous ethanol, however, since trace amounts of acidic impurities are usually present.) If Figs. 13-21 and 13-22 are compared, it will be seen that there are distinct differences in the splitting patterns for the hydroxyl proton signal (in the region of 4.6 to 4.7 τ) and for the methylene proton signal (in the region of 6.4 τ). Whereas the hydroxyl proton signal is a triplet when ethanol is very pure (Fig. 13-21), it is a singlet when traces of acidic impurities are present (Fig. 13-22); and whereas the signal for the methylene protons is essentially a higher-order octet when the ethanol is very pure (Fig. 13-21), it appears as a first-order quartet when the ethanol contains traces of acidic impurities (Fig. 13-22). The apparent simplification of the ethanol spectrum when acidic impurities are present is attributed to rapid *chemical exchange* of the hydroxyl proton in the presence of trace amounts of acid catalysts. In other words, in the presence of a catalyst, a hydroxyl proton is rapidly exchanged between different ethanol molecules. One effect of this rapid exchange is that the hydroxyl proton is in almost constant contact with all possible spin arrangements of the methylene group, and therefore the magnetic effects corresponding to *three* possible spin arrangements of the methylenic protons are *averaged*; instead of a triplet a *single* sharp absorption

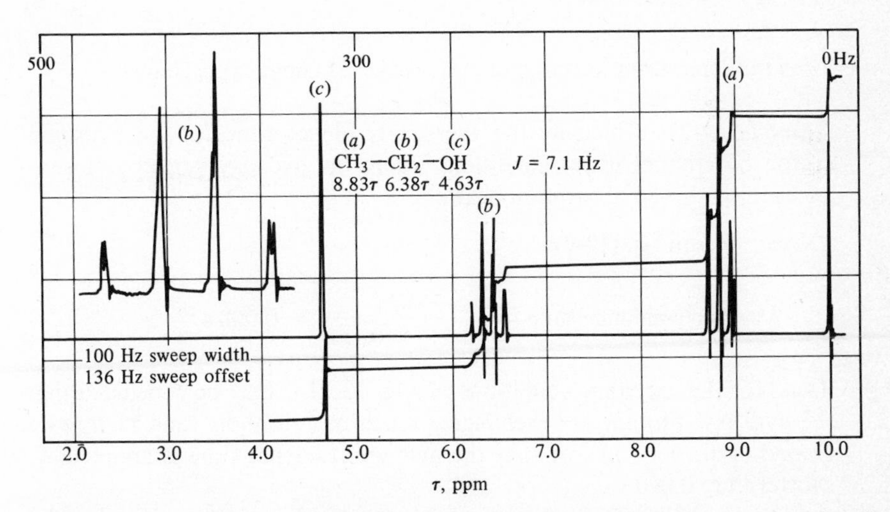

FIGURE 13-22 Nuclear magnetic resonance spectrum of ethanol containing a trace of acidic impurities. Coupling constant measured from spectrum, $J = 7.1$ Hz. Sample was reagent-grade ethanol containing a trace of hydrochloric acid; no solvent was used; i.e., a neat sample; Varian A-60 spectrometer, 500-Hz sweep width, except for upper tracing, which represents a fivefold scale expansion of the sweep axis in the region of 6.4 τ. (*From J. R. Dyer, "Applications of Absorption Spectroscopy of Organic Compounds,"* p. 67, *Prentice-Hall, Inc., Englewood Cliffs, N.J.,* 1965, *by permission.*)

line is observed. Similarly, the methylene protons no longer experience the magnetic effects of two different spin arrangements of the hydroxyl proton and instead simply experience an average, nonsplitting field from the hydroxyl proton. Therefore, the only significant splitting of the methylene group comes from the methyl group, and a first-order quartet results. In a case like this, rapid chemical change is said to cause *spin decoupling* of the hydroxylic and methylenic protons.

The criterion for spin decoupling by chemical exchange is that the *frequency* of exchange must be substantially faster than the frequency *separation* of the components of the multiplet. If the rate of chemical exchange is of the same order of magnitude as the frequency separation, the instantaneous spin orientations are only partially averaged and a broad adsorption peak results. If, on the other hand, the rate of chemical exchange is much slower than the frequency separation, the full multiplicity will be observed. In the case of the acidified ethanol shown in Fig. 13-22 the frequency of exchange of the hydroxyl proton in the acidified ethanol is faster than the J_{AM} coupling constant in ethanol, 5.0 Hz. When the frequency of chemical exchange is known, e.g., when the conditions for exchange are such that the multiplet just coalesces into a broad absorption peak, the average residence *time* (in seconds) for a proton at one site can be calculated from

$$\text{Average residence time} = \frac{\sqrt{2}}{\pi J} \tag{13-9}$$

where J is the separation of peaks in the absence of chemical exchange.

Example 13-25 Calculate the average residence time for the hydroxyl proton on ethanol under conditions where the hydroxyl triplet just coalesces into a broad absorption peak.

Answer From Eq. (13-9),

$$\text{Average residence time} = \frac{\sqrt{2}}{(3.14)\,(5.0\ \text{Hz})} \approx 0.090\ \text{s}$$

Thus, for the exchange conditions of Fig. 13-22 it may be concluded that the hydroxyl protons are exchanging much more rapidly than 11 times a second ($1/0.09 \approx 11$), or that the average residence time is appreciably shorter than 0.090 s.

As one might expect, the rate of chemical exchange increases with increasing temperature, and thus spin decoupling can sometimes be achieved by raising the temperature. Alternatively, some systems that are normally uncoupled at room temperature can be coupled by lowering the temperature. Spectra of methanol at various temperatures are shown in Fig. 13-23. At

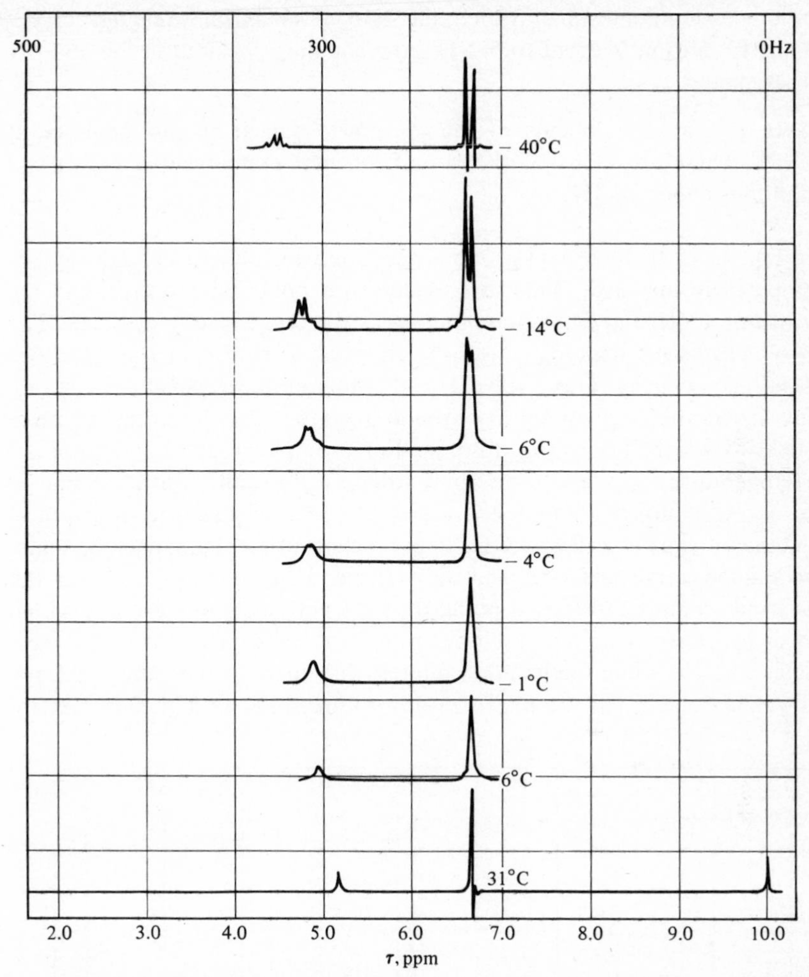

FIGURE 13-23 Nuclear magnetic resonance spectra of methanol at various temperatures. A neat sample of reagent-grade methanol was used, along with a Varian A-60 spectrometer set for a 500-Hz sweep width. The coupling constant for methanol is $J = 5.2$ Hz. (*From J. R. Dyer, "Applications of Absorption Spectroscopy of Organic Compounds," p. 72, Prentice-Hall, Inc., Englewood Cliffs, N.J., 1965, by permission.*)

room temperature (31°C) the spectrum of methanol shows sharp singlets for both the hydroxyl and methyl protons, indicating decoupling between the various protons. As the temperature is lowered, successive broadening and then splitting of the peaks occurs, indicating that coupling starts when the rate of chemical exchange of the hydroxyl proton becomes sufficiently slow. At −40°C the hydroxyl band is clearly split into a first-order quartet, and the methyl band appears as a first-order doublet.

Example 13-26 From the methanol spectra shown in Fig. 13-23, estimate the temperature at which the rate of chemical exchange of the hydroxyl proton is equal to 5.2 Hz, the coupling constant in the absence of exchange.

Answer The components of the multiplets appear to just coalesce at $-4°C$. Therefore, at $-4°C$ the rate of chemical exchange of the hydroxyl proton is about 5.2 Hz.

In Fig. 13-23 the hydroxyl-group absorption band shifts to higher field as the temperature increases. This shift is due to a progressive decrease in the extent of intermolecular *hydrogen bonding* as the temperature increases. The presence of hydrogen bonding causes the chemical shift of hydroxylic protons to depend upon concentration, solvent, and temperature. In general, hydrogen bonding lowers the electron density around a proton, thus lowering the magnetic field at which the proton absorbs. For example, Fig. 13-24 shows the room-temperature spectrum of an approximately 7% solution of methanol in $CDCl_3$, and by comparing this spectrum to the room-temperature spectrum of pure methanol shown in Fig. 13-23, it can be seen that the hydroxyl singlet has undergone a marked chemical shift (from 5.2τ to 8.6τ), whereas the methyl band remains constant at about 6.6τ. In pure methanol there is a high degree of hydrogen bonding, resulting in hydroxyl absorption at a relatively low field, whereas when methanol is diluted with inert solvent, the hydrogen bonding is disrupted and the hydroxyl absorption shifts to high field. At in-

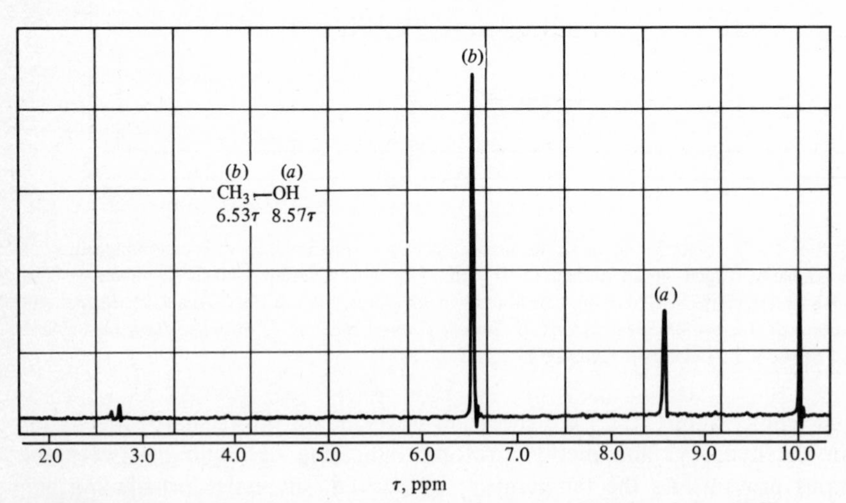

FIGURE 13-24 Nuclear magnetic resonance spectrum of methanol in $CDCl_3$ at room temperature. Sample was approximately 7% solution of methanol in $CDCl_3$; instrument was a Varian A-60 spectrometer, set for a 500-Hz sweep width. (*From Varian Associates, "High Resolution NMR Spectra Catalogue," spectrum* 1, *National Press, Palo Alto, Calif.*, 1962, 1963, *by permission.*)

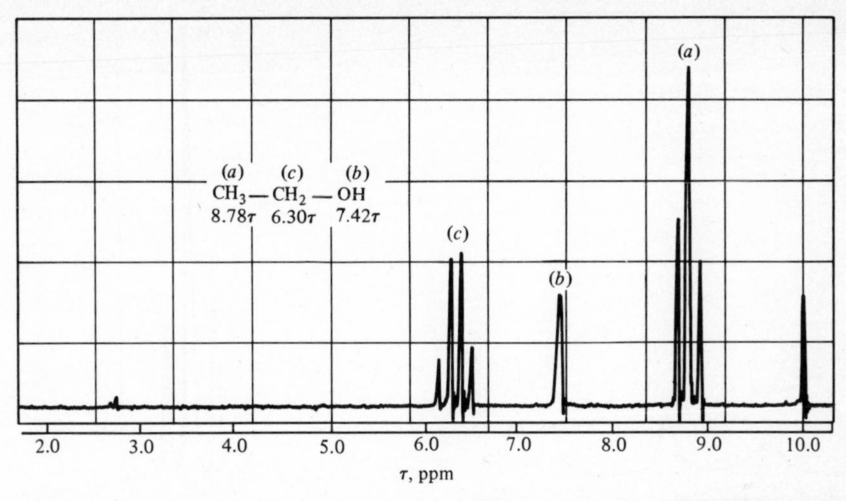

FIGURE 13-25 Nuclear magnetic resonance spectrum of ethanol in $CDCl_3$. Sample was approximately 7% solution of ethanol in $CDCl_3$; instrument was a Varian A-60 spectrometer, set for a 500-Hz sweep width. (*From Varian Associates. "High Resolution NMR Spectra Catalogue," spectrum* 14, *National Press, Palo Alto, Calif.*, 1962, 1963, *by permission.*)

finite dilution or in the vapor phase the hydroxyl group shifts up to about 9.5τ.

Another example of the effect of hydrogen bonding can be seen in the spectrum for a 7 percent solution of ethanol in $CDCl_3$, shown in Fig. 13-25. In Fig. 13-25 the hydroxyl proton absorbs at about 7.4τ, whereas in neat ethanol the hydroxyl proton absorbs at about 4.7τ (Fig. 13-21).

If two different hydroxylic species are present in the same solution, rapid chemical exchange can occur between them, and depending on the particular conditions and extent of chemical exchange, both the chemical shifts and the splitting patterns can be affected. For example, Fig. 13-26 shows the nmr spectrum of an ethanol-water mixture which is free of any acid or base catalysts. The protons of the water can be seen to give a peak separate from that of the hydroxylic protons of ethanol (the resolution of this spectrum is not sufficiently high to reveal whether the hydroxylic protons of ethanol are split into a triplet). If a trace of hydrochloric acid is added as a catalyst, as in Fig. 13-27, the two types of hydroxyl groups merge to give a single absorption band. Rapid proton exchange between water and ethanol averages the shielding of each environment. The conditions for observing the merging of lines of two different compounds into a common singlet are the same as for spin decoupling by chemical exchange within a single compound; in other words, the frequency of proton exchange between the two compounds must be faster than the frequency separation of the two lines observed in the absence of exchange. With these principles it is possible to adjust conditions until the hydroxyl proton lines of two different compounds just coalesce and in this way

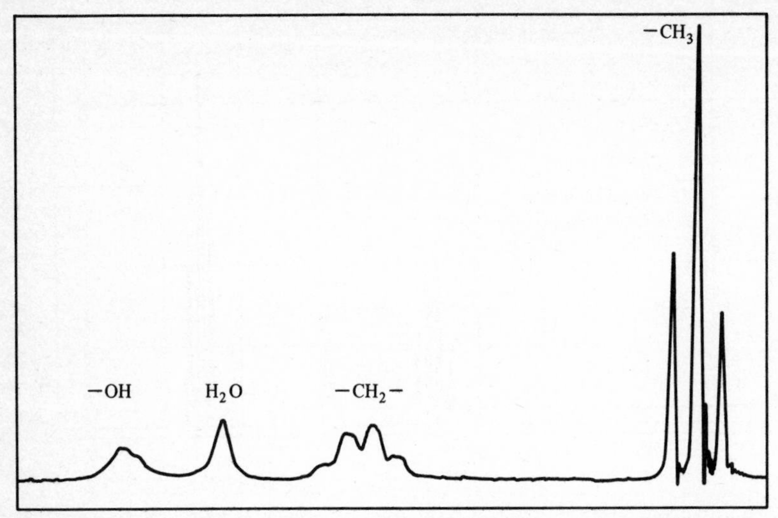

FIGURE 13-26 Nuclear magnetic resonance spectrum of an ethanol-water mixture free of acid or base catalysts. (*From L. M. Jackman, "Applications of Nuclear Magnetic Resonance Spectroscopy in Organic Chemistry," p.* 27, *Pergamon Publishing Company, New York and Oxford,* 1959, *by permission.*)

to measure the rates of exchange between the compounds, even though the rates may be extremely fast.

Chemical exchange and hydrogen bonding are not restricted to protons residing on oxygen, though these effects are particularly pronounced with hydroxylic groups (especially with alcohols and to a somewhat lesser extent

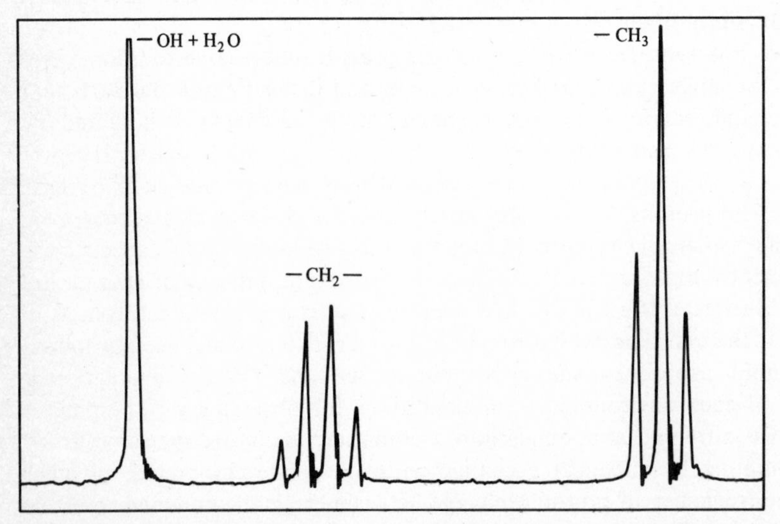

FIGURE 13-27 Nuclear magnetic resonance spectrum of an ethanol-water mixture containing a trace of hydrochloric acid. (From L. M. Jackman, "Applications of Nuclear Magnetic Resonance Spectroscopy in Organic Chemistry," p. 27, *Pergamon Publishing Company, New York and Oxford,* 1959, *by permission.*)

with phenols, enols, and carboxylic acids). Other heteroatoms which may contain an exchangeable proton include nitrogen, sulfur, and the halogens. Spectral complications that can arise with these heteroatoms are discussed in Refs. 10 and 15. Most complications can be removed or identified by shaking the sample solution with a few drops of deuterium oxide and repeating the spectrum. Exchangeable protons will be replaced by deuterium, removing those proton peaks from the spectrum.

13-1F Spin Decoupling (Multiple Irradiation)

A remarkably powerful tool for simplifying complex spectra involves irradiating the sample with two or more fixed radio-frequency signals while gradually changing the magnetic field strength (this is in contrast to the usual nmr experiment, in which only a single radio-frequency signal is imposed on the sample while the magnetic field strength is changed). In theory, any number of radio frequencies can be used, but in practice only double or triple irradiation is used. For adequate spectral interpretation there is little need for any more than triple irradiation, and instrumental complications become acute as the number of signals increase. Double or triple irradiation is often called double or triple *resonance* or more simply *spin decoupling*.

The phenomenon of double resonance is very similar to spin decoupling through rapid chemical exchange. For example, if a very pure ethanol sample is subject to double irradiation, whereby the sample is irradiated with a second and very strong radio-frequency signal at the resonance frequency of the hydroxyl group, the protons of the hydroxyl group will be *saturated* and no longer absorb, and instead of the higher-order octet being observed for the methylene group (see Figs. 13-21 and 13-28*a*), a first-order quartet will result (Fig. 13-28*b*). Likewise, when protons of the methyl group are irradiated by double resonance (Fig. 13-28*c*), the methyl-group absorption band (normally at about 8.8 τ) disappears and the methylene group (at about 6.3 τ) becomes a doublet because it is coupled only to the hydroxyl proton. The greatest simplification of all occurs when the middle (methylene) group of pure ethanol is irradiated by double resonance (Fig. 13-28*d*). There is now no significant spin coupling in ethanol, and only a hydroxyl singlet and a methyl singlet are observed.

The similarities and the differences between spin decoupling by chemical exchange and by double irradiation should be clearly understood. In both cases a simpler spectrum always results. In both cases the decoupling is brought about by averaging out the differences in magnetic environment that neighboring protons experience, with a resulting decrease in the multiplicity of neighboring groups. However, in chemical exchange the exchanging proton still absorbs, although its multiplicity is reduced to a singlet (see, for example, the hydroxyl-group absorption at 4.6 τ in Fig. 13-22), whereas in double irradiation the proton (or group of protons) being irradiated ceases to absorb

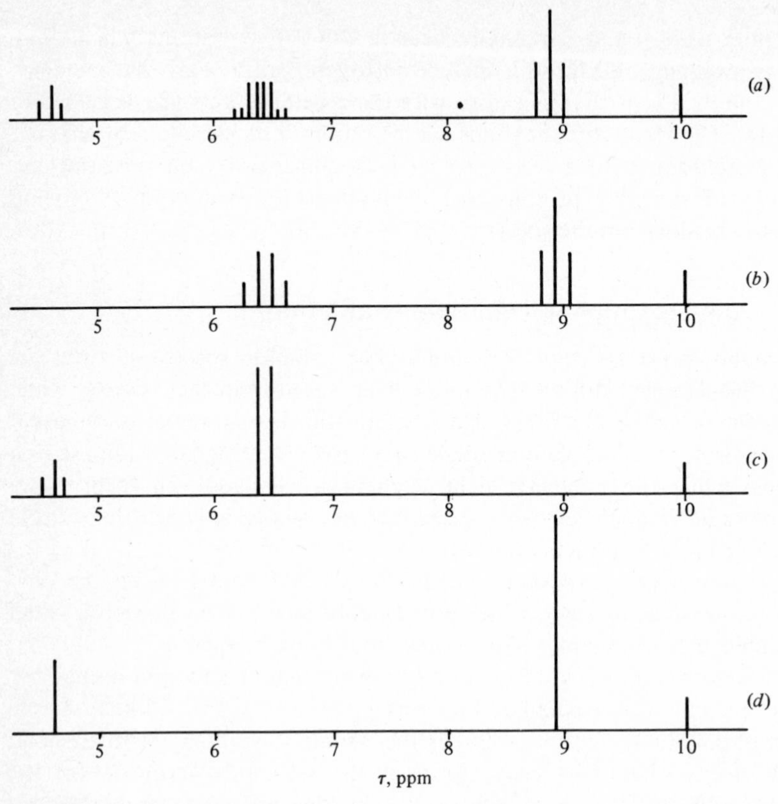

FIGURE 13-28 Effect of double resonance on nuclear magnetic resonance spectrum of very pure ethyl alcohol: (*a*) ordinary, single-resonance spectrum; (*b*) irradiation of hydroxyl group (double resonance); (*c*) irradiation of methyl group (double resonance); (*d*) irradiation of methylene group (double resonance).

(note the absence of a hydroxyl peak at 4.6 τ in Fig. 13-28*b*). Decoupling by chemical exchange is limited to protons on heteroatoms, whereas decoupling by double resonance is a much more unrestricted and useful tool. A few restrictions of decoupling by multiple irradiation will now be discussed.

An inherent requirement of the nmr experiment is that the power level of the radio-frequency signal used for resonance absorption be low, in order to optimize resolution and prevent saturation (Sec. 13-1*B*). Conversely, however, any additional radio-frequency signals used for decoupling must be relatively powerful, in order to ensure saturation of the particular protons we wish to decouple and remove from the spectrum. Because of the frequency distribution of any electromagnetic-signal source there is a finite limitation on how close together in absorption frequency two protons to be decoupled can be. As a rough *rule of thumb,* two protons must differ by at least 0.2 ppm in chemical shift (12-Hz separation at 60 MHz, or 20-Hz separation at 100

MHz) in order to consider decoupling them by double irradiation. For this and other reasons it turns out that complete decoupling of higher-order systems, such as the *AB* type, cannot be achieved by double irradiation, and attempts at this often lead to even more complex patterns. As another example, however, a system of the type *ABX* can be simplified by irradiating the *X* nucleus, thereby reducing the spectrum to a much simpler *AB* type, from which coupling constants and other assignments can be obtained with much greater ease.

A technique known as *tickling* is sometimes useful when the chemical shift between two protons is too small to use conventional double resonance.† This technique consists of using a second *low-power* radio-frequency signal to irradiate a neighboring proton weakly while otherwise scanning the spectrum as in a conventional double-resonance experiment. The effect of this second weak irradiation is to make it possible for certain complex higher-order systems (such as an *ABC* system) to be interpreted more conclusively by introducing a controlled amount of splitting of groups adjacent to the group being tickled.

In summary, spin decoupling by multiple irradiation is a very useful aid in the interpretation of complex spectra. In general, a simpler spectrum always results. Sometimes when bands involving spin-spin coupling overlap and features are hidden in the spectrum, chemical shifts that would otherwise be unmeasurable can be determined by experimentally varying the difference in the irradiating frequencies until decoupling is observed. Furthermore, decoupling is not limited to protons. In fact, the first double-resonance experiments involved heteronuclear (in contrast to proton-proton, or homonuclear) irradiation of nuclei such as ^{14}N, ^{31}P, or ^{19}F. These latter three isotopes have spins of 1, ½, and ½, respectively, allowing them to spin-couple with protons. Heteronuclear double irradiation is relatively easy to carry out because of the very large frequency differences from the proton resonance.

Finally, it should be mentioned that heteronuclear double irradiation is sometimes useful in conjunction with chemical-exchange experiments with deuterium oxide. Whereas deuterium oxide can be used to identify exchangeable protons and simplify the spectrum by removing those proton peaks from the spectrum (deuterium itself would not absorb on scale, having a resonance frequency of about 51 MHz using the magnetic field of a 60-MHz spectrometer), nonetheless deuterium nuclei (with a spin of 1) couple weakly with protons (deuterium-proton coupling constants are about one-seventh the value of proton-proton coupling constants). Therefore, by irradiating deuterium in a double-resonance technique, the pattern can be simplified because the small couplings to deuterium nuclei (giving partial splittings or very broad bands) can be removed.

† R. Freeman and W. A. Anderson, *J. Chem. Phys.*, **37**:2053 (1962).

13-1G NMR Studies of Nuclei Other than Protons

Although about 90 percent or more of all nmr studies are concerned with proton resonance, there is a significant and increasing interest in a number of other nuclei. Since the applications of these other resonance studies tend to be somewhat specialized, this section will present only the basic factors which determine whether a given isotope can be studied successfully by high-resolution nmr. This is followed by a brief summary of the potentialities of a few of the most widely studied isotopes. Further details will be found in the references.

For an isotope to show nmr it must have a nuclear spin other than zero. A few common isotopes with spin I equal to zero are ^{12}C, ^{16}O, and ^{32}S, all of which have an even number of protons and neutrons in their nuclei. Most isotopes have a nuclear spin (and thus a magnetic moment), and to date the magnetic moments have been measured for about 100 isotopes. Table 13-1 gave the magnetic moment for about a dozen of the more common isotopes. Several factors affect the ease of using a given isotope for nmr studies.

1 The magnetic moment of the nucleus must by large, because the sensitivity of detection at constant field strength is proportional to the cube of the magnetic moment μ^3. Both ^1H and ^{19}F have large moments ($\mu_H = 2.793$ and $\mu_F = 2.627$ nuclear magnetons; see Table 13-1), and they are two of the nuclei which are most sensitive to detection. ^{31}P, ^{13}C, and ^{14}N, of the fairly commonly studied elements, have successively smaller magnetic moments, and ^{14}N can be detected with only about 0.1 percent of the sensitivity of the proton using the same magnetic field strength.

2 The feasibility of studying a given isotope is governed by the natural abundance of that isotope (see Table 13-1). This becomes the critical factor, for example, with an isotope like ^{13}C having only a 1.11 percent natural abundance. This can be compensated for to some extent by isotope enrichment of samples, by increasing the size of the sample, or by using a computer averaging technique (CAT), in which the spectrum of a weak signal is repeatedly scanned and a computer is used to sum electronically any coherent signals while disregarding the incoherent (random) noise signals. The use of a CAT requires an extremely stable magnetic field. Some loss of resolution results from all three of these methods for increasing sensitivity.

3 The success of studying a given isotope is also affected by the spin number I. For high-resolution studies the spin number should be $\frac{1}{2}$, as it is, for example, with ^1H, ^{19}F, ^{31}P, and ^{13}C. Isotopes with higher spin numbers have *quadrupole moments* associated with them, which result in excessive line broadening because the quadrupole moments interact with fluctuating electric field gradients in the molecule and provide an efficient

relaxation mechanism for the nuclei. For example, ^{2}H ($I = 1$), ^{11}B ($I = 3/2$), and ^{14}N ($I = 1$) all possess quadrupole moments.

4 Another factor affecting the feasibility of studying a given isotope is the inherent relaxation time of the isotope. The relaxation time must be short to avoid saturation when suitable power levels and sweep rates are used. For example, relaxation times for ^{19}F are often longer than for ^{1}H, although for both cases they are sufficiently short to allow the nuclei to be studied easily. The relaxation times for ^{13}C are so long that usually it is necessary to examine the nucleus under conditions known as *rapid passage in the dispersion mode.* Dispersion mode means that a phase sensitive detector is used and tuned so that a derivative (dispersion) curve is obtained. This dispersion mode is much more stable than the conventional absorption mode during rapid sweep rates (typically 50 to 100 mG/s) and permits the application of sufficient power to get moderately strong signals without saturation, although the resolution is less under these conditions than under conventional conditions.

Some of the potentialities of nmr studies of ^{19}F, ^{31}P, ^{13}C, and ^{14}N will be summarized.

FLUORINE 19

Whereas chemical shifts for protons cover only about 15 ppm, fluorine chemical shifts cover about 500 ppm, and this gives ^{19}F nmr an advantage. For example, in certain conformational studies the different arrangements of atoms in space may give only minute differences in proton shielding but large and easily measured differences in ^{19}F shielding. Numerous ^{19}F conformational studies of substituted fluoroethanes and fluorocyclohexanes have been reported, and compilations of ^{19}F chemical-shift data are available [13]. As a reference compound for chemical-shift measurements, CF_3COOH is commonly used (external reference). When a proton nmr spectrometer with a 14,092-G field is used (the size used in proton nmr spectrometers that operate at 60 MHz), generally all that is needed for ^{19}F studies is a different radio-frequency source (56.6 MHz) and amplifier.

PHOSPHORUS 31

^{31}P chemical shifts have the advantage of covering a range of about 500 ppm, like that of ^{19}F, but its sensitivity is lower than that of ^{1}H and ^{19}F because of a smaller magnetic moment (see Table 13-1). As a result, samples must be relatively high in phosphorus concentration (generally 1 *M* or more in phosphorus), and this limits most of its usefulness to solid samples. The reference material usually used is 85 percent aqueous phosphorous acid (external reference). Phosphorus shows large coupling constants, for example, 500 to 700 Hz, when directly bonded to hydrogen or fluorine, and this permits very complex splitting patterns to be well resolved. Tables of ^{31}P chemical shifts and coupling constants are available [13]. To convert a proton nmr spectrometer with a 14,092-G field to ^{31}P studies, a separate sample probe is required, as well as a different radio-frequency source (24.3 MHz), amplifier, and bridge controls.

CARBON 13

^{13}C is the only naturally occurring isotope of carbon having a magnetic moment (spin $I = \frac{1}{2}$). Despite its low natural abundance (1.1 percent), long relaxation times, and low natural sensitivity to nmr detection (about 2 percent of the sensitivity of the proton using the same magnetic field strength), many successful ^{13}C resonance studies have been reported, and the potential for future studies looks great. To overcome the sensitivity problem associated with low natural abundance, large sample tubes (up to 15 mm OD) and as high a power level as possible are used. To avoid saturation a dispersion mode and rapid sweep rates are used.

One of the exciting features of ^{13}C high-resolution nmr studies is that direct information on carbon atoms not attached to any hydrogen atoms can be obtained. For example, a direct probing of carbonyl groups is possible, and Fig. 13-29 shows a typical ^{13}C resonance spectrum of acetic acid. The large peak, arbitrarily labeled 0 ppm, is due to the carbon in the carboxyl group, while the quartet centered at about 160 ppm is due to the carbon in the methyl group. The spin-spin interaction with the three methyl protons has split the methyl carbon resonance into four components, but there is no observable long-range coupling of either the methyl or the hydroxyl protons with the carboxyl carbon.

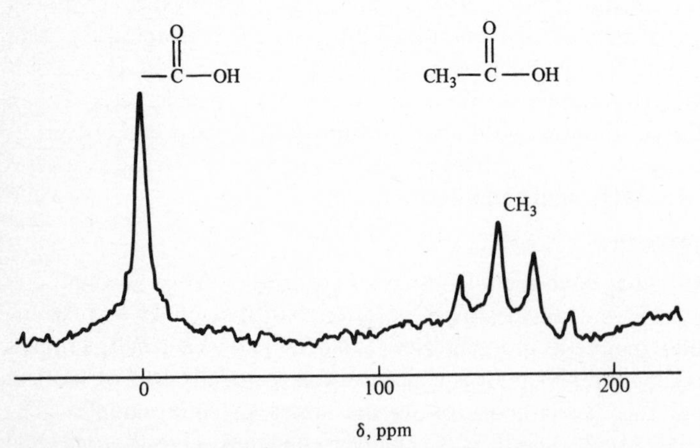

FIGURE 13-29 A ^{13}C nmr spectrum of acetic acid. (*After D. Chapman and P. D. Magnus, "Introduction to Practical High Resolution Nuclear Magnetic Resonance Spectroscopy," p.* 50, *Academic Press Inc., New York,* 1966, *by permission.*)

Tables of ^{13}C chemical shifts and coupling constants are becoming available [8, 13], and it seems certain that organic chemists will devote increasing attention to this nucleus in the future.

NITROGEN 14

The nitrogen nucleus possesses a quadrupole moment ($I = 1$), and therefore its resonance signals are broad. Chemical shifts, however, have been measured for the ^{14}N nucleus in numerous molecular environments [8, 13] and range over about 600 ppm. Nitrate ion in aqueous solution appears to be the most suitable reference. When a proton is bonded to a nitrogen atom and proton resonance is being observed, the proton signals are usually broad and sometimes barely detectable because the quadrupole moment of nitrogen generates a strong relaxation mechanism. Another distinct limitation of ^{14}N for resonance studies is its relatively small magnetic moment and low natural sensitivity (about 0.1 percent of the sensitivity of hydrogen at a constant field strength). Although ^{15}N, the natural companion to ^{14}N, has a low natural abundance (about 0.37 percent), it has a spin of $\frac{1}{2}$, and there is some future promise for ^{15}N studies using ^{15}N-enriched compounds. At present there are few applications of ^{14}N spectra to structure determinations.

13-2 INSTRUMENTATION

An nmr spectrometer is similar in principle to a more conventional absorption spectrometer. The three main components are a source, a dispersion device, and a detector. The *source* is monochromatic radiation produced by a radio-frequency oscillator, and the *detector* is simply a radio-frequency receiver. The *dispersion device* is a magnet with a strong homogeneous field. This field produces the energy-level splitting which allows nuclear transitions. Usually a spectrum is scanned by varying the strength of the magnetic field over a small range by means of magnetic sweep coils (field sweep), keeping the source frequency constant. Alternately, the spectrum may be scanned by maintaining the magnetic field constant and varying the source frequency (frequency sweep).

13-2A The Magnetic Field

There are several essential requirements for the magnet used in high-resolution nmr experiments. It must be very *powerful* (of the order of 10 kG or more) to produce splitting of the nuclear-spin energy levels sufficient for absorption of radio-frequency energy to be detected [recall, from Sec. 13-1 and Eqs. (13-2) and (13-4), that the differences in population of nuclear magnetic levels are directly proportional to the strength of the applied magnetic field]. In addition to being very powerful, the magnetic field must be very *stable* and *homogeneous* in order to be able to resolve small differences in chemical shifts and observe spin-spin splittings. Most high-resolution nmr spectrometers can distinguish between lines which are only 0.1 mG or 0.5 Hz apart, which, at an applied field of 10 kG or frequency of 60 MHz, represents a resolving power of 1 part in 10^8 or better; no other spectroscopic technique approaches this precision. This requires that the magnetic field be homogeneous to at least 1 part in 10^8 over the whole sample volume and for at least as long as it takes to scan the spectrum. Typically the sample area is about 0.5 cm wide and about 2 cm deep. While it is possible to scan the spectrum in about 10 s or less, it is desirable that the stability be maintainable for a period of days, in order to assure long-term reproducibility and precision. These requirements are very stringent. Techniques used to achieve good homogeneity include (1) the use of a large magnet with pole pieces typically about 20 cm (about 8 in) in diameter, allowing the sample to be placed in a small volume near the center; (2) polishing the pole faces to optical flatness; (3) using electrical shims around the magnet through which small currents are passed to offset any departure from homogeneity; and (4) spinning the whole sample about an axis at right angles to the field direction, thus effectively averaging the field experienced by each part of the sample. Stability of the magnetic field is mainly achieved by taking elaborate precautions to stabilize the current from the power supply. However, the magnetic field can also be affected by movement of metallic masses and by stray fields in the laboratory. Such *transient fluctuations* are compensated for by providing sensing coils wound around the pole pieces which feed back a signal to a control circuit.

The devices discussed cannot rule out *slow drifts* of the magnetic field. Drifts of only a few parts per *billion* could cause serious reproducibility errors. Fortunately, however, the condition for resonance does not require either the magnetic field or the oscillator frequency to remain constant but only their *ratio*, which may be seen by combining Eqs. (13-2) and (13-3) to give

$$\nu = \frac{\gamma H}{2\pi} \tag{13-10}$$

Thus, it is only necessary to keep the ratio of frequency to magnetic field strength constant at a value of $\gamma/2\pi$ in order to achieve the desired long-term stability. Some means for accomplishing this will be discussed in Sec. 13-2*C*.

Three types of magnets are in commercial use, permanent magnets, electromagnets, and superconducting magnets, each with advantages and disadvantages. *Permanent magnets* have advantages in economy, stability, and ease of operation but are more limited in field strengths obtainable (about 4.1 kG is the upper limit), and it is not possible to vary the field strength in order to study different nuclei. Different nuclei can be studied only by a change in the operating *frequency* of the instrument, and thus a separate crystal oscillator is required for each nucleus of interest. The temperature of permanent magnets is usually controlled to very fine limits, but this is far easier to do than with electromagnets, which dissipate large amounts of heat in the magnet windings.

Examples of commercial instruments employing a permanent magnet are the Perkin-Elmer models R12 and R-20A and the Varian model T-60.

Electromagnets have the advantage of greater flexibility, making it possible to change the field strength in a matter of minutes for the study of different nuclei or even to determine the spectrum of the same isotope at different field strengths without altering the radio-frequency oscillator. The disadvantage of electromagnets is the cost of the elaborate equipment necessary to obtain sufficient stability in terms of resolution and field strength. Nonetheless, the great majority of commercial instruments employ electromagnets. Electromagnets are commonly C-shaped, the pole pieces being 15 to 30 cm in diameter and separated by a gap of approximately 2.5 to 5.0 cm. The strength of the field is approximately inversely proportional to this gap, provided magnetic saturation does not become limiting. The magnet must be held by a massive yoke since the force between the pole pieces at high fields may be nearly 10 tons.

The maximum operating field for normal electromagnets is about 25 kG, but a recent advance, the *superconducting magnet,* has almost no limit to the field strengths possible. There are two advantages to increasing the field strength: the intensity of the signal will increase with the applied field, making it possible to obtain greater sensitivity or use smaller samples or both; more importantly, however, chemical shifts increase, which means that

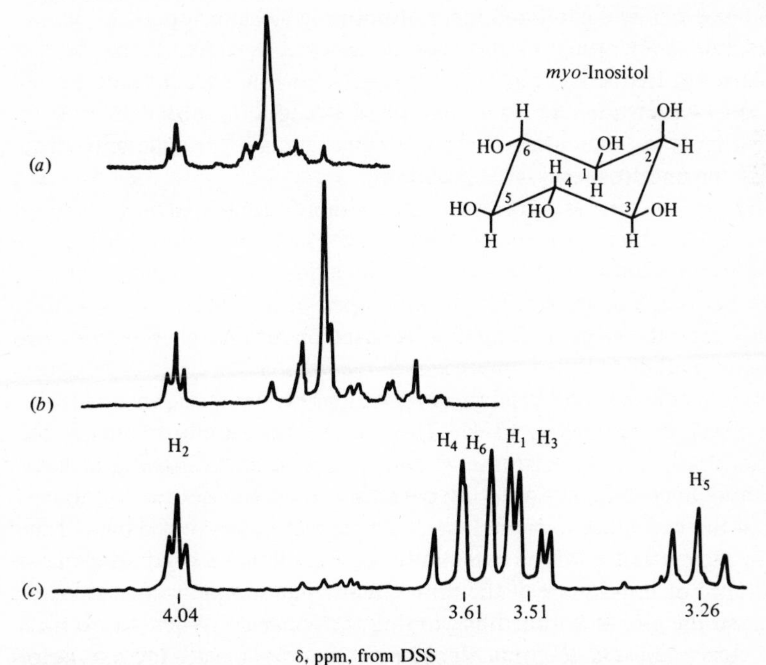

FIGURE 13-30 Effect of increasing magnetic field (and increasing oscillator frequency) on the resolution of nmr spectra. Nmr spectra of the myo diastereomer of inositol is shown at (*a*) 60, (*b*) 100, and (*c*) 220 MHz. (*Varian Associates.*)

clearer and more detailed spectra become possible. Figure 13-30 gives the nmr proton spectra of *myo*-inositol at 60, 100, and 220 MHz. Note the greater detail obtained with increasing frequency (increasing field strength).

The principle behind a superconducting magnet is that certain metals which are cooled to temperatures of less than a degree above absolute zero have an electric resistance which falls to zero. A current once started in a coil of such material will continue to circulate until the metal is allowed to warm up. An example of a commercially available nmr spectrometer that employs a superconducting magnet is the Varian model HR-220.

For *wide-line* nuclear magnetic resonance work, field homogeneity of the order of 1 part in 10^4 or 10^5 is adequate, since line widths in this case are of the order of several gauss. This degree of homogeneity and stability is relatively easy to obtain with permanent or electromagnets.

13-2B Other Basic Components

The *source* of electromagnetic radiation in nmr spectrometry is essentially a carefully tuned crystal oscillator. The frequency generated must be maintained constant to at least the constancy of the magnetic field, which means to at least 1 part in 10^8. At 60 MHz, for example, the frequency must be constant to at least 0.6 Hz and preferably to about 0.2 Hz. This is a stringent requirement and requires careful control of both the voltage applied to the oscillator and the temperature of the crystal. Since it is easier to produce a source of constant frequency than one accurately tunable over a range of frequencies, most spectrometers vary the field strength in order to scan a spectrum, although an increasing number of spectrometers provide both magnetic-field sweep and frequency-sweep options.

The *detector* in nmr spectrometers is a simple radio-frequency receiver coil, but it may be either the same coil which serves as the transmitter (single-coil method) or a separate coil placed at right angles to the transmitter coil (two-coil or crossed-coil method). The principles behind these two methods are quite different; the single-coil method is based on nmr *absorption*, whereas the double-coil method is based on *nuclear induction*.

The principle of the *single-coil* (absorption) *method* can be explained with the simple circuit given in Fig. 13-31. The source S is a radio-frequency oscillator which feeds a high resistance R and a tank circuit consisting of a capacitor C and a coil L arranged in parallel. Since the source voltage is constant and the resistance is made large with respect to the impedance of the tank circuit, the current I will be essentially constant in spite of small changes in the resistance or impedance of the tank circuit. The sample is placed inside the coil L, and the circuit is tuned by varying the capacity. When an external magnetic field is adjusted to bring the sample into resonance, the energy or power absorbed by the sample causes an apparent increase of resistance in the coil, and hence an increase in the voltage drop E. The voltage drop is amplified and observed on an oscilloscope screen or recorder.

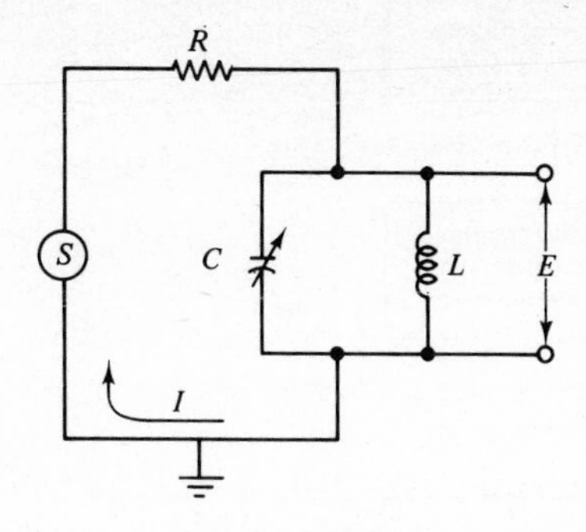

FIGURE 13-31 Schematic diagram of a single-coil circuit, which is the basis of nmr absorption detection; S = radio-frequency oscillator source, R = resistance, C = variable capacitance, L = inductance coil, E = voltage, I = current.

In practice, higher sensitivity can be achieved with a bridge circuit, and Fig. 13-32 illustrates an elementary form of bridge circuit which can be used. The bridge functions much like a Wheatstone bridge. It is balanced when no resonance is occurring, and absorption of energy at resonance causes an unbalance of the bridge which is amplified and recorded. With a fixed-frequency oscillator the magnetic field is varied as a function of time (a linearly increasing current is applied to the field-sweep coils), whereas with a fixed magnetic field the frequency of the oscillator must be linearly varied with time. Examples of instruments using this single-coil (absorption) method are the Jeol model JNM-4H-100, the Perkin-Elmer model R-20A, and the Varian models T-60 and A60D.

The arrangement for the *double-coil* (nuclear-induction) *method* of de-

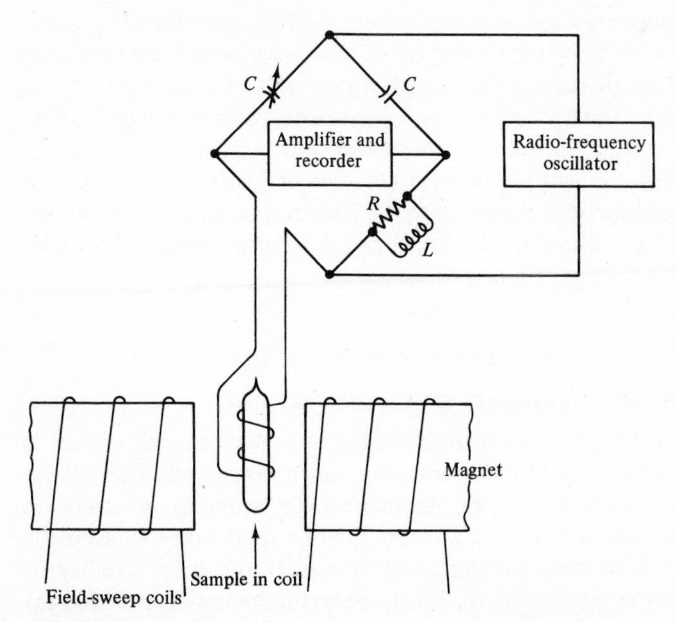

FIGURE 13-32 Elementary bridge circuit for single-coil nmr detection; R = resistance, C = capacitance, and L = inductance coil.

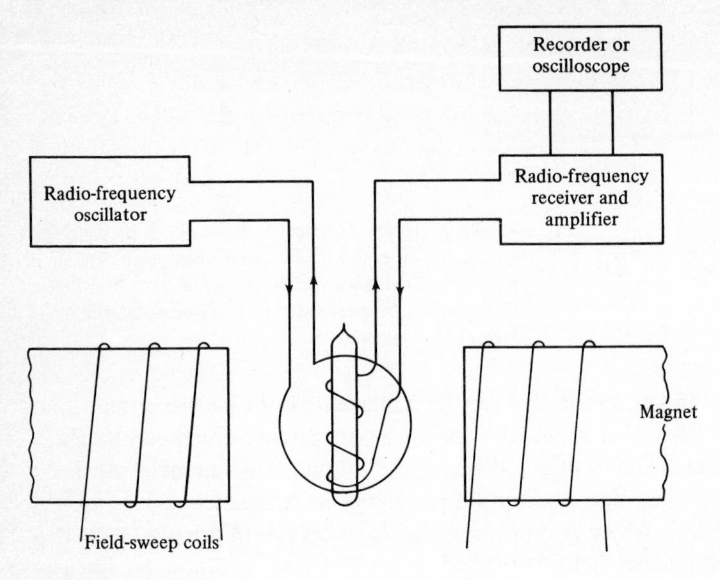

FIGURE 13-33 Block diagram showing arrangement of coils in double-coil method of nmr detection.

tection is illustrated in Fig. 13-33. The oscillator coil is oriented with its axis perpendicular to the magnetic field, and the receiver coil, wound around the sample tube, has its axis perpendicular to both the magnetic field and the axis of the oscillator coil. The radio-frequency receiver is tuned to the oscillator frequency, but in the absence of resonance there is little or no direct coupling between the oscillator and receiver coils. When resonance occurs, however, the induced magnetization of the sample causes a current to be generated in the receiver coil, which is then amplified and recorded. Thus, the signal arises from an indirect coupling between the oscillator and receiver coils, with the coupling produced by the sample itself. Examples of instruments using this double-coil (nuclear-induction) method are the Bruker and models HX-60, 90, and HFX-10, the Perkin-Elmer model R-12, and the Varian models HA-100D and HR-220.

13-2C Other Features of Commercial Instruments

METHODS OF MAINTAINING BASE-LINE STABILITY

It is essential in high-resolution nmr studies that the *base line* (which is conventionally *expressed* in frequency units but which may in *practice* involve scanning either the frequency or the magnetic field strength) be stable to within 1 part in 10^8 and for a time at least as long as it takes to scan the spectrum. For long-term reproducibility it is desirable that this stability be maintained for hours or even days, and it is becoming common practice for instrument manufacturers to specify the resolution drift or resolution stability in units like hertz per day. A typical resolution stability might be 0.5 Hz/d, or even less. This implies that *both* the magnetic field and the oscillator frequency are independently stable to at least the specified value (say 1 part in 10^8) or that the *ratio* of field to frequency is stable to those limits.

There are two fundamental ways of achieving base-line stability: (1) by *separately* stabilizing both the magnetic field and the oscillator frequency and (2) by controlling the *ratio* of field to frequency. Methods for separately stabilizing the magnetic field and the oscillator frequency were discussed in Secs. 13-2*A* and 13-2*B*, and examples of commercial instruments based on this are Varian models T-60 and HR-220.

There are several ways of controlling the *ratio* of field to frequency, including (1) forcing the magnetic field strength to follow any fluctuations in the oscillator frequency, (2) forcing the oscillator frequency to follow any fluctuations in the magnetic field strength, and (3) special sideband-modulation techniques, which may do both. The first two approaches were used in early instrument designs, but almost all modern spectrometers now use *sideband-modulation* techniques to achieve stability. In sideband modulation, one or two audio-frequency oscillators (typically in the range of about 1 to 8 kHz) are used in addition to the main radio-frequency oscillator, and the superimposed sidebands are used in various ways to maintain a resonance condition with a reference compound, regardless of fluctuations in the magnetic field strength or the main oscillator frequency. In effect, by monitoring the resonance of a reference compound and continuously correcting for any fluctuations in conditions that the reference compound sees, the unknown sample is automatically subjected to corrected and unfluctuating field-frequency conditions.

Two sideband-modulation techniques are common. The first uses a separate, *external* control sample (usually water), and this technique is usually referred to as *external-lock* stabilization. The second uses a reference compound (often TMS) which has been added directly to the analytical sample, and this technique is usually referred to as *internal-lock* stabilization.

Figure 13-34 is a block diagram of the Varian model A-60 spectrometer, which uses *external-lock stabilization,* and Fig. 13-35 shows details of the probe. A modulation frequency of about 5 kHz is generated by a sideband oscillator (called a field modulator in Fig. 13-34), and is applied to the *modulating coils* in the sample probe. When the magnetic field is thus modulated, a spectral line appears not as a single peak but as a *center band* with *sidebands* spaced at 5-kHz intervals, as shown in Fig. 13-36. The frequency of the center band is the same as the main *transmitter* frequency, in this case 60 MHz. The intensity of the sidebands relative to the center band depends on the power of the audio signal used. Of the various sidebands, the first upper sideband is used for the observation of the spectrum of the analytical sample. From Fig. 13-34, note that the dc field sweep is applied to the analytical sample only, allowing the control sample to remain in constant resonance at the unswept-

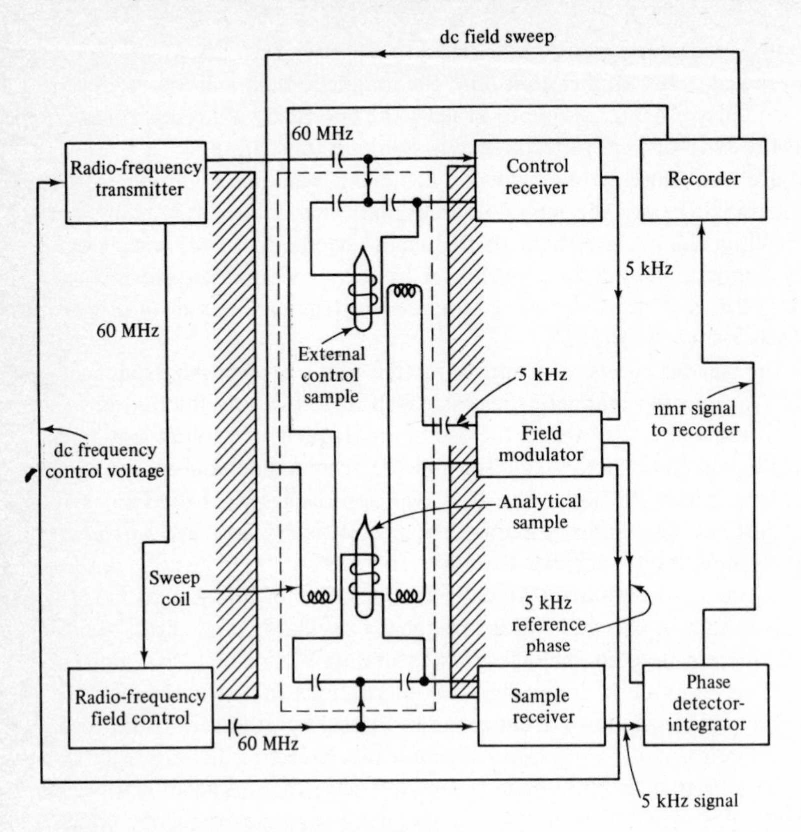

FIGURE 13-34 Block diagram of Varian model A-60 nmr spectrometer, showing external-lock arrangement. (*Varian Associates.*)

field strength. On the other hand, the 5-kHz modulation frequency is applied to both the analytical sample and the control sample, so that any corrections (required, for example, by fluctuations in the magnetic field strength) necessary at the control sample are also applied to the analytical sample. At the *control receiver* the 60-MHz carrier frequency from the *transmitter* is demodulated; i.e., the center band is rejected, and only the upper 5-kHz sideband remains. This frequency (which will be 5 kHz plus or minus whatever frequency correction was necessary to keep the control sample in resonance) is then applied to the *input* of the field modulator, which tends to reinforce and amplify the corrected 5-kHz signal. The corrected 5-kHz signal is reapplied to the modulating coils, completing the control loop.

To understand how this system stabilizes the base line, it is useful to invoke the Larmor resonance equation

$$\omega = \gamma H \tag{13-1}$$

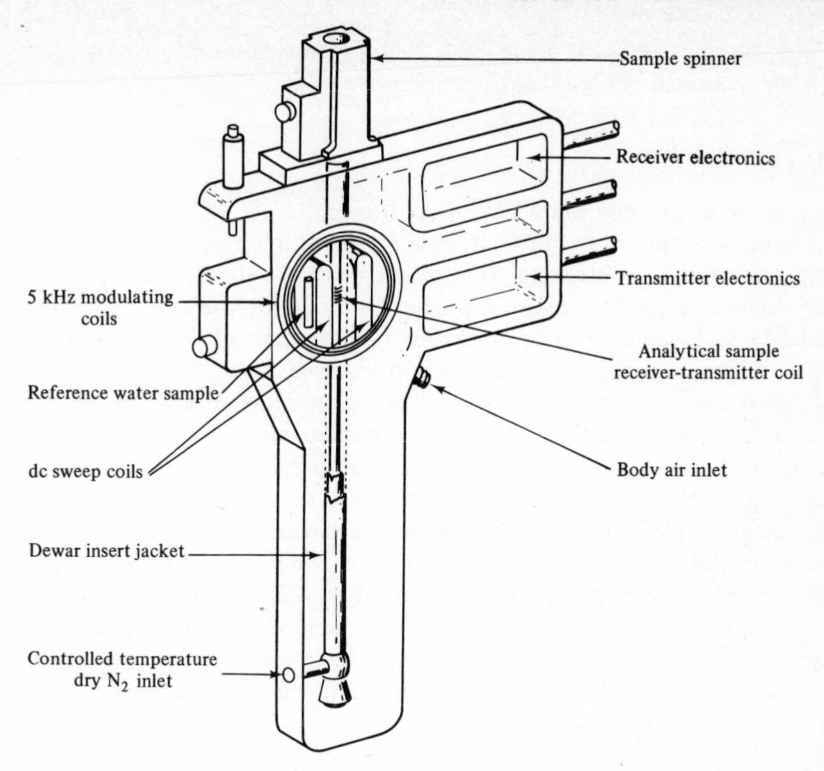

FIGURE 13-35 Cutaway drawing of the probe used in the Varian model A-60 nmr spectrometer. (*Varian Associates.*)

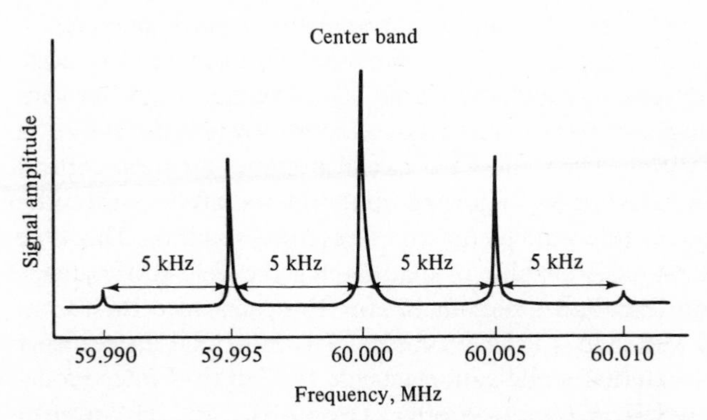

FIGURE 13-36 Effect of 5-kHz modulation of the magnetic field on a resonance line. (*After F. A. Bovey, "Nuclear Magnetic Resonance Spectroscopy," p. 36, Academic Press Inc., New York, 1969, by permission.*)

When a sideband oscillator is being used, the total applied oscillator frequency ω should be separated into two parts, as follows:

$$\omega_t + \omega_m = \gamma H \tag{13-11}$$

where ω_t is the fixed radio frequency of the transmitter (60 MHz in this case) and ω_m is the modulation frequency (about 5 kHz in this case). Suppose now that the magnetic field strength H were to *increase* slightly. The *control receiver* would attempt to remain in oscillation (maintain the control sample in resonance) and would do so by *increasing* the frequency of the sideband oscillator ω_m. Thus, the upper sideband would increase to 60 MHz plus a frequency of somewhat *more* than 5 kHz. Any resonance signal we may have been observing from the analytical sample (regardless of what particular gyromagnetic ratio γ and time we are at with respect to the dc field sweep) would likewise stay in resonance, because the increase in magnetic field strength at the analytical sample would also have been precisely compensated for by the increase in modulator frequency ω_m. Likewise, if the magnetic field strength were to *decrease,* an appropriate *decrease* in the modulation frequency would occur to maintain resonance. Similarly, if the main oscillator frequency ω_t were to change, an appropriate change in the modulation frequency would occur to maintain resonance. Thus, this sideband-modulation technique compensates for any change in *either* the oscillator frequency or magnetic field strength.

There is one additional consideration for optimum operation of the sideband oscillator. Sideband oscillators are inherently more stable and reliable if they have to operate only over a relatively narrow frequency range, say from 4 to 6 kHz. Long-term drifts in the magnetic field strength could, if uncorrected for, force the sideband oscillator to go outside this 4 to 6 kHz range in order to keep the control sample in resonance. To prevent this from happening an automatic-frequency-control (AFC) circuit is provided which senses any large excursions from 5-kHz frequency and sends a signal to the magnet power supply which pulls the magnetic field into a more correct relationship with the frequency of the transmitter. Thus, the AFC circuit provides a coarse control over the field-frequency relationship, allowing optimum operation of the sideband oscillator and precise fine-tuning control of base-line stability. This type of external-lock stabilization is capable of maintaining the field-to-frequency ratio constant to within less than 1 part in 10^8 for 16 h, provided the room-temperature change is less than $\pm 1°C$. Examples of commercial instruments employing this type of external stabilization include the Perkin-Elmer model R-20A and Varian models A-60, A-56/60D, DA-60-EL, and HA-60-EL. Most Bruker and Jeol Ltd. models have external-lock stabilization in addition to internal-lock stabilization or as an option.

Internal-lock stabilization can be used to achieve even more precise field-to-frequency control, often approaching 3 parts in 10^9 over a 24-h period.

Internal-lock stabilization can be more precise than external lock because in internal lock the reference compound used to control the field-to-frequency ratio is experiencing exactly the same magnetic field as the analytical sample, whereas in the external-lock system the control sample is monitoring the magnetic field a short distance away from the analytical sample, and the field at slightly different locations may change in slightly different ways with time. Examples of commercial instruments using internal-lock stabilization include most models of Bruker and Jeol Ltd., the Perkin-Elmer model R-12, and the Varian model HA-100; internal lock is optionally available on certain other Varian models.

WIDE-LINE NMR

Wide-line nmr differs from high-resolution nmr in that resonance lines are very broad instead of being sharp. Early nmr experiments tended to give wide-line results, but this broadening was due to limitations in the instrumentation then available, i.e., limitations in the homogeneity of the magnetic field. Modern wide-line nmr experiments measure lines that are broad due to the sample environment and not instrumentation limitations. Wide-line nmr spectra may be defined as those spectra in which the observed (and *true*) width of the resonance line is as large or larger than the resonance shifts caused by differences in the chemical environment. This broadening is usually observed at low temperatures and/or with solids, although it is applicable to liquids and even gases. Most forms of line broadening are due to restricted motion of atoms or molecules in the sample which causes the nucleus being observed to experience a heterogeneous pattern of local magnetic fields coming from the surrounding atoms. In these cases the line width usually becomes narrower as the temperature is raised, since the increased molecular motion causes the heterogeneity of the local magnetic fields to be reduced by an averaging effect. Other causes of line broadening include paramagnetic properties of the sample, electric quadrupole effects, and critical collision frequencies. These latter effects often cause the lines to be excessively broadened, resulting in poor detection.

From the definition of wide-line nmr spectra, it follows that no information about the chemical environment of the nucleus can be obtained. However, these spectra provide valuable information about the *physical* environment of the observed nucleus, including a measure of the amount and kind of molecular motion present at a given temperature, the degree of crystallinity of polymers, etc. Furthermore, these spectra are uniquely suitable for *quantitative analysis,* since the area under the absorption band is directly proportional to the number of atoms of the observed isotope, independent of the chemical or physical environment of the sample. Only a single standard is required to calibrate each isotope, eliminating the need for different standards to represent various compounds or environments. The technique is applicable to all isotopes observable by nmr, which includes isotopes having an odd number of protons in their nuclei plus a few isotopes, like beryllium, which have an even number of protons and an odd number of neutrons.

Several factors influence the instrumentation requirements in wide-line nmr. First, line widths being measured are often in the order of 1 G, so that the magnetic field homogeneity need only be of the order of 1 part in 10^4 or 10^5, thereby greatly simplifying the instrumentation requirements. On the other hand, as any absorption line broadens, the signal-to-noise ratio deteriorates, and thus sensitivity becomes even more of a critical problem in wide-line nmr than in high-resolution nmr. Furthermore, most applications of wide-line nmr require a quantitative measurement of peak shape and width. For these reasons most wide-line nmr instruments measure the *derivative* of the absorption band.

The signal-to-noise ratio and thus the sensitivity can be improved by modulating the magnetic field at a low frequency so that the resonance condition and hence the output signal vary at the same frequency. This signal can be further amplified by a narrow-band or lock-in amplifier which selectively accepts the tuned frequency and rejects noise contributions. If the amplitude of the modulation signal is *less* than the width of the observed line, the low-frequency signal which is detected will be the first derivative of the absorption curve, as shown by Fig. 13-37. At points 1 and 3 in Fig. 13-37*a*, for example, the ac signal being *detected* will be essentially zero, since there will be little difference in the signal strength throughout the magnetic field fluctuation. At point 2 in Fig. 13-37*a*, which is the point of maximum slope, a maximum ac signal will flow during the magnetic field modulation (see Fig. 13-37*b*). In practice, a modulation frequency in the range of about 30 to 100 Hz is generally used.

Since *electron spin resonance* (esr) spectrometers use low-frequency modulation and lock-in amplifiers and the magnetic field homogeneity generally used is more than adequate (usually of the order of 1 part in 10^6 or better), it is fairly common to convert esr spectrometers into wide-line nmr spectrometers. The main addition required is one or more oscillators of the proper frequency. For example, most of the various Varian esr spectrometers of the general classification V-4502 can be converted into wide-line nmr spectrometers with the addition of an oscillator and three probes, covering the range of 2 to 16 MHz. This allows wide-line studies of approximately 85 different isotopes to be made at a field strength of about 3400 G.

A second approach to obtaining wide-line nmr instrumentation is to modify a high-resolution nmr spectrometer by the addition of a low-frequency modulator and a lock-in amplifier. This route is more expensive than the modification of an esr spectrometer but has the decided advantage of yielding higher sensitivity by virtue of the much stronger magnetic field strengths provided. The pole gap of the magnet being used should be large enough to accommodate a large sample (a 15-mm-OD sample tube typically being used), so as to optimize the signal-to-noise ratio. Examples of high-resolution esr

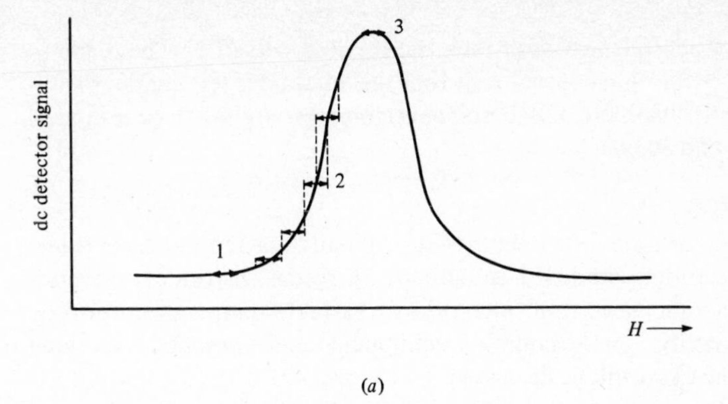

(a)

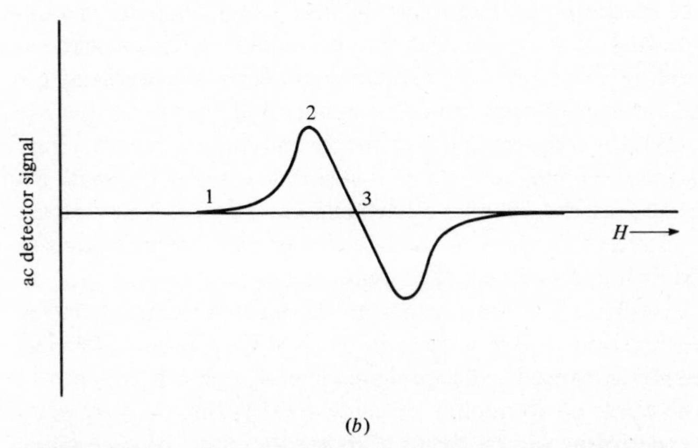

(b)

FIGURE 13-37 Use of low-frequency modulation of the magnetic field to produce the first derivative of the absorption curve. (*a*) Absorbance curve; (*b*) amplified first-derivative curve. Horizontal arrows in (*a*) show the amplitude of the modulation signal superimposed on the magnetic field sweep; vertical dotted lines in (*a*) show the magnitude of the ac signal detected, forming the basis for curve (*b*).

instruments which can readily be converted to wide-line nmr spectrometers include almost all the various Bruker models, Jeol model JNM-4H-100, and the Varian model DA-60.

PULSED NMR

Pulse techniques can be used for accurate measurements of relaxation times and other kinetic processes such as exchange effects in liquids. Here the radio-frequency field is applied in *pulses* instead of continuously, and the transient nuclear induction signals that result both during the application of the pulse and after its termination yield the kinetic information desired. Pulses from the high-frequency oscillator are best obtained by applying a programmed "gate" to the output of an ordinary, continuously oscillating generator. Alternately, but less desirably, the oscillator itself can be made to pulsate instead of oscillating continuously, but this method is less stable. Examples of commercially available pulsed nmr spectrometers are the Bruker models B-KR 301s, 302s, and 303s.

ACCESSORIES

Many features of nmr instruments that originally started out as optional accessories are becoming standard equipment on recent instruments. Examples include electronic spectrum integrators, variable-temperature probes, multinucleus capability, spin-decoupling equipment, and computer-averaging devices. Only the last two will be discussed.

Spin-decoupling equipment is used for decoupling both *homonuclear* and *heteronuclear* spin interactions. Examples of the former include *proton-proton* and *fluorine-fluorine* decoupling. Examples of the latter include *proton-fluorine* and *proton-phosphorus* decoupling. In either homonuclear or heteronuclear decoupling there are two fundamental ways of sweeping the spectrum, and the advantages of each should be understood. In the *field-sweep* method, the main oscillator frequency ν_1 and the decoupling frequency ν_2 are each kept constant, and the magnetic field H is swept. Generally the main oscillator frequency ν_1, called the *observing frequency,* is kept at a low power level, while the decoupling frequency ν_2 is at a power level high enough to saturate the nucleus being decoupled. The main limitation of this type of decoupling is that a series of separate experiments must be carried out to uncouple the entire spectrum. For example, in an AMX spectrum (see Sec. 13-1D for definitions), if ν_2 is used to decouple nucleus A, one experiment will be required to see the effect of decoupling on nucleus M (using $\nu_1 - \nu_2 = \nu_M - \nu_A$) and another experiment will be required to see the effect of decoupling on nucleus X (by making $\nu_1 - \nu_2 = \nu_X - \nu_A$). This mode of decoupling has advantages of simplicity of equipment and ease of maintaining high stability, since only fixed-frequency oscillators are used. With the *frequency-sweep* method, the applied field H_0 and decoupling field at frequency ν_2 are maintained constant and the spectrum is recorded by varying ν_1. Thus, in one experiment nucleus A is decoupled from all other nuclei, resulting in a much more efficient method of decoupling.

For specialized applications a modification of the frequency-sweep method of spin decoupling called INDOR (for internuclear double resonance) is used. In the INDOR technique the *observing* frequency ν_1, which is at the usual low observing power, is kept constant, and the *decoupling* frequency ν_2, which is at a higher power level than ν_1 but not high enough to cause saturation, is swept. The net effect is to change the intensity (peak height) of the line which is being observed at frequency ν_1. This information is particularly useful in determining the relative order of energy levels.

Computer-averaging devices (commonly referred to as CAT or sometimes as TAD) are available for use with high-resolution nmr instruments in order to enhance the signal-to-noise ratio. These units are based on the principle that noise is random and therefore on repetitive scans this signal tends to average toward zero, whereas the signal of interest, being reproducible, tends to add linearly. Statistical analysis reveals that the signal-to-noise ratio will be increased in proportion to the square root of the number of repetitive scans. Up to 100 scans or more are often made when it is necessary to clear up a particularly noisy (low signal-to-noise) spectrum.

13-3 SAMPLE HANDLING

The optimum size and shape of sample to be used for high-resolution nmr measurements involve a compromise between a number of factors. The most important factor is the homogeneity of the magnetic field, and if possible the entire volume of sample (at least that which is within the radio-frequency measuring coil of the probe) should be in a region of homogeneous magnetic field. This condition is more nearly approached as the sample volume is reduced, but this in turn tends to decrease the signal strength. It should be noted, however, that there is no direct proportion between signal intensity and sample volume, since smaller samples usually give greater field homogeneity, which in turn gives sharper signals. In practice the usual sample volume is about 0.2 to 0.5 ml, though capillary cells holding as little as 0.003 ml have been used. For wide-line studies, where field homogeneity is much less critical, sample volumes up to 50 ml have been used, resulting in considerably increased sensitivity.

It is difficult to generalize about the actual *concentration* of solute molecules that must be present in a sample in order to detect the signal adequately, since it depends on the sensitivity (signal-to-noise ratio) of the particular instrument being used, the multiplicity and sharpness of the observed lines, the number of equivalent protons in the molecule, etc. However, as a rough rule of thumb, the usual solute concentration required for proton spectra with a 60-MHz (14-kG) instrument is about 5 to 20 weight per volume percent, and a 100-MHz (23-kG) instrument would be roughly twice as sensitive. The amount of compound required for a proton spectrum is only about 5 to 50 mg, and this, coupled with the fact that nmr is nondestructive, is an advantage when it comes to studying expensive samples. The sensitivity of nmr for other isotopes, relative to the sensitivity for protons, was given in Table 13-1. A time-averaging computer can be used to increase the sensitivity.

For high-resolution studies the sample must be in a low-viscosity liquid state, either as a pure liquid (neat state) or as a dilute solution in an inert solvent. Theoretically, a gaseous sample could also be used, but to obtain a large enough number of nuclei in the sample tube a pressure of several atmospheres would be required. It is usually preferable to examine a sample as a dilute solution rather than as a neat liquid, especially if the sample has highly polar groups, since intermolecular attractions between molecules can appreciably affect nuclear shielding (chemical shifts). A solid sample must be melted or dissolved in a suitable solvent, and a viscious liquid must be heated or diluted with a suitable solvent. Slurries of organic solids should not be used, because any system which restricts motion or gives incomplete averaging of dipole-dipole interactions will lead to field inhomogeneities and broadening of spectral lines.

The solvent chosen must be capable of dissolving the sample to the extent of about 5 to 20 percent and should also be chemically inert, magnetically isotropic, and devoid of the nuclear species being observed (usually hydrogen). *Carbon tetrachloride* is the ideal solvent for compounds that will dissolve in it, and *carbon disulfide, deuterium oxide,* and *deuterochloroform* are other widely used solvents. Deuterochloroform, $CDCl_3$, is probably the single most widely used solvent, despite its relatively high cost. Other solvents occasionally used are chloroform (with a proton resonance at about 2.7τ with respect to TMS), pyridine (1.5 to 3.1τ), trifluoroacetic acid (0.17τ), benzene (2.7τ), dioxane (6.3τ), acetone (7.8τ), acetonitrite (8.0τ), and dimethyl sulfoxide (7.4τ). (The τ values noted may vary slightly with the solute.)

In order to have a reference line included in each spectrum from which chemical shifts can be measured (and to provide the reference for instruments using internal-lock stabilization; see Sec. 13-2*C*), it is customary to add TMS directly to the sample solution, forming an *internal reference.* A concentration of about 0.2 percent is suitable for most purposes, and it is convenient to add the TMS directly to the solvent. When water or deuterium oxide is the solvent, TMS can be used as an *external reference* by sealing the TMS in a capillary immersed in the solution. However, with external references it is necessary to apply a correction factor to the chemical shifts to compensate for the difference in magnetic susceptibility of the sample and reference solutions, and thus in polar solvents an *internal* reference like DSS is usually preferred.

Finally nmr samples should be free of small amounts of foreign matter that could lower the resolution or contribute impurity lines. Often, for example, it is important to remove dissolved oxygen, which might otherwise broaden line widths by affecting the spin-lattice relaxation time through paramagnetic interactions. Oxygen can be removed by distilling the sample in a vacuum or by repeatedly freezing and evacuating the sample. Once oxygen is removed, the sample should be sealed in its glass cylindrical sample tube. The concentration of other impurities which can be tolerated depends upon where in the spectrum the impurity lines lie and the relative molecular weights of the impurity compound and sample compound. If the impurity has a much lower molecular weight than the sample compound, a small amount of impurity may give very intense lines in the spectrum, especially if the impurity contains many identical protons or other nuclei being observed.

13-4 APPLICATIONS

The most important applications of nmr spectroscopy include structure elucidation, qualitative analysis, kinetic studies, property studies, and quantitative analysis.

13-4A Structure Elucidation and Qualitative Analysis

These two applications will be discussed together since they both use a similar approach. For these purposes, nmr data should be used in conjunction with data obtained by other techniques, such as infrared and ultraviolet spectroscopy and mass spectrometry. Fundamental data such as melting and boiling points are likewise useful. However, in contrast to other spectroscopic techniques, an nmr spectrum by itself conveys a great deal of unambiguous structural information, and it is not at all uncommon to be able to locate and assign every band in an nmr spectrum, whereas this is rarely possible with other spectroscopic techniques. The definitive nature of nmr spectra is due mainly to the fact that only one type of isotope, e.g., protons, is measured at a time, and thus other nuclei present in a molecule are largely transparent, though of course the other nuclei may modify the signal being measured. Another consequence of this, however, is that an nmr spectrum is usually not as unique a fingerprint of a molecule as, say, an infrared spectrum, where absorption is determined by vibrations of the whole molecule; a possible exception occurs with high-molecular-weight (in excess of 300) molecules, which may give very complex nmr spectra, entirely suitable for use as fingerprints.

Recall from Sec. 13-1 that three main types of information can be obtained from an nmr spectrum. (1) The position of a line, or its *chemical-shift value,* provides information about the molecular environment of the proton. Protons in similar chemical environments in different molecules show similar chemical-shift values, so that a knowledge of such shifts is useful in identifying particular protons in chemical compounds, just as group vibrational frequencies in infrared and Raman spectroscopy are an aid in identifying certain groups in a molecule. (2) The *intensity* of a peak within a spectrum is directly proportional to the number of protons contributing to the peak, and thus the relative number of protons in different chemical environments can be determined. (3) The fine structure, or *splitting pattern,* arising from spin-spin coupling often gives unambiguous information on the kind and number of neighboring nuclei. Use of all three kinds of information provides a powerful tool for identifying molecular structures.

Further information can be obtained by observing the nmr spectra of other magnetic nuclei in the molecule (Sec. 13-1*G*). Also, for spectra complicated by extensive spin-spin coupling, spin-decoupling techniques may greatly simplify the interpretation. In addition, the use of different solvents and isotopic substitutions may further clarify the interpretation (Sec. 13-1*E*).

13-4B Kinetic Studies

The ability of nmr to indicate rapidly and quantitatively the changes that may be occurring in key functional groups during chemical reactions makes it very suitable for the study of reaction kinetics. It is desirable that the system being studied have unique resonances for key reactants and products, but in certain cases it may be possible to resolve systems giving overlapping bands. The control of temperature, atmosphere (especially in a sealed system), and light is reasonably convenient with equipment now available.

Fairly fast reactions (half-times of the order of seconds or minutes) can be studied by adding the last key component while the sample tube is in the spectrometer and the system is being scanned. Many spectrometers permit scans to be made as rapidly as once every 10 s, allowing complete spectral information to be obtained for reactions lasting a minute or so.

Very fast reactions, e.g., a conformational interchange or a proton exchange in an organic acid, can often be quantitatively studied. For example, in the case of a reversible hydrogen exchange from one environment to another, the particular nmr spectrum observed depends on the rate of exchange and the chemical-shift differences (frequency separation) between the separate environments. If the rate of exchange is slow compared with the frequency separation, separate resonance bands will be observed. As the rate of exchange approaches the frequency separation, the bands broaden and coalesce into a single broad band. At higher exchange rates, e.g., at higher temperatures, the band may sharpen into a single line at a position determined by the relative lifetimes of the proton in each of the two environments. This information can often be used to determine rates of exchange or thermodynamic information like activation energies and potential-barrier energies.

13-4C Property Studies

These applications are too numerous to discuss in detail, but a few of the properties that can be studied include isomerism, tautomeric equilibria, relaxation processes, hydrogen bonding, solvation effects, molecular conformations, electronic structure of molecules, nuclear moments, electrical quadrupole interactions, and the type and length of bonds.

13-4D Quantitative Analysis

Nuclear magnetic resonance is not often thought of as a tool for quantitative analysis, and certainly it is not directly competitive with most other types of spectroscopic techniques in this respect. The requirement of a relatively high concentration (Sec. 13-3) is one limitation. There are a few cases for which nmr is particularly well suited or where it may have definite advantages over other techniques. The unique advantages of wide-line nmr have already been mentioned in Sec. 13-2*C*. In high-resolution nmr there are advantages in cases where a mixture is to be analyzed and the components are unknown or where the components are known but pure compounds are not available for the calibrations necessary with other types of spectroscopic instruments. With nmr, the absolute proportionality that exists between line intensity and the number of atoms is a powerful tool in accounting for the presence or absence of more than one compound and also for directly giving the relative amounts of the various compounds present. Unlike other spectroscopic methods, the measurement of extinction coefficients is not necessary in nmr determinations.

13-5 ELECTRON SPIN RESONANCE

Electron spin resonance (esr), also called electron paramagnetic resonance (epr) or electron magnetic resonance (emr), is similar to nuclear magnetic resonance except that the reorientation of the magnetic moment of an *unpaired electron*, rather than a nucleus, in a magnetic field is observed. Electron spin resonance is applicable only to *unpaired* electrons, since the magnetic moments of paired electrons are canceled. Unpaired electrons occur in free radicals (unstable paramagnetic materials formed either as intermediates in a chemical reaction or by irradiation of a "normal" molecule) and in certain naturally occurring substances such as NO, O_2, NO_2, and transition-metal species, all of which are important to the chemist. Since free radicals frequently exist only fleetingly and in very low concentrations during the course of a chemical reaction, one of the primary requirements of any method for studying them is the ability to measure them accurately in very low concentrations and at high speeds. Electron spin resonance is far more sensitive and rapid than any other available technique sensitive to free radicals, and thus it is considered the technique of choice for such studies. Radicals with lifetimes greater than about 1 μs can routinely be studied with esr, and shorter-lived species can often be studied by going to lower temperatures in the solid state (matrix techniques), since this increase their lifetimes.

13-5A Theory

Four properties of esr spectral lines are important, their *intensity, width, position,* and *multiplet structure.*

INTENSITY OF SPECTRAL LINES

The intensity of an esr absorption line is proportional to the concentration of the free radical or paramagnetic material present, and a quantitative measurement of the number of unpaired electrons in the sample can be made by comparing peak areas of the unknown with a standard having the same general line shape as the unknown and a known number of unpaired electrons. A common standard is diphenylpicrylhydrazyl (DPPH), a relatively stable free radical having 1.53×10^{21} unpaired electrons per gram. At typical operating conditions (a magnetic field of about 3400 G and an oscillator frequency of about 9500 MHz, which is in the microwave region), it is possible to detect less than 1 ng of DPPH, which means that as few at 10^{12} spins are detectable. This extremely high sensitivity, superior to nmr sensitivities by several orders of magnitude, is due to the large magnetic moment of the electron (the magnetic moment μ of an electron is about 1836, compared with about 2.8 for a proton; see Table 13-1). The large magnetic moment causes a large energy separation between spin states in a given magnetic field, which may be seen from

$$\Delta E = h\nu = \frac{\mu H}{I} \tag{13-2}$$

It will be recalled from Sec. 13-1*B* that in nmr spectroscopy the ratio of protons in the upper nuclear spin state and the lower spin state is typically only 0.999993, as given by the Maxwell-Boltzmann expression

$$\frac{n_i}{n_j} = e^{-\Delta E/kT} \tag{13-4}$$

Under typical esr operating conditions, the ratio of electrons in the upper spin state n_i to the lower spin state n_j is about 0.9984 at room temperature, and this much greater population difference is what gives esr its superior sensitivity, as explained in Sec. 13-1*B*. Equation (13-4) also indicates that greater sensitivity can be obtained by working at low temperatures, since as T decreases, the difference between n_i and n_j increases, giving a larger net absorption. Whereas samples must be in the liquid state for high-resolution nmr studies, there is no such restriction on esr, and solids down to 4 K have been studied.

Equations (13-2) and (13-4) likewise indicate that greater sensitivity can be achieved by working at high magnetic field strength H and oscillator frequency ν. However, practical considerations set an upper limit on the field strength and frequency. First, most free-radical studies require a sample volume of the order of 1 ml, and it is hard to design a magnet much in excess of 10 kG which will produce a uniform magnetic field over such a relatively large volume. Second, there are few klystrons or other microwave sources that have high output power at frequencies greater than about 40 GHz. In practice, the upper limit of operation for most electron resonance studies is a frequency of about 36 GHz and a field strength of about 13 kG, though sufficient sensitivity for most studies is obtained at 9500 MHz and 3400 G. Another consideration is that at the upper-frequency limit the wavelength and waveguide cavity are only 8 mm, whereas at the lower frequency the wavelength and cavity are 3 cm, thereby permitting larger samples. Typically liquid samples are 0.05 to 0.16 ml, and the Pyrex tubing holding them often occupies a large fraction of the wavelength cavity.

In esr spectroscopy, concentrations down to 10^{-5} *M* or lower can be analyzed, whereas nmr is limited to about 0.1 to 1 *M* (concentrations down to about 10^{-2} *M* are sometimes possible with a 100-MHz instrument).

WIDTH OF SPECTRAL LINES

The width of an esr line depends on the *relaxation time* of the electronic spin state, completely analogous to nmr lines. As with nmr, the two relaxation processes that are important are *spin-lattice* interactions (where the energy of the upper spin state is converted into thermal vibrations of the solid as a whole) and *spin-spin* interactions (where the energy of the upper spin state is dissipated by interactions with local magnetic nuclei or unpaired electrons). In general, electron spin relaxations are much more efficient than those of nuclei, and relaxation times of the order of 10^{-7} or 10^{-8} s are common (for comparison, nuclear relaxation times are often of the order of 1 s for typical liquids).

The *Heisenberg uncertainty principle* can be used to estimate line widths from relaxation times. It states that the uncertainty in energy times the uncertainty in time is a constant, equal to $h/2\pi$:

$$\Delta E\ \Delta t = \frac{h}{2\pi} \tag{13-12}$$

Example 13-27 Estimate the usual line widths (in frequency units) in esr spectra.

Answer The Planck equation states that

$$\Delta E = h\nu \tag{1-5}$$

The *uncertainty* in ΔE is given by $d\Delta E$, or just ΔE, and the derivative of Eq. (1-5) yields

$$\Delta E = h\ \Delta\nu \tag{13-13}$$

Dividing both sides of Eq. (13-12) through by $h\Delta t$ and substituting Eq. (13-13) gives

$$\frac{\Delta E}{h} = \frac{1}{2\pi\ \Delta t} = \Delta\nu$$

An order-of-magnitude estimate of $\Delta\nu$ can be obtained if we approximate 2π as about 10, giving

$$\Delta\nu \approx \frac{0.1}{\Delta t}$$

Therefore, if $\Delta t = 10^{-7}$ s,

$$\Delta\nu \approx 10^6\ \text{Hz}$$

If $\Delta t = 10^{-8}$ s,

$$\Delta\nu \approx 10^7\ \text{Hz}$$

As Example 13-27 indicates, line widths in esr spectra are usually of the order of 1 to 10 MHz. In contrast, theoretical line widths in nmr are usually of the order of 0.1 Hz, although the resolving power of most nmr spectrometers is limited to about 1.0 Hz. By comparison, therefore, it might seem that esr lines are extremely wide. On the other hand, it must be remembered that esr uses very high frequencies (in the order of 10 GHz, whereas nmr uses about 100 MHz), and thus, relative to the applied frequency, esr line widths are of the order of 10 to 100 ppm, which still give sharp, well-defined spectral lines. In fact, in esr work such lines are called narrow, to distinguish them from broadened lines which appear when relaxation times are shortened.

POSITION OF SPECTRAL LINES

The frequency of absorption ν will depend on the strength of magnetic field H that is used [see Eq. (13-2)]. Figure 13-38 illustrates how the energy levels of an electron are split by the presence of an external magnetic field (the *Zeeman effect*). In the absence of a magnetic field a collection of spinning electrons are aligned randomly, but in the presence of an externally applied field there will be a preferred direction (designated in Fig. 13-38 as the lower energy level with a spin quantum number of $-\frac{1}{2}$). Quantum-mechanical restrictions allow only two spin orientations, one in which the field generated by the spinning magnet is aligned *with* the direction of the external field (spin $-\frac{1}{2}$) and one in which the generated field *opposes* the external field (spin $+\frac{1}{2}$). Most electrons will be in the lower energy level, but a significant number will always have sufficient energy to be in the upper energy level, with the ratio of electrons in the two states being predictable from the Maxwell-Boltzmann dis-

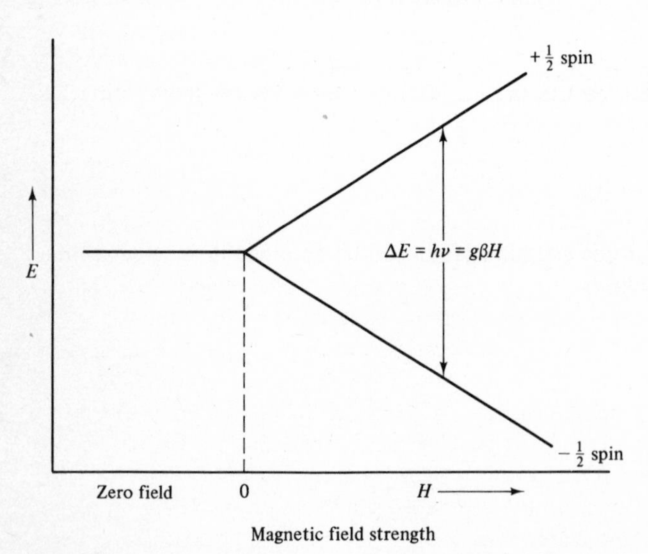

FIGURE 13-38 The effect of an external magnetic field on the energy levels of an unpaired electron.

tribution [Eq. (13-4)]. The fact that the difference in energy between the two spin states increases with increasing magnetic field strength is easy to understand by analogy with an ordinary bar magnet: it requires more energy to hold a bar magnet opposed to a *strong* field than it does to hold one opposed to a *weaker* field. It follows that to reverse the spin of an electron we need to supply energy ΔE, and more energy is required in high than in low magnetic fields.

The basic resonance energy equation in esr spectroscopy is

$$\Delta E = h\nu = g\beta H \tag{13-14}$$

where g is the Landé, or spectroscopic, splitting factor and β is the Bohr magneton.† The Bohr magneton is a true constant equal to $eh/4\pi m_e c$ (where m_e is the mass of an electron) and has a value of 9.273×10^{-21} erg/G. The g factor is a dimensionless constant which is a physical property of the electron and is conventionally taken as the parameter which governs the position of the resonance absorption in esr studies. The g factor is sometimes thought of as the esr counterpart of the nmr chemical shift, but the analogy is not particularly good because among most paramagnetic species there is little variation in the g factor, whereas, of course, chemical shifts vary markedly from compound to compound. Classical analysis indicates that the g value of any free-spinning electron should be exactly 2, but when a relativity correction is applied, an actual free-spin g value of 2.0023 is obtained. Virtually all free radicals and even some ionic crystals have g values which differ from the free-electron value by only ± 0.003 unit, a difference so small that it is often difficult to avoid overlapping lines. The reason for this constancy is essentially that in free radicals the electron can move about more or less freely over an orbital encompassing the whole molecule and is not confined to a localized orbital between just two of the atoms in the molecule. In this sense the electron behaves much the same as an electron in free space.

On the other hand, transition-metal and rare-earth ions with partially filled d and f shells have unpaired electrons which are much more localized (belonging to a particular atom), and thus their g values depart drastically from the free-electron value of 2.0023, varying from about 1 to 8. Systems with such widely varying g values obviously require an extremely wide search to find the resonance condition, and thus it is a requirement of the esr spectrometer that the applied field H and/or frequency ν be variable over a wide range.

The main purpose of determining g values is to characterize the resonance condition for the system being measured, independent of the particular frequency ν and magnetic field strength H being used. It can be seen from Eq. (13-14) that g is equal to a constant h/β times the *ratio* of ν to H. To measure the g value for the system under study, it is necessary to standardize the esr

† The derivation of this equation will be found in several books, e.g., Ref. 26, pp. 2–7.

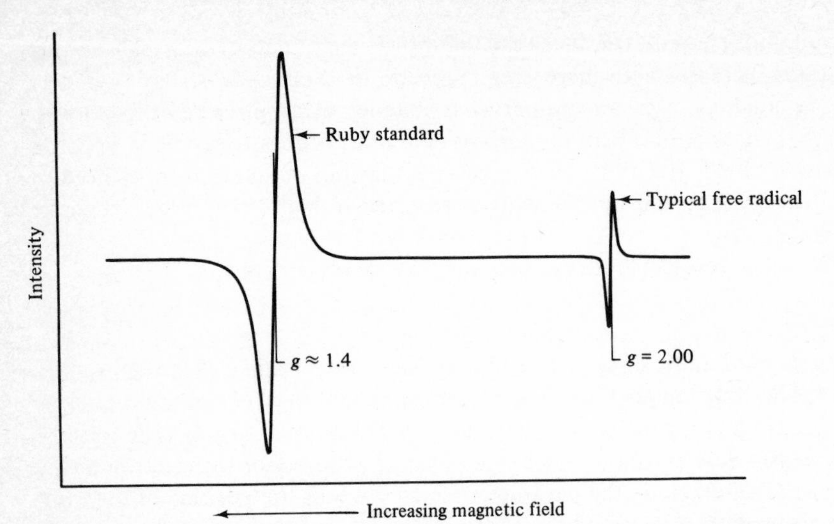

FIGURE 13-39 Electron spin resonance spectrum of ruby internal standard and typical free radical. [*After N. F. Chamberlain, p.* 1950, *in I. M. Kolthoff et al. (eds.), "Treatise on Analytical Chemistry," pt.* I, *vol.* 4, *John Wiley & Sons, Inc., New York,* 1963, *by permission.*]

spectrometer with a sample of known g value. DPPH, with a g of 2.0036, is a favorite for this purpose, although the g values of most free radicals that might be under study are so similar that it is usually necessary to obtain separate spectra for the standard and the unknown, rather than attempt to measure both resonances during the same scan and on the same chart. To avoid the problem of overlapping spectral lines and a separate calibration scan, an *internal standard* having a resonance well separated from that of the unknown is sometimes used. For example, a very small piece of ruby crystal ($g \approx 1.4$), glued to the side of the guide tube into which the sample fits, has been successfully used as a permanent reference standard. Figure 13-39 shows a spectrum for ruby on the same chart with a typical free-radical resonance (for simplicity, the free-radical resonance shown has only one line, rather than being split into multiplets as is usually the case; see next section).† For precise calibration of the base-line linearity, two reference materials with accurately known and well-separated g factors should be used; more often, nuclear resonance or a fluxmeter is used.

Finally, it should be pointed out that although the g values of organic radicals are all close to the value for a free electron (2.0023), the deviations from this value tend to follow reasonably predictable trends. In aromatic radicals, for example, the g value tends to increase with increasing atomic number of the atoms in the radical molecule. More specifically, pure-

† It should be noted that esr absorption bands are commonly recorded as first-derivative curves (Fig. 13-37 compares integral and differential curves), since they can be measured with greater sensitivity. For quantitative measurements it is necessary to obtain the area under the derivative curve.

hydrocarbon radicals tend to have lower g values than aromatic radicals containing nitrogen or oxygen, and these, in turn, have lower g values than aromatic radicals containing halogen atoms. A thorough discussion of g values, along with tables of typical values, can be found in Ref. 26.

THE MULTIPLET STRUCTURE OF ESR SPECTRA

It is necessary to distinguish between two distinct kinds of multiplet structures. The first is called *fine structure,* generally involving large energy separations between the multiplet lines, but this type is of limited importance because it has been observed only in solids and then only in a few isolated cases where the paramagnetic material had more than one unpaired electron. The second kind of multiplet structure is called *hyperfine structure,* involving much smaller energy separations between the multiplet lines. This type of splitting is very common, occurring both in crystals and in free radicals, and provides useful information about electronic structures. Hyperfine interactions are analogous to spin-spin coupling in nmr.

Theoretically, *fine structure* can occur whenever a paramagnetic material has two or more unpaired electrons in the same molecule, since then the total spin quantum number S can have more than one value. Thus for two electrons the total spin S can be either 1 (the two electrons have *parallel* spins, giving a total spin of $S = \frac{1}{2} + \frac{1}{2} = 1$) or 0 (the two electrons have *opposed* spins, giving a total or net spin of zero). The interaction of the two different spin states with the internal crystal field and the externally applied magnetic field leads to two distinct esr lines, as shown in Fig. 13-40. On the left side of Fig. 13-40*a* is shown the splitting between the $S = 1$ and $S = 0$ spin states caused by the *crystal field,* which is always present in the solid state (the so-called zero-field splitting). On the right side of Fig. 13-40*a* the effect of gradually increasing the external magnetic field H_z is shown; the $S = 0$ state is essentially diamagnetic and is unaffected, whereas the $S = 1$ state splits into upper and lower ($S_z = +1$ and -1, respectively) levels. If now the system is irradiated at a constant microwave frequency ν, that is, at constant energy ΔE, and the magnetic field H_z is swept over the range necessary for resonance absorption, two distinct resonance lines will be found, as shown in Fig. 13-40*b*, corresponding to the two transitions allowed under the selection rule $\Delta S_z = \pm 1$ (see Fig. 13-40*a*). In general, n unpaired spins should result in a fine-structure spectrum of n lines.

In practice, however, such fine structure is rarely observed. If, for example, two unpaired electrons occupy sites which are well separated in the molecule, as is usually the case in *organic biradicals,* little or no interaction occurs between the two electrons, and each gives an electron resonance signal as though it belonged to a separate monoradical; i.e., no splitting occurs, and a singlet peak is observed. If, on the other hand, the two electrons are sufficiently close, as in molecules in the triplet state, interactions become im-

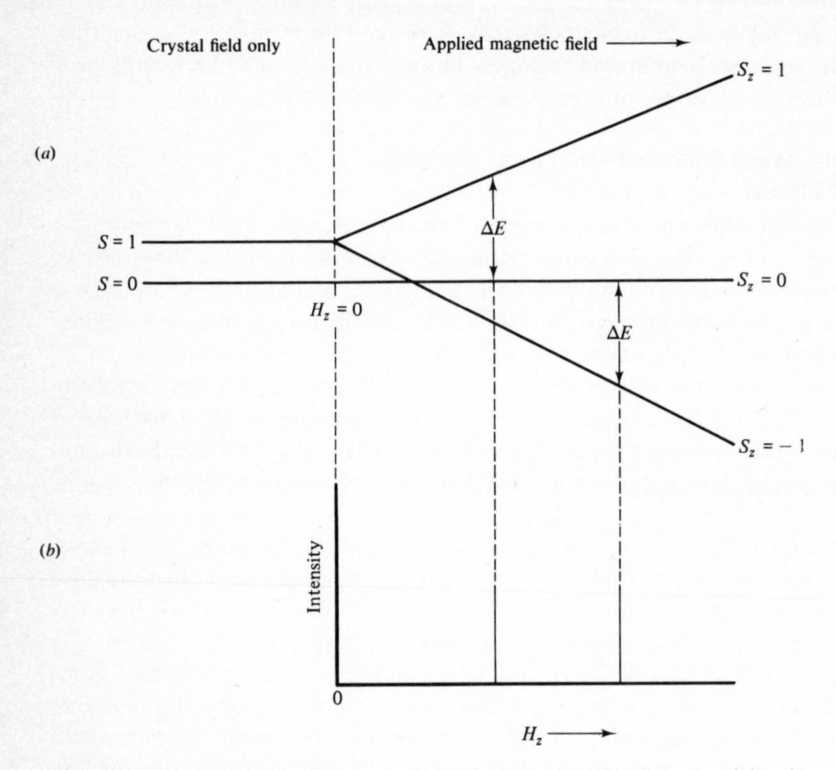

FIGURE 13-40 Electron spin resonance fine structure (a doublet) caused by the presence of two unpaired electrons in the same molecule. (*a*) Effect of crystal field and magnetic field on energy of spin states; (*b*) esr spectrum obtained at constant ΔE, increasing H_z.

portant, but often the interaction between electrons is so strong that it overshadows the effect of the externally applied magnetic field. For example, molecular oxygen is the best-known inorganic free radical, having two unpaired electrons in its most stable ground state; oxygen is thus a stable triplet-state molecule.† However, the unpaired electrons in O_2 are so close together that spin-spin splitting amounts to about 13 kG, which is larger than the external magnetic fields ordinarily applied. This essentially means that the electrons will precess about the *molecular axis* rather than about H_z, the externally applied magnetic field. In short, fine splitting will not be observed with oxygen under normal resonance conditions.

One case where fine structure has been observed involves photoexcited naphthalene in the solid state. This triplet-state molecule gives fine structure

† The molecule has a total spin S of 1, and it therefore has *three* magnetic quantum numbers, 1, 0, or −1, giving it *triplet* status.

with lines separated in the order of hundreds of gauss, for example, 2100 G in experiments in durene.† Alternately, in a fixed field of 3 kG, the multiplet lines for naphthalene in its triplet state are separated by about 6 GHz [3].

Hyperfine structure, on the other hand, arises through coupling of an unpaired electron with neighboring *nuclear* spins in much the same way that coupling between nuclear spins occurs in nmr. However, whereas spin-spin coupling in nmr gives separations between lines of the order of a few hertz, the hyperfine structure of esr spectra have line separations of some 50 MHz (or about 1 to 20 G), which is larger by a factor of about 10^6 than nucleus-nucleus coupling. Electron-nucleus coupling is so much stronger than nucleus-nucleus coupling because an electron can approach a nucleus closer than another nucleus can.

When an unpaired electron interacts with the small magnetic field coming from a nucleus, the energy levels of the electron will be *split* by a small amount. Consider first the particular case where the unpaired electron interacts with a proton (or with any other single nucleus having a spin $I = \frac{1}{2}$). Any one proton will be lined up either parallel or antiparallel to the applied field and the electron spin, and in a collection of molecules the protons will presumably be about equally distributed between the two orientations. Assume that a constant magnetic field H is applied and the frequency ν of microwave radiation is varied through the region of possible resonances.‡ Figure 13-41 illustrates how the spin energy levels of an unpaired electron are split, starting at the left with no external applied field, and then moving to the right, successively adding the effects of (1) a constant field H, (2) one interacting proton, and finally (3) a second interacting proton. With no applied field, electrons with spin of $+\frac{1}{2}$ and $-\frac{1}{2}$ will be degenerate (equal energy), whereas the external applied field H will split the energy levels. If there is one proton which can interact with the spinning electron, the magnetic moment of the proton will produce a small additional field at the electron, and this will either add to or subtract from the effect of the external field, according as the proton is quantized with its spin parallel or antiparallel to the field. Accordingly, the proton will bring about the additional splitting shown in Fig. 13-41*b*. Quantum mechanics requires that allowed esr transitions obey the *selection rules* $\Delta M_s = \pm 1$, $\Delta M_I = 0$ (see, for example, the derivation in Ref. 9, pp. 21–22). Thus, whereas a singlet absorption line is observed with no nuclear interaction (see lower section of Fig. 13-41*a*), the presence of one proton splits the line into a doublet of equal (1:1) intensity (Fig. 13-41*b*). The separation between lines gives the *hyperfine splitting constant a*. (Note that the hyperfine coupling constant is completely analogous to the nmr coupling constant.)

† C. A. Hutchison, Jr., and B. W. Mangum, *J. Chem. Phys.*, **29**:952 (1958); **34**:908 (1961).

‡ The alternative experiment, giving equivalent results, is to use a *constant frequency* and vary the magnetic field through the region of resonances; how this type of experiment affects the spin energy levels and the prediction of resonance lines is left as an exercise (see Exercise 13–23).

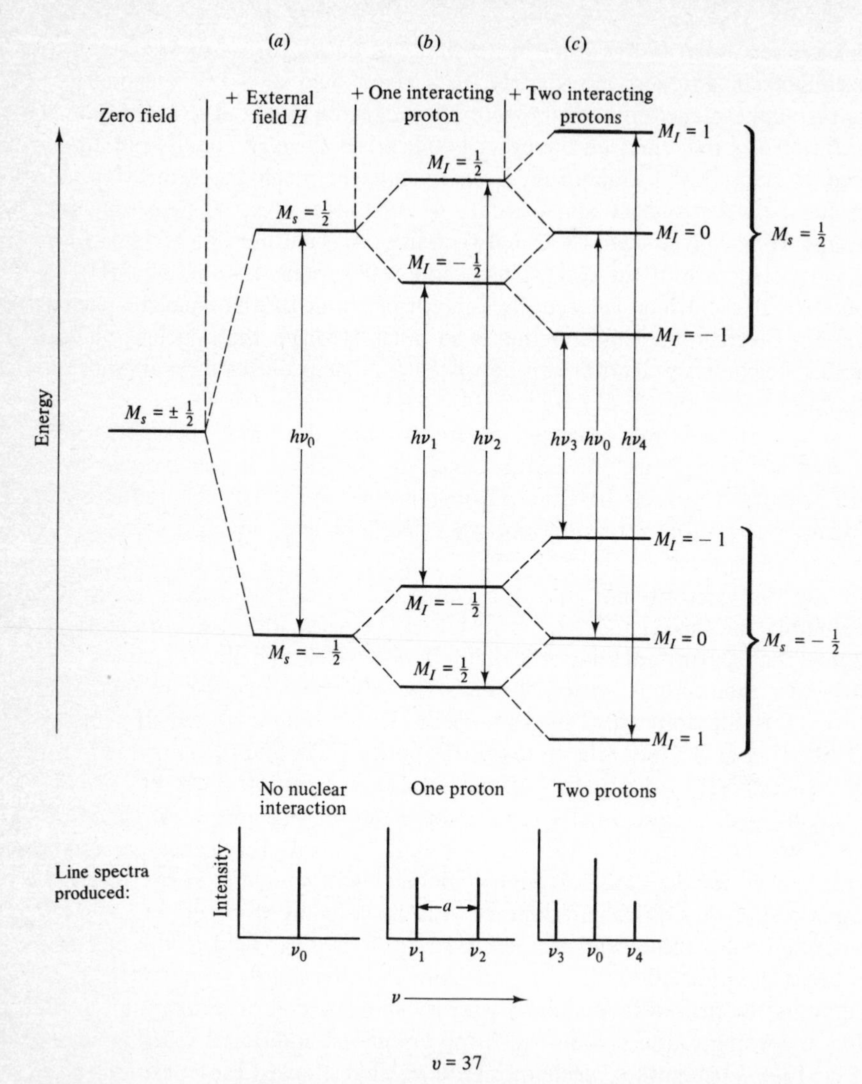

$v = 37$

FIGURE 13-41 The splitting of electron spin energy levels by the successive addition of (*a*) an external magnetic field *H*; (*b*) one interacting protin, and (*c*) a second interacting proton. The line spectra produced as shown immediately below the splitting diagram; all three spectra are at a constant field *H*. M_s = electron spin quantum number; M_I = nuclear spin quantum number.

If now a second proton is present in the molecule, the spectrum of the unpaired electron appears as a triplet (see lower section of Fig. 13-41*c*). The relative intensity of the three lines is 1:2:1, consistent with the fact that the $M_I = 0$ level is twice as probable as the $M_I = 1$ or $M_I = -1$ levels (Fig. 13-41*c*). In general, for *n* equivalent protons† there will be $n + 1$ lines, and the relative intensities of the lines will be proportional to the coefficients in a binomial expansion of order *n*; that is, $(1 + X)^n$.‡

Example 13-28 Predict the type of esr spectrum to be obtained for 2,3-dichlorobenzoquinone

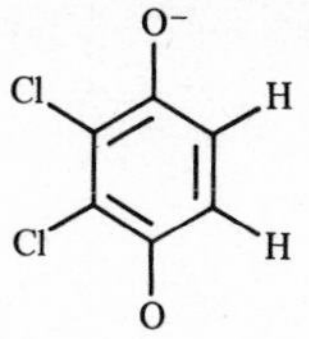

Cl, Cl, H, H, O⁻, O

Answer Since the only nuclei of nonzero spin are the two protons, and since they are equivalent, the esr spectrum should be a triplet, of relative intensity 1:2:1. This is, in fact, experimentally observed.

When there are *nonequivalent* nuclei in a molecule which interact with an unpaired electron, the esr spectra become more complicated in a way analogous to the second-order splitting of nmr spectra. For two different types of protons, for example, it can be predicted that there will be a total of $(n_1 + 1)(n_2 + 1)$ lines. The magnitude of the interaction of the electron with each type of nucleus will be proportional to the hyperfine splitting constant for each type of nucleus, which in turn depends on the product of the electron density at the nucleus in question and some interaction constant which is proportional to the nuclear moment. Often very complex spectra can be interpreted by starting from a known or postulated structure and predicting the number of lines to be observed for each type of equivalent nuclei. One example will be given.

Example 13-29 When diacetyl is condensed in the presence of base to form 2,5-dimethylbenzoquinone, the esr spectrum shown in Fig. 13-42 is obtained during the course of the reaction. Show how this spectrum is consistent with the postulation of a semiquinone free-radical intermediate, as follows:

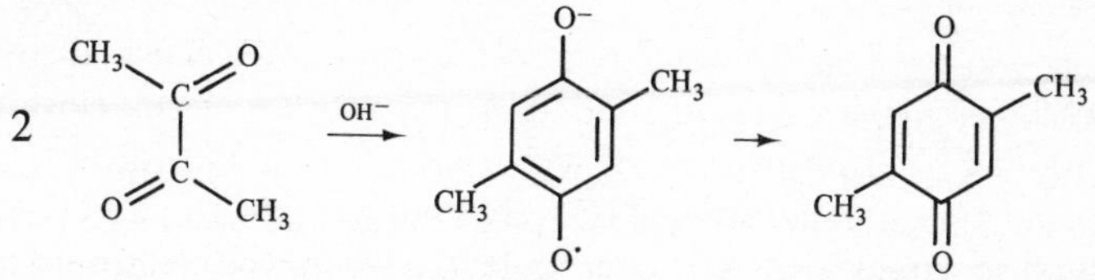

$$2\ CH_3COCOCH_3 \xrightarrow{OH^-} \text{(2,5-dimethylsemiquinone radical anion: } O^-, O^\cdot, CH_3, CH_3) \rightarrow \text{(2,5-dimethylbenzoquinone: } O, O, CH_3, CH_3)$$

Answer The only nuclei with nonzero spin present in the radical intermediate are protons, and there are six equivalent methyl protons and two equivalent ring protons. Therefore, we might expect the resonance of the unpaired electron to be split into seven, that is, $n + 1$, lines from the

---

† Equivalent protons in esr are those with the same hyperfine splitting constant as a result of symmetry of the molecule. Examples of equivalent protons are given in subsequent illustrations (Examples 13-28 and 13-29).

‡ This is the same rule as developed in Sec. 13-1*D* for nmr spectra (see Example 13-17).

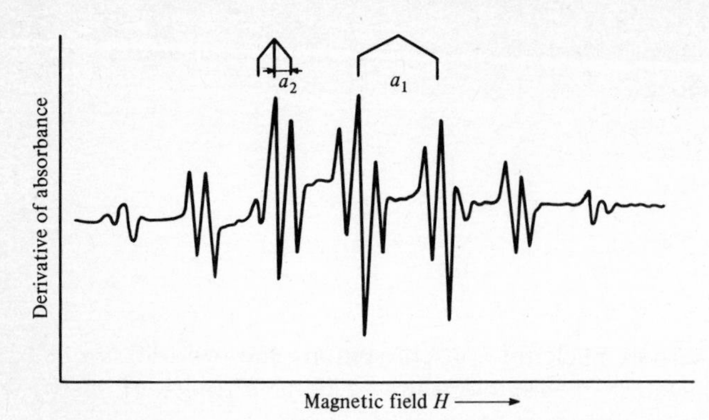

FIGURE 13-42 Electron spin resonance spectrum obtained during reaction of diacetyl condensing to form 2,5-dimethylbenzoquinone. [*Reprinted by permission from Varian Staff, "NMR and EPR Spectroscopy,"* p. 208, *Pergamon Publishing Company New York*. 1960 *(out of print)*.]

methyl protons; each of these lines, in turn, should be split three times by the ring protons. Looking closely at the spectrum, we can identify an intense septet which roughly satisfies the predicted intensity ratio of 1:6:15:20:15:6:1, and indeed each of those major lines is split into a triplet, which, though badly distorted, gives reasonable agreement with the 1:2:1 intensity ratios predicted. The identification of the major and minor lines is aided by the assignment of the hyperfine splitting constants a_1 and a_2, labeled on the spectrum. The relative magnitudes of the hyperfine splitting constants give evidence that the unpaired electron interacts strongly with the methyl protons, i.e., spends a large amount of time in their vicinity, and interacts only weakly with the ring protons. Without the esr spectrum this might not have been expected. It is left as an exercise to predict the spectrum if the interaction had been otherwise (see Exercise 13-22).

Electron spin resonance data can be used to measure electron distribution within a molecule, giving values in good agreement with molecular-orbital and valence-bond theory. Information of this type is of great value in understanding chemical reactions, since positively charged (electrophilic) reactants obviously tend to attack that part of a molecule where the electron density is the greatest, and vice versa.

13-5B Instrumentation

The instrumentation for esr is quite different from that for nmr. Most of the differences are due to the much higher magnetic moment of an electron than a proton (1836 compared with 2.79 nuclear magneton units; see Table 13-1). This affects the frequency and magnetic field strength used to accomplish resonance [see Eq. (13-2)] and also the sensitivity, which hinges on the Boltzmann distribution of spin states [Eq. (13-4)]. In addition, the characteristically wide esr lines allow the use of less homogeneous magnets, while the large splitting constants necessitate a larger sweep of the magnetic field than is required in nmr.

CHOICE OF FREQUENCY AND MAGNETIC FIELD STRENGTH

The effect of the magnetic moment of the electron on the choice of applied frequency and magnetic field strength can be seen from Eq. (13-2) after substitution of the spin of an electron (½), giving

$$\Delta E = h\nu = 2\mu H \tag{13-15}$$

Since μ = 1836 nuclear magneton units for a free electron and each nuclear magneton unit† has a value of 5.051×10^{-24} erg/G,

$$\nu = \frac{2\mu H}{h} = \frac{(2)(1836)(5.051 \times 10^{-24}\ \text{erg/G})H}{6.62 \times 10^{-27}\ \text{erg-s}}$$

$$\nu(\text{Hz}) = (2.803 \times 10^{6}\ \text{s-G}^{-1})H \tag{13-16}$$

From Eq. (13-16) it is readily calculated that if a typical nmr magnet with a field strength of 14,092 G is used, the resonance frequency for a free electron will be 39.5 GHz, which is in the microwave region of the electromagnetic spectrum (see Fig. 1-4). In practice, the magnetic fields and frequencies used in esr studies are seldom, if ever, this high, the upper limit for most electron resonance studies being 13 kG and 36 GHz (which corresponds to a wavelength of 8 mm). The upper limit on the magnetic field strength is dictated by the desirability of using relatively large (1-ml) samples, which in turn necessitates a relatively large pole gap, and it is difficult to design a magnet much in excess of 10 kG which maintains homogeneity over such a large sample volume. By the same token, microwave sources such as klystrons perform poorly in the region of 40 GHz or higher. The nature of microwave energy further restricts the upper energy limit, since rectangular tubes called *waveguides* are required to transport the radiation, and the dimensions of the waveguide and sample cavity must be of the same magnitude as the wavelength of radiation used. Thus, a 36-GHz source frequency needs a waveguide and sample cavity of only 8 mm, requiring samples that are even smaller than this and making the alignment of all the microwave components fairly critical. Therefore, for all these reasons, the overwhelming majority of commercial esr spectrometers operate at about 9.5 GHz (a wavelength of about 3 cm), corresponding to a magnetic field of about 3400 G. An intermediate frequency of 24 GHz (8600 G) is sometimes used. These three frequency regions in the microwave region (9.5, 24, and 36 GHz) are often called the *X*, *K*, and *Q* bands of the microwave region, respectively.

† The nuclear magneton unit is calculated from $eh/4\pi Mc$, where M is the mass of a proton and $h = 6.62 \times 10^{-27}$ erg-s.

THE SENSITIVITY OF ESR SPECTROMETERS

The fact that the magnetic moment of an electron is much larger than that for a proton means that esr has an inherently higher sensitivity than nmr, which can be understood from Eq. (13-15) and the Boltzmann distribution equation

$$\frac{n_i}{n_j} = e^{-\Delta E/kT} \tag{13-4}$$

From Eq. (13-4) it can be shown that there is only about a 7-ppm difference in upper- and lower-level spin-state populations in nmr experiments, whereas there is about a 1600-ppm population imbalance in esr experiments, giving esr spectrometers about a 230-fold sensitivity advantage. Energy-level splitting ΔE is directly proportional to the magnetic field strength [see Eq. (13-15)], and thus greater sensitivity could be expected at higher field strengths and resonance frequencies, but the practical considerations mentioned in the last paragraph restrict this means of increasing the sensitivity. Equation (13-4) indicates that lowering the temperature will likewise increase the sensitivity, but considerations such as the effect of temperature on molecular motion usually limit this approach.

MAGNETIC FIELD HOMOGENEITY REQUIREMENTS

The homogeneity of the field required for most esr studies is generally less critical than it is for high-resolution nmr studies, since esr line widths are relatively broader (often several gauss, or more, which is in the same order of broadness of many wide-line nmr peaks). Field homogeneity and stability of 1 part in 10^6 are adequate for most esr studies, especially free-radical studies in solids, but higher homogeneity and stability are desirable for high-resolution studies of solutions.

MAGNETIC FIELD SWEEP REQUIREMENTS

The large splitting that occurs in esr spectroscopy (of the order of 1 to 20 G, compared with splittings of the order of 1 mG or less in nmr spectroscopy) requires a very large change in magnetic field sweep in order to scan a complete spectrum. Furthermore, although the spectroscopic splitting factor [g in Eq. (13-14)] is fairly constant for many free radicals, it varies considerably for transition-metal ions. Thus, to be able to use the same instrument for a variety of sample materials, it is desirable to be able to vary the magnetic field for a typical 9.5-GHz instrument over a field range of from about 50 to 5500 G. These large variations in magnetic field strength require that an *electromagnet* rather than a permanent magnet be used, since such large sweeps of a permanent magnet, even if possible, would rapidly ruin its performance through hysteresis.

Whereas the design of nmr spectrometers allows the option of varying *either* the magnetic field H *or* the oscillator frequency ν in order to scan a spectrum in search of resonances, it is only feasible to vary the magnetic field in esr spectrometers. This is mainly because esr spectrometers employ a sample resonant cavity to amplify the microwave signal, and the cavity would not remain at resonance if the microwave frequency were varied. A second, but less important, reason is that it is difficult to vary the frequency of a klystron oscillator linearly and reproducibly.

PRINCIPLES OF INSTRUMENT DESIGN

A simple form of esr spectrometer is shown in Fig. 13-43. The usual microwave source is a *klystron* oscillator, delivering 30 to 300 mW of power. The energy is transmitted by means of a *waveguide,* a rectangular tube made of copper or brass with dimensions appropriate to the wavelength of the radiation. The sample is placed inside a *resonant cavity,* which, in its simplest form, is a blanked-off section of waveguide with a hole in each end wall to transmit power in and out. The purpose of the resonant cavity is to concentrate the source energy for the sample by multiple reflections of the traveling microwave from the two end walls. Detection is generally accomplished with a semiconducting crystal detector which acts as a *rectifier,* converting the microwave power into a direct current. The magnetic field supplied by an electromagnet is arranged perpendicular to the field of the microwave radiation.

In operation the magnetic field strength is slowly varied while a constant microwave source frequency is maintained. As the magnetic field strength approaches the value corresponding to resonance for the sample being studied, power is absorbed by the sample and the power transmitted through the cavity to the detector is reduced. The decrease in measured current represents the esr signal.

A spectrometer of the type just described would be very insensitive, since

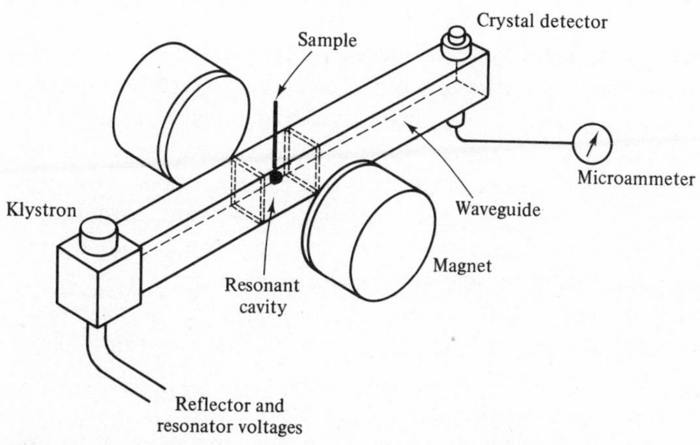

FIGURE 13-43 Simple form of esr spectrometer. (*From P. B. Ayscough, "Electron Spin Resonance in Chemistry,"* p. 137, *Methuen & Co., Ltd., London,* 1967, *by permission.*)

it would be necessary to look for a relatively small decrease in the dc crystal current from a level that is normally fairly high. It is possible, however, to introduce a comparatively small modification into this simple design and increase the sensitivity greatly. If, instead of measuring the direct current from the detector, the *derivative* of the absorption curve is taken, an ac signal results, which will consist of a zero signal off resonance and a finite signal at resonance; this ac signal is readily amplified. The derivative of the absorption curve can be obtained by superimposing a small *oscillating* magnetic field on top of the large dc field, a process called *modulation.* The ac signal is detected with an ac amplifier in the detector circuit. Early instruments modulated the magnetic field at a frequency around 60 Hz, whereas modern instruments often use about 100 kHz. In any case, the *amplitude* of the modulation signal must be smaller than the resonance-line width (10 percent or less) in order to obtain a true derivative signal. The instrument shown in Fig. 13-43, modified only by the addition of modulation, could be used for demonstration purposes, but further redesign is necessary to obtain the high sensitivity available in modern esr spectrometers, a sensitivity increase which may sometimes approach a factor of 10^8 over that of the instrument just described.

A block diagram of a typical modern esr spectrometer is shown in Fig. 13-44. The major difference between this instrument and the one diagramed in Fig. 13-43 is that instead of using a *transmission cavity* and placing the sample directly in the path between the source and the detector, the sample is in a *reflection cavity* located at right angles to the source and detector. This arrangement necessitates some means of transmitting the microwave power into the sample cavity and thence to the detector. Several types of *divider* devices are available, including a *hybrid tee* (shown in Fig. 13-44), a *circulator,* and a *directional coupler,* but only the hybrid tee, which is probably most used, will be described here (see Ref. 23, p. 54, for a discussion of the other two).

The hybrid tee is a device which will not allow microwave power to pass in a straight line from one arm to the opposite. Instead, it "turns the corner," as shown by the arrows in Fig. 13-44. Thus, power coming from arm 1 is equally divided between arms 2 and 3. Arm 3 usually contains a resistor, which forms one arm of a balanced Wheatstone bridge circuit, and any change in the impedance of arms 2 or 3 will unbalance the bridge, resulting in a signal reaching the crystal detector. In practice, when resonance conditions for the sample are not satisfied, power will be absorbed in the sample *cavity* and the impedance of arm 2 will balance that of arm 3. At resonance, however, the *sample* absorbs energy from the microwave field, and this energy *change* unbalances the impedance of the cavity, which is reflected in the change in signal received by the crystal detector (arm 4).

This reflection-cavity arrangement has two main advantages over the simple transmission cavity: (1) the reflection cavity is more efficient in building up (amplifying) microwave power, and in fact gives more than twice

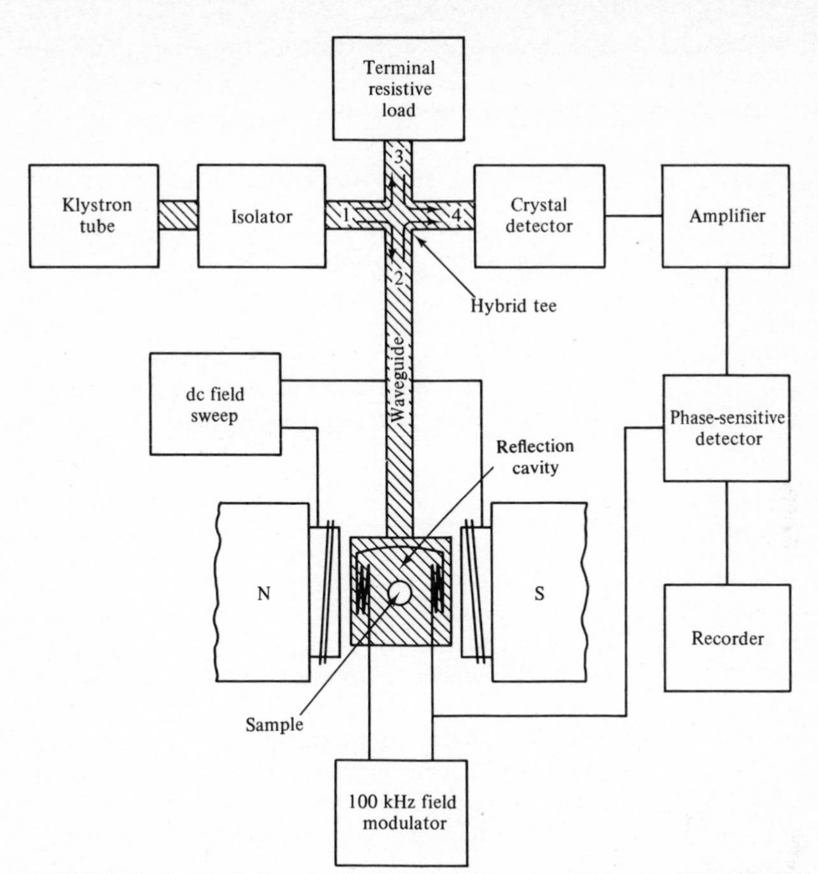

FIGURE 13-44 Block diagram of a typical modern esr spectrometer. (*After H. G. Hecht. "Magnetic Resonance Spectroscopy," p.* 122, *John Wiley & Sons, Inc., New York,* 1967, *by permission.*)

the sensitivity of a transmission cavity; (2) this bridge circuitry gives a positive signal only in the region of a resonance peak, no signal being detected off resonance, whereas the transmission-cavity system must measure resonance signals which are only a small *change* from a normally large nonresonance signal. With both arrangements, it is standard practice to modulate the dc magnetic field (see dc field-sweep coils in Fig. 13-44) with a small-amplitude ac generator (see 100-kHz field modulator coils in Fig. 13-44), and only the ac component of the detector signal is amplified. With ac field modulation, amplification of the detector signal can be achieved, and the derivative of the resonance absorption peak is recorded. The phase-sensitive detector shown in Fig. 13-44 keeps the recorder scan in phase with the 100-kHz field modulator.

The isolator diagramed in Fig. 13-44 allows transmission of microwave radiation in only one direction, from the *klystron tube* to the *hybrid tee,* and inhibits the reverse transmission. In this way there is no danger that the klystron will become detuned through reflective feedback from the sample cell at resonance.

Klystron oscillator tubes and crystal detectors were discussed in Sec. 7-2*A*.

ESR SAMPLES

Solid, liquid, or gaseous samples can be used in esr spectroscopy, the first two being more common. Gaseous samples normally give signals that are too weak to be detected accurately. Samples are usually enclosed in cylindrical *quartz* tubes, since glass often contains a sufficient quantity of paramagnetic impurities to interfere. Typical sample volume for liquids is 0.05 to 0.16 ml, analyzed in tubes of about 3 mm ID. With aqueous solutions the sample volume should be restricted since there is a large dielectric power loss in water.

13-5C Applications

Electron spin resonance spectrometers can respond only to systems containing unpaired electrons, such as: (*a*) free radicals; (*b*) compounds whose normal bonds have been modified or broken, say by irradiation or electrolysis; (*c*) atoms and molecules having an odd number of electrons, e.g., atomic hydrogen or nitrogen, or molecular NO; (*d*) molecules like O_2 which are paramagnetic despite an even number of electrons; (*e*) transition elements with partly filled inner electron shells; and (*f*) metals and semiconductors where the paramagnetism is caused by conducting electrons. The following discussions will emphasize only the first two systems, which give the greatest number of applications in chemistry.

The detection of esr in a sample demonstrates unequivocally the presence of unpaired electrons. The spectroscopic splitting factor g and especially the hyperfine coupling (splitting between lines) produced by the interaction of the unpaired electron with neighboring magnetic nuclei gives strong evidence not only of the presence and identity of *free radicals* but yields information about the relative unpaired-electron density around the different magnetic nuclei in the molecule (see Sec. 13-5*A*). By following the appearance or disappearance of free radicals during the course of a reaction, the role radicals play can be determined. Short-lived or transient radicals which occur as reaction intermediates can often be observed in rapid-flow mixing chambers. Steady-state concentrations of radicals can be observed in a sample cell which allows the esr signal to be recorded as the samples mix and pass through the cavity. Such observations, for example, have provided strong support for the theory that many biochemical reactions proceed by way of free-radical intermediates [5]. Other esr studies have shown that free radicals are involved in vinyl polymerization† and in the adsorption of aromatic hydrocarbons on cracking catalysts.‡

The study of irradiation damage and observations of photochemical reactions are among the most widely used applications of esr spectroscopy. If a solid organic material such as a protein, polypeptide, or other polymer is irradiated with high-energy ionizing radiation, a free radical is produced in almost every case. The stability of these radicals depends upon the lattice in which they are formed. Although many are short-lived, in some cases they are stable for *years*. Irradiation sources that have been commonly used include x-ray, gamma-ray, and high-energy electron beams. Radiation sources in the visible and particularly the ultraviolet region are widely used to follow photochemical reactions. Sample cells have been constructed which allow all these forms of radiation to be used while simultaneously monitoring the esr signal. In addition, some sample cells allow monitoring the esr spectra of species produced during *electrolytic* oxidation or reduction.

In a technique which essentially *combines* esr and nmr spectroscopy, called *electron nuclear double resonance* (ENDOR), an esr signal is monitored while inducing nmr transitions. If the *microwave* intensity and frequency are adjusted to saturate a particular electronic transition and the *radio-frequency* (nmr) field is swept slowly, a deflection of the esr signal is achieved when the nmr resonance condition is fulfilled. The techniques essentially provides a greatly enhanced sensitivity for the study of *nuclear resonances*, taking advantage of the much larger magnetic moment that the electron has over that of the nucleus. The most important applications of this double-resonance technique so far have been to resolve hyperfine structure which otherwise might have been lost in the electron-resonance line width. A large gain in *resolution* of the esr spectrum, often by a factor of 10^4, is obtained, since the limiting line width is that of the *nuclear* resonance rather than that of the electron resonance. Thus, some of the favorable features of both nmr and esr are combined into one technique, and it can be expected that ENDOR will become increasingly useful.

EXERCISES

13-1 What causes chemical shifts in nmr spectroscopy?

13-2 The nmr spectrum of diethyl ether shows a band A which is a triplet and a band B which is a quartet. What is the area of band A relative to band B?

† G. K. Fraenkel, J. M. Hirshon, and C. Walling, *J. Am. Chem. Soc.*, **76**:3606 (1954).
‡ J. J. Rooney and R. C. Pink, *Proc. Chem. Soc.*, **1961**:70.

13-3 The fundamental processes responsible for nmr and ultraviolet and visible spectra have certain similarities and differences. Outline them by answering the following questions. (*a*) What is responsible for the energy-level separations in each case? (*b*) What is the magnitude of the energy-level separations in each case? (*c*) What is the relative population level of the lower and higher energy levels in each case? (*d*) What are typical lifetimes of the excited state in each case, and what, if any, effect is there on spectral line widths? (*e*) How can energy-level separations be modified experimentally?

13-4 In proton nmr spectroscopy, typical relaxation times are in the order of 1 s.
(*a*) Estimate the theoretical line width (in frequency units) to be expected in proton nmr. *Ans:* 0.1 Hz
(*b*) By referring to various nmr spectra, estimate typical line widths (in frequency units) experimentally observed, and explain any difference from the theoretical value calculated in part (*a*).
(*c*) Express the line widths found in parts (*a*) and (*b*) in units of magnetic field strength, e.g., milligauss. *Ans:* for part (*a*) approximately 0.02 mG

13-5 Calculate the frequency of radiation required to interact with ^{19}F nuclei in a field of 14,092 G. *Ans:* 56.446 MHz

13-6 Calculate the magnetic field strength required to give resonance absorption by ^{31}P nuclei using a 100-MHz nmr spectrometer.

13-7 Sketch the approximate shape of proton magnetic resonance spectra (show number of peaks and their sharpness) to be expected for (*a*) NH_3, where N is the ^{15}N isotope ($I = \frac{1}{2}$); (*b*) NH_3, where N is ^{14}N ($I = 1$).

13-8 Derive the relationship between magnetic moment, spin number, and magnetogyric ratio [Eq. (13-3)] using the Larmor equation and Eq. (13-2).

13-9 Calculate the magnetogyric ratio for ^{11}B nuclei, and use this to calculate the frequency of radiation that would be required for resonance with ^{11}B in a magnetic field of 10 kG. *Ans:* 13.66 MHz

13-10 Using the Boltzmann distribution law, calculate the relative population of high- and low-energy rotational states (*a*) at room temperature and (*b*) at 1000 K. Assume a typical rotational energy-level spacing of 1.24×10^{-4} eV. *Ans:* (*a*) 0.9952; (*b*) 0.99855

13-11 Some types of bonds give *shielding* along the axis of the bond and *deshielding* perpendicular to the bond axis, whereas with other types of bonds the effects are reversed. Summarize the shielding-deshielding effects for (*a*) double bonds, (*b*) triple bonds, and (*c*) conjugated double bonds in an aromatic ring.

13-12 The inductive effect has been used to explain the relative chemical shifts of methane and its mono-, di-, and trisubstituted phenyl derivatives (see Example 13-10). How else can these shifts be explained? Which effect is probably the larger contributor?

Ans: Deshielding due to phenyl ring currents. Ring currents are undoubtedly the most significant effect, as evidenced, for example, by the very large deshielding of protons on the benzene ring ($2.7\,\tau$).

13-13 In the external-lock stabilization method, why is it desirable to use a sideband oscillator rather than simply lock onto the reference sample with the main transmitter frequency?

13-14 The proton nmr spectrum of ethyl acetate is given in Fig. 13-45. (*a*) Assign each of the bands; (*b*) classify the protons in ethyl acetate according to the AB_2X system.

13-15 (*a*) Draw the line spectrum expected for 1,1-dibromoethane, $CHBr_2CH_3$. (*b*) Classify the protons in 1,1-dibromoethane according to the ABX convention. *Ans:* A_3X

13-16 Classify each of the following molecules according to the type of spin pattern (AB_2X, etc.) it should have (note that ^{19}F has a spin of ½, like ^1H). (*a*) CH_2O, (*b*) CH_3F, (*c*) $CH_3CH_2CH_3$, (*d*) $CH_2{=}CHCl$, (*e*) $CH_2{=}CF_2$, (*f*) C_6H_5F.

13-17 A certain organic hydrocarbon contains ^{19}F. (*a*) Do you expect an absorption peak due to ^{19}F in the proton nmr spectrum obtained with a Varian

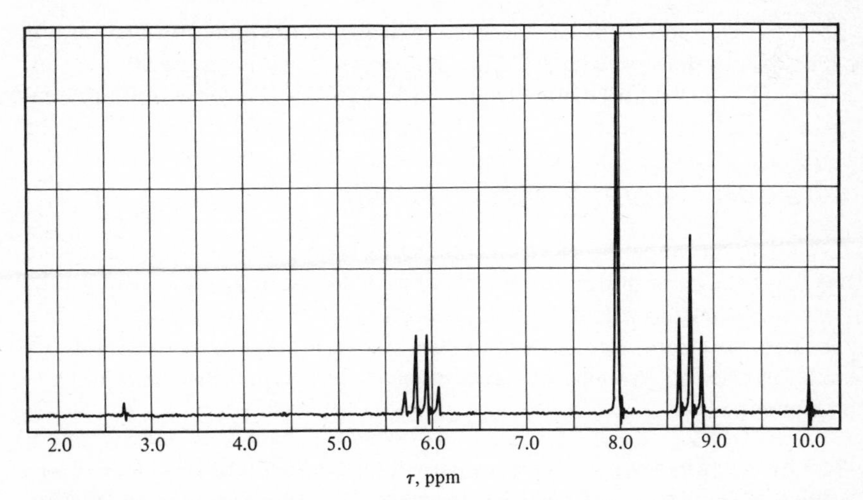

FIGURE 13-45 Proton nmr spectrum of ethyl acetate. (*From Varian Associates, "High Resolution NMR Spectra Catalogue," spectrum* 79, *National Press, Palo Alto. Calif.*, 1962, 1963, *by permission.*)

model A-60 spectrometer? (*b*) Do you expect that the presence of ^{19}F will have any effect on the proton spectrum?

13-18 A certain compound is obtained in very poor yield (though high purity) with the result that the nmr spectrum obtained with a Varian A-60 has peaks too low in intensity to interpret adequately. What can be done to increase the signal-to-noise ratio?

13-19 In wide-line nmr spectroscopy, a typical line width is 1 G. Estimate the degree of homogeneity a magnetic field should have for wide-line proton resonance studies at 60 MHz. *Ans:* about 30 ppm or better

13-20 From the weight per volume percent sample composition typically used in nmr experiments, estimate the molar concentration range for which nmr is sensitive. *Ans:* about 0.1 to 2 *M*

13-21 Predict the type of esr spectrum (number of lines and relative intensities) to be obtained for *p*-benzosemiquinone ion

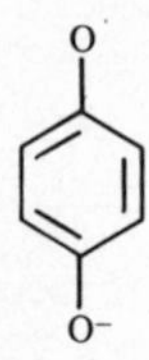

13-22 The esr spectrum of the free-radical intermediate formed in the condensation of diacetyl with 2,5-dimethylbenzoquinone is essentially a septet, with second-order splitting into triplets (see Fig. 13-42). Draw the approximate line spectrum you would expect to get for the radical intermediate if the unpaired electron intereacted mainly with the *ring* protons and less with the methyl protons, instead of the other way around.

13-23 Given an unpaired electron interacting with a single proton, draw an energy-level diagram depicting the splitting of the electron spin states by the nuclear spin, and show how the energy levels are affected by a magnetic field which increases from zero to some value H. Assuming that a constant frequency ν is used to irradiate this system, show how a multiplet esr spectrum will be obtained.

13-24 In esr spectroscopy, typical line widths are theoretically of the order of 1 to 10 MHz (see Example 13-27). Express these line widths in units of applied field (gauss). Check your answer against typical esr spectra, and explain any differences.

REFERENCES

MAGNETIC RESONANCE

Introductory

1 SCHWARZ, J. C. P. (ed.): "Physical Methods in Organic Chemistry," Oliver & Boyd, Edinburgh, 1964.

2 BARROW, G. M.: "Physical Chemistry," 2d ed., McGraw-Hill, New York, 1966.

3 WHIFFEN, D. H.: "Spectroscopy," Wiley, New York, 1966.

4 DAVIS, J. C., JR.: "Advanced Physical Chemistry," Ronald Press, New York, 1965.

5 VARIAN STAFF: "NMR and EPR Spectroscopy," Pergamon, New York, 1960.

6 CHAMBERLAIN, N. F.: Nuclear Magnetic Resonance and Electron Paramagnetic Resonance, chap. 39 in I. M. Kolthoff, P. J. Elving, and E. P. Sandell (eds.), "Treatise on Analytical Chemistry," pt. I, vol. 4, Wiley-Interscience, New York, 1963.

Intermediate and Advanced

7 WALKER, S., and H. STRAW: "Spectroscopy," vol. 1, Chapman & Hall, London, 1961 (reprinted with amendments, 1966). Chapters 4 and 5 treat esr and nmr, respectively.

8 NACHOD, F. C., and W. D. PHILLIPS: "Determination of Organic Structures by Physical Methods," vol. 2, Academic, New York, 1962.

9 CARRINGTON, A., and A. D. MCLACHLAN: "Introduction to Magnetic Resonance," Harper & Row, New York, 1967.

NMR

10 BIBLE, R. H., JR.: "Interpretation of NMR Spectra," Plenum, New York, 1965.

11 BIBLE, R. H., JR.: "Guide to the NMR Empirical Method: A Workbook," Plenum, New York, 1967.

12 JACKMAN, L. M.: "Applications of Nuclear Magnetic Resonance Spectroscopy in Organic Chemistry," Pergamon, New York, 1959.

13 MATHIESON, D. W. (ed.): "Nuclear Magnetic Resonance for Organic Chemists," Academic, New York, 1967.

14 MATHIESON, D. W. (ed.): "Interpretation of Organic Spectra," Academic, New York, 1965.

15 SILVERSTEIN, R. M., and G. C. BASSLER: "Spectrometric Identification of Organic Compounds," 2d ed., Wiley, New York, 1967.

16 FREEMAN, S. K. (ed.): "Interpretive Spectroscopy," Reinhold, New York, 1965.

17 BRAND, J. C. D., and G. EGLINTON: "Applications of Spectroscopy to Organic Chemistry," Oldbourne, London, 1965.

18 BOVEY, F. A.: "Nuclear Magnetic Resonance Spectroscopy," Academic, New York, 1969.
19 POPLE, J. A., W. G. SCHNEIDER, and H. J. BERNSTEIN: "High Resolution Nuclear Magnetic Resonance," McGraw-Hill, New York, 1959.

Spectra Catalogs

20 VARIAN ASSOCIATES: "High Resolution NMR Spectra Catalogue," National Press, Palo Alto, Calif., 1962, 1963.
21 WIBERG, K. B., and B. J. NIST: "The Interpretation of NMR Spectra," Benjamin, New York, 1962. A computerized printout of spectra.

ESR

22 SQUIRE, T. L.: "An Introduction to Electron Spin Resonance," Academic, New York, 1964.
23 ASSENHEIM, H. M.: "Introduction to Electron Spin Resonance," Plenum, New York, 1967.
24 Ayscough, P. B.: "Electron Spin Resonance in Chemistry," Methuen, London, 1967.
25 WILMHURST, T. H.: "Electron Spin Resonance Spectrometers," Plenum, New York, 1967.
26 BERSOHN, M., and J. C. BAIRD: "An Introduction to Electron Paramagnetic Resonance," Benjamin, New York, 1966.
27 INGRAM, D. J. E.: "Free Radicals as Studied by Electron Spin Resonance," Butterworths, London, 1958.
28 POOLE, C. P., JR.: "Electron Spin Resonance," Wiley-Interscience, New York, 1967.

14

Gamma-Ray Spectroscopy and Mössbauer Spectroscopy

Gamma-ray spectroscopy and Mössbauer spectroscopy have in common that gamma rays are being measured, but the techniques and applications of the two methods differ. Gamma-ray spectroscopy generally involves a liquid or solid scintillator as the *detector,* and indeed another common name for gamma-ray spectroscopy is scintillation spectrometry. Gamma-ray spectroscopy is used for qualitative and quantitative analysis of radioactive samples in a way that is similar to emission spectrophotometry in other regions of the electromagnetic spectrum. Qualitative analysis is accomplished by measuring the energy (or wavelength) of the emitted gamma rays, and quantitative analysis is accomplished by measuring the intensity (usually number of counts per minute) of the gamma ray of interest.

On the other hand, in Mössbauer spectroscopy a gamma-emitting isotope is used as a *source* in a Mössbauer spectrometer, completely analogous to the use of a hollow-cathode tube as a source in an atomic absorption spectrometer (see Sec. 5-1). The absorption of the monochromatic gamma ray by a sample containing the same element as the emitting isotope is measured. Mössbauer spectroscopy often gives the same type of molecular information as magnetic resonance spectroscopy, while in other cases, e.g., those dealing with electric quadrupole interactions, it provides a unique type of information. The most important applications of Mössbauer spectroscopy thus far have been in iron chemistry, and this has been of vital importance, both in metallurgy and in organometallic chemistry.

14-1 GAMMA-RAY SPECTROSCOPY

Although the main topic of this section is gamma-ray spectroscopy, the pertinent characteristics of alpha, beta, and gamma radiation will first be reviewed so that the effects of one type of radiation on the measurement of another can be understood; this will also serve to clarify why gamma rays are the most advantageous of the three types for measurement. A short discussion of beta-ray spectroscopy with liquid scintillators is included, since this type of counting is related to gamma-ray spectroscopy and is of vital importance for biological studies, where the low-energy betas of ^{3}H and ^{14}C are of great interest.

14-1A Theory

Alpha particles are doubly ionized helium particles ejected with very high energy from the nuclei of certain radioactive isotopes. In general, all heavy radioactive elements above bismuth in the periodic table are alpha emitters, and there are a few other alpha emitters scattered through the periodic table (mainly the heavier elements). It is important to understand that alpha particles have only slight penetrating power through matter, being stopped by paper-thin sheets of solid materials and penetrating only a few centimeters (5 to 7 cm) of air. Their energies, however, are generally very high, often of the order of 5 or 6 MeV and sometimes in excess of 10 MeV. As a result of its relatively high mass (4), high charge (+2), and high energy, the ionizing power of an alpha particle is very great, and large numbers of ion pairs are produced along the linear path traversed by an alpha particle. Because of the high ionizing power, alphas are relatively easy to distinguish from betas and gammas using a proportional counter and measuring the pulse amplitude. Or alpha rays can simply be excluded from entering a given counting device by using a thin absorber, e.g., a thin aluminum sheet, while allowing the betas or gammas to pass through with little or no attenuation. In a typical gamma-ray spectrometer, this means of eliminating interference from alpha radiation is generally used; i.e., the alphas are prevented from entering the scintillation crystal by a thin aluminum cover on the scintillation crystal.

Beta particles are energetic electrons emitted from certain unstable nuclei. The most common type of beta particle is negatively charged, but a few radioactive nuclei emit positively charged electrons known as *positrons*. Positrons are not observed as such, since after leaving the nucleus they quickly interact with one of the extranuclear electrons; the resulting *annihilation* process results in the creation of gamma radiation of about 0.51 MeV. Thus, the observation of gamma rays with an energy of 0.51 MeV is indirect evidence of positron emission by the nucleus.

The most important thing to understand about negative† beta particles is

† Normally, the adjective "negative" is not used in referring to the usual kind of beta particle.

that they are emitted from the nucleus with a *continuous* spread of energies, in contrast to the *discrete* energy with which alphas and gammas are emitted. Figure 14-1 gives a typical beta-particle spectrum. A detailed explanation of why a beta spectrum is a continuum where other emissions give line spectra is beyond the scope of this book. When beta particles are emitted, a *neutrino* is emitted simultaneously, and the neutrino, with zero charge and very small mass, carries off random amounts of kinetic energy, leaving a collection of betas with a random distribution of energy. The *maximum* energy of the ejected betas (identified on Fig. 14-1) is characteristic of the emitting substance and thus can be used for qualitative analysis, but the wide bands of beta energy makes beta-ray spectrometry (the differentiation of a mixture of betas) difficult. Generally in counting betas one must be satisfied with the gross beta activity, instead of attempting selectively to count one type of beta emitter in the presence of a mixture of others. Where the E_{max} of two or more betas differ greatly, it may be possible to use variable amounts of shielding or absorbers to increase the selectivity. Specialized references on nuclear counters should be consulted for details.

Gamma rays are high-energy photons emitted by excited nuclei of atoms. When the excited nucleus returns to the ground state, the difference in energy is given off as gamma radiation. Since only definite, discrete energy levels are possible in the nucleus, gamma emissions have discrete, well-defined energies. The penetrating power of gamma radiation is much greater than that of either alpha or beta particles, but the ionizing power is less.

It is possible, with a scintillation spectrometer, to select conditions so as to count alpha, beta, or gamma radiation. However, counting alpha radiation has only limited application, since relatively few isotopes are alpha emitters and the penetrating power of alpha radiation is very poor. Similarly, the counting of beta radiation is relatively limited, except for liquid-scintillation counting, where only one or two beta emitters are present, e.g., biological studies with ^{3}H- or ^{14}C-tagged compounds. But scintillation spectrometry is

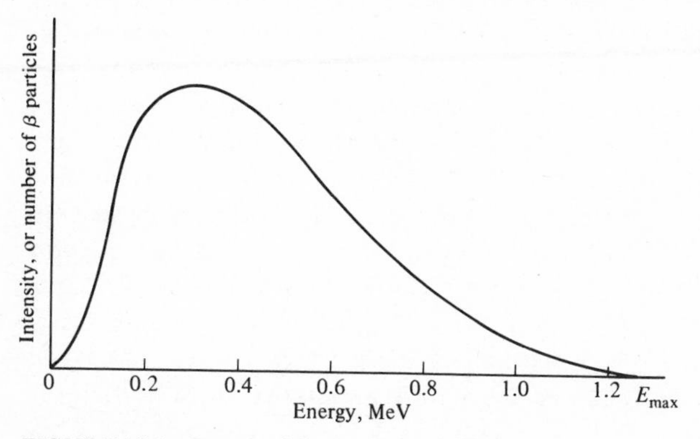

FIGURE 14-1 Beta-particle spectrum for ^{210}Bi.

ideally suited for counting gamma rays, since scintillator materials are available for detecting gamma rays with very high efficiency, and the discrete-line spectra of gamma radiation allows mixtures of gamma emitters to be analyzed. A gamma-ray spectrum allows qualitative and quantitative analysis to be performed, since the energy at which a peak occurs is characteristic of the isotope which emitted the gamma ray and the intensity of the peak is proportional to the concentration of that isotope.

14-1B Instrumentation

A block diagram of a single-channel scintillation spectrometer is shown in Fig. 14-2. The instrument diagramed is considered a single-channel analyzer because it measures only one energy region of the spectrum at a time and has only one scaler or recorder. Multichannel spectrometers simultaneously monitor more than one energy region of the spectrum and require multiple scalers or recorders for readout.

To understand how the scintillation spectrometer works, assume that a sample of some radioactive isotope which emits only one energy of gamma radiation is placed in the test-tube well of the scintillation crystal. As each gamma ray enters the scintillation crystal, the crystal scintillates, or produces visible-light photons; the number of photons produced is directly proportional to the energy of the incoming gamma radiation. For example, a 1-MeV gamma might produce 1000 photons, whereas a 2-MeV gamma would produce 2000 photons.† These photons then strike the *photocathode* of the

† Actually, for a NaI thallium-activated scintillation crystal, a 1-MeV gamma will produce about 17,000 photons, with energy in the blue region of the visible spectrum.

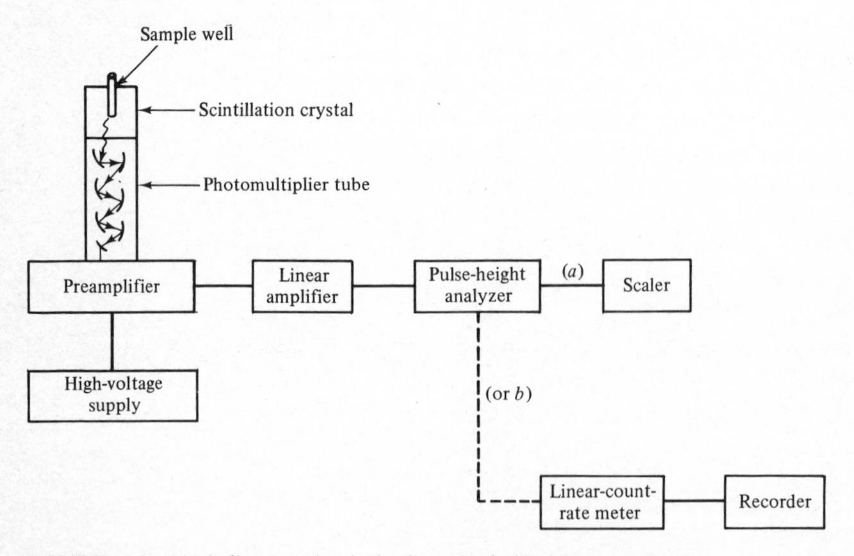

FIGURE 14-2 Block diagram of a single-channel scintillation spectrometer.

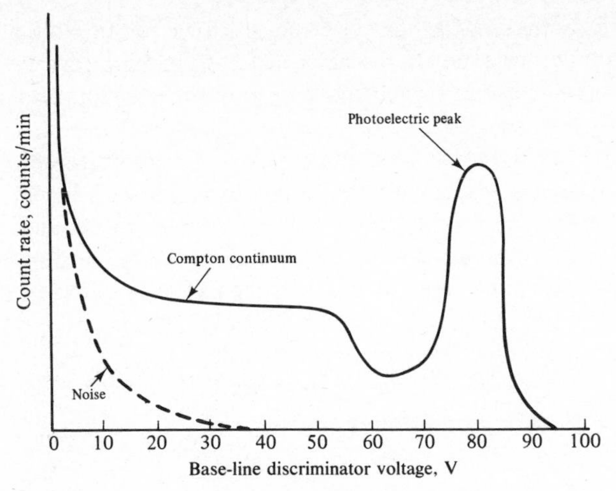

FIGURE 14-3 Gamma-ray spectrum for a typical gamma-emitting isotope (differential scan).

photomultiplier tube, and a certain proportion of those photons (about 10 percent) eject electrons at the photocathode. The ejected electrons are accelerated and multiplied in number as they move from dynode to dynode through the photomultiplier, and the output is about 1 million times as many electrons as entered (see Sec. 2-2*B*). The electrons leaving the photomultiplier, which represent an electric current, are fed through a load resistor, and the voltage drop across the resistor represents a voltage pulse that is directly proportional to the energy of the original gamma ray.

The voltage pulse produced at the input of the *preamplifier* circuit (see Fig. 14-2) is then amplified by a fixed amount in the preamplifier and *linear amplifier* and fed into a discriminator circuit in the *pulse-height analyzer,* where only a narrow range in size of voltage pulses is accepted. If the voltage pulse is of the proper size to be measured, the *scaler* is tripped or the pulse activates a *count-rate meter* and *recorder.*

To obtain a gamma-ray spectrum with a single-channel scintillation spectrometer it is first necessary to set the *high-voltage supply* and *gain* of the linear amplifier to fixed, constant values.† Next the *voltage-acceptance slit* in the pulse-height analyzer is set at a constant value, e.g., a width of 1 V. (This voltage slit width is often referred to as the voltage *window.*) Finally, the gamma-ray spectrum is scanned by continuous or stepwise changing of the lower-voltage bias in the pulse-height analyzer. The lower-voltage bias is the lowest voltage pulse that the discriminator circuit in the pulse-height analyzer will accept.

Figure 14-3 shows the type of spectrum which would be observed for a typical gamma-emitting isotope. The abscissa in Fig. 14-3 is given in units of

† The high-voltage supply determines the gain of the photomultiplier tube.

base-line discriminator voltage and represents a sweep of the lower-voltage bias from 0 to 100 V, in a continuous scan. One or two isotopes with gamma rays of known energy (in megaelectronvolts) can be used to calibrate the base line into energy units.

Figure 14-3 represents a *differential scan,* since only a narrow region of the spectrum (that which passes through the voltage slit, or window) is being monitored at one time. Conversely, it is possible to open the voltage slit and monitor all signals above a certain minimum, i.e., all voltage pulses exceeding the lower-voltage bias, with no upper bias at all, in which case an *integral* spectrum like that shown in Fig. 14-4 is obtained.† Differential spectra are much more useful, in general, being completely analogous to the type of spectra observed in other regions of the electromagnetic spectrum, but in certain counting situations it may be advantageous to remove the voltage window and observe the gross (or integral) count rate above a certain minimum voltage value.

To understand a gamma-ray spectrum it is useful to divide the differential spectrum given in Fig. 14-3 into two characteristic parts. The sharp symmetrical peak appearing in Fig. 14-3 at 70 to 90 V is known as the *photoelectric* or *full-energy peak.* The continuous curve appearing at lower voltages is mainly due to Compton interactions in the scintillation crystal, and is referred to as the *Compton continuum.*

In the *photoelectric effect,* an incoming gamma ray strikes an orbital electron in the scintillation crystal head on and knocks the electron out of its

† The relationship of Figs. 14-3 and 14-4 to each other should be evident from the calculus; if the curve in Fig. 14-4 is differentiated, Fig. 14-3 will result; conversely, if Fig. 14-3 is integrated, Fig. 14-4 will result.

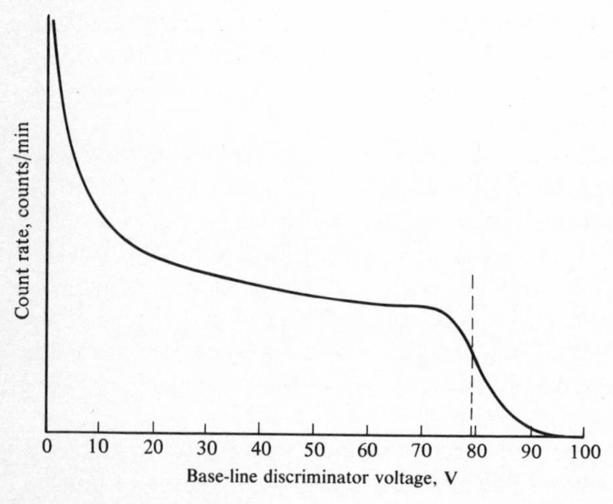

FIGURE 14-4 Integral gramma-ray spectrum for a typical gamma-emitting isotope.

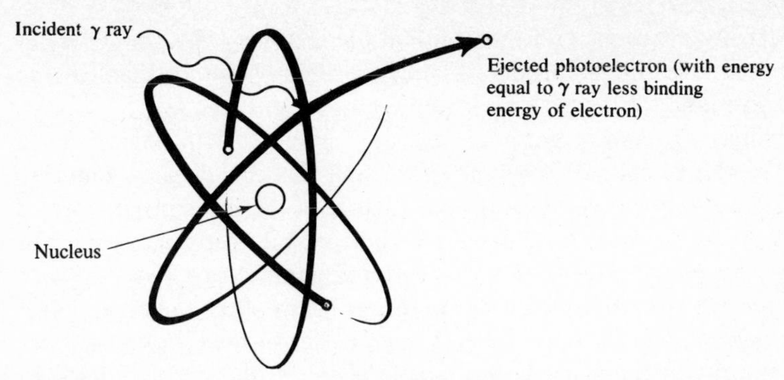

FIGURE 14-5 Simplified representation of the interaction between a gamma ray and an orbital electron in the photoelectric effect.

orbital, and all the energy of the gamma ray is given to the electron. Figure 14-5 gives a simplified picture of the interaction that takes place in the photoelectric effect. The ejected photoelectron travels through the scintillation crystal and causes ionization and excitation, producing visible photons. The greater the energy of the photoelectron, the further it can travel before its energy is all used up and the more light photons will be produced (see Fig. 14-6). The greater the number of light photons formed, the larger the voltage pulse produced at the output of the photomultiplier tube will be. Therefore,

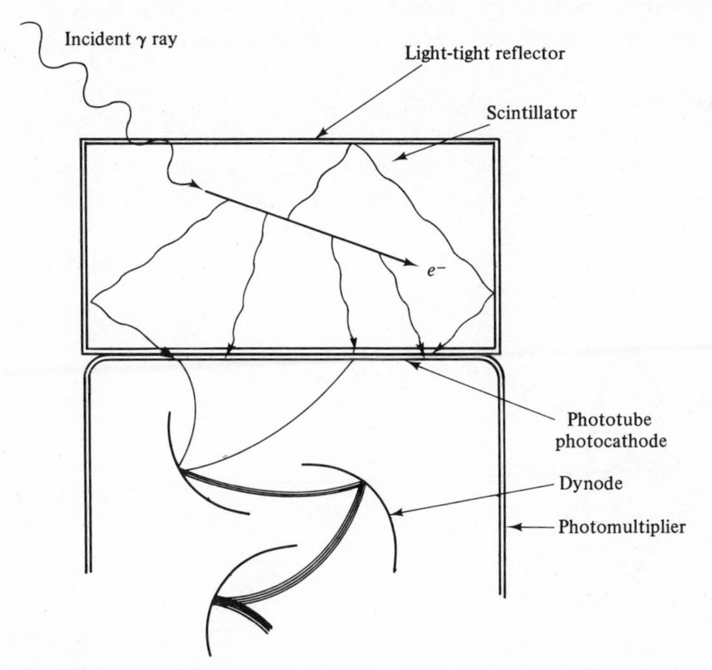

FIGURE 14-6 Schematic representation of the production of light photons in a scintillation detector.

when a gamma ray interacts with the scintillation crystal by the photoelectric effect, a voltage pulse will be produced which is directly proportional to the energy of the gamma.

Unfortunately, however, gamma rays can interact with matter in a second way, by the so-called *Compton effect,* and this complicates counting the gammas. Figure 14-7 gives a simplified picture of the interaction. Here a gamma ray can be pictured as glancing off an orbital electron, giving the electron only *part* of its energy, and the scattered gamma ray, now of lower energy, may go on to interact with other atoms or (more often the case) travel *out* of the scintillation crystal and be lost. Just as the ejected *photoelectrons* cause light photons in the scintillation crystal (Fig. 14-6), so do the ejected *Compton* electrons produce light photons; but since the Compton electrons have a wide range of energies (all lower than the energy of the original gamma), the Compton effect results in the wide range of voltage pulses that make up the Compton continuum. The minimum in the Compton continuum just below the photoelectric peak is predicted by quantum theory as an edge where Compton interactions stop and beyond which the only interaction that can take place, if any, is a direct (photoelectric) hit on the electron. In general, the Compton continuum is an undesirable background and must be corrected for in quantitative measurements on mixtures of gamma-emitting isotopes. An additional contributor to the background is thermal-electron *noise,* caused by thermal electrons spontaneously being emitted from the photocathode of the photomultiplier tube. This thermal-electron noise becomes serious only when gammas of very low energy are being measured.

From a detailed theory of the photoelectric effect and Compton effect it would be predicted that an ideal gamma-ray detector would give a gamma-ray spectrum like that in Fig. 14-8 for an isotope emitting a gamma ray of only one energy. In other words, the photoelectric peak would be a sharp line (rather than a symmetrical band), and the Compton continuum would gradually increase in intensity up to an edge, after which no voltage pulse

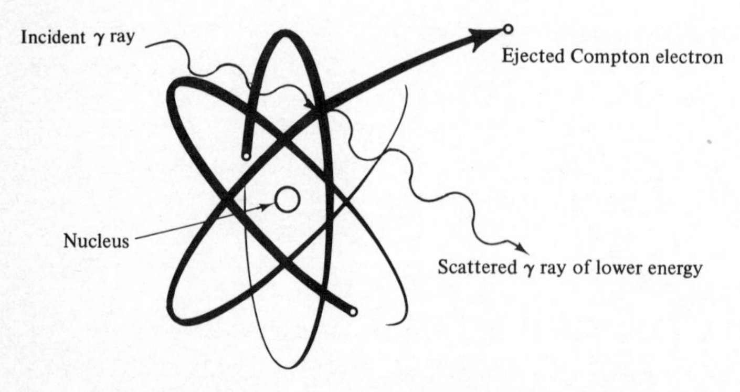

FIGURE 14-7 Simplified representation of the interaction between a gamma ray and an orbital electron in the Compton effect.

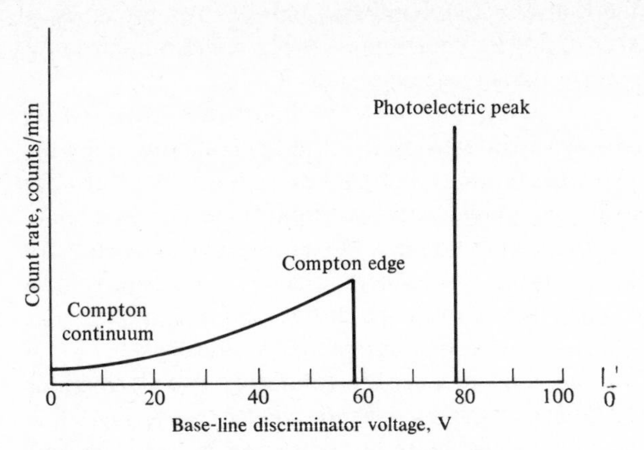

FIGURE 14-8 Idealized gamma-ray spectrum for an isotope emitting a gamma ray of one energy (ideal detector system).

would be observed until the photoelectric peak is reached. The reason why an actual, observed spectrum is more like Fig. 14-3 than Fig. 14-8 has to do with inherent features of the instrumentation used for the measurement.

The photoelectric peak is always observed as a symmetrical band rather than as a sharp line, for two reasons. (1) Only a fraction of the light photons produced in the scintillation crystal actually result in electrons being ejected from the photocathode of the photomultiplier tube (see Fig. 14-6). As a rough rule of thumb, about 10 percent of the photons striking the photocathode eject photoelectrons, and in fact the percentage fluctuates *randomly* about the median percentage. It follows that this random fluctuation in the efficiency of ejecting electrons in the photomultiplier will manifest itself in a random range of voltage pulses at the output of the photomultiplier tube. This process thus contributes to a spread of voltage pulses about the median or true voltage of the photoelectric peak, and the spread is symmetrical, fitting a normal error curve as expected for a random or statistical process. (2) Small fluctuations occur in the high-voltage supply which controls the gain of the photomultiplier tube and of the preamplifier and linear amplifer (see Fig. 14-2). The high-voltage supply must be engineered to be exceptionally stable, since even small fluctuations in the voltage at the various dynodes of the photomultiplier tube will greatly affect the gain, thus causing a variation in the output voltage of the photomultiplier. A good spectrometer has a high-voltage supply that is stable to within 0.1 percent or less over a 24-h period, but even the smallest fluctuations in the high-voltage supply contribute to the width of the photoelectric peak.

The Compton continuum should theoretically be more like that shown in Fig. 14-8 than that in Fig. 14-3. The differences in the background radiation in the two spectra are due mainly to noise in the photomultiplier tube. Even at room temperature a photocathode will spontaneously emit small but measurable numbers of thermal electrons, and this noise shows up as background pulses. Thermal-electron noise contributes to the background mainly at low discriminator voltages (see the dotted noise curve in Fig. 14-3).

Additional contributions to background are *backscattered radiation,* or gamma radiation which has been scattered by materials surrounding the scintillation crystal, and *bremsstrahlung,* or x- and gamma radiation produced by the deceleration of beta particles coming from the sample or surrounding areas. Neither backscattering nor bremsstrahlung is sufficiently important to merit further discussion here, as long as it is remembered that the background must be kept constant and reproducible and must be corrected for if quantitative analyses are to be performed on mixtures of gamma-emitted isotopes.

To conclude the discussion of instrumentation, two remaining components, the *linear amplifier* and the *pulse-height analyzer* (see Fig. 14-2) must be discussed. The *linear amplifier* must be stable and have a relatively high and variable gain; most important, the gain must be *linear* over a wide range of voltage pulses. The degree of linearity of the amplifier will determine the linearity of the abscissa of the gamma-ray spectrum. If the amplifier is sufficiently linear, it may be possible to calibrate the base line with only one gamma ray of known energy and determine other gamma energies by interpolation. On the other hand, if the amplifier is nonlinear, it may be necessary to use a number of gamma emitters of known energy to calibrate the base line; interpolation between known energy points may then result in significant errors.

The *pulse-height analyzer* is the monochromator of a gamma-ray spectrometer. Without detailing the electronics involved, it will suffice to say that the analyzer consists of two electronic discriminator circuits, which in turn feed an anticoincidence circuit. The block diagram in Fig. 14-9*a* illustrates how the circuits are connected. Each discriminator circuit is biased to block any signal pulse from the linear amplifier that has an amplitude below a set minimum height but will pass all others. For example, suppose the lower-level discriminator is set to pass pulses of amplitude greater than 30 V. If a 10-V channel (window) is wanted, the upper-level discriminator would be set at 40 V (see Fig. 14-9). The anticoincidence circuit which follows the two discriminator circuits will transmit a pulse *only* if a pulse gets through *one* of the discriminator circuits but not the other. If the anticoincidence circuit sees the same kind of signal coming from *both* discriminator circuits, it will block it, and no signal will be transmitted out of the pulse-height analyzer.

To understand how the pulse-height analyzer works, consider, one at a time, three different-sized voltage pulses entering a pulse-height analyzer. First, consider a 25-V input signal, as shown in Fig. 14-9*a*. Since this voltage pulse is too small to get through either discriminator circuit, both discriminator circuits block the signal, the anticoincidence circuit sees the same kind of signal (which may be thought of as two negative signals) from each discriminator circuit, and no signal is transmitted beyond the anticoincidence circuit.

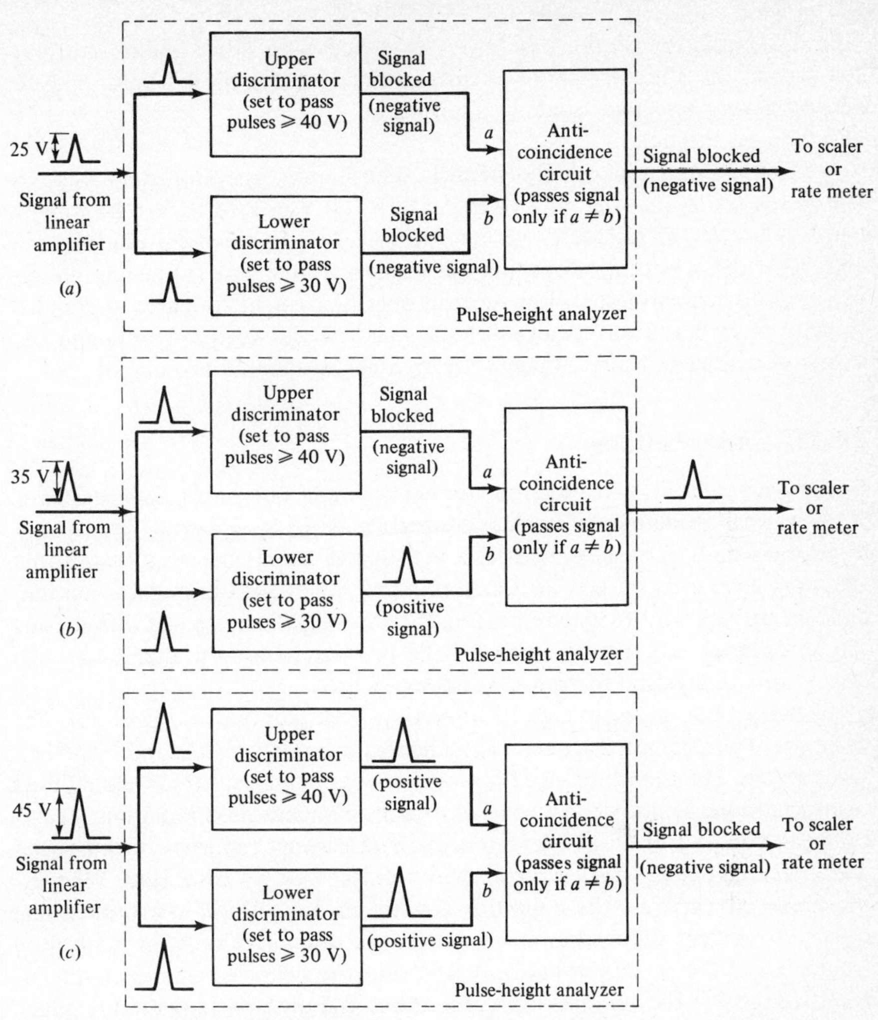

FIGURE 14-9 Schematic representation of a pulse-height analyzer: (*a*) 25-V input signal; (*b*) 35-V input signal; (*c*) 45-V input signal.

However, now consider the case of a 35-V signal entering the pulse-height analyzer (see Fig. 14-9*b*). This voltage pulse will pass through the lower-level discriminator but will be blocked by the upper-level discriminator. The anticoincidence circuit now sees a pulse coming from only *one* of the discriminator circuits (in other words, a plus signal from one discriminator and a negative signal from the other). These out-of-phase signals are permitted to pass through the anticoincidence circuit, and a pulse is transmitted to the scaler or count-rate meter. Finally, consider a 45-V signal entering the pulse-height analyzer. This voltage pulse will pass through *both* discriminators, but since the anticoincidence circuit sees the same kind of signal coming from both discriminators (in this case, two plus signals), no signal will be transmitted beyond the anticoincidence circuit. In this way the pulse-height analyzer acts as a very selective window, allowing only pulses of a preselected magnitude to get through.

Most commercial single-channel scintillation spectrometers amplify voltage pulses so that the range is either 0 to 100 V or 0 to 10 V. The window width is usually continuously variable from 0 to 10 percent of the full-scale voltage, with an option of opening the window all the way (removing the upper-level discrminator) to allow integral operation. A high degree of stability is required in the discriminators of the pulse-height analyzer, although this stability is easier to achieve than in the high-voltage supply and amplifiers.

14-1C Applications

Gamma-ray scintillation spectrometers have found numerous applications in studying such fundamental nuclear properties as decay schemes, interactions of photons and high-energy particles with matter, and shielding studies, but the most widespread use is as a counting tool when both high detection efficiency and high selectivity are needed. This is best illustrated by considering the advantages and disadvantages of the techniques for the qualitative and quantitative analysis of mixtures of radioisotopes.

Figure 14-10 gives a typical three-component gamma-ray spectrum. The three sharp photopeaks are due to the gammas from ^{51}Cr, ^{103}Ru, and ^{95}Zr-^{95}Nb, respectively. The peak labeled "Pb x-ray" results from the lead shielding used to surround the scintillation crystal to minimize background radiation. The x-ray peak results from high-energy cosmic and gamma radiation striking lead and knocking out K-shell electrons, with resultant x-rays from lead. Whereas the observed curve is the solid-line tracing in Fig. 14-10, the broken-line tracings represent the background contribution (mainly Compton scattering) from each of the three gamma rays, and thus the observed tracing is the cumulative total of the gammas. The procedures and problems involved in qualitative and quantitative analysis in gamma-ray scintillation spectrometry may be understood by reference to Fig. 14-10.

In *qualitative analysis,* the number of different photopeaks that can be clearly identified depends on ρ, the resolution, defined in Fig. 14-10 as the width of a photopeak at half its maximum height. The magnitude of ρ depends on the particular instrument being used and its operating parameters and also on where in the energy spectrum the peak appears, as emphasized by

$$\rho = k(E_\gamma)^{1/2} \tag{14-1}$$

where E_γ is the energy of the gamma ray and k is a proportionality constant. The value of k depends on many variables, such as the size of the scintillation

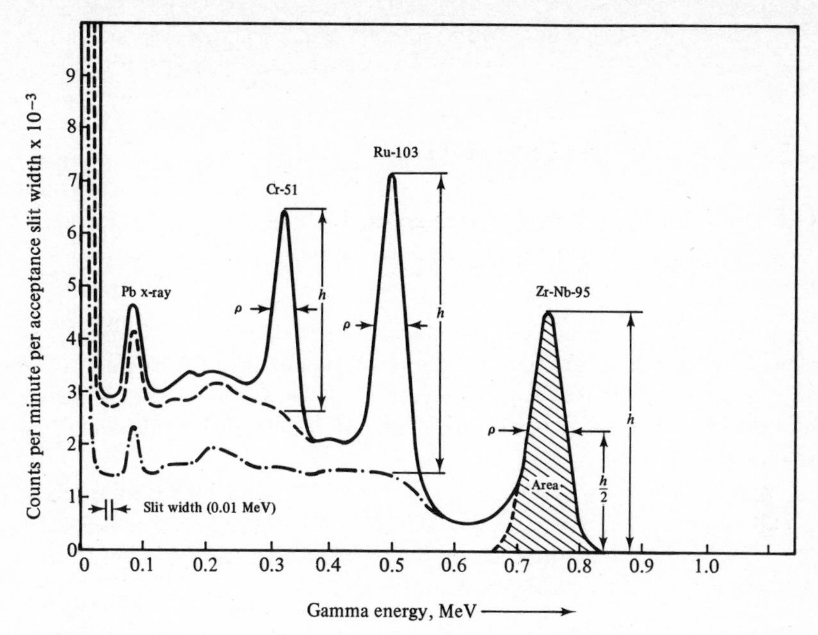

FIGURE 14-10 Typical three-component gamma-ray spectrum. Area = 1.07 ρh, where ρ = peak width at one-half maximum in units of slit width. [*Reprinted with permission from R. E. Connally, Anal. Chem.*, **28**:1847 (1956). *Copyright by the American Chemical Society.*]

crystal, the stability of the high-voltage supply, etc., but the important thing is to keep these variables constant. The numerical value of k can be determined by experiment. The avantage of Eq. (14-1) is that once k has been determined for a given spectrometer, the resolution ρ that can be expected for any gamma energy can be calculated. As a practical rule of thumb, two gammas should differ in energy by more than ρ if it is hoped to clearly resolve and identify each gamma ray in a given gamma-ray spectrum.

Example 14-1 A certain gamma-ray scintillation spectrometer is calibrated with the 0.663-MeV gamma ray of ^{137}Cs. The width of the cesium photopeak at half the maximum peak height ρ is found to be 0.093 MeV. (*a*) Estimate the resolution ρ that the instrument will show at a gamma energy of 0.77 MeV. (*b*) Can this instrument be expected to resolve the photopeaks of ^{95}Zr and ^{95}Nb? (^{95}Zr has a 0.71-MeV gamma, and ^{95}Nb has a 0.77-MeV gamma.) (*c*) Can this instrument be expected to resolve the photopeaks of ^{95}Nb and ^{137}Cs?

Answer (*a*)

$$\rho = k(E_\gamma)^{1/2} \qquad (14\text{-}1)$$

For ^{137}Cs, $E_\gamma = 0.663$ MeV, and $\rho = 0.093$ MeV. Therefore,

$$0.093 = k(0.663)^{1/2}$$

$$k = \frac{0.093}{0.663^{1/2}} = \frac{0.093}{0.814} = 0.114$$

At $E_\gamma = 0.77$ MeV, the resolution expected will be

$$\rho = (0.114)(0.77)^{1/2} = 0.100 \text{ MeV}$$

(*b*) The difference in energy between the ^{95}Nb and ^{95}Zr gamma rays is 0.77 − 0.71 = 0.06 MeV. Since the resolution of the instrument at 0.77 MeV is only 0.100 MeV [part (*a*)], it is to be expected that the two photopeaks will overlap badly and will *not* be clearly resolved by this instrument.

(*c*) The difference in energy between the ^{95}Nb and ^{137}Cs gamma rays is 0.77 − 0.66 = 0.11 MeV. Since the resolution of the instrument is 0.10 MeV at 0.77 MeV [part (*a*)], it can be expected that the two photopeaks *will* be clearly resolved by this instrument.

For *quantitative analysis* it is the net area under each photopeak which is directly proportional to the concentration (or number of atoms) of the isotope emitting the gamma ray being measured. Since the photopeaks are very nearly gaussian in shape, it is possible to calculate the area very accurately from

$$\text{Area under photopeak} = 1.07\,\rho h \tag{14-2}$$

The factor 1.07 is a normalizing factor from gaussian statistics. To relate the area under a photopeak to the concentration of the element or isotope whose gamma ray is being measured it is necessary to take one or more known samples of each gamma emitter and experimentally determine the proportionality constant involved. The proportionality constant will be a characteristic of the instrument being used (size of the scintillation crystal, etc.) and will depend somewhat on the particular counting conditions, i.e., whether the sample is a solid or in solution, whether it is being counted in a test tube or a planchet, etc.

The biggest problem in quantitative analyses of mixtures of gamma-emitted isotopes is in accounting for the Compton background contribution from each istope. Since each full-energy (photoelectric) peak will have its own Compton background contribution at lower energies, it is necessary to start with the most energetic photopeak (the ^{95}Zr-^{95}Nb photopeak in Fig. 14-10), where the Compton background will be zero, and systematically subtract out its Compton contribution to the area of each photopeak appearing at lower energy. The Compton contribution at all energies below the photopeak can be determined as a fraction of the photopeak area simply by running a spectrum of each isotope by itself. Assuming that instrumental conditions remain constant, the relative Compton contribution will remain constant. Thus, in Fig. 14-10, after the ^{95}Zr-^{95}Nb Compton background is subtracted from the photopeak of ^{103}Ru, the area of the ^{103}Ru photopeak can be accurately calculated. Then the Compton contribution from both ^{103}Ru and ^{95}Zr-^{95}Nb must be subtracted from the ^{51}Cr photopeak so that its area can be determined. In this way, starting from the most energetic photopeak and working to lower and lower energy peaks, a very complex gamma-ray spectrum can be resolved and quantitative analyses performed.

The precision of measurement of each peak becomes less as successive Compton continuum corrections have to be made. In practice a five-component mixture is about the maximum which can be analyzed without the assistance of radioactive decay or chemical separation. The quantitative accuracy which can be expected is in the order of 1 to 2 percent for the most energetic photopeak, falling to roughly 10 to 20 percent for the third most energetic photopeak, depending of course on the relative amounts of the various isotopes. Thus, minor constituents falling at very low energies may be determinable only with order-of-magnitude (± 50 or 100 percent) accuracy. While this accuracy may seem intolerable, it must be remembered that the measurement of radioactivity is an extremely sensitive method of measurment, since every count is a direct measurement of an individual atom. Therefore, an order-of-magnitude estimate of an extremely low concentration of any substance must be viewed as a very successful achievement, particularly in view of the relative ease with which such measurements are made.

14-1D Liquid-Scintillation Counting

Liquid-scintillation counting has become vitally important in certain biological studies, where weak beta-emitting isotopes such as ^{3}H and ^{14}C are used as tracers. The theory of liquid-scintillation counting ties in very closely with gamma-ray scintillation spectrometry, but the instrumentation and experimental approach are different.

Where gamma-ray scintillation counting involves placing a solid, liquid, or gaseous sample, suitably contained, *next to* a solid scintillation crystal, in liquid-scintillation counting the compound containing the radionuclide must be *dissolved in* a liquid scintillator because very weak betas are largely unable to penetrate the walls of even the thinnest-walled containers and those which do get through would probably be scattered at the surface of a scintillation crystal. Conventionally, the radioactive compound is dissolved in toluene or xylene which contains 1 to 10 g/l of a primary scintillator plus 0.1 g/l of a secondary scintillator (the latter serving to increase the pulse height, usually by acting as a wavelength shifter). Combinations frequently used are *p*-

terphenyl or 2,5-diphenyloxazole (PPO) plus 1,4-bis-[2-(5-phenyloxazolyl)]-benzene (POPOP), or *p*-terphenyl plus 1,6-diphenyl-1,3,5-hexatriene (DPH). Aqueous solutions are counted in alcohol- or dioxane-based "cocktails" to facilitate solubilization of the sample.

A formidable problem in counting of low-energy beta radiation is competition and interference from thermal-electron noise in a photomultiplier tube. This problem is fairly serious with carbon 14, which has a 0.155-MeV beta, but it is critical with tritium, whose beta is only 0.0176 MeV. To minimize the interference from thermal electrons it is standard practice to use *two* photomultiplier detectors connected through a *coincidence* circuit. Since a given beta ray will produce a large number of light photons, all within a very short time, the two photomultipliers will see these photons simultaneously, whereas background pulses arising from thermal noise within each tube will be independently random. The coincidence circuit will reject the random noise pulses and accept only the coincident pulses, thereby selectively measuring the beta emissions and rejecting most of the noise. Until recently, it was also necessary to refrigerate the photomultiplier tubes to keep the noise pulses below the level that affects tritium counting, but some photomultiplier tubes no longer require refrigeration.

Finally, it is necessary to make a quench correction in liquid-scintillation counting of betas. Whereas in gamma-scintillation counting the magnitude of the voltage pulse generated by a photomultiplier tube is directly proportional to the energy of the gamma ray, resulting in constant counting efficiency, there are many more variables in liquid-scintillation counting of betas that make it impractical to maintain a constant proportionality between beta energy and magnitude of the voltage pulse generated. The many variables are collectively termed *quenching*. Causes of quenching include dilution, e.g., a variable quantity of primary or secondary fluor; color present in the sample cocktail (which may cause light photons to be absorbed, decreasing efficiency); chemical interferences, e.g., an oxidizing agent that may react with the fluor, decreasing efficiency; and self-absorption (where the sample is not dissolved in the cocktail, and the presence of two phases causes variable and unpredictable counting efficiency). There are various ways of preparing a quench-correction curve. One of the most common is to use a series of samples with known activity and take counts in two different channels while varying the amount of a quenching agent, e.g., chloroform. The quench curve is a plot of the measured efficiency (ratio of measured count rate to known disintegration rate) vs. the ratio of counts in the two channels (channels-ratio method). By measuring the ratio of counts in the two channels for the unknown sample, the efficiency of counting can be obtained from the quench curve. Newer instrumentation employs the automatic external standardization (AES) method almost exclusively for quench correction. The AES is a high-energy gamma source which can be positioned automatically (by means of a pneumatic system) close to the sample vials. The count rate obtained with the

AES depends on the degree of quenching in the sample. A quench-correction curve can be constructed as described for the channels-ratio technique. Neither the channels-ratio technique nor the AES technique will correct for quenching due to self-absorption.

In summary, liquid-scintillation counting is a valuable means of counting weak beta-emitting radioisotopes. By dissolving the sample directly in the liquid scintillator the problem of blocking the betas by window absorption or self-absorption is eliminated. The only serious disadvantages are the relatively high cost of the counting apparatus and the fact that the method is limited to samples that are soluble or readily suspended in the liquid scintillator.

14-2 MÖSSBAUER SPECTROSCOPY

The Mössbauer effect was discovered in 1958, and by approximately 1962, chemists began to realize its potential. Chemists have applied Mössbauer spectroscopy to such studies as chemical bonding, crystal structure, electron density, and the determination of magnetic properties. Although complementary to nmr, esr, and nuclear quadrupole resonance in many of its applications, it is more limited in scope. For example, qualitative elemental analysis is not feasible by Mössbauer spectroscopy, since it is necessary first to choose as a source a gamma-emitting isotope of the element to be studied, thereby precluding the use of the instrument for qualitative analysis. Similarly, its use for elemental quantitative analysis is rather limited. But as a nondestructive method of analysis for studying reaction mechanisms and solution chemistry, it has a bright future.

14-2A Theory

A number of radioactive isotopes undergo *isomeric transitions,* whereby nuclei in an excited state emit gamma rays and end up in the ground state without having undergone a change in mass number or charge. Only the *energy* of the isotope changes. If these gamma rays are allowed to interact with a similar nucleus in the ground state, the gamma ray may be reabsorbed in a *resonance-capture* process. Although the possibility of this process had been recognized for some time, it was not considered feasible because the act of emitting gamma rays imparted significant amounts of *recoil energy* to the emitting nucleus, thereby decreasing the energy of the gamma ray by the amount of the recoil energy. Thus, the emitted gamma ray always contained insufficient energy to be absorbed by similar nuclei in the ground state. Furthermore, the momentum balance of the reabsorption process required that the newly excited nucleus possess excess kinetic energy after the absorption process. Consequently, the emitted and absorbed gamma would differ by *twice* the kinetic energy (or recoil energy) involved, and thus the conditions for resonant absorption were not met. However, in 1958 Mössbauer showed that resonance capture of the gamma ray can occur with moderate efficiency if the source and the absorber are both contained in solid lattices. The conditions for resonance absorption appeared to be met, since any nucleus which is rigidly locked in a crystalline lattice should neither recoil during emission nor take on kinetic energy during absorption.

This discussion is oversimplified in several respects. First, it is not true that a gamma ray emitted from a nucleus locked in a solid lattice loses *zero* recoil energy; instead the recoil losses are very small. Conservation of energy and momentum requires *some* recoil and kinetic energy to be involved, but it will be reduced to very small values because in both emission and absorption the momentum is taken up by the entire crystal, rather than by the individual nuclei. This may be understood as follows. Since kinetic energy (KE) is given by

$$\mathrm{KE} = \tfrac{1}{2}mv^2 \qquad (14\text{-}3)$$

and momentum p by

$$p = mv \qquad (14\text{-}4)$$

it follows that

$$\mathrm{KE} = \frac{1}{2}\frac{p^2}{m} \qquad (14\text{-}5)$$

In other words, for a given momentum the kinetic energy is inversely proportional to the mass. By going from individual nuclei to a small crystal lattice, the mass will be raised by a factor of about 10^{18}, and thus the kinetic energy involved in the emission and absorption process will become negligible.

A further requirement for resonance absorption is that the recoil energy not be sufficient to excite any optical modes of vibration in the crystalline lattice. Vibrational excitation would heat the lattice and modulate the energy of the emitted gamma ray, thereby nullifying the conditions for resonance absorption. The practical consequence of this is that only gamma rays with an energy of less than about 150 keV (0.150 MeV) are usable.

Finally, it must be understood that all emission and absorption lines have finite width and are never "infinitely thin." If gamma-ray emission and absorption lines were infinitely thin, the Mössbauer effect would be impossible, since even the smallest loss of recoil energy would cause the emission energy to differ from the absorption energy and no absorption would be possible. This is illustrated in Fig. 14-11*a*, where E_T is the actual energy-level separation in the nucleus involved (transition energy), $E_T - E_R$ is the energy of the emitted gamma ray (where E_R is the recoil energy), and $E_T + E_{\mathrm{KE}}$ is the energy *required* for resonance absorption (where E_{KE} is the kinetic-energy requirement of the absorber). Since the energy of the emitted gamma ($E_T - E_R$) and the energy required for resonant absorption ($E_T + E_{\mathrm{KE}}$) are different and in no way overlap, resonance absorption is impossible.

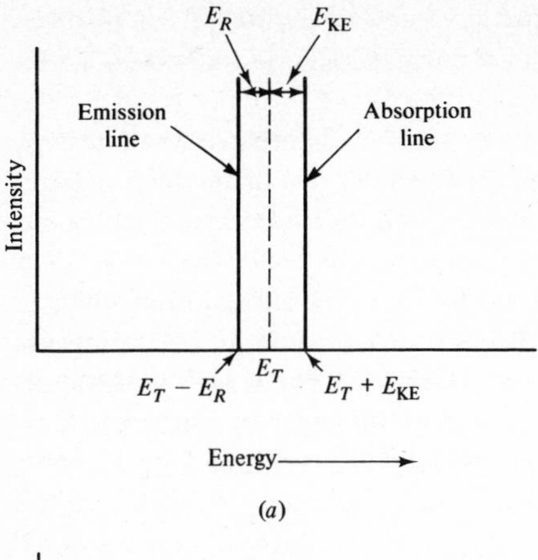

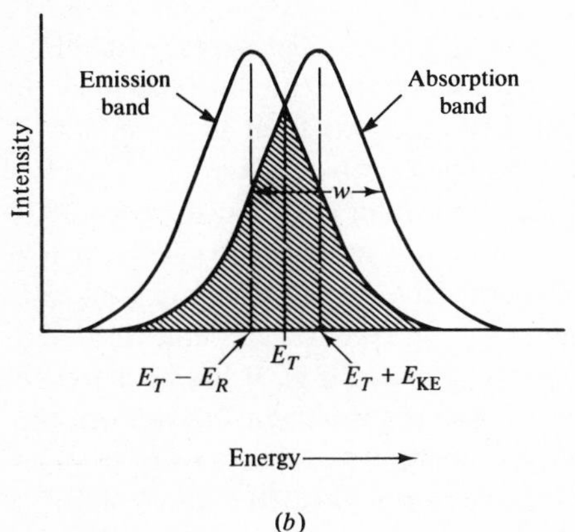

FIGURE 14-11 Schematic diagram of the relationships between line width, transition energy, and recoil energy. (*a*) Emission and absorption lines with zero line width (hypothetical); (*b*) emission and absorption lines with finite width w. E_R = recoil energy, E_{KE} = kinetic-energy requirement of the absorber, E_T = transition energy; w = the line width at half maximum peak height.

In practice, however, all absorption and emission lines have a finite theoretical width that can be calculated from the lifetime of the nuclear excited state using the Heisenberg uncertainty principle (see Sec. 13-5*A*).† These line widths are of the order of 10^{-7} eV (about 10 MHz) for most Mössbauer radioisotopes. Line widths of this magnitude are extremely narrow. Assuming that a typical Mössbauer gamma has an energy of 100 keV, a line width of 10^{-7} eV is only 10^{-12} times the energy of the gamma ray, which is equivalent to saying that the energy of the gamma ray is defined to 1 part in 10^{12}, making it

† The uncertainty principle states that the uncertainty in energy ΔE times the uncertainty in time Δt is a constant:

$$\Delta E \, \Delta t = \frac{h}{2\pi} \tag{13-12}$$

The more short-lived a given high-energy quantum state, the smaller the uncertainty in time Δt and therefore the larger the uncertainty in the energy ΔE of the emitted radiation. Conversely the more stable (long-lived) the quantum state, the larger the uncertainty in when the system will decay Δt and the sharper the emitted energy, i.e., the smaller the ΔE.

the most accurately defined electromagnetic radiation available for physical measurements. Yet the recoil energy E_R of a nucleus locked in a crystal lattice is very much smaller than the line width (a typical recoil energy for a crystal might be 10^{-19} eV, making E_R much smaller than w). Therefore, the emission band not only overlaps the absorption band, as shown by the shaded area of Fig. 14-11*b*, but is virtually superimposed on it, rendering the effect of the recoil energy negligible.

The 14.4-keV gamma-emitting isomer of ^{57}Fe is the most important and widely used nucleus for Mössbauer experiments, and the energy-level diagram for ^{57}Fe and its precursor ^{57}Co are shown in Fig. 14-12. The stability of the excited state is the determining factor as far as the line width is concerned, and for the ^{57}Fe excited state with a spin of $3/2$ the half-life is about 10^{-7} s. From the uncertainty principle the line width can be calculated as 1.6 MHz. This is the same order of magnitude of, or less than, the energy typically encountered in nmr spectroscopy. Thus, this line width is sufficiently narrow to allow measurement of very small changes in the energy level of the nucleus. For example, it is possible to observe hyperfine interactions such as the effect of the extranuclear electron field upon low-lying nuclear energy levels.

In order to utilize the Mössbauer effect to obtain information about different nuclear orientations or the interaction of nuclei with surrounding electrons, some means must be found to modify the energy of the gamma-ray source systematically by a few megahertz, because if the emitting and absorbing nuclei were in different nuclear orientations, e.g., if the nuclei were in different chemical compounds, an exact energy match would not be achieved and resonance absorption would not occur. The most convenient way of making these energy adjustments is through the *doppler* effect.

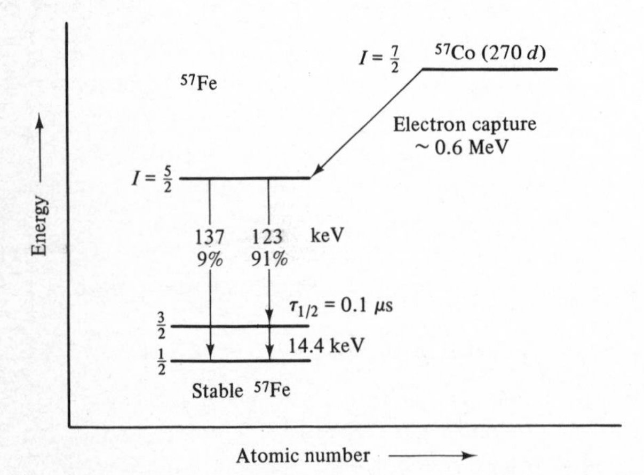

FIGURE 14-12 Energy-level diagram for the decay of ^{57}Co to ^{57}Fe, showing the 14.4-keV isomeric transition of ^{57}Fe. (*After G. K. Wertheim, "Mössbauer Effect: Principles and Applications,"* p. 14, *Academic Press Inc., New York,* 1964, *by permission.*)

USE OF THE DOPPLER EFFECT TO VARY THE GAMMA-RAY ENERGY

Mössbauer was the first to utilize the doppler effect to modify the energy of the gamma-ray source to match the difference in energy between the ground state and the excited state of the absorber, and this principle has found almost universal acceptance, although the mechanisms now employed bear little resemblance to those originally used. Other methods of varying the energy, e.g., utilizing the thermal red shift or ultrasonically produced sidebands, have been proposed for special applications but have not been adopted for general use.

In the doppler effect the gamma-ray source is moved with a velocity v relative to the absorber. The modified frequency of the gamma ray ν will be related to its original frequency ν_0 by

$$\nu = \nu_0\left(1 + \frac{vc}{c^2}\right) = \nu_0\left(1 + \frac{v}{c}\cos\theta\right) \tag{14-6}$$

where θ is the angle between the axis of symmetry and the optical axis. Since energy E is directly proportional to the frequency ($E = h\nu$), the significance of Eq. (14-6) is best visualized in terms of the energy. If an absorber and a gamma-ray source are traveling directly toward each other with a velocity v, the energy will seem to be $E_0\,(1 + v/c)$, where E_0 is the inherent energy of the emitted gamma ray and $\cos\theta = 1$. On the other hand, if the absorber and the gamma-ray source are moving directly away from each other ($\cos\theta = -1$), the energy will seem to be $E_0\,(1 - v/c)$. Clearly then, the doppler effect affords a means of varying the energy of the gamma ray. To estimate the magnitude of the doppler shift $\Delta\nu$ which is possible with ordinary velocity changes, the following equation can be used:

$$\Delta\nu = \nu_0\frac{v}{c} \tag{14-7}$$

For example, with the 14.4-keV gamma ray of ^{57}Fe, a velocity of 1 cm/s would give a frequency shift of 116 MHz [Eq. (14-7)]. Since the line width of ^{57}Fe is about 1.6 MHz (calculated earlier from the uncertainty principle), a frequency scan of 116 MHz is more than adequate to scan an entire Mössbauer spectrum. A relative velocity of 1 cm/s is easily produced in the laboratory, and in fact, velocities considerably less than that are sufficient for most studies. Mechanical motion devices based on gears and cams have met with only limited success because wear and bearing vibrations tend to cause problems. Most of the devices now used to move the source relative to the absorber are electromechanical feedback systems, e.g., a microphone coil fed by some type of waveform generator.

INTERPRETATION OF MÖSSBAUER SPECTRA

The raw data of a Mössbauer spectrum are the relative number of gamma rays per second (counts per second) being transmitted through the absorbing sample as a function of the relative velocity of the source and sample. A typical Mössbauer spectrum is shown in Fig. 14-13. Note that the ordinate has been transcribed to percent absorption units, where the maximum count rate is taken to be 100 percent transmittance or 0 percent absorption, and a zero (or background) count rate would be taken as 0 percent transmittance or 100 percent absorption. The standard convention in presenting the abscissa is to represent motion of the source toward the absorber by positive velocities and motion of the source away from the absorber by negative velocities. The parameters of a Mössbauer spectrum of greatest interest to a chemist are the size of the absorption peak ϵ, the line width w, the isomer shift (IS), quadrupole splitting (QS), magnetic hyperfine structure (MHFS), and the temperature coefficients of these parameters.

Size of the Absorption Peak ϵ This obviously depends on the number of absorber nuclei in the optical path, but at present it is very difficult to make quantitative use of this fact because a variety of interference effects and background-scattering contributions from the environment near the system contribute to the counting rate. In most experiments the nuclear transition of interest is accompanied by other radiations which may constitute an interference. With ^{57}Fe, for example, the 14.4-keV gamma ray is of primary interest (see Fig. 14-12), but for every 14.4-keV gamma there is an equal number of 123-keV precursor events, as well as 9 percent of 137-keV radiation and numerous low-energy x-rays both from Auger effects and scattering from the environment. Despite careful energy resolution, some of this radiation will reach the detector and contribute to the overall counting rate, thus rendering the absolute value of the Mössbauer peak height variable and unpredictable.

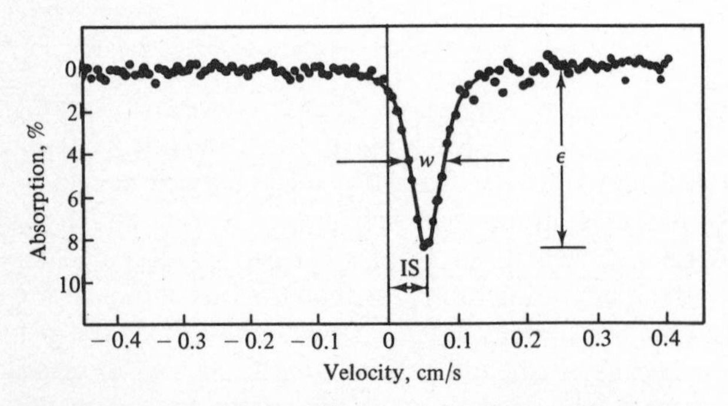

FIGURE 14-13 Typical Mössbauer spectrum (ferricinium bromide at 20 K using a ^{57}Fe source). (*After G. K. Wertheim, "Mössbauer Effect: Principles and Applications," p.* 51, *Academic Press Inc., New York,* 1964, *by permission.*)

A further complication is that the Mössbauer resonance is with a particular isotope among many isotopes that may be present, and the problem of a low or variable percentage abundance of that isotope can lead to large relative errors in attempts to reproduce the value of the peak height.

Line Width *w* This depends theoretically on the stability of the excited state and the Heisenberg uncertainty principle, as already noted. In practice, however, the observed line width is always wider than the theoretical value because of experimental complications. Experimental imperfections in the velocity drive, the presence of solid-state defects and impurities, thermal effects, and the finite thickness of both the source and the absorber all tend to broaden the resonance line. The most important use of the resonance line width is as a measure of the quality of a given Mössbauer source. When a standard absorber is used, the magnitude of w for two or more sources can be compared, and the source giving the smallest w should be selected.

Isomer Shift (IS) This is the Mössbauer parameter which yields the greatest amount of chemical information. A few authors refer to it as the *chemical shift* but isomer shift is the term most widely used, since the effect depends on the difference in the nuclear radius of the ground and excited (isomeric) states. Experimentally the isomer shift is defined as the displacement of the resonance maximum from zero velocity, as shown in Fig. 14-13.

The isomer shift arises from the fact that the nucleus of an atom occupies a finite volume and interacts as a region of charge with its electronic environment. There are two fundamental requirements for the creation of an isomer shift: (1) the radii of the nuclear excited state and the ground state must differ, in order that each state interact differently with the electronic environment; (2) the absorber and source must be in chemically different environments, e.g., in two different oxidation states, so that the electron density at the nucleus will be different for the absorber and source. According as the excited-state radius is larger or smaller than the ground state, the isomer shift will be positive or negative. For example, with ^{57}Fe the excited state is physically smaller than the ground state (a negative-radius effect), and thus an increase in electron density in the absorber will lead to a negative isomer shift. Thus, for example, if the absorber is changed from Fe^{2+} to Fe^{3+}, the charge density at the nucleus *increases* and a sizable negative isomer shift results. The absolute value of the isomer shift is not too significant; it is the *change* in isomer shift with change in conditions that is most important. The magnitude of the shift depends on the chemical environment, e.g., whether the cation is combined with chloride, bromide, or some other anion.

A specific example of the use of the isomer shift is the determination of the type of bonding and the oxidation state of iron in various heme compounds.†

† A. J. Bearden, T. H. Moss, W. S. Caughey, and C. A. Beaudreau, *Proc. Natl. Acad. Sci. U.S.*, **53**:1246 (1965).

Quadrupole Splitting (QS) This phenomenon is often more sensitive to the nature of the chemical bond, its ionic character, for example, than the isomer shift. Quadrupole splitting results from a nonsymmetrical electric field gradient around the nucleus. This might be caused by a nonspherical (oblate, or flattened) nucleus or (of more interest in Mössbauer spectroscopy) by a distortion of the normally spherical electron distribution of the filled electronic shells by the valence-electron configuration. This nonsymmetrical field has the effect of removing the degeneracy of the spin states in the excited state of nuclei, as illustrated in Fig. 14-14. In Fig. 14-14*a* the energy levels for free nuclei are shown on the left. Transitions between these states represent resonance absorption at zero relative velocity. In the center of Fig. 14-14*a* the displacement due to an isomer shift is shown. The displacement for the excited state is different from that for the ground state because the chemical environment is affecting the two states differently. At the right of Fig. 14-14*a* is shown the effect of an asymmetric electrostatic field gradient at the nucleus. The quadrupole coupling between the electric field and the spin of the nucleus

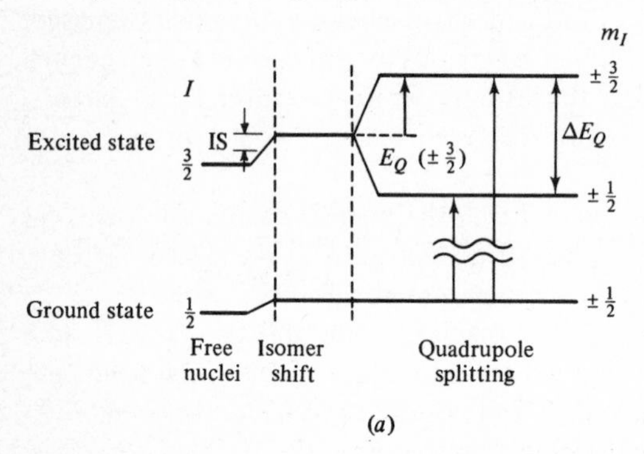

(*a*)

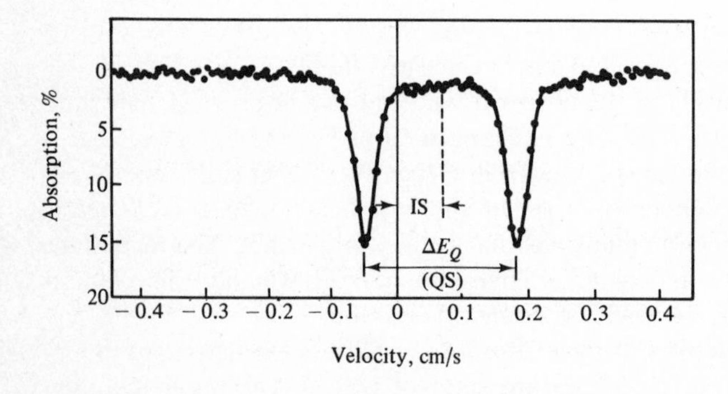

(*b*)

FIGURE 14-14 The origin of quadrupole splitting interaction. (*After G. K. Wertheim, p.* 62, *Academic Press Inc., New York,* 1964, *by permission.*)

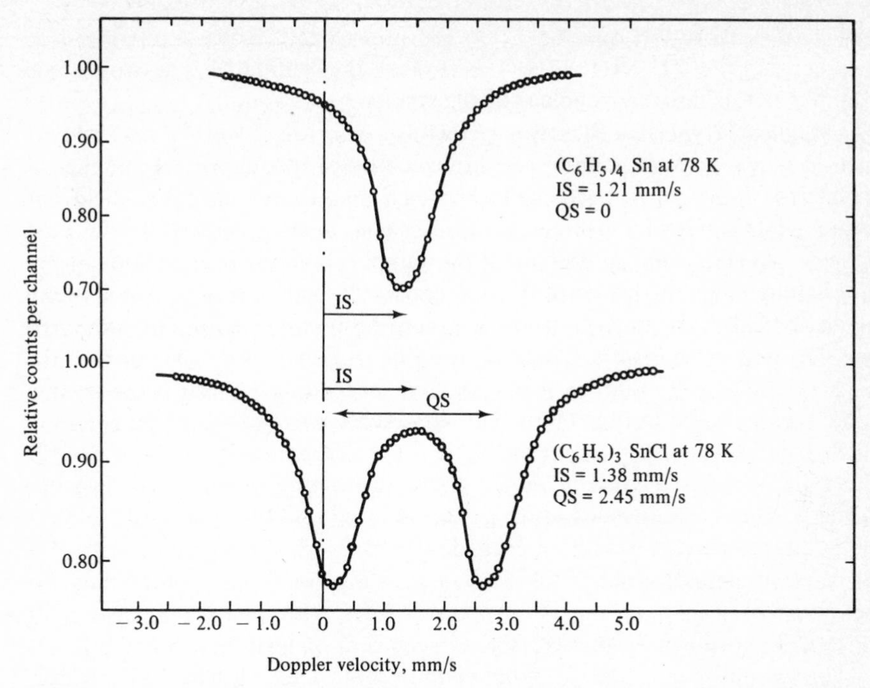

FIGURE 14-15 Mössbauer spectra of $(C_6H_5)_4Sn$ and $(C_6H_5)_3SnCl$. Substitution of a halogen ligand for a C_6H_5 group produces quadrupole splitting. (*From R. H. Herber, "The Mössbauer and Its Application in Chemistry," p.* 15 *Advan. Chem. Ser.* 68, 1967, *by permission.*)

causes a splitting of the upper spin state (one is raised by an energy E_Q, and one is lowered by the same amount). Selection rules indicate that transitions to the two new states should be equally probable. Hence, a Mössbauer spectrum of an element which is subject to such quadrupole splitting will show *two* lines of equal intensity. This is illustrated with the spectrum of biferrocenyl, shown in Fig. 14-14*b*. Note from Fig. 14-14*b* that the isomer shift is measured from the zero velocity point to the exact midpoint between the two peaks, since in the absence of the quadrupole field the singlet peak would appear at that point.

Quadrupole splitting is a sensitive and powerful means of monitoring the symmetry of a given compound. An example is shown in Fig. 14-15. The upper spectrum is that of tetraphenyltin, with tetrahedral symmetry about the nucleus, and because of this symmetry, no quadrupole splitting occurs. However, if a phenyl group is replaced with a chlorine atom, as in triphenyltin chloride, quadrupole splitting is observed (see the bottom spectrum of Fig. 14-15).

Numerous examples of such studies have been made with iron compounds. For example, compounds like $K_3[Fe(CN)_6]$ and $K_4[Fe(CN)_6]$, with octahedral symmetry, and K_2FeO_4, with tetrahedral symmetry, show zero or very small quadrupole splittings. On the other hand, when a nitroprusside absorber, $Na_2[Fe(CN)_5NO] \cdot 2H_2O$, is studied, the symmetry is destroyed and a doublet caused by quadrupole splitting results.

Magnetic Hyperfine Structure (MHFS) This arises when an external or internal *magnetic field* is imposed in the vicinity of the nucleus. A number of interesting studies have been made in which an *external* magnetic field has been used [9], but of more general Mössbauer interest are cases where an *internal* magnetic field is present in the sample. For example, paramagnetic materials having outer electrons with unpaired spins will generate a small magnetic field at the nucleus, and even more pronounced effects are observed from ferromagnetic materials such as metallic iron or Fe_2O_3. The effect of the magnetic field is to remove the remaining degeneracy of the various spin states, as illustrated in Fig. 14-16. This represents a nuclear Zeeman effect, in which each nuclear level is split into $2|M_I| + 1$ components; i.e., each nuclear spin state becomes aligned in the magnetic field, either with the field or opposing it. The selection rules which govern allowed transitions are $\Delta M_I = 0$ or ± 1, so that an absorber with an excited state of $I = 3/2$ and a ground state of $1/2$ will have six possible transitions, as shown in Fig. 14-16. The probability for each type of transition results in spectral lines with relative intensities of 3:2:1:1:2:3, as shown by the Mössbauer spectrum of metallic iron in Fig. 14-17. The two resonance lines at either end are the $-1/2$ to $-3/2$ and $1/2$ to $3/2$ transitions, respectively, and the separation between these two can be directly related to the magnitude of the internal magnetic field. The separation between lines 2 and 4 (or between lines 3 and 5; see Fig. 14-16) is just the ground-state splitting. This splitting, which amounts to 3.924 mm/s in metallic iron at room temperature (see Fig. 14-17), can also be measured by nmr techniques, and this affords an independent (non-Mössbauer) calibration of the doppler velocity scale of a Mössbauer spectrometer.

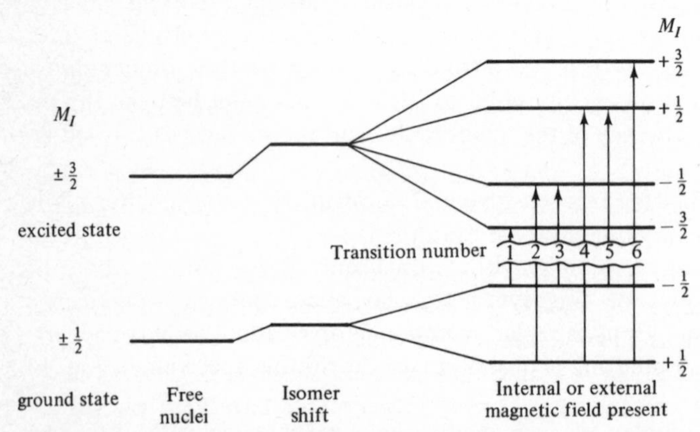

FIGURE 14-16 Origin of magnetic hyperfine structure. (*After G. K. Wertheim, "Mössbauer Effect: Principles and Applications," p. 73, Academic Press Inc., New York, 1964, by permission.*)

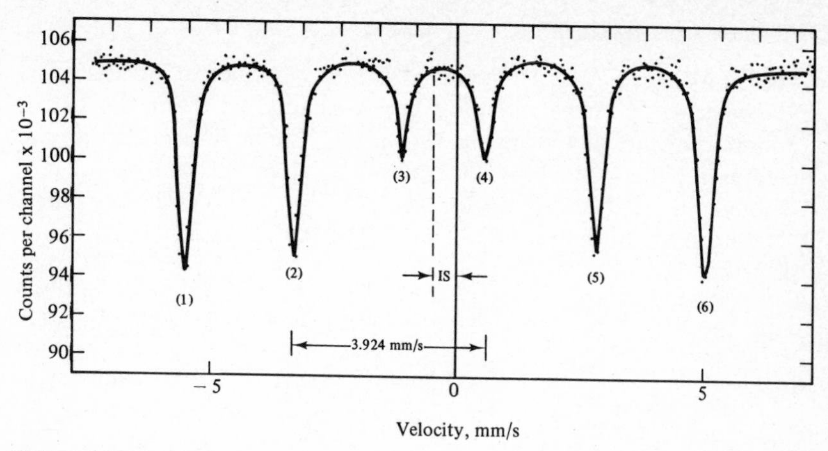

FIGURE 14-17 Mössbauer spectrum of metallic iron at room temperature. (*After R. H. Herber, "The Mössbauer and Its Application in Chemistry,"* p. 16 *Advan. Chem. Ser.* 68, 1967, *by permission.*)

Mössbauer spectroscopy sometimes has an advantage over nmr spectroscopy for measuring ground-state splitting and other hyperfine interactions because bulk metallic samples often cause trouble in nmr measurements because of their electrical conductivity. If thin samples are used in nmr studies in an attempt to minimize the interference from electrical conductivity, the ferromagnetism of the sample may be altered.

It is sometimes possible to make correlations between Mössbauer and esr data. For example, the paramagnetism in some samples give Mössbauer magnetic hyperfine splittings which are related to the hyperfine coupling constants measured in esr experiments.

Simultaneous magnetic and electric quadrupole coupling may occur. For example, in certain unsymmetrical iron compounds like Fe_2O_3, the ferromagnetism leads to a magnetic hyperfine splitting into six lines, much like Fig. 14-17, but the asymmetry of the chemical compound can superimpose an electric quadrupole coupling which generally changes the regularity of the spacing between the six lines [8, 9].

Temperature Coefficients of Various Mössbauer Parameters These coefficients can yield a considerable amount of chemical information, particularly with respect to the solid-state properties of the host lattice. This is primarily because the fraction of recoil-free emissions depends on the state of excitation of the lattice vibrations. Temperature studies are also of importance in the study of ferromagnetic and antiferromagnetic materials and second-order doppler effects. Temperature studies may also yield useful information about crystal-field effects and relaxation times. Experiments have been carried out in which just the sample temperature is controlled, allowing the gamma rays to pass through the walls of an isothermal enclosure; in other cases both the source and sample are temperature-controlled.

14-2B Instrumentation

Many types of Mössbauer spectrometers have been described in the literature, but the fundamental principles of most spectrometers, including the half a dozen or more commercially available instruments, can be illustrated with the block diagram in Fig. 14-18. The *source* is usually a thin foil, perhaps 0.1 mm thick, containing the radioactive nuclei. The *absorber,* which is the sample of interest, is used in the form of a film of similar thickness covering the window of the detector. The *detector* is generally a NaI (thallium-activated) scintillation crystal, although it is advantageous to use a proportional counter for gamma energies below 20 keV because of the high background with scintillation systems at low energies (see Sec. 14-1*B*). To minimize the time required for measuring a spectrum, as high a count rate as possible is desirable. Therefore, a fast-preamplifier single-channel analyzer (labeled "energy analyzer" in Fig. 14-18) and *multichannel scaler* capable of counting at 10 MHz are very useful. Even with these components and a highly radioactive source it generally takes 2 to 4 h to obtain a good Mössbauer spectrum because the random error in counting radioactivity is proportional to the square root of the number of counts. If, for example, only 10^4 counts were made at each velocity along the abscissa, the precision in the transmission data would only be 1 percent of the count. This is scarcely precise enough since the Mössbauer absorption is only 3 percent or so at the peak of the resonance.

In practice the source is generally attached to a moving *microphone coil,* whose motion is controlled by a *waveform generator.* Generally a sawtooth variation of velocity with time is used, and the entire velocity range is scanned several times a second. The counts at each velocity are fed, over equal time intervals, into separate channels of a *multichannel analyzer,* and the channel (or address) for each velocity is determined by an *address signal* appropriately

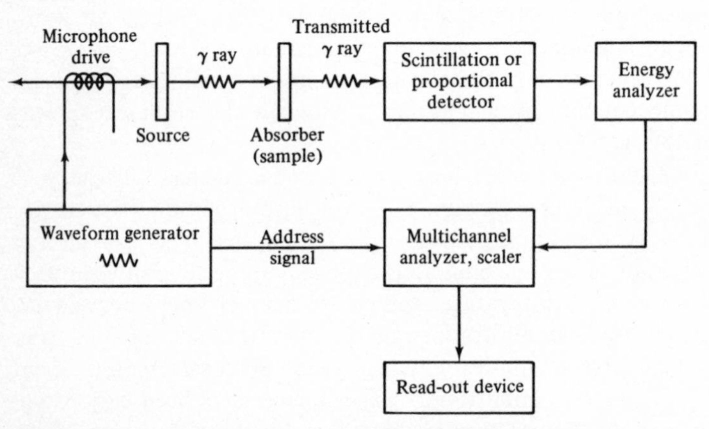

FIGURE 14-18 Block diagram of a typical Mössbauer spectrometer. (*After D. H. Whiffen, "Spectroscopy," p. 69, John Wiley & Sons, Inc., New York,* 1966, *by permission.*)

TABLE 14-1 Some Isotopes in Which the Mössbauer Effect Has Been Observed, in Order of Decreasing Frequency of Use†

Isotope	Gamma energy, keV	Half-life of excited state, ns
^{57}Fe	14.4	98
^{119}Sn	23.9	18
^{161}Dy	25.7	29
^{160}Dy	86.8	2.0
^{197}Au	77.3	1.8
^{169}Tm	8.4	3.9
^{129}I	28	16
^{40}K	29.4	3.9
^{61}Ni	67.4	5.3
^{67}Zn	93	9400
^{73}Ge	67	1.6
^{83}Kr	9.3	147
^{99}Ru	90	20
^{107}Ag	93	44×10^9
^{121}Sb	37	3.5
^{125}Te	36	1.5
^{129}Xe	40	1.0
^{133}Cs	81	6.3
^{177}Hf	113	50
^{181}Ta	6.3	6800
^{182}W	100	1.4
^{183}W	47	0.2
	99	0.6
^{187}Re	134	0.01
^{186}Os	137	0.8
^{191}Ir	129	0.1
^{195}Pt	99	0.2
	129	0.6
^{197}Au	77	1.8
Nine rare earths (Pr, Sm, Eu, Gd, Tb, Dy, Er, Tm, Yb)		
^{237}Np	59	63

† From A. H. Muir, K. J. Ando, and H. M. Coogan, "Mössbauer Effect Data Index, 1958–1965," p. xvii, John Wiley & Sons, Inc., New York, 1966, by permission.

derived from the waveform generator. Each channel represents one velocity on the abscissa of the spectrum. After a 2 to 4 h counting period each channel will usually have accumulated enough counts to allow an accurate readout of the gamma-ray transmittance at that velocity.

14-2C Applications

Numerous applications of Mössbauer spectroscopy have already been mentioned. A complete listing will be found in the reviews of Mössbauer spectroscopy appearing every two years in *Analytical Chemistry* [11].

Table 14-1 lists some isotopes in which the Mössbauer effect has been observed. Note that all the isotopes listed have gammas with energies below the 150-keV limit. The half-life of the excited state is of interest because it determines the inherent line width of the gamma. Note that ^{107}Ag and ^{67}Zn have the longest half-lives, and thus the narrowest line widths, of all the isotopes listed in Table 14-1.

Isotopes are given in Table 14-1 roughly in the order of the frequency with which they have been used. Thus, in the majority of the studies thus far (slightly over 50 of the publications) ^{57}Fe has been used, ^{119}Sn being the second most widely used source (about 20 percent of the publications). Isotopes of the remaining approximately 30 elements have found much more limited use thus far. Unfortunately, not all of the elements in the periodic table have isotopes with suitable nuclear transitions, thereby limiting the usefulness of the Mössbauer effect. On the other hand, for the elements for which suitable isotopes are available, Mössbauer spectroscopy should prove to be increasingly useful.

EXERCISES

GAMMA-RAY SPECTROSCOPY

14-1 A certain single-channel scintillation spectrometer is calibrated with the 0.663-MeV gamma of ^{137}Cs, and it is found that the cesium photopeak has a width of 0.102 meV at half the maximum photopeak height. Calculate whether this instrument has sufficient resolution to resolve the following pairs of gamma-emitting isotopes:

(*a*)	^{95}Zr	0.71 MeV	and	^{147}Nd	0.53 MeV
(*b*)	^{131}I	0.364 MeV	and	^{147}Nd	0.53 MeV
(*c*)	^{95}Zr	0.71 MeV	and	^{95}Nb	0.77 MeV
(*d*)	^{144}Ce-^{144}Pr	1.25 MeV	and	^{195}Te	0.80 MeV
(*e*)	^{195}Te	0.80 MeV	and	^{95}Zr	0.71 MeV

Ans: (*c*) and (*e*) no; the rest yes

14-2 (*a*) How can the linearity of a gamma-ray spectrometer be tested? (*b*)

Explain how accurate values of gamma-ray energies can be measured on a nonlinear spectrometer.

14-3 What effect would an unstable high-voltage supply in a gamma-ray spectrometer have on the spectrum observed?

14-4 Explain how a pulse-height analyzer works.

14-5 (*a*) What are the essential components of a multichannel scintillation spectrometer? (*b*) Draw a block diagram of a three-channel analyzer.

MÖSSBAUER SPECTROSCOPY

14-6 Explain the cause of an isomer shift in Mössbauer spectroscopy, and give two examples.

14-7 In what way, if any, does recoil energy contribute to an observed isomer shift?

14-8 Explain the cause of quadrupole coupling in Mössbauer spectroscopy, and indicate what chemical information can be gained from it.

14-9 Explain the origin of magnetic hyperfine structure in Mössbauer spectroscopy, and indicate what useful information can be gained from it.

14-10 The isotope ^{127}I has a 60-keV gamma ray and an excited state having a half-life of 2.7 ns. Does ^{127}I appear to have suitable properties for use as a Mössbauer source? Speculate on any advantages or disadvantages it may have compared with ^{129}I, whose properties are listed in Table 14-1.

14-11 The 60-keV gamma ray from ^{127}I involves a transition from a spin state of $\pm 7/2$ to a ground state of $\pm 5/2$. If a magnetic field were applied to an ^{127}I nucleus, (*a*) into how many energy levels would the $I = 7/2$ state be split? (*b*) Into how many energy levels would the $I = 5/2$ state be split? (*c*) How many lines (MHFS) might be expected in the Mössbauer spectrum?

Ans: (*a*) 8; (*b*) 6

14-12 The Mössbauer spectrum of ^{57}Fe in ferrocene, $Fe(C_5H_5)_2$, at 20 K consists of two lines at -0.50 and $+1.88$ mm/s. (The ^{57}Fe was encapsulated in a stainless-steel source.) Calculate the isomer shift and the quadrupole splitting for the excited nucleus in millimeters per second.

Ans: IS = 0.69 mm/s; QS = 2.38 mm/s

REFERENCES

GAMMA-RAY SPECTROSCOPY

1 CONNALLY, R. E.: *Anal. Chem.*, **28**:1847 (1956). An excellent discussion of gamma-ray spectrometry and its applications.

2 CONNALLY, R. E., and M. B. LEBOEUF: *Anal. Chem.*, **25**:1095 (1953). Principles of gamma-ray spectroscopy.
3 WILLARD, H. H., L. L. MERRITT, JR., and J. A. DEAN: "Instrumental Methods of Analysis," 4th ed., Van Nostrand, New York, 1965.
4 BIRKS, J. B.: "The Theory and Practice of Scintillation Counting," Pergamon, New York, 1964.
5 SIEGBAHN, K. (ed.): "Alpha-, Beta-, and Gamma-Ray Spectroscopy," vol. 1, North-Holland, Amsterdam, 1968.
6 BELL, C. G., JR., and F. N. HAYES: "Liquid Scintillation Counting," Pergamon, New York, 1961.

MÖSSBAUER SPECTROSCOPY

7 WHIFFEN, D. H.: "Spectroscopy," chap. 6, Wiley, New York, 1966.
8 WERTHEIM, G. K.: "Mössbauer Effect: Principles and Applications," Academic, New York, 1964.
9 HERBER, R. H.: "The Mössbauer Effect and Its Application in Chemistry," *Adv. Chem. Ser.* 68, 1967.
10 GOL'DANSKII, V. I.: "The Mössbauer Effect and Its Applications in Chemistry," trans. from Russian, Consultants Bureau, New York, 1964.
11 STEVENS, J. G., J. C. TRAVIS, and J. R. DEVOE: *Anal. Chem.*, **44**:384R (1972). The fourth biannual review of fundamentals and applications that has appeared in *Analytical Chemistry*.
12 MUIR, A. H., JR., K. J. ANDO, and H. M. COOGAN: "Mössbauer Effect Data Index," Wiley-Interscience, New York, 1960.

appendix A Supplementary Definitions and Principles

A-1 DIFFRACTION BY A SLIT (SEE SEC. 1-4E)

Fraunhofer Diffraction: Derivation of Equation for Angle of Diffraction

Figure 1-18 shows the angle θ at which the minimum P_1 occurs. The calculation of angle θ can be made from

$$\sin \theta = \frac{\lambda}{b} \tag{1-19}$$

where λ is the wavelength and b is the slit width. Equation (1-19) can be derived from an expanded view of Fig. 1-18, shown in Fig. A-1. It should be emphasized that the presence of the focusing lens has the effect of making the screen an infinite distance from the slit, which is a necessary condition of Fraunhofer diffraction, and allows the diffracted rays to be considered exactly parallel. The condition for a minimum at P_1 is that the wavelets from the *extremes* of the slit arrive just one wavelength out of phase or, as shown in Fig. A-1, that the distance RP_1 be one wavelength longer than the distance SP_1. Thus, the secondary wavelet from point R will travel approximately $\lambda/2$ farther than the wavelet from point Q in the center of the slit, and thus the wavelets from R and Q will arrive at P_1 180° out of phase, giving *destructive interference*. Similarly the wavelet originating from the next point below R will cancel with that from the next point below Q, and we can continue this

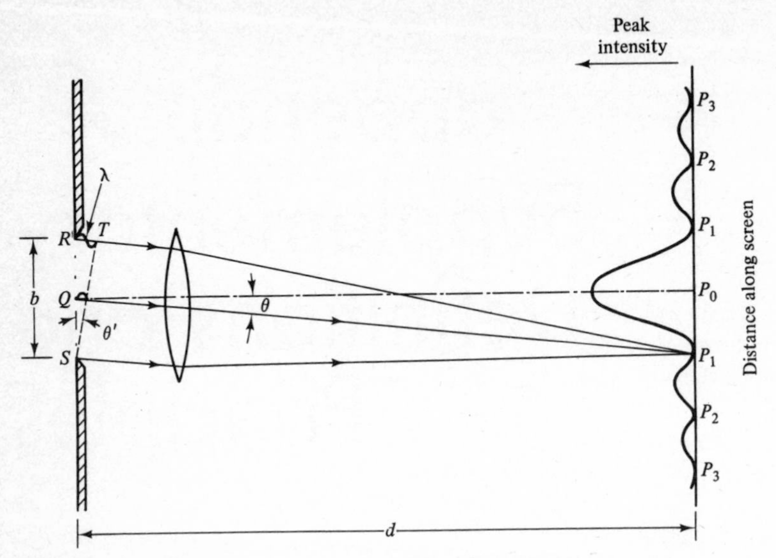

FIGURE A-1 Expanded view of Fraunhofer diffraction pattern of a single slit.

pairing off to include all points in the wavefront, so that the resultant intensity at P_1 is zero.

Although it is not our purpose to do, it could be shown in a similar fashion that at P_2 the path difference from the extremes of the slit is 2λ.

Assuming that ST is drawn perpendicular to RP_1 in Fig. A-1, it follows that

$$\sin \theta' = \frac{RT}{RS} = \frac{\lambda}{b} \tag{A-1}$$

where b is the slit width and the distance $RT = \lambda$. Furthermore, the effective parallel arrangement of wavelets RP_1, QP_1, and SP_1 makes the angles θ and θ' exactly equal, and thus it follows that

$$\sin \theta = \frac{\lambda}{b} \tag{1-19}$$

A-2 DISPERSION (SEE SEC. 1-4F)

Anomalous Dispersion

The relationship between anomalous dispersion and absorption bands has been theoretically explained by Sellmeier† in terms of incident radiation inducing oscillations in the bound atoms in the material through which it passes. Sellmeier's derivation assumed that atoms are bound by elastic forces and are capable of vibrating with a certain definite frequency ν^0. This is the so-called

† See F. A. Jenkins and H. E. White, "Fundamentals of Optics," 3d ed., p. 472, McGraw-Hill, New York, 1957.

natural frequency, and if radiation of this frequency is incident upon the substance, it will cause vibrations and be absorbed. If the frequency ν of the incident light does not agree with ν^0, the vibrations will be forced vibrations of relatively small amplitude and frequency ν. A simple form of Sellmeier's equation is

$$n^2 = 1 + \frac{A\lambda^2}{\lambda^2 - \lambda_0^2} \tag{A-2}$$

where n = refractive index
A = constant proportional to number of electrons per cubic centimeter
λ_0 = wavelength in vacuum corresponding to the natural frequency ν^0
λ = wavelength of incident radiation

To allow for the possibility of several different natural frequencies, the equation can be written with a series of terms

$$n^2 = 1 + \frac{A_0\lambda^2}{\lambda^2 - \lambda_0^2} + \frac{A_1\lambda^2}{\lambda^2 - \lambda_1^2} + \cdots \tag{A-3}$$

where $\lambda_0, \lambda_1, \ldots$ correspond to various natural frequencies. Note that Eq. (A-3) predicts many of the features of Fig. 1-27. (1) It predicts the regions of anomalous dispersion in absorption-band regions. As λ approaches λ_0 or λ_1 on the short-wavelength side, n may have values less than 1, whereas n is always greater than 1 when approaching λ_0 or λ_1 on the long-wavelength side. And when λ equals λ_0 or λ_1, n becomes indeterminate. (2) Equation (A-3) predicts that n approaches unity as λ approaches zero, which Fig. 1-27 shows does occur in the short x-ray and gamma-ray regions. (3) As λ approaches infinity, n levels out at some value higher than 1, since n^2 approaches the value of $1 + A_0 + A_1 + \cdots$.

The existence of refractive indexes less than unity might appear to violate the theory of relativity, since n is defined by Eq. (1-14) as c/v, where c is the speed of light in a vacuum and v is the velocity in the medium of interest. There is actually no contradiction, however, since relativity applies to the velocity with which *energy* is transmitted and this is always less than c. Since the induced oscillators radiate secondary radiation, there can be constructive interference which appears to give a vector velocity greater than c, but the overall *group* velocity which carries the energy travels with a velocity less than c. This point is discussed in texts on optics.†

Normal Dispersion

If Eq. (A-3) is expanded by the binomial theorem and considered only in regions of normal dispersion, where λ is appreciably greater than λ_0, the dif-

† For example, ibid., pp. 477 and 484.

ferentiated equation becomes

$$\frac{dn}{d\lambda} = -\frac{2A'}{\lambda^3} \tag{A-4}$$

where A' is directly related to the A terms in Eq. (A-3). This equation is useful because it shows the dependence of the dispersion of a prism on wavelength in regions of the wavelength spectrum where prisms are used. Dispersion varies approximately as the inverse cube of the wavelength, which predicts that it will be about 8 times as large at 400 nm as at 800 nm. The minus sign corresponds to the usual negative slope of the dispersion curve.

A-3 PROOF THAT BEER'S LAW FAILS FOR POLYCHROMATIC RADIATION (SEE SEC. 1-5C)

Consider an absorbing system of fixed path b illuminated by radiation that is a mixture of two different monochromatic radiations of wavelength λ_1 and λ_2. The wavelengths λ_1 and λ_2 might represent, for example, the extremes of wavelength out of a band of wavelengths being used to illuminate a given sample cell at some fixed (nominal) wavelength setting of the instrument. For simplicity in our derivation, however, we shall ignore the wavelengths in between λ_1 and λ_2, which will not harm our arguments. Assume further that Beer's law is obeyed at each individual wavelength and that the absorptivities of the system at these wavelengths are a_1 and a_2, respectively. (The case of Beer's law *failing* at either or both wavelengths is trivial, leading to obvious failures.) Figure A-2 schematically pictures this situation, where $I_{0,1}$ and $I_{0,2}$ represent the incident intensities at wavelengths 1 and 2 and I_1 and I_2 represent the transmitted intensities at the two wavelengths. Under these conditions the total light entering the sample is given by

$$I_{0,\text{tot}} = I_{0,1} + I_{0,2} \tag{A-5}$$

and the total light transmitted (which is what is *measured*) is given by

$$I_{\text{tot}} = I_1 + I_2 \tag{A-6}$$

If Beer's law is used in its exponential form, the incident intensities at wave-

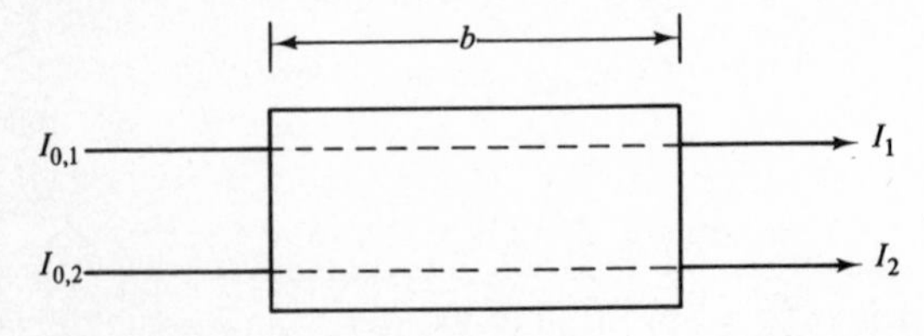

FIGURE A-2 Transmittance of two wavelengths after passing through an absorbing sample.

lengths 1 and 2 can be expressed as

$$I_{0,1} = I_1 \times 10^{a_1bC} \tag{A-7}$$

and

$$I_{0,2} = I_2 \times 10^{a_2bC} \tag{A-8}$$

Combining Eqs. (A-5), (A-7), and (A-8) and factoring out 10^{a_1bC} gives

$$I_{0,\text{tot}} = 10^{a_1bC}(I_1 + I_2 \times 10^{(a_2-a_1)bC}) \tag{A-9}$$

Dividing Eq. (A-9) by Eq. (A-6) gives the final equation

$$\frac{I_{0,\text{tot}}}{I_{\text{tot}}} = \frac{10^{a_1bC}(I_1 + I_2 \times 10^{(a_2-a_1)bC})}{I_1 + I_2} \tag{A-10}$$

The message of Eq. (A-10) will be clearer if we convert back to the logarithmic form of Beer's law, giving

$$A = \log \frac{I_{0,\text{tot}}}{I_{\text{tot}}} = a_1bC + \log(I_1 + I_2 \times 10^{(a_2-a_1)bC}) - \log(I_1 + I_2) \tag{A-11}$$

The consequences of Eq. (A-11) can be seen if we take two cases: (1) where $a_1 \neq a_2$ and (2) where $a_1 = a_2$. In case (1), Beer's law fails, since A is not linearly proportional to C but depends on C in a complex, exponential way. On the other hand, in case (2), when $a_1 = a_2$, Beer's law holds, since Eq. (A-11) reduces to

$$A = a_1bC \tag{A-12}$$

These results are of great interest since they serve to guide us in selecting the proper wavelength at which to carry out an analysis. (See examples in Sec. 1-5*C*.)

A-4 PROOF THAT THERE IS A REGION OF OPTIMUM TRANSMITTANCE FOR MINIMUM READING ERROR (SEE SEC. 1-5D)

In the following derivation, three assumptions are made: (1) that Beer's law holds for the system, (2) that the instrument used is adjusted to read 100 percent T with the reference (solvent) solution, and (3) that the instrument used gives a readout that is linear in percent T; that is, the output is directly proportional to the intensity of light falling on the detector. The last two assumptions are in accordance with the most usual spectrophotometric conditions, and small deviations from assumption 1 will not seriously affect our conclusions. The more complicated case of instruments giving readout that is linear in absorbance A will be commented upon briefly at the conclusion of this derivation.

It is convenient to use Beer's law in the form

$$abC = -\log T \tag{A-13}$$

where all terms were defined in Sec. 1-5*B*. If we let dT be the random error in reading the transmittance scale (which may include associated random errors such as slight nonlinearity in the readout device and variations in stray light) and dC be the corresponding uncertainty in concentration (or absorbance) which is generated, these quantities can be evaluated by differentiating Eq. (A-13):

$$ab\,dC = -\log e \frac{1}{T}\,dT \tag{A-14}$$

Since $\log e = 0.434$ and $ab = -(\log T)/C$, these terms can be substituted into Eq. (A-14) to give

$$\frac{dC}{C} = \frac{0.434}{T(\log T)}\,dT \tag{A-15}$$

It should be noted that it is the *relative* error in concentration dC/C which is of the greatest analytical interest, and this can be evaluated from Eq. (A-15). To visualize this error function, Fig. A-3 plots the percent relative error in the concentration ($100\ dC/C$) vs. percent transmittance for a fairly typical reading error dT of ± 0.005 (or ± 0.5 percent transmittance) as well as for a fairly high reading error of ± 0.01. As can be seen from Fig. A-3, the optimum transmittance range is about 20 to 60 percent (corresponding to an absorbance range of about 0.2 to 0.7), the error becoming very appreciable

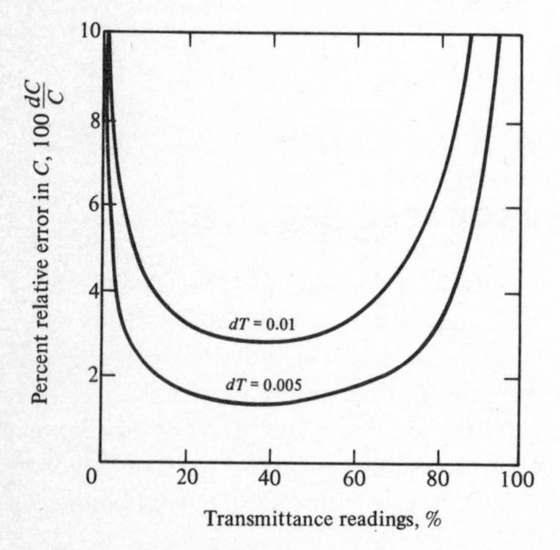

FIGURE A-3 Relative error in concentration due to a transmittance reading error.

outside this range. In particular, if a sample transmittance falls below about 10 percent or above about 70 percent, the sample concentration or cell path should be adjusted to fall within the optimum range. Since a reading error dT of 0.005 is fairly typical (about 0.002 is the *minimum* error in ordinary work), Fig. A-3 shows that a relative concentration error of about 1 percent is typical when working within the optimum transmittance interval.

To find the point of minimum relative error, Eq. (A-15) can be differentiated again and the derivative set equal to zero, yielding the solution that the optimum transmittance point is 36.8 percent T, or $A = 0.434$.

If the instrument used gives readout which is linear in absorbance instead of transmittance, the optimum absorbance range is less easy to deduce but appears to fall in the range of 0.4 to 1.4 absorbance units, with a minimum error at about 0.87 absorbance unit.† The reader should not confuse instruments giving linear absorbance readout with those which merely provide a nonlinear absorbance-scale calibration in addition to a linear transmittance scale; readings on this latter instrument should be maintained between 0.2 and 0.7 absorbance unit, as discussed earlier.

A-5 DERIVATION OF STRAY-LIGHT EQUATION (SEE SEC. 1-5D)

Equation (1-49) can be derived as follows. Let $I_{0,s}$ be the intensity of the unwanted stray radiation striking the detector and I_0 and I be the usual incident and transmitted intensities. The relationship of these quantities is pictured in Fig. A-4. From Fig. A-4, the observed transmittance will be given by

$$T_{\text{obs}} = \frac{I + I_{0,s}}{I_0 + I_{0,s}} \tag{A-16}$$

The fraction of stray light ρ is defined by the equation

$$\rho = \frac{I_{0,s}}{I_0} \tag{A-17}$$

Equation (A-17) can be solved for $I_{0,s}$ and substituted into Eq. (A-16), giving

$$T_{\text{obs}} = \frac{I + \rho I_0}{I_0 + \rho I_0} \tag{A-18}$$

The transmittance that would be observed if no stray light were present can

† H. K. Hughes, *Appl. Opt.*, 2:937 (1963).

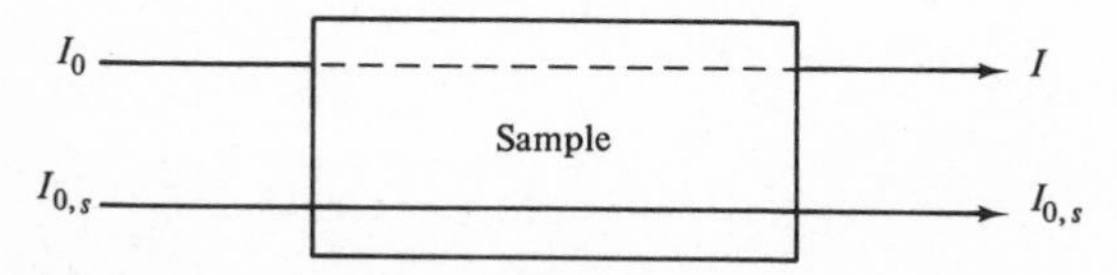

FIGURE A-4 Schematic illustration of stray radition $I_{0,s}$ adding to radiation which is absorbed.

be designated as T_{true} and defined by the equation

$$T_{\text{true}} = \frac{I}{I_0} \tag{A-19}$$

Thus, if the numerator and denominator of Eq. (A-18) are divided through by I_0, Eq. (A-19) can be inserted to give

$$T_{\text{obs}} = \frac{T_{\text{true}} + \rho}{1 + \rho} \tag{1-49}$$

which is the final stray-light equation given in Sec. 1-5*D*.

appendix B Tables

TABLE B-1 Metric-System Prefixes

Value	Prefix	Abbreviation	Value	Prefix	Abbreviation
10^{12}	tera	T	10^{-2}	centi	c
10^{9}	giga	G	10^{-3}	milli	m
10^{6}	mega	M	10^{-6}	micro	μ
10^{3}	kilo	k	10^{-9}	nano	n
10^{2}	hecto	h	10^{-12}	pico	p
10	deka	da	10^{-15}	femto	f
10^{-1}	deci	d	10^{-18}	atto	a

TABLE B-2 Constants

π	pi = 3.1416
e	base of natural logarithms = 2.7183
	ln 10 = 2.3026
	log e = 0.4343
c	velocity of light in vacuum = 2.9979×10^{10} cm/s
h	Planck's constant = 6.6256×10^{-27} erg-s
e	electronic charge = 4.8029×10^{-10} esu
e_m	electron mass = 9.1085×10^{-28} g
g	acceleration due to gravity = 980.665 cm/s^2
N	Avogadro's number = 6.0247×10^{23} mol^{-1}
k	Boltzmann constant = 1.3805×10^{-16} erg/(K) (molecule)

TABLE B-3 Conversion Factors

1 angstrom (Å) = 10^{-8} cm = 10^{-10} m
1 hertz (Hz) = 1 cycle/s
1 absolute joule (J) = 10^7 ergs = 10^7 dyn-cm = 1 W-s
1 thermochemical calorie (cal) = 4.184 J = 4.184×10^7 g-cm^2/s^2 (ergs)
1 electronvolt (eV) = 1.6021×10^{-19} J

$$1 \text{ radian (rad)} = \frac{360°}{2\pi} = 57.2958°$$

1° = 60 minutes = 0.01745 rad
1 lb = 453.59 g
1 kg = 2.2046 lb
1 wave number (cm^{-1}) = 2.8591 cal/mol
= 1.2398×10^{-4} eV/molecule
= 1.9863×10^{-16} erg/molecule

To interconvert wavelength in micrometers and wave numbers use

$$\text{Wavelength } (\mu\text{m}) = \frac{1}{\text{wave number (cm}^{-1})} \times 10^4$$

To interconvert wavelength in nanometers and energy in electronvolts use

$$\text{Wavelength (nm)} = \frac{1240}{\text{energy (eV)}}$$

Index

Page numbers in **boldface** indicate figures.